AF358630

Advances in Anatomy and Its History

Advances in Anatomy and Its History

Editors

Gianfranco Natale
Francesco Fornai

Basel • Beijing • Wuhan • Barcelona • Belgrade • Novi Sad • Cluj • Manchester

Editors
Gianfranco Natale   Francesco Fornai
University of Pisa   University of Pisa
Pisa   Pisa
Italy   Italy

Editorial Office
MDPI
St. Alban-Anlage 66
4052 Basel, Switzerland

This is a reprint of articles from the Special Issue published online in the open access journal *Anatomia* (ISSN 2813-0545) (available at: https://www.mdpi.com/journal/anatomia/special_issues/Advances_in_Anatomy_History).

For citation purposes, cite each article independently as indicated on the article page online and as indicated below:

Lastname, A.A.; Lastname, B.B. Article Title. *Journal Name* **Year**, *Volume Number*, Page Range.

ISBN 978-3-7258-1247-9 (Hbk)
ISBN 978-3-7258-1248-6 (PDF)
doi.org/10.3390/books978-3-7258-1248-6

Cover image courtesy of Gianfranco Natale

Contents

About the Editors

Gianfranco Natale

Gianfranco Natale graduated in Medicine and Surgery and specialized in Clinical Pharmacology at the University of Pisa. He is a Professor of Human Anatomy at the School of Medicine of the University of Pisa, with teaching activity in several university courses of the medical area, including doctorate courses, master's degrees, and winter and summer schools. He is the Director of the Museum of Human Anatomy "Filippo Civinini". He is a past coordinator of international relationships in the medical field, including the Erasmus Program, and a member of several international projects, including the Uzhelth Project (cooperation with Uzbekistan; 2016–2017) and the Tuning Asia South East (TA-SE) Project (2017–2019). His main research interests are the morphometry, pharmacology, and physiopathology of the experimental peptic ulcer, as well as morphological and functional alterations of the digestive tract in inflammatory and neurodegenerative diseases. Another interest encompasses the history of medicine, with special attention to anatomy. He is a member of the Editorial Board of the scientific journal *Clinical Cancer Drugs*, and of the web magazine *University Heritage*. He is a member of the Italian Society of Anatomy and Histology, and the Italian Society of History of Medicine.

Francesco Fornai

Francesco Fornai is currently a Full Professor and the Chair of Human Anatomy at the University of Pisa, as well as the Team Leader of Neurobiology of Movement Disorders at IRCCS Neuromed. He received his training at the University of Pisa, Italy, at Georgetown University, Washington (D.C.), and at the University of Washington, Seattle (WA). At the University of Pisa, he completed his M.D. course and received a Ph.D. degree in Medical Sciences at SSSUP S. Anna. He worked as a Post- Doctoral Fellow while finishing his residency in Clinical Pharmacology. He has expertise in neuroanatomy and neurodegeneration, witnessed by over 350 peer-reviewed publications, encompassing the alteration of autophagy in motor neuron disease. He provided innovative methods for the exploration of the morphology of the spinal cord and the molecular mechanisms of neurodegeneration. In collaboration with the Nobel Prize Winner Dr. Thomas Sudhof, he demonstrated the detrimental effects of the synaptic protein alpha synuclein in Parkinsonism.

Preface

This Special Issue deals with one of the most important disciplines in the medical area: anatomy. The morphological investigation represents the most ancient approach to the study of living beings. The purpose of this issue is to examine the great value of this long history. At the same time, it also considers the importance of evaluating the role of this discipline in modern medicine. In this respect, new technological approaches are challenging the classic morphological methods for studying anatomical structures. In line with this, anatomical education is changing, as well, and new tools are needed to face modern demands. This collective work is addressed to all people who are interested in a new vision of anatomy: physicians, surgeons, technicians, nurses, historians, teachers, and students. A special acknowledgement must be addressed to the numerous contributors to this Special Issue.

Gianfranco Natale and Francesco Fornai
Editors

 *anatomia*

Editorial

Advances in Anatomy and Its History

Gianfranco Natale [1,2,*] and Francesco Fornai [1,3]

[1] Department of Translational Research and New Technologies in Medicine and Surgery, University of Pisa, 56126 Pisa, Italy; francesco.fornai@unipi.it
[2] Museum of Human Anatomy "Filippo Civinini", University of Pisa, 56126 Pisa, Italy
[3] Istituti di Ricovero e Cura a Carattere Scientifico (I.R.C.C.S.) Neuromed, 86007 Pozzilli, Italy
* Correspondence: gianfranco.natale@unipi.it

1. Introduction

Anatomy is still considered the most ancient and basic discipline in the medical sciences. Since the work of Andreas Vesalius (1514–1564), anatomy has played a strong role in both medical research and education, being the main axis of medical research over the centuries, to foster novel findings about human bodies and shape the education provided at medical schools. In the second half of the last century, this trend was slightly modified while gross anatomical observations reduced, and education became the primary purpose of anatomy institutes [1,2]. There is also a conservative tendency in anatomic descriptions, with a seeming reluctance among acknowledged anatomists to explore new theories by returning to direct studies of gross anatomy. Although this discipline is considered an undisputed core component of the medical curriculum, anatomical research and education have been reduced in scope and status to accommodate other disciplines [3].

This occurred despite the fields of microscopic and ultramicroscopic anatomy undergoing expansion to explore the micro-world at the molecular level. This has happened despite gross anatomy, which is still recognized as a fundamental discipline, appearing quite obsolete. Such a scenario emerges when shallow perspectives are applied; in fact, in the last few years, some gross anatomical discoveries have been reported, which overturned the long existing dogmas about morpho-functional aspects. For instance, the anatomy of a previously unrecognized, though widespread, macroscopic, fluid-filled space within and between tissues represents a widened and in-depth focus on the classic concept of the human interstitium. The mesenteric continuity, as well as the novel concept of the glymphatic system, have been reported [4,5]. The gross anatomy of myofascia is terminologically evolving through the work of the Fascia Nomenclature Committee [6]. A new field of research is exploring how gut microbiota acts as an "invisible organ" to modulate gastrointestinal functions, as well as the general homeostasis of the body [7]. Apart from the sensational discovery of "new organs", gross anatomy research remains an important tool to correct or redefine classic concepts and ascertain anatomical varieties. Independently of the approach used (*ex vivo* dissection or in vivo imagining techniques), it represents a unique way to explore the human body. Cadaveric dissection has several advantages: (i) gross anatomy research; (ii) the development of new dissection and preservation techniques; (iii) forensic medicine research; (iv) planning virtual anatomy software; (v) surgical training and simulation; (vi) the enrichment of anatomic museums with specific preparations (third mission of the university).

This was the opinion of the Irish anatomist Alexander Macalister (1844–1919): Practical anatomy was considered a discipline of incalculable value, for not only was the knowledge acquired an end in itself, but the method through which it was gained was the most valuable part of a student's entire medical training. Anatomy occupied a unique position in the curriculum. In all other departments, the student had no alternative but to accept the dictum of their teacher or textbook. Here, the student was trained to use their eyes and

Citation: Natale, G.; Fornai, F. Advances in Anatomy and Its History. *Anatomia* **2024**, *3*, 50–56. https://doi.org/10.3390/anatomia3010005

Received: 27 February 2024
Accepted: 12 March 2024
Published: 14 March 2024

hands, as well as educated to translate the impression made by the objects of study on their senses into words. This enabled students to check the descriptions in their book or those taught in the lecture room by comparing practical experience with the verbal accounts given by teachers. In the dissecting room, every student could be an investigator [8].

In this respect, the Special Issue *Advances in Anatomy and Its History* represents the best way to inaugurate the new scientific journal *Anatomia* as it provides a broad view on this topic and considers both holistic and humanistic perspectives, taking into account the innumerous connections between the history of anatomy and its recent advances.

All multidisciplinary approaches related to anatomy besides morphological techniques lie within the scope of this Special Issue, including anatomical education. Comparative and veterinary anatomy were also included. At the same time, this Special Issue looks back to the past of this discipline and focuses its attention to the history of anatomy to rediscover a long tradition of morphological investigation from an integrated perspective.

2. An Overview of Published Articles

This Special Issue includes 12 original research articles, 9 articles dealing with anatomical variations, and 8 review articles. This Special Issue captures the diversity of the studies that focus on the anatomical field. The diversity of articles, the depth of the topics, and the relative geographical reach of the authors in this Special Issue confirmed the continued interest of researchers in anatomy.

It is not the purpose of this Editorial to elaborate on each of the articles but rather to encourage the reader to read them. These articles encompass a diverse range of study methods and topics, elucidating the richness of this research field.

Original *Research Articles* represent the most numerous group included in this Special Issue. Four articles consider the important topic of anatomical education. In one *Communication*, a study provides preliminary data regarding the use of immersive virtual reality with 3D holography to improve learning of the anatomy of the heart, concluding that anatomical classes that make use of this technology, although attractive for medical students, are not superior to traditional learning in terms of knowledge retention if such classes are not precisely designed or strictly supervised by academic teachers and the students do not become familiar with the use of virtual reality goggles before anatomical classes (Czaja's article, contribution 22). In another attempt to improve anatomy teaching and learning, the use of virtual anatomy and histology and small portable autostereoscopic 3D screens was beneficial for anatomical education, and the increasing availability of such systems and reduced costs will allow this technology to have a significant impact in the coming years (Otoo's article, contribution 19). In a further study, the educational value and the students' perceptions of the effectiveness and usefulness of surgical and bioengineering technologies' integration into the anatomy course are evaluated, and the authors show that students appreciate in-depth surgical integration much more than in-depth engineering integration, even if those who prefer in-depth surgical studies also appreciate in-depth engineering studies (Miglietta's article, contribution 17). Finally, one study is dedicated to the relationship between neuroanatomy and the evolution of human brain atlases, as well as the impacts of these atlases on the understanding, presentation, and advancement of neuroanatomy, demonstrating that they are excellent ways to represent, disseminate, and support this discipline (Nowinski's article, contribution 16).

Two works deal with animal anatomy. The rostral epidural rete mirabile structure in giraffes (*Giraffa camelopardalis*) is recognized as an essential blood flow regulatory mechanism balancing physiological difficulties arising due to the extensive heart-to-head distance, and it might fulfill the same function in other long-necked artiodactyls (van der Walt's article, contribution 22). In one *Communication*, the transition from wolf to dog is observed through the study of skeletal features, accompanied by a series of parameters useful for understanding this evolutionary shift. A diminution in size occurs early in the process of animal domestication; this phenomenon is characterized in many animal species rather than just dogs. This study suggests that in order to differentiate between ancient wolf skeletal specimens of *Canis lupus* L. and similar specimens of *Canis familiaris* L., it is necessary to carry out several morphological studies, since there is not one single significant parameter that can be used alone for this investigation (Coli's article, contribution 18).

The pathological investigation is the topic of other two studies. This first study evaluates ultra-low-dose computed tomography protocols regarding the detectability of pulmonary nodules. The impacts of tube current settings, kernels, strength levels of third-generation iterative reconstruction algorithms, and pitch are investigated. A chest phantom with artificial spherical nodules of different densities and diameters is examined using a third-generation dual-source CT. Scanning and post-processing protocols, tube current levels, and ultra-high and non-high pitch modes are applied. This chest phantom study shows no difference in nodule detection when using ultra-high pitch protocols compared to a regular pitch mode (Leitzig's article, contribution 15). The second study investigates the diagnostic and prognostic value of different quantitative analysis methods assessing adrenal gland parameters on contrast-enhanced CT scans in patients bearing septic shock conditions, demonstrating that there is no additional diagnostic value in performing time-consuming semi-automated whole-organ adrenal gland segmentation analyses (Milberg's article, contribution 1).

The other three relevant studies refer to pure anatomical studies. One study details the morphology of the incisive or naso-palatine canal, showing that there are different anatomical features, including the absence of this canal (Iamandoiu's article, contribution 5). Another study examines, via a cross-sectional analysis, the development of hippocampal dentation in healthy children and adolescents, considering that it is associated with episodic memory in healthy adults. This is the first study indicating wide variability in the degree of hippocampal dentation (Beattie's article, contribution 3). The third study investigates three-dimensional variations in the shape of the volar rim of the distal radius with the aim of proposing a personalized implant shape design corresponding to the individual morphology for the treatment of fractures. The authors find individual differences in the shape of the radius epiphysis and suggest the development of several types of semi-custom plates with correlated sagittal and transverse bending angles. This allows for osteosynthesis with minimal soft tissue invasion while accommodating the anatomical diversity of the volar rim (Yoneda's article, contribution 10).

Moreover, in a historical article, the medical pictorial cycle devised in the annex of the villa of Flaminio Rota, a 16th century anatomist who worked the University of Bologna, is examined, revealing an interesting cultural dimension that provides new information about Rota's academic life (Armocida's article, contribution 14).

Nine original articles are dedicated to anatomical variations. Seven of them are *Case Reports*. Three articles explore blood vessel variations in greater depth: a rare bilateral variation in the branches of the internal thoracic artery (Hawi's article, contribution 27), a duplicated inferior vena cava (Klansek's article, contribution 21), and a rare configuration origin of the superior thyroid, lingual, and facial arteries in a pentafurcated common carotid artery (Zaccheo's article, contribution 12). These studies emphasize that such unexpected variations may be asymptomatic, as well as associated with severe symptoms and complicated surgical procedures.

The nervous tissue is the topic of two articles: they cover a very rare anatomical variation, namely the anastomosis between the median and musculo-cutaneous nerves (Barone's article, contribution 4), and an unusual morphologic variant related to the clinical pathological features of neuroblastoma-like schwannoma (Mortazavi's article, contribution 13).

Another two articles focus their attention on the muscular tissue: the presumed very uncommon presence of extensor indicis and digiti medii communis muscles (Penkwitz's article, contribution 20) and the distribution and appearance of the arrector pili muscle in the skin of the faces of Rhesus monkeys, showing regional differences in the distribution of hair follicles and associated arrector pili muscles (May's article, contribution 2). A similar matter is discussed in an *Opinion* article, where the authors propose a new classification of the variant cleido-occipital muscle (Silawal's article, contribution 7).

In a *Technical Note*, the author identifies and correlates oro-facial anatomy discrepancies as biometric data and the impact of its rehabilitation as an educational, forensic approach for human identification (Corte-Real's article, contribution 9).

Finally, eight *Review* articles complete the issue. Four reviews provide informative biographies of eminent figures in the field of anatomy. The first article delves into the life and accomplishments of the Spanish anatomist Juan Valverde de Amusco (1525–1587), with a special focus on his masterpiece titled *Historia de la composición del cuerpo humano*, written in the Castilian Spanish language This book provides the first accurate description of pulmonary circulation, the vomer bone, and four extraocular rectus muscles and corrects several of Vesalius' anatomical observations. In this respect, Juan Valverde de Amusco pioneers the implementation and transfer of post-Vesalian anatomical scientific knowledge (Arráez-Aybar's article, contribution 29). Another article focuses on Jean Cruveilhier (1791–1874) and his book titled *Anatomie descriptive*, which was a great success during the author's lifetime. The masterpiece deals with the anatomy of the human body in a way that can be applied by both students and medical professionals. Due to his scientific rigor, Cruveilhier always investigates the anatomic changes linked to pathological processes, and for this reason, he could certainly be considered a predecessor of evidence-based medicine (Arráez-Aybar's article, contribution 25). Ludwik Henryk Bojanus (1776–1827), the anatomist and co-founder of Polish veterinary education, is also discussed. His most important achievements include the organization and launch of the first veterinary school in Polish lands in 1823, the compilation of a monograph about the anatomy of the European pond turtle "Anatome testudinis europaeae" between 1819 and 1821, and the development of methods for dealing with infectious diseases in animals. He is considered one of the most important evolutionists before Darwin (Sobolewski's article, contribution 26). The last biography refers to the Swiss naturalist René-Édouard Claparède (1832–1871), a pioneering protozoologist and comparative anatomist. He made important contributions to diverse areas of natural science, biology, and comparative anatomy, including the structures of infusoria, annelids, and earthworms; the evolution of arthropods; the embryology of spiders; and observations about marine invertebrates (Kollarou's article, contribution 24).

Two of the review articles concern vascular anatomy. The first article considers the anatomy of cerebral arteries with clinical aspects in patients with ischemic stroke. Computed tomography angiography is the main method for the initial evaluation of cerebral circulation in acute stroke, and a comprehensive examination allows for the identification of most abnormalities and normal variants. Anatomical knowledge of the presence of any normal variants, such as fenestration, duplications, and persistent fetal arteries, plays a crucial role in the diagnosis and therapeutic management of acute stroke. Sometimes, it is the clinical picture that allows us to weigh the relevance of an alteration (Barbato's article, contribution 8). The second article explores common anatomical variations in neurovascular canals and foramina that are relevant to oral surgeons. In this respect, all health professionals must continually challenge themselves to learn more about the different anatomical variations that the human body presents and how these may variations affect clinical practice to perform a surgery that is as safe and minimally invasive as possible (Sferlazza's article, contribution 6).

Another review discusses the recent topic of microbiota and refutes the notion that it should be considered an organ, given that an organ comprises tissue of similar or different embryological origins, while the microbiota is a pool of different microbial species originating individually from single replications rather than a common ancestral cellular element. Conversely, the authors present the muco-microbiotic layer as a novel morphofunctional structure (Fucarino's article, contribution 11).

Finally, a very long review in the form of a mini-book addresses the role of chemical communication in mammals, giving special attention to the vomero-nasal system in pheromone-mediated interactions. This intricate sensory system profoundly influences several aspects of social and sexual interactions, encompassing reproductive processes, the formation of hierarchies, maternal bonding, and intraspecies recognition. Furthermore, it is evident that there is a noticeable range of variability resulting from evolutionary adaptations within this particular system, surpassing the documented variants in the olfactory system. This study opens the doors to the use of innovative approaches for enhancing the well-being and health of both animals and humans (Torres's article, contribution 28).

3. Conclusions

The present Special Issue *Advances in Anatomy and Its History* aimed at reinforcing the attention paid to this basic medical discipline. Indeed, anatomy is still considered a fundamental pillar to understand other medical fields, including pathology, physiology, and surgery. Furthermore, morphological notions are essential for better interpreting the origin, development and resolution of diseases. For these reasons, clinical and functional anatomy have great importance [9]. The advancements in imaging techniques and the availability of modern digitalized devices allow researchers to revive anatomy in living beings. These strategies represent an opportunity to innovate and integrate classic dissection and improve educational anatomy [10–12].

This Special Issue includes several articles that cover the main aspects of the anatomical discipline. A large body of work deals with classic anatomical issues in humans and animals. Cadaver dissection or novel in vivo imaging techniques provide useful original information about anatomical variations that must be taken into account during the clinical and surgical activities. Another group of articles considers the need to propose new educational strategies. Modern technological devices and digitalized data will allow us to improve anatomical teaching. The integration of medical imaging into anatomical education provides advantages in understanding and learning about the human body.

Finally, this Special Issue represents a constructive incitement to preserve the memory of this ancient discipline and, at same time, stimulate innovation and improve strategies to connect anatomy to new fields of research. It is also important to continue teaching anatomy as part of the modern medical curriculum. The integration of novel teaching modalities and modern technologies will encourage interest in and the retention of anatomical knowledge and its clinical relevance.

Author Contributions: The authors contributed to the manuscript equally. All authors have read and agreed to the published version of the manuscript.

Conflicts of Interest: The authors declare no conflicts of interest.

List of Contributions

1. Milberg, M.; Kindt, A.; Luft, L.; Hoffmann, U.; Behnes, M.; Schoenberg, S.; Janssen, S. Diagnostic and Prognostic Value of Quantitative Computed Tomography Parameters of Adrenal Glands in Patients from Internistled ICU with Sepsis and Septic Shock. *Anatomia* **2022**, *1*, 14–32; https://doi.org/10.3390/anatomia1010003.
2. May, I.; Mätz-Rensing, K.; May, C. Distribution and Appearance of Arrector Pili Muscle in the Skin of the Rhesus Monkey Face. *Anatomia* **2022**, *1*, 33–40; https://doi.org/10.3390/anatomia1010004.
3. Beattie, J.; Martin, R.; Cook, E.; Thompson, M.; Kana, R.; Jacobs, R.; Correya, T.; Ramaniharan, A.; Ver Hoef, L. Hippocampal Dentation in Children and Adolescents: A Cross-Sectional Analysis from Birth to 18 Years Old. *Anatomia* **2022**, *1*, 41–53; https://doi.org/10.3390/anatomia1010005.
4. Barone, R.; D'Amico, A.; Di Lorenzo, N.; Di Grado, G.; Matranga, E.; Spinoso, G.; Bavuso, L.; Marino Gammazza, A.; Rappa, F.; Bucchieri, F.; Cappello, F.; Piotrowska, W.; Spodnik, J.; Spodnik, E.; Wójcik, S. Anastomosis between Median and Musculocutaneous Nerve: Presentation of a Very Rare Anatomical Variation in Comparison to Classical Divisions. *Anatomia* **2022**, *1*, 68–74; https://doi.org/10.3390/anatomia1010007.
5. Iamandoiu, A.; Mureşan, A.; Rusu, M. Detailed Morphology of the Incisive or Nasopalatine Canal. *Anatomia* **2022**, *1*, 75–85; https://doi.org/10.3390/anatomia1010008.
6. Sferlazza, L.; Zaccheo, F.; Campogrande, M.; Petroni, G.; Cicconetti, A. Common Anatomical Variations of Neurovascular Canals and Foramina Relevant to Oral Surgeons: A Review. *Anatomia* **2022**, *1*, 91–106; https://doi.org/10.3390/anatomia1010010.

7.   Silawal, S.; Pandey, S.; Schulze-Tanzil, G. Introduction of a New Classification of the Cleidooc-cipital Muscle. *Anatomia* **2022**, *1*, 148–151; https://doi.org/10.3390/anatomia1020015.

8.   Barbato, F.; Allocca, R.; Bosso, G.; Numis, F. Anatomy of Cerebral Arteries with Clinical Aspects in Patients with Ischemic Stroke. *Anatomia* **2022**, *1*, 152–169; https://doi.org/10.3390/anatomia1 020016.

9.   Corte-Real, A. Orofacial Anatomy Discrepancies and Human Identification—An Education Forensic Approach. *Anatomia* **2022**, *1*, 170–176; https://doi.org/10.3390/anatomia1020017.

10. Yoneda, H.; Iwatsuki, K.; Saeki, M.; Yamamoto, M.; Tatebe, M. Do Anatomical Differences of the Volar Rim of the Distal Radius Affect Implant Design? A Three-Dimensional Analysis of Its Anatomy and Need for Personalized Medicine. *Anatomia* **2022**, *1*, 177–185; https://doi.org/10.3 390/anatomia1020018.

11. Fucarino, A.; Burgio, S.; Paladino, L.; Caruso Bavisotto, C.; Pitruzzella, A.; Bucchieri, F.; Cappello, F. The Microbiota Is Not an Organ: Introducing the Muco-Microbiotic Layer as a Novel Morphofunctional Structure. *Anatomia* **2022**, *1*, 186–203; https://doi.org/10.3390/ anatomia1020019.

12. Zaccheo, F.; Mariotti, F.; Guttadauro, A.; Passaretti, A.; Campogrande, M.; Petroni, G.; Cicconetti, A. A Rare Configuration origin of the Superior Thyroid, Lingual and Facial Arteries in a Pentafurcated Common Carotid Artery. *Anatomia* **2022**, *1*, 204–209; https://doi.org/10.3390/ anatomia1020020.

13. Mortazavi, S.; Kamyab Hesari, K.; Khorsand, A.; Ardalan, M. Clinicopathologic Features of Neuroblastoma-like Schwannoma: A Case Report of Unusual Morphologic Variant. *Anatomia* **2022**, *1*, 217–221; https://doi.org/10.3390/anatomia1020022.

14. Armocida, E.; Fornai, F.; Natale, G. Flaminio Rota, 16th Century Anatomist at the University of Bologna: A Biography on the Walls. *Anatomia* **2023**, *2*, 1–14; https://doi.org/10.3390/anatomia2 010001.

15. Leitzig, N.; Janssen, S.; Kayed, H.; Schönberg, S.; Scheffel, H. Observer Sensitivity for Detection of Pulmonary Nodules in Ultra-Low Dose Computed Tomography Protocols Using a Third-Generation Dual-Source CT with Ultra-High Pitch—A Phantom Study. *Anatomia* **2023**, *2*, 15–26; https://doi.org/10.3390/anatomia2010002.

16. Nowinski, W. Advances in Neuroanatomy through Brain Atlasing. *Anatomia* **2023**, *2*, 28–42; https://doi.org/10.3390/anatomia2010004.

17. Miglietta, S.; Familiari, G.; Relucenti, M.; Basili, S.; Bini, F.; Bove, G.; Barbaranelli, C.; Familiari, P. Surgical and Bioengineering Integration in the Anatomy Course of Medicine and Surgery High Technology: Knowledge and Perception of Anatomy. *Anatomia* **2023**, *2*, 63–77; https://doi.org/10.3390/anatomia2010006.

18. Coli, A.; Prinetto, D.; Giannessi, E. Wolf and Dog: What Differences Exist? *Anatomia* **2023**, *2*, 78–87; https://doi.org/10.3390/anatomia2010007.

19. Otoo, E.; Leibowitz, H.; Wong, O.; Rhode, K. Using a Portable Autostereoscopic Screen to Improve Anatomy Teaching and Learning. *Anatomia* **2023**, *2*, 88–98; https://doi.org/10.3390/ anatomia2010008.

20. Penkwitz, I.; Wind, G.; Maynes, E.; Leighton, M.; Granite, G. Presumed Presence of Extensor Indicis et Digiti Medii Communis Muscle in a 70-Year-Old White Male Donor. *Anatomia* **2023**, *2*, 109–116; https://doi.org/10.3390/anatomia2010010.

21. Klansek, J.; Meshida, K.; Maynes, E.; Leighton, M.; Wind, G.; Granite, G. Duplicated Inferior Vena Cava in a 69-Year-Old White Female Donor. *Anatomia* **2023**, *2*, 117–123; https://doi.org/10.3390/anatomia2020011.

22. van der Walt, M.; Daffue, W.; Goedhals, J.; van der Merwe, S.; Deacon, F. The Rostral Epidural Rete Mirabile: Functional Significance in Blood Flow Regulatory Mechanisms in Giraffe (*Giraffa camelopardalis*). *Anatomia* **2023**, *2*, 138–155; https://doi.org/10.3390/anatomia2020013.

23. Czaja, J.; Skuła, M.; Kowalczyk, D.; Redelbach, W.; Hobot, J.; Nowak, M.; Halaba, Z.; Simka, M. Does Immersive Virtual Reality with the Use of 3D Holography Improve Learning the Anatomy of the Heart?: Results of a Preliminary Study. *Anatomia* **2023**, *2*, 156–164; https://doi.org/10.3390/anatomia2020014.

24. Kollarou, P.; Triarhou, L. René-Édouard Claparède (1832–1871), Pioneer Protozoologist and Comparative Anatomist. *Anatomia* **2023**, *2*, 165–175; https://doi.org/10.3390/anatomia2020015.

25. Arráez-Aybar, L.; Fuentes-Redondo, T.; Bueno-López, J.; Romero-Reverón, R. Jean Cruveilhier (1791–1874), a Predecessor of Evidence-Based Medicine. *Anatomia* **2023**, *2*, 206–221; https://doi.org/10.3390/anatomia2030019.

26. Sobolewski, J.; Zdun, M. Anatomist and Co-Founder of Polish Veterinary Education—Ludwik Henryk Bojanus (1776–1827). *Anatomia* **2023**, *2*, 261–270; https://doi.org/10.3390/anatomia2030024.
27. Hawi, J.; Jurjus, R.; Daouk, H.; Ghazi, M.; Basset, C.; Cappello, F.; Hajj Hussein, I.; Leone, A.; Jurjus, A. A Rare Bilateral Variation in the Branches of the Internal Thoracic Artery: A Case Report. *Anatomia* **2023**, *2*, 320–327; https://doi.org/10.3390/anatomia2040028.
28. Torres, M.; Ortiz-Leal, I.; Sanchez-Quinteiro, P. Pheromone Sensing in Mammals: A Review of the Vomeronasal System. *Anatomia* **2023**, *2*, 346–413; https://doi.org/10.3390/anatomia2040031.
29. Arráez-Aybar, L.; Reblet, C.; Bueno-López, J. Juan Valverde de Amusco: Pioneering the Transfer of Post-Vesalian Anatomy. *Anatomia* **2023**, *2*, 450–471; https://doi.org/10.3390/anatomia2040033.

References

1. Trelease, R.B. From chalkboard, slides, and paper to e-learning: How computing technologies have transformed anatomical sciences education. *Anat. Sci. Educ.* **2016**, *9*, 583–602. [CrossRef] [PubMed]
2. Govender, S.; Cronjé, J.Y.; Keough, N.; Oberholster, A.J.; van Schoor, A.N.; de Jager, E.J.; Naicker, J. Emerging Imaging Techniques in Anatomy: For Teaching, Research and Clinical Practice. *Adv. Exp. Med. Biol.* **2023**, *1392*, 19–42. [CrossRef] [PubMed]
3. Lassek, A.M. *Human Dissection Its Drama and Struggle*; Thomas Nelson: Springfield, IL, USA, 1958.
4. Kumar, A.; Ghosh, S.K.; Faiq, M.A.; Deshmukh, V.R.; Kumari, C.; Pareek, V. A brief review of recent discoveries in human anatomy. *QJM* **2019**, *112*, 567–573. [CrossRef] [PubMed]
5. Natale, G.; Limanaqi, F.; Busceti, C.L.; Mastroiacovo, F.; Nicoletti, F.; Puglisi-Allegra, S.; Fornai, F. Glymphatic System as a Gateway to Connect Neurodegeneration from Periphery to CNS. *Front. Neurosci.* **2021**, *15*, 639140. [CrossRef] [PubMed]
6. Schleip, R.; Hedley, G.; Yucesoy, C.A. Fascial nomenclature: Update on related consensus process. *Clin. Anat.* **2019**, *32*, 929–933. [CrossRef] [PubMed]
7. Li, X.; Liu, L.; Cao, Z.; Li, W.; Li, H.; Lu, C.; Yang, X.; Liu, Y. Gut microbiota as an "invisible organ" that modulates the function of drugs. *Biomed. Pharmacother.* **2020**, *121*, 109653. [CrossRef] [PubMed]
8. Tubbs, R.S. Anatomy occupies a unique position in the curriculum. *Clin. Anat.* **2023**, *36*, 847. [CrossRef] [PubMed]
9. Tubbs, R.S. What is clinical anatomy? *Clin. Anat.* **2017**, *30*, 285. [CrossRef] [PubMed]
10. Sugand, K.; Abrahams, P.; Khurana, A. The anatomy of anatomy: A review for its modernization. *Anat. Sci. Educ.* **2010**, *3*, 83–93. [CrossRef] [PubMed]
11. Binder, J.; Krautz, C.; Engel, K.; Grützmann, R.; Fellner, F.A.; Burger, P.H.M.; Scholz, M. Leveraging medical imaging for medical education—A cinematic rendering-featured lecture. *Ann. Anat.* **2019**, *222*, 159–165. [CrossRef] [PubMed]
12. Pettersson, A.; Karlgren, K.; Hjelmqvist, H.; Meister, B.; Silén, C. An exploration of students' use of digital resources for self-study in anatomy: A survey study. *BMC Med. Educ.* **2024**, *24*, 45. [CrossRef] [PubMed]

Communication

Does Immersive Virtual Reality with the Use of 3D Holography Improve Learning the Anatomy of the Heart?: Results of a Preliminary Study

Joanna Czaja, Marcin Skuła, Dariusz Kowalczyk, Wojciech Redelbach, Jacek Hobot, Marta Nowak, Zenon Halaba and Marian Simka *

Institute of Medical Sciences, University of Opole, 45-040 Opole, Poland
* Correspondence: msimka@uni.opole.pl

Abstract: Immersive virtual reality with the use of 3D holography is a new method that is being currently introduced for teaching anatomy, yet the actual educational benefits associated with its use remain unclear. Here, we present our preliminary observations and conclusions after the pilot phase of the study on a 3D holographic human heart. The study was conducted on a group of 96 students of medical faculty. Students were randomly divided into two groups: 57 students who were taught anatomy using traditional methods (plastinated human hearts, anatomical models, and atlases) and 39 students who were taught using 3D holographic hearts. Assessment of knowledge retention of the heart anatomy comprised 3 tests, which were performed 1 week and 3 and 6 months after the classes on heart anatomy. We have found that although anatomical classes with the use of immersive virtual reality were attractive for students; still, unsupervised teaching with the use of 3D holograms was not superior to traditional medical education. Differences between the groups in terms of anatomical knowledge retention were not statistically significant. Results of this pilot study suggest that in order to achieve better knowledge retention and understanding of the anatomy of the heart, classes should be precisely planned and strictly supervised by academic teachers. Moreover, students should get familiar with the use of virtual reality goggles before the classes.

Keywords: heart; teaching of anatomy; virtual reality; 3D holography

Citation: Czaja, J.; Skuła, M.; Kowalczyk, D.; Redelbach, W.; Hobot, J.; Nowak, M.; Halaba, Z.; Simka, M. Does Immersive Virtual Reality with the Use of 3D Holography Improve Learning the Anatomy of the Heart?: Results of a Preliminary Study. *Anatomia* **2023**, *2*, 156–164. https://doi.org/10.3390/anatomia2020014

Academic Editors: Gianfranco Natale and Francesco Fornai

Received: 1 February 2023
Revised: 26 April 2023
Accepted: 17 May 2023
Published: 21 May 2023

1. Introduction

Although cadaveric dissections are still regarded as the basis of teaching anatomy at medical universities, nowadays this method is increasingly displaced by other educational tools. This is primarily associated with less time allocated by current curricula to teaching gross anatomy, in favor of clinically oriented education. The virtual immersive reality, which is a new and rapidly progressing educational method, has already been introduced with some success into postgraduate medical education, primarily associated with invasive procedures [1–7]. Similarly, this new method is currently being introduced for teaching anatomy at some medical universities throughout the world. Attractiveness for medical students, as well as no need for cadavers, appears to be the main benefit of such a teaching method. However, at the moment, virtual immersive reality as an educational tool is at its experimental stage and the benefits associated with the use of 3D holography for teaching anatomy remain unclear [8–11], even if preliminary reports are encouraging [9,12–16]. Our digital project was aimed at the evaluation of the educational value of an immersive virtual reality with the use of 3D holography of the human heart. In this paper, we present our preliminary observations and conclusions after the pilot phase of this study.

2. Materials and Methods

2.1. Digital Design of the Heart Holograms

Unlike typical 3D medical holograms that utilize anatomical 3D images that are derived from artist's drawings, or—in the case of 3D holograms of the brain—basing it on MRI images of this organ, in our project 3D images of the heart were created using a genuine porcine heart as a digital framework. Fresh animal hearts were scanned as whole organs and also after trans-sections at different levels. Thereafter, scans of the heart were digitally modified in order to create a more human appearance of the organ (Figure 1), although morphologically human and porcine hearts are very similar. Large vessels, such as the aorta and the pulmonary trunk, were digitally added and movements of heart chambers and heart valves were generated by a special software. Then, anatomical descriptions of the particular parts of the heart were embedded into the final 3D hologram. The final digital product comprised a 3D hologram of a beating heart, which exhibited a real-life appearance. Our digital heart consisted of the musculature of all four heart chambers, of the heart valves and also of the proximal part of the aorta with its major branches, the pulmonary trunk with proximal parts of the pulmonary arteries, the superior and inferior vena cava, and proximal parts of the pulmonary veins. Three-dimensional holograms of the heart could be displayed using special goggles for immersive virtual reality. These 3D holograms of the heart could be enlarged, moved, rotated, or seen from the inside of heart chambers. Additionally, the digital heart could be virtually cross-sectioned at different levels in order to better visualize its structures, particularly those situated inside the ventricles (Figures 2 and 3). Using special commands on the goggles, the descriptions of anatomical details could be displayed on the side of a 3D hologram. Simultaneously, these details were marked with color dots and could also be heard by a student through the headphones of the goggles (Figure 4).

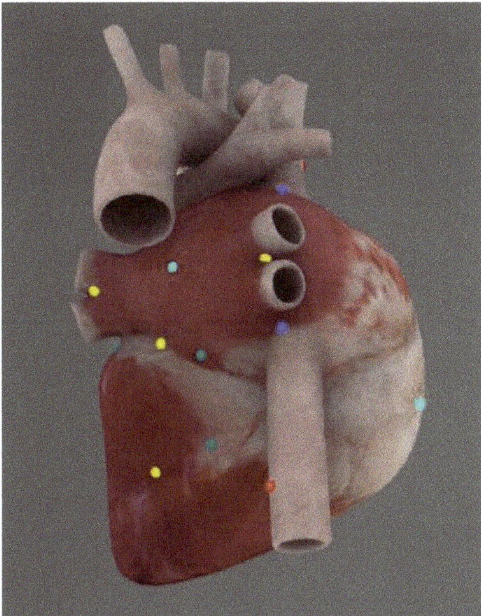

Figure 1. Our holographic 3-dimensional heart. This figure shows 2-dimensional representation of the view. Actually, with the use of goggles, it was possible to see 3-dimensional and moving heart. Color dots point to anatomical structures of the heart and their names can be heard through the headphones of goggles when these virtual structures are "touched" by student's sight.

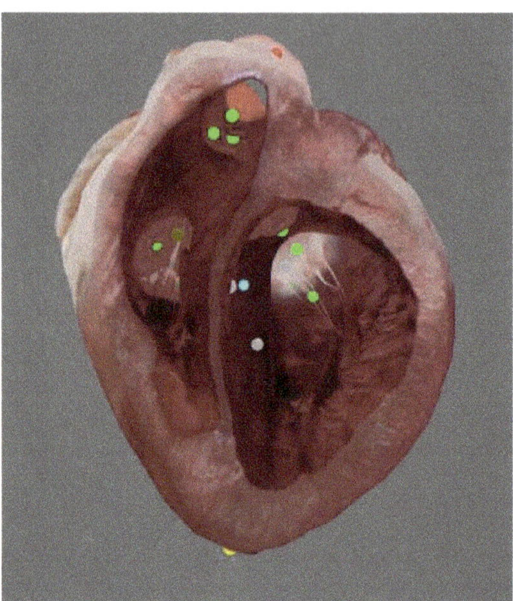

Figure 2. Three-dimensional hologram of the longitudinally cross-sectioned heart. Similarly to Figure 1, all structures were 3-dimensional, were moving, could be rotated around any axis, and could be cross-sectioned through several preset planes. Color dots point to structures located inside heart chambers.

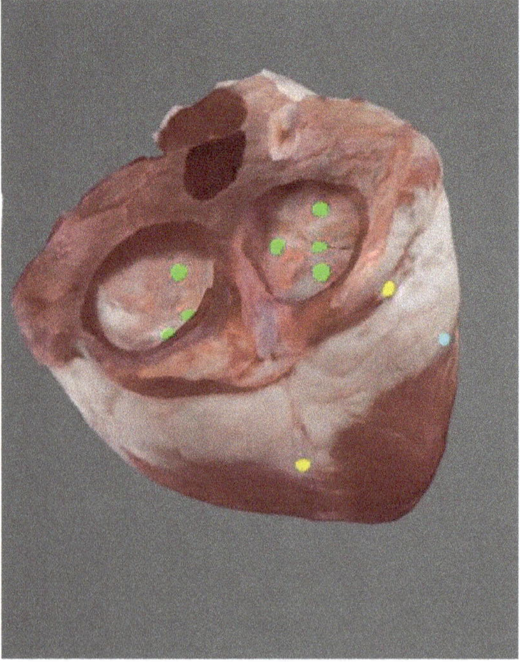

Figure 3. Three-dimensional hologram of the transversally cross-sectioned heart.

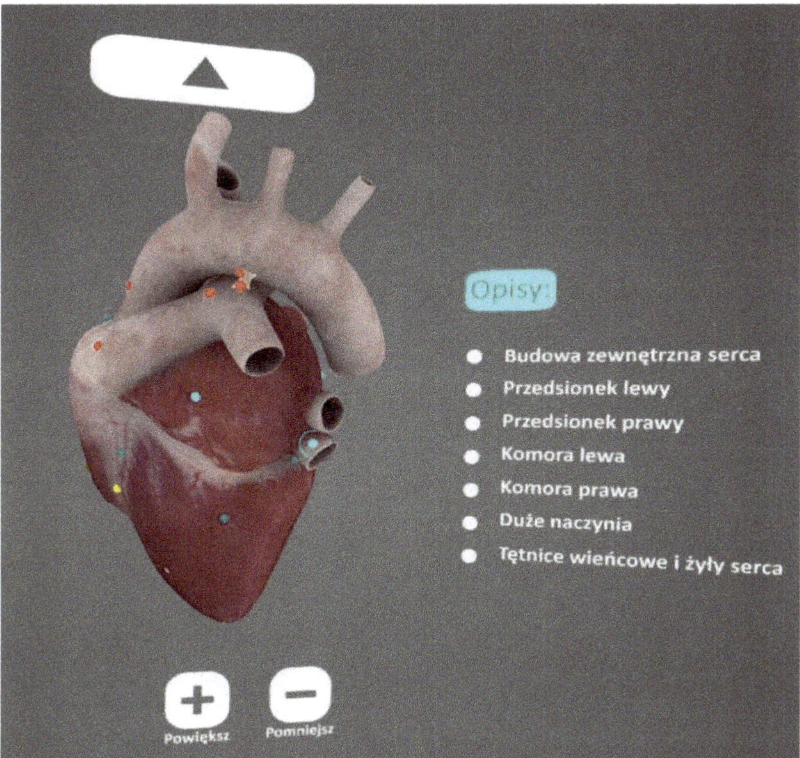

Figure 4. Using special commands on the goggles, the descriptions of anatomical details could be displayed on the side of the hologram. Manipulation of the images and anatomical descriptions is either using the sight of the user (looking at the detail activates the function) or using the remote control. Similarly to the previous figures, these written descriptions are also 3-dimensional; thus, in this 2-dimensional figure, they seem distorted. In this figure anatomical terms in the Polish language are visible. With the buttons (+) and (−), it is possible to enlarge or reduce the hologram; button (Δ) leads to the previous menu.

2.2. Study Design

The study was conducted on a group of 96 students of the medical faculty. The primary endpoint of this educational study was the assessment of the value of this e-learning tool in terms of:

- Improvement of short-term anatomical knowledge retention;
- Improvement of long-term anatomical knowledge retention.
- Secondary endpoints of this study comprised:
- Prevalence of adverse events associated with an immersive virtual reality (such as headache, vertigo, or nausea);
- Attractiveness of anatomical classes with the use of 3D holography;
- Identification of problems associated with teaching anatomy using this particular method.

For the purpose of this study, we used the Samsung Gear VR goggles for immersive virtual reality (Samsung Electronics, Suwon, Republic of Korea) and our digital heart software. The entire study has been approved by the Committee on Ethics of Scientific Investigations of our university.

Students were randomly allocated into two groups. Group 1 consisted of 57 students, who were taught the anatomy of the heart using traditional methods (plastinated human hearts, anatomical models of the heart, and anatomical atlases). Academic teachers supervising the classes in Group 1, similarly to their students, had no previous contact with the 3D holographic heart. Group 2 comprised 39 students who were taught using our 3D holographic heart. In both groups, the classes on heart anatomy were performed in 8–10 person subgroups and lasted 3 h. It should be emphasized, however, that in Group 2 the use of goggles was not strictly supervised by academic teachers, and students were allowed to study 3D holograms on their own, while the teachers helped them to use the goggles properly, but did not focus on anatomical aspects of the holograms. This primarily resulted from the fact that this e-learning tool was new, both for the students and academic teachers, and at that time it remained unclear how such teaching should be conducted to achieve the desired educational effects. After the classes students from Group 2 were asked if there were any unpleasant sensations associated with the use of goggles. Of note is that the above-described classes on heart anatomy, both with the use of 3D holography and the traditional ones, were performed a few weeks after standard anatomical classes on this topic and were seen as additional ones. At our medical university, the approach to anatomical curriculum is the regional one. The anatomy of the heart is taught within the block dedicated to chest anatomy.

An assessment of anatomical knowledge retention by the students consisted of four standard anatomical tests; each of them consisted of 40 questions that evaluated students' knowledge. The first test regarded anatomical knowledge not associated with the heart (anatomy of the extremities). It was conducted about one month before the classes on the anatomy of the heart and served the purpose of comparing the groups. Then, there were three tests assessing knowledge of heart anatomy. The first one was conducted one week after the classes and served the purpose of the assessment of short-term knowledge retention. Other tests, utilizing the same questions, were performed 3 and 6 months after the classes on heart anatomy. They were aimed at the assessment of long-term anatomical knowledge retention. After 6 months the students from both groups were also asked about their opinions on the benefits and obstacles associated with the use of 3D holograms.

For the assessment of the results of this study, another test evaluating students' knowledge, which initially was not a part of the protocol, was also utilized. It was a test comprising 100 questions regarding the entire human anatomy, which was performed about 3 months after the classes on heart anatomy. The reason why the results of this test were also taken into account will be discussed in the Results section of this paper.

2.3. Statistical Analysis

The two-sample t-test was used to test the null hypothesis that the anatomical knowledge retention in both groups of students was equal, against the alternative hypothesis that there were significant differences between study groups. In addition, the F-test was used to find out whether there was a significant difference between variances within the groups. The significance of the p values was set at $p < 0.05$. This statistical analysis was performed using the PAST data analysis package (version 2.09; University of Oslo, Oslo, Norway).

3. Results

The results of the tests performed are presented in Table 1. While all students (i.e., 57 students from Group 1 and 39 students from Group 2) were present during the first test on heart anatomy, for different reasons some students were missing during other tests. Still, the number of missing students was not high (see: Table 1). The internal consistency of the test on heart anatomy was assessed with Cronbach's Alpha index, and the average value of the three examinations performed was 0.89, indicating its good consistency.

Table 1. Results of anatomical tests; *p* values considered statistically significant if <0.05.

		Group 1 (Traditional Teaching)	Group 2 (3D Holography)	*p* Value (Two-Sample t-Test)	*p* Value (F-Test)
Test not associated with heart anatomy	number of students	56	36	-	-
	mean result	27.8	27.3	0.72	-
	95% confidence interval	26.0–29.7	24.9–29.7	-	0.95
Test on heart anatomy after 1 week	number of students	57	39	-	-
	mean result	25.1	20.4	0.02	-
	95% confidence interval	22.5–27.8	17.6–23.1	-	0.34
Test on heart anatomy after 3 months	number of students	55	38	-	-
	mean result	24.5	20.6	0.02	
	95% confidence interval	22.8–26.3	17.2–23.9	-	0.002
Test on heart anatomy after 6 months	number of students	46	34	-	-
	mean result	22.4	18.5	0.01	-
	95% confidence interval	20.1–24.7	16.7–20.4	-	0.03

An initial test on the anatomy of the extremities revealed similar anatomical knowledge in both groups (Table 1), which indicated that the groups were comparable. Regarding further tests, unexpectedly, the students who were taught using 3D holography performed significantly worse than those who had their classes on heart anatomy with the use of traditional educational tools. These worse results were seen during all three tests on heart anatomy. However, statistical analysis of the results with the F-test has also revealed significant differences between the groups regarding variances. Consequently, the results of another test that was performed about 3 months after the classes on heart anatomy were also taken into account. This test, consisting of 100 questions on the entire human anatomy, revealed a trend ($p = 0.06$) towards worse results in Group 2. The mean difference in students' performance during this test was at the level of 10%, in favor of Group 1. Since this statistical analysis suggested that the groups were not actually equal (on average, there were stronger students in Group 1 and weaker ones in Group 2), we adjusted the results of the tests on heart anatomy, taking into account the difference revealed by the test on entire human anatomy. After this adjustment, there were no statistically significant differences between the groups (Table 2).

Analysis of the secondary endpoints of this study revealed that only one student from the Group 2 (2.6%) complained of a headache during the use of goggles. This symptom, although mild, was likely to be associated with immersive virtual reality. There we no other adverse events reported by students. All students found the 3D holography as an attractive educational method. The possibility to study the organ in three dimensions, to rotate it, or to cross-sect the heart, was particularly seen as an advantage of this learning tool. However, some students found the navigation of holograms difficult. It was especially troubling for those presenting with mixed astigmatism, where the proper use of immersive virtual reality goggles was not possible. In addition, students felt that in order to fully benefit from this new educational method more time than just a 3 h class was needed.

Table 2. Results of the tests on heart anatomy after adjustment according to the results of the test on the entire human anatomy.

		Group 1	Group 2 (Adjusted Results)	p Value (Two-Sample t-Test)	p Value (F-Test)
Test on heart anatomy after 1 week	number of students	57	39	-	-
	mean result	25.1	22.4	0.17	-
	95% confidence interval	22.5–27.8	19.4–25.4	-	0.75
Test on heart anatomy after 3 months	number of students	55	38	-	-
	mean result	24.5	22.6	0.29	
	95% confidence interval	22.8–26.3	18.9–26.3	-	0.0002
Test on heart anatomy after 6 months	number of students	46	34	-	-
	mean result	22.4	20.4	0.20	-
	95% confidence interval	20.1–24.7	18.3–22.4	-	0.09

4. Discussion

In this pilot study, it has been demonstrated that although anatomical classes with the use of immersive virtual reality can be attractive for students, unsupervised teaching with the use of 3D holograms was still not superior to traditional medical education. We have also found that this educational method was relatively safe, but in some individuals, adverse events, such as headaches, can occur. In addition, some students with astigmatism could not fully benefit from this method, at least using currently available virtual reality goggles.

In spite of the possible great potential of immersive virtual reality in the medical curriculum, currently, this educational method is used by a minority of medical universities. It is primarily related to the lack of evidence demonstrating its educational efficacy. Unfortunately, at the moment only a few studies comparing traditional teaching with new 3D modalities have been published [13,17]. Moreover, the results of these studies are not congruent. Moro et al., who compared virtual reality with augmented reality and 3D tablet application for teaching the anatomy of the skull, did not reveal significant differences between study groups in terms of anatomical knowledge retention, although students found virtual reality to be a very attractive method [18]. Additionally, these authors found quite a high prevalence of adverse events (headache, drowsiness, fatigue and general discomfort) associated with the use of virtual reality. Similarly, Stepan et al. found the 3D virtual reality to be equally effective as the traditional methods during learning neuroanatomy [19]. Hackett et al. demonstrated a better approach to learning heart anatomy after the classes with the use of 3D holograms, in comparison with monoscopic 3D visualizations and 2D printed images of the heart [12]. This study, however, was conducted on a group of nursing students. Thus, detailed knowledge of heart anatomy (at the level required from future doctors) was probably not evaluated. In another study, Weinman et al. found that traditional teaching of the female pelvis with the use of physical anatomical models was better than 3D visualizations [20]. Yet, in this study 3D images were displayed on computer screens; thus, it was not a real 3D immersive virtual reality method. Of note is that it has already been revealed that 3D computer models projected on flat screens are actually perceived by students as two-dimensional images. Consequently, learning anatomy using such 3D models can be inferior to learning by utilizing anatomical models or cadaver specimens, which are really three-dimensional [14]. In this context, it remains unclear whether 3D holograms are actually perceived by students as 3D objects since holograms cannot be touched, but touch is probably an important part of learning. An interesting observation comes from the study performed by Miller. He found that the learning benefits associated with the use of 3D holography were primarily regarding weak students, while there was no additional benefit of this method among strong students [21]. He suggested that this educational tool can be of particular value for those medical students who are challenged to learn a lot of material using traditional methods, and consequently, are performing worse in comparison with their better-skilled peers.

Our preliminary educational study on the use of our 3D holographic heart for teaching anatomy indicates that in order to achieve better knowledge retention and understanding of the anatomy of this organ by students, anatomical classes should probably be strictly supervised by academic teachers. The same conclusion comes from the pilot study by Fairén et al. [11]. They found that the real-time interplay between medical students and teachers is of crucial importance while studying anatomy using this new educational tool. Consequently, in the context of the study by Miller [21], perhaps this novel didactic modality should be primarily offered to weaker students, though not for them to perform at home, but rather under the supervision of the academic teachers.

Of note is that students should get familiar with the use of goggles for immersive virtual reality before anatomical classes. They should learn how to move, rotate and increase objects, or activate special functions of the virtual application. Testing a simpler virtual anatomical application could probably serve this purpose. Additionally, classes on heart anatomy with the use of virtual reality should be precisely planned. Anatomical problems that could be easier explained with 3D holograms in comparison with traditional educational methods should be identified, including, for example, the shape and topography of coronary arteries in relation to the topography of heart chambers, or blood supply provided by a particular coronary branch in relation to its topography. Hopefully, such designed anatomical classes, with the participation of students who are familiar with the goggles, would result in better learning of human anatomy. Yet, whether such a goal is actually achievable should be validated by the next phase of our study.

We acknowledge that there are some limitations to our study. Firstly, the number of our students was not very high. Secondly, in our study, students' anatomical knowledge was assessed through tests. A practical examination of gross anatomy, as well as the radiological anatomy of the heart, would undoubtedly provide more information on the actual didactic value of 3D holography. Thirdly, in this study, the use of 3D holography was not supervised and moderated by academic teachers. Such a moderation of a new didactic modality is desirable and should be included in future studies on this method.

5. Conclusions

Anatomical classes with the use of 3D immersive virtual reality, although attractive for medical students, are not superior to traditional learning in terms of knowledge retention, if such classes are not precisely designed, not strictly supervised by academic teachers and the students did not get familiar with the use of virtual reality goggles before anatomical classes.

Author Contributions: Conceptualization, Z.H. and M.S. (Marian Simka); methodology, J.C., J.H., and M.S. (Marcin Skuła); software, Z.H.; validation, J.C., Z.H., M.S. (Marian Simka) and M.N.; formal analysis, M.S. (Marian Simka); investigation, J.C., J.H., D.K., W.R. and M.S. (Marcin Skuła); resources, Z.H.; data curation, M.S. (Marian Simka), J.C., and M.N.; writing—original draft preparation, M.S. (Marian Simka); writing—review and editing, M.S. (Marcin Skuła) and M.S. (Marian Simka); visualization, Z.H. and M.S. (Marian Simka); supervision, M.S. (Marian Simka); project administration, Z.H. and M.S. (Marian Simka). All authors have read and agreed to the published version of the manuscript.

Funding: This research received no external funding.

Institutional Review Board Statement: Not applicable.

Informed Consent Statement: Not applicable.

Data Availability Statement: The data presented in this study are available on request from the corresponding author. The data are not publicly available due to privacy restrictions.

Acknowledgments: The authors wish to thank Marcin Adamczyk, the IT specialist, for his help in preparing the software and 3D holography goggles for the classes. The authors wish to thank our collaborators: the Professor Zbigniew Religa Foundation of Cardiac Surgery Development (Zabrze, Poland) and the Farm 51 Group SA (Gliwice, Poland), a technological company specializing in virtual reality applications.

Conflicts of Interest: The authors declare no conflict of interest.

References

1. Condino, S.; Turini, G.; Parchi, P.D.; Viglialoro, R.M.; Piolanti, N.; Gesi, M.; Ferrari, M.; Ferrari, V. How to build a patient-specific hybrid simulator for orthopaedic open surgery: Benefits and limits of mixed-reality using the Microsoft HoloLens. *J. Healthc. Eng.* **2018**, *2018*, 5435097. [CrossRef] [PubMed]
2. García-Vázquez, V.; von Haxthausen, F.; Jäckle, S.; Schumann, C.; Kuhlemann, I.; Bouchagiar, J.; Höfer, A.-C.; Matysiak, F.; Hüttmann, G.; Goltz, J.P.; et al. Navigation and visualisation with HoloLens in endovascular aortic repair. *Innov. Surg. Sci.* **2018**, *3*, 167–177. [CrossRef]
3. Hanna, M.G.; Ahmed, I.; Nine, J.; Prajapati, S.; Pantanowitz, L. Augmented reality technology using Microsoft HoloLens in anatomic pathology. *Arch. Pathol. Lab. Med.* **2018**, *142*, 638–644. [CrossRef]
4. Jang, J.; Tschabrunn, C.M.; Barkagan, M.; Anter, E.; Menze, B.; Nezafat, R. Three-dimensional holographic visualization of high-resolution myocardial scar on HoloLens. *PLoS ONE* **2018**, *13*, e0205188. [CrossRef] [PubMed]
5. Siff, L.N.; Mehta, N. An interactive holographic curriculum for urogynecologic surgery. *Obstet. Gynecol.* **2018**, *132* (Suppl. 1), 27S–32S. [CrossRef] [PubMed]
6. Tepper, O.M.; Rudy, H.L.; Lefkowitz, A.; Weimer, K.A.; Marks, S.M.; Stern, C.S.; Garfein, E.S. Mixed reality with HoloLens: Where virtual reality meets augmented reality in the operating room. *Plast. Reconstr. Surg.* **2017**, *140*, 1066–1070. [CrossRef]
7. Wong, K.; Yee, H.M.; Xavier, B.A.; Grillone, G.A. Applications of augmented reality in otolaryngology: A systematic review. *Otolaryngol. Head Neck Surg.* **2018**, *159*, 956–967. [CrossRef] [PubMed]
8. Azer, S.A.; Azer, S. 3D anatomy models and impact on learning: A review of the quality of the literature. *Health Prof. Educ.* **2016**, *2*, 80–98. [CrossRef]
9. Montes, W.B.; Gómez, M.G. Implementar la realidad virtual en la enseñanza de anatomía una necesidad en la formación de profesionales de la salud. *Morfolia* **2021**, *13*, 11–18.
10. Chytas, D.; Johnson, E.O.; Piagkou, M.; Mazarakis, A.; Babis, G.C.; Chronopoulos, E.; Nikolaou, V.S.; Lazaridis, N.; Natsis, K. The role of augmented reality in anatomical education: An overview. *Ann. Anat.* **2020**, *229*, 151463. [CrossRef]
11. Fairén, M.; Moyés, J.; Insa, E. VR4Health: Personalized teaching and learning anatomy using VR. *J. Med. Syst.* **2020**, *44*, 94. [CrossRef]
12. Hackett, M.; Proctor, M. The effect of autostereoscopic holograms on anatomical knowledge: A randomised trial. *Med. Educ.* **2018**, *52*, 1147–1155. [CrossRef]
13. Hackett, M.; Proctor, M. Three-dimensional display technologies for anatomical education: A literature review. *J. Sci. Educ. Technol.* **2016**, *25*, 641–854. [CrossRef]
14. Romaniuk, M.; Lamb, J.; Bayer, J.; Bayer, I.; Wainman, B. The promise of mixed reality in anatomy education. *FASEB J.* **2017**, *31*, 736.6.
15. Yong, V.; Sridharan, P.; Ali, S.A.; Tingle, G.; Enterline, R.; Ulrey, L.; Tan, L.; Eastman, H.; Gotschall, R.; Henninger, E.; et al. Cadaver vs. Microsoft HoloLens: A comparison of educational outcomes of a breast anatomy module. *FASEB J.* **2018**, *32*, 635.6. [CrossRef]
16. Chen, S.; Zhu, J.; Cheng, C.; Pan, Z.; Liu, L.; Du, J.; Shen, X.; Shen, Z.; Zhu, H.; Liu, J.; et al. Can virtual reality improve traditional anatomy education programmes? A mixed-methods study on the use of a 3D skull model. *BMC Med. Educ.* **2020**, *20*, 395. [CrossRef] [PubMed]
17. Preim, B.; Saalfeld, P. A survey of virtual human anatomy educational systems. *Comput. Graph.* **2018**, *71*, 132–153. [CrossRef]
18. Moro, C.; Štromberga, Z.; Raikos, A.; Stirling, A. The effectiveness of virtual and augmented reality in health sciences and medical anatomy. *Anat. Sci. Educ.* **2017**, *10*, 549–559. [CrossRef]
19. Stepan, K.; Zeiger, J.; Hanchuk, S.; Del Signore, A.; Shrivastava, R.; Govindaraj, S.; Iloreta, A. Immersive virtual reality as a teaching tool for neuroanatomy. *Int. Forum Allergy Rhinol.* **2017**, *7*, 1006–1013. [CrossRef] [PubMed]
20. Weinman, B.; Wolak, L.; Pukas, G.; Zheng, E.; Norman, G.R. The superiority of three-dimensional physical models to two-dimensional computer presentations in anatomy learning. *Med. Educ.* **2018**, *52*, 1138–1146. [CrossRef] [PubMed]
21. Miller, M. Use of computer-aided holographic models improves performance in a cadaver dissection-based course in gross anatomy. *Clin. Anat.* **2016**, *29*, 917–924. [CrossRef] [PubMed]

 **anatomia**

Article

The Rostral Epidural Rete Mirabile: Functional Significance in Blood Flow Regulatory Mechanisms in Giraffe (*Giraffa camelopardalis*)

Marna S. van der Walt [1], Willem Daffue [2], Jacqueline Goedhals [3], Sean van der Merwe [4] and Francois Deacon [1,*

[1] Department of Animal-, Wildlife- and Grassland Sciences, Bloemfontein 9301, South Africa
[2] Kroonstad Dierehospitaal, Kroonstad 9500, South Africa
[3] Department of Anatomical Pathology and National Health Laboratory Services, Bloemfontein 9301, South Africa
[4] Department of Mathematical Statistics and Actuarial Science, Bloemfontein 9301, South Africa
* Correspondence: deaconf@ufs.ac.za

Citation: van der Walt, M.S.; Daffue, W.; Goedhals, J.; van der Merwe, S.; Deacon, F. The Rostral Epidural Rete Mirabile: Functional Significance in Blood Flow Regulatory Mechanisms in Giraffe (*Giraffa camelopardalis*). *Anatomia* **2023**, *2*, 138–155. https://doi.org/10.3390/anatomia2020013

Academic Editor: Francesco Cappello

Received: 28 January 2023
Revised: 19 March 2023
Accepted: 20 March 2023
Published: 6 May 2023

Abstract: The distinctive long neck of the giraffe (*Giraffa camelopardalis*) entails functional difficulties brought about by the extended distance between the heart and the head. Blood must be circulated over 2 m from the heart to the brain against gravitational force. The natural movement of the head to ground level would result in a large volume of blood moving toward the brain with the force of gravity. Large blood volumes also rush to the brain during bulls' fighting (necking), rendering the giraffe susceptible to possible brain damage. The natural movement of the head from ground level to fully erect would result in blood moving away from the brain with gravitational force. The lack of blood perfusing the brain can cause fainting. The giraffe, however, suffers neither brain damage nor fainting. What adaptations do giraffes have to counteract these challenges? The aim of this study was to investigate the functionality of the rostral epidural rete mirabile situated just beneath the brain and its possible contribution to successful circulation in long-necked giraffes. The unique rostral epidural rete mirabile structure significantly contributes to counteract physiological challenges. Turns and bends characterize this structural arterial meshwork and subsequently an increased artery length through which blood flow must proceed before entrance into the brain, exerting resistance to blood racing to the brain when the head is lowered to the ground. The brain is supplied mainly by the maxillary artery through the carotid rete, with a rudimentary basilar artery not contributing to the brain's blood supply. The resistance to blood flow due to the structure and position of the rostral epidural rete mirabile when the head is in the upright position is counteracted by the unique carotid-vertebral anastomosis allowing immediate cerebral blood supply. The rostral epidural rete mirabile structure in giraffes is an essential feature balancing physiological difficulties arising due to the extensive heart-to-head distance and might fulfill the same function in other long-necked artiodactyls.

Keywords: giraffe; cerebral blood supply; rostral epidural rete mirabile

1. Introduction

Pragmatically, we can assume giraffes (*Giraffa camelopardalis*) experience several physiological challenges due to their extraordinary build. Detailed tests on captive giraffes showed that these animals have a very high blood pressure of approximately 200–400 mmHg [1–3]. At this pressure, certain key physiological issues arise. For instance, the heart must pump blood to the brain against gravitational pressure and would encounter vascular friction towards the brain, situated approximately 2 m above the heart. Some studies have investigated possible adaptations to the physiological challenges of giraffes [2,4,5]. Few studies describe the unified function of how giraffes successfully circulate blood, avoid fainting and blackouts, escape brain damage and prevent oedema in the legs. A unique anatomical structural differentiation called the rostral epidural rete mirabile [4,6–9] is present in Cetartiodactyla, including giraffe. The rostral epidural rete mirabile is absent in Perissodactyls, but cats and dogs also have a rostral epidural rete mirabile [10–15]. The rostral epidural rete mirabile is described as a complex network of arteries and veins situated at the base of the cranium [8,9,16,17]. The entire rete in Artiodactyls is situated intracranially, within the cavernous venous sinus, which is different from other species, such as cats and certain primates that also have a rostral epidural rete mirabile [8,15,16,18]. Alteration of the intracranial segment of the internal carotid artery developing into the rostral epidural rete mirabile network structure occurs in the embryo. In Artiodactyls, the common carotid artery transitions into the maxillary artery, from where it enters the skull through the carotid foramen and transitions into the network of the rostral epidural rete mirabile. The brain blood supply is thus different from the supply via the internal carotid and basilar artery, as in other species [6,8], to the maxillary artery connecting to the carotid arterial network in giraffes. The rostral cerebral artery and the caudal communicating artery are responsible for the main supply to the cerebral arterial circle. In giraffes, the basilar artery is undeveloped and therefore does not form part of the pathways through which blood can be supplied to the brain [7,8,19]. The basilar artery in sperm whales (*Physeter macrocephalus*) is similarly non-functional concerning the blood supply to the brain [12]. The basilar artery is prominent in dromedary camels (*Camelus dromedarius*) and other ruminants and services the medulla oblongata, pons, and cerebellum [20]. The maxillary artery is dominant in its blood supply to the circle of Willis. Still, the basilar artery in these species is directly connected to the circle of Willis, with the course of flow directed caudally and not rostral [21]. Extending from the retial meshwork, the intracranial maxillary artery branch configures into the brain's arterial circle, also known as the circle of Willis [6,17,19]. The intracranial maxillary artery segment emerging from the rete forms a split with the caudal intracranial carotid artery [15]. The arterial circle of the brain receives blood predominantly from the maxillary artery with a non-functional basilar artery that cannot be utilized to supply any substantial amount of blood to the cranium in giraffes [8,15,19,21,22]. The study of [7] on Old and New World camelids showed that the cerebral arterial blood supply originates directly from the rostral epidural rete mirabile and the basilar artery. Additionally, in contrast to an artiodactyl's, the rostral epidural rete mirabile of camelids anastomose with branches of both the maxillary and internal carotid artery. [18] have suggested three functions of the rostral epidural rete mirabile: the regulation of blood flow and pressure to encephalic circulation, temperature regulation and the movement of pheromones from venous blood originating from the nasal mucosa to the hypophysis. [23] focused on the rostral epidural rete mirabile functioning to regulate blood flow in the average bodily environment. [10,14,24] described the rete to allow adequate brain perfusion in instances of break-in blood flowing towards the brain as a direct result of the various anastomoses between the two sides of the rete. The rete thus acts as a protective measure preventing high perfusion pressure with the facilitation of increased blood flow during unusual conditions. Furthermore, the rostral epidural rete mirabile allows for adequate blood flow necessary for brain function concurrently with reducing high systolic blood pressure, influencing mean perfusion pressure minimally [24], and the lack of valves within the cranial cavity support venous blood resuming from the brain. The morphology and structural function of the rostral epidural rete mirabile may vary in different species, with a distinction in the main blood suppliers to this meshwork [25].

This study points out the structure of the rostral epidural rete mirabile and its function in the long-necked giraffe. Furthermore, this study is part of a more significant study that aims to describe the combined features that allow giraffes to successfully circulate blood to the brain, avoid fainting, escape brain damage and prevent oedema in the legs.

2. Materials and Methods

Ethics approval nr. UFS-AED2020/0083. Approval was obtained from the Animal Ethics Research Committee at the UFS, SPCA and DESTEA.

2.1. Experimental Model and Subject Details

Samples were attained from animals culled by several nature reserves in South Africa and thus on an ad hoc basis. Unfortunately, the giraffe is culled by these reserves due to overstocking concentrations and diet shortages. The study samples were opportunistically gathered when these culling operations were conducted. The researchers were not involved in organizing or carrying out the culling procedure. As part of sustainable habitat use, most game ranches manage the number of animals conferring to the farm's carrying capacity (available diet) that gets evaluated yearly. This annual management procedure includes live trade, translocations or culling the excess animals. This process is a well-thought-out recurrent management method used in the Southern Africa Wildlife Industry to maintain herbivore numbers applicable to the carrying capacity of the farm or reserve and to evade death due to starvation [26].

2.2. Animal Sources

Five giraffes (*G. camelopardalis*), female n = 3 and male n = 2, were included in this study. Giraffe males weigh around 1200 kg with a total height of approximately 5.5 m. The female giraffe weighs around 830 kg with a total height of about 4.5 m [27]. Sample animals were obtained opportunistically on an ad hoc basis.

2.3. Method Details

A giraffe head severed between C3 and C4 cervical vertebrae was obtained. The vascular system was physically washed out with warm water. The specimen samples were raised with the nose pointing down and washed until the water was clear.

2.4. Dissection

Four giraffe heads were used for latex infusion. The latex mixture combinations used for the arterial and venous systems varied. The basic ingredients included: Barium sulphate ($BaSO_4$) (X-ray grade, Kyron Powder, Kyron, SA, and Latex moulding rubber (A. Shak (Pty) Ltd., Durban, South Africa). Additionally, to distinguish and differentiate the arteries from the veins, a red (Stamp ink, Office Mate, South Africa) pigment was used to infuse arteries, and a blue pigment (Print ink, Treeline, SA) was used in combination to infuse the veins. Arteries were infused with $BaSO_4$ powder at a forty percent volume, sixty percent latex volume and two percent volume red ink. Veins were infused with $BaSO_4$ powder of a twenty percent volume, eighty percent latex volume and two percent volume of blue ink. The vascular system was thoroughly rinsed with lukewarm water and subsequently filled with the different latex mixtures. A tube (10-mm PVW piping) was positioned 2 m above the vascular system to be able to use gravitation force to aid adequate latex infusion. Sutures (Catgut 3 Kyron, South Africa) and strings were utilized to keep the tubing in place. The common carotid artery was infused with the red combination latex, and the jugular vein with the blue combination latex. The infused arteries and veins were closed with catgut (Catgut 3 Kyron, SA). To encourage latex to set, the specimens were positioned at a 30° angle for 24 h. After the latex had been set, the specimens were separated at the midpoint between the ossicones with a Recipro saw (Makita). Care was taken to divide the specimens into identical halves through the skull and neck vertebra. Dissection originated at the C3 cervical vertebra trailing the arteries and veins up to the rete mirabile and the brain. $BaSO_4$ was included in the latex mixture for future radiographic imaging, as part of a dissertation.

2.5. Brain Endocast

Two giraffe heads were used to determine brain volume. Most of the skin and meat were removed from the skulls. The skulls were then buried and left to decay naturally, allowing a completely clean skull. The two halves of each head were filled with latex molding rubber (A. Shak (Pty) Ltd., SA) and allowed to set. Each latex cast was removed and paired with the correct other halves. Brain volume was determined by submerging the endocast in a measurement jug filled with water and measuring the amount of water displaced.

2.6. Histology Sample

One giraffe head was used for dissection to obtain histological samples of the carotid rete, the artery at the origin of the carotid rete and the artery at the exit of the carotid rete. The head was divided into two symmetrical halves, cut along the midline from the middle point between the two ossicones through the skull and neck vertebra with a Makita Recipro saw. The rostral epidural rete mirabile was located above the brain plate in the cranium and then photographed with a Canon 7D camera and a 24–105 mm lens. A sample of the rostral epidural rete mirabile was taken, as well as a sample at the origin before branching into the rostral epidural rete mirabile network. A sample was also taken at the exit of the rostral epidural rete mirabile network.

The samples were then placed in a 10% buffered formalin (10% aqueous solution and 4% formaldehyde). The formalin-fixed tissue was processed overnight in a VIP6 tissue processor and was embedded manually in wax blocks. Following embedding, 2–4 µm thick sections were cut, placed on glass slides, and stained with haemotoxylin and eosin (H&E) and Masson's Trichrome (for muscle and collagen fibers) and Verhoeff–Van Gieson (for elastic fibers) using standard methods. The stained slides were then covered, slipped and evaluated. Pictures were taken from the section pre-carotid rete, post-carotid rete and sections within the rostral epidural rete mirabile at a 2.5× magnification with a Leica DM750 microscope with a Leica ICC50 W camera. The Tunica media of the arteries were measured using the microscope.

2.7. Pressure and Flow Tempo Experiment

We mimicked a basic rete network to determine what influence the number of turns or spirals in the rostral epidural rete mirabile structure might have on flow tempo. The serial setup of the experimental flow test, even though a limitation to the actual anatomy of the carotid rete, is utilized to represent the influence of the rete structure. A rostral epidural rete mirabile network was mimicked using 4-mm diameter plastic fish tank air pump tubing. The tubing was measured into 0.5 m, 1 m, 2 m, 3 m, 4 m, 5 m and 6-m lengths and then winded around a standard 25-mm diameter plastic pipe to imitate the rostral epidural rete mirabile (Figure 1a,b). Plastic fish piping was fixed against a wall, giving a length of 4750 mm. Water, at room temperature, was used. Connections were constructed in a manner so that water could be opened with a switch and for the water to be pumped through the rostral epidural rete mirabile and upwards to the height of 4750 mm. The water flow was opened for one minute. The amount of water that ran through the system was measured. The change in water level before and after the rete mirabile was measured after the minute (Figure 2A). The experiment was repeated five times for each of the different lengths that simulated the carotid rete.

Flow per minute was calculated by averaging the five measurements for each rostral epidural rete mirabile length. The pressure was calculated by using the same experimental setup. A pressure meter (Yoto PG802C-3-10KPA,2017101312,0-10KPA,0.25%FS) was inserted after the rostral epidural rete mirabile to measure pressure. The pressure difference was calculated using the pressure measured before the flow entered the rostral epidural rete mirabile network and after the flow through the rete mirabile network. The pressure measured post flow through the rostral epidural rete mirabile was subtracted from the pressure measured prior to entering the rete. The pressure before 4750 mmH$_2$O measured height gives the water column in meters, from where the pressure was calculated and expressed in mmHg. The five pressure measurements were again averaged for each rostral epidural rete mirabile length (Figure 2B).

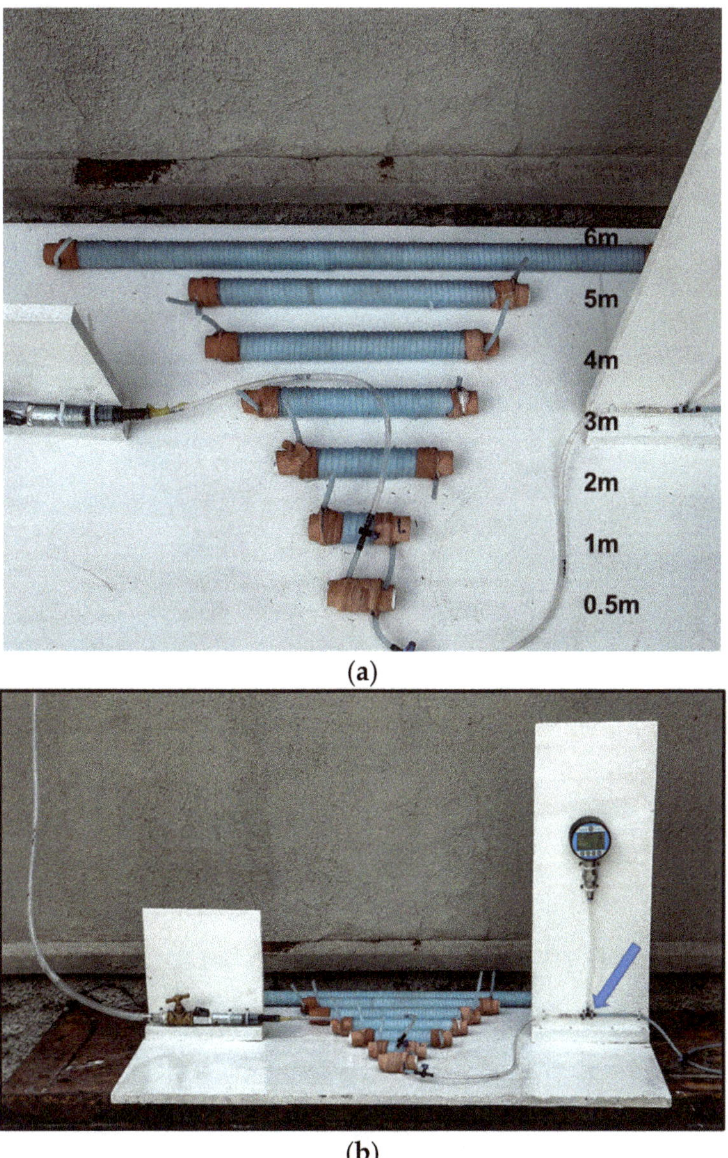

(a)

(b)

Figure 1. (**a**). Mimicked rostral epidural rete mirabile structures. (**b**). Mimicked rostral epidural rete mirabile network with pressure meter attached. Pressure is measured at the point indicated by the blue arrow.

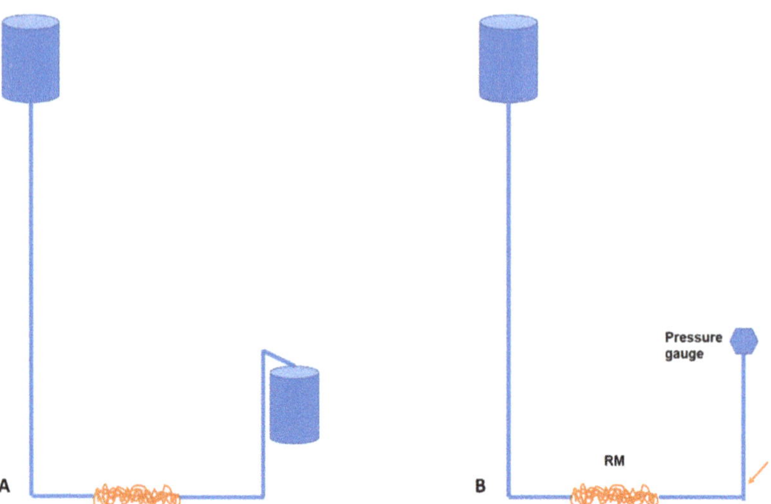

Figure 2. Schematic illustration of water flow (**A**) and pressure (**B**) experiment. Pressure is measured at the point indicated by the orange arrow.

3. Results

3.1. Dissection

During the dissection procedure, we first identified the exact location of the rostral epidural rete mirabile within the giraffe's skull. The location of the rostral epidural rete mirabile is in the cavernous venous sinus, above the brain plate within the cranium (Figure 3a,b).

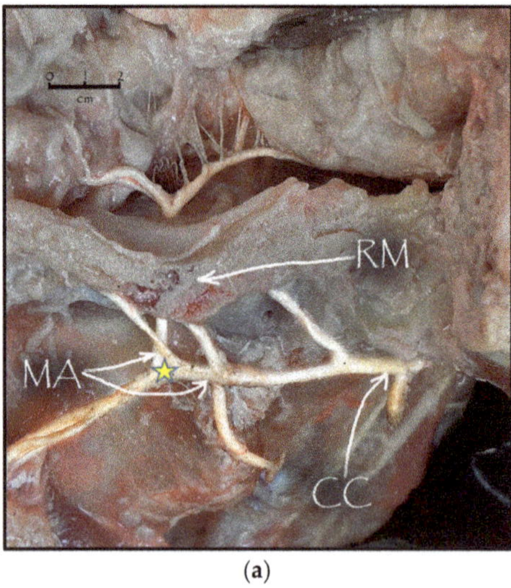

(**a**)

Figure 3. *Cont.*

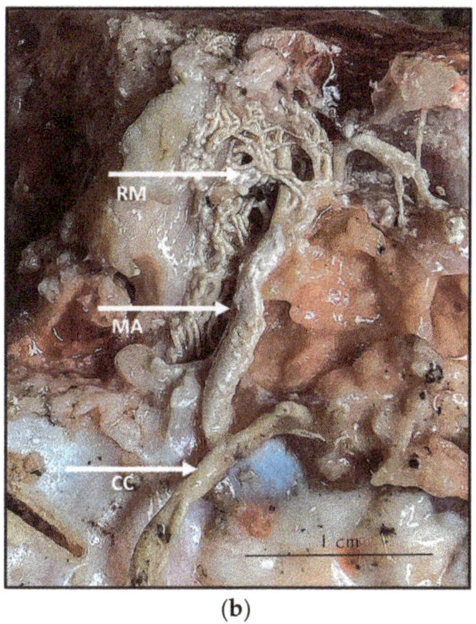

(**b**)

Figure 3. (**a**). The rostral epidural rete mirabile network is located in the cavernous venous sinus, above the brain plate within the cranium illustrating the rostral epidural rete mirabile (RM), the maxillary artery (MA) and the common carotid (CC). Notable bend towards the RM indicated by a yellow star. (**b**). Image illustrating rostral epidural rete mirabile (RM) structure below the giraffe brain, the maxillary artery (MA) and the common carotid (CC).

Upon dissection of the latex-filled arteries, the arterial pathway was followed from the C3 cervical vertebra to where the maxillary arteries enter the brain via the carotid foramina. The internal carotid artery forms the rostral epidural rete mirabile, the external carotid artery transitions into the maxillary artery. The pathway of the common carotid artery, with its transition to the external carotid artery and transition to the maxillary artery, is observed as a relatively straight pathway without bends. An interesting sharp bend is present before the entrance of the maxillary artery into the rostral epidural rete mirabile network (Figure 3a). Even though the latex injection did not perfuse much of the rostral epidural rete mirabile network, the network could clearly be identified, irrespective of size limitation, as visible in Figure 3b. Having access to the complete giraffe, we observed that the heart is situated anteriorly in the chest compared to the heart of other ruminants, such as elephants and sable antelope (Figure 4). The pathway of the blood from the heart to the brain is straight to allow for the easy and quick transport of blood without the resistance exerted by a curve or bend in the pathway.

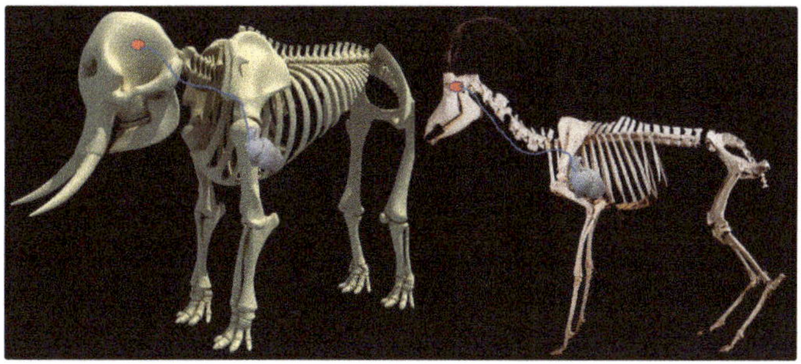

Figure 4. *Cont.*

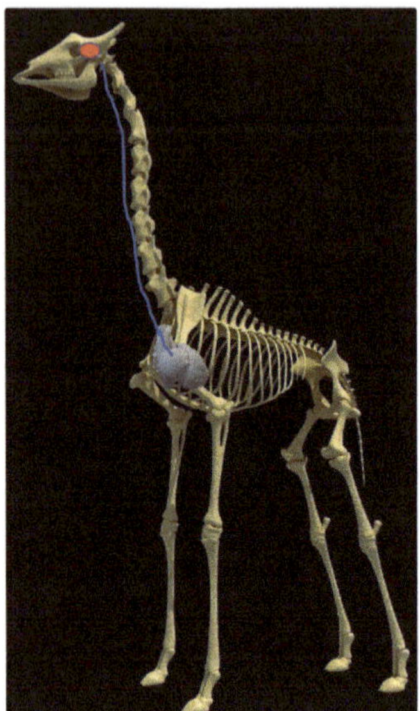

Figure 4. Position of the heart in a sable antelope (*Hippotragus niger*), elephant (*Loxodonta africana*) and giraffe (*G. camelopardalis*), with the blood flow pathway illustrated by the blue arrows, emphasizing the giraffe heart situated anteriorly in the chest.

We also confirmed that the basilar artery in the giraffe, unlike in other mammals, does not offer a second supply of blood to the brain [8,15,19,22]. The maxillary artery supplies the circle of Willis via the rostral epidural rete mirabile in the giraffe.

3.2. Brain Volume Endocast

The brain endocasts revealed that the male giraffe's volume was 900 mL and the female giraffe's 700 mL. In most mammals, the total volume of blood is 5.5–8% of body mass [28]. For our calculations, we will use an average of 6.75%. In humans, the total blood volume in the body is 5.7 L of body mass in males and 4.3 L in females on average [29]. The human brain weighs about 1500 g with 100–130 mL of blood present at any moment in the brain [30]. We used the average of 115 mL of blood at a specific time in a 1500-g brain for our calculations. Blood flow through the brain is thus 60 mL of blood/100 g/min, with a total of 900 mL of blood per minute in the brain of humans. In humans, blood needed in the brain equates to 2.02% of the total blood volume.

In this study, we found that for giraffes, the males have 69 mL of blood, and the females have 43 mL of blood present at any moment in the brain. This is notably a small volume compared to humans and other mammals. However, blood flow is relatively high, with 540 mL of blood per minute in the brain of a male giraffe and 420 mL of blood per minute in the brain of a female giraffe. If the brain's blood percentage is calculated in male (0.089%) and female (0.077%) giraffes, it equates to less than 0.01% of the total blood volume. Giraffes thus need an average of 24.46 times less blood in the brain, as a proportion of the total supply, compared to humans. Therefore, only a small amount of blood, 0.01%, must be present in the giraffe's brain at a specific time. Consequently, the physiological stress of large amounts of blood that need to be moved over the long head-to-heart distance to supply to the brain in the giraffe is reduced.

3.3. Histology

In Figure 5, an illustration of the section of the pre-carotid rete, within the carotid rete, and post-rostral epidural rete mirabile is shown. Pictures were taken from the Verhoeff–Van Gieson-stained sections at a 2.5× magnification with a Leica DM750 microscope with a

Leica ICC50 W camera. Measurements with the Leica DM750 microscope of the Tunica media of each of the sections are shown in Table 1.

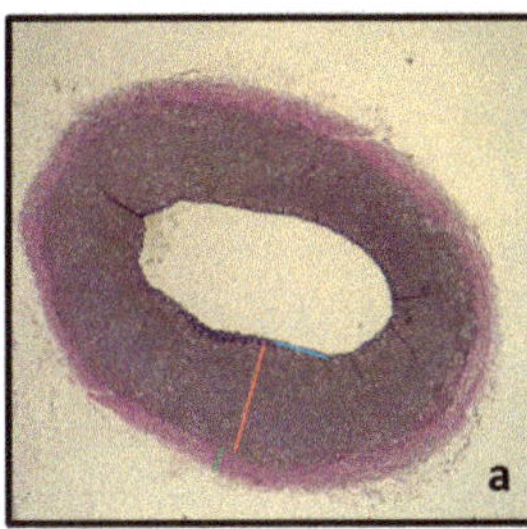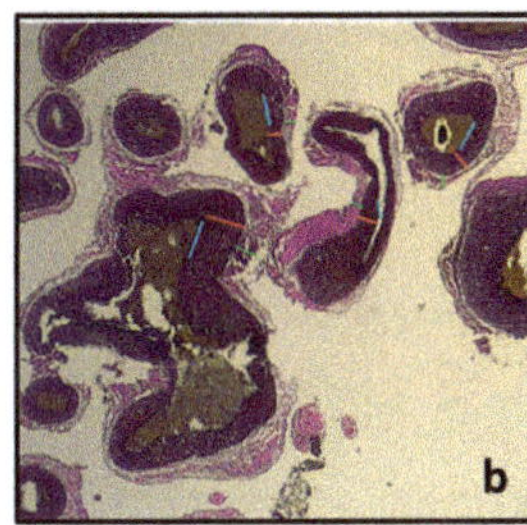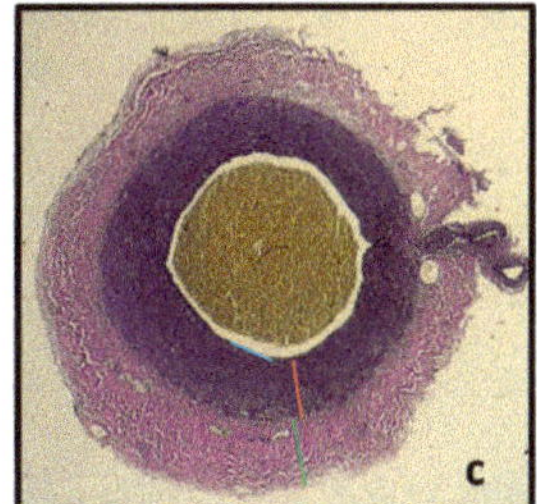

Figure 5. Section of the maxillary artery before entrance into the rostral epidural rete mirabile (**a**) showing the rostral epidural rete mirabile arteries (**b**) and the intracranial maxillary artery (**c**) with the T. intima (blue line), T. media (red line) and T. adventitia (green line).

Table 1. Measurements of the Tunica media, T. intima and T. adventitia of the sections before the carotid rete, in the carotid rete and after the rostral epidural rete mirabile and the thickness of the T. media as a percentage of the total thickness of the vessel.

	T. Media	T. Media + T. Intima	T. Media + T. Intima + T. Adventitia	T. Media Thickness: Total Thickness (%)
Pre-carotid rete	0.53–0.60 mm	0.55–0.63 mm	0.63–0.73 mm	83%
Carotid rete	0.09–0.18 mm	0.10–0.19 mm	0.15–0.28 mm	62%
Post-carotid rete	0.28–0.33 mm	0.30–0.35 mm	0.43–0.68 mm	54%

From Table 1, it is evident that the Tunica media of the maxillary artery before it enters the rostral epidural rete mirabile and the tunica media of the intracranial maxillary artery at the exit of the rostral epidural rete mirabile are approximately three times thicker in comparison to the T. media of the rostral epidural rete mirabile arteries. We expected that the pre-carotid vessel's tunica media would be larger than the T. media of the rete for more regulation pre-rete.

3.4. Pressure and Flow Experiment

Flow rate:

The average flow rate of water through each different length of simulated rostral epidural rete mirabile is shown in Table 2. This table clearly illustrates that the flow rate decreased when the rostral epidural rete mirabile length increased. Table 3 confirms this by linear regression statistics through various lengths of the carotid rete network.

Table 2. In one minute, the average flow rate through the simulated rostral epidural rete mirabile network of varying lengths expresses as milliliters per minute.

CR Length (m)	mL Per Minute (mL/min)
6	275.00
5	400.00
4	400.40
3	428.00
2	510.40
1	575.00
0.5	647.80

Table 3. Linear regression statistics for flow rate through variable lengths of rostral epidural rete mirabile.

Term	Estimate	Std. Error	*t*-Statistic	*p*-Value
Intercept	643.347	9.682	66.450	$p < 0.001$
Length	−58.922	2.682	−21.97	$p < 0.001$

Linear regression was performed through the measurements. The fit of the measurements is illustrated in Figure 6.

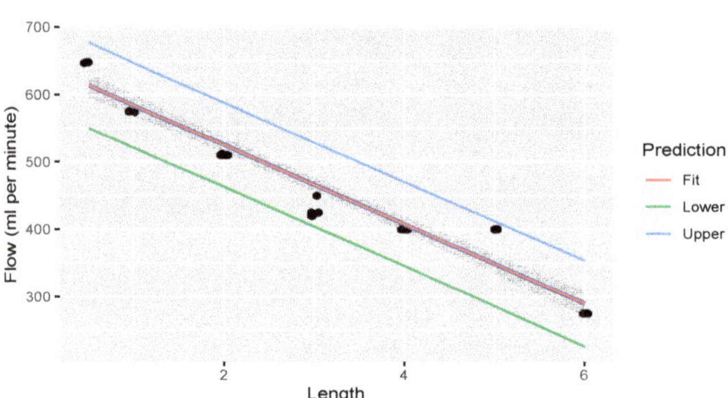

Figure 6. Linear regression through measurements illustrating flow rate decreasing with an increase in rostral epidural rete mirabile length. The shaded area indicates the 95% confidence region of the regression line, and the outer lines indicate the 95% prediction interval for observations.

The average pressure of water through each different length of rostral epidural rete mirabile is shown in Table 4. Pressure decreased as the rete length increased.

Table 4. Average pressure measured post the rostral epidural rete mirabile network of varying lengths in one minute, expressed as millimeters of mercury, at the head-up and head-down positions.

Rete Length (m)	Average Pressure (mmHg) Head-Up	Average Pressure (mmHg) Head-Down
6	313.53	157.51
5	298.52	150.01
4	286.52	142.51
3	277.52	136.51
2	264.02	133.51
1	256.52	127.51
0.5	238.52	120.01

Linear regression was performed through the measurements. The fit of the measurements is illustrated in Figure 7a,b. Table 5 illustrates linear regression statistic of pressure at variable lengths of rostral epidural rete mirabile, for the head-up position, with pressure starting at 330.027 mmHg. Table 6 illustrate linear regression statistic of pressure at variable lengths of rostral epidural rete mirabile, for the head-down position, with pressure starting at 180.015 mmHg.

From the data obtained in the experiment, we found that both the flow rate and pressure were indirectly proportional to rete length. The meshwork structure of numerous arteries of the rostral epidural rete mirabile reduces the flow rate of blood and blood pressure as it moves through the network to subsequently enter the brain.

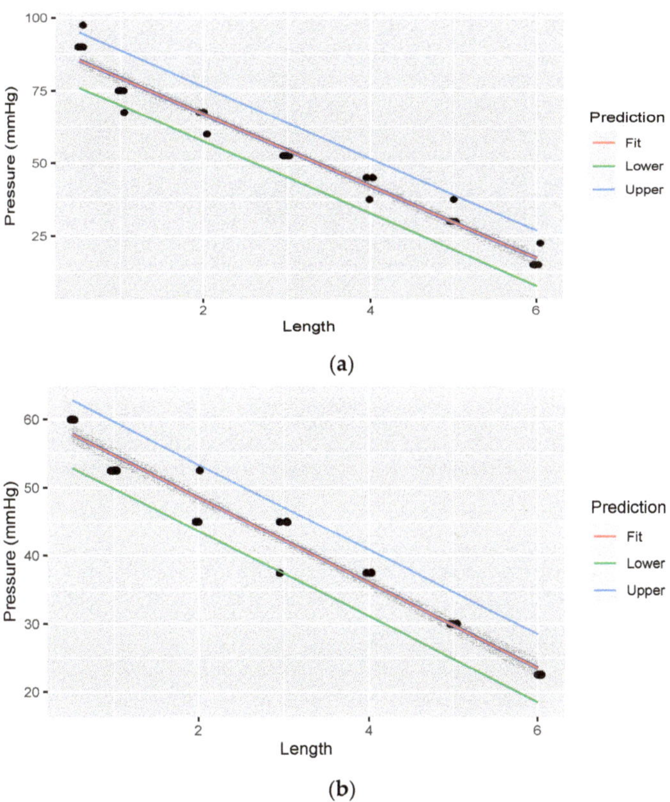

Figure 7. (**a**). Linear regression through measurements illustrating pressure decreasing with an increase in rostral epidural rete mirabile length, with head-up, with pressure measured post the rostral epidural rete structure. The shaded area indicates the 95% confidence region of the regression line, and the outer lines indicate the 95% prediction interval for observations. (**b**). Linear regression through measurements illustrating pressure decreasing with an increase in rostral epidural rete mirabile length, with head-down, with pressure measured post the rostral epidural rete structure. The shaded area indicates the 95% confidence region of the regression line, and the outer lines indicate the 95% prediction interval for observations.

Table 5. Linear regression statistic of pressure at variable lengths of rostral epidural rete mirabile, for the head-up position, with pressure starting at 330.027 mmHg.

Term	Estimate	Std. Error	t-Statistic	p-Value
Intercept	91.650	1.451	63.184	$p < 0.001$
Length	−12.396	0.402	−30.856	$p < 0.001$

Table 6. Linear regression statistic of pressure at variable lengths of rostral epidural rete mirabile, for the head-down position, with pressure starting at 180.015 mmHg.

Term	Estimate	Std. Error	t-Statistic	p-Value
Intercept	60.963	0.756	80.668	$p < 0.001$
Length	−6.243	0.209	−29.825	$p < 0.001$

The implication is that the structure of the rete only functions when blood flows. The pressure before, within, and after the rete is equal if there is no flow. Pressure decrease occurs due to (a) a decrease in the diameter of the vessel it flows through and (b) resistance against the inside area of the vessel it flows through. The resistance depends on the (a) change in vessel wall diameter and (c) the change in momentum of laminar flow.

In the rete, the resistance does not change due to a decrease in vessel wall diameter, as shown in Table 1. An example of a pressure drop due to a decrease in vessel wall diameter is in the tibialis artery, where the pressure drops due to the reduction in flow. We calculated the area of a 10-mm vessel (Table 7) by the following equation:

$$A = \pi r^2$$

where:

A = Area
π = 3.14
r = radius

Table 7. Area of a 10 mm-vessel with a constant radius.

Radius	Area (mm^2)	Number of Vessels	Total Area (mm^2)
5	78.54	1×10	78.54
2.5	19.64	2×5	39.28
2.5	19.64	4×5	78.56

We can confidently say that the rete arteries bifurcate in such a way that the total area of each vessel increases; the pressure drop and then the volume or flow does not change (Figure 8a,b).

The following equation was used to calculate the circumference of the artery (Table 8):

$$C = 2\pi r^2$$

where:

C = circumference
π = 3.14
r = radius

Table 8. The circumference of a 10-mm vessel with a constant radius.

Radius	Circumference	Number of Vessels	Total Circumference
5	31.42	$\times 1$	31.42
2.5	15.70	$\times 4$	62.80

These data suggest that, consequently, the resistance due to the friction of the vessel wall is double in the bifurcation. An area offers resistance in the vessel through which blood flows, doubles and thus gives the friction coefficient, resulting in a pressure drop.

This, in turn, would mean that the resistance in the rete is influenced by the viscosity of the blood or fluid and the resistance exerted by the vessel wall.

The rostral epidural rete mirabile structure, with its numerous bends and turns, influences pressure by the law of laminar flow. The following factors all affect the flow and thus the pressure in a vessel:

$$\Delta P = \frac{1}{2}f_s pu^2 \frac{\pi R b}{D}\, \theta/180° + \frac{1}{2}Kbpu^2$$

where:

f_s = Moody friction factor
p = density
u = mean flow velocity

Rb = bend radius
D = tube diameter
Θ = bend angle
Kb = bend loss coefficient

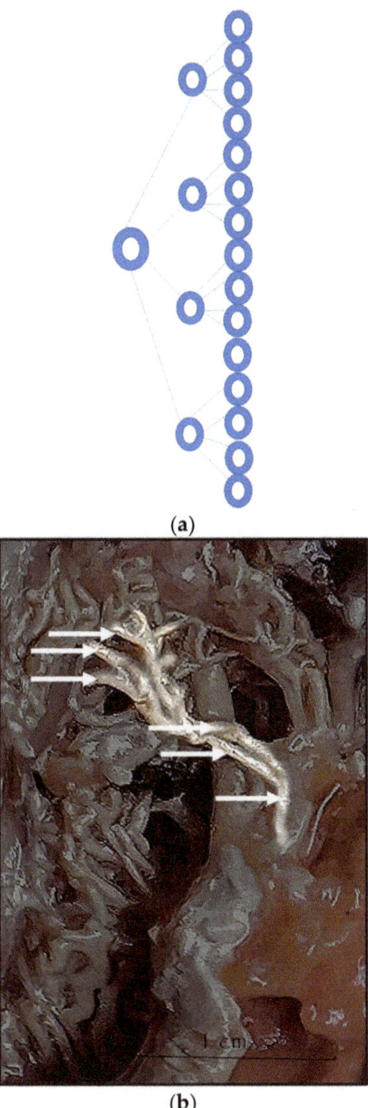

(a)

(b)

Figure 8. (**a**). Schematic presentation of bifurcation of the rostral epidural rete mirabile arteries. (**b**). Highlighted arteries of the rostral epidural rete mirabile indicating bifurcation (marked by white arrows) with arteries splitting into several arteries remaining similar in size.

Centripetal force is the force a rotating object experiences. The path of the centripetal force is in the direction of the pivot point or inward-directed to the middle of the circle. The giraffe moves its head from ground level to fully direct it in a circular path (Figure 9), similar to an object moving in a circle from the bottom to the top of the circle. The head experiences a centripetal force directed caudally along the atlanto-occipital joint. Blood that moves towards the head as it is moved to ground level experiences a centripetal force away from the head. The blood is forced to remain towards the head area, instead of flowing back towards that heart as expected due to gravity.

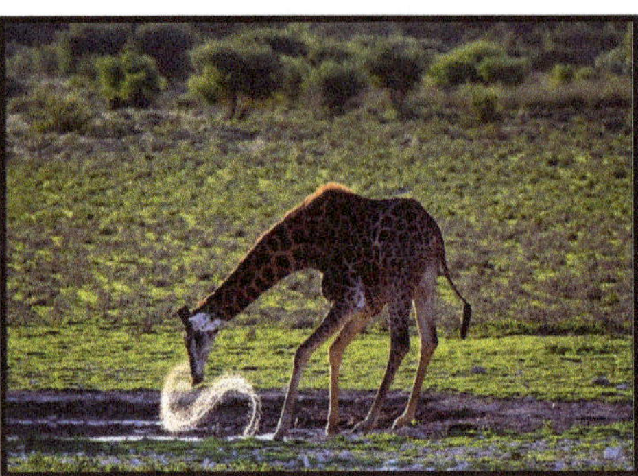

Figure 9. Illustration of movement of giraffe head from ground level to erect in a circular path towards the head area, instead of flowing back towards that heart as expected due to gravity. Centripetal force is calculated as per the below formula (Photo: F. Deacon).

The speed at which the giraffe moves its head from ground level to fully erect is currently unknown and, unfortunately, not documented anywhere. Therefore, we can only speculate on the additional effect of the centripetal force on the cerebral blood supply in the giraffe.

4. Discussion

Our study on the blood supply from the heart to the brain in the physiologically challenged giraffe (*G. camelopardalis*) reveals the unique functional significance of the rostral epidural rete mirabile.

Ref. [31] showed that the rostral epidural rete mirabile arteries in dromedary camels (*C. dromedarius*) measured 305 ± 9.7 cm, indicating the importance of the length and increased surface area for influencing the regulation of blood flow and, subsequently, blood pressure. Unfortunately, the latex mixture in our study did not fill the rete structure. Due to its extremely fine and interconnected network structure, we could not dissect the rostral epidural rete mirabile and measure its length. However, observing the rete structure closely revealed the network of numerous bends, suggesting a definite influence on blood flow through resistance exerted by bends in arterial walls [32,33]. The interconnected meshwork of arteries adds an overall significant enlargement of the cross-sectional area in vessels, resulting in reduced resistance as conferred using Poiseuille's Law [9,32]. The structure will subsequently also reduce pressure [34]. In a study on the cerebral arterial blood flow in the Yak (*Bos grunniens*), the rostral epidural rete mirabile was found to be highly developed compared to the rostral epidural rete mirabile in Chinese cattle (*Bos taurus*). In the Yak, the rete enables more efficient blood exchange and fulfills the critical function of a blood reservoir for blood supply to the brain [14]. Yak are well known to reside at high altitudes with characteristically low oxygen availability. This species' rostral epidural rete mirabile is essential to efficiently shield, transport and stabilize blood flow and oxygen to the brain [14].

Similarly, the alpaca (*Vicugna pacos*) resides at high altitudes with low oxygen availability and endure extreme ambient temperature variability. It is speculated that the rostral epidural rete mirabile in the alpaca might aid in coping with hypoxia and extreme environmental temperature fluctuations [9]. This brings us back to our initial research question: Does the rostral epidural rete mirabile similarly fulfill this function in the giraffe, with its extreme head-to-heart distance aiding as a reservoir for brain perfusion and buffering blood rushing to the brain at the different head positions? This deduction was verified by our experiment, showing that rostral epidural rete mirabile length is indirectly proportional to both blood flow and pressure. The rostral epidural rete mirabile is, therefore, automatically altering the hemodynamics of cerebral blood flow using its structure.

Studies conducted on selective brain cooling in artiodactyls by [17,21,35] describe the inimitable position of the rostral epidural rete mirabile where it lies in the cavernous sinus, at the base of the brain; indeed, it is positioned below the brain, within the cranium [4,36]. The position of the rostral epidural rete mirabile in the brain is significant regarding brain cooling and supplying blood efficiently to the brain [14,17,36]. Correspondingly, this position in the giraffe further supports the idea that the rostral epidural rete mirabile influences blood flow and pressure to and from the brain, which is especially important with the quick movement and position change of the head to ground level and again to a fully erect position. In a giraffe, the alternative pathway of blood to the brain via the vertebral arteries and the basilar artery that enters the circle of Willis in other animals does not exist. The basilar artery in the giraffe does not assist in the blood supply to the brain [8,19,21]. The maxillary artery through the rostral epidural rete mirabile is the dominant vessel supplying blood to the brain in giraffes.

Another observation in a giraffe that differs from other Artiodactyls and Perissodactyls is the heart's position concerning the brain. The pathway from the heart to the brain is notably straight, lowering resistance to flow and relieving pressure from the heart pumping 2 m against gravity to the brain. In a study on giraffes, [5] measured carotid blood flow and calculated the resistance to flow at head-up and head-down positions. Their results showed that when the head is raised, extracranial vasoconstriction takes place to prevent fainting. When the head is down, they suggest that blood flow to the brain is unrestricted. When looking at the position of the heart and the pathway from the heart to the brain, a notable straight path can be observed in comparison to other animals in which the heart is situated at an angle to the brain. The angle creates a bent path for blood flow from the heart to the brain, increasing arterial resistance [32,37,38]. The 2-m head-to-heart pathway of blood in giraffes has less resistance than, for example, sables and elephants' pathways due to the straight pathway. This supports the idea of little or no control of blood flow to the brain in the head-down position. The observed notable bending of the maxillary artery before entering the rostral epidural rete mirabile in an otherwise straight pathway creates additional resistance to blood flow rushing to the brain when the head is lowered. The twisted arteries of the rostral epidural rete mirabile, the significant length of the rete arteries and the notable bend at the maxillary before entrance into the rete all in conjunction provide resistance to flow and therefore aid in the prevention of brain damage with the quick bending down from a fully erect to a ground position.

Histologically, the maxillary artery prior to entering the rostral epidural rete mirabile structure and the intracranial maxillary artery post the rete are characteristically thicker-walled than the rete arteries. The muscular content of the rostral epidural rete mirabile arteries enables a greater extent of control. Again, resistance against blood flow is exerted by structural features, with blood moving from a thicker maxillary to the thinner rete arteries and again through the thicker intracranial maxillary into the circle of Willis. The rostral epidural rete mirabile arteries contain dense, smooth muscle indicating its ability to contract and expand. Expansion of the rostral epidural rete mirabile arteries will be restricted due to the limited space where it lies within the cavernous sinus, enclosed by bone. However, contraction of the rostral epidural rete mirabile network arteries will be possible due to the high muscular content. The ability of a large thick artery to contract in a limited space is restricted; however, the contraction ability of a network of various small arteries is more remarkable [39].

We suggest that the pool of blood present within the rostral epidural rete mirabile meshwork, situated just beneath the brain, helps prevent the giraffe from fainting together with the carotid-vertebral anastomotic connection when the head is moved in a quick motion from ground level to fully erect. It is essential that the brain receives a continuous flow of blood and thus oxygen [10,24,33]. The blood in the rete can be pushed via the contraction of the rostral epidural rete mirabile arteries into the circle of Willis. Additionally, when the head is moved from ground level to fully erect, the force of gravitation pulls blood away from the brain, and the carotid artery enlarges with a parallel drop in pressure. The vertebral arteries are neighbored by connective tissue and back muscles as it courses through the transverse foramina and through the axis and atlas, avoiding enlargement. At this point, the vertebral arteries connect to the anastomotic artery that subsequently connects to the carotid artery. During the loss of pressure in the carotid artery, blood can be moved from the vertebral arteries via the anastomotic artery into the carotid artery that transitions into the maxillary artery, pushing blood into the rostral epidural rete mirabile and the circle of Willis. This pathway requires less pressure to be generated by the heart to supply adequate blood quickly to the brain during the movement of the head to a fully erect position. Fainting due to a loss of continuous blood flow to the brain is negated by the immediate source of blood to the cerebral arterial circle originating from the vertebral artery via the anastomotic artery into the carotid artery, as well as a small volume of blood contained within the rete arterial structure. The retained blood volume in the rete and the vertebral blood supply contributes to and ensures the crucial continuous flow of blood and thus oxygen to the brain to prevent fainting or even an ischemic cascade, owing to too little blood present in the brain [40]. This pathway of cerebral blood supply aid in temporary relief during the split seconds time-lapse, allowing the nervous system to instigate the heart to pump blood against gravity to reach the brain. Even though nervous stimulation is instant, a time-lapse of a few seconds occurs before the heart can pump blood to the brain in a fully erect position. A study by [40] on blood flow circulation in intracranial networks showed that due to the elasticity and subsequent cross-sectional area of arterial blood vessels, a relative blood volume is partly retained, slowing the pulsatile flow. The retained blood is released using blood vessel contraction [40]. The centripetal force exerted on the blood accumulated towards the head also adds to blood availability to the brain during the movement of the head from ground level to fully erect, at a critical moment, i.e., milliseconds.

Additionally, [35] investigated selective brain cooling and the influence of the rostral epidural rete mirabile in sheep. They concluded that brain blood flow and brain metabolic heat production differed contrary to the expected close relationship, as previously verified by [41]. Can this variation to the desired close relationship be due to the effect exerted by the rostral epidural rete mirabile structure on blood flow?

The combination of the results in our study gives a clear indication of the functional significance and efficiency of the rostral epidural rete mirabile in giraffes, as follows: (a) the structural arterial meshwork with characteristic turns and bends, and subsequently increased artery length through which blood flow must proceed before entrance into the brain, exert resistance to blood racing to the brain when the head is dropped to the ground; (b) the rostral epidural rete mirabile structure is positioned just beneath the brain, retaining a minimal amount of blood; (c) the brain is supplied mainly by the maxillary artery through the rostral epidural rete mirabile, with a rudimentary basilar artery not contributing to the brain blood supply and (d) the resistance to blood flow due to the structure and position of the rete when the head is in the upright position is counteracted by the unique carotid-vertebral anastomosis allowing an immediate cerebral blood supply.

5. Conclusions

The structure, function and position of the rostral epidural rete enable adequate circulation to the brain in the long-necked giraffe. Future research can focus on determining its functionality in other long-necked artiodactyls such as the llama (*Lama glama*), vicuna (*Vicugna vicugna*) and alpaca (*V. pacos*). Additionally, do fossil giraffids, although with characteristically shorter necks, also have a rostral epidural rete mirabile to assist with physiological challenges associated with a long head-to-heart distance? The rostral epidural rete mirabile in the giraffe is one of a collection of features that, in combination, enable efficient brain perfusion and prevention of brain aneurisms despite its peculiar build resulting in numerous physiological challenges.

Author Contributions: Conceptualisation: M.S.v.d.W., W.D. and F.D.; methodology: M.S.v.d.W.; formal analysis: S.v.d.M.; investigation: M.S.v.d.W.; writing—original draft: M.S.v.d.W.; Writing—Review and editing: F.D., W.D., J.G. and S.v.d.M. All authors have read and agreed to the published version of the manuscript.

Funding: This project was funded by the Kroonstad Dierehospitaal and Midlands Veterinary Wholesalers Pty Ltd. This work is based on the research supported wholly/partially by the National Research Foundation (NRF) of South Africa (Grant Number: RA201126576714).

Institutional Review Board Statement: Ethics approval nr. UFS-AED2020/0083. Approval was obtained from the Animal Ethics Research Committee at the UFS, SPCA and DESTEA.

Informed Consent Statement: Not applicable.

Data Availability Statement: All data will be available from the main author.

Acknowledgments: Kroonstad Dierehospitaal for the use of the animal hospital premises, and their staff, especially Ben Koloti and Daniel Mthembu, for assistance with experimental work. Midlands Veterinary Wholesaler Pty Ltd. for assistance in the funding of the project. This work is supported on the research wholly / in part by the National Research Foundation (NRF) of South Africa (Grant Number: RA201126576714).

Conflicts of Interest: The authors declare no conflict of interest.

References

1. Hargens, A.R.; Gershuni, D.H.; Danzig, L.A.; Millard, R.W.; Pettersson, K. Tissue Adaptations to gravitational stress: Newborn versus Adult Giraffes. *Physiologist* **1988**, *31*, S110–S113.
2. Ostergaard, K.H.; Bertelsen, M.F.; Brondum, E.T.; Alkjaer, C.; Hasenkam, J.M.; Smerup, M.; Wang, T.; Nyengaard, J.R.; Baandrup, U. Pressure profile and morphology of the arteries along the giraffe limb. *J. Comp. Physiol. B* **2011**, *181*, 691–698. [CrossRef]
3. Peterson, K.K.; Horlyck, A.; Ostergaard, K.H.; Andresen, J.; Skovgaard, N.; Telinius, N.; Laher, I.; Bertelsel, M.F.; Grondahl, C.; Smerup, M.; et al. Protection against high intravascular pressure in giraffe legs. *Am. J. Physiol. Regul. Integr.* **2013**, *305*, R1021–R1030. [CrossRef] [PubMed]
4. Aalkjaer, C.; Wang, T. The Remarkable Cardiovascular System of Giraffes. *Annu. Rev. Physiol.* **2021**, *83*, 1–15. [CrossRef] [PubMed]
5. Mitchell, G.; Skinner, J.D. How Giraffe adapt to their extraordinary shape. *Trans. R. Soc. S. Afr.* **1993**, *48*, 207–2108. [CrossRef]
6. Fukuta, L.; Kudo, H.; Sasaki, M.; Kiura, J.; Ismail, D.; Endo, H. Absence of rostral epidural rete mirabile in small tropical ruminants: Implication for the evolution of the arterial system in artiodactyls. *Anatomy* **2007**, *210*, 112–116. [CrossRef]
7. Kieltyka-Kurc, A.; Frackowiak, H.; Zdun, M.; Nabzdyk, M.; Kowalczyk, K.; Tokacz, M. The arteries on the base of the brain in the camelids (Camelidae). *Ital. J. Zool.* **2014**, *81*, 215–220.
8. O'Brien, H.D.; Gignac, P.M.; Hieronymus, T.L.; Witmer, L.M. A comparison of postnatal patterns in a growth series of giraffe (Artiodactyla: *Giraffa Camelopardalis*). *Dev. Biol. Evol. Stud. Zool.* **2016**, *4*, e1696.
9. O'Brien, H.D. Cranial arterial patterns of the alpaca (Camelidae: *Vicugna pacos*). *R. Soc. Open Sci.* **2017**, *4*, 160967. [CrossRef]
10. Daniel, P.M.; Dawes, J.D.K.; Prichard, M.M.L. Studies of the rostral epidural rete mirabile and its associated arteries. *Philos. Trans. R. Soc. Lond. Ser. B Biol. Sci.* **1953**, *237*, 173–208.
11. Dieguez, G.; Garcia-Villalon, A.L.; Gomez, B.; Lluch, L. Hemodynamic significance of the rostral epidural rete mirabile during changes in arterial blood pressure. *Am. Physiol. Soc.* **1988**, *254*, R770–R775.
12. Melnikov, V.V. The Arterial System of the Sperm Whale (*Physeter macrocephalus*). *J. Morphol.* **1997**, *234*, 37–50. [CrossRef]
13. Rommel, S.A.; Costidis, A.M.; Fernandez, A.; Jepson, P.D.; Pabst, D.A.; McLellan, W.A.; Houser, D.S.; Cranford, T.W.; Van Helden, A.L.; Allen, D.M.; et al. Elements of beaked whale anatomy and diving physiology and some hypothetical causes of sonar-related stranding. *J. Cetacean Res. Manag.* **2006**, *7*, 189–209. [CrossRef]
14. Wang, X.R.; Liu, Y.; Guo, T.F.; Wu, J.P. Anatomic peculiarities of cerebral arterial system and blood supply in Yak (*Bos grunniens*). *J. Anim. Vet. Adv.* **2012**, *11*, 2533–2539.
15. O'Brien, H.D.; Bourke, J. Physical and computational fluid dynamics models for the hemodynamics of the artiodactyl carotid rete. *J. Theor. Biol.* **2015**, *386*, 122–131. [CrossRef] [PubMed]
16. Frackowiak, H.; Godynicki, S. Brain basal arteries in various species of Felidae. *Pol. J. Vet. Sci.* **2003**, *6*, 195–200. [PubMed]
17. Jessen, C. Selective brain cooling in mammals and birds. *Jpn. J. Physiol.* **2001**, *51*, 291–301. [CrossRef]
18. Mitchell, G.; Lust, A. The carotid rete and artiodactyl success. *Biol. Lett.* **2008**, *4*, 415–418. [CrossRef]
19. Frackowiak, H.; Jakubowski, H. Arterial vascularization in the giraffe brain. *Ann. Zool. Fenn.* **2008**, *45*, 353–359. [CrossRef]

20. Aiyan, A.A.; Menan, P.; AlDarwich, A.; Almuhairi, F.; Alnuaimi, S.; Bulshawareb, A.; Qablan, M.; Shebab, S. Descriptive Analysis of Cerebral Arterial Vascular Architecture in Dromedary Camel (*Camelus dromedarius*). *Front. Neuroanat.* **2019**, *13*, 1–13. [CrossRef] [PubMed]

21. Jerbi, H.; Vazquez, N.; Perez, W. Morphological Configuration and Topography of the brain arterial supply of the One-humped Camel (*Camelus dromedarius*, Linnaeus 1758). *Int. J. Morphol.* **2019**, *37*, 1095–1100. [CrossRef]

22. Oliveira, J.C.D.; Campos, R. Systematic study of brain-based arteries in the wild boar (*Sus scrofa scrofa*). *Anat. Histol. Embriol.* **2005**, *34*, 232–239. [CrossRef] [PubMed]

23. Khamas, W.A.; Goshal, N.G.; Bal, H.S. Histomorphologic structure of the carotid rete-cavernous sinus complex and its functional importance in sheep (*Ovis aries*). *Am. J. Vet. Res.* **1983**, *45*, 156–158.

24. Edelman, N.H.; Epstein, P.; Cherniak, N.S.; Fishman, A.P. Control of cerebral blood flow in the goat; role of the carotid rete. *Am. J. Physiol.* **1972**, *223*, 615–619. [CrossRef]

25. O'Brien, H.D. From Anomalous arteries to Selective Brain Cooling: Parallel Evolution of the Artiodactyl Carotid Rete. *Anat. Rec.* **2020**, *303*, 208–317. [CrossRef]

26. Deacon, F.; Tutchings, A. The South African giraffe *Giraffa camelopardalis* giraffa: A conservation success story. *Oryx* **2019**, *53*, 45–48. [CrossRef]

27. Kock, M.D.; Burroughs, R. *Chemical and Physical Restraint of Wild Animals: A Training and Field Manual for African Species*, 2nd ed.; IWVS (Africa): Greyton, South Africa, 2012.

28. Lindstedt, S.L.; Schaeffer, P.J. Use of allometry in predicting anatomical and physiological parameters of mammals. *Lab. Anim.* **2002**, *26*, 1–19. [CrossRef]

29. Sharma, R.; Sharma, S. Physiology, Blood Volume. In *StatPearls, Treasure Island (F.L.)*; StatPearls: Tampa, FL, USA, 2021; Available online: https://europepmc.org/article/nbk/nbk526077#free-full-text (accessed on 16 October 2022).

30. Parent, A.; Carpenter, M.B. *Carpenters Human Neuroanatomy*, 9th ed.; Williams and Wilkins Publisher: Philadelphia, PA, USA, 1995.

31. Jerbi, H.; Khaldi, S.; Perez, W. Morphometric study of the rostral epidural rete mirabile in the dromedary (*Camelus dromedarius*). *Int. J. Morphol.* **2016**, *34*, 1429–1435. [CrossRef]

32. Dormer, K.J.; Denn, M.J.; Stone, J.L. Cerebral blood flow in the sea lion (*Zalophus californianus*) during voluntary dives. *Comp. Biochem. Physiol.* **1977**, *58*, 11–18. [CrossRef]

33. Blix, A.S.; Walloe, L.; Messelt, E.B. On how whales avoid decompression sickness and why they sometimes strand. *J. Exp. Biol.* **2013**, *216*, 3385–3387. [CrossRef]

34. Kieltyka-Kurc, A.; Frackowiak, H.; Brudnicki, W. The arteries of brain base in species of the cervid family. *Anat. Rec.* **2015**, *298*, 735–740. [CrossRef] [PubMed]

35. Maloney, S.K.; Mitchell, D.; Blache, D. The contribution of rostral epidural rete mirabilevariability to brain temperature variability in sheep in a thermoneutral environment. *Am. J. Physiol. Regul. Integr. Comp. Physiol.* **2007**, *292*, R1298–R1305. [CrossRef] [PubMed]

36. De Salles, A.A.F.; Solberg, T.D.; Mischel, P.; Massoud, T.F.; Plasencia, A.; Goetsch, S.; De Souza, E.; Vinuela, F. Arteriovenous Malformation Animal Model for Radiosurgery: The Rete Mirabile. *Am. J. Neuroradiol.* **1996**, *17*, 1451–1458. [PubMed]

37. Jayanti, S. *Bends, Flow and Pressure Drop in*; Begell House: Danbury, CT, USA, 2011. [CrossRef]

38. Vogl, A.W.; Fisher, H.D. Arterial retia related to supply of the cerebral nervous system in two small toothed whales–narwhal (*Monodon monoceros*) and beluga (*Delphinapterus leucas*). *J. Morphol.* **1982**, *174*, 41–56. [CrossRef]

39. Barnett, C.H.; Marsden, C.D. Functions of the Mammalian Rostral epidural rete mirabile. *Nature* **1961**, *191*, 88–89. [CrossRef]

40. Grinberg, L.; Cheever, E.; Anor, T.; Madsen, J.R.; Karniadakis, G.E. Modelling Blood Flow Circulation in Intracranial Arterial Networks: A Comparative 3D/1D simulation study. *Ann. Biomed. Eng.* **2011**, *39*, 297–309. [CrossRef] [PubMed]

41. Zhu, M.; Ackerman, J.J.H.; Sukstanskii, A.L.; Yablonskiy, D.A. How the body controls brain temperature: The temperature shielding effect of cerebral blood flow. *J. Appl. Physiol.* **2006**, *101*, 1481–1488. [CrossRef] [PubMed]

Article

Using a Portable Autostereoscopic Screen to Improve Anatomy Teaching and Learning

Elsa-Marie A. Otoo *, Hannah Leibowitz, Oliver Wong and Kawal Rhode

Department of Biomedical Engineering, School of Biomedical Engineering and Imaging Sciences, King's College London, St. Thomas' Hospital, Westminster Bridge Road, London SE1 7EH, UK
* Correspondence: elsa-marie.otoo@kcl.ac.uk

Abstract: Conventional anatomical models and cadaveric specimens can be time-consuming and resource intensive for any anatomical institute. In recent years, there has been a push for more flexible and varied approaches to teaching, including problem-based and computer-aided learning, which includes web-based anatomical models or the use of three-dimensional visualization technology. With advances in hardware, autostereoscopic (AS) 3D screens have become more affordable, portable, and accessible to individuals, not just institutes. At King's College London (KCL), we developed the Virtual Anatomy and Histology (VAH) platform—an online resource which focuses on perspective volumetric 3D viewing of medical scan data and 3D models to facilitate the online teaching and learning of anatomy. This paper presents the features of VAH and details a study that was conducted in 2022, to evaluate the VAH 3D AS viewer configured with The Looking Glass Portrait (TLG) (Looking Glass, New York, NY, USA) 8-inch AS display. We tested the hypothesis that using an AS display can improve spatial understanding of cardiovascular anatomy. A cardiovascular 3D textured model was used from our gallery to carry out a spatial test. Twenty current healthcare students at King's participated in the study and completed a structured questionnaire. Results showed that 47.6% and 52.4% of participants agreed and strongly agreed, respectively, that identifying anatomical structures was easier in 3D compared to 2D. Qualitative feedback was positive as most students found King's VAH and TLG display "useful for people who need help with spatial understanding" and that "it was a good tool to test your anatomical knowledge". In conclusion, based on the quantitative results and feedback, we are optimistic that King's VAH and portable AS displays can be beneficial in anatomy education. With the increasing availability of such systems and competitive pricing, this technology is likely to have a significant impact in education in coming years.

Keywords: anatomy education; 3D model; medical imaging; 3D autostereoscopic display; online learning

Citation: Otoo, E.-M.A.; Leibowitz, H.; Wong, O.; Rhode, K. Using a Portable Autostereoscopic Screen to Improve Anatomy Teaching and Learning. *Anatomia* **2023**, *2*, 88–98. https://doi.org/10.3390/anatomia2010008

Academic Editors: Gianfranco Natale and Francesco Fornai

Received: 29 November 2022
Revised: 25 January 2023
Accepted: 6 February 2023
Published: 14 February 2023

1. Introduction

There is a consensus that knowledge of gross anatomy is fundamental in medical education and professions [1]. Understanding the relationship between the structure and function of the human body under normal and pathologic circumstances creates a solid foundation in all healthcare specialties. Moreover, with the growing prominence of medical imaging and minimally invasive therapy, the need to interpret images and anatomical pathways is increasingly important [2].

The long-standing format of many institutions' anatomical education can be categorized into lectures, practical sessions, and assessments. However, with the rapid development of technology, and in turn medical imaging, there has been a significant paradigm shift in medical education from passive and informative lectures centered around the teacher to a more active and clinical-based student approach. Therefore, the demand for practical sessions has gained more precedence in anatomical education. Practical examinations were identified as the most recommended method of assessment in anatomical education in the Rowland et al. (2011) study by medical students (59.1%), trainees (all stages combined; 54.2%), and specialists (51.7%). This further supports the need for practical educational sessions [3].

1.1. Issues Facing Current Implementation of Practical Anatomy

Historically, cadaveric dissection has been the most significant aid in teaching gross anatomy in conjunction with textbooks. A study conducted by Azer and Eizenberg (2007) found that some students feel at a disadvantage without dissection classes [4]. Their study also found that cadaveric dissection was twice as useful as a teaching resource for learning gross anatomy compared to textbooks alone. Cadaveric dissections are impractical at times due to their time constraints, because there is a set number of accessible hours in a dissection theater. Additionally, cadavers are in short supply. Both the UK and the US have been increasing medical student placements to match population demands, but donation numbers are not increasing proportionately [5–7]. Cadavers have high costs too; in the UK, cadavers are not freely bought or sold, but institutes must pay for transportation, embalming, and maintaining dissection labs that store the donated cadavers, which all increase cost. In the US, donated human bodies can be sold for around $3000 to $5000, though prices sometimes reach $10,000 [8–10].

The coronavirus (COVID-19) pandemic was an unprecedented emergency which had massive economic and social repercussions, and the education sector was not immune to its ramifications [11]. The implementation of social distancing and isolation protocols to curb rising COVID-19 infections within the population meant that more than 1.5 billion students worldwide were affected by closures of schools and universities [12]. Educational institutions around the world were faced with the challenge of teaching and learning in a distant manner.

For teaching, this had instant consequences, which included the cessation of or limited access to conventional anatomy education pedagogies, such as cadaveric dissection and access to non-digitized textbooks. Therefore, many medical institutions were forced to adopt a blended learning or a full e-learning approach. Blended learning involves the integration of online learning and face-to face teaching, which includes online lectures, pre-recorded dissection, and computer-based 3D platforms. While e-learning is fully online, where students interact with learning materials and receive feedback on an online platform, designing an online platform that replicates the visuo-spatial and social interaction of traditional teaching is a unique and growing challenge for educational developers and facilitators [13].

A recurring obstacle among some medical students is the capability to grasp the necessary spatial understanding of 3D anatomy when learning from static 2D illustrations in textbooks and medical imaging [14,15]. Visuo-spatial or spatial ability (SA) is the cognitive capacity to understand and perform mental manipulation, such as object rotation and folding to imagine different perspectives of objects and memorize a spatial representation of the object that includes the relationships among their parts and surroundings [16–18]. Preece et al. (2013) and Marks (2000) drew similar conclusions that a poor visuo-spatial ability can compromise students' abilities to keep up with advancing medical imaging in clinical practice [19,20].

One way to reduce some of the visuo-spatial information lost when one is unable to perform in-person cadaveric practical dissection is to use 3D visualization technologies that have gained momentum in the technological boom in the last couple of decades.

1.2. 3D Displays

3D display technologies are split into two categories. First, there are stereoscopic display technologies that require special glasses or headwear to create the 3D sensation. Second are autostereoscopic 3D display technologies that are glasses-free 3D displays and can be viewed with the naked eye. Stereoscopic and autostereoscopic (AS) displays are groups of 3D displays that only use binocular disparity out of all of the possible physical depth cues.

Stereoscopic displays require 3D glasses to be worn, which can make them a more solitary learning style as they immerse but also isolate the viewer from the world around them. AS displays, on the other hand, can cater to other learning styles, such as social learning, and do not require a headset that can be uncomfortable to wear.

Some AS displays work by using a lenticular lens, such as The Looking Glass (TLG) (Looking Glass, New York, NY, USA) displays, shown in Figure 1a below. In these devices, an array of lenticular lenses is attached to the top of the screen. The lenticular array distributes the pixels of the display to multiple viewpoints. Each lens behaves like a tiny magnifying glass, inflating one of the sub-pixels and hiding the remaining ones. However, depending on the angle of view, each inflated sub-pixel is different. Since both eyes watch the screen from slightly offset angles, they both see different sub-pixels in any given location of the screen, creating binocular disparity, as shown in Figure 1b [21]. Figure 1c showcases the TLG's simultaneous 45 viewpoints/images rendered to create its 3D visualization.

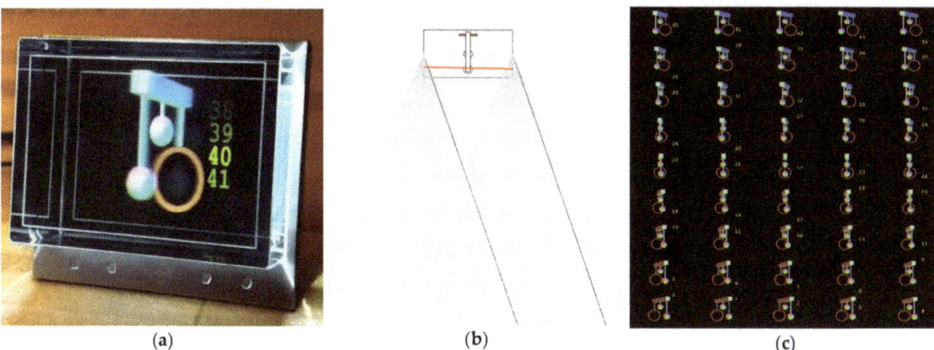

(a)   (b)   (c)

Figure 1. Diagrams demonstrating how the TLG display works: (**a**) is an image that shows what is viewed on the TLG display at different positions. The numbers in the scene showcase the different views/perspective of the scene as the viewer moves from one edge to the other [22]. (**b**) is a top-down ray diagram of the parallax used to create the 3D visualization at different positions. The bold lines show the eye positions needed to see the viewpoint shown in (**a**) [22]. It also shows the zero-parallax plane (the thin red line) where objects are the sharpest in the scene. (**c**) is known as the 'quilt', which displays the 45 views simultaneously on TLG [22].

Dedicated holographic devices are markedly expensive. Depending on size of the display, a standard device used to price at a minimum of $3000. However, with recent advances in hardware capabilities and a growing market, now more affordable, portable, and accessible systems are available, such as TLG Portrait (displayed in Figure 2) which is $399, and the Leia Inc. (Leia Inc., Menlo Park, CA, USA) Lume pad tablet, which is $650 [23,24]. This accessibility opens up more opportunities for educators and students to use these devices in anatomical teaching and learning.

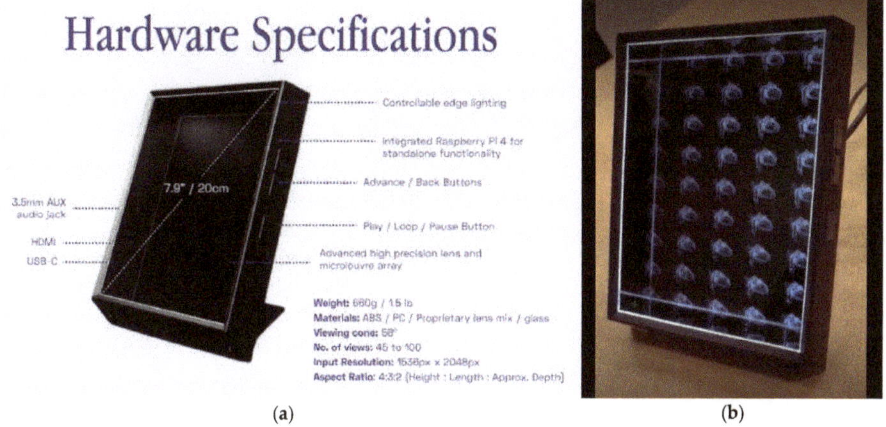

(a)   (b)

Figure 2. (**a**) The hardware specification of TLG portrait [25]. (**b**) An image of TLG portrait in quilt mode showing an X-ray textured model [23].

1.3. 3D Displays in Anatomical Education

There is some evidence in anatomical education literature showing the application of autostereoscopic displays in the medical field. In a few studies, AS has been used for training surgeons. Narita et al. (2014) studied twenty neurosurgeons and asked them to identify blood vessels using a 2D screen and an autostereoscopic screen. They found significantly more correct answers were attained using the 3D autostereoscopic screens, with 91.7% correct responses compared to 56.7% with the 2D screen [26]. This shows that AS gives the viewer a higher sense of realism and accuracy, therefore enhancing their performance in identifying key anatomical structures. Silvestri et al. (2011) also found that AS enhanced the viewers' performance. They found that the fifteen participants, who were physicians, were able to carry out the five assigned tasks faster on the autostereoscopic screens compared to the 2D monitor [27].

The benefits of this system and the studies carried out on the training of surgeons mean that there is some promise in using AS to teach anatomy. Luursema et al. (2011) studied stereoscopic screens that utilized glasses and showed that students with low visuospatial ability benefitted more from making use of the screen than students with higher spatial abilities, as students were able to visualize abdominal structures more easily than with 2D models [28]. Although this study was conducted with a stereoscopic screen, the results are likely to be replicated with AS, as they are more user-friendly. Di Natale et al. (2020) have also identified that using immersive VR in teaching promoted experiential learning, intrinsic motivation, and engagement, and supported the transfer of knowledge [29]. Sinha et al. (2022) evaluated a 42-inch AS display for anatomical education and found that the AS display promoted 3D stereoscopic viewing in a group setting for collaborative learning that a single-user VR headset could not [30]. This type of social learning and the other promoted attributes mentioned are all key psychological factors that support deep learning and teaching. Therefore, in this study, we will explore the use of the portable 3D AS display we configured to be compatible with our online learning resource.

2. Materials and Methods

We proposed a 3D visualization system that incorporates volumetric 3D medical data and 3D models with the capabilities of displaying these on 3D displays.

2.1. System Overview

Virtual Anatomy and Histology (VAH) is a university-based online learning web-based application, shown in Figure 3 below, and was developed to allow students to study and interact with anatomy through 3D models, histological images, and 3D imaging scans that the different King's College London (KCL) departments and museums possessed. The departments and museums had numerous subscriptions for anatomical software. It was decided that an in-house all-in-one online anatomical learning platform or tool could reduce costs, as the new cost would go towards cloud and web app hosting. It also provides a more bespoke solution, for example, by allowing educators and students to design and save their own 3D annotated anatomical scenes for lessons and private study. VAH has volumetric datasets, 3D anatomical models, and histology slides from a cloud-based gallery which can be accessed by users to view and interact with.

The TLG Portrait 8-inch AS display was configured to be compatible with VAH's model viewer. From this point onwards, we will be focusing on the VAH TLG viewer, which creates a popup window that can be dragged onto the TLG's screen, to allow models to be viewed in 3D mode or 2D mode. The viewer can load 3D models from the site catalogue or from the user's personal computer files. The VAH TLG viewer was now ready for usability testing.

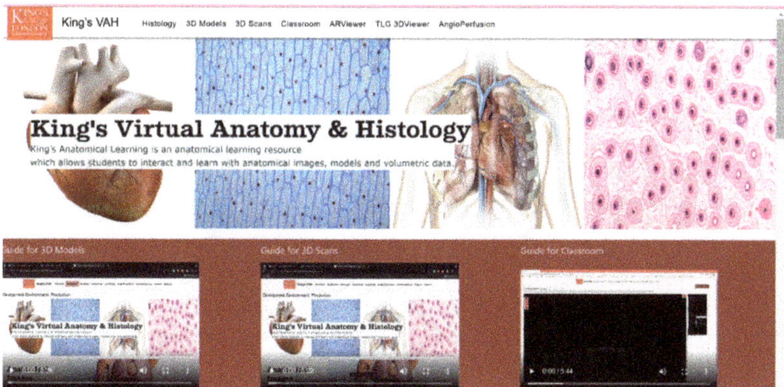

Figure 3. Shows King's VAH Home Page with its different features, such as Histology (slide) viewer, 3D model viewer, 3D scans (volumetric) viewer, augmented reality (AR) viewer, and TLG Viewer.

2.2. *Study Design*

2.2.1. Participants

Ethics approval was obtained via the Institutional Review Board (or Ethics Committee) of the KCL Research Ethics Office (MRSP-20/21-21158). Participation in the study was anonymous and voluntary. Information sheets about the study were provided during recruitment via cohort mailing lists, as required by the ethics committee. As explained in the information sheet and discussed in person, students who completed and submitted the survey, had provided informed consent for their data to be used. Inclusion criteria included being a current healthcare student in the Faculty of Life Sciences and Medicine. Exclusion criteria included if the user had photosensitive epilepsy or cybersickness. Ultimately, twenty current healthcare students at KCL were recruited to participate in the study via this strategy. Their level of education ranged from 1st year undergraduates to Ph.D. candidates, with 10 undergraduates (UG) and 10 postgraduates (PG).

2.2.2. Procedure

The study was conducted in 2022, with the aim to assess the use of VAH in anatomical education and to test the hypothesis that visuospatial understanding of cardiovascular anatomy was improved by using TLG display. The cardiovascular 3D textured model from our gallery that was used in the spatial test can be seen in Figure 4a.

The assessment used a repeated measures study design, where participants were asked to describe the spatial relationship between two anatomical structures in a heart model in both 3D and 2D. The two anatomical structures were selected randomly when the spatial test button was pressed, with their names also on display. The rest of the model structures were textured with an "X-ray"-style shader to highlight the positions of the selected structures, as shown in Figure 4b. The spatial relationship had to be described using anatomical orientations and directional terms with the use of the reference diagram shown in Figure 4c. For example, in the case of Figure 4b, the papillary muscles and the aortic valve are the selected/highlighted test objects 1 and 2. Therefore, participants were asked to describe the relationship of object 2 with respect to object 1 with the diagram Figure 4c as a reference while they interact with the viewer through orbiting, rotating, translation, and zooming. There were 4 parts to an answer:

- Superior or inferior or in the same transversal plane;
- Anterior or posterior or in the same coronal plane;
- Left or right or in the same sagittal plane;
- Proximal and distal or lateral and medial or on the same medial line.

Each correct response was worth 0.25, amounting to a total mark of 1 for a perfect score. We rated the performance by comparing our version, which we made from analyzing the anatomical model we used in the assessment. We calculated the centroid and the bounding box of each of the objects and compared them in the different planes. Therefore, someone who scored 0 was interpreted as having little spatial understanding of the physical relationship between the anatomical objects when using 3D or 2D display mode. On the other hand, someone who scored 1 was interpreted as having a high understanding of the spatial relationship between the anatomical objects.

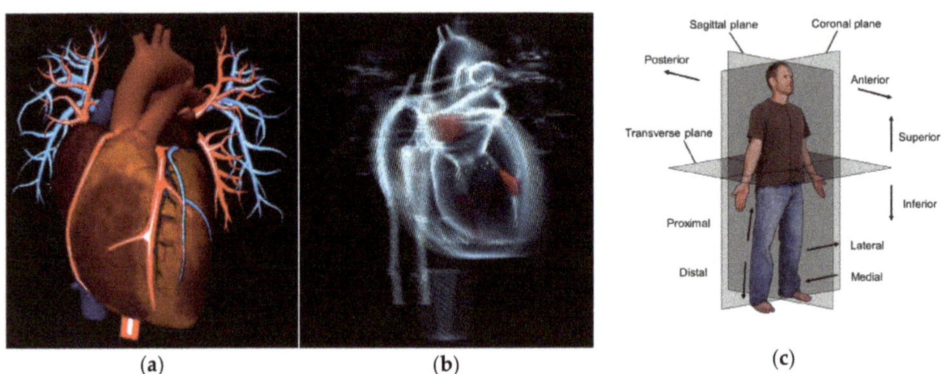

Figure 4. Shows some of the scenes throughout the experiment. (**a**) Shows the normal textured heart model used in the assessment through a 2D display. (**b**) Shows one of the randomized spatial test scenes in 3D mode. (**c**) Shows the anatomical orientation and the reference diagram participants used [31].

Volunteers performed the activity three times, with an initial attempt in 3D mode using the 3D display. The initial attempt was used as practice to get used to VAH and to confirm that they were able to perceive the 3D visuals of the screen, and to allow the users to adjust the settings of the viewer to best enhance the 3D visualizations (i.e., background color, directional lighting, model size, and model position). We allowed participants autonomy to choose the order they wanted to complete their assessed 2D and 3D tries, which were all conducted on the TLG display. All results were filled in and collected using an online Microsoft form (Microsoft, Redmond, WA, USA), for the possibility of remote participants. However, in the end, all the 3D TLG assessments ended up being completed in person on campus in a classroom, and not remotely. Consequently, there was an additional in-person supervisor who provided initial assistance when participants used the 3D screen for the first time. A paired *t*-test analysis was performed comparing the overall 2D mode ("before") vs 3D mode ("after") test scores, to establish statistical significance at $p = 0.05$. Participants were also asked to complete a questionnaire on the form containing a series of Likert scale user experience questions and statements such as "How likely are you to recommend King's VAH to someone else?", or grading their agreement with the statement "Identifying anatomical structures was easier in 3D compared to 2D". There were also some additional open-ended questions, such as "Based on today's session, to what extent do you think King's VAH might impact your future anatomical learning?" and "Do you have any final feedback regarding the King's VAH website?". All these helped to collect feedback about the site, the 3D display, and the tools which the volunteers found particularly useful.

3. Results

Figure 5 shows the anatomical spatial test scores for 3D and 2D viewing.

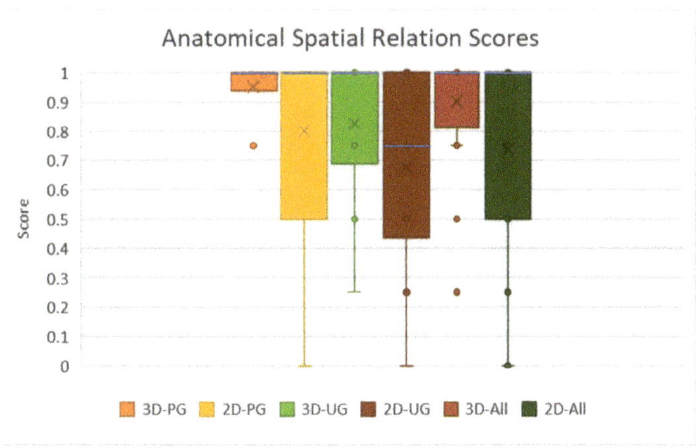

Figure 5. Shows box and whisker plots of the anatomical spatial relation scores different level of education (undergraduates—UG, postgraduates—PG and All). The cross indicates the mean, and the medium line has been highlighted blue.

The 3D-All score had a first quartile (Q1 = 0.875) and a third quartile (Q3 = 1) with an interquartile range (IQR = 0.125) and a median = 1. While the 2D-All score had a first quartile (Q1 = 0.5) and a third quartile (Q3 = 1) with an interquartile range (IQR = 0.5) and a median = 1. In terms of the mean (M) and standard deviation (SD), the 2D display had a score of (M = 0.73, SD = 0.34) and the 3D display one of (M = 0.90, SD = 0.20). For the paired *t*-test, *t* and *p*-values were $t(19) = 2.37$ and $p = 0.028$, which is less than a = 0.05. The 3D-All mean score was higher than the mean score of the 2D-All, and the results of the paired *t*-test indicated there was a significant difference between the two groups. If we separate the groups into level of education, UG and PG scores, we still see similar trends between 2D and 3D groups. The 3D groups have lower IGRs than 2D, 3D-UG = 0.313 and 3D-PG = 0.062 vs. 2D-UG = 0.563 and 2D-PG = 0.500. In terms of medians, both 3D-PG = 1 and 2D-PG = 1 were the same, but for undergraduates, 3D-UG = 1 had a higher median than 2D-UG = 0.5. The 2D-UG score had the most evenly distributed test scores, while 3D-UG had a more positively skewed score.

Figure 6 shows the agreement level of participants based on the Likert scale for the statement "Identifying anatomical structures was easier in 3D compared to 2D".

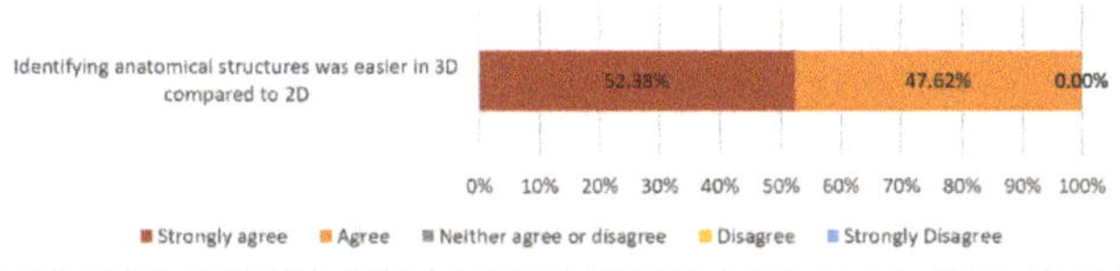

Figure 6. A stacked bar graph showing the Likert scale results for participant responses to the statement comparing 2D and 3D.

From Figure 6, we can see that 47.6% and 52.4% of participants agreed and strongly agreed, respectively, that identifying anatomical structures was easier in 3D compared to 2D. The results show an extreme positive skew.

Figure 7 displays participants' responses to two further questions asked in the questionnaire. Both results are skewed to more positive responses.

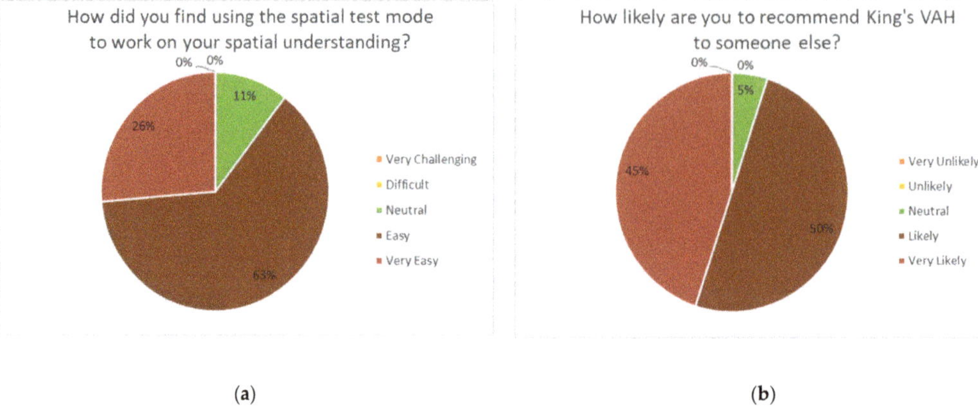

(a)   (b)

Figure 7. Pie charts of two user experience Likert scale results. (**a**) The figure shows a pie chart of the results to the question "How did you find using the spatial test mode to work on your spatial understanding?". (**b**) The figure shows a pie chart of the results to the question "How likely are you to recommend King's VAH to someone else?".

Qualitative Results

The most common themes that appeared in the responses to the question "Based on today's session, to what extent do you think King's VAH might impact your future anatomical learning?" were 'helpful' or 'useful'. Visualized in a word cloud in Figure 8, the words helpful or useful appeared in 45% of people's responses.

Figure 8. Shows a collection of responses to the question "Based on today's session, to what extent do you think King's VAH might impact your future anatomical learning?".

When asked the question "Do you have any final feedback regarding the King's VAH website?", 30% of users suggested they would have liked an anatomical position reset button in the spatial test.

4. Discussion and Conclusions

Most participants that used the 3D visualization system with the 3D display performed better than in the 2D display. Both the 2D and 3D test scores had a median of 1, but the interquartile range of 0.5 for the 2D was 4× larger than the 0.125 of the 3D, shown in Figure 5. This shows that both sets of data are left skewed, but 2D has a higher variability. The positive skew of both datasets could be related to the freedom of VAH as an interactive viewer to allow for explorative spatial understanding, through orbiting the model. From the less variability of the 3D results, we can infer that the AS screen further enhances anatomical spatial understanding through the visualization of depth information in a natural way for participants. This is further reinforced by looking at UG scores alone, as 2D-UG had the most normal distribution. However, when participants used the 3D display, 3D-UG scores were more positively skewed. We can also analyze the different group means, where 2D-All had a lower mean (M = 0.73, SD = 0.34) than 3D-All (M = 0.90, SD = 0.20). In addition, the paired *t*-test results indicate that the mean difference between 3D and 2D datasets were different enough to support the hypothesis that "*visuospatial understanding of cardiovascular anatomy was improved by using the TLG display*". The hypothesis is further supported by the fact that 47.6% and 52.4% of participants agree and strongly agree, respectively, that identifying anatomical structures was easier in 3D compared to 2D. This is particularly positive, because it suggests that portable and affordable AS screens can provide the necessary stereo depth to assist with spatial understanding, which could enhance remote learning. However, $n = 20$ is a small sample size for an educational assessment and in future work we would like to increase the cohort size.

An observation throughout the study was that all participants chose to complete the 3D assessment first. A probable reason may have been the excitement related to a new technology. As all participants ended up choosing 2D mode display last, we can ask whether there was any brain fatigue that affected the scores. It may have been better to have randomized the order of the tests.

A limitation of the spatial test was that pairing selection of the organ structures were chosen at random and therefore, a few combinations of test objects had easier spatial relationships to describe than others. The supervisors during the experiment did allow for additional randomizer selection tries, if the relationship was too easy, to balance out the difficulty. However, anatomical names generally do provide some prior information of the anatomical position of an object, for example, middle cardiac vein. Therefore, part of the spatial relationship between structures could be deciphered before visualization. Nevertheless, as the focus was on learning, we still provided the names of the anatomical parts, to help them match names and characteristics of model structures, such as positional information, which is good for learning and is one of our aims.

One of the main critiques and feedback provided was the need for an anatomical reset button. This is very reasonable as the 3D scene itself is rotatable because we can orbit around the model even in 2D. We anticipated that if resolved, this would reduce the time taken to manually reset the orientation. Since the writing of this paper, the button has been added. In addition, the toolbar was also mentioned as something to improve.

Helpful or useful appeared in 45% of people's responses when describing the future impact of VAH in anatomical learning. We can infer that VAH has a place in the future of anatomical learning, especially as 95% of participants would likely or very likely recommend VAH to someone else.

In conclusion, based on the quantitative and qualitative results and feedback, we are optimistic that King's VAH and small portable AS displays can be beneficial in anatomical education. There are improvements to be made to the controls and additional features. With the increasing availability of such systems and reducing costs, this technology is likely to have a significant impact in education in coming years.

Author Contributions: Conceptualization, K.R.; Investigation, O.W. and E.-M.A.O.; Methodology, E.-M.A.O.; Software, E.-M.A.O.; Validation, H.L. and E.-M.A.O.; Supervision, K.R. All authors have read and agreed to the published version of the manuscript.

Funding: This work is funded by the King's College London and Imperial College London EPSRC Centre for Doctoral Training in Medical Imaging (EP/L015226/1) and EPSRC National Productivity Investment Fund (NPIF) (EP/R512552/1). This research was supported by the National Institute for Health Research (NIHR) Biomedical Research Centre award to Guy's and St Thomas' NHS Foundation Trust in partnership with King's College London, and by the NIHR Healthcare Technology Co-operative for Cardiovascular Disease at Guy's and St Thomas' NHS Foundation Trust, Project 651120.

Institutional Review Board Statement: The study was conducted in accordance with the Declaration of Helsinki, and approved by the Institutional Review Board (or Ethics Committee) of KCL Research Ethics Office (MRSP-20/21-21158).

Informed Consent Statement: Informed consent was obtained from all subjects involved in the study.

Data Availability Statement: Not applicable.

Acknowledgments: The Heart texture model was provided by YingLiang Ma.

Conflicts of Interest: The authors declare no conflict of interest.

References

1. Böckers, A.; Jerg-Bretzke, L.; Lamp, C.; Brinkmann, A.; Traue, H.C.; Böckers, T.M. The Gross Anatomy Course: An Analysis of Its Importance. *Anat. Sci. Educ.* **2010**, *3*, 3–11. [CrossRef] [PubMed]
2. McCuskey, R.S.; Carmichael, S.W.; Kirch, D.G. The Importance of Anatomy in Health Professions Education and the Shortage of Qualified Educators. *Acad. Med.* **2005**, *80*, 349–351. [CrossRef] [PubMed]
3. Rowland, S.; Ahmed, K.; Davies, D.C.; Ashrafian, H.; Patel, V.; Darzi, A.; Paraskeva, P.A.; Athanasiou, T. Assessment of Anatomical Knowledge for Clinical Practice: Perceptions of Clinicians and Students. *Surg. Radiol. Anat.* **2011**, *33*, 263–269. [CrossRef] [PubMed]
4. Azer, S.A.; Eizenberg, N. Do We Need Dissection in an Integrated Problem-Based Learning Medical Course? Perceptions of First- and Second-Year Students. *Surg. Radiol. Anat.* **2007**, *29*, 173–180. [CrossRef] [PubMed]
5. *The Expansion of Medical Student Numbers in the United Kingdom Medical Schools Council Position Paper*; Medical Schools Council: London, UK, 2021. Available online: https://www.medschools.ac.uk/media/2899/the-expansion-of-medical-student-numbers-in-the-united-kingdom-msc-position-paper-october-2021.pdf (accessed on 29 November 2022).
6. AAMC. U.S. Medical School Enrollment Rises 30%. Available online: https://www.aamc.org/news-insights/us-medical-school-enrollment-rises-30 (accessed on 12 January 2023).
7. Habicht, J.L.; Kiessling, C.; Winkelmann, A. Bodies for Anatomy Education in Medical Schools: An Overview of the Sources of Cadavers Worldwide. *Acad. Med.* **2018**, *93*, 1293–1300. [CrossRef] [PubMed]
8. Singh, V.; Kharb, P. A Paradigm Shift from Teaching to Learning Gross Anatomy: Meta-Analysis of Implications for Instructional Methods. *J. Anat. Soc. India* **2013**, *62*, 84–89. [CrossRef]
9. In the U.S. Market for Human Bodies, Anyone Can Sell the Donated Dead. Available online: https://www.reuters.com/investigates/special-report/usa-bodies-brokers/ (accessed on 12 January 2023).
10. Heizenrader. Why Virtual Cadavers Are a Wise Investment for Medical Schools. Available online: https://heizenrader.com/why-virtual-cadavers-are-a-wise-investment-for-medical-schools/ (accessed on 12 January 2023).
11. Ayittey, F.K.; Ayittey, M.K.; Chiwero, N.B.; Kamasah, J.S.; Dzuvor, C. Economic Impacts of Wuhan 2019-NCoV on China and the World. *J. Med. Virol.* **2020**, *92*, 473–475. [CrossRef] [PubMed]
12. UNESCO's Education Response to COVID-19. Available online: https://en.unesco.org/covid19/educationresponse/support (accessed on 8 August 2022).
13. Tóth, Á.; Pentelényi, P.; Tóth, P. Virtual Learning Aspects of Curriculum Development in Technical Teacher Training. In Proceedings of the 2006 International Conference on Intelligent Engineering Systems, London, UK, 26–28 June 2006; pp. 308–313. [CrossRef]
14. Berney, S.; Bétrancourt, M.; Molinari, G.; Hoyek, N. How Spatial Abilities and Dynamic Visualizations Interplay When Learning Functional Anatomy with 3D Anatomical Models. *Anat. Sci. Educ.* **2015**, *8*, 452–462. [CrossRef] [PubMed]
15. Pedersen, K. Supporting Students with Varied Spatial Reasoning Abilities in the Anatomy Classroom. *Teach. Innov. Proj.* **2012**, *2*. Available online: https://ojs.lib.uwo.ca/index.php/tips/article/view/3576 (accessed on 29 November 2022).
16. Bogomolova, K.; Hierck, B.P.; van der Hage, J.A.; Hovius, S.E.R. Anatomy Dissection Course Improves the Initially Lower Levels of Visual-Spatial Abilities of Medical Undergraduates. *Anat. Sci. Educ.* **2020**, *13*, 333–342. [CrossRef] [PubMed]
17. Gonzales, R.A.; Ferns, G.; Vorstenbosch, M.A.T.M.; Smith, C.F. Does Spatial Awareness Training Affect Anatomy Learning in Medical Students? *Anat. Sci. Educ.* **2020**, *13*, 707–720. [CrossRef] [PubMed]
18. Hegarty, M.; Waller, D. A Dissociation between Mental Rotation and Perspective-Taking Spatial Abilities. *Intelligence* **2004**, *32*, 175–191. [CrossRef]
19. Preece, D.; Williams, S.B.; Lam, R.; Weller, R. "Let's Get Physical": Advantages of a Physical Model over 3D Computer Models and Textbooks in Learning Imaging Anatomy. *Anat. Sci. Educ.* **2013**, *6*, 216–224. [CrossRef] [PubMed]
20. Marks, S.C., Jr. The Role of Three-Dimensional Information in Health Care and Medical Education: The Implications for Anatomy and Dissection. *Clin. Anat.* **2000**, *13*, 448–452. [CrossRef] [PubMed]
21. Alioscopy. How Does It Work? Available online: https://www.alioscopy.com/en/principles.php (accessed on 30 September 2022).
22. Looking Glass Documentation. How the Looking Glass Works. Available online: https://docs.lookingglassfactory.com/keyconcepts/how-it-works (accessed on 30 September 2022).
23. Looking Glass Portrait. Available online: https://lookingglassfactory.com/looking-glass-portrait (accessed on 4 November 2022).
24. CNET. Lume Pad Brings Back Glasses-Free 3D—This Time on an Android Tablet. Available online: https://www.cnet.com/tech/computing/lume-pad-brings-glasses-free-3d-back-again-on-an-android-tablet/ (accessed on 4 November 2022).
25. Kickstarter. Looking Glass Portrait by Looking Glass. Available online: https://www.kickstarter.com/projects/lookingglass/looking-glass-portrait/description (accessed on 30 September 2022).
26. Narita, Y.; Tsukagoshi, S.; Suzuki, M.; Miyakita, Y.; Ohno, M.; Arita, H.; Saito, Y.; Kokojima, Y.; Watanabe, N.; Moriyama, N.; et al. Usefulness of a Glass-Free Medical Three-Dimensional Autostereoscopic Display in Neurosurgery. *Int. J. Comput. Assist. Radiol. Surg.* **2014**, *9*, 905–911. [CrossRef] [PubMed]
27. Silvestri, M.; Ranzani, T.; Argiolas, A.; Vatteroni, M.; Menciassi, A. A Multi-Point of View 3D Camera System for Minimally Invasive Surgery. *Procedia Eng.* **2012**, *47*, 1211–1214. [CrossRef]
28. Luursema, J.-M.; Verwey, W.B.; Editor, A.; Kumar Tripathi, A. The Contribution of Dynamic Exploration to Virtual Anatomical Learning. *Adv. Hum. Comput. Interact.* **2011**, *2011*, 965342. [CrossRef]

29. Di Natale, A.F.; Repetto, C.; Riva, G.; Villani, D. Immersive Virtual Reality in K-12 and Higher Education: A 10-Year Systematic Review of Empirical Research. *Br. J. Educ. Technol.* **2020**, *51*, 2006–2033. [CrossRef]
30. Sinha, S.; DeYoung, V.; Chan, S.; Ives, R.; Lohit, S.; Reis, I.; Touliopoulos, E.; Nehru, A.; Mitchell, J.P.; Brewer-Deluce, D.; et al. Evaluating Autostereoscopy (AlioscopyTM) Use for Anatomy Education. *FASEB J.* **2022**, *36*. [CrossRef]
31. Quizlet. Directional Terms Anatomy and Physiology Diagram. Available online: https://quizlet.com/424696628/directional-terms-anatomy-and-physiology-diagram/ (accessed on 27 November 2022).

Communication

Wolf and Dog: What Differences Exist?

Alessandra Coli [1,*], **Davide Prinetto** [2] **and Elisabetta Giannessi** [1]

[1] Department of Veterinary Sciences, University of Pisa, 56124 Pisa, Italy
[2] Veterinary Anatomy Museum, University of Pisa, 56124 Pisa, Italy
* Correspondence: alessandra.coli@unipi.it; Tel.: +39-050-221-3856

Abstract: A morphological study of the skeletal specimen of *Canis lupus* L. from an archeological dig of Agnano (Pisa) (Fauna Laboratory, Department of Archaeological Sciences, University of Pisa, Italy) that is chronologically placed in the Wurm period (last glaciation) was done to perform an anatomical comparison between this wild ancestor and osteological specimens of *Canis familiaris* L. present in the Veterinary Anatomy Museum (University of Pisa). Marked morphological differences in the splanchnocranium (nasal bone, zygomatic arch and orbital angle), neurocranium (sagittal crest) and temporomandibular joint (due to different developments of the masticatory muscles) are highlighted on the wolf specimen compared to those in the domestic dog specimens present in Museum. The appendicular skeletal bones of the wolf show anatomical features similar to those of dog bone specimens, confirming their belonging to the same family (*Canidae*). This result confirms that domestication has almost exclusively affected the anatomical features of the skull that have changed due to the difference in dietary approach between wolves and dogs.

Keywords: domestication; skeleton; morphological study; *Canidae*

Citation: Coli, A.; Prinetto, D.; Giannessi, E. Wolf and Dog: What Differences Exist? *Anatomia* **2023**, *2*, 78–87. https://doi.org/10.3390/anatomia2010007

Academic Editors: Gianfranco Natale and Francesco Fornai

Received: 13 December 2022
Revised: 23 January 2023
Accepted: 7 February 2023
Published: 9 February 2023

1. Introduction

The wolf (*Canis lupus* L., 1758) belongs to the *Canidae* family, whose main morphological characteristics are a long dental row, a large number of teeth (42), a long tail, digitigrade limbs and four fingers in the hind limb. The wolf is considered to be the wild ancestor of the domestic dog (*Canis lupus familiaris* L., 1758). The anatomical similarity between wild and domesticated species is actually considered important in the study and monitoring of infectious diseases' spread [1]; for instance, the domestic dog is actually considered to be the main reservoir of the Canine Distemper Virus (CDV) [2], thanks to epidemiological monitoring using Geographical Information System (GIS) [3].

In Italy, Altobello [4] highlighted characteristics to distinguish the Italian wolf from the populations of other European wolves, regarding it a subspecies of the gray wolf (*Canis lupus* L., 1758) and calling it "Apennine wolf". Recent genetic investigations have confirmed this statement by elevating the Apennine wolf to subspecies (*Canis lupus italicus*), thus distinguishing it, by morphological and genetic characteristics, from the remaining populations of European wolves [5]. From a genetic point of view, the gray wolf represents the original dog line in Asia and Europe. A case of hybridization between wolf and dog was reported in Europe, and its occurrences were well analyzed by genetic analysis of the wolf–dog hybrids in several European countries [6]. The Apennine wolf represents a wolf colony that recolonized the western Italian Alps, as reported [7] by analyzing the DNA extracted from Apennine wolf tissue samples and genotyping at 12 microsatellite loci. From this genetic study, it can be stated that the Apennine wolf has a significantly higher heterozygosity than the those in the wolves from the Alps. Currently the wolf distribution affects the whole Apennine chain with branches in Lazio and Tuscany, which has a total population estimated at 400–500 individuals [8]. The explanation for where and when this animal domestication took place remains surprisingly inaccurate [9]. Scientific studies show that it was not man who sought the wolf, transforming it into a dog, but the exact opposite; the wolf approached human settlements to eat the remains of meals, losing fear of man over time and making itself tamable [10]. It is therefore a process of "self-domestication", wherein wolves and men shared an ecological niche, allowing the wolf to become domesticable and showing changes in the behavioral and morphological features [11]. In fact, the bones can show the morphological and structural transformations due to the domestication process that occurred over 14,000 ears ago, for instance, in the bones of the primitive *Canis lupus familiaris* that were found in a Pleistocene archaeological excavation in a human burial. However, results based on the study of mitochondrial DNA have suggested that the domestication process can be traced back as early as 40,000 years ago [12].

From the literature, the study of the wolf skeleton was particularly focused on the morphological features of the skull; it reached a length of about 23–27 cm and a width of about 15–18 cm, while the dog skull has a different length and width in relation to the breed.

A long splanchnocranium, large zygomatic arches and a developed external sagittal crest are particularly developed in the wolf skull [13]. The angle formed by the intersection between the straight-line tangent to the top of the skull and the tangent line to the zygomatic arch, or "orbital angle", is a parameter showing distinction between the wolf skull and the dog skull, especially in dogs of similar morphology (e.g., German Shepherd dog). This angle is taken into account for the distinction between the two types of animals; it measures 39–46° in the wolf and 49–55° in dogs [14,15] (Figure 1).

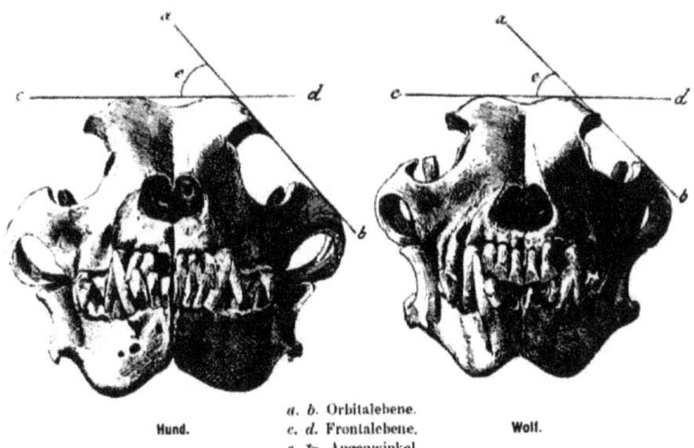

Figure 1. Orbital angle as depicted in the original Studer publication (1901).

The orbital angle justifies the different development of chewing musculature in the wolf, which has its own attachment between the considered bones.

The angle between the nasal and frontal bones, or "frontal stop", is a parameter that helps in the identification of the skull specimens; in the wolf, a flatter frontal angle than that in the dog is present (Figure 2).

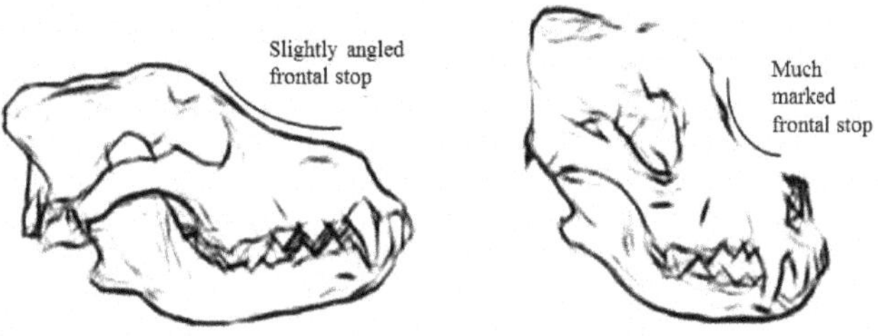

Figure 2. Drawing of "frontal stop" of the wolf skull (**left**) and the dog skull (**right**). Copyright © Davide Prinetto.

The wolf adult dental formula is the same as that of the dog, but the canines and "feral" or "carnassial" teeth (PM4/M1) stand out in respect in comparison to those of the dog [16]. An earlier review showed that [17], compared to a wolf of the same size, the dog shows a lighter skull, smaller teeth, wider palate but larger neurocranium. These propositions can be explained by artificial selection.

Currently, there is no motivation related to domestication regarding the shape of the coronoid process of the jaw (on which the temporal and masseter muscles are inserted), which is curved backwards along the ascendant branch in the dog skull (Figure 3).

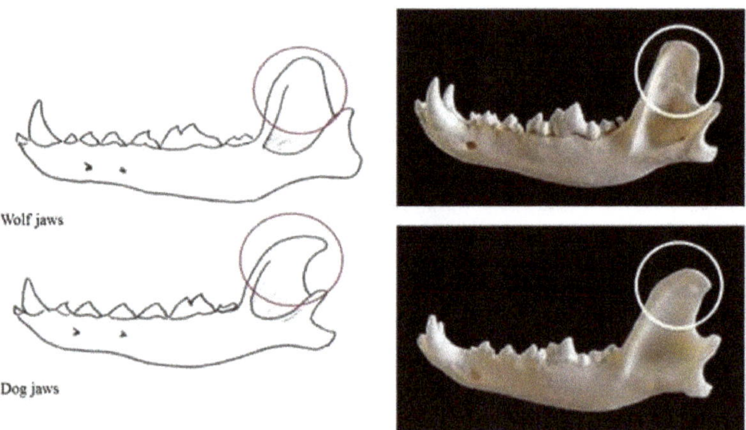

Figure 3. The coronoid process of the wolf jaw and the dog jaw. (**left**), Drawing Copyright © Davide Prinetto; (**right**), bone findings (Veterinary Anatomy Museum, University of Pisa).

This study has the purpose of carrying out a morphological investigation of bone specimens of the ancient Italian wolf and dog bone specimens from the Veterinary Anatomy Museum (University of Pisa) to check for significant variables in the two types of bone specimens, as described in the literature.

2. Materials and Methods

The authors performed a comparative study on the bone specimens of *Canis lupus* L., chronologically placed in the Wurm period (last glaciation) that were recovered from the archeological dig of Agnano (Pisa) and four osteological skeletal specimens and ten skulls of *Canis lupus familiaris* L. that were kept in the Veterinary Anatomy Museum (University of Pisa), datable around 1850 and of which, in the museum archives, there is no documentation regarding the breed.

The wolf specimen included the skull with the jaw, segments of thoracic and pelvic limbs with a part of the coxal. In the skull, the profile of the nasal bone, the development and profile of the sagittal crest, the measure of the orbital angle, the position of the orbital cavity with respect to the median plane of the skull, the extension of the zygomatic arch, the length of the cranial cavity and the jaw bone processes (jaw body and coronoid process), the width of the cranial cavity (widest interparietal distance) and the depth of the masseteric pit were studied and were compared with the same bone processes of the dog specimens. The morphology of the wolf teeth was compared to that of the dog specimens.

The bone segments relating to the appendicular skeleton of the wolf include the fore-limb bones (humerus, radius and ulna, carpal bones, five metacarpal bones and phalanges of the hand) and the hindlimb bones (tibia, fibula, tarsal bones, five metatarsal bones and phalanges of the foot) and were comparatively evaluated with the same topographical findings of the four dog skeletons kept in the Museum.

3. Results

The structure of the wolf skull shows the typical features of a predator; the sagittal crest is very well developed to allow a broad attachment of the temporal muscle, which is more developed in carnivores (Figure 4).

Figure 4. The wolf skull (Department of Archeological Sciences, University of Pisa), lateral view.

The dog skulls found housed in the Museum show sagittal crests that are less developed than that of the wolf, and they tend to decline gradually in mesaticephalic breeds and disappear in brachycephalic ones (Figure 5).

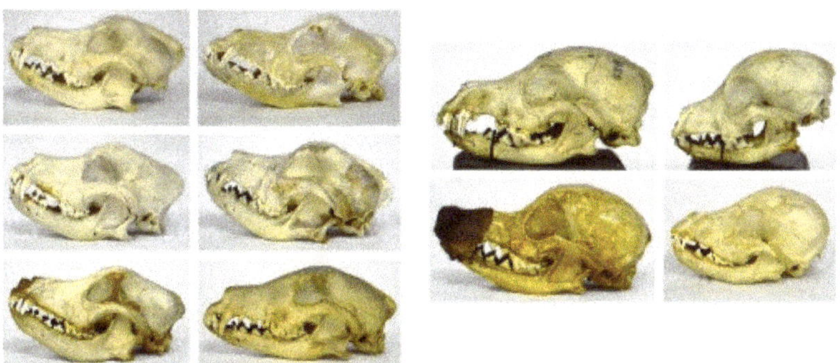

Figure 5. Dog skulls (Veterinary Anatomy Museum, University of Pisa).

Due to this, the temporal muscle has a lower efficiency of contraction in some dog breeds due to domestication.

The wolf nasal bone is long and wide and is continuous with the frontal bone, compressed dorso-ventrally (Figure 6).

With the progressive shortening of the splanchnocranium, due to domestication, the dog nasal bone shows an obtuse angle with the frontal bone (called "stop") in dolichocephalic breeds, and it tends to change to an acute angle in mesaticephalic and brachycephalic breeds (Figure 7). The sagittal crest disappears in brachycephalic breeds.

In the wolf skull, the acute orbital angle, between the tangent line to the top of the skull and the zygomatic orbital line, measures 40°. The orbital cavity is more open in the frontal position because the ventral zygomatic margin of the orbit, which is almost straight, is seen higher than and close to the zygomatic process of the frontal bone (Figure 8).

Figure 6. The wolf skull (Department of Archeological Sciences, University of Pisa), dorsal view.

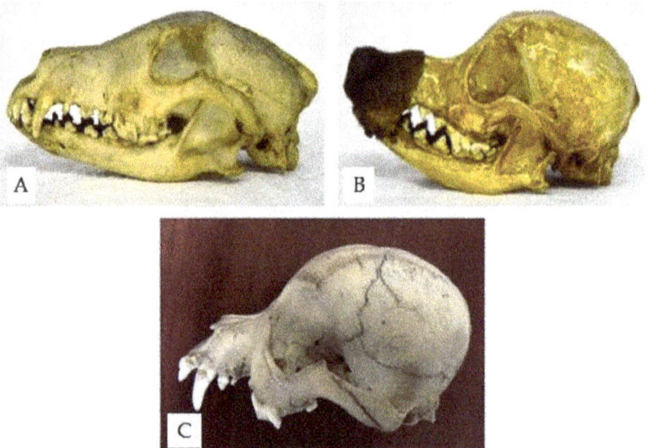

Figure 7. Angle between the nasal and frontal bones ("stop") in dolichocephalic (**A**), mesaticephalic (**B**) and brachycephalic breeds (**C**) (Veterinary Anatomy Museum, University of Pisa).

The zygomatic process of the frontal bone expands on the lateral plane. The cranial cavity, measured by a line between the orbital cavity and the occipital bone, is 125 mm in length and 60 mm in width.

In the dog skull, the orbital cavity has a more rounded shape since the ventral zygomatic edge has a more concave profile, allowing a position that is more lateral to the eyeball (field of view less than 180°). The zygomatic process of the frontal bone progressively reduces until it disappears in brachycephalic breeds (Figure 9).

The orbital angle is measured to be between 53° and 60° in the examined bone specimens. The cranial cavity's length and width are 97–99 mm and 53–56 mm for brachycephalic breeds, 101–103 mm and 56–57 mm for mesaticephalic breeds and 123–124 mm and 57–62 mm for dolichocephalic breeds, respectively.

In the wolf jaw, the masseteric pit (point of insertion of the masseteric muscle) is deep and the profile of the jaw body is linear with the coronoid process (point of insertion of the temporalis muscle); it is broad and rounded at its apex, which diverges laterally, and is in connection with the breadth of the zygomatic arch. Due to shortening of the splanchnocranium during domestication, the profile of the dog jaw body becomes progressively more convex, with the maximum degree present in brachycephalic breeds (Figure 10).

Figure 8. Ventral zygomatic margin of the wolf orbit (inserted line) (Department of Archeological Sciences, University of Pisa).

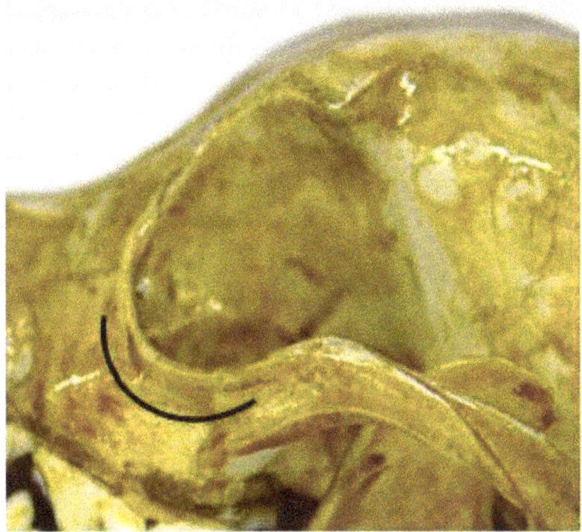

Figure 9. Ventral zygomatic margin of the dog orbit (inserted line) (Veterinary Anatomy Museum, University of Pisa).

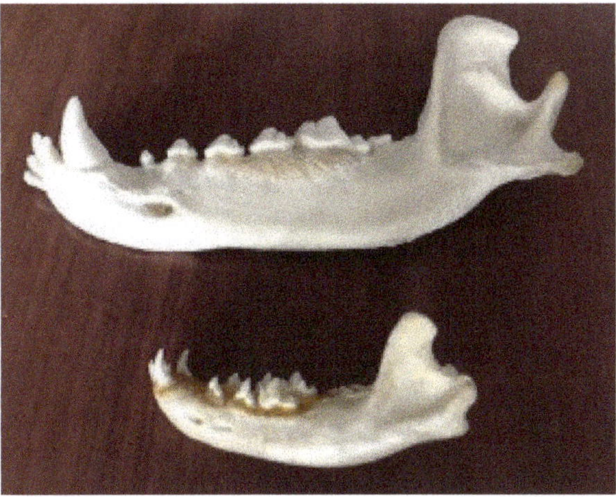

Figure 10. Dog jaws in dolichocephalic (**up**) and brachycephalic (**down**) skulls (Veterinary Anatomy Museum, University of Pisa).

The masseteric pit is shallower than that in the wolf, and the coronoid process is more slender and caudally curved in the profile of the jaw body.

The wolf teeth are typical of a carnivore, with very well developed canine teeth and "feral" or "carnassial" teeth, suitable for slicing and keeping the prey firmly in the mouth. The process of domestication has not resulted in profound changes in the dog teeth; they are typical of a carnivore, with a lateral overlap of the maxillary teeth over the mandibular ones during occlusion.

A deviation in the placement of the teeth results as the sizes of the teeth do not decrease proportionately with a reduction in the length of the jaw; as a result of the shortening of the two bone arches in brachycephalic breeds, the teeth appear closer together, reducing the sizes of the canines and feral teeth, due to a different food supply compared to the wild progenitor [18] (Figure 11).

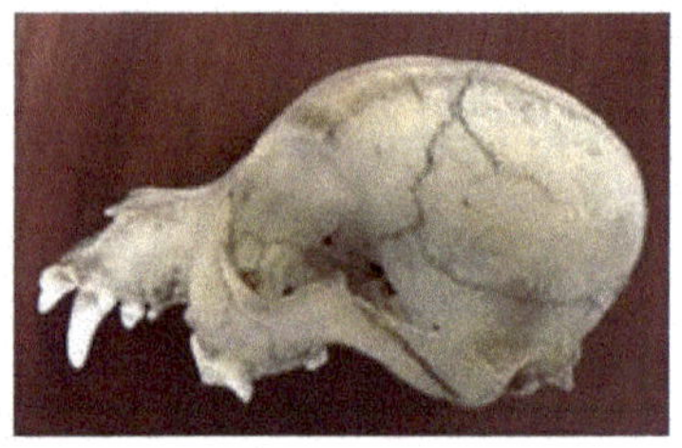

Figure 11. Wolf teeth (**left**) (Department of Archeological Sciences, University of Pisa) and brachycephalic dog teeth (**right**) (Veterinary Anatomy Museum, University of Pisa).

The appendicular skeleton, which includes the wolf forelimb and hindlimb, shows bone segments similar to those in dog skeletons of comparable size; the epiphyses and diaphyses of long bones and the hand and foot bones have the same anatomical features for muscle and ligament attachment and for the joint surfaces for diarthrosis (Figures 12 and 13).

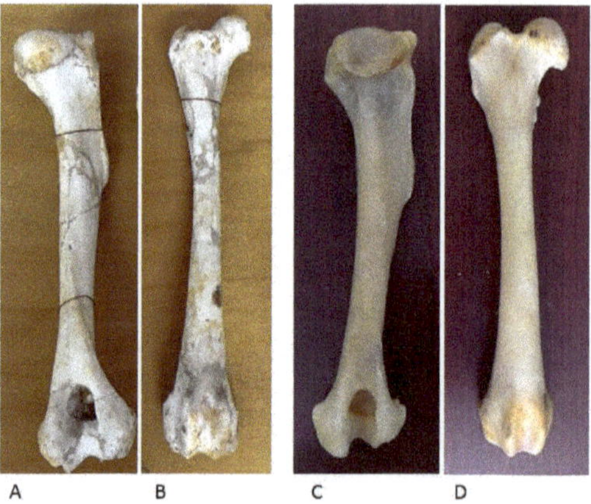

Figure 12. Wolf humerus (**A**) and femur (**B**) (Department of Archeological Sciences, University of Pisa). Dog humerus (**C**) and femur (**D**) (Veterinary Anatomy Museum, University of Pisa).

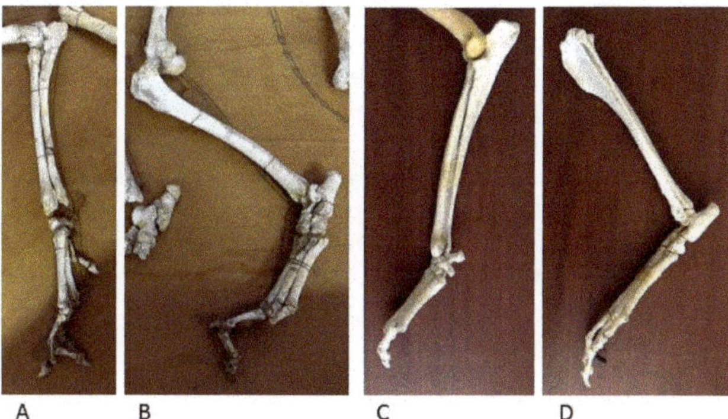

Figure 13. Wolf distal forelimb (**A**) and hindlimb (**B**) (Department of Archeological Sciences, University of Pisa). Dog distal forelimb (**C**) and hindlimb (**D**) (Veterinary Anatomy Museum, University of Pisa).

4. Discussion

The transition from wolf to dog is observed by the study of skeletal features, accompanied by a series of parameters useful in understanding this biological passage. A diminution in size occurs early in the process of animal domestication; this phenomenon is characterized in many animal species, not just the dog.

This work suggests that, in order to differentiate ancient wolf skeletal specimens of *Canis lupus* L. and similar specimens of *Canis familiaris* L., it is necessary to carry out a series of morphological investigations, since there is not one single significant parameter that can be used alone for this investigation.

In the skeletal study, the skull features of the splanchnocranium and neurocranium bones are more interesting than the skeletal features in relation to the appendicular skeleton detected in the lupine subject of the archaeological excavation. In the wolf skull, almost all of the studied anatomical parameters show different features with respect to the same features in the dog specimen. In particular, the morphology of the sagittal crest and the nasal bone, the position of the orbital cavity and the measurement of the orbital angle and the morphology of the masseteric pit proved to be valuable features in differentiating the wolf skull from the dog skull.

The similar features in the wolf and dog teeth indicate that the domestication process did not change the approach to food, which remains typical of a carnivorous animal even after the transformation of the dog into an omnivorous animal. The teeth are therefore more conservative and remain large. Crowding of the teeth and overlapping of cheek teeth is diagnostic of the early domesticated dog compared to wolves [18].

Different developments of the previous parameters consequently indicate a modification in the development of the neurocranium. The analysis of cranial length and width, in the different types of breeds of the analyzed dog skulls, is a useful parameter for discriminating the wolf and dog skulls; indeed it is pointed out that in a dog that is also dolicocephalic, the neurocranium is generally wider and shorter in size, in accordance with some studies in the literature [19] that refer to this feature as being related to the reduction in the development of the limbic system and rhinencephalon during domestication. The morphological differences highlighted could also be attributed to the phenomenon of neotenic pedomorphism, i.e., the conservation in adult dogs of morphological and behavioral traits typical only of different juvenile stages of wolf development, as a result of the selection processes following the domestication process.

Instead, the anatomy of the appendicular skeleton does not vary in the organization of the long or short bones that characterize it; beyond the variable length and morphology of the diaphysis in different dog breeds, domestication has not led to structural variations.

The lack of evidence of remarkable changes in skeletal anatomy from Pleistocene (the geological time to which the examined wolf skeleton belongs) to Upper Paleolithic suggests that no dependent relationship has yet been established between wolves and humans. Casual association must have occurred if a wolf puppy would sometimes be kept by a human family [18].

The only skull differences found might derive from the different types of lives of wolves that would have approached humans; an animal integrated into the world of humans does not need to kill for autonomous survival, leading to a reduced brain capacity and therefore to a change in the morphology of the skull. Since these conclusions are valid only for the small group of subjects included in this work, further studies are needed to validate these conclusions.

Author Contributions: Conceptualization, A.C. and E.G.; methodology, A.C.; investigation, D.P.; resources, A.C.; writing–original draft preparation, A.C.; writing–review and editing, A.C.; supervision, E.G. All authors have read and agreed to the published version of the manuscript.

Funding: This research received no external funding.

Institutional Review Board Statement: Ethical review and approval were waived for this study due to the fact that for the anatomical observations in a wolf skeleton (from archeological site) compared with dog skeletons (preserved in the Veterinary Anatomy Museum of the University of Pisa), no animal manipulation was carried out on living subjects. The study does not fall within the typology for which it is necessary to comply with the indications of the European legislation that regulates its use (Directive 2010/63/EU on the protection of animals used for scientific purposes).

Informed Consent Statement: Not applicable.

Data Availability Statement: Not applicable.

Acknowledgments: All individuals included in this section have consented to the acknowledgement of Department of Archaeological Sciences, University of Pisa.

Conflicts of Interest: The authors declare no conflict of interest.

References

1. Orusa, T.; Orusa, R.; Viani, A.; Carella, E.; Borgogno Mondino, E. Geomatics and EO data to support wildlife diseases assessment at landscapleve: A pilot experience to map infectious keratoconjunctivitis in Chamois and phenological trends in Aosta valley (NW Italy). *Remote Sens.* **2020**, *12*, 3542. [CrossRef]
2. Carella, E.; Orusa, T.; Viani, A.; Meloni, D.; Borgogno Mondino, E. An integrated, tentative remote-sensing approach based on NDVI entropy to model Canine Distemper Virus in wildlife and to prompt science-based management policies. *Animals* **2022**, *12*, 1049. [CrossRef] [PubMed]
3. Norstrom, M. Geographical Information System (GIS) as a tool in surveillance and monitoring of animal diseases. *Acta Vet. Scand.* **2001**, *42*, 79–85. [CrossRef] [PubMed]
4. Altobello, G. Fauna dell'abruzzo e del molise. In *Mammiferi*, 4th ed.; Colitti: Campobasso, Italy, 1921; pp. 1–61.
5. Ciucani, M.M.; Palumbo, D.; Galaverni, M.; Serventi, P.; Fabbri, E.; Ravegnini, G.; Angelini, S.; Maini, E.; Persico, D.; Caniglia, R.; et al. Old wild wolves: Ancient DNA survey unveils population dynamics in Late Pleistocene and Holocene Italian remains. *Peer J.* **2019**, *7*, e6424. [CrossRef] [PubMed]
6. Salvatori, V.; Donfrancesco, V.; Trouwborst, A.; Boitani, L.; Linnel, J.D.C.; Alvares, F.; Akesson, M.; Bally, V.; Blanco, J.C.; Chiriac, S.; et al. European agreements for nature conservation need to explicitly address wolf-dog hybridization. *Biol. Conserv.* **2020**, *248*, 108525. [CrossRef]
7. Fabbri, E.; Miquel, C.; Lucchini, V.; Santini, A.; Caniglia, R.; Duchamp, C.; Weber, J.M.; Lequette, B.; Marucco, F.; Boitani, L.; et al. From the Apennines to the Alps: Colonization genetics of the naturally expanding Italian wolf (*Canis lupus*) population. *Mol. Ecol.* **2007**, *16*, 1661–1671. [CrossRef] [PubMed]
8. Apollonio, M.; Mattioli, L.; Scandura, M.; Mauri, L.; Gazzola, A.; Avanzinelli, E. Wolves in the Casentinesi Forests: Insight for wolf conservation in Italy from a protected area with a rich wild prey community. *Biol. Conserv.* **2004**, *120*, 253–264. [CrossRef]
9. Larson, G.; Karlsson, E.K.; Perri, A.; Webster, M.T.; Ho, S.Y.W.; Peters, J.; Stahl, P.W.; Piper, P.J.; Lingaas, F.; Fredholm, M.; et al. Rethinking dog domestication by integrating genetics, archeology, and biogeography. *Proc. Natl. Acad. Sci. USA* **2012**, *109*, 8878–8883. [CrossRef] [PubMed]
10. Gazzano, A. A nostra immagine e somiglianza, la creazione del cane e delle sue doti mentali e emotive. In *Emotività Animali. Ricerche e Discipline a Confronto*; LED Edizioni Universitarie: Milano, Italy, 2013; pp. 147–154. [CrossRef]
11. Driscoll, C.A.; Macdonald, D.W. Top dogs: Wolf domestication and wealth. *J. Biol.* **2010**, *9*, 10–16. [CrossRef] [PubMed]

12. Ding, Z.L.; Oskarsson, M.; Ardalan, A.; Angleby, H.; Dahlgren, L.G.; Tepeli, C.; Kirkness, E.; Savolainen, P.; Zhang, Y.P. Origins of domestic dog in southern East Asia is supported by analysis of Y-chromosome DNA. *Heredity* **2012**, *108*, 507–514. [CrossRef]

13. Ciucci, P.; Boitani, L. Il lupo: Elementi di biologia, gestione, ricerca. In *Istituto Nazionale per la Fauna Selvatica "Alessandro Ghigi"*; Documenti Tecnici; Ozzano dell'Emilia: Bologna, Italy, 1998; Volume 23, pp. 1–116.

14. Studer, T. *Die Prähistorischen Hunde in Ihrer Beziehung zu den Gegenwärtig Lebenden Rassen*; Zurcher und Furrer ed.: Zurich, Switzerland, 1901; pp. 1–154. Available online: https://publikationen.ub.uni-frankfurt.de/opus4/frontdoor/deliver/index/docId/5947/file/E000013978.pdf (accessed on 13 December 2022).

15. Janssens, L.; Spanoghe, I.; Miller, R.; Van Donden, S. Can orbital angle morphology distinguish dogs from wolves? *Zoomorphology* **2016**, *135*, 149–158. [CrossRef] [PubMed]

16. Severtsov, A.S.; Kormylitsin, A.A.; Severtsova, E.A.; Yatsuk, I.A. Functional Differentiation of Teeth in the Wolf (Canis lupus, Canidae, Carnivora). *Biol. Bull.* **2016**, *43*, 1271–1280. [CrossRef]

17. Olsen, J.S.; Olsen, J.W. The Chinese Wolf, Ancestor of New World Dogs. *Science* **1977**, *197*, 533–535. [CrossRef] [PubMed]

18. Evans, H.E. *Miller' Anatomy of the Dog*, 3rd ed.; Elsevier: Amsterdam, The Netherlands, 2006.

19. Zeder, M.A. The Domestication of Animals. *J. Anthropol. Res.* **1982**, *9*, 321–327. [CrossRef]

Article

Surgical and Bioengineering Integration in the Anatomy Course of Medicine and Surgery High Technology: Knowledge and Perception of Anatomy

Selenia Miglietta [1], Giuseppe Familiari [1], Michela Relucenti [1,*], Stefania Basili [2], Fabiano Bini [3], Gabriele Bove [4], Claudio Barbaranelli [5] and Pietro Familiari [6]

[1] Department of Anatomy, Histology, Forensic Medicine and Orthopedics, Sapienza University of Rome, 00161 Rome, Italy
[2] Department of Translational and Precision Medicine, Sapienza University of Rome, 00185 Rome, Italy
[3] Department of Mechanical and Aerospace Engineering, Sapienza University of Rome, 00184 Rome, Italy
[4] I.N.I. Group, Orthopaedic Unit, Grottaferrata, 00046 Rome, Italy
[5] Department of Psychology, Sapienza University of Rome, 00185 Rome, Italy
[6] Division of Neurosurgery, Department of Human Neurosciences, Policlinico Umberto I, Sapienza University of Rome, 00185 Rome, Italy
* Correspondence: michela.relucenti@uniroma1.it

Abstract: The Locomotor System Anatomy (LSA) course, placed in the first semester of the first year of the new Master's degree in Medicine and Surgery High Technology (MSHT) at the Sapienza University of Rome, was integrated with surgical and bioengineering content. This study investigated the educational value and the students' perceptions of the effectiveness of these two types of integration, comparing surgical integration (SI) with engineering integration (EI). Anatomy knowledge and students' opinions attending the LSA course in MSHT degree (n = 30) were compared with those of students (n = 32) attending another medical and surgery course not comprising EI. Data show that students in the MSHT course like in-depth SI much more than in-depth EI. However, those who like in-depth SI also like in-depth EI. Significant differences were in anatomy knowledge between the two groups in the three sections of the test. There was no significant correlation between the three test scores and the levels of liking, while there was a significant correlation between students liking SI and those liking EI. A statistically significant correlation was also found in students who correctly responded to questions on the head and trunk, with students responding correctly to questions on the upper limbs. This study will be important in optimizing the deepening of SI and EI in the LSA course.

Keywords: human anatomy; anatomical sciences education; gross anatomy teaching; locomotor system; neurosurgery; orthopedics; surgical integration; bioengineering integration; technical physician; technical medicine

Citation: Miglietta, S.; Familiari, G.; Relucenti, M.; Basili, S.; Bini, F.; Bove, G.; Barbaranelli, C.; Familiari, P. Surgical and Bioengineering Integration in the Anatomy Course of Medicine and Surgery High Technology: Knowledge and Perception of Anatomy. *Anatomia* **2023**, *2*, 63–77. https://doi.org/10.3390/anatomia2010006

Academic Editors: Gianfranco Natale and Francesco Fornai

Received: 29 November 2022
Revised: 16 December 2022
Accepted: 28 January 2023
Published: 1 February 2023

1. Introduction

The extreme dynamism and continuous introduction of new technologies that improve patient care and preventive health care characterize modern medicine; thus, the implementation of the educational curriculum of medicine and surgery with bioengineering contents is fundamental. Such educational curriculum modernization is aimed at creating future physicians with skills for digital health and quality improvement, a mindset for precision and personalized medicine. Health professionals, informed about the advantages of artificial intelligence, machine learning, medical robotics, network medicine, big data analysis, genomics, omics, and all other technologies related to bioengineering, bioinformatics, and bioelectronics will be introduced in this work [1–4]. The first Master's degree course in Medicine and Surgery High Technology (HT) in the Italian State University system was activated at the Sapienza University of Rome in the academic year 2020–2021 [5,6]. Strong vertical integration of basic and clinical sciences characterizes this degree course [5,7];

thus, the Human Anatomy course (placed in the first and second years) is integrated with surgical and bioengineering content. In particular, the Human Anatomy 1 module (focused on the locomotor system) is articulated with anatomy lessons interspersed with surgical lessons (given by orthopedists and neurosurgeons, who collaborate in subsequent modules of the anatomy course [8]) and is further implemented with a module on biological systems mechanics and biomechanics of the locomotor system (this module is carried out by an industrial bioengineering teacher). While the integration of anatomy with the clinical sciences is now well established in anatomy teaching [9,10], the integration with a bioengineering discipline is a major pedagogical innovation. Therefore, in this study, we investigated the educational value and the students' perceptions of the effectiveness and usefulness of this new type of integration compared to surgical integration.

2. Materials and Methods

2.1. The Undergraduate Course of Medicine and Surgery HT at Rome Sapienza University's Medical School

The undergraduate course is held at the Policlinico Umberto I Hospital associated with the faculty; the biomedical-technological training program consists of a 6-year curriculum, designed in collaboration with the Faculty of Medicine and Engineering, and trains students in both technological and bioengineering skills [6].

This new undergraduate curriculum was activated in the academic year 2020–2021 and was divided into 12 semesters and included 36 integrated courses with related exams. Bioengineering sciences are added to the vertical and horizontal integration of basic and clinical sciences during the six years of the degree program. This curriculum organization reduces the emphasis on teacher-centered lectures and focuses on a more student-centered learning model. For this purpose, activities, including practical integrated experiences and tutorials or seminars, were introduced since the first-year course [5,6,8].

2.2. The Course of the Locomotor System of Medicine and Surgery HT at the Sapienza University of Rome

The locomotor system course is a part of the initial stage of the human anatomy curriculum taught during the first-year's first semester of the undergraduate course. It comprises face-to-face lessons as well as practicals when students in small groups use plastic models, histology slides, and interactive multimedia tools proposed by the teacher.

Clinically-integrated lectures were organized as a presentation of clinical cases by an orthopedic surgeon or a neurosurgeon in the presence of an anatomy teacher, who actively contextualizes the presentation of surgical cases.

Surgical integrated lessons consisted of the clinical case presentation: the orthopedic surgeon illustrated cases of the shoulder, hip, and knee joint surgery, emphasizing the technical aspects of the operative procedures, including robotic surgery (examples of the orthopedic surgeon's teaching activities are shown in Figure 1).

The neurosurgeon presented the cases of brain and spinal cord surgery, describing the particular operative techniques of intervention, such as the different types of craniotomies and the different surgical approaches to the spine. Special emphasis was given to illustrating the advantages of innovative neuronavigation techniques concerning traditional approaches (examples of neurosurgeon's activities are shown in Figure 2).

As a further activity in addition to the lectures, small groups of students were organized to train their technical and surgical skills by simulating both orthopedic robotic surgery (Mako robotics–Stryke®, Kalamazoo, MI, USA) and neurosurgery with the aid of 3D simulation and neuronavigation tools (Brain-Lab®, Munich, Germany).

The bioengineering-integrated lectures, given by a teacher in industrial bioengineering, were intended to provide fundamental knowledge about the study of motion and stresses in biological systems to forces caused by biomechanical phenomena. The basic principles of computational biomechanical analysis of a multi-link model of the human body were also explained. Frontal teaching was devoted to the reference system for biomechanical analysis, degrees of freedom of joints, elementary movements and kinematic models of limbs, and geometry of masses. The hands-on exercises used specific platforms to illustrate the principles of numerical and computational methods for biomechanical analysis and were devoted to the modeling of the limb to analyze and simulate muscle actions and exchanged forces. Examples of the bioengineering teacher's teaching activities are shown in Figure 3.

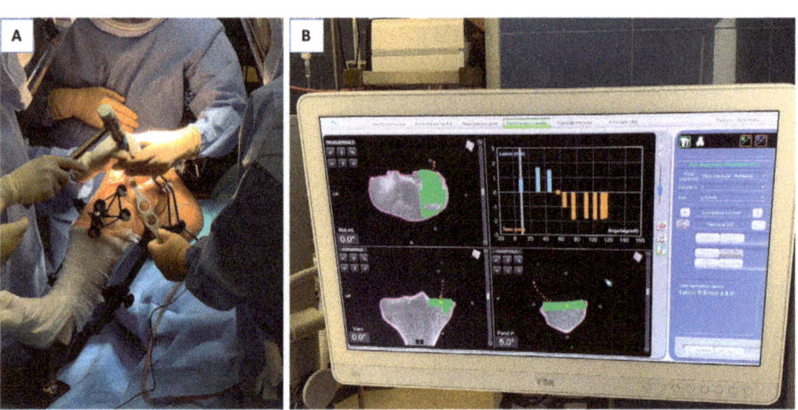

Figure 1. (**A**) Knee joint prosthetic implant surgery using the Mako robotics–Stryker® robotic system. Note the presence of the navigation system applied to the patient's leg. (**B**) Control monitor of Mako robotics–Stryker® robotic system. During knee joint implant surgery, the surgeon can observe the removed bone surface while performing the procedure and placing the prosthesis.

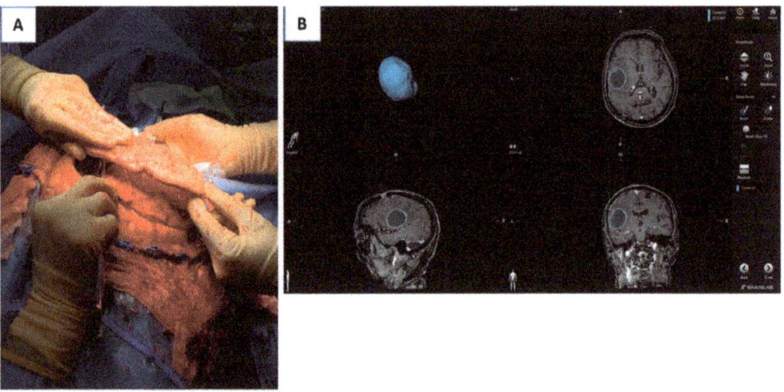

Figure 2. (**A**) Separation of the craniotomy bone operculum from the dura mater during decompressive craniotomy surgery for the evacuation of acute subdural hematoma. (**B**) BrainLab® neuronavigation monitor during neurooncological surgery. The 3D reconstruction of the intra-axial neoplastic lesion: axial, sagittal, and coronal of the tumor lesion inside the skull.

2.3. Student Sampling

Students (n = 62) who attended at least 67% of the mandatory locomotor system anatomy course during the academic year 2021–2022 were the subject of this study. The ages of the students ranged between 19 and 20; 65.6% of them were female, and 34.4% were male. Students (n = 30) from the first year of the course degree in medicine and surgery HT (Faculty of Medicine and Dentistry) were considered the study group, while students (n = 32) attending at least 67% of the mandatory anatomy course of the locomotor system and belonging from the first year of another degree in medicine and surgery course degree not comprising engineering integration (Faculty of Medicine and Psychology), acted as the control group. The anatomy and clinical teachers were the same in both the HT Medicine course at the Faculty of Medicine and Dentistry and the Medicine and Surgery course at the Faculty of Medicine and Psychology.

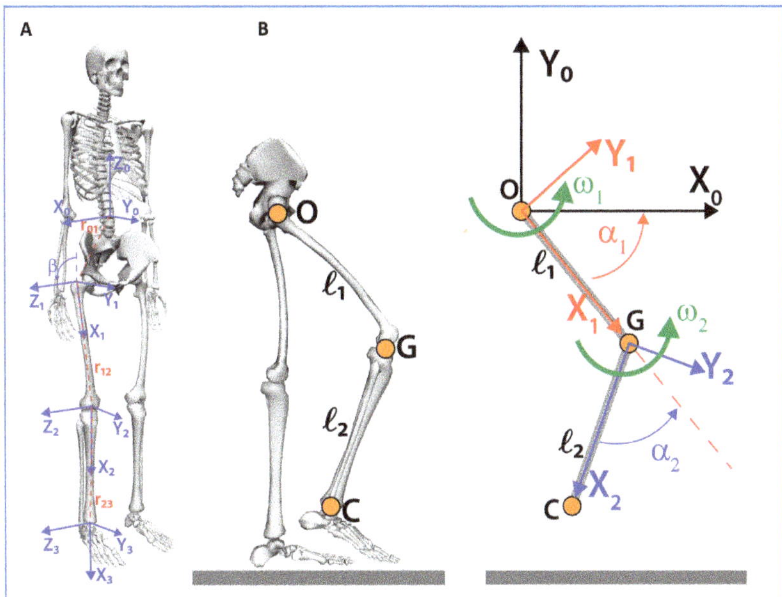

Figure 3. (**A**) Illustration of the coordinate systems of human skeletal structures: torso coordinate system X0Y0Z0 with origin coincident with the barycentre of the body, hip joint coordinate system $X_1Y_1Z_1$, knee joint coordinate system $X_2Y_2Z_2$, and ankle joint coordinate system $X_3Y_3Z_3$. Vectors r_{01}, r_{12}, and r_{23} connect the coordinate system [i-1] with the system [i], where i = 1,2,3. The coordinate system (1) is obtained from the translation of the coordinate system(0)in the center of the hip joint and the subsequent rotation around the Y_1 axis by an angle β. (**B**) Kinematic analysis of the lower limb using a simplified model composed of two segments of length ℓ_1, and ℓ_2, respectively, knowing the system configuration through the angles α and the joint angular velocities ω.

2.4. Students' Views

The questionnaire used in this study was adapted from a questionnaire previously used in our study dealing with the integration of neurosurgery in neuroanatomy, where the Cronbach alpha was 0.9707 [8]. It was administered after the last lecture on the locomotor system course to all Medicine and Surgery HT students (n = 30) who attended the entire cycle of lessons. The students were asked to fill out the questionnaire anonymously. Informed consent to participate was obtained from each student after explaining the objective and purpose of the study.

The questionnaire was divided into two sections, A and B.

Section A of the questionnaire was designed to gather data concerning the didactic usefulness of clinically-integrated or engineering-integrated learning. Subsection A1 evaluated the usefulness or the uselessness of the surgical integration, whereas Subsection A2 evaluated the usefulness or the uselessness of the engineering integration.

Section B of the questionnaire was designed to collect data on the didactic usefulness of clinical case presentations or principles of computational biomechanical analysis modeling. Subsection B1 evaluated the usefulness or the uselessness of the lessons, including orthopedic and neurosurgical case presentations, while Subsection B2 evaluated the usefulness or the uselessness of the lessons in which modeling computational biomechanical analysis principles have been presented.

Each Subsection, A1/A2 and B1/B2, contained four topic-related items. Two of these items were positive, while the other two were negative, in agreement with a twofold cross-check. A Likert scale was applicated to each item to test the agreement of the students. Respondents were invited to indicate their agreement on a five-point Likert scale (1 = strongly disagree, 2 = disagree, 3 = neither agree nor disagree, 4 = agree, 5 = strongly agree).

2.5. Data Analysis of Students' Views

Individual responses obtained using the Likert scale were treated as variables measured at the level of equivalent intervals, and then mean standard deviations were computed being an adequate statistic for this measurement level [11] (regarding treating Likert scales at the level of equivalent intervals, see also [12]). Pairwise comparisons of sections A1 vs. A2 and B1 vs. B2 were performed using the t-student test. Statistical significance was established at $p \leq 0.05$. The internal consistency of the data was assessed using Cronbach's Alpha calculations. A Cronbach's Alpha value higher than 0.70 is considered adequate for the internal consistency of the questionnaire [13]. Data were analyzed using IBM-SPSS 27.

2.6. Assessment of Students' Knowledge of Locomotor System Anatomy

The students' knowledge of the functional anatomy of the locomotor system was assessed using a test containing 30 questions on the functional anatomy of the locomotor system taken from a human anatomy text containing a collection of multiple-choice questions for testing and self-testing of learning used by Italian students [14]. The test was structured into 30 questions, out of which 11 concerned the functions of the muscles of the head and neck, another 11 pertained to the functions of the muscles of the upper limb, and, finally, eight questions were inherent to the actions of the muscles of the lower limb. The same test was administered, after the last lecture on the locomotor system anatomy course, to students belonging to the Medicine and Surgery HT degree (study group, n = 30) and to students belonging to another degree in medicine and surgery at Sapienza University of Rome (control group, n = 32). The 30 test questions are provided as material in Appendix A of this scientific article.

2.7. Data Analysis of the Assessment of Students' Knowledge of Locomotor System Anatomy

The sum of the correct answers in the three sections of the administered test questions on locomotor system anatomy was considered a measure of learning. Differences between traditional and HT students in these three scores were analyzed using the ANOVA test. Statistical significance was established at $p \leq 0.05$. Data were analyzed using IBM-SPSS 27.

2.8. Correlation Analysis between Student Opinions and Anatomy Knowledge Results

Correlation analysis among students' opinions on the usefulness of clinical integration or engineering integration and results obtained in the assessment test of students' functional anatomy knowledge was performed using the Bravais–Pearson coefficient. Statistical significance was established at $p \leq 0.05$. Data were analyzed using IBM-SPSS 27.

3. Results

3.1. Students' Opinions

3.1.1. Evaluation of the Usefulness of Clinically-Integrated versus Engineering-Integrated Lessons in Learning Functional Anatomy

Data are presented in Table 1 in two distinct subsections: Subsection A1 contains data concerning the usefulness or the uselessness of surgical integration, whereas Subsection A2 presents data on the usefulness or the uselessness of the bioengineering integration.

Table 1. Student view of didactic usefulness of surgical vs. bioengineering integrated class usefulness in learning functional anatomy.

Subsection A1 Surgical Integration:	Mean ($\pm$SD) [a] Min–Max	Subsection A2 Bioengineering Integration:	Mean ($\pm$SD) [a] Min–Max
It is useful to improve general knowledge of locomotor system functional anatomy (A1.1)	4.33 ($\pm$0.884) 2–5	It is useful to improve one's general knowledge of locomotor system functional anatomy (A2.1)	3.70 ($\pm$0.750) 2–5
Makes the lessons dynamic and interesting (A1.2)	4.50 ($\pm$0.572) 3–5	Makes the lessons dynamic and interesting (A2.2)	3.20 ($\pm$0.925) 1–5
It is totally useless for improving one's general knowledge of the locomotor system functional anatomy (A1.3)	1.70 ($\pm$1.022) 1–5	It is completely useless for improving general knowledge of locomotor system functional anatomy (A2.3)	2.07 ($\pm$0.785) 1–4
Makes lessons hard and boring (A1.4)	1.43 ($\pm$0.568) 1–3	Makes lessons hard and boring (A2.4)	2.60 ($\pm$0.894) 1–4

[a] Likert scale: 1 = strongly disagree, 2 = disagree, 3 = neither agree nor disagree, 4 = agree, 5 = strongly agree; Cronbach's alpha = 0.876.

As shown in Table 1, the answers to statement A1.1 (opposite A1.3) showed that a very high number of students stated that clinical integration of the locomotor system was useful for improving one's general knowledge of locomotor system functional anatomy. Even responses to statement A2.1 (opposite A2.3) showed that a high number of students stated that the engineering integration of the lessons from the locomotor system lessons was useful for improving general knowledge of locomotor system functional anatomy.

In Table 2, the results of the paired *t*-test with comparisons between statements of Section A1 and Section A2 are presented. Data revealed that student satisfaction was higher for surgical integration compared to satisfaction for engineering integration.

Table 2. Paired *t*-test results with comparisons between statements of Sections A1 and A2.

Statements	Mean Difference	T
A1.1 vs. A2.1 positive	0.633	3471 **
A1.2 vs. A2.2 positive	1300	7208 ***
A1.3 vs. A2.3 negative	−0.367	−1690
A1.4 vs. A2.4 negative	−1167	−7000 ***

The degrees of freedom are 29 for all tests. ** $p \leq 0.01$, *** $p \leq 0.001$.

Furthermore, the answers to statement A1.2 (opposite to A1.4) revealed that a very high number of students stated that the surgical integration of the locomotor system makes the lessons dynamic and interesting. The answers to Statement A2.2 also reflected students' positive assessment of engineering integration, although this evaluation was significantly lower than the evaluation of surgical integration evaluation.

However, the responses to the questions about the perceived lack of usefulness or boringness of these lessons show very low values, although, in this case, there is a statistical difference in the difficulty of engineering integration, which prevails over the difficulty found in surgical integration. In contrast, there were no statistically significant differences in perceived low usefulness for surgical or engineering integration, both of which show very low values.

3.1.2. Evaluation of Clinically-Integrated or Engineering-Integrated Lessons' Usefulness in Learning Morphological Anatomy

Data are presented in Table 3 in two distinct subsections: Subsection B1 contains data concerning the usefulness or the uselessness of lessons, including surgical cases, whereas Subsection B2 contains data on the usefulness or the uselessness of the lessons, including modeling computational biomechanical analysis principles.

Table 3. Students' views of the didactic usefulness of surgical case presentations or principles of modeling the computational biomechanical analysis.

Subsection B1 Lessons Including Surgical Cases:	Mean (±SD) [a] Min–Max	Subsection B2 Lessons Including Modeling Computational Biomechanical Analysis Principles:	Mean (±SD) [a] Min–Max
B1.1 improve one's general knowledge of locomotor system morphodynamics	4.30 (±0.794) 3–5	B2.1 improve one's general knowledge of locomotor system morphodynamics	3.90 (±0.662) 3–5
B1.2 are totally useless for improving one's general knowledge of locomotor system morphodynamics	1.53 (±0.629) 1–3	B2.2 are totally useless for improving one's general knowledge of locomotor system morphodynamics	1.97 (±0.669) 1–3
B1.3 are dynamic and interesting	4.33 (±0.711) 3–5	B2.3 are dynamic and interesting	3.23 (±1.135) 1–5
B1.4 are hard and boring	1.60 (±0.675) 1–3	B1.4 are hard and boring	2.37 (±1.066) 1–4

[a] Likert scale: 1 = strongly disagree, 2 = disagree, 3 = neither agree nor disagree, 4 = agree, 5 = strongly agree; Cronbach's alpha = 0.876.

As shown in Table 3, responses to statement B1.1 (opposite B1.2) showed that a very high number of students stated that lessons, including clinical cases related to orthopedics and neurosurgery, improve general knowledge of locomotor system morphodynamics. Even answers to statement B2.1 (opposite B2.2) showed that a high number of students stated that lessons incorporating modeling computational biomechanical analysis principles improve one's general knowledge of locomotor system morphodynamics.

In Table 4, the results of the paired *t*-test with comparisons between statements of section B1 and section B2 are presented. Data revealed that student satisfaction was higher for the lessons that included clinical cases compared to satisfaction for the lessons that included computational biomechanical analysis principles.

Table 4. Paired *t*-test results with comparisons between statements of Sections B1 and B2.

Statements	Mean Difference	T
B1.1 vs. B2.1 positive	0.149	2.693 *
B1.2 vs. B2.2 negative	0.141	−3.067 **
B1.3 vs. B2.3 positive	0.216	5.086 ***
B1.4 vs. B2.4 negative	0.207	−3.699 ***

The degrees of freedom are 29 for all tests. * $p \leq 0.05$; ** $p \leq 0.01$, *** $p \leq 0.001$.

Additionally, the answers to statement B1.3 (opposite B1.4) revealed that a very high number of students stated that the lessons, including clinical cases, were dynamic and interesting. A positive student lesson evaluation of the lessons that included modeling computational biomechanical analysis principles was also reflected by statement B2.3, even if this evaluation was lower than the evaluation of the surgical presentation.

However, the responses to the questions about the perceived low usefulness or boringness of these lectures show very low values, although, in this case, there is a statistical difference in the difficulty of the engineering lectures, which prevails over the difficulty found in the clinical integration lectures. There are statistically significant differences regarding perceived low usefulness for the clinical or engineering classes. Both show very low values but are significantly different from each other.

The Cronbach's alpha index value (0.876) revealed a good internal consistency for Sections A and B of the questionnaire.

3.2. Assessment of Student Knowledge of Locomotor System Anatomy

As shown in Table 5, results obtained in the assessment test of students' knowledge of the functional anatomy of the locomotor system showed differences between the two groups of students in the three sections of the test.

Table 5. Results obtained in the assessment test of students' knowledge of the functional anatomy of the locomotor system (Ns = number of students; Nq = number of questions).

Test	Students	Ns	Nq	Mean (±SD)	95% CI	Min	Max	ANOVA
Head and Trunk	HT	30	11	7.0667 (±2.0833)	6.2887–7.8446	2	11	$F = 7.911$ **
	Control	32	11	5.4688 (±2.3689)	4.6147–6.3228	1	10	
Upper Limb	HT	30	11	5.1000 (±1.5833)	4.5088–5.6912	0	8	$F = 3.834$
	Control	32	11	4.2188 (±1.9299)	3.5229–4.9146	0	8	
Lower Limb	HT	30	8	2.5000 (±1.3582)	1.9928–3.0072	0	5	$F = 9.454$ **
	Control	32	8	3.7188 (±1.7271)	3.0961–4.3414	0	7	

** $p \leq 0.01$.

In fact, in the 11 questions related to the functions of the head and trunk muscles, students in the HT course obtained significantly better results than students in the control course in terms of the number of correct answers.

In the 11 questions related to upper limb muscle functions, HT students also obtained better results, bordering on statistical significance, than control group students in terms of the number of correct answers (note that the significance level is only very slightly above the significance level of 0.05).

In contrast, the opposite result was obtained in the responses to eight questions related to lower limb muscle function. In fact, in this case, there was a significant prevalence of correct answers in the control medicine and surgery students compared to the correct answers given to the HT medicine and surgery students.

3.3. Correlation Analysis between Students' Views and Anatomy Knowledge Results

As shown in Table 6, there is a significant correlation between the evaluation of the usefulness of the two different types of clinical and engineering integration. Instead, these two values did not show any significant correlation with the three subtests measuring knowledge of the topic thought. A significant positive correlation emerged between the "head and neck" and "upper limb" subtests.

Table 6. Pearson correlation coefficients between the usefulness of surgical integration or bioengineering integration and the results obtained in the assessment test of students' functional anatomy knowledge (head and trunk, upper limb, lower limb).

	Surgical Integration	Bioengineering Integration	Head and Trunk	Upper Limb
Bioengeneering Integration	0.491 **			
Head and Trunk	−0.197	−0.168		
Upper Limb	−0.980	0.038	0.408 **	
Lower Limb	−0.039	0.302	0.171	0.032

N = 30. ** Correlation significance ≤ 0.01 (two-tails).

4. Discussion

4.1. Curricular Integration in Medical Schools

The educational curricula of Italian medical schools are characterized by the presence of horizontal and vertical integration through the use of different interactive and multidisciplinary pedagogical approaches [5,7,9,15]. These innovative curricula can be represented by a Z shape or as an inverted triangle structure, where students are introduced to clinical sciences at the beginning of the curriculum [5,7,9,15]. From a broader perspective, Italian curricula are now adapting to a general definition of vertical integration, allowing for a gradual involvement of students and young physicians in the professional community. This happens through a gradual increase in responsibilities in patient care, crossing the boundaries among undergraduate, postgraduate, and continuing medical education [5,16,17]. In the new Medicine and Surgery HT courses in international universities [1–3] and Italy, at Sapienza University of Rome [6], the curriculum is fully embedded with the disciplinary scientific field of biomedical engineering. This new curriculum needs new forms of integration with the human anatomy course, involving, for the first time, biomedical engineering.

4.2. New Curricular Integration in the Human Anatomy Course: Students' Views

In this general context, the surgical integration of human anatomy lectures, which usually occur in the first and second years of the degree program, as well as the presence of in-depth human anatomy lectures in later years within integrated clinical courses, represent long-recommended goals for improving human anatomy teaching and learning, contextualizing its learning concerning the procedures and clinicaßl skills to be acquired by the student [9,10,18–22]. Therefore, the clinical integration of anatomy is not new but has been established for many years in medicine and surgery Master's degree courses.

Anatomical knowledge is the basis of clinical procedures and the focus of the anatomy program should be centered on being propaedeutic to the clinical sciences [19]. The current trend is also to increase clinical integration in the human anatomy course to provide students with a stronger motivation to learn and understand every single topic [9,23,24]. In addition, this integration with technology provides students with a guiding principle for the problem-solving process that underlies the reasoning of a physician with technological expertise. The expertise that must underlie the problem-solving of this technological physician is precise and involves an integrated analysis of anatomy, physiology, pathophysiology, and technology [1]. The aim of integrating biomechanic skills into the human anatomy of the locomotor system course is to help students understand and apply robotic technologies related to joint surgery or spinal surgery, robotic surgery being an increasingly used technique in current surgical practice [25–29].

Our data showed that students like in-depth surgical integration much more than in-depth engineering integration. However, those who like in-depth surgical studies also like in-depth engineering studies. There was no significant correlation between the three test scores and levels of liking, while there was a significant correlation between students who liked clinical integration and those who liked engineering integration. A statistically significant correlation was also found in students who correctly answered questions on the head and trunk, with students answering correctly to questions on the upper limb.

The international literature lacks reports on this type of transdisciplinary teaching integration; our data are the first from an Italian HT Master's degree program in medicine and surgery and provide interesting insights. The greater preference for clinical integration vs. engineering integration is not surprising. The lectures given by the orthopedic surgeon and neurosurgeon showed clinical cases easily understandable even by the first-year students. In the surgically-integrated lectures, the anatomical aspects were emphasized, creating a strong link between surgery and anatomy. This had two effects: on the one hand, the students were motivated; on the other, they were allowed to contextualize the complex topics of the locomotor system anatomy. This was important to increase the otherwise misperceived usefulness of some topics in the anatomy program. Engineering-integrated lectures were also appreciated, even if to a lesser extent, being perceived as more complex lectures of the first year; this was due to a deeper knowledge of mathematics and calculus required to fully understand the usefulness of this type of lecture. Based on the results obtained in this study, the engineering-integrated lectures will be calibrated in the next year of the course to delve into aspects of biomechanics at a level more appropriate for medical students.

4.3. New Curricular Integration in the Human Anatomy Course: Students' Knowledge of Functional Anatomy

Data obtained in the test of knowledge of the functional anatomy of the locomotor system showed significant differences in the higher number of correct answers given by the students in the HT course, compared with the students in the control group, in the head and trunk and upper limb sections of the test, while students in the control group significantly better answered the lower limb questions. The administered test contained no specific questions on the engineering topics covered in class, partly because the same test was administered to students from the other medical and surgical courses at Sapienza, where this type of integration was not present, and the questions were standard questions taken from a text used by all Italian medical students [14]. The addition of questions more relevant to engineering topics would probably have resulted in a better learning outcome but introduced a bias in the analysis on the actual usefulness of this type of integration.

The difference in results found in the different sections (or sub-tests) of the test may be because, in the HT medicine course, four fewer hours of lecture on the lower limb were provided (12 percent less than the total hours), compared to the hours of the lecture provided to the students in the control group, who thus had the opportunity to study this part of the program in greater depth. Another possible cause of these differences may be due to the personal preference in studying a specific topic, but unfortunately, preference for a specific topic was not asked about, so this could be an object of new studies. Further analysis on this point will help to specifically define the actual usefulness of engineering integration within the anatomy of the locomotor system course in the HT Master's degree program in medicine and surgery.

4.4. Study Limitations and Strengths

The limited number of students who participated in the study represents a limitation of the research, even though this new degree program has fewer students enrolled, and almost all of them have been involved in the study itself. Further analysis is needed to explain the differences in locomotor system anatomy test knowledge between the two groups of students analyzed.

4.5. Conclusions

This study, and the other in-depth analyses that will follow, are important in recalibrating and optimizing new modalities for curricular integration in human anatomy courses of the new Master's degree courses of Medicine and Surgery HT recently activated in Italy for the training of physicians with technological skills.

Author Contributions: Conceptualization, S.M., G.F. and P.F.; methodology, G.F., S.B., C.B., S.M. and P.F.; software, C.B.; validation, G.F., S.B., S.M., M.R. and P.F.; formal analysis, S.M., P.F. and M.R.; investigation, P.F., S.M., F.B., G.B. and C.B.; resources, M.R.; data curation, S.M. and P.F.; writing—original draft preparation, S.M., G.F. and P.F.; writing—review and editing, C.B., F.B. and S.B.; visualization, S.M. and P.F.; supervision, M.R., S.M. and P.F.; project administration, S.M. and P.F.; funding acquisition, S.M. All authors have read and agreed to the published version of the manuscript.

Funding: This research was funded by Sapienza University of Rome Ateneo 2021 grant number B83C22000960005.

Institutional Review Board Statement: Ethical review and approval were waived for this study due to its observance of the Italian University System's evaluation regulations administered by the National Agency for the Evaluation of the University System and Research (ANVUR-https://www.anvur.it/, accesed on 28 November 2022) and due to the observance of Sapienza University rules for anonymous students' satisfaction evaluation and Teaching Quality Assessment Questionnaire.

Informed Consent Statement: Informed consent was obtained from all subjects involved in the study.

Conflicts of Interest: The authors declare no conflict of interest.

Appendix A

Table A1. Anatomy test questions. Locomotor system Anatomy—Functional Anatomy.

The interspinal muscles' actions:	(a) flex the cervical spine (b) extend the spine (c) head on the neck rotation (d) flex the cervical and lumbar spine (e) contributes to respiratory movements
External oblique muscle action:	(a) bilaterally lowers the ribs and increases abdominal pressure (b) unilaterally flexes and rotates the trunk homolaterally (c) unilaterally rotates the thorax on the same side (d) bilaterally raises the ribs and increases abdominal pressure (e) bilaterally rotates the thorax on the opposite side
The coracobrachialis muscle:	(a) rotates the shoulder joint medially (b) extends the shoulder joint (c) adducts the shoulder joint (d) adducts the shoulder joint (e) laterally rotates the shoulder joint
Which of these statements is correct?	(a) the deltoid muscle is the main flexor of the arm (b) the deltoid muscle is the main adductor of the arm (c) the deltoid muscle is the main lateral rotator of the arm (d) the deltoid muscle is the main medial rotator of the arm (e) the deltoid muscle is the main abductor of the arm.
These statements are all true for the muscles of the anterior ligament of the leg except for one. Which one?	(a) they are responsible for the extension of the toes (b) they are responsible for the dorsal flexion of the ankle (c) they are innervated by the deep peroneal nerve (d) they are innervated by the femoral and saphenous nerves (e) they are responsible for the inversion/eversion of the ankle
Which of these statements is incorrect?	(a) the quadratus femoris muscle is considered an adductor muscle (b) the gracilis muscle is considered an adductor muscle (c) the pectineus muscle is considered an adductor muscle (d) the long adductor muscle is considered an adductor muscle (e) the short adductor muscle is considered an adductor muscle
Which of these statements is incorrect?	(a) the coracobrachialis muscle adducts the shoulder joint (b) the coracobrachialis muscle flexes the shoulder joint (c) the great round muscle rotates the arm medially (d) the deltoid muscle flexes the arm (e) the infraspinatus muscle rotates the arm medially
The semimembranosus muscle is	(a) lateral rotator of the thigh (b) flexor of the leg and extensor of the thigh (c) thigh flexor (d) extensors of the leg (e) adductor of the thigh
Which of these statements is incorrect?	(a) the great round muscle flexes the arm (b) the coracobrachialis muscle flexes the shoulder joint (c) the coracobrachialis muscle adducts the shoulder joint (d) the infraspinatus muscle rotates the arm medially (e) the great round muscle rotates the arm medially

Table A1. *Cont.*

Which of these muscles are the external rotators of the hip?	(a) the external obturator muscle (b) the gluteus medius muscle (c) the great gluteus muscle (d) the gluteus minimus muscle (e) the tensor fascia lata muscle
Which of these muscles are the internal rotators of the hip?	(a) the great gluteus muscle (b) the quadratus femoris muscle (c) the internal obturator muscle (d) the gluteus medius muscle (e) the external obturator muscle
The long head of the triceps brachii muscle is a:	(a) supinator of the forearm (b) extensor of the elbow (c) flexor of the elbow (d) pronator of the forearm (e) abductor of the shoulder
Which of these statements is correct?	(a) the teres minor muscle rotates the arm laterally. (b) the teres minor muscle adducts the arm (c) the teres minor muscle adducts the arm (d) the teres minor muscle flexes the arm (e) the teres minor muscle rotates the arm medially
The brachialis muscle is a:	(a) supinator of the hand (b) elbow flexor (c) pronator of the hand (d) extensor of the elbow (e) wrist abductor
Bilateral contraction of the splenius cervicis muscles:	(a) flexes the neck (b) tilts the neck laterally (c) tilts the neck laterally from the opposite side (d) extends the neck (e) rotates the neck
The action of the medial rectus muscle results in the following:	(a) abduction of the eye (b) adduction of the eyes (c) looking up and medially at (d) looking down and laterally (e) looking down and medially
The rectus abdominis muscle:	(a) flexes the trunk forward (b) with a fixed point on the trunk, acting individually, rotates the pelvis (c) with a fixed point on the trunk flexes the lower limb on the pelvis (d) acting individually, flexes the trunk laterally (e) contributes to inhalation
The coccygeal muscle determines	(a) elevation and support of the pelvic floor (b) extension of the coccygeal joints (c) retroposition of the anal canal (d) depression of the pelvic floor (e) nutation of the sacrum
Which of these statements is correct?	(a) the gastrocnemius muscle extends the leg (b) the gastrocnemius muscle is an extensor muscle (c) the gastrocnemius muscle extends the ankle (d) the gastrocnemius muscle is one of the extrinsic muscles of the foot (e) the gastrocnemius muscle is one of the intrinsic muscles of the foot

Table A1. *Cont.*

The levator muscle of the upper eyelid determines	(a) closing of the eyes (b) lifting of the eyebrow (c) elevation of the lower eyelid (d) elevation of the upper eyelid (e) wrinkling of the forehead
Which of these statements is correct?	(a) the iliopsoas muscle flexes the lumbar spine (b) the iliopsoas muscle adducts the lumbar spine (c) the iliopsoas muscle extrude the lumbar spine (d) the iliopsoas muscle adducts the lumbar spine (e) the iliopsoas muscle is an internal rotator of the lumbar spine
Which of these statements is correct?	(a) the piriformis muscle adducts the lumbar spine (b) the psoas major muscle flexes the lumbar spine (c) the iliopsoas muscle intrudes the lumbar spine (d) the long adductor muscle extends the lumbar spine (e) the iliopsoas muscle adducts the lumbar spine
Which of these muscles extends the femur	(a) the gluteus minimus muscle (b) the tensor fascia lata muscle (c) the piriformis muscle (d) the gluteus maximus muscle (e) the gluteus medius muscle
Which of these statements is correct?	(a) the subscapularis muscle adducts the arm (b) the subscapularis muscle flexes the arm (c) the subscapularis muscle rotates the arm laterally (d) the subscapularis muscle adducts the arm (e) the subscapularis muscle rotates the arm medially
The medial head of the triceps brachii muscle is a:	(a) supinator of the forearm (b) extensor of the elbow (c) elbow flexor (d) pronator of the forearm (e) abductor of the shoulder
Which of these statements is incorrect?	(a) the coracobrachialis muscle flexes the shoulder joint (b) the coracobrachialis muscle adducts the shoulder joint (c) the teres major rotates the arm medially (d) the subscapularis muscle rotates the arm medially (e) the coracobrachialis muscle adducts the shoulder joint
The action of the palatine veil elevator and palatine veil tensor muscles is to:	(a) elevates the soft palate (b) constrict the pharynx (c) elevate the pharynx (d) constrict the isthmus of the jaws (e) elevate the larynx
The piriform muscle:	(a) extends the femur (b) intrarotates the hip (c) adducts the hip (d) flexes the lumbar spine (e) flexes the femur
The extensor digitorum communis muscle:	(a) flexes all fingers (b) has a common origin with the extensor carpi radialis brevis muscle (c) flexes the wrist (d) flexes the thumb (e) extends the thumb
The external anal sphincter muscle:	(a) raises the anal canal (b) narrows and lowers the anal canal (c) closes the anal canal and the anus (d) narrows and lifts the anal canal (e) provides involuntary control for defecation

References

1. Groenier, M.; Pieters, J.M.; Miedema, H.A.T. Technical medicine: Designing medical technological solutions for improved health care. *Med. Sci. Educ.* **2017**, *27*, 621–631. [CrossRef]
2. Han, E.R.; Yeo, S.; Kim, M.J.; Lee, Y.H.; Park, K.H.; Roh, H. Medical Education trends for future physicians in the era of advanced technology and artificial intelligence: An integrative review. *BMC Med. Educ.* **2019**, *19*, 460. [CrossRef] [PubMed]
3. Rambukwella, M.; Balamurugan, A.; Klapholz, H.; Beninger, P. The application of engineering principles and practices to medical education: Preparing the next generation of physicians. *Med. Sci. Educ.* **2021**, *31*, 897–904. [CrossRef] [PubMed]
4. Wang, J.J.; Singh, R.K.; Miselis, H.H.; Stapleton, S.N. Technology literacy in undergraduate medical education: Review and survey of the US medical school innovation and technology programs. *JMIR Med. Educ.* **2022**, *8*, e32183. [CrossRef] [PubMed]
5. Consorti, F.; Familiari, G.; Lotti, A.; Torre, D. Medical education in Italy: Challenges and opportunities. *Med. Teach.* **2021**, *43*, 1242–1248. [CrossRef] [PubMed]
6. Basili, S.; Familiari, G.; Del Prete, Z.; Farina, L.; Stefanini, L.; Polimeni, A. A new biomedical-technological training program at Sapienza University of Rome. In Proceedings of the Virtual Conference AMEE 2021, Dundee, UK, 30 August 2021.
7. Snelgrove, H.; Familiari, G.; Gallo, P.; Gaudio, E.; Lenzi, A.; Ziparo, V.; Frati, L. The challenge of reform: 10 years of curricula change in Italian medical schools. *Med. Teach.* **2009**, *31*, 1047–1055. [CrossRef]
8. Familiari, G.; Relucenti, M.; Heyn, R.; Baldini, R.; D'Andrea, G.; Familiari, P.; Bozzao, A.; Raco, A. The value of neurosurgical and intraoperative magnetic resonance imaging and diffusion tensor imaging tractography in Clinically integrated neuroanatomy modules: Across sectional study. *Anat. Sci. Educ.* **2013**, *6*, 294–306. [CrossRef]
9. Chan, L.K.; Pawlina, W. *Teaching Anatomy: A Practical Guide*; Springer International Publishing: New York, NY, USA, 2015; pp. 1–403.
10. Sadeqi, H.; Valiani, A.; Avizhgan, M.; Ebrahimi, S.A.; Manteghinejad, A.; Miralai, P.; Omid, A. The effect of teaching integrated course of physical examination and radiological anatomy in practical limb anatomy on medical students' learning outcomes. *BMC Med. Educ.* **2021**, *21*, 461–467. [CrossRef]
11. Carifio, J.; Perla, R. Resolving the 50-year debate around using and misusing Likert scales. *Med. Educ.* **2008**, *42*, 1150–1152. [CrossRef]
12. Little, T.D. *Longitudinal Structural Equation Modelling*; Guilford Press: New York, NY, USA, 2013; pp. 1–386.
13. Bland, J.M.; Altman, D.G. Cronbach's Alpha. *BMJ* **1997**, *314*, 572. [CrossRef]
14. Bandiera, P.; Bucchieri, F.; Carpino, G.; Castaldo, C.; Conconi, M.T.; Consalez, G.; Cremona, O.; Cusella De Angelis, M.G.; De Luca, A.; Di Meglio, F.; et al. Anatomia Umana. In *Raccolta di Quesiti a Risposta Multipla per la Verifica e L'autoverifica Degli Apprendimenti SSD Bio-16*; EdiSES Srl: Napoli, Italy, 2017; pp. 1–853.
15. Brauer, D.G.; Ferguson, K.J. The integrated curriculum in medical education. AMEE Guide n. 96. *Med. Teach.* **2015**, *37*, 312–322. [CrossRef] [PubMed]
16. Rosenthal, D.; Worley, P.S.; Mugford, B.; Stagg, P. Vertical integration of medical education: Riverland experience, south Australia. *Rural. Remote Health* **2004**, *4*, 1–11. [CrossRef]
17. Wijnen-Meijer, M.; van der Broek, S.; Koens, F.; ten Cate, O. Vertical integration in medical education: The broader perspective. *BMC Med. Educ.* **2020**, *20*, 509–513. [CrossRef] [PubMed]
18. Bergman, E.M.; Prince, K.J.; Drukker, J.; van der Vleuten, C.P.; Scherpbier, A.J. How much anatomy is enough? *Anat. Sci. Educ.* **2008**, *1*, 184–188. [CrossRef]
19. Louw, G.; Eizenberg, N.; Carmichael, W. The place of anatomy in medical education. AMEE Guide n. 41. *Med. Teach.* **2009**, *31*, 373–386. [CrossRef]
20. Bergman, E.M.; van der Vleuten, C.P.; Scherpbier, A.J. Why don't they know enough about anatomy? A narrative review. *Med. Teach.* **2011**, *33*, 403–409. [CrossRef]
21. Sbayeh, A.; Choo, M.A.Q.; Quane, K.A.; Finucane, P.; McGrath, D.; O'Flynn, S.; O'Mahony, S.M.; O'Tuathaigh, C.M.P. Relevance of anatomy to medical education and clinical practice: Perspectives of medical students, clinicians, and educators. *Perspect. Med. Educ.* **2016**, *5*, 338–346. [CrossRef]
22. Barry, D.S.; Dent, J.M.; Hankin, M.; Moyer, D.; Shah, N.L.; Tuskey, A.; Soukoulis, V. The clinical anatomy and imaging laboratory: Vertical integration in the preclerkship curriculum. *MedEDPORTAL* **2019**, *15*, 10824. [CrossRef]
23. Pabst, R.; Westermann, J.; Lippert, H. Integration of clinical problems in teaching gross anatomy: Living anatomy, X-ray anatomy, patient presentations, and film depicting clinical problems. *Anat. Rec.* **1986**, *215*, 92–94. [CrossRef]
24. Rizzolo, L.J.; Rando, W.C.; O'Brien, M.K.; Haims, A.H.; Abrahams, J.J.; Stewart, W.B. Design, implementation, and evaluation of an innovative anatomy course. *Anat. Sci. Educ.* **2010**, *3*, 102–120. [CrossRef]
25. Innocenti, B.; Bori, E. Robotics in orthopaedic surgery: Why, what and how? *Arch. Orthop. Trauma Surg.* **2021**, *141*, 2035–2042. [CrossRef]
26. Callaghan, J.J.; DeMik, D.E.; Carender, C.N.; Bedard, N.A. Analysis of New Orthopaedic Technologies in Large Database Research. *J. Bone Jt. Surg.* **2022**, *10* (Suppl. 3), 47–50. [CrossRef] [PubMed]
27. Perfetti, D.C.; Kisinde, S.; Rogers-LaVanne, M.P.; Satin, A.M.; Lieberman, I.H. Robotic Spine Surgery: Past, Present, and Future. *Spine* **2022**, *47*, 909–921. [CrossRef] [PubMed]

28. Tovar, M.A.; Dowlati, E.; Zhao, D.Y.; Khan, Z.; Pasko, K.B.D.; Sandhu, F.A.; Voyadzis, J.M. Robot-assisted and augmented reality-assisted spinal instrumentation: A systematic review and meta-analysis of screw accuracy and outcomes over the last decade. *J. Neurosurg. Spine* **2022**, *25*, 1–16. [CrossRef] [PubMed]
29. Sielatycki, J.A.; Mitchell, K.; Leung, E.; Lehman, R.A. State of the art review of new technologies in spine deformity surgery-robotics and navigation. *Spine Deform.* **2022**, *10*, 5–17. [CrossRef]

Article

Advances in Neuroanatomy through Brain Atlasing

Wieslaw L. Nowinski

Nowinski Brain Foundation, Warsaw West County, P.O. Box 56, 05-092 Lomianki, Poland; info@nowinbrain.org

Abstract: Human brain atlases are tools to gather, present, use, and discover knowledge about the human brain. The developments in brain atlases parallel the advances in neuroanatomy. The brain atlas evolution has been from hand-drawn cortical maps to print atlases to digital platforms which, thanks to tremendous advancements in acquisition techniques and computing, has enabled progress in neuroanatomy from gross (macro) to meso-, micro-, and nano-neuroanatomy. Advances in neuroanatomy have been feasible because of introducing new modalities, from the initial cadaveric dissections, morphology, light microscopy imaging and neuroelectrophysiology to non-invasive in vivo imaging, connectivity, electron microscopy imaging, genomics, proteomics, transcriptomics, and epigenomics. Presently, large and long-term brain projects along with big data drive the development in micro- and nano-neuroanatomy. The goal of this work is to address the relationship between neuroanatomy and human brain atlases and, particularly, the impact of these atlases on the understanding, presentation, and advancement of neuroanatomy. To better illustrate this relationship, a brief outline on the evolution of the human brain atlas concept, creation of brain atlases, atlas-based applications, and future brain-related developments is also presented. In conclusion, human brain atlases are excellent means to represent, present, disseminate, and support neuroanatomy.

Keywords: neuroanatomy; human brain atlases; neuroeducation; brain research; brain atlases in clinics; large brain projects; big brain data

Citation: Nowinski, W.L. Advances in Neuroanatomy through Brain Atlasing. *Anatomia* **2023**, 2, 28–42. https://doi.org/10.3390/anatomia2010004

Academic Editor: Lars Ove Brandenburg

Received: 11 November 2022
Revised: 4 January 2023
Accepted: 14 January 2023
Published: 19 January 2023

1. Introduction

For centuries, the human brain has been an enormous challenge for scientists and an abundant inspiration for artists. However, the great importance of the brain has not always been fully understood. In Ancient Egypt, for instance, the brain was considered a rather useless organ with no need to be mummified. In Ancient Greece, Herodotus advising on the mummification process recommended removing as much of the brain as possible and mixing any remains of it with drugs, implying the brain was toxic. One of the greatest philosophers of Antiquity, Aristotle, who also substantially contributed to natural sciences, viewed the brain as a cooling mechanism for blood, while the heart was the seat of intelligence. Toward the end of Antiquity, St. Augustine, considered the father of psychology, demonstrated a better understanding of the brain by dividing it into three compartments, the environment with the senses, the movement environment, and the seat of memory. Then, after one thousand years of stagnation, Leonardo da Vinci created beautiful images, though not always anatomically correct, of the brain capturing its anatomy, by bridging art and science. It was however Vesalius, universally considered to be the most important anatomist and the founder of modern anatomy, who started a new era of anatomical investigation ending its dependence on Greek and Arabic authorities, often erroneous and based upon animal rather than human studies [1]. Vesalius also made a substantial contribution to neuroanatomy by providing the first description of the human corpus callosum linking two halves of the brain, putamen, globus pallidus, caudate nucleus, pulvinar, midbrain, pineal body, and internal capsule, among others. Willis introduced a new level of neuroanatomical accuracy and reclassified the cranial nerves. Neuroanatomy advancements through brain gross dissections were accomplished by 19th-century neuroanatomists including Arnold, Burdach, Foville, Gratiolet, Mayo, and Reil as it was illustrated and reviewed by Schmahmann and Pandya [2]. One of the first maps of the human cortical surface based on cytoarchitectonics was created in 1909 by a German neurologist named Korbinian Brodmann [3]. Brodmann postulated that areas differing in structure perform different functions. Brodmann's areas are still in use today in neuroeducation and research.

Since then, there has been a tremendous development of human brain maps and atlases in terms of concept, content, functionality, applications, and availability. I have earlier distinguished four generations of brain atlases: early cortical maps, print stereotactic atlases, early digital atlases, and advanced brain atlas platforms [4].

Neuroanatomy, as the study of the structure and organization of the nervous system, and human brain atlases, as tools to gather, present, use, and discover knowledge about the human brain, are obviously linked. The goal of this work is to address the relationship between neuroanatomy and human brain atlases and, particularly, the impact of these atlases on the understanding, presentation, and advancement of neuroanatomy. To better illustrate this impact, a brief outline about the evolution of the human brain atlas concept, creation of brain atlases, atlas-based applications, and future brain-related developments is also presented.

2. Evolution of Brain Atlas Concept

The concept of the brain atlas has been evolving together with the tremendous progress in neuroanatomy thanks to imaging and computing. It should be noted that various authors consider or define the brain atlas differently as briefly overviewed below. Traditionally, the brain atlas is considered a collection of brain maps or a database. Here, there are a few examples. Roland and Zilles define brain atlases as collections of micrographs or schematic drawings of brain sections with identified anatomic structures [5]. Evans et al. treat brain atlases as large-scale neuroimaging databases providing the mean and variance in the population [6]. Mori et al. consider the brain atlas a tool for image structurization via atlas-based image subdivision to exploit a great amount of imaging information offered by medical systems [7]. Amunts et al. regard brain atlases as central for integrating diversified information about various aspects of the brain [8]. Kuan et al. consider the brain atlas a tool aiming to integrate diverse information, understand complex brain anatomy, localize experimental data, and plan experiments [9]. Costa et al. consider the atlases the means able to produce specific, testable hypotheses about circuit organization and connectivity [10]. Chon et al. find anatomical atlases in standard coordinates to be necessary for the interpretation and integration of research findings in a common spatial context [11]. Hence, despite some minor differences, what is common for all these approaches is that they mainly reflect a research usefulness of brain atlases in human and/or animal studies.

I proposed a different concept of the human brain atlas by extending its standard imaging content with a knowledge database, tools for content processing and analysis, and means to broaden this content with the user's data [12]. This concept has been customized to stereotactic and functional neurosurgery as a population-based, self-growing, and structural-functional multi-atlas. Subsequently, based on the atlas evolution review [4] and considering various perspectives and applications, my latest definition of the human brain atlas has evolved as follows: "the reference human brain atlas is a vehicle to gather, present, use, and discover knowledge about the human brain with a highly organized content, tools enabling a wide range of its applications, massive and heterogeneous knowledge database, and means for content and knowledge updating and growing by its user" [13]. Correspondingly, an architecture embodying such a brain atlas is proposed along with a method of its implementation [13].

3. Creation of Human Brain Maps and Atlases

The evolution of brain fixation techniques combined with optical microscopy enabled neuroanatomy advancement beyond gross anatomy toward microanatomy. Several early cortical maps were created from microscopy in the first three decades of the 20th century encapsulating new knowledge about the human brain. Early brain mappers include Brodmann [3], Campbell [14], Flechsig [15], Vogt and Vogt [16], and Von Economo and Koskinas [17]. Their maps were made for a single modality, cytoarchitectonics [3,17] or myeloarchitectonics [15,16], and varied in the number of parcellated cortical areas. This development was a substantial step forward in comparison to examining gross neuroanatomy from cadaveric studies.

To localize cerebral structures in neurosurgery in the pre-tomographic imaging era, stereotactic brain atlases were developed. These, initially print, atlases represented a significant step forward in atlas development both in terms of atlas content and concept. In the 1950s, stereotactic brain atlases were created by Speigel and Wycis in 1952 [18], Talairach et al. in 1957 [19], and Schaltenbrand and Bailey in 1959 [20], followed by Andrew and Watkins in 1969 [21], Van Buren and Borke in 1972 [22], Schaltenbrand and Wahren in 1977 [23], Afshar et al. in 1978 [24], and Talairach and Tournoux in 1988 [25] and 1993 [26]. The contents of these atlases vary covering deep gray nuclei (by Talairach et al., 1957), the thalamus and adjacent structures (by Andrew and Watkins, 1969), variations and connections of the thalamus (by Van Buren and Borke, 1972), deep structures and the whole brain (by Schaltenbrand and Wahren, 1977), the brainstem and cerebellar nuclei (by Afshar et al., 1978), the whole brain (by Talairach and Tournoux (1988), and brain connections (by Talairach and Tournoux, 1993).

Besides stereotactic, other print atlases were published for neuroradiology, neurosurgery, neuroscience, and neuroeducation, including a brain atlas for computed tomography [27], an atlas of the hippocampus [28], an atlas of the cerebral sulci [29], an atlas of brain function [30], an atlas of the brainstem and cerebellum [31], an atlas of morphology and functional neuroanatomy [32], an atlas of the brainstem and cerebellum with magnetic resonance 9.4 Tesla (T) images [33], and the Netter's atlas of neuroscience [34].

As print atlases had several limitations, including static content, sparseness of image plates, limited functionality, and difficulty in mapping into patients' scans, electronic and interactive brain atlases have been developed. Initially, these were digitalized versions of the stereotactic print atlases followed by their enhancements and extensions as reviewed in [4,35].

In particular, two stereotactic brain atlases are of great importance, "Atlas of Stereotaxy of the Human Brain" by Schaltenbrand and Wahren [23] and "Co-Planar Stereotactic Atlas of the Human Brain" by Talairach and Tournoux" [25]. The Schaltenbrand and Wahren atlas is based on 111 brains and comprises photographic plates of macroscopic and microscopic sections through the hemispheres and the brainstem. The macroscopic plates provide the extent of variation in the brain structures. The microscopic myelin-stained sections demonstrate in great detail cerebral deep structures which usually are not well visible on brain scans. This atlas is available in most surgical workstations. The Talairach and Tournoux atlas presents the cerebral structures as colored drawings through axial, coronal, and sagittal sections of a single, normal brain specimen. It is applied in neurosurgery and brain research reaching over 22,000 citations.

Because of the importance of these two brain atlases, we have developed their enhanced and extended electronic versions, and the applied processing was explained in detail in [36]. These electronic atlases are fully parcellated which enables their automatic labeling. This parcellation is by unique coloring and closed contouring (a contour representation is additionally useful for atlas-to-data registration as the contours do not block the actual patient data); see Figure 1. These electronic atlases have been embedded into atlas-assisted stand-alone applications [37–40] and plug-in libraries licensed to 13 companies and integrated with major surgical workstations [41].

Figure 1. Electronic brain atlases: (**left**) Talairach and Tournoux axial fully color-coded plate 4 mm above the intercommissural plane; (**right**) Schaltenbrand and Wahren coronal microscopic plate in contour representation 4 mm behind the posterior commissure (note that all the contours are closed).

Enormous advancements in imaging, brain mapping, and computing drive the development of human brain atlas platforms. I have specified 23 directions in the evolution of brain atlas content development grouped into eight categories by employing various criteria, including scope, parcellation, plurality, modality, scale, ab/normality, ethnicity, and a combination of them [4]. I briefly overview these brain atlas categories and provide some examples of brain atlases from numerous centers.

The scope of brain atlases ranges from structural neuroanatomy [42–45] to connectional neuroanatomy [46–51] to vascular neuroanatomy [52,53] including cerebral variants [54] to cranial nerves and nuclei [55] to gene expression [56] including gene expression in brain development [57].

In general, the human brain can be parcellated into numerous anatomically and/or functionally distinct cortical regions and subcortical structures based on macrostructural, microstructural, functional, and/or connectional features. The parcellation category represents novel and/or finer parcellations of brain structures and surfaces based on various modalities and approaches. The developments here are from classic gross anatomy, cytoarchitecture, and myeloarchitecture to functional magnetic resonance imaging (fMRI) exploiting resting-state and task-based sequences [58], chemoarchitecture [59], vascular territories [60], anatomic connectivity based on diffusion tensor imaging [48] and diffusion spectrum imaging [61], anatomic-functional connectivity based on diffusion and resting-state MRI [62], electroencephalography [63], (multi)receptor architecture [64], and/or multiplicity of them [50,65]. Both the size and the number of the parcellated regions can be variable; for instance, a multi-modal MRI-based parcellation of the cerebral cortex results in 180 variable-size areas per hemisphere [65], the *Brainnetome* atlas is parcellated into 210 various cortical areas and 36 subcortical regions [62], and the *Yale Brain Atlas* consists of 690 same-size one-square centimeter parcels [63].

Parcellation not only introduces subdivision but also enables systematization, localization, and comparison, ideally making the brain "addressable". Parcellated regions can be named based on some existing nomenclatures, such as *Terminologia Anatomica* [66] which is an international standard for the whole body or *Terminologia Neuroanatomica* targeting the central nervous system, peripheral nervous system, and sensory organs [67]. Several nomenclatures have been introduced for research applications, such as *NeuroNames* supporting synonyms and multiple languages [68], *Uberon* [69] supporting single- and cross-species queries, *Foundation Model of Anatomy* (FMA) providing a structure-based template from the molecular to the macroscopic levels for representing biological functions of the human body [70], and *Common Coordinate Framework* (CCF) ontology to define positions in the body down to individual cells [71]. Alternatively, parcellation-related identifiers are used, such as numbers in naming Brodmann's areas [3] or parcel unique names with a gyrus code and a letter indicating the parcel position within the gyrus in the *Yale Brain Atlas* [63].

Within the plurality category, probabilistic brain atlases provide novel neuroanatomical information in terms of statistical distributions of the studied entities. For instance, these atlases may contain the mean values, standard deviations, moments, and other quantifiers of volumes (e.g., for the entire brain [72], white matter [73], cerebellum [74], or subcortical structures [75]), areas (such as cortical surface regions [76]) or distances (e.g., the thickness of the cortical mantle). Multi-atlases can illustrate neuroanatomy over the lifespan. For instance, a mega multi-atlas [77] comprises 90 component brain atlases with the brain specimens ranging from 4 to 82 years of age.

In the modality category, the major advancement has been from postmortem to in vivo data enabled by neuroimaging allowing to accomplish a "living neuroanatomy". Furthermore, more detailed neuroanatomical images with better quality are feasible in brain atlasing due to the increased teslage of the acquired MRI neuroimages, namely, from 1.5T [78] to 3T [45,53] to 7T [52,79–83] to 9.4T [84].

The scale category includes brain atlases with various temporal, spatial, and combined spatiotemporal scales. Several temporal scale-related brain atlases aggregate age-dependent neuroanatomical changes ranging from pediatric to geriatric populations [85–87]. Other relevant works include a dynamic 4D atlas of the developing brain [88] and a temporal cell atlas of gene expression in brain development [57].

The spatial scale of brain atlases ranges from macro- to meso- to micro- to nano-scale, including the integration of atlas data across multiple scales. The developments in this area include the *BigBrain* with a 20-micrometer resolution [89], a comprehensive cellular-resolution (of 1 μm/pixel) atlas linking macroscopic anatomical and microscopic cytoarchitectural parcellations [90], a whole-brain cell atlas integrating anatomical, physiological and molecular annotations for a complete characterization of neuronal cell types, their distributions, and patterns of connectivity [91], a genomics brain atlas [56], an atlas of brain transcriptome [92], an atlas of serotonin [93], and a proteomic brain atlas [94].

Several disease-specific atlases have been created, e.g., for Alzheimer's disease [95], dementia [96], stroke [97,98], brain tumors [99], and epilepsy [100]. Some of them enable the quantification of brain structural deficits in epilepsy, depression, schizophrenia, Alzheimer's disease, autism, and bipolar disorders [101]; others include the *Probabilistic Stroke Atlas* [98] which facilitates outcome prediction, the *Virtual Epileptic Patient* atlas which provides an automated brain region parcellation and labeling for epileptology and functional neurosurgery [100], and the *Probabilistic Atlas of Diffuse WHO Grade II Glioma Locations* which identifies the preferential locations of these gliomas in the brain [99]. A different way of atlas use is presented in [102] to investigate genetic correlations between brain phenotypes (attained as cortical surface area and thickness) and psychiatric/neurological disorders by means of genetically informed brain atlases. This study revealed the association between global surface and fronto-parietal thickness with attention-deficit hyperactivity disorder, temporal area with schizophrenia and autism spectrum disorder, and fronto-occipital morphology with neurological disorders.

Ethnicity-based brain atlases enable comparison of neuroanatomy between various populations, such as Chinese and Caucasian [103] and Indian with Chinese and Caucasian [104].

The design, development, and validation of a human brain atlas is a painstaking and time-consuming process that requires high attention to detail. The design principles of a holistic and reference brain atlas are formulated in [105], computational methods employed in brain atlas development are addressed in [106], visualization and interaction are discussed in [107], and a user-centric and application-balanced architecture cum implementation of a reference human brain atlas is proposed in [13].

4. Brain Atlas-Assisted Applications

The human brain atlases are employed across education, research, and clinics [4]. In neuroeducation, the brain atlas assists students and educators as a visual and interactive tool with parcellated and labeled virtual brain models, equipped with an intuitive and friendly user interface, able to communicate cerebral complexity in a more convenient and comprehensible manner. In research, brain atlases focus predominantly on how to integrate and openly share massive amounts of heterogeneous experimental data in a common reference atlas space and to relate these data across scales. In clinics, brain atlases are valuable computer-aided tools to support and enhance screening, diagnosis, treatment, and prediction.

4.1. Education

The history of neuroanatomy over the centuries has been linked to the teaching methods employed, including cadaveric dissection, plastination, observation of live models, live surgery, animal dissection, synthetic models, bibliographic sources, radiology, and audiovisual virtual reality including stereoscopy [108]. Electronic and interactive brain atlases may be embedded in synthetic models, radiology, audiovisual virtual reality, and computer-aided live surgery.

Several standard neuroeducational brain atlases have been developed, such as *Digital Anatomist* [109], *A.D.A.M.* [110], *The Electronic Clinical Brain Atlas* [37], *Voxel-man* [78], *The Cerefy Atlas of Brain Anatomy* [39], Primal's *Interactive Head and Neck* [111], and *The Cerefy Clinical Brain Atlas* [40].

In comparison to the standard brain atlases, advanced atlases provide novel features in neuroeducation facilitating brain exploration and understanding. Examples of such atlases are *The Cerefy Atlas of Cerebral Vasculature* [53], *The Human Brain in 1492 Pieces* [43], *The Human Brain in 1969 Pieces: Structure, Vasculature, Tracts, Cranial Nerves, Systems, Head Muscles, and Glands* [44], and *The Human Brain, Head and Neck in 2953 Pieces* [81]. These novel features include continuous navigation and exploration, free composing and decomposing of a 3D explorable scene (see Figure 2), joint surface and sectional anatomy, presentation in context, correlation of anatomy and terminology, simultaneous presentation of multiple systems, wide scope of presentations (from local to global neuroanatomy), virtual dissections, quantification, and generation of teaching materials [112,113] as well as automatic testing and assessment of neuroanatomy knowledge [114] available, e.g., in *The Cerefy Atlas of Cerebral Vasculature* [53].

Technology advancements open new avenues in brain atlasing, although on the other hand, they may cause an increased cost and decreased accessibility of brain atlas applications, especially for users in less privileged countries. To address this issue, I have created the *NOWinBRAIN* 3D neuroimage public repository at www.nowinbrain.org. *NOWinBRAIN* is a large (the largest so far), systematic, comprehensive, extendable, spatially consistent, easy to use, long-lasting, and beautiful repository of 3D reconstructed images of a living human brain extended to the head and neck populated with over 7800 images (version 3.1) organized in 10 galleries. The design, development, and content of the primary and multi-tissue galleries are addressed in [115], the combined planar–surface gallery in [116], the dissection gallery in [117]; and the gallery of dual white matter–cortical surfaces with the cerebral sulci in [118]. Note that despite the tremendous development of various brain-related resources, such a repository is not yet available. This systematically designed repository is empowered with many novel features, such as multi-tissue galleries, the use of various spatially co-registered image sequences, and unique image-naming syntax. It is freely available and easily accessible as a web resource without any password or registration. These features make *NOWinBRAIN* valuable for neuroeducators, medical students, neuroscientists, and clinicians, especially, in less privileged countries. The current users are from over 75 countries on six continents. Most users are from Europe and the United States including the technologically advanced Silicon Valley. Frequent users are from India, China, and Egypt. There are also visitors from Nepal, Afghanistan, Sudan, Tanzania, Brazil, Argentine, and Peru.

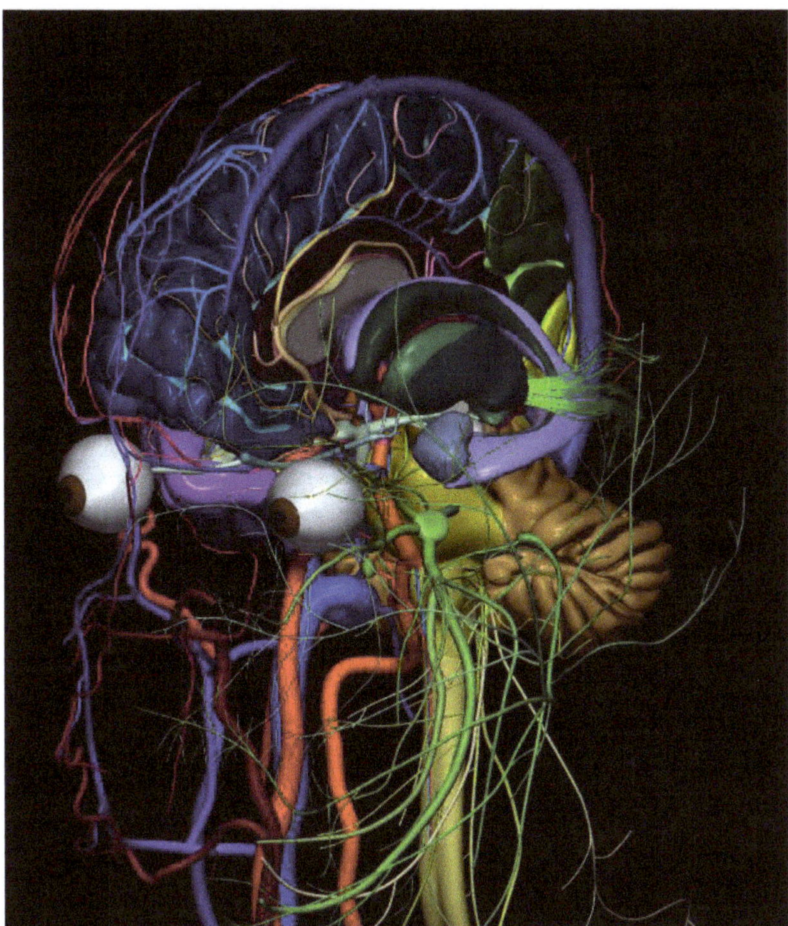

Figure 2. Neuroanatomy composed of 3D pieces (such as Lego blocks) and parcellated by unique color coding. The composed 3D scene contains the brain with the left hemisphere removed and the right hemisphere parcellated into gyri and sulci, cervical spine, deep gray nuclei, cerebral ventricles, intracranial and extracranial vasculature on the right, cranial nerves on the left, and the visual system (an antero-left lateral view).

4.2. Research

Brain atlases are widely applied in research for various purposes and play a key role in modern neuroimage analysis [119]. One of the main areas of brain atlas applications is human brain mapping. Then, the brain atlases, such as the *BrainMap* [120] or the *Brain Atlas for Functional Imaging* [38], provide the underlying neuroanatomy enabling the activation loci in functional images to be automatically labeled with cortical areas and stereotactic coordinates. Brain atlases are widely applicable for fast, automatic, and robust segmentation of neuroimages [121–126]. Brain atlases are central tools for data integration [127] enabling combining various brain-related information, such as micro- and macrostructural parcellation, connectivity, temporal dynamics, and regional functional specialization [8]. The brain atlas also serves as a tool for localizing experimental data and planning experiments [9] as well as to generate hypotheses about brain organization [10]. In addition, brain atlases enable knowledge discovery; for instance, Makowski et al. employed genetically informed brain atlases to determine the impact of genetic variants on the brain in genome-wide association studies of regional cortical surface area and thickness in about 40,000 adults and 9000 children [102]. These studies uncovered 440 genome-wide significant loci (largely acquired in childhood) related to early neurodevelopment and associated with neuropsychiatric risk.

4.3. Clinics

The first clinical application of human brain atlases has been stereotactic and functional neurosurgery. Initially, a digital atlas, such as *The Electronic Clinical Brain Atlas* [37], was employed offline in the operating room to aid neurosurgery. Subsequently, the brain atlas libraries derived from our brain atlas database [36] were directly incorporated into several surgical workstations, including the *StealthStation* (Medtronic) [41], to assist neurosurgery. In general, the brain atlas provides pre-, intra-, and post-operative support [128]. Pre-operatively, the atlas assists to plan the target and trajectory as well as provides a list of structures intersected by the trajectory. The usage of multiple brain atlases improves the planning quality and surgeon's confidence [129,130]. Intra-operatively, the brain atlas specifies the structures already traversed by the electrode, identifies the actual structure where the electrode tip is located, measures distances to important structures, and provides the neuroanatomic and vascular context [130]. Post-operatively, the atlas enables the examination of the precision of placement of the stimulating electrode or a permanent lesion. Other atlas-assisted applications in neurosurgery include atlas-guided do-it-yourself neurosurgery [41] and an atlas-enhanced operating room for the future [131].

Several brain atlas-aided proofs of concepts (prototypes) have been developed in some other areas. Namely, in neuroradiology, brain atlases can assist in neuroimage interpretation by segmenting and labeling brain scans including pathological, template-based reporting, dealing with data explosion by facilitating processing multi-detector (especially 320-raw computed tomography) scans, and communication for both doctor-to-doctor and especially doctor-to-patient [132]. Multiple brain atlases have the potential in stroke management including prediction, diagnosis, and treatment by providing automated processes ensuring fast decisions [60,98,133]. In neurology, the *3D Atlas of Neurologic Disorders* [134] demonstrates various locations of brain damage, including local neuroanatomy, cranial nerves, and cerebrovasculature, along with the resulting neurologic deficits, bridging in this way neuroanatomy, neuroradiology, and neurology [135]. Finally in psychiatry, a brain atlas allows for the automatic generation of neuroanatomic volumes of interest for statistical analysis, e.g., to study schizophrenic patients and controls [136].

5. Future Developments

There has been an enormous explosion of human brain-related endeavors in the last few years. These are advanced, big, government-led, and/or well-funded projects, initiatives, and/or national brain programs, such as *The Human Connectome Project* to map structural and functional connections to investigate the relationship between brain circuits and behavior [51]; *The Allen Brain Atlas* to map gene expression [56]; *The Big Brain* to acquire ultra-high resolution neuroimages [89]; *The CONNECT* project combining macro- and micro-structure [137]; the *Brainnetome* project to understand the brain and its disorders, develop methods for multi-scale brain network analysis, and create the *Brainnetome* atlas [138]; *The BRAIN Initiative* (*Brain Research through Advancing Innovate Neurotechnologies*) [139] to develop technology to advance neuroscience discovery [140]; *The Blue Brain Project* to simulate neocortical micro-circuitry [141]; *The Human Brain Project* to create a research infrastructure to decipher the human brain, reconstruct its multiscale organization, and develop brain-inspired information technology [142]; the *Chinese Color Nest Project* to study human connectomics across the life span [87]; the Japanese *Brain/MINDS (Brain Mapping by Integrating Neurotechnologies for Disease Studies)* project to better understand the human brain and neuropsychiatric disorders through "translatable" biomarkers [143]; and *SYNAPSE (Synchrotron for Neuroscience—an Asia-Pacific Strategic Enterprise)* to map the entire human brain at sub-cellular level by employing synchrotron tomography [144]—a proposal of how to build a corresponding human brain atlas I have recently presented at the SYNAPSE 2022 meeting; https://www.slri.or.th/th/index.php?option=com_attachments& task=download&id=4493 (28 December 2022).

These and other efforts have resulted in the acquisition of big data and the development of diverse brain-related databases, such as *BigBrain, Allen Brain Atlas, HCP (Human Connectome Project) database* and *HCP Young Adult Data, BIRN (Biomedical Informatics Research Network) MRI and fMRI data, OpenNEURO, OASIS (Open Access Series of Imaging Studies) Brains Project, ABCD (Adolescent Brain Cognitive Development) Data Repository, BCP (Baby Connectome Project) database, BP* (bipolar disorder) *neuroimaging database*, and the *Alzheimer's Disease Neuroimaging Initiative (ADNI)* as overviewed in [145]. Moreover, the *BRAIN Initiative* resulted in the development of the *Neuroscience Multi-Omic Archive* repository containing transcriptomic and epigenomic data from over 50 million brain cells [146]. In addition, the online community repository *NeuroMorpho.Org* contains more than 140,000 neural reconstructions (including glia) consisting of 3D representations of branch geometry and connectivity in a standardized format, and for each reconstruction, a set of morphometric features is extracted [147].

The abovementioned large-scale endeavors and big data empowered with high-performance computing at peta- and exascale will enormously increase our knowledge and understanding of the human brain at various scales and will propel the development of novel and more powerful brain atlases.

6. Summary and Conclusions

Neuroanatomy, as the study of the structure and organization of the nervous system, and human electronic brain atlases, as tools to gather, present, use, and discover knowledge about the human brain, are naturally linked. Consequently, this work addresses this human brain atlas–neuroanatomy mutual relationship.

Brain atlasing has progressed from the initial brain drawings and hand-drawn cortical maps to advanced brain atlas platforms. Presently, human electronic brain atlases have been advancing tremendously in terms of content, functionality, and applications. The advancement is empowered by software engineering methods and tools, such as databases, image processing, computer graphics, and virtual and augmented reality. This advancement spreads in multiple directions which can be grouped with respect to scope, parcellation, plurality, modality, scale, ab/normality, ethnicity, and combination of them.

Neuroanatomy has also been transformed enormously. From gross neuroanatomy facilitated by cadaveric dissections to micro-neuroanatomy enabled by brain fixation techniques combined with optical microscopy to nano-neuroanatomy empowered by modern electron microscopy, genomics, proteomics, transcriptomics, and epigenomics, and also from cadaveric neuroanatomy to living neuroanatomy enabled by modern imaging of structure, function, vasculature, structural and functional connectivity, and molecular processes. Moreover, imaging offers new acquisition methods, ever-increasing spatial and temporal resolutions, a better quality of images, and shorter acquisition times, all supported by artificial intelligence.

This ever-growing neuroanatomical knowledge enables the creation of human electronic brain atlases. These atlases mirror the advances in neuroanatomy capturing the dramatically increasing knowledge about the human brain in health and disease. Numerous centers contribute to neuroanatomy and brain atlasing advancements from various perspectives as briefly outlined here.

Furthermore, reciprocally, the developments in brain atlasing impact neuroanatomy enabling the use, presentation, mining, dissemination, and growth of this knowledge as well as facilitating learning, understanding, exploring, researching, diagnosing, screening, decision making, outcome prediction, and treatment of the human brain. In addition, because of remarkable progress in brain atlasing, these atlases are able to more accurately, realistically, and completely represent and present this neuroanatomical knowledge and better disseminate and use it. In my opinion, human brain atlases are the best means to represent, present, disseminate, and support neuroanatomy.

Finally, the impact on neuroanatomy and brain atlasing by the ongoing large brain projects and acquired big data may be expected to be enormous.

Funding: This research received no external funding.

Institutional Review Board Statement: Not applicable.

Informed Consent Statement: Not applicable.

Data Availability Statement: *NOWinBRAIN* 3D neuroimage repository is publically available at www.nowinbrain.org.

Conflicts of Interest: The author declares no conflict of interest.

References

1. Polyak, S. *The Vertebrate Visual System*; Kluever, K., Ed.; University of Chicago Press: Chicago, IL, USA, 1957.
2. Schmahmann, J.D.; Pandya, D.N. *Fiber Pathways of the Brain*; Oxford University Press: Oxford, UK, 2006.
3. Brodmann, K. *Vergleichende Lokalisationslehre der Grosshirnrinde in ihren Prinzipien dargestellt auf Grund des Zellenbaues*; Barth JA: Leipzig, Germany, 1909.
4. Nowinski, W.L. Evolution of human brain atlases in terms of content, applications, functionality, and availability. *Neuroinformatics* **2021**, *19*, 1–22. [CrossRef] [PubMed]
5. Roland, P.E.; Zilles, K. Brain atlases–a new research tool. *Trends Neurosci.* **1994**, *17*, 458–467. [CrossRef] [PubMed]
6. Evans, A.C.; Janke, A.L.; Collins, D.L.; Baillet, S. Brain templates and atlases. *Neuroimage* **2012**, *62*, 911–922. [CrossRef] [PubMed]
7. Mori, S.; Oishi, K.; Faria, A.V.; Miller, M.I. Atlas-based neuroinformatics via MRI: Harnessing information from past clinical cases and quantitative image analysis for patient care. *Annu. Rev. Biomed. Eng.* **2013**, *15*, 71–92. [CrossRef] [PubMed]
8. Amunts, K.; Hawrylycz, M.J.; Van Essen, D.C.; Van Horn, J.D.; Harel, N.; Poline, J.B.; De Martino, F.; Bjaalie, J.G.; Dehaene-Lambertz, G.; Dehaene, S.; et al. Interoperable atlases of the human brain. *Neuroimage* **2014**, *99*, 525–532. [CrossRef] [PubMed]
9. Kuan, L.; Li, Y.; Lau, C.; Feng, D.; Bernard, A.; Sunkin, S.M.; Zeng, H.; Dang, C.; Hawrylycz, M.; Ng, L. Neuroinformatics of the Allen Mouse Brain Connectivity Atlas. *Methods* **2015**, *73*, 4–17. [CrossRef] [PubMed]
10. Costa, M.; Manton, J.D.; Ostrovsky, A.D.; Prohaska, S.; Jefferis, G.S. NBLAST: Rapid, sensitive comparison of neuronal structure and construction of neuron family databases. *Neuron* **2016**, *91*, 293–311. [CrossRef] [PubMed]
11. Chon, U.; Vanselow, D.J.; Cheng, K.C.; Kim, Y. Enhanced and unified anatomical labeling for a common mouse brain atlas. *Nat. Commun.* **2019**, *10*, 5067. [CrossRef]
12. Nowinski, W.L. Towards constructing an ideal stereotactic brain atlas. *Acta Neurochir.* **2008**, *150*, 1–14. [CrossRef]
13. Nowinski, W.L. Towards an architecture of a multi-purpose, user-extendable reference human brain atlas. *Neuroinformatics* **2022**, *20*, 405–426. [CrossRef]
14. Campbell, A.W. *Histological Studies on the Localisation of Cerebral Function*; Cambridge University Press: Cambridge, UK, 1905.
15. Flechsig, P. *Anatomie des Menschlichen Gehirns und Rückenmarks auf Myelogenetischer Grundlage*; Thieme: Leipzig, Germany, 1920.
16. Vogt, C.; Vogt, O. Allgemeinere Ergebnisse unserer Hirnforschung (English Translation: Results of our brain research in a broader context). *J. Psychol. Neurol.* **1919**, *25*, 292–398.
17. Von Economo, C.; Koskinas, G.N. *Die Cytoarchitektonik der Hirnrinde des Erwachsenen Menschen*; Springer: Berlin, Germany, 1925.
18. Speigel, E.A.; Wycis, H.T. *Stereoencephalotomy: Part I. Methods and Stereotactic Atlas of the Human Brain*; Grune and Stratton: New York, NY, USA, 1952.
19. Talairach, J.; David, M.; Tournoux, P. *Atlas d'Anatomie Stereotaxique des Noyaux Gris Centraux*; Masson: Paris, France, 1957.
20. Schaltenbrand, G.; Bailey, W. *Atlas of Stereotaxy of the Human Brain*; Georg Thieme Verlag: Stuttgart, Germany, 1959.
21. Andrew, J.; Watkins, E.S. *A Stereotaxic Atlas of the Human Thalamus and Adjacent Structures. A Variability Study*; Williams and Wilkins: Baltimore, MD, USA, 1969.
22. Van Buren, J.M.; Borke, R.C. *Variations and Connections of the Human Thalamus*; Springer: Berlin, Germany, 1972.
23. Schaltenbrand, G.; Wahren, W. *Atlas of Stereotaxy of the Human Brain*; Georg Thieme Verlag: Stuttgart, Germany, 1977.
24. Afshar, E.; Watkins, E.S.; Yap, J.C. *Stereotactic Atlas of the Human Brainstem and Cerebellar Nuclei*; Raven Press: New York, NY, USA, 1978.
25. Talairach, J.; Tournoux, P. *Co-Planar Stereotactic Atlas of the Human Brain*; Thieme: Stuttgart, Germany; New York, NY, USA, 1988.
26. Talairach, J.; Tournoux, P. *Referentially Oriented Cerebral MRI Anatomy: Atlas of Stereotaxic Anatomical Correlations for Gray and White Matter*; Thieme: Stuttgart, Germany, 1993.

27. Takayoshi, M.; Hirano, A. *Atlas of the Human Brain for Computerized Tomography*; Igaku Shoin Medical Publishers: New York, NY, USA, 1978.

28. Duvernoy, H.M. *The Human Hippocampus: Atlas of Applied Anatomy*; Bergman: Munich, Germany, 1988.

29. Ono, M.; Kubik, S.; Abernathey, C.D. *Atlas of the Cerebral Sulci*; Georg Thieme Verlag/Thieme Medical Publishers: Stuttgart, Germany; New York, NY, USA, 1990.

30. Orrison, W.W., Jr. *Atlas of Brain Function*; Thieme: New York, NY, USA, 1995.

31. Duvernoy, H.M. *The Human Brain Stem and Cerebellum. Surface, Structure, Vascularization, and Three-Dimensional Sectional Anatomy, with MRI*; Springer: Wien, Austria; New York, NY, USA, 1995.

32. Scarabino, T.; Salvolini, U.; DiSalle, F.; Duvernoy, H.; Rabischong, P. (Eds.) *Atlas of Morphology and Functional Anatomy of the Brain*; Springer: Berlin, Germany, 2006.

33. Naidich, T.h.P.; Duvernoy, H.M.; Delman, B.N.; Sorensen, A.G.; Kollias, S.S.; Haacke, E.M. *Duvernoy's Atlas of the Human Brain Stem and Cerebellum*; Springer: Wien, Austria; New York, NY, USA, 2009.

34. Felten, D.L.; O'Banion, M.K.; Maida, M.E. *Netter's Atlas of Neuroscience*, 3rd ed.; Elsevier: Amsterdam, The Netherlands, 2015.

35. Alho, E.J.L.; Grinberg, L.; Heinsen, H. Review of printed and electronic stereotactic atlases of the human brain. In *Neuroimaging for Clinicians: Combining Research and Practice*; Peres, J.F.P., Ed.; InTech: Rijeka, Croatia, 2011; pp. 145–172.

36. Nowinski, W.L.; Fang, A.; Nguyen, B.T.; Raphel, J.K.; Jagannathan, L.; Raghavan, R.; Bryan, R.N.; Miller, G. Multiple brain atlas database and atlas-based neuroimaging system. *Comput. Aided Surg.* **1997**, *2*, 42–66. [CrossRef] [PubMed]

37. Nowinski, W.L.; Bryan, R.N.; Raghavan, R. *The Electronic Clinical Brain Atlas. Multiplanar Navigation of the Human Brain*; Thieme: New York, NY, USA, 1997.

38. Nowinski, W.L.; Thirunavuukarasuu, A.; Kennedy, D.N. *Brain Atlas for Functional Imaging. Clinical and Research Applications*; Thieme: New York, NY, USA, 2000.

39. Nowinski, W.L.; Thirunavuukarasuu, A.; Bryan, R.N. *The Cerefy Atlas of Brain Anatomy. An Introduction to Reading Radiological Scans for Students, Teachers, and Researchers*; Thieme: New York, NY, USA, 2002.

40. Nowinski, W.L.; Thirunavuukarasuu, A. *The Cerefy Clinical Brain Atlas on CD-ROM*; Thieme: New York, NY, USA, 2004.

41. Nowinski, W.L. Anatomical and probabilistic functional atlases in stereotactic and functional neurosurgery. In *Textbook of Stereotactic and Functional Neurosurgery*, 2nd ed.; Lozano, A., Gildenberg, P., Tasker, R., Eds.; Springer: Berlin, Germany, 2009; pp. 395–441.

42. Mandal, P.K.; Mahajan, R.; Dinov, I.D. Structural brain atlases: Design, rationale, and applications in normal and pathological cohorts. *J. Alzheimers Dis.* **2012**, *31* (Suppl. 3), S169–S188. [CrossRef] [PubMed]

43. Nowinski, W.L.; Chua, B.C.; Qian, G.Y.; Marchenko, Y.; Puspitasari, F.; Nowinska, N.G.; Knopp, M.V. *The Human Brain in 1492 Pieces: Structure, Vasculature, and Tracts*; Thieme: New York, NY, USA, 2011.

44. Nowinski, W.L.; Chua, B.C. *The Human Brain in 1969 Pieces: Structure, Vasculature, Tracts, Cranial Nerves, Systems, Head Muscles, and Glands (Version 2.0)*; Thieme: New York, NY, USA, 2014.

45. Rohlfing, T.; Zahr, N.M.; Sullivan, E.V.; Pfefferbaum, A. The SRI24 multichannel atlas of normal adult human brain structure. *Hum. Brain Mapp.* **2010**, *31*, 798–819. [CrossRef]

46. Baker, C.M.; Burks, J.D.; Briggs, R.G.; Conner, A.K.; Glenn, C.A.; Sali, G.; McCoy, T.M.; Battiste, J.D.; O'Donoghue, D.L.; Sughrue, M.E. A connectomic atlas of the human cerebrum chapter 1, introduction, methods, and significance. *Oper. Neurosurg.* **2018**, *15*, S1–S9. [CrossRef]

47. Briggs, R.G.; Conner, A.K.; Baker, C.M.; Burks, J.D.; Glenn, C.A.; Sali, G.; Battiste, J.D.; O'Donoghue, D.L.; Sughrue, M.E. A connectomic atlas of the human cerebrum-Chapter 18, The Connectional Anatomy of Human Brain Networks. *Oper. Neurosurg.* **2018**, *15* (Suppl. 1), S470–S480. [CrossRef]

48. Mori, S.; Wakana, S.; Nagae-Poetscher, L.M.; van Zijl, P.C. *MRI Atlas of Human White Matter*; Elsevier: Amsterdam, The Netherlands.

49. Nowinski, W.L.; Chua, B.C.; Yang, G.L.; Qian, G.Y. Three-dimensional interactive human brain atlas of white matter tracts. *Neuroinformatics* **2012**, *10*, 33–55. [CrossRef] [PubMed]

50. Van Essen, D.C. Cartography and connectomes. *Neuron* **2013**, *80*, 775–790. [CrossRef]

51. Van Essen, D.C.; Smith, S.M.; Barch, D.M.; Behrens, T.E.J.; Yacoub, E.; Ugurbil, K. The WU-Minn Human Connectome Project: An overview. *NeuroImage* **2013**, *80*, 62–79. [CrossRef]

52. Huck, J.; Wanner, Y.; Fan, A.P.; Jäger, A.T.; Grahl, S.; Schneider, U.; Villringer, A.; Steele, C.J.; Tardif, C.L.; Bazin, P.L.; et al. High resolution atlas of the venous brain vasculature from 7 T quantitative susceptibility maps. *Brain Struct. Funct.* **2019**, *224*, 2467–2485. [CrossRef]

53. Nowinski, W.L.; Thirunavuukarasuu, A.; Volkau, I.; Marchenko, Y.; Runge, V.M. *The Cerefy Atlas of Cerebral Vasculature*; Thieme: New York, NY, USA, 2009.

54. Nowinski WL, A. Thirunnavuukarasuu, Volkau I, Marchenko Y, Aminah B, Puspitasaari F, Runge VM. A three-dimensional interactive atlas of cerebral arterial variants. *Neuroinformatics* **2009**, *7*, 255–264. [CrossRef] [PubMed]

55. Nowinski, W.L.; Johnson, A.; Chua, B.C.; Nowinska, N.G. Three-dimensional interactive and stereotactic atlas of cranial nerves and nuclei correlated with surface neuroanatomy, vasculature and magnetic resonance imaging. *J. Neurosci. Methods* **2012**, *206*, 205–216. [CrossRef] [PubMed]

56. Sunkin, S.M.; Ng, L.; Lau, C.; Dolbeare, T.; Gilbert, T.L.; Thompson, C.L.; Hawrylycz, M.; Dang, C. Allen Brain Atlas: An integrated spatio-temporal portal for exploring the central nervous system. *Nucleic Acids Res.* **2013**, *41*, D996–D1008. [CrossRef] [PubMed]

57. Kanton, S.; Boyle, M.J.; He, Z.; Santel, M.; Weigert, A.; Sanchís-Calleja, F.; Guijarro, P.; Sidow, L.; Fleck, J.S.; Han, D.; et al. Organoid single-cell genomic atlas uncovers human-specific features of brain development. *Nature* **2019**, *574*, 418–422. [CrossRef]

58. James, G.A.; Hazaroglu, O.; Bush, K.A. A human brain atlas derived via n-cut parcellation of resting-state and task-based fMRI data. *Magn. Reason. Imaging* **2016**, *34*, 209–218. [CrossRef]

59. Yelnik, J.; Bardinet, E.; Dormont, D.; Malandain, G.; Ourselin, S.; Tandé, D.; Karachi, C.; Ayache, N.; Cornu, P.; Agid, Y. A three-dimensional, histological and deformable atlas of human basal ganglia. I. Atlas construction based on immunohistochemical and MRI data. *NeuroImage* **2007**, *34*, 618–638. [CrossRef]

60. Nowinski, W.L.; Qian, G.; Bhanu Prakash, K.N.; Thirunavuukarasuu, A.; Hu, Q.M.; Ivanov, N.; Parimal, A.S.; Runge, V.M.; Beauchamp, N.J. Analysis of ischemic stroke MR images by means of brain atlases of anatomy and blood supply territories. *Acad. Radiol.* **2006**, *13*, 1025–1034. [CrossRef]

61. Arsiwalla, X.D.; Zucca, R.; Betella, A.; Martinez, E.; Dalmazzo, D.; Omedas, P.; Deco, G.; Verschure, P.F. Network dynamics with BrainX(3): A large-scale simulation of the human brain network with real-time interaction. *Front. Neuroinform.* **2015**, *9*, 2. [CrossRef]

62. Fan, L.; Li, H.; Zhuo, J.; Zhang, Y.; Wang, J.; Chen, L.; Yang, Z.; Chu, C.; Xie, S.; Laird, A.R.; et al. The human brainnetome atlas: A new brain atlas based on connectional architecture. *Cereb. Cortex* **2016**, *26*, 3508–3526. [CrossRef]

63. McGrath, H.; Zaveri, H.P.; Collins, E.; Jafar, T.; Chishti, O.; Obaid, S.; Ksendzovsky, A.; Wu, K.; Papademetris, X.; Spencer, D.D. High-resolution cortical parcellation based on conserved brain landmarks for localization of multimodal data to the nearest centimeter. *Sci. Rep.* **2022**, *12*, 18778. [CrossRef]

64. Amunts, K.; Lenzen, M.; Friederici, A.D.; Schleicher, A.; Morosan, P.; Palomero-Gallagher, N.; Zilles, K. Broca's region: Novel organizational principles and multiple receptor mapping. *PLoS Biol.* **2010**, *8*, e1000489. [CrossRef] [PubMed]

65. Glasser, M.F.; Coalson, T.S.; Robinson, E.C.; Hacker, C.D.; Harwell, J.; Yacoub, E.; Ugurbil, K.; Andersson, J.; Beckmann, C.F.; Jenkinson, M.; et al. A multi-modal parcellation of human cerebral cortex. *Nature* **2016**, *536*, 171–178. [CrossRef] [PubMed]

66. Federative Committee on Anatomical Terminology (FCAT). *Terminologia Anatomica, International Anatomical Terminology*; Thieme: Stuttgart, Germany; New York, NY, USA, 1988.

67. Federative International Programme for Anatomical Terminology. Terminologia Neuroanatomica. 2017. Available online: https://fipat.library.dal.ca (accessed on 28 December 2022).

68. Bowden, D.M.; Song, E.; Kosheleva, J.; Dubach, M.F. NeuroNames: An ontology for the BrainInfo portal to neuroscience on the web. *Neuroinformatics* **2012**, *10*, 97–114. [CrossRef] [PubMed]

69. Haendel, M.A.; Balhoff, J.P.; Bastian, F.B.; Blackburn, D.C.; Blake, J.A.; Bradford, Y.; Comte, A.; Dahdul, W.M.; Dececchi, T.A.; Druzinsky, R.E.; et al. Unification of multi-species vertebrate anatomy ontologies for comparative biology in Uberon. *J. Biomed. Semant.* **2014**, *5*, 21. [CrossRef]

70. Rosse, C.; Mejino, J.L., Jr. A reference ontology for biomedical informatics: The Foundational Model of Anatomy. *J. Biomed. Inform.* **2003**, *36*, 478–500. [CrossRef]

71. Börner, K.; Teichmann, S.A.; Quardokus, E.M.; Gee, J.C.; Browne, K.; Osumi-Sutherland, D.; Herr BW 2nd Bueckle, A.; Paul, H.; Haniffa, M.; Jardine, L.; et al. Anatomical structures, cell types and biomarkers of the Human Reference Atlas. *Nat. Cell Biol.* **2021**, *23*, 1117–1128. [CrossRef]

72. Liang, P.; Shi, L.; Chen, N.; Luo, Y.; Wang, X.; Liu, K.; Mok, V.C.; Chu, W.C.; Wang, D.; Li, K. Construction of brain atlases based on a multi-center MRI dataset of 2020 Chinese adults. *Sci. Rep.* **2015**, *5*, 18216. [CrossRef]

73. Figley, T.D.; Mortazavi Moghadam, B.; Bhullar, N.; Kornelsen, J.; Courtney, S.M.; Figley, C.R. Probabilistic white matter atlases of human auditory, basal ganglia, language, precuneus, sensorimotor, visual and visuospatial networks. *Front. Hum. Neurosci.* **2017**, *11*, 306. [CrossRef]

74. Diedrichsen, J.; Balsters, J.H.; Flavell, J.; Cussans, E.; Ramnani, N. A probabilistic MR atlas of the human cerebellum. *NeuroImage* **2009**, *46*, 39–46. [CrossRef]

75. Pauli, W.M.; Nili, A.N.; Tyszka, J.M. A high-resolution probabilistic in vivo atlas of human subcortical brain nuclei. *Sci. Data* **2018**, *5*, 180063. [CrossRef]

76. Shattuck, D.W.; Mirza, M.; Adisetiyo, V.; Hojatkashani, C.; Salamon, G.; Narr, K.L.; Poldrack, R.A.; Bilder, R.M. Toga AW Construction of a 3D probabilistic atlas of human cortical structures. *NeuroImage* **2008**, *39*, 1064–1080. [CrossRef] [PubMed]

77. Wu, D.; Ma, T.; Ceritoglu, C.; Li, Y.; Chotiyanonta, J.; Hou, Z.; Hsu, J.; Xu, X.; Brown, T.; Miller, M.I.; et al. Resource atlases for multi-atlas brain segmentations with multiple ontology levels based on T1-weighted MRI. *Neuroimage* **2016**, *125*, 120–130. [CrossRef] [PubMed]

78. Hoehne, K.H. *VOXEL-MAN, Part 1, Brain and Skull*; Version 2.0.; Springer: Berlin/Heidelberg, Germany, 2001.

79. Cho, Z.H.; Kim, Y.B.; Han, J.Y.; Min, H.K.; Kim, K.N.; Choi, S.H.; Veklerov, E.; Shepp, L.A. New brain atlas—Mapping the human brain in vivo with 7.0 T MRI and comparison with postmortem histology: Will these images change modern medicine? *Int. J. Imaging Syst. Technol.* **2008**, *18*, 2–8. [CrossRef]

80. Liu, Y.; D'Haese, P.F.; Newton, A.T.; Dawant, B.M. Generation of human thalamus atlases from 7 T data and application to intrathalamic nuclei segmentation in clinical 3 T T1-weighted images. *Magn. Reason. Imaging* **2020**, *65*, 114–128. [CrossRef]

81. Nowinski, W.L.; Chua, B.C.; Thaung, T.S.L.; Wut Yi, S.H. *The Human Brain, Head and Neck in 2953 Pieces*; Thieme: New York, NY, USA, 2015. Available online: http://www.thieme.com/nowinski/ (accessed on 2 January 2023).

82. Saygin, Z.M.; Kliemann, D.; Iglesias, J.E.; van der Kouwe, A.J.W.; Boyd, E.; Reuter, M.; Stevens, A.; Van Leemput, K.; McKee, A.; Frosch, M.P.; et al. High-resolution magnetic resonance imaging reveals nuclei of the human amygdala: Manual segmentation to automatic atlas. *Neuroimage* **2017**, *155*, 370–382. [CrossRef] [PubMed]

83. Schira, M.M.; Isherwood, Z.J.; Kassem, M.; Barth, M.; Shaw, T.B.; Roberts, M.M.; Paxinos, G. HumanBrainAtlas: An in vivo MRI dataset for detailed segmentations. *bioRxiv* **2022**. [CrossRef]

84. Yushkevich, P.A.; Avants, B.B.; Pluta, J.; Das, S.; Minkoff, D.; Mechanic-Hamilton, D.; Glynn, S.; Pickup, S.; Liu, W.; Gee, J.C.; et al. A high-resolution computational atlas of the human hippocampus from postmortem magnetic resonance imaging at 9.4. T. *NeuroImage* **2009**, *44*, 385–398. [CrossRef]

85. Oishi, K.; Linda Chang, L.; Huang, H. Baby brain atlases. *NeuroImage* **2019**, *185*, 865–880. [CrossRef]

86. Zhang, Y.; Wei, H.; Cronin, M.J.; He, N.; Yan, F.; Liu, C. Longitudinal atlas for normative human brain development and aging over the lifespan using quantitative susceptibility mapping. *Neuroimage* **2018**, *171*, 176–189. [CrossRef]

87. Zuo, X.N.; He, Y.; Betzel, R.F.; Colcombe, S.; Sporns, O.; Milham, M.P. Human connectomics across the life span. *Trends Cogn. Sci.* **2017**, *21*, 32–45. [CrossRef]

88. Kuklisova-Murgasova, M.; Aljabar, P.; Srinivasan, L.; Counsell, S.J.; Doria, V.; Serag, A.; Gousias, I.S.; Boardman, J.P.; Rutherford, M.A.; Edwards, A.D.; et al. A dynamic 4D probabilistic atlas of the developing brain. *Neuroimage* **2011**, *54*, 2750–2763. [CrossRef] [PubMed]

89. Amunts, K.; Lepage, C.; Borgeat, L.; Mohlberg, H.; Dickscheid, T.; Rousseau, M.É.; Bludau, S.; Bazin, P.L.; Lewis, L.B.; Oros-Peusquens, A.M.; et al. Bigbrain: An ultrahigh-resolution 3D human brain model. *Science* **2013**, *340*, 1472–1475. [CrossRef] [PubMed]

90. Ding, S.L.; Royall, J.J.; Sunkin, S.M.; Ng, L.; Facer, B.A.; Lesnar, P.; Guillozet-Bongaarts, A.; McMurray, B.; Szafer, A.; Dolbeare, T.A.; et al. Comprehensive cellular-resolution atlas of the adult human brain. *J. Comp. Neurol.* **2016**, *524*, 3127–3481. [CrossRef] [PubMed]

91. Ecker, J.R.; Geschwind, D.H.; Kriegstein, A.R.; Ngai, J.; Osten, P.; Polioudakis, D.; Regev, A.; Sestan, N.; Wickersham, I.R.; Zeng, H. The BRAIN Initiative Cell Census Consortium: Lessons learned toward generating a Comprehensive Brain Cell Atlas. *Neuron* **2017**, *96*, 542–557. [CrossRef] [PubMed]

92. Hawrylycz, M.J.; Lein, E.S.; Guillozet-Bongaarts, A.L.; Shen, E.H.; Ng, L.; Miller, J.A.; van de Lagemaat, L.N.; Smith, K.A.; Ebbert, A.; Riley, Z.L.; et al. An anatomically comprehensive atlas of the adult human brain transcriptome. *Nature* **2012**, *489*, 391–399. [CrossRef]

93. Beliveau, V.; Ganz, M.; Feng, L.; Ozenne, B.; Højgaard, L.; Fisher, P.M.; Svarer, C.; Greve, D.N.; Knudsen, G.M. A high-resolution in vivo atlas of the human brain's serotonin system. *J. Neurosci.* **2017**, *37*, 120–128. [CrossRef]

94. McKetney, J.; Runde, R.M.; Hebert, A.S.; Salamat, S.; Roy, S.; Coon, J.J. Proteomic atlas of the human brain in Alzheimer's Disease. *J. Proteome Res.* **2019**, *18*, 1380–1391. [CrossRef]

95. Thompson, P.M.; Mega, M.S.; Woods, R.P.; Zoumalan, C.I.; Lindshield, C.J.; Blanton, R.E.; Moussai, J.; Holmes, C.J.; Cummings, J.L.; Toga, A.W. Cortical change in Alzheimer's disease detected with a disease-specific population-based brain atlas. *Cereb. Cortex* **2001**, *11*, 1–16. [CrossRef]

96. Mega, M.S.; Dinov, I.D.; Mazziotta, J.C.; Manese, M.; Thompson, P.M.; Lindshield, C.; Moussai, J.; Tran, N.; Olsen, K.; Zoumalan, C.I.; et al. Automated brain tissue assessment in the elderly and demented population: Construction and validation of a sub-volume probabilistic brain atlas. *NeuroImage* **2005**, *26*, 1009–1018. [CrossRef]

97. de Haan, B.; Karnath, H.O. 'Whose atlas I use, his song I sing?'–The impact of anatomical atlases on fiber tract contributions to cognitive deficits after stroke. *Neuroimage* **2017**, *163*, 301–309. [CrossRef]

98. Nowinski, W.L.; Gupta, V.; Qian, G.; Ambrosius, W.; Kazmierski, R. Population-based stroke atlas for outcome prediction: Method and preliminary results for ischemic stroke from CT. *PLoS ONE* **2014**, *9*, e102048. [CrossRef] [PubMed]

99. Parisot, S.; Darlix, A.; Baumann, C.; Zouaoui, S.; Yordanova, Y.; Blonski, M.; Rigau, V.; Chemouny, S.; Taillandier, L.; Bauchet, L.; et al. A Probabilistic Atlas of Diffuse WHO Grade II Glioma Locations in the Brain. *PLoS ONE* **2021**, *11*, e0144200. [CrossRef] [PubMed]

100. Wang, H.E.; Scholly, J.; Triebkorn, P.; Sip, V.; Medina Villalon, S.; Woodman, M.M.; Le Troter, A.; Guye, M.; Bartolomei, F.; Jirsa, V. VEP atlas: An anatomic and functional human brain atlas dedicated to epilepsy patients. *J. Neurosci. Methods* **2021**, *348*, 108983. [CrossRef] [PubMed]

101. Toga, A.W.; Thompson, P.M. Brain atlases of normal and diseased populations. *Int. Rev. Neurobiol.* **2005**, *66*, 1–54. [PubMed]

102. Makowski, C.; van der Meer, D.; Dong, W.; Wang, H.; Wu, Y.; Zou, J.; Liu, C.; Rosenthal, S.B.; Hagler, D.J., Jr.; Fan, C.C.; et al. Discovery of genomic loci of the human cerebral cortex using genetically informed brain atlases. *Science* **2022**, *375*, 522–528. [CrossRef]

103. Tang, Y.; Hojatkashani, C.; Dinov, I.D.; Sun, B.; Fan, L.; Lin, X.; Qi, H.; Hua, X.; Liu, S.; Toga, A.W. The construction of a Chinese MRI brain atlas: A morphometric comparison study between Chinese and Caucasian cohorts. *Neuroimage* **2010**, *51*, 33–41. [CrossRef]

104. Bhalerao, G.V.; Parlikar, R.; Agrawal, R.; Shivakumar, V.; Kalmady, S.V.; Rao, N.P.; Agarwal, S.M.; Narayanaswamy, J.C.; Reddy, Y.C.J.; Venkatasubramanian, G. Construction of population-specific Indian MRI brain template: Morphometric comparison with Chinese and Caucasian templates. *Asian J. Psychiatr.* **2018**, *35*, 93–100. [CrossRef]

105. Nowinski, W.L. Towards the holistic, reference and extendable atlas of the human brain, head and neck. *Brain Inform.* **2015**, *2*, 65–76. [CrossRef]

106. Nowinski, W.L. Computational and mathematical methods in brain atlasing. *Neuroradiol. J.* **2017**, *30*, 520–534. [CrossRef]

107. Nowinski, W.L. Visualization and interaction in the atlas of the human brain, head and neck. *Mach. Graph. Vis.* **2014**, *23*, 3–10. [CrossRef]

108. Abarca-Olivas, J.; González-López, P.; Fernández-Cornejo, V.; Verdú-Martínez, I.; Martorell-Llobregat, C.; Baldoncini, M.; Campero, A. 3D stereoscopic view in neurosurgical anatomy: Compilation of basic methods. *World Neurosurg.* **2022**, *163*, e593–e609. [CrossRef] [PubMed]

109. Sundsten, J.W.; Brinkley, J.F.; Eno, K.; Prothero, J. *The Digital Anatomist. Interactive Brain Atlas. CD ROM for the Macintosh*; University of Washington: Seattle, WA, USA, 1994.
110. A.D.A.M. *A.D.A.M Animated Dissection of Anatomy for Medicine. User's Guide*; A.D.A.M. Inc.: Atlanta, GA, USA, 1996.
111. Berkovitz, B.; Kirsch, C.; Moxham, B.; Alusi, G.; Cheeseman, T. *Interactive Head & Neck.*; Primal Pictures Ltd.: London, UK, 2003.
112. Nowinski, W.L. 3D atlas of the brain, head and neck in 2953 pieces. *Neuroinformatics* **2017**, *15*, 395–400. [CrossRef] [PubMed]
113. Nowinski, W.L.; Chua, B.C.; Qian, G.Y.; Nowinska, N.G. The human brain in 1700 pieces: Design and development of a three-dimensional, interactive and reference atlas. *J. Neurosci. Methods* **2012**, *204*, 44–60. [CrossRef] [PubMed]
114. Nowinski, W.L.; AThirunavuukarasuu, A.; Ananthasubramaniam, A.; Chua, B.C.; Qian, G.; Nowinska, N.G.; Marchenko, Y.; Volkau, I. Automatic testing and assessment of neuroanatomy using a digital brain atlas: Method and development of computer- and mobile-based applications. *Anat. Sci. Educ.* **2009**, *2*, 244–252. [CrossRef]
115. Nowinski, W.L. NOWinBRAIN: A large, systematic, and extendable repository of 3D reconstructed images of a living human brain cum head and neck. *J. Digit. Imaging* **2022**, *35*, 98–114. [CrossRef]
116. Nowinski, W.L. Bridging neuroradiology and neuroanatomy: NOWinBRAIN–a repository with sequences of correlated and labeled planar-surface neuroimages. *Neuroradiol. J.* **2022**. Available online: https://pubmed.ncbi.nlm.nih.gov/35702757/ (accessed on 2 January 2023). [CrossRef]
117. Nowinski, W.L. NOWinBRAIN 3D neuroimage repository: Exploring the human brain via systematic and stereotactic dissections. *Neurosci. Inform.* **2022**, *2*, 100085. [CrossRef]
118. Nowinski, W.L. On the definition, construction, and presentation of the human cerebral sulci: A morphology-based approach. *J. Anat.* **2022**. [CrossRef]
119. Hess, A.; Hinz, R.; Keliris, G.A.; Boehm-Sturm, P. On the usage of brain atlases in neuroimaging research. *Mol. Imaging Biol.* **2018**, *20*, 742–749. [CrossRef]
120. Lancaster, J.L.; Woldorff, M.G.; Parsons, L.M.; Liotti, M.; Freitas, C.S.; Rainey, L.; Kochunov, P.V.; Nickerson, D.; Mikiten, S.A.; Fox, P.T. Automated Talairach atlas labels for functional brain mapping. *Hum. Brain Mapp.* **2000**, *10*, 120–131. [CrossRef]
121. Aljabar, P.; Heckemann, R.A.; Hammers, A.; Hajnal, J.V.; Rueckert, D. Multi-atlas based segmentation of brain images: Atlas selection and its effect on accuracy. *Neuroimage* **2009**, *46*, 726–738. [CrossRef] [PubMed]
122. Artaechevarria, X.; Munoz-Barrutia, A.; Ortiz-de Solorzano, C. Combination strategies in multi-atlas image segmentation: Application to brain MR data. *IEEE Trans. Med. Imaging* **2009**, *28*, 1266–1277. [CrossRef] [PubMed]
123. Labra, N.; Guevara, P.; Duclap, D.; Houenou, J.; Poupon, C.; Mangin, J.F.; Figueroa, M. Fast automatic segmentation of white matter streamlines based on a multi-subject bundle atlas. *Neuroinformatics* **2017**, *15*, 71–86. [CrossRef] [PubMed]
124. Li, X.; Chen, L.; Kutten, K.; Ceritoglu, C.; Li, Y.; Kang, N.; Hsu, J.T.; Qiao, Y.; Wei, H.; Liu, C.; et al. Multi-atlas tool for automated segmentation of brain gray matter nuclei and quantification of their magnetic susceptibility. *Neuroimage* **2019**, *191*, 337–349. [CrossRef] [PubMed]
125. Lötjönen, J.M.; Wolz, R.; Koikkalainen, J.R.; Thurfjell, L.; Waldemar, G.; Soininen, H.; Rueckert, D.; Alzheimer's Disease Neuroimaging Initiative. Fast and robust multi-atlas segmentation of brain magnetic resonance images. *NeuroImage* **2010**, *49*, 2352–2365. [CrossRef]
126. Zaffino, P.; Ciardo, D.; Raudaschl, P.; Fritscher, K.; Ricotti, R.; Alterio, D.; Marvaso, G.; Fodor, C.; Baroni, G.; Amato, F.; et al. Multi atlas based segmentation: Should we prefer the best atlas group over the group of best atlases? *Phys. Med. Biol.* **2018**, *63*, 12NT01. [CrossRef]
127. Bjerke, I.E.; Øvsthus, M.; Papp, E.A.; Yates, S.C.; Silvestri, L.; Fiorilli, J.; Pennartz, C.M.A.; Pavone, F.S.; Puchades, M.A.; Leergaard, T.B.; et al. Data integration through brain atlasing: Human Brain Project tools and strategies. *Eur. Psychiatry* **2018**, *50*, 70–76. [CrossRef]
128. Nowinski, W.L. Computerized brain atlases for surgery of movement disorders. *Semin. Neurosurg.* **2001**, *12*, 183–194. [CrossRef]
129. Nowinski, W.L.; Yang, G.L.; Yeo, T.T. Computer-aided stereotactic functional neurosurgery enhanced by the use of the multiple brain atlas database. *IEEE Trans. Med. Imaging* **2000**, *19*, 62–69. [CrossRef]
130. Nowinski, W.L.; Chua, B.C.; Volkau, I.; Puspitasari, F.; Marchenko, Y.; Runge, V.M.; Knopp, M.V. Simulation and assessment of cerebrovascular damage in deep brain stimulation using a stereotactic atlas of vasculature and structure derived from multiple 3T and 7T scans. *J. Neurosurg.* **2010**, *113*, 1234–1241. [CrossRef]
131. Benabid, A.L.; Nowinski, W.L. Intraoperative robotics for the practice of neurosurgery: A surgeon's perspective. In *The Operating Room for the 21th Century*; Apuzzo, M.L., Ed.; American Association of Neurological Surgeons: Rolling Meadows, IL, USA, 2003; pp. 103–118.
132. Nowinski, W.L. Usefulness of brain atlases in neuroradiology: Current status and future potential. *Neuroradiol. J.* **2016**, *29*, 260–268. [CrossRef] [PubMed]
133. Nowinski, W.L. Human brain atlases in stroke management. *Neuroinformatics* **2020**, *18*, 549–567. [CrossRef]
134. Nowinski, W.L.; Chua, B.C.; Wut Yi, S.H. *3D Atlas of Neurologic Disorders*; Thieme: New York, NY, USA, 2014.
135. Nowinski, W.L.; Chua, B.C. Bridging neuroanatomy, neuroradiology and neurology: Three-dimensional interactive atlas of neurological disorders. *Neuroradiol. J.* **2013**, *26*, 252–262. [CrossRef]
136. Sim, K.; Yang, G.L.; Loh, D.; Poon, L.Y.; Sitoh, Y.Y.; Verma, S.; Keefe, R.; Collinson, S.; Chong, S.A.; Heckers, S.; et al. White matter abnormalities and neurocognitive deficits associated with the passivity phenomenon in schizophrenia: A diffusion tensor imaging study. *Psychiatry Res.* **2009**, *172*, 121–127. [CrossRef]
137. Assaf, Y.; Alexander, D.C.; Jones, D.K.; Bizzi, A.; Behrens, T.E.; Clark, C.A.; Cohen, Y.; Dyrby, T.B.; Huppi, P.S.; Knoesche, T.R.; et al. The CONNECT project: Combining macro- and micro-structure. *Neuroimage* **2013**, *80*, 273–282. [CrossRef]

138. Jiang, T. Brainnetome: A new -ome to understand the brain and its disorders. *Neuroimage* **2013**, *80*, 263–272. [CrossRef]
139. BRAIN Working Group. 2014. *BRAIN 2025. A Scientific Vision*. NIH. Available online: https://www.braininitiative.nih.gov/pdf/BRAIN2025_508C.pdf (accessed on 2 January 2023).
140. Jorgenson, L.A.; Newsome, W.T.; Anderson, D.J.; Bargmann, C.I.; Brown, E.N.; Deisseroth, K.; Donoghue, J.P.; Hudson, K.L.; Ling, G.S.; MacLeish, P.R.; et al. The BRAINInitiative: Developing technology to catalyse neuroscience discovery. *Philos. Trans. R. Soc. B Biol. Sci.* **2015**, *370*, 20140164. [CrossRef]
141. Markram, H.; Muller, E.; Ramaswamy, S.; Reimann, M.W.; Abdellah, M.; Sanchez, C.A.; Ailamaki, A.; Alonso-Nanclares, L.; Antille, N.; Arsever, S.; et al. Reconstruction and simulation of neocortical microcircuitry. *Cell* **2015**, *163*, 456–492. [CrossRef]
142. Amunts, K.; Ebell, C.; Muller, J.; Telefont, M.; Knoll, A.; Lippert, T. The Human Brain Project: Creating a European research infrastructure to decode the human brain. *Neuron* **2016**, *92*, 574–581. [CrossRef]
143. Sadato, N.; Morita, K.; Kasai, K.; Fukushi, T.; Nakamura, K.; Nakazawa, E.; Okano, H.; Okabe, S. Neuroethical issues of the Brain/MINDS Project of Japan. *Neuron* **2019**, *101*, 385–389. [CrossRef]
144. Chin, A.L.; Yang, S.M.; Chen, H.H.; Li, M.T.; Lee, T.T.; Chen, Y.J.; Lee, T.K.; Petibois, C.; Cai, X.; Low, C.M.; et al. A synchrotron X-ray imaging strategy to map large animal brains. *Chin. J. Phys.* **2020**, *65*, 24–32. [CrossRef]
145. Chen, S.; He, Z.; Han, X.; He, X.; Li, R.; Zhu, H.; Zhao, D.; Dai, C.; Zhang, Y.; Lu, Z.; et al. How big data and high-performance computing drive brain science. *Genom. Proteom. Bioinform.* **2019**, *17*, 381–392. [CrossRef] [PubMed]
146. Ament, S.A.; Adkins, R.S.; Carter, R.; Chrysostomou, E.; Colantuoni, C.; Crabtree, J.; Creasy, H.H.; Degatano, K.; Felix, V.; Gandt, P.; et al. The Neuroscience Multi-Omic Archive: A BRAIN Initiative resource for single-cell transcriptomic and epigenomic data from the mammalian brain. *Nucleic Acids Res.* **2022**, *9*, 2022. [CrossRef] [PubMed]
147. Akram, M.A.; Ljungquist, B.; Ascoli, G.A. Efficient metadata mining of web-accessible neural morphologies. *Prog. Biophys. Mol. Biol.* **2022**, *168*, 94–102. [CrossRef] [PubMed]

Article

Observer Sensitivity for Detection of Pulmonary Nodules in Ultra-Low Dose Computed Tomography Protocols Using a Third-Generation Dual-Source CT with Ultra-High Pitch—A Phantom Study

Natascha Leitzig [1,*], Sonja Janssen [2], Hany Kayed [1], Stefan O. Schönberg [1] and Hans Scheffel [3]

[1] Clinic of Radiology and Nuclear Medicine, University Medical Centre Mannheim, Heidelberg University, Theodor-Kutzer-Ufer 1-3, 68167 Mannheim, Germany
[2] MVZ Radiologie Westpfalz GmbH, Lutrinastraße 27, 67655 Kaiserslautern, Germany
[3] Institute of Radiology and Nuclear Medicine, Diakonie Hospital, Speyerer Straße 91, 68163 Mannheim, Germany
* Correspondence: natascha.leitzig@gmail.com; Tel.: +49-176-72661397

Abstract: This study evaluates ultra-low-dose computed tomography (ULDCT) protocols concerning the detectability of pulmonary nodules. The influence of tube current settings, kernels, strength levels of third-generation iterative reconstruction algorithms, and pitch was investigated. A chest phantom with artificial spherical nodules of different densities and diameters was examined with a third-generation dual-source CT. Scanning and post-processing protocols, tube current levels, and ultra-high and non-high pitch modes were applied. Images were reconstructed with filtered back-projection (FBP) or advanced model-based iterative reconstruction (ADMIRE) algorithms. Sharp (Bl57) or medium-soft (Br36) convolution kernels were applied. The reading was performed by an experienced and an inexperienced reader. The highest observer sensitivity was found using a non-high pitch protocol at tube currents of 120 mAs and 90 mAs with the sharp kernel and iterative reconstruction level of 5. Non-high pitch protocols showed better detectability of solid nodules. Combinations with the medium-soft kernel achieved slightly higher observer sensitivity than with the sharp kernel. False positives (FP) occurred more often for subsolid nodules, at a tube current level of 120 mAs, and with the sharp kernel. A tube current level of 90 mAs combined with the highest iterative reconstruction level achieved the highest accuracy in lung nodule detection regardless of size, density, and reader experience.

Keywords: anthropomorphic chest phantom; detectability; iterative reconstruction; pulmonary nodules; ultra-low-dose CT

Citation: Leitzig, N.; Janssen, S.; Kayed, H.; Schönberg, S.O.; Scheffel, H. Observer Sensitivity for Detection of Pulmonary Nodules in Ultra-Low Dose Computed Tomography Protocols Using a Third-Generation Dual-Source CT with Ultra-High Pitch—A Phantom Study. *Anatomia* **2023**, *2*, 15–26. https://doi.org/10.3390/anatomia2010002

Academic Editors: Gianfranco Natale and Francesco Fornai

Received: 2 November 2022
Revised: 7 December 2022
Accepted: 6 January 2023
Published: 13 January 2023

1. Introduction

Lung cancer has been an important topic in medicine for a long time and is still one of the biggest challenges for medical doctors and scientists worldwide [1–3].

In Europe, a total of 4,042,263 new cases were diagnosed in 2020 and 1,942,552 deaths were attributed to lung cancer. It is the second most common cancer in men and the third most common cancer in women. It also ranks first as the cause of cancer deaths for men and ranks second for women. For both sexes and all ages, the mortality rate is 54.2, which is by far the highest mortality rate of all cancers in European countries [2].

Treatment outcome is significantly better in the early stages of the disease [4]. However, three-quarters of lung cancer diseases are only diagnosed after they reached a point beyond curative treatment [5].

Chest X-ray has been the method of choice to investigate lung cancer in patients, but due to its limited sensitivity, it could not be used for screening. In previous studies using X-rays as a screening method, lung cancer was diagnosed more often, but mortality was not affected [6]. In the randomized PLCO (Prostate, Lung, Colorectal, and Ovarian) screening trial, lung cancer screening involved four annual chest radiographs for up to four years. There was no evidence of a shift in stage at diagnosis, nor of significantly lower lung cancer mortality [7].

In 2011, the National Lung Screening Trial (NLST) was the first to show a significant 20% reduction in lung cancer mortality through annual computed tomography (CT) screening of high-risk patients compared to chest radiography resulting in earlier detection of cancer lesions [8].

In 2012, the Lung Cancer Screening Panel of the National Comprehensive Cancer Network in the USA was the first to officially recommend annual LDCT screening for high-risk individuals, and in 2013, the American Cancer society published lung cancer guidelines including LDCT screening [9].

In 2017, a European interdisciplinary consensus statement named CT the best method available to date for lung cancer screening and recommended planning the implementation of lung cancer screening in high-risk individuals throughout Europe at certified multidisciplinary medical centers [9].

When The Nederlands-Leuvens Longkanker Screening Onderzoek (NELSON) results were published in 2020, they reported a 26% reduction in lung cancer mortality in men and a 36% reduction in mortality in women after a 10-year follow-up. The screening was found to result in a systematic shift and earlier diagnosis at a lower stage [10,11].

To date, however, there is no established low-dose CT screening for lung cancer in many European countries. Critical points are a high rate of over-diagnosis, a bad risk–benefit ratio concerning radiation exposure of asymptomatic individuals in the screening situation, and missing guidelines and standardized protocols for the different CT scanners available [11–13].

Due to the constant and fast technological advances in the field of radiology, the possibilities for screening are also constantly improving. As screening is implemented in more countries, radiologists will need data to rely on when optimizing their setup for screening. There have been few approaches to optimizing CT protocols so far.

In 2020, a study evaluated nodule visibility using different effective mAs levels. They concluded that a high detection rate can be achieved at low mAs if the other detection parameters are set as best as possible [14].

We raised the general question about which settings contribute best to improving screening. In this study, we investigated dedicated ultra-low dose CT (ULDCT) protocols using a third-generation dual-source CT with ultra-high pitch and different post-processing settings to evaluate their impact on the observer sensitivity for the detection of pulmonary nodules at different reader experience levels using an anthropomorphic phantom.

2. Materials and Methods

2.1. Phantom and Setup

The anthropomorphic chest phantom used in this study was life-sized and equipped with an artificial thoracic wall, heart, mediastinum, diaphragm, and lung with pulmonary vessels (Figure 1). The soft tissue was made of polyurethane resin, and for bone structures, epoxy resin was used.

Artificial spherical nodules with a density of 100 Hounsfield units (HU) were made of polyurethane resin to represent the more common but also often benign solid pulmonary nodules. To represent the rare, but often malignant, subsolid nodes, nodules with densities of $-630\,\mathrm{HU}$ and $-800\,\mathrm{HU}$ were made of polyurethane foam resin. There were five different diameters: 3, 5, 8, 10, and 12 mm (Figure 2).

A third researcher who was not involved in the readout randomly distributed the nodules in both lobes of the lung so that their position would remain unknown to the readers. The artificial nodules adhered to the artificial bronchovascular structures inside the phantom. This was repeated 13 times to generate 13 different settings.

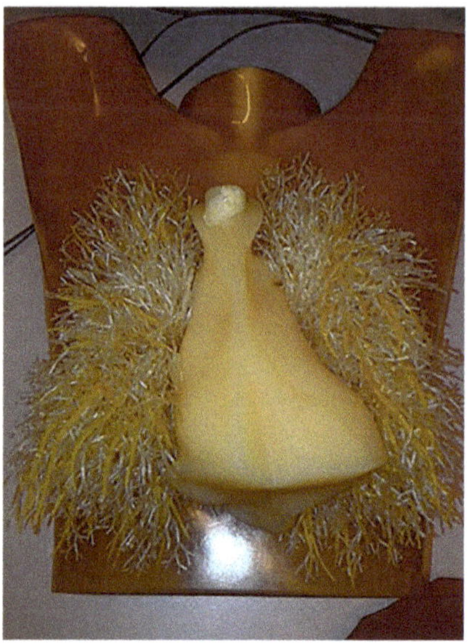

Figure 1. Anthropomorphic chest phantom.

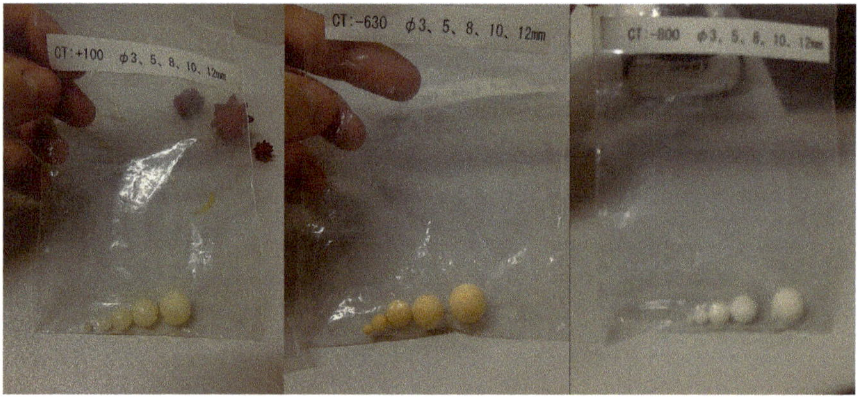

Figure 2. Artificial nodules with diameters of 3, 5, 8, 10 and 12 mm in 3 densities (from left to right solid +100, subsolid −630 and −800).

2.2. Data Acquisition and Post-Processing

The phantom was scanned with different ultra-low-dose protocol combinations using a third-generation dual-source CT scanner system (2 × 192 slices; Somatom Force, Siemens Healthineers, Forchheim, Germany). The protocol parameters were a tube voltage of 100 kVp with a tin filter (100 kVp/Sn), four different tube current levels of 30 mAs, 60 mAs, 90 mAs, and 120 mAs, a rotation time of 0.25 s, detector collimation of 192 × 0.6 mm, and a matrix size of 512 × 512.

Each of the setups created was scanned with a non-high pitch of 1.2 and an ultra-high pitch of 3.2. The slice thickness was constantly kept at 1.5 mm, along with an increment of 1.0 mm throughout all study post-processing.

Filtered back-projection reconstruction (FBP) and advanced model-based iterative reconstruction (ADMIRE) at the strength levels of 1, 3, and 5 were applied. All scans were reconstructed once with a lung kernel Bl57 ("sharp") and once with a medium-soft kernel

Br36 ("soft") for the study read-out (Table 1). Due to the use of an artificial chest phantom, IRB approval was not required for this study.

Table 1. Reading overview.

Pitch	mAs	Reconstruction Kernel (3rd Generation Iterative Reconstruction Strength)
non-high pitch (1.2)	120	Br36(3)
ultra-high pitch (3.2)	90	Br36(5)
	60	Bl57 (FBP)
	30	Bl57(1)
		Bl57(3)
		Bl57(5)

2.3. Data Analysis

The reconstructed images were archived in the hospital's research Picture Archiving and Communication System (PACS). For image analysis, dedicated software was used (Osirix DICOM viewer Version v.11.0 64-bit, Pixmeo, Geneva, Switzerland). The maximum intensity projection (MIP) was set to 10.

A total of 221 datasets were evaluated by two readers. One was an experienced radiologist with over 15 years of experience in thoracic CT imaging. The other one was an inexperienced reader—with no relevant experience in chest CT evaluation so far—who received basic training before the evaluation. Both readers were blinded to the number, location, and size of nodules within each of the nodule settings. To check the visibility of the nodules a control scan outside the chest phantom was performed for all protocols (Figures 3 and 4). Records were kept on the number of nodules found in every setup and their measured density.

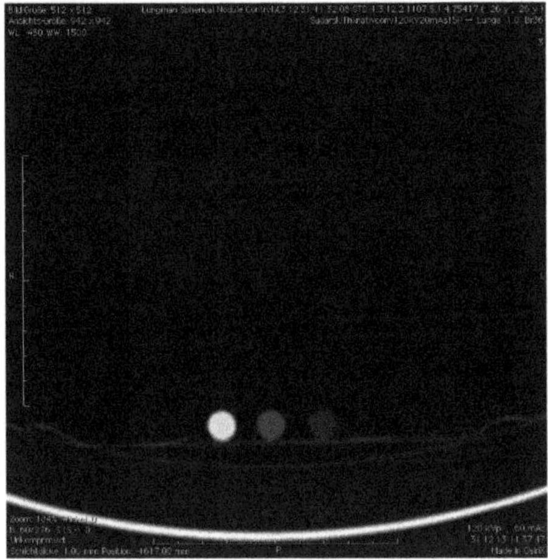

Figure 3. Test scan, 12 mm nodules of all densities.

Figure 4. Test scan, 3 mm nodules of all densities.

2.4. Statistical Analysis

The statistical analysis and the creation of graphs were performed by a professional statistician from the Department of Medical Statistics, Biomathematics, and Information Processing of the University Medical Centre Mannheim. The dedicated software used was SAS v9.4. 64-bit, SAS, Cary, NC, USA.

Observer detection sensitivity was calculated per setup as a percentage of true positives out of all findings. The Kruskal–Wallis test and the Mann–Whitney U-test for unconnected samples were performed.

Results are presented as the mean ± standard deviation. False positives (FPs) are presented as an absolute number and relative percentage.

3. Results

3.1. Observer Sensitivity

The highest observer sensitivity for both readers irrespective of nodule density was found in the non-high pitch protocols ADMIRE 5 sharp-120 mAs and ADMIRE 5 sharp–90 mAs with an observer sensitivity of 76.7% each (Table 2). The lowest observer sensitivity of 54.7% was achieved with protocol FBP sharp-30 mAs using an ultra-high pitch. The readers achieved a mean overall observer sensitivity of 68.5% ± 20.6 ($p = 0.02$).

Table 2. Observer sensitivity of both readers, overall densities.

	Reconstruction Kernel (3rd Generation Iterative Reconstruction Strength)	mAs 120	mAs 90	mAs 60	mAs 30
non-high pitch	Br36(3)	66.7	60.7	69.4	68.0
	Br36(5)	70.7	60.0	72.0	68.7
	Bl57 (FBP)	69.4	66.7	60.7	59.4
	Bl57(1)	71.3	67.4	62.7	62.0
	Bl57(3)	74.7	73.4	71.4	68.7
	Bl57(5)	76.7	76.7	75.4	71.4
ultra-high pitch	Br36(3)	68.0	70.7	68.7	67.4
	Br36(5)	67.4	69.4	70.0	66.0
	Bl57 (FBP)	66.7	65.4	63.4	54.7
	Bl57(1)	65.4	66.0	64.7	56.0
	Bl57(3)	66.0	69.4	66.0	56.0
	Bl57(5)	72.0	72.7	66.7	62.7

For nodule densities of 100 HU, the highest observer sensitivity was measured for ADMIRE 3 sharp-30 mAs with 90% at the non-high pitch. The lowest sensitivity was 68% and was achieved in the protocols ADMIRE 5 soft-120 mAs and FBP sharp-30 mAs with ultra-high pitch and ADMIRE 3 soft and ADMIRE 5 soft at 90 mAs with non-high pitch.

For nodule densities of −630 HU, the best results in detection were achieved with ADMIRE 5 sharp-120 mAs with an 80% observer sensitivity at the non-high pitch. The highest results were obtained with ADMIRE 1 sharp-120 mAs at the ultra-high pitch and FBP sharp at 60 mAs and 30 mAs at non-high pitch (58%).

For nodule densities of −800 HU, ADMIRE 5 sharp-120 mAs showed a sensitivity of 72% at the non-high pitch. Only 34% was achieved with ADMIRE 1 sharp-30 mAs and ADMIRE 3 sharp-30 mAs at the ultra-high pitch and FBP sharp-30 mAs at the non-high pitch.

3.2. Pitch

Non-high pitch protocols showed significantly better results for the nodule densities of 100 HU and overall compared to ultra-high pitch protocols. The experienced reader achieved a mean observer sensitivity of 77.8% ± 35.8 with non-high pitch vs. 59.4% ± 40.2 with ultra-high pitch at 100 HU ($p < 0.01$). For all nodule densities, he reached 70.0% ± 20.4 with non-high pitch and 62.1% ± 22.5 with ultra-high pitch ($p < 0.01$).

The highest observer sensitivity for both readers combined was achieved with the non-high pitch for nodule densities of 100 HU with 80.4% ± 34.5 ($p < 0.01$). The lowest was achieved for nodule densities of −800 HU at non-high pitch (58.5% ± 36.3).

3.3. Tube Current Level

From 30 mAs–120 mAs, the observer sensitivity increased steadily. The lowest observer sensitivity was obtained at 30 mAs (64.6% ± 21.2).

The inexperienced reader obtained the highest accuracy over all densities at the highest tube current setting at 120 mAs (73.5% ± 17.6; $p = 0.03$) (Figure 5), while the experienced reader achieved the highest observer sensitivity with 68.8% ± 20.5 at 90 mAs and 67.9% ± 25.4 at 120 mAs ($p = 0.07$) (Figure 6).

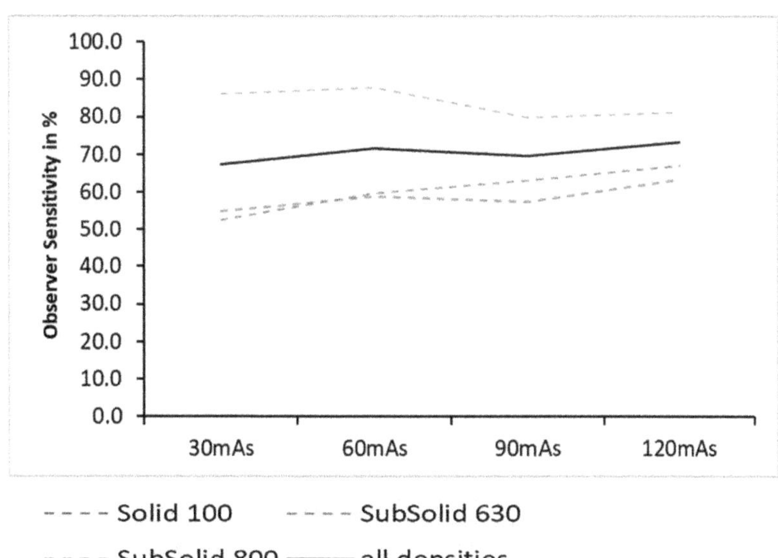

Figure 5. Observer sensitivity vs. mAs, all densities—inexperienced reader.

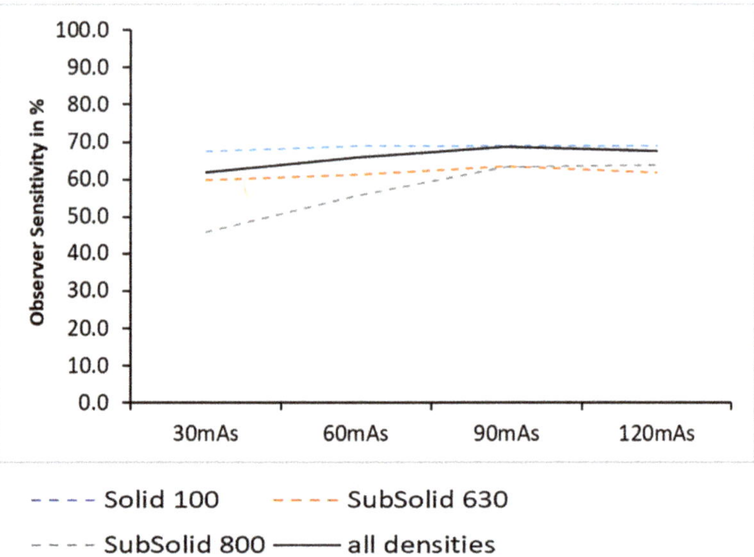

Figure 6. Observer sensitivity vs. mAs, all densities—experienced reader.

For nodule densities of −800 HU, both readers combined achieved the best observer sensitivity at 120 mAs (65.6% ± 35.29) and the lowest results at 30 mAs (49.2% ± 35.9).

The dose decreased from a DLP of 14.31 at 120 mAs to 11.00 at 90 mAs and 7.15 at 60 mAs to 3.69 at 30 mAs. The conversion factor from DLP to mSv was 0.015.

3.4. Kernel

There was a tendency for the highest observer sensitivity to occur for all images combined with the soft kernel for both readers together (68.8% ± 20.8, $p = 0.60$) as well as for each alone (71.0% ± 18.9, $p = 0.80$ and 66.6% ± 22.3, $p = 0.64$). However, these results were not significant. Nevertheless, the results of both kernels were close to each other (Figures 7 and 8).

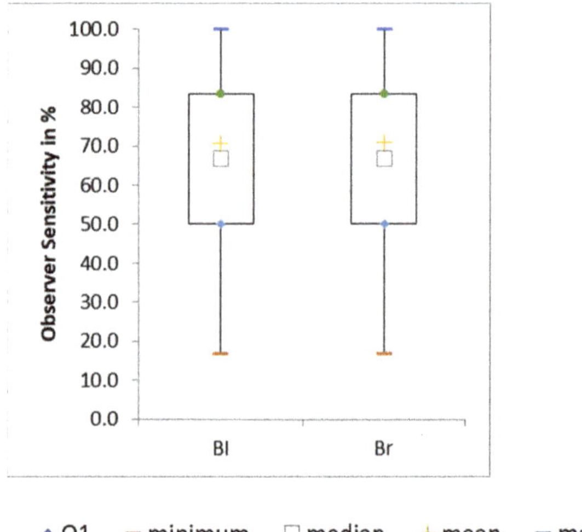

Figure 7. Observer sensitivity vs. kernel, all densities—inexperienced reader.

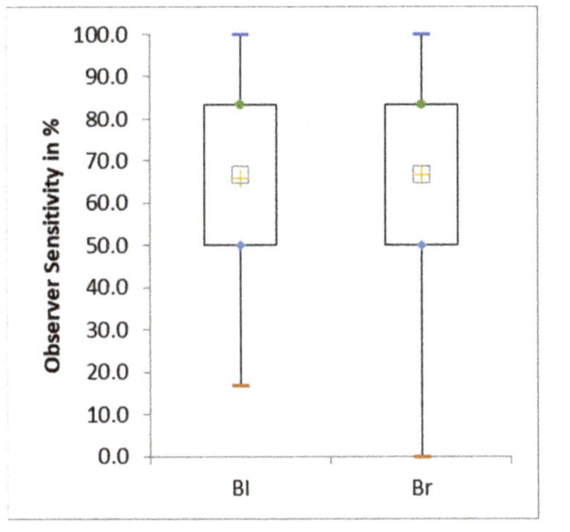

Figure 8. Observer sensitivity vs. kernel, all densities—experienced reader.

There was an improvement in observer sensitivity with increasing nodule density for both kernels and both readers. The highest observer sensitivity was achieved with the sharp lung kernel at −800 HU (58.5% ± 36.5).

3.5. False Positives

The protocol with the most FPs was ADMIRE 3 soft-120 mAs at non-high pitch (4 of 40 FP; 10%). For nodule densities of 100 HU, there were only 6 FPs (15%), for −630 HU, there were 15 FPs (38%), and the most false positives occurred at −800 HU (19 FPs, equal to 48%).

Considering the mAs level, the most FPs occurred for a mAs setting of 120 (15; 38%).

Furthermore, 24 FPs (60%) occurred with the sharp kernel and 16 (40%) with the soft kernel.

4. Discussion

There have not been many trials using third-generation dual-source CT for lung examinations. A study from 2019 compared a dual-source CT of this generation with a 16-row Light-speed CT in the examination of pediatric oncology patients. They unsurprisingly achieved significantly lower tube voltage and tube current with better image quality with the dual-source scanner. Moreover, a reduction of motion artefacts was apparent [15].

In 2020, another group investigated the sensitivity of multidetector CT for the detection of interstitial lung disease. They measured a high sensitivity of 91.4% [16].

In our study, dual-source CT images reconstructed with ADMIRE 5 were found to give better results than lower ADMIRE settings or FBP.

While ADMIRE 5 sharp-30 mAs at non-high pitch reached an observer sensitivity of 71.4%, a tube current of 60 mAs was needed to reach the same value as ADMIRE 3 sharp at the non-high pitch. ADMIRE 1 sharp required 120 mAs to achieve 71.3%, and FBP sharp only reached 69.4%, even at 120 mAs (Table 2). This shows how iterative reconstruction algorithms can reduce the dose while obtaining high observer sensitivity rates using the manual readout.

Several trials support this evidence. The previously mentioned study from 2019 proved similar to another iterative reconstruction mode, adaptive statistical iterative reconstruction (ASIR-V, GE Healthcare, Chicago, IL, USA), which was also able to improve image quality in ULDCT compared to FBP in a lung phantom [17].

As early as 2011, another study showed a dose reduction of 40% for second-generation dual-source CT while applying iterative reconstruction in images space (IRIS, Siemens Healthineers, Erlangen, Germany) to result in the same noise as FBP at a 100% dose in a phantom [18].

While the efficacy of iterative reconstruction is not unheard of, it is of interest for the clinical routine to know what level of reconstruction might achieve the best results. In 2015, a study compared FBP to ADMIRE 3 and 5 while reducing the radiation dose in a similar setting using a chest phantom and artificial lung nodules on a third-generation dual-source CT. They noted significantly lower image noise and fewer false positives of nodules using ADMIRE 5 [19].

In line with that, our findings showed ADMIRE 5 as the best choice for all reconstruction modes for the detection of lung nodules.

Furthermore, ultra-high pitch and non-high pitch were compared. For the inexperienced reader, there was no significant difference ($p = 0.06$). The experienced reader showed a significantly higher observer sensitivity for non-high pitch protocols in the U-Test for the densities of 100 HU ($p < 0.01$) and overall ($p < 0.01$). This result indicates that radiological practices that only have standard equipment can also offer screening with their equipment as many scanners only provide pitch levels up to 1.5. The experience of the reader seems to be of greater importance here than the advantage of the pitch.

From our point of view, there is no apparent explanation for these results. However, a research group evaluated the influence of high pitch on the accuracy of volumetry for solid pulmonary nodules using a second-generation CT in a similar setup in 2015 [20]. They also concluded that the accuracy was comparable to those of conventional pitch but also found high-pitch protocols to be less accurate for the volumetry of smaller nodules. This might be dependent on the use of an artificial phantom thorax with no breathing. In reality, ultra-high pitch is primarily a method to reduce radiation dose and motion artefacts. Especially in screening, the patient should be exposed to as little harmful radiation as possible so that the benefit outweighs the risk of the examination. Therefore, ultra-high pitch, which is expected to be of great benefit under real conditions, might not provide an advantage here. In 2018, another group of researchers demonstrated on a second-generation dual-source CT that ultra-high pitch improves image quality and lowers radiation exposure at the same time when used for coronary angiography [21]. With this, image quality was mainly determined by motion artefacts.

Keeping this in mind, the setting should still be useful for lung screening in a clinical routine, when the patient is breathing, and radiation dose needs to be considered.

Results show that a higher tube current equals a higher observer sensitivity. In 2010, a study showed a diagnostically relevant decrease in quality below 60 mAs in a dose simulation study of lung parenchyma at a high-resolution kernel [22]. This lines up with the findings of our study. Especially with low-density nodules, choosing the right tube current can improve the reader's detection rate. With high-density nodules, only the inexperienced reader had a significant benefit.

The number of false-positive nodules was at its maximum across all densities at the mAs setting of 120 (15; 38%), while 90 mAs showed the lowest number (4; 10%) and 60 mAs showed 6 FPs (15%). This could be an indication that 90 mAs would be the best compromise between observer sensitivity and false positives. The cause of this difference should be evaluated in more detail in further studies.

Another outstanding result is that the kernels do not make that much of a difference. The soft kernel is normally applied to soft tissue and is often excluded from studies on the lung. Although the best observer sensitivity for both readers combined was achieved with a sharp kernel at 100 HU (76.6% $\pm$ 36.6), there has been a tendency toward better detectability with the soft kernel across all nodule densities combined, as well as for the subsolid nodules ($p = 0.60$).

Additionally, the sharp kernel showed more false positives across all densities than the soft kernel (24; 60% vs. 16; 40%). In total, it not only shows more false positives but also insignificantly worse results than the soft kernel. Considering this, the habit of including only the sharp kernel in studies and its status as the standard lung kernel might need to be reconsidered, especially given the number of FPs occurring.

With decreasing nodule density, the number of FPs increased. As mentioned before, at 100 HU, only 15% of all FPs occurred, while at −630 HU, it was already 38%. Almost half of the false positives occurred at −800 HU (48%). This illustrates the often-mentioned issue of overdiagnosis. The NELSON trial showed an overdiagnosis rate of 19.7% in real patients within ten years and 8.9% after eleven years, claiming that a longer screening interval can reduce false positives [10]. This was not possible to include in our phantom study. However, it can be seen clearly that the overdiagnosis rate is much lower for solid nodules and increases with diminishing density.

Limitations

The use of a phantom causes some limitations. First, it does not breathe. Second, it shows factory-caused irregularities, which might feign nodules. Furthermore, the nodules inevitably adhere to the structures of the phantom and cannot be located freely in the parenchyma. This makes measurement more difficult for the readers and might have increased the number of false positives.

Another limiting factor was the use of spherical nodules only. There is no information on whether these correlations also apply to spiculated or lobulated nodules.

There was no semi-automated software involved in the assessment. A complementary phantom study focusing on the possible advantages of using computer-aided detection (CAD) can be found by the same institution with an identical chest phantom setting and similar scan parameters [23].

Furthermore, it cannot be ruled out that the inexperienced reader benefits even from a small training effect. To minimize this influence, a two-week briefing and practice time was allowed to become accustomed to the data analysis course.

Thirteen different setups were scanned for each protocol. Both assessors were completely blinded to the number and location of nodules. Moreover, they had no information if the enumeration of setups was changed in between. This ought to avoid a recognition effect. The tube voltage and current levels, kernels, and reconstruction settings were mixed so that they would not become familiar with one setting.

As can be seen, a high standard deviation was determined. To increase the sharpness of the results, the number of assessors and readings should be expanded further in future studies.

5. Conclusions

Overall, this chest phantom study showed no difference in nodule detection when using ultra-high pitch protocols compared to a regular pitch mode. The same applies to the sharp kernel and soft kernel. We found the highest nodule detectability with an increasing tube current and at high iterative reconstruction levels. A protocol containing third-generation iterative recon algorithms at high strength and 90 mAs appears to provide the best balance between observer sensitivity and false positives in the detection of solid and subsolid nodules from 3–12 mm.

Author Contributions: Conceptualization, S.J.; methodology, S.J.; software, N.L. and H.S.; validation, H.S.; formal analysis, N.L.; investigation, N.L. and H.S.; resources, S.O.S.; data curation, N.L.; writing—original draft preparation, N.L.; writing—review and editing, S.J., H.K. and H.S.; visualization, N.L. and H.S.; supervision, S.O.S.; project administration, S.J. and H.S. All authors have read and agreed to the published version of the manuscript.

Funding: This research received no external funding.

Institutional Review Board Statement: Not applicable.

Informed Consent Statement: Not applicable.

Data Availability Statement: The data presented in this study are available upon request from the corresponding author.

Acknowledgments: The authors would like to thank Rozemarijn Vliegenthart from the "Centre for Medical Imaging", University of Groningen, for the support and advice to carry out this study and the expert of the Department of Medical Statistics, Biomathematics and Information Processing of the University Medical Centre Mannheim for her statistical advice and the performance of statistical tests. Furthermore, the authors would like to thank Sina Grund for adapting the image data.

Conflicts of Interest: The authors declare no conflict of interest.

References

1. Cancer Research U.K. Lung Cancer Statistics. Available online: https://www.cancerresearchuk.org/health-professional/cancer-statistics/statistics-by-cancer-type/lung-cancer#heading-Four (accessed on 15 August 2021).
2. European Union. ECIS—European Cancer Information System—Incidence and Mortality Europe. Available online: https://ecis.jrc.ec.europa.eu/explorer.php?$0-0$1-AEE$4-1,2$3-All$6-0,85$5-2008,2008$7-7,8$CEstByCancer$X0_8-3$CEstRelativeCanc$X1_8-3$X1_9-AE27$CEstBySexByCancer$X2_8-3$X2_-1-1 (accessed on 16 August 2021).
3. US Preventive Services Task Force. Final Recommendation Statement. Lung Cancer: Screening. Available online: http://www.uspreventiveservicestaskforce.org/Page/Document/UpdateSummaryFinal/lung-cancer-screening (accessed on 24 August 2021).
4. Guerrera, F.; Errico, L.; Evangelista, A.; Filosso, P.L.; Ruffini, E.; Lisi, E.; Bora, G.; Asteggiano, E.; Olivetti, S.; Lausi, P.; et al. Exploring Stage I non-small-cell lung cancer: Development of a prognostic model predicting 5-year survival after surgical resection. *Eur. J. Cardio-Thorac. Surg. Off. J. Eur. Assoc. Cardio-Thorac. Surg.* **2015**, *47*, 1037–1043. [CrossRef] [PubMed]
5. Cancer Research U.K. Lung Cancer Diagnosis Statistics. Available online: https://www.cancerresearchuk.org/health-professional/lung-cancer-diagnosis-statistics#ref-5 (accessed on 22 April 2021).
6. Finigan, J.H.; Kern, J.A. Lung cancer screening: Past, present and future. *Clin. Chest Med.* **2013**, *34*, 365–371. [CrossRef] [PubMed]
7. Oken, M.M.; Hocking, W.G.; Kvale, P.A.; Andriole, G.L.; Buys, S.S.; Church, T.R.; Crawford, E.D.; Fouad, M.N.; Isaacs, C.; Reding, D.J.; et al. Screening by chest radiograph and lung cancer mortality: The Prostate, Lung, Colorectal, and Ovarian (PLCO) randomized trial. *JAMA* **2011**, *306*, 1865–1873. [CrossRef] [PubMed]
8. Aberle, D.R.; Adams, A.M.; Berg, C.D.; Black, W.C.; Clapp, J.D.; Fagerstrom, R.M.; Gareen, I.F.; Gatsonis, C.; Marcus, P.M.; Sicks, J.D. Reduced lung-cancer mortality with low-dose computed tomographic screening. *N. Engl. J. Med.* **2011**, *365*, 395–409. [CrossRef] [PubMed]
9. Kauczor, H.U.; Bonomo, L.; Gaga, M.; Nackaerts, K.; Peled, N.; Prokop, M.; Remy-Jardin, M.; von Stackelberg, O.; Sculier, J.P. ESR/ERS white paper on lung cancer screening. *Eur. Respir. J.* **2015**, *46*, 28–39. [CrossRef] [PubMed]
10. De Koning, H.J.; van der Aalst, C.M.; de Jong, P.A.; Scholten, E.T.; Nackaerts, K.; Heuvelmans, M.A.; Lammers, J.J.; Weenink, C.; Yousaf-Khan, U.; Horeweg, N.; et al. Reduced Lung-Cancer Mortality with Volume CT Screening in a Randomized Trial. *N. Engl. J. Med.* **2020**, *382*, 503–513. [CrossRef] [PubMed]
11. Sudarshan, M.; Murthy, S. Computed tomography screening for lung cancer. *Gen. Thorac. Cardiovasc. Surg.* **2020**, *68*, 660–664. [CrossRef] [PubMed]
12. Kalinke, L.; Thakrar, R.; Janes, S.M. The promises and challenges of early non-small cell lung cancer detection: Patient perceptions, low-dose CT screening, bronchoscopy and biomarkers. *Mol. Oncol.* **2020**, *15*, 2544–2564. [CrossRef]
13. Prosch, H.; Ebner, L. Lung cancer screening. *Der Radiol.* **2019**, *59*, 19–22. [CrossRef] [PubMed]
14. Tugwell-Allsup, J.; Owen, B.W.; England, A. Low-dose chest CT and the impact on nodule visibility. *Radiography* **2021**, *27*, 24–30. [CrossRef] [PubMed]
15. Agostini, A.; Mari, A.; Lanza, C.; Schicchi, N.; Borgheresi, A.; Maggi, S.; Giovagnoni, A. Trends in radiation dose and image quality for pediatric patients with a multidetector CT and a third-generation dual-source dual-energy CT. *La Radiol. Med.* **2019**, *124*, 745–752. [CrossRef]
16. Ahmed, S.; Khanduri, S.; Husain, M.; Khan, A.U.; Singh, A.; Rajurkar, M.; Abbas, S.Z.; Khan, N. Diagnostic Accuracy of Multidetector CT in Detection of Early Interstitial Lung Disease with Its Role in Characterization. *Cureus* **2020**, *12*, e8253. [CrossRef] [PubMed]
17. Afadzi, M.; Lysvik, E.K.; Andersen, H.K.; Martinsen, A.C.T. Ultra-low dose chest computed tomography: Effect of iterative reconstruction levels on image quality. *Eur. J. Radiol.* **2019**, *114*, 62–68. [CrossRef] [PubMed]
18. Hu, X.H.; Ding, X.F.; Wu, R.Z.; Zhang, M.M. Radiation dose of non-enhanced chest CT can be reduced 40% by using iterative reconstruction in image space. *Clin. Radiol.* **2011**, *66*, 1023–1029. [CrossRef]
19. Martini, K.; Higashigaito, K.; Barth, B.K.; Baumueller, S.; Alkadhi, H.; Frauenfelder, T. Ultralow-dose CT with tin filtration for detection of solid and sub solid pulmonary nodules: A phantom study. *Br. J. Radiol.* **2015**, *88*, 20150389. [CrossRef]
20. Hwang, S.H.; Oh, Y.W.; Ham, S.Y.; Kang, E.Y.; Lee, K.Y. Effect of the high-pitch mode in dual-source computed tomography on the accuracy of three-dimensional volumetry of solid pulmonary nodules: A phantom study. *Korean J. Radiol. Off. J. Korean Radiol. Soc.* **2015**, *16*, 641–647. [CrossRef] [PubMed]

21. Smettei, O.A.; Sayed, S.; Al Habib, M.A.; Alharbi, F.; Abazid, R.M. Ultra-fast, low dose high-pitch (FLASH) versus prospectively-gated coronary computed tomography angiography: Comparison of image quality and patient radiation exposure. *J. Saudi Heart Assoc.* **2018**, *30*, 165–171. [CrossRef] [PubMed]
22. Ley-Zaporozhan, J.; Ley, S.; Krummenauer, F.; Ohno, Y.; Hatabu, H.; Kauczor, H.U. Low dose multi-detector CT of the chest (iLEAD Study): Visual ranking of different simulated mAs levels. *Eur. J. Radiol.* **2010**, *73*, 428–433. [CrossRef]
23. Janssen, S.; Overhoff, D.; Froelich, M.F.; Schoenberg, S.O.; Rathmann, N. Detectability of Lung Nodules in Ultra-low Dose CT. *Anticancer Res.* **2021**, *41*, 5053–5058. [CrossRef] [PubMed]

Article

Flaminio Rota, 16th Century Anatomist at the University of Bologna: A Biography on the Walls

Emanuele Armocida [1], Francesco Fornai [2,3] and Gianfranco Natale [2,4,*]

[1] Department of Medicine and Surgery, University of Parma, 43126 Parma, Italy
[2] Department of Translational Research and New Technologies in Medicine and Surgery, University of Pisa, 56126 Pisa, Italy
[3] I.R.C.C.S. Neuromed, 86007 Pozzilli, Italy
[4] Museum of Human Anatomy "Filippo Civinini", University of Pisa, 56126 Pisa, Italy
*   Correspondence: gianfranco.natale@unipi.it

Abstract: Flaminio Rota was a 16th century anatomist and medical figure at the University of Bologna. He was highly praised, despite his poor scientific production. As a matter of fact, Rota competed with scientific activities in different anatomical arguments, but he did not publish any important research. Nevertheless, we know the principal results of his scientific activity because indirect information can be found in other publications, where some of his studies were emphasized by his contemporary colleagues. Henning Witte even mentioned Rota as a very famous Italian medical figure, together with Galilei and Santorio. On the other hand, Rota was a highly esteemed teacher. The best evidence of his recognition is well-documented in the Palace of Archiginnasio in Bologna, where Rota's teaching activity was praised with six memorial epigraphs. In the south-eastern outskirts of Bologna, there is an 18th century villa, including a more ancient annex, that belonged to Rota. At this location, the upper parts of the walls and the ceiling are decorated with a pictorial cycle illustrating medical scenes. In this paper, we theorize regarding his scientific thinking by analyzing the pictorial cycle he commissioned.

Keywords: Flaminio Rota; University of Bologna; anatomy; university teaching; university research

Citation: Armocida, E.; Fornai, F.; Natale, G. Flaminio Rota, 16th Century Anatomist at the University of Bologna: A Biography on the Walls. *Anatomia* **2023**, 2, 1–14. https://doi.org/10.3390/anatomia2010001

Academic Editors: Francesco Cappello and Rafael Coveñas Rodríguez

Received: 21 November 2022
Revised: 7 December 2022
Accepted: 29 December 2022
Published: 6 January 2023

1. Introduction

Flaminio Rota (1555–1611) was a 16th century anatomist and medical figure at the University of Bologna. Rota's career was influenced by the fame of his father, Giovanni Francesco Rota (1520–1558). From 1546–1547, Giovanni Francesco graduated in Philosophy and Medicine at the University of Bologna, becoming a professor of anatomy and surgery. His works were particularly appreciated, including his most important book, *De bellicorum tormentorum vulnerum natura et curatione liber* (*A Book on the Nature and Treatment of Artillery Wounds in War*), written on the basis of his experience as a military surgeon, describing the effects of contusions and shell explosions. Gunshot wounds were a very new field of research at that time. A bust was placed inside the famous anatomical theater of the Archiginnasio of Bologna in his memory [1].

Flaminio followed the career of his father. He graduated in Philosophy and Medicine at the University of Bologna on 8 March 1577, and he enrolled in all of the most important medical institutions in the city. Since 1579, he was a lecturer (Ad lecturam chirurgiae), and from 1589–1590, he also taught anatomy (Ad anathomiam), until his death (16 January 1611). Although the chairs of anatomy and surgery were separated at the behest of Aranzio in 1570, there was certainly still a strong affinity between the two disciplines [2–5]. Lectures in surgery were based on Galen's tripartite work: (1) *De tumoribus praeter naturam* (*Unnatural Tumors*); (2) *De ulceribus* (*Ulcers*); and (3) *De vulneribus* (*Wounds*) [6]. Rota's famous colleagues were Giulio Cesare Aranzio (1530–1589), pupil of the famous Flemish anatomist Andreas Vesalius (Andreas van Wesel; 1514–1564), Gaspare Tagliacozzi (1545–1599), a pioneer in plastic surgery, Girolamo Mercuriale (1530–1606), who renewed the therapy of syphilis, and Giovanni Battista Cortesi (c. 1553–1634), known for his pharmacopeia and anatomical studies on the central nervous system [1]. In 1595, Rota enrolled in the prestigious *Collegio di Filosofia e Medicina*, an important status symbol in Bologna, which allowed special contact between political and economic circles [7].

Rota was also an appreciated surgeon and physician. From 1576 to 1580, he was a student assistant (a so-called astante) at the Hospital of *Santa Maria della Morte* in Bologna. The student assistant fed the sick and poor according to the dietetic prescriptions indicated by the physicians. Rota also needed to visit the sick several times a day, checking medications and reporting on patient pain levels. From 1578 until his death, Rota was active as a surgeon at the Hospital of Saint Job of the Incurable, one of the first hospitals specialized in venereal diseases. In 1585, he was named "supernumerary surgeon" at the Hospital of *Santa Maria della Vita* [6,8].

Medici [3] evidenced that some historians neglected Rota's memory, very likely because he never published his scientific research. In the past centuries, the modern aphorism, *"publish or perish,"* was not so compulsory and determinant. In the time of Rota, only the lower-status learned surgeons were really moved to publish their anatomical and surgical practical activities to gain advancement and to establish their names. An example of such a situation can be found at the University of Bologna, where five surgeons had different backgrounds. Giulio Cesare Aranzio and Gaspare Tagliacozzi came from the artisanal class, whereas Giovanni Battista Cortesi was born into poor conditions. On the contrary, Rota and Angelo Michele Sacchi (1538–1611) were part of medical families of collegiate physicians. Therefore, not surprisingly, both Rota and Sacchi were the only ones to not publish. As a matter of fact, Rota competed with scientific activities in different anatomical arguments, but he did not publish any important research. Nevertheless, we know the principal results of his scientific activity because indirect information can be found in other publications. Some of his studies were admirably mentioned by his contemporary colleagues, including Ulisse Aldrovandi (1522–1605), Vincenzo Alsario dalla Croce (1576 ca–1632), Giovanni Battista Codronchi (1547–1628), Daniel Sennert (1572–1637), and Giovanni Zecchi (1533–1601) [1]. In *Memoriae medicorum nostri seculi clarissimorum renovatae decas prima* (*Memoirs of the Most Famous Physicians of our Century, Renewed in the First Decade*), Henning Witte (1634–1696) even mentioned Rota as a very famous Italian medical figure, together with Galileo Galilei (1564–1642) and Santorio Santorio (1561–1636) [9].

The memory of Flaminio Rota was guaranteed by his teaching ability. Thus, "Rota built his career upon his ability as a teacher, both in private and in public, and by exploiting the social prestige he inherited from his family tradition" [8]. Indeed, Rota was undoubtedly a praised teacher. The Palace of Archiginnasio in Bologna preserves the best evidence of the recognition of his teachings [3,10,11]. This palace was the seat of Bologna University, where today, visitors can still appreciate a remarkable collection of coats of arms referring to the different colleges, as well as marble epigraphs, in memory of professors particularly praised by students. In this context, when alive, Rota was celebrated with six memorial epigraphs. To have an idea of the value of this recognition, the previously mentioned anatomist, Aranzio, deserved eight epigraphs. Rota's success was due to the radical innovation he provided in anatomical teaching, with interactive lectures and open discussions, and, in the course of public anatomy lectures, he was quick to provide solutions to "the very difficult objections raised extemporaneously by distinguished scholars" [11]. It is solely for this reason that he is regarded as a celebrated anatomist [1].

Other than his teaching ability, what do we know about Rota? In this paper, we try to theorize regarding his scientific thinking by analyzing the pictorial cycle he commissioned in his Bolognese country house.

2. Rota's Life on the Walls

Although he never published his scientific works, Flaminio Rota's fame has become known to us thanks to the epigraphs that his students left on the walls of the Archiginnasio, which are still visible to everyone, to celebrate his teaching skills. Less visible is the direct testimony from Rota that would likely represent his scientific thoughts.

In the south-eastern outskirts of Bologna, there is an 18th century villa realized by the architect Angelo Venturoli (1749–1821). It includes a more ancient annex that is likely to be the country estate of Flaminio Rota—the presence of his heraldic crest makes this hypothesis very plausible—whose intended use has not yet been recognized with certainty. After many years of neglect, the villa, known as Villa Rivalta or Villa Negri, was restored in the 2000s and is currently a privately inhabited property (Figure 1).

Figure 1. Villa Rivalta today. On the left, covered by trees, is Rota's ancient annex building that is decorated with a pictorial cycle illustrating medical scenes.

Inside the annex, the upper parts of the walls and the ceiling are decorated with a pictorial cycle illustrating medical scenes. Here, archives witness that Rota hosted students, as it was the custom at that time. For example, the official biographies of Cesare Magati claim this [3,12]. Thus, it is likely that the building was intended both for professional and educational activity. The frescoes are attributable to the school of Cesare Baglione (1550–1615), who was very active in Bologna at the end of the 16th century. The pictorial cycle may have been made between 1580 and 1610. A stone architrave placed under the frescoes, engraved with MDCIX (1609), suggests that this year corresponds with the pictorial realization. The work was likely commissioned before Rota's second marriage in 1610 [13].

The decoration of the ceiling (Figure 2), implanted on very curled cartouches and grotesques, with figures of birds, presents a human figure at the center of each background, delimited by the joists. Among these, pagan divinities (Mars, Venus, Diana, Minerva, Mercury, and another disappeared due to the collapse of a portion of the canopy that supports the painted surface) (Figure 3) alternate with putti. The heraldic symbol of the wheel (in old-fashion Italian, Rota means wheel) appears in badges that accompany the mythological figures, and it is also found in the hand of one of the putti, who is playing with it.

Figure 2. General view of the decoration of the ceiling of Rota's annex.

Figure 3. Ceiling details of Rota's annex. From **left** to **right**: Mars, Mercury, Venus, and Diana.

For the purposes of this study, the most interesting images are undoubtedly those painted on the upper parts of the walls of the room. In fact, the room is frescoed with a series of oval paintings, two on each wall, which depict scenes that recall the fundamental disciplines of medical practice and well-known characters in the history of medicine. This was likely to be considered the most relevant by the client. In each oval painting, explicit writings identify the discipline and the characters depicted, leaving no interpretative doubts.

It is very likely that the pictorial cycle was inspired by the title page of the *Hippocratis Coi* opera, that is, the publication of the writings of Hippocrates in Venice by Giunti printers in 1588 and supervised by Girolamo Mercuriale [14]. It was engraved with a burin by the

Italian engraver and publisher by Giacomo Franco (1550–1620), as attested by the signature "Iacobus Francus f." affixed to the lower right corner. It was later used until 1609 by the same publishers for the editions of scientific works in folio format (for example, Avicenna's *Canon*) [15] (Figure 4).

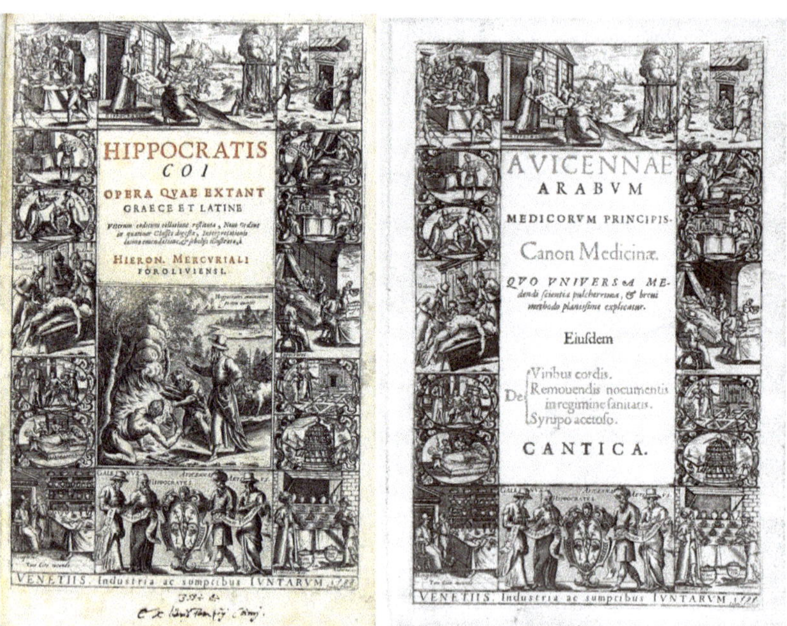

Figure 4. Frontispieces of Hippocrates' [14] and Avicenna's [15] works, published by Giunti. Permission was granted by the Ministero dei Beni Culturali (commons license CC BY-NC-SA 3.0 IT).

Mercuriale is remembered as one of the fathers of dermatology. His *De morbis cutaneis et omnibus corporis humani excrementis* (*Diseases of the Skin, and All the Wastes of the Human Body*) in five volumes, collected around 1572, is considered the first treatise on dermatology. In 1587, Mercuriale left the chair of Padua to move to Bologna. At that time, Rota was already a lecturer in anatomy and surgery and a surgeon at the Hospital of Saint Job, set up for the treatment of syphilitics. It is conceivable to think that the two physicians met and confronted each other on scientific issues.

3. About the Frescoes

On the eastern wall, the first oval painting is dedicated to CHIRVRGIA (surgery), where a scene of the trephination of the skull is depicted. The operation is being performed by Podalirius and Machaon (the names of the two surgeons appear near their feet as "PODALIRIVS" and "MACHAON"), who are supported by two assistants. Podalirius and Machaon were doctors from Greek mythology. Tradition has it that they were sons of Asclepius (god of medicine) and Epione (princess of Kos). They learned their healing arts from their father and the centaur Chiron [16,17]. This image is also found in the title page of Giunti's edition, supervised by Mercuriale (Figure 5).

Figure 5. Left: Oval painting dedicated to surgery, where Podalirius and Machaon are depicted. **Right**: A similar detail taken from the frontispiece of Giunti's edition.

The next oval painting represents the "distillery" (at the bottom, there is the inscription DESTILATORIA), where a figure with the explanatory inscription MESVES approaches a colossal support full of stills. The distillery also appears in the frontispiece but with no characters nearby (Figure 6).

Figure 6. Left: The oval painting represents the "distillery," where Mesue is the protagonist of the scene. **Right**: The distillery in the frontispiece of Giunti's edition.

Distillery, and in particular, the use of alcohol, is attributed to the Arab tradition. However, it is curious to note that to represent this practice, Rota did not use the much better-known Avicenna (represented in another oval painting) or Rhazes, remembered in the history of medicine for his pioneering alchemical studies based on distillation. Mesue is known as Mesue senior, that is, Yuhanna Ibn Masawaih (777–857). He became a knowledgeable physician and anatomist, and he was appointed as director of a hospital in Baghdad. He was the personal physician to four caliphs. He composed medical treatises on a number of topics, including ophthalmology, fevers, leprosy, headache, melancholia, dietetics, and medical aphorisms. One of Mesue's treatises concerns aromatics, entitled *On Simple Aromatic Substances*. It was reported that Mesue regularly held an assembly of some sort, where he consulted with patients and discussed subjects with pupils. He apparently attracted considerable audiences, having acquired a reputation for repartee. He translated various Greek medical works into Syriac, but wrote his own works in Arabic.

Apes were supplied to him by the caliph al-Mu'tasim for dissection. Many anatomical and medical writings are credited to him, notably the *Disorder of the Eye* (Daghal al-'ain), which is the earliest systematic treatise on ophthalmology extant in Arabic, and *The Aphorisms*, the Latin translation of which became very popular in the Middle Ages [18]. Mesue was representative of a transition period in Arabic medicine, when physicians no longer limited themselves to translations but also began to develop original medical studies and practices [19]. In his works, there were many innovations that would provide the basis for the theory and practice of pharmacy for centuries and arguably formed part of the artisanal epistemological influence on the scientific revolution [20].

In the first oval painting of the southern wall, there is a scene where a sick person, naked on a bed, is visited by Aulus Cornelius Celsus and Nicola Fiorentino (as reported in the captions CORNELIVS CELSVS and NICOLAVS FLORENTINVS). At the bottom, there is the inscription TUTO CITO IVCVNDE (Figure 7). Cito tuto iucunde (meaning to treat patients "swiftly, safely, and sweetly") was a motto of Asclepiades of Bithynia (c. 130–40 BC). He opposed the humoral conception of Hippocrates and affirmed the atomistic doctrine, believing that all the phenomena of life must be attributed to mechanical laws, considering that diseases reside in the atoms, which live in continuous movement among them and are joined by channels, the so-called pores. As a skilled surgeon, he was certainly the first doctor who came to Rome to practice medicine based on scientific concepts [17,21].

Figure 7. Left: Oval painting on the southern wall depicting Aulus Cornelius Celsus and Nicolò Falcucci at work. **Middle**: Herophilos seated in front of a table on which the dissection instruments are exhibited; on the floor of the room where he sits is the inscription Tuto Cito Iucunde (frontispiece of Giunti's edition). **Right**: Galen intent on dissecting a corpse under the gaze of Eudemus and Alessandro Damasceno (frontispiece of Giunti's edition).

However, this fresco can raise doubts concerning its interpretation [13]. Indeed, in Giunti's edition, the motto is associated with another scene, which depicts the anatomist, Herophilos, seated in front of a table with dissection instruments. It is evident that the artist's intentions were to depict an anatomical dissection exercise. In fact, on the frontispiece, this vignette is in the right column, where surgery scenes are depicted. Furthermore, Giunti's frontispiece features Galen (ca. 129–ca. 200) intent on dissecting a corpse under the gazes of the philosopher Eudemus, who benefited from his care, and Alessandro Damasceno, who, according to the tradition, was skeptical of the experiments carried out by Galen [22,23]. On the other hand, in the oval painting, the three figures were reinterpreted to feature Aulus Cornelius Celsus (c. 25 BC–c. 50 AD) and Nicolò Falcucci (also known as Nicolò Nicolai), who was a Florentine doctor who died in 1412, while the figure in the middle is just an unnamed bystander.

The Roman Celsus, a profound connoisseur of Hippocrates, certainly had contacts with Alexandrian medicine and with some Greek doctors transferred to Rome. He wrote a general encyclopedia (*De Artibus*), including some medical arguments (*De Medicina*), and two books of surgery. Following Hippocrates' tradition, it can be considered the most significant medical work, and it had a great diffusion in Roman medical teachings until the arrival of Galen. Forgotten for several centuries, it was rediscovered in 1443 by Pope Nicolas V, and it was the first medical and surgical book to be printed. It is a matter of debate as to whether Celsus actually practiced as a surgeon or simply collected the medical knowledge available at that time [24].

Even the fame of Nicolò Falcucci, more than to the practice of the medical profession, is linked to his remarkable work (*Sermones medicinales*), which, according to many historians, was for several decades an obligatory point of reference for students in medicine. This was a handbook and a summary of the medical science of the time. He provided detailed indications on how medicine could be taught to medical students, and the book was to be an educational tool to improve the skills and knowledge of physicians. In this formation, theoretical and doctrinal competence and operative skill must support each other. This peculiar feature of medicine implied that the education of a physician must follow a unified course of study but according to a composite pattern of competence, in which "to know" and "to do" are integrated. In his book, Falcucci also introduced the good student and the good teacher of medicine. He prescribed that the student of medicine must be willing to learn (disciplinabilis), must not show off, and should attend the Studia. Above all, the pupil must love his teacher as a father, praise him, and defend his fame. To the master, on the contrary, he assigns more specific tasks. He affirmed, indeed, that the physician who is doctoratus must transform himself into a teacher and transmit whatever he has learned. We recognize here, translated in terms of educational relations and forms of writing, the two main divisions of medical knowledge: the theoretic part, with the techniques of transmission of commentary, and the practical part, which is better transmitted in the form of compendia and collections. We can also recognize three institutional characteristics imposed on medicine, as on every other discipline, by university structure: 1) The importance of the written text; 2) the role of authoritative texts in defining the scope and the identity of a discipline; and 3) the general need that what has been acquired by experience should be checked by comparison with authoritative texts and should be transformed by means of writing into doctrine that is thus rendered valid and transmissible [25].

The second oval painting of the southern wall depicts a scene that takes place completely outdoors: in a countryside not far from a city, two men are burning infected materials according to the directives of a doctor, whose name can be read below: HIPPOCRATES. In the cartouche below, the inscription PESTIS, FVGA clearly explains the meaning of the representation. It should be remembered that Hippocrates acquired great fame in ancient times by eradicating the great plague of Athens in 429 BC [26]. In Giunti's frontispiece, there is a similar vignette to that of the oval painting just described (Figure 8).

Figure 8. Left: The second oval painting of the southern wall, showing two men who are burning infected materials according to the directives of Hippocrates. **Right**: Vignette of Giunti's edition that can be associated with the oval painting.

On the western wall, the first painting shows the two figures of ARISTOTELES and AVICENAS in the act of scrutinizing the skies, the air, and the countryside. The caption below reports CAVSARVM INVESTIGATIO (Figure 9). In Giunti's frontispiece, there are no vignettes similar to that of the oval painting just described. Rota entrusted the search for the causes of the diseases to two philosophers, Aristotle and Avicenna, respectively, exponents of Greek and Arab culture.

Figure 9. The oval painting depicting Aristotle and Avicenna in the act of scrutinizing the skies.

Avicenna reworked and updated all the Aristotelian sciences (with the addition of mathematics) to conform them to the criteria of demonstration that Aristotle outlined in the *Posterior Analytics*. In the Avicenna's system of science, metaphysics has to provide the foundation of all special sciences [27].

The next depiction of the western wall is incomplete, due to a window being open on the wall after the completion of the fresco. The oval painting shows a man holding a bunch of freshly picked medicinal herbs. At the man's feet, there is the following inscription: DIOSCORIDES. In Giunti's frontispiece, there is a similar scene, where Dioscorides is in the company of Theophrastus in a botanical garden (Figure 10). Pedanius Dioscorides (c. 40–90 AD) was a Greek surgeon and military physician for the emperor Nero; he was also a pharmacologist, botanist, and the author of *De materia medica*, an encyclopedia about herbal medicine and related medicinal substances. More than 600 plants and 1000 drugs were described in this pharmacopeia, which remained the standard medical text until the 17th century. His notes on the plants include their habitats, methods of preparation, and medicinal uses of the drugs they contained [17,28]. With regard to Western materia medica, through the early modern period, Dioscorides' text eclipsed the *Hippocratic corpus*.

Figure 10. Left: The oval painting shows Dioscorides holding a bunch of freshly picked medicinal herbs. **Right**: Vignette of Giunti's edition that can be associated with the oval painting, where Dioscorides is in the company of Theophrastus in a botanical garden.

On the fourth and last wall, the northern one, the first oval painting is dedicated to dietetics: a messenger wrapped in a large cloak receives a book entitled DIETETICA from the hands of GALENVS, while HEROPHILVS and ORIBASIVS are present behind him. It is interesting to note that a very similar image is present in Giunti's frontispiece, but the book is in the hands of Hippocrates (Figure 11).

Figure 11. Left: In the oval painting, a messenger receives the book entitled *Dietetica* from the hands of Galen, while Herophilos and Oribasius are present behind him. **Right**: Vignette of Giunti's edition that can be associated with the oval painting, but the messenger receives the book from the hands of Hippocrates.

Hippocrates (5th century BC), the father of medicine, emphasized the concept that the diet is the best way to treat a disease. In the following centuries, Herophilos (3rd century BC), Galen (2nd century AD), and Oribasius (4th century AD) produced many writings on dietetics [17,29].

Herophilos (335–280 BC) was a Greek physician regarded as one of the earliest anatomists. He was the founder of the great medical school of Alexandria in Egypt and the first scientist to systematically perform scientific dissections of human cadavers. Herophilos believed that exercise and a healthy diet were integral to an individual's quality of life [17,30].

The aforementioned Galen was a Greek physician, surgeon, and philosopher in the Roman Empire. He is considered one of the most accomplished of all medical researchers of antiquity, and his doctrine in medicine and science became law for the following twelve centuries. He was one of the believers in diet, stating that health depended chiefly on the choice of food [17,31].

Oribasius (c. 320–403) was a medical writer and the personal physician of the Roman emperor Julian. He was the most famous doctor of Byzantium. In his works, he discussed hygiene and dietetics. According to Oribasius, the physician must pave the way for new research by drawing lessons from the clinical observations and experimental methods of the great masters of the classical era. This figure also represents an important step on the path of Galenism, since he was the first to attribute a relevant part to the work of Galen, considering it as fundamental for the progress of medicine [17,32].

The second oval painting of the northern wall, the last of the series, depicts the interior of a pharmaceutical workshop with a large number of jars of various shapes and sizes, a mortar, a scale, and other objects; in the foreground, two large vessels bear the name of the medicaments they contain: THERIACA MAGNA and MITRIDATVS. Near these jars, on the sides of a counter cluttered with preparations, are the inventors of the two remedies. In this scene, however, the painter was the victim of a misunderstanding because under the figure of Mithridates, clearly recognizable by the dress and the crown he wears, he wrote ANDROMACVS REX, and under the figure of Andromachus, he wrote MITRIDATVS, thus exchanging one character with another. Below is the title of the representation: PHARMACEVTICA. A very similar scene is depicted in Giunti's frontispiece (Figure 12).

Figure 12. Left: The oval painting depicts Mithridates and Andromachus in a pharmaceutical workshop. **Right**: A very similar scene is present in Giunti's frontispiece.

Mithridates, king of Pontus in the first century BC, was famous for taking poisons in small doses to be immunized against their effects, and the property of a universal antidote against poisoning was attributed to the drug that bears his name [17].

Andromachus lived during the first century and was Nero's physician. He came from the island of Crete, where herbs were gathered by "botanical persons" in the service of the emperor and placed in knitted pots, which were sent not only to Rome but also to other countries. The vast knowledge of botany by Andromachus helped him "provide the requisite medicines for mankind" [33]. Andromachus took a great deal of Mithridates' ingredients and produced a new antidote. The most important alteration of Andromachus to the Mithridatium antidote was the replacement of an African lizard, called skink, with the viper, which created a new antidote: the Theriac. This remedy continued to be used in Western medicine until the 17th century [17,33]. The composition of the Theriac was a controversial topic in 16th century Bologna, and the college of doctors was also consulted a few decades before Rota joined [34].

4. Conclusions

This paper examined the medical pictorial cycle realized in the annex of the villa of Flaminio Rota and revealed a very interesting cultural aspect that added new information to a previously published article dealing with the biographical knowledge and the academic life of this intriguing 16th-century anatomist [1]. In detail, this additional research allowed us to reach some considerations.

The first consideration is of an artistic nature. It is clear that the pictorial cycle attributed to the school of Cesare Baglione painted in the annex that belonged to Rota was inspired by the frontispieces of *Hippocratis Coi* and Avicenna's *Canon* published by Giunti and not vice versa.

By analyzing the differences rather than the similarities, between the pictorial cycle and Giunti's editions, it is possible to outline the intellectual and scientific profile of the Bolognese anatomist. In fact, it is assumed that the differences are the result of the client's preferences.

For example, in the frescoes, there are no scenes dealing with medical treatments (present in Giunti's editions), an activity not practiced by Rota. In the oval painting representing dietetics, Rota preferred Galen surrounded by the greatest exponents of the Alexandrian school (Herophilos and the Byzantine Oribasius), rather than Hippocrates. Galen paved the way for progress in medicine by combining the anatomo–physiological method of the school of Alexandria in Egypt with the Hippocratic tradition. Similarly, in another oval painting, Rota decided to support Aristotle with Avicenna to represent the passing of the baton of medical knowledge directed towards progress [35].

The scene dedicated to anatomy from the analysis is perhaps the most significant, considering Rota's career. At the dissecting table, he decided to put Celsus and Nicolò Falcucci. They are not famous anatomists, such as Herophilos or Galen, but didactic innovators or, better said, clarifiers and interpreters of already existing theories. Furthermore, in the figure of Celsus, there is the intention to honor the memory of his father, Giovanni Francesco Rota. Celsus attended valetudinarians, was a classifier of skin pathologies, and became appreciated as an elegant writer. Giovanni Francesco was a doctor of the papal army, professor of anatomy, a scholar of gunshot burns, and was remembered as an elegant writer. Even the choice of enhancing the figure of Mesue is perhaps attributable to his skills as an anatomist and teacher.

In summary, from the pictorial cycle, it is evident that Flaminio Rota was a doctor open to scientific progress and a supporter of experimental research based on the study of anatomy, which is capable of improving knowledge of Hippocratic origin. From the university point of view, it can be thought that Rota was attentive when teaching, which he carried out in a clear and concise but, at the same time, exhaustive and elegant way. Furthermore, his tendency to combine academic teaching with practical teaching clearly emerged. This tendency can be confirmed by the documentary biographical information known so far.

At the same time, thanks to the biographical information we know about Rota, it is possible to recognize him as the client of the pictorial cycle and to consolidate the hypothesis that the annex belonged to him.

Finally, regardless of the artistic value of the pictorial cycle and the biography of Flaminio Rota, questioning the meaning of the proposed images allows the reader to have an overview of the medical knowledge in the late 16th century.

Author Contributions: Conceptualization, E.A.; writing—original draft preparation, E.A. and G.N.; writing—review and editing, G.N.; supervision, F.F. All authors have read and agreed to the published version of the manuscript.

Funding: This research received no external funding.

Institutional Review Board Statement: Not applicable.

Informed Consent Statement: Not applicable.

Data Availability Statement: Not applicable.

Acknowledgments: The authors wish to thank the owners of Villa Rivalta, Marco Degidi and Francesca Baldi, for their permission to examine and disseminate the pictorial cycle.

Conflicts of Interest: The authors declare no conflict of interest.

References

1. Natale, G.; Soldani, P.; Gesi, M.; Armocida, E. Flaminio Rota: Fame and Glory of a 16th Century Anatomist without Scientific Publications. *Int. J. Environ. Res. Public. Health.* **2021**, *18*, 8772. [CrossRef] [PubMed]
2. Di Galeotti, B. *Trattato de gli Huomini Illustri di Bologna*; Appresso Vittorio Baldini: Ferrara, Italy, 1590; p. 111.
3. Medici, M. *Compendio Storico della Scuola Anatomica di Bologna dal Rinascimento delle Scienze e delle Lettere a Tutto il Secolo XVIII con un Paragone fra la sua Antichità e Quella delle Scuole di Salerno e di Padova*; Tip. Government della Volpe e del Sassi: Bologna, Italy, 1857.
4. Dallari, U. *I Rotuli dei Lettori Legisti e Artisti dello Studio Bolognese dal 1384 al 1799*; Regia Tip. dei Fratelli Merlani: Bologna, Italy, 1889; Volume 2, pp. 203–307.
5. Ascanelli, P. *I Fascicoli Personali dei Lettori Artisti della Assunteria di Studio all'Archivio di Stato di Bologna (Archivio dell'università): Studio Documentario e Biobibliografico*; Tipografia Valbonesi: Forlì, Italy, 1968.
6. Savoia, P. Skills, Knowledge, and Status: The Career of an Early Modern Italian Surgeon. *Bull. Hist. Med.* **2019**, *93*, 27–54. [CrossRef]
7. Duranti, T. Il collegio dei dottori di medicina di Bologna: Università, professioni e ruolo sociale in un organismo oligarchico della fine del medioevo. *Ann. Stor. Univ. Ital.* **2017**, *21*, 151–178.
8. Majocchi, D. I primi vestigi dell'insegnamento della Dermatologia nello Studio di Bologna e la fondazione della Cattedra di specialità Dermosifilopatica nella nostra Università. *Arch. Ital. Dermatol. Sifilogr. Venereol.* **1930**, *5*, 49–105.
9. Witte, H. *Memoriae Medicorum Nostri Seculi Clarissimorum Renovatae Decas Prima*; Apud Martinum Hallervord: Francofurt, Germany, 1676; Volume 1, p. 56.
10. Malagola, C. *Monografie Storiche Sullo Studio Bolognese*; Nicola Zanichelli: Bologna, Italy, 1888; p. 63.
11. Ferrari, G. Public anatomy lessons and the carnival: The anatomy theatre of Bologna. *Past Present.* **1987**, *117*, 50–106. [CrossRef] [PubMed]
12. Putti, V. *Biografie di Chirurghi del XVI e XIX Secolo*; L. Cappelli Editore: Bologna, Italy, 1941; p. 9.
13. Fanti, M. I resti della dimora di campagna di un medico del Cinquecento. *Strenna Storica Bolognese* **1982**, *32*, 203–221.
14. Hippocrates. *Hippocratis Coi Opera quae Extant Graece et Latine Veterum Codicum Collatione Restituta, Nouo Ordine in Quattuor Classes Digesta, Interpretationis Latinae Emendatione, & Scholijs Illustrata, à Hieron. Mercuriali Foroliuiensi*; Industria ac Sumptibus Iuntarum: Venetiis, Italy, 1588.
15. Avicenna. *Avicennae Arabum Medicorum Principis Canon Medicinae*; Industria ac Sumptibus Iuntarum: Venetiis, Italy, 1595.
16. Koutserimpas, C.; Samonis, G. Machaon: The first trauma surgeon in Western history? *J. Wound Care.* **2018**, *27*, 659–661. [CrossRef] [PubMed]
17. Natale, G.; Zampieri, A. Part 1: From the origins to the Middle Ages. E-book. Edited by Franco Mosca. In *Pain and Its Therapy in Western Medicine. From the Origins to the Mid-Nineteenth Century*; Preface by Andrea Bocelli; Pisa University Press s.r.l.: Pisa, Italy, 2017; ISBN 978-88-6741-800-8.
18. Sarton, G. From Homer to Omar Khayyam. In *Introduction to the History of Science*; Carnegie Institution of Washington: Baltimore, MD, USA, 1927; Volume I, p. 574.
19. Shoja, M.M.; Tubbs, R.S. The history of anatomy in Persia. *J. Anat.* **2007**, *210*, 359–378. [CrossRef]
20. De Vos, P. The "Prince of Medicine": Yūhannā ibn Māsawayh and the foundations of the western pharmaceutical tradition. *Isis* **2013**, *104*, 667–712. [CrossRef]
21. Santacroce, L.; Bottalico, L.; Charitos, I.A. Greek Medicine Practice at Ancient Rome: The Physician Molecularist Asclepiades. *Medicines* **2017**, *4*, 92. [CrossRef]
22. Walsh, J. Galen's Discovery and Promulgation of the Function of the Recurrent Laryngeal Nerve. *Ann. Med. Hist.* **1926**, *8*, 176–184. [PubMed]
23. Malloch, A. Galen. *Ann. Med. Hist.* **1926**, *8*, 61–68. [PubMed]
24. Köckerling, F.; Köckerling, D.; Lomas, C. Cornelius Celsus—Ancient encyclopedist, surgeon-scientist, or master of surgery? *Langenbecks Arch. Surg.* **2013**, *398*, 609–616. [CrossRef] [PubMed]
25. Crisciani, C. Teachers and learners in scholastic medicine: Some images and metaphors. *Hist. Univ.* **1999**, *15*, 75–101.
26. Pinault, J.R. How Hippocrates cured the plague. *J. Hist. Med. Allied Sci.* **1986**, *41*, 52–75. [CrossRef]
27. Alpina, T. Exercising impartiality to favor Aristotle: Avicenna and "The Accomplished Anatomists" (Aṣḥāb Al-Tašrīḥ Al-Muḥaṣṣilūna). *Arab. Sci. Philos.* **2022**, *32*, 137–178. [CrossRef]
28. Haas, L.F. Pedanius Dioscorides (born about AD40, died about AD90). *J. Neurol. Neurosurg. Psychiatry* **1996**, *60*, 427. [CrossRef] [PubMed]
29. Hwalla, N.; Koleilat, M. Dietetic practice: The past, present and future. *East Mediterr. Health. J.* **2004**, *10*, 716–730. [CrossRef]
30. Dobson, J.F. Herophilus of Alexandria. *Proc. R. Soc. Med.* **1925**, *18*, 19–32. [CrossRef]

31. Gill, C.; Whitmarsh, T.; Wilkins, J. (Eds.) *Galen and the World of Knowledge*; Cambridge University Press: Cambridge, MA, USA, 2009. [CrossRef]
32. Browning, R.; Nutton, V. Oribasius, Greek medical writer, c. 320–c. 400 CE. In *Oxford Classical Dictionary*; Oxford University Press: Oxford, UK, 2016.
33. Aleem, M.; Khan, M.I.; Danish, M.; Ahmad, A. History and Traditional uses of Tiryaq (Theriac): An important formulation in Unani medicine. *J. Phytopharm.* **2020**, *9*, 429–432. [CrossRef]
34. Cevolani, E.; Buscaroli, G. Dispute sulla teriaca tra gli speziali e Ulisse Androvandi nella Bologna del XVI secolo. *Atti Memorie Rivista Storia Farmacia* **2018**, *1*, 39–50.
35. Gutas, D. *Avicenna and the Aristotelian Tradition. Introduction to Reading Avicenna's Philosophical Works*, 2nd ed.; E.J. Brill: Leiden, The Netherlands, 2014.

Article

Do Anatomical Differences of the Volar Rim of the Distal Radius Affect Implant Design? A Three-Dimensional Analysis of Its Anatomy and Need for Personalized Medicine

Hidemasa Yoneda *, Katsuyuki Iwatsuki, Masaomi Saeki, Michiro Yamamoto and Masahiro Tatebe

Department of Human Enhancement and Hand Surgery, Nagoya University, 65 Tsurumai-cho, Showa-ku, Nagoya 466-8560, Japan
* Correspondence: yoneda@med.nagoya-u.ac.jp; Tel.: +81-52-744-2957

Abstract: The distal radius, one of the frequent sites of upper extremity fractures, includes unique anatomy referred to as the volar rim. Few studies have addressed its interindividual differences. Additionally, implants for osteosynthesis must match the anatomical structures to prevent soft tissue invasion, but no implants have focused on that so far. In this study, three-dimensional surface models were created from CT images of 101 cases. Analysis of the distal radius, including the volar rim anatomy, was performed to design plates to minimize the discrepancy between the bone anatomy and the implant. The results showed that there were considerable interindividual differences in the morphology of the distal radius, particularly in the degree of palmar protrusion of the volar rim. A moderate correlation between the width of the distal radius and the shape of the volar rim was demonstrated. Considering that variations in plate width are available for treatment of normal distal radius fractures and that the shape of the volar rim changes in correlation with the width of the bone, we infer that simply adding volar rim shape information to the current variations should suffice in preventing complications. We conclude that individualized design according to the shape of the volar rim is unnecessary in fracture surgery.

Keywords: interindividual difference; distal radius fracture; tree-dimensional analysis; osteosynthesis with anatomical plate

Citation: Yoneda, H.; Iwatsuki, K.; Saeki, M.; Yamamoto, M.; Tatebe, M. Do Anatomical Differences of the Volar Rim of the Distal Radius Affect Implant Design? A Three-Dimensional Analysis of Its Anatomy and Need for Personalized Medicine. *Anatomia* **2022**, *1*, 177–185. https://doi.org/10.3390/anatomia1020018

Academic Editors: Gianfranco Natale and Francesco Fornai

Received: 19 October 2022
Accepted: 7 November 2022
Published: 10 November 2022

Publisher's Note: MDPI stays neutral with regard to jurisdictional claims in published maps and institutional affiliations.

1. Introduction

The anatomical morphology of bones is determined by various parameters such as height, weight, sex, age, nutritional status, and systemic diseases. The diaphysis is generally similar in shape [1], although there are variations in width and length [2], and the epiphysis and metaphysis near the joints show variations in three-dimensional shapes between individuals.

The volar rim, a characteristic morphology of the distal radius, is a structure with significant individual anatomical variation [2]. The volar rim projects palmarly, shifting the load-bearing axis from the lunate in the sagittal plane volarly, thus, transmitting the load of the hand to the elbow. This occurs largely through the palmar cortex, where the bone strength is high (Figure 1) [3]. The proximal part of the volar rim is covered with the pronator quadratus and the distal part is covered with fibrous tissue, forming the sliding floor of the flexor tendon located just above the rim [4,5].

The metaphyseal and epiphyseal areas of the radius, including this volar rim, are the sites where fractures frequently occur [6]. It is the area where strong forces are exerted during load transmission. Hence, a strong plate is used to maintain the restored position of the fracture. Since most plates used are ≥2 mm thick and are placed between the flexor tendon and bone where there is not much space, the influence on the flexor tendon and other soft tissues on the palmar side is substantial. Plates placed distally across the watershed line, the most prominent part of the volar rim, and plates that protrude more than 2 mm palmarly in the sagittal plane have been reported to be at risk for flexor tendon injury [7,8].

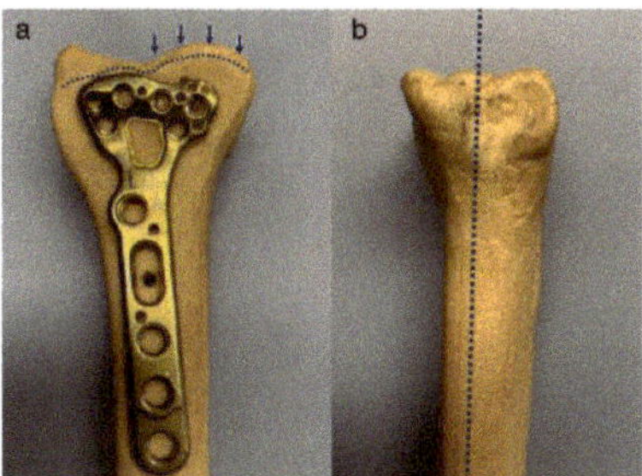

Figure 1. (**a**) Metaphyseal anatomy of the distal radius: The ulnar part of the distal radius, including the lunate fossa, constitutes the volar rim; the most prominent part is referred to as the watershed line (arrow). (**b**) In lateral view, the distal end of the bone, including the volar rim, is shifted palmarly, and the load from the lunate is mainly received by the palmar cortex of the bone (dotted line) rather than the center of the bony axis.

In recent years, plates designed to have a smaller palmar protrusion than the volar rim protrusion and ones with three-dimensional bending to match the shape of the volar rim have been introduced to reduce this effect on the palmar soft tissue (Figure 2) [4]. However, the only option for plates for osteosynthesis is bone width, which does not correspond to the anatomical variation of the volar rim.

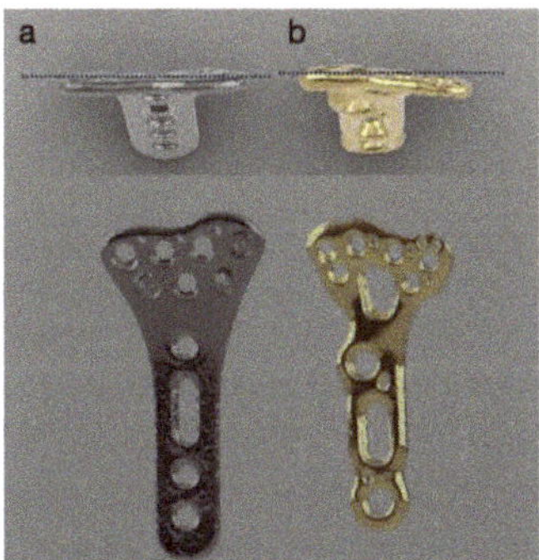

Figure 2. A plate with flat distal portion (**a**) and a plate bent to match the rim shape (**b**).

Few studies have investigated individual variations in the volar rim. Sometimes plates with shapes significantly different from the anatomical shape are placed and can cause the rupture of the flexor tendon [2,9–12]. In this study, we examined three-dimensional variations in the shape of the volar rim. Additionally, we discussed whether a personalized implant shape design corresponding to the individual morphology of the volar rim is necessary for the treatment of fractures.

2. Materials and Methods

This study was performed after the approval of the institutional review board of our hospital. The patient age and gender were extracted from the medical records. We used computed tomography (CT) images of the wrist taken at our hospital over the past 10 years. The CT scanner was an Acquilion (Canon Medical Systems, Tokyo, Japan) and the image data were outputted with a slice thickness of 0.4 mm, segmented using Mimics 21.0 (Materialise, Leuven, Belgium), and then again outputted as a 3D surface model (Standard

Triangulated Language, STL file) of the bone. Patients with inflammatory diseases such as rheumatoid arthritis or osteoarthritis of the hand and patients with a history of fractures were excluded from the analysis. Patients younger than 20 years of age were excluded. The surface models with less than 7 cm were excluded.

Measurements were performed using the 3D analysis software Rhinoceros 6.0 (Robert McNeel & Associates, Seattle, WA, USA). Concentric circles were created at 4 cm and 6 cm from the articular surface and the line connecting the centers of the circles was defined as the Z-axis (Figure 3). If the Z-axis was apparently different from the long axis of the radius, the Z-axis was adjusted to match its long axis. The distal radioulnar joint was identified in a cross-section perpendicular to the Z-axis and included the most proximal part of the radiocarpal joint. In the joint, the X-axis was defined as the line connecting the two points on the palmar and dorsal sides of the articular surface. The Y-axis is the axis perpendicular to it in the cross-section. The planes perpendicular to the X-, Y-, and Z-axes were defined as coronal, sagittal, and transverse planes, respectively.

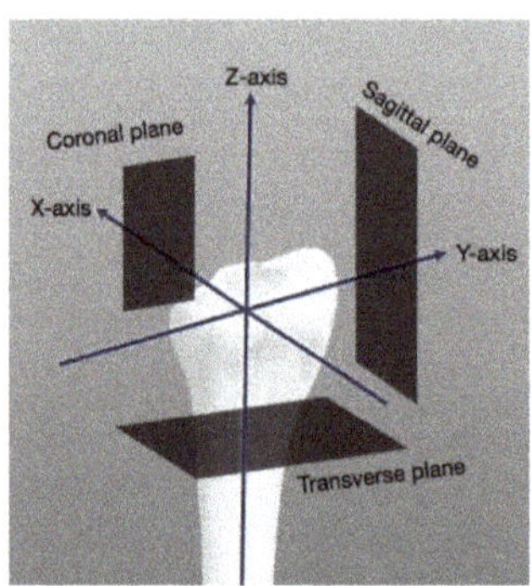

Figure 3. Schemas of measurement axes (X-, Y-, and Z-axes) and coronal, sagittal, and transverse planes.

The tangent line of the bone cortex at the most prominent part of the palmar side in a transverse section, 4 cm from the articular surface, was set as the U-axis (Figure 4). The T-axis was then defined as a straight line passing through the most prominent part of the palmar volar rim and the prominent part of the palmar cortex of the scaphoid fossa in a transverse section passing through the most prominent part of the volar rim (Figure 4). The angle between the T-axis and the U-axis was measured as the angle of rotation of the metaphysis relative to the diaphysis [13].

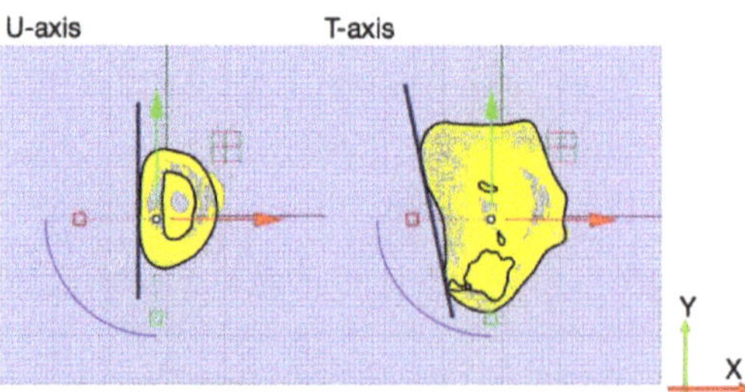

Figure 4. Measuring of the angle of rotation in the metaphysis: The angle of rotation of the metaphysis was measured as the angle between the tangent line (U-axis) of the most prominent part of the palmar cortex in a transverse section, 4 cm from the articular surface, and the line connecting the volar rim at its central level of it and the prominent part of the palmar cortex of the scaphoid fossa (T-axis).

Measurements for the quantitative analyses were performed in the transverse and sagittal planes. In the transverse plane, we measured the angle of metaphysis rotation as described above and the bone width of the radius in the S plane. We also measured the angle between the Y-axis and the tangent line of the most prominent part (axial angle) and the prominent distance from the concave surface of the palmar side to the most prominent point of the volar rim (Figure 5). In the sagittal plane, as per the method of Yoneda et al., the angle of the inflection point on the cortex was taken as the teardrop inclination angle and the distance to the inflection point as the teardrop length (Figure 6) [2]. The teardrop inclination angle cannot be used for designing plates because the position of the inflection point varies in each case (Figure 7). To design the plate, the distance between the watershed line and inflection point was set to the average of the teardrop length and the sagittal angle, which corresponds to the bending angle of the plate in the sagittal plane, was measured. Finally, the transverse section was tilted based on the sagittal angle and the axial angle was measured as the modified axial angle.

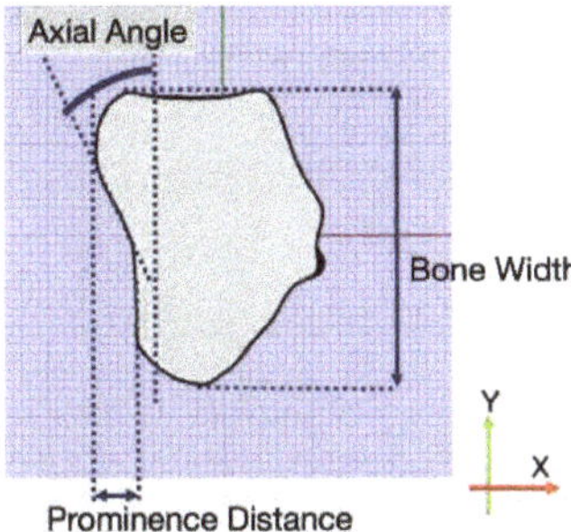

Figure 5. Measurements of bone width, axial angle, and prominence distance: They were measured in transverse section including the base of the subchondral bone forming the radiocarpal joint. The bone width was the longest length parallel to Y-axis and the axial angle was the angle between the Y-axis and the tangent line through the palmar projection of the volar rim. The prominence distance was the protrusion of the lunate fossa relative to the scaphoid fossa, measured on the X-axis.

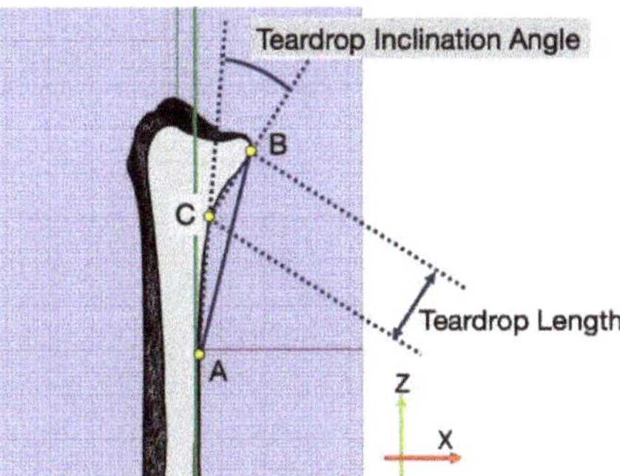

Figure 6. Measurement of teardrop inclination angle and teardrop length: In the sagittal plane passing through the center of the volar rim, draw a shaft line through the palmar aspect of the radius and parallel to the bony axis. Drag a line segment AB connecting point A, where the palmar cortex separates from the shaft line, and the inflection point B of the cortex. Then, determine the point C on the ulnar cortex farthest from AB. The teardrop inclination angle is measured as the external angle created by the two lines, AC and BC, and the length of BC is the teardrop length.

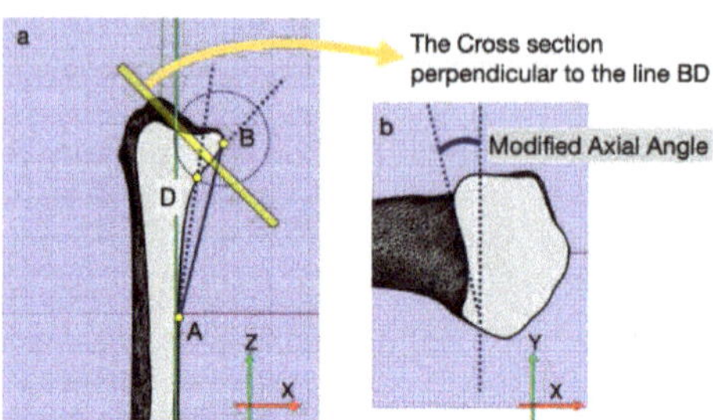

Figure 7. Measurement of sagittal angle and modified axial angle: (**a**) The same method was used to determine points A and B, as in Figure 6. Point D on the palmar cortex is plotted in the sagittal plane so that the length of D is the determined value of the teardrop length. The angle between AD and BD is measured as the sagittal angle. (**b**) The plane perpendicular to BD is the modified transverse plane and the axial angle is measured on that plane.

3. Results

The number of cases for image analysis was 108, of which seven were excluded based on the exclusion criteria. Ultimately, surface models for the analysis were created using CT images of 101 cases (50 males and 51 females). A summary of the quantitative measurements is shown in Table 1. In the transverse section, the bone width averaged 27.1 mm (SD 2.8), the rotation angle averaged 8.8° (SD 6.0), and the axial angle was 16.4 (SD 8.3). The mean teardrop inclination angle was 22.8 (SD 6.2) and the mean teardrop length was 12.1 (SD 2.7). Then, when the teardrop length was set to 12 mm, the average sagittal angle was 26.6° (SD 6.5) and the modified axial angle was 12.0° (SD 2.8). The coefficients of variation were more than 0.2 except for the bone width, suggesting individual differences in the anatomical morphology of the radius.

Table 1. Summary of the quantitative measurements.

	Average ± SD (CV)		
	Male	**Female**	**Total**
Number of patients	50	51	101
Age [y]	58.9 ± 14.0	61.1 ± 12.0	60.0 ± 13.0
Bone width [mm] *	29.1 ± 1.6 (0.05)	25.2 ± 2.3 (0.09)	27.1 ± 2.8 (0.10)
Rotation of metaphysis [°]	8.8 ± 6.1 (0.69)	8.8 ± 5.9 (0.67)	8.8 ± 6.0 (0.68)
Teardrop inclination angle [°]	25.2 ± 4.7 (0.19)	20.4 ± 6.7 (0.33)	22.8 ± 6.2 (0.27)
Teardrop length [mm] *	11.7 ± 2.7 (0.23)	12.5 ± 2.7 (0.22)	12.1 ± 2.7 (0.23)
Sagittal angle [°] *	28.9 ± 4.5 (0.16)	24.4 ± 7.4 (0.30)	26.6 ± 6.5 (0.24)
Prominence distance [mm] *	6.2 ± 1.4 (0.47)	5.3 ± 1.6 (0.30)	5.76 ± 1.6 (0.27)
Axial angle [°]	17.9 ± 8.4 (0.22)	14.9 ± 8.0 (0.10)	16.4 ± 8.3 (0.51)
Modified axial angle [°]	12.0 ± 2.9 (0.24)	11.8 ± 2.7 (0.23)	12.0 ± 2.8 (0.23)

SD, standard deviation; CV, coefficient of variation. * Significant difference between males and females by Welch test ($p < 0.05$).

Among the measured items, the mean values of bone width, teardrop inclination angle, sagittal angle, and prominence distance differed between males and females (Table 1). No correlation with age was observed for any of the items. There was no correlation between bone width and angle of rotation, but moderate correlations were found between the sagittal angle, teardrop inclination angle, and axial angle (Table 2). Moderate correlations were also observed between modified axial angles.

Table 2. Correlations between measured parameters.

Parameters	Coefficient of Correlation	
Age—Bone width	−0.27	
Age—Rotation of metaphysis	−0.18	
Age—Teardrop inclination angle	−0.13	
Age—Teardrop length	0.09	
Age—Sagittal angle	−0.09	
Age—Axial angle	−0.23	
Age—Prominence distance	−0.28	
Age—Modified axial angle	−0.19	
Bone width—Rotation of metaphysis	0.16	
Bone width—Sagittal angle	0.46	*
Sagittal angle—Axial angle	0.53	*
Teardrop inclination angle—Axial angle	0.53	*
Sagittal angle—Modified axial angle	0.46	*
Teardrop inclination angle—Modified axial angle	0.44	*

* $p < 0.05$.

The subgroup analysis was performed by dividing all models into three groups based on the bone width: lower 1/3 (range from 21.3 to 25.6 mm; small), middle 1/3 (range from 25.7 to 28.0 mm; medium), and upper 1/3 (range from 28.1 to 35.3 mm; large) with 33, 35, and 33 cases in the small, medium, and large groups, respectively. In the small and medium groups, a moderate correlation was observed between the sagittal and modified axial angles. However, the correlation between the sagittal and modified axial angles was lower in the large group (Figure 8). These results indicate that in bones with moderate or smaller widths, it is possible to create implants that match the shape of the rim. This can be achieved by preparing several variations correlating the angle of bending in the sagittal plane with the angle in the transverse plane. Conversely, in bones with large widths, the variation of the volar rim in the transverse and sagittal planes is large, causing it to be difficult to design an implant matching the shape of the volar rim even if multiple variations can be prepared. Possibly, the ulnar distal part of the plate should be adjustable to match the anatomy of the individual volar rim.

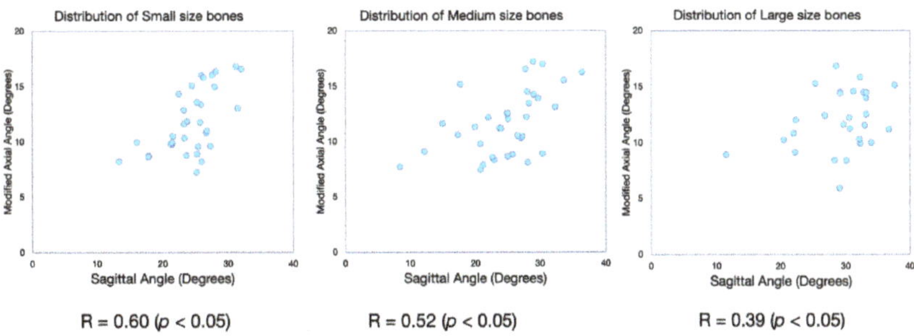

Figure 8. Distribution of sagittal angle and modified axial angle in subgroup analysis.

4. Discussion

It is widely known that there are individual differences in the morphology of the volar rim that cannot be explained by gender, extension, or weight alone. Several quantitative reports have been published previously and implants have been developed based on these findings. Oura et al. found that the amount of palmar protrusion in the transverse section was maximized at 4–6 mm from the articular surface, corresponding to the volar rim area [13]. In an analysis of reconstructed images from CT images, Andarmahr et al. found that the volar rim protrusion averaged 6 mm in the sagittal plane and Yoneda et al. showed that the angle between the protruding portion in the sagittal plane and the bone axis was

29° [2,14]. Kwon et al. found that the slope of the volar rim in the sagittal plane is steeper in males than in females [9].

A variation in the amount of palmar projection of the volar rim is one of the factors that concerns surgeons during plate placement in distal radius fracture surgery. Suppose the palmar protrusion of the volar rim is extensive, in that case, the plate may protrude similarly, preventing reconstruction of the flexor tendon gliding floor, and cause irritation of the flexor tendons or the median nerve. Therefore, the current treatment consensus requires that the plate not be placed distal to the watershed line and that it should not protrude palmarly more than 2 mm beyond the palmar aspect of the volar rim in the sagittal plane. Although it is desirable to use implants designed for individual geometries, current fracture implants offer only options for width and length and few implants offer possibilities for the bending angle of the plate that considers the shape of the volar rim.

The use of angularly stable locking plates to assemble fractures using implants as templates has become common [15,16]. In contrast, osteotomies for malunions are being performed based on preoperative planning using 3D image analysis technology and implants have been introduced that are placed according to the individual geometry of the osteotomy [17]. For fracture treatment, the use of implants that match the individual anatomy is ideal because a large discrepancy between the anatomical shape to be restored and the shape of the plate can cause malunion. However, because fractures are operated on a semi-emergency basis, individualized treatment is usually not feasible. Therefore, it is realistic to prepare multiple variations of implants and semi-customize the selection at the time of surgery and it is necessary to clarify the shape of the implant to accommodate these variations.

This study exhibited that gender is a factor that can affect bone width and volar rim anatomy, but age is unrelated to it. Previous studies have reported an effect of gender on fracture plate fit and the findings of this study support this [18]. The sagittal and transverse protrusions of the volar rim are moderately correlated. Furthermore, the individual differences in volar rim shape were found to be significant for bones with large widths. These results suggest that for bones with bone widths approximately less than 28 mm, plates should be designed with bending angles in the transverse plane that correlate to the bending angle in the sagittal plane. In bones with large bone widths, the variation in the shape of the volar rim tends to be significant, so it is desirable to be able to individually adjust the shape of the plate to match the rim shape. We conclude that for most bones, a plate with minimal soft tissue invasion can be designed using the above information and that custom-made plates are not necessary for treating fractures.

Next, the same investigation should be performed for other metaphyseal sites, such as distal femoral, proximal tibia, and distal tibia fractures. Similar to the distal radius fractures, the fixation strength of fractures with implants has been generally satisfactory in these sites and the next issue to be solved is the prevention of soft tissue problems caused by discrepancies between bone anatomy and implant shape [19,20]. The need for semi-custom implant design for fracture treatment based on individual bone anatomy for the treatment of any epiphyseal fractures needs to be clarified as well.

5. Limitation

This study had some limitations. First, because the analysis was limited to the bones of Asians, the external validity of the results for other races is uncertain. Second, the number of bones examined was small and multivariate analysis could not be performed to identify the factors affecting shape. Third, the thickness of the plate was not considered in this study, so the invasion of soft tissue could not be accurately predicted.

6. Conclusions

In conclusion, there are individual differences in the shape of the radius epiphysis and for bones of moderate or lower grade the size of the volar rim has a moderate correlation with the bone width in both the sagittal and transverse planes. As per the results of this study, we propose the development of several types of semi-custom plates with correlated sagittal and transverse bending angles. This allows for osteosynthesis with minimal soft tissue invasion while accommodating for the anatomical diversity of the volar rim.

Author Contributions: Conceptualization, H.Y.; methodology, H.Y.; analysis, H.Y. and M.S.; writing—original draft preparation, H.Y.; writing—review and editing, K.I. and M.Y.; supervision, M.T.; project administration, H.Y.; funding acquisition, H.Y. All authors have read and agreed to the published version of the manuscript.

Funding: This work was supported by the Japan Society for the Promotion of Science KAKENHI Grant Number 21H03288.

Institutional Review Board Statement: The study was conducted in accordance with the Declaration of Helsinki and approved by the Institutional Review Board of Nagoya University Hospital (code 2014-0326 and approved on 2022-03-28).

Informed Consent Statement: Patient consent was waived due to the retrospective nature of the study.

Data Availability Statement: The data that support the findings of this study are available on request from the corresponding author, [HY]. The data are not publicly available due to restrictions their containing information that could compromise the privacy of research participants.

Conflicts of Interest: The authors declare no conflict of interest.

References

1. Gosman, J.H.; Hubbell, Z.R.; Shaw, C.N.; Ryan, T.M. Development of Cortical Bone Geometry in the Human Femoral and Tibial Diaphysis. *Anat. Rec.* **2013**, *296*, 774–787. [CrossRef] [PubMed]
2. Yoneda, H.; Iwatsuki, K.; Hara, T.; Kurimoto, S.; Yamamoto, M.; Hirata, H. Interindividual anatomical variations affect the plate-to-bone fit during osteosynthesis of distal radius fractures. *J. Orthop. Res.* **2016**, *34*, 953–960. [CrossRef]
3. Heifner, J.J.; Orbay, J.L. Assessment and Management of Acute Volar Rim Fractures. *J. Wrist Surg.* **2022**, *11*, 214–218. [CrossRef]
4. Limthongthang, R.; Bachoura, A.; Jacoby, S.M.; Osterman, A.L. Distal Radius Volar Locking Plate Design and Associated Vulnerability of the Flexor Pollicis Longus. *J. Hand Surg.* **2014**, *39*, 852–860. [CrossRef] [PubMed]
5. Imatani, J.; Akita, K.; Yamaguchi, K.; Shimizu, H.; Kondou, H.; Ozaki, T. An Anatomical Study of the Watershed Line on the Volar, Distal Aspect of the Radius: Implications for Plate Placement and Avoidance of Tendon Ruptures. *J. Hand Surg.* **2012**, *37*, 1550–1554. [CrossRef] [PubMed]
6. Baumbach, S.F.; Schmidt, R.; Varga, P.; Heinz, T.; Vécsei, V.; Zysset, P.K. Where is the distal fracture line location of dorsally displaced distal radius fractures? *J. Orthop. Res.* **2011**, *29*, 489–494. [CrossRef] [PubMed]
7. Soong, M.; Earp, B.E.; Bishop, G.; Leung, A.; Blazar, P. Volar Locking Plate Implant Prominence and Flexor Tendon Rupture. *J. Bone Jt. Surg.* **2011**, *93*, 328–335. [CrossRef] [PubMed]
8. Kitay, A.; Swanstrom, M.; Schreiber, J.J.; Carlson, M.G.; Nguyen, J.T.; Weiland, A.J.; Daluiski, A. Volar Plate Position and Flexor Tendon Rupture Following Distal Radius Fracture Fixation. *J. Hand Surg.* **2013**, *38*, 1091–1096. [CrossRef]
9. Kwon, B.C.; Lee, J.K.; Lee, S.Y.; Hwang, J.Y.; Seo, J.-H. Morphometric Variations in the Volar Aspect of the Distal Radius. *Clin. Orthop. Surg.* **2018**, *10*, 462–467. [CrossRef] [PubMed]
10. Imatani, J.; Akita, K. Volar Distal Radius Anatomy Applied to the Treatment of Distal Radius Fracture. *J. Wrist Surg.* **2017**, *6*, 174–177. [CrossRef]
11. Mishra, P.K.; Nagar, M.; Gaur, S.C.; Gupta, A. Morphometry of distal end radius in the Indian population: A radiological study. *Indian J. Orthop.* **2016**, *50*, 610–615. [CrossRef]
12. Kumar, A.; Passey, J.; Chouhan, D.; Saini, M.; Narang, A. CT Based Characterization of Volar Surface of Distal Radius: Can an Ideal Volar Plate for Fixation of Distal Radial Fractures be Designed? *J. Hand Surg. (Asian-Pac. Vol.)* **2021**, *26*, 77–83. [CrossRef] [PubMed]
13. Oura, K.; Oka, K.; Kawanishi, Y.; Sugamoto, K.; Yoshikawa, H.; Murase, T. Volar morphology of the distal radius in axial planes: A quantitative analysis. *J. Orthop. Res.* **2015**, *33*, 496–503. [CrossRef] [PubMed]
14. Andermahr, J.; Lozano-Calderon, S.; Trafton, T.; Crisco, J.J.; Ring, D. The Volar Extension of the Lunate Facet of the Distal Radius: A Quantitative Anatomic Study. *J. Hand Surg.* **2006**, *31*, 892–895. [CrossRef] [PubMed]
15. Toros, T.; Sügün, T.S.; Özaksar, K. Complications of distal radius locking plates. *Injury* **2013**, *44*, 336–339. [CrossRef] [PubMed]
16. Buzzell, J.E.; Weikert, D.R.; Watson, J.T.; Lee, D.H. Precontoured Fixed-Angle Volar Distal Radius Plates: A Comparison of Anatomic Fit. *J. Hand Surg.* **2008**, *33*, 1144–1152. [CrossRef] [PubMed]
17. Oka, K.; Shigi, A.; Tanaka, H.; Moritomo, H.; Arimitsu, S.; Murase, T. Intra-articular corrective osteotomy for intra-articular malunion of distal radius fracture using three-dimensional surgical computer simulation and patient-matched instrument. *J. Orthop. Sci.* **2020**, *25*, 847–853. [CrossRef] [PubMed]

18. Perrin, M.; Badre, A.; Suh, N.; Lalone, E.A. Analysis of Three-Dimensional Anatomical Variance and Fit of the Distal Radius to Current Volar Locking Plate Designs. *J. Hand Surg. Glob. Online* **2020**, *2*, 277–285. [CrossRef] [PubMed]
19. Mb, O.; Aksan, T.; Ertekin, C.; Tezcan, M. Coverage of exposed bone and hardware of the medial malleolus with tibialis posterior artery perforator flap after ankle fracture surgery complications. *Int. Wound J.* **2020**, *17*, 429–435. [CrossRef] [PubMed]
20. Chou, Y.-C.; Wu, C.-C.; Chan, Y.-S.; Chang, C.-H.; Hsu, Y.-H.; Huang, Y.-C. Medial Gastrocnemius Muscle Flap for Treating Wound Complications After Double-Plate Fixation via Two-Incision Approach for Complex Tibial Plateau Fractures. *J. Trauma Acute Care Surg.* **2010**, *68*, 138–145. [CrossRef] [PubMed]

Article

Detailed Morphology of the Incisive or Nasopalatine Canal

Andrei Valentin Iamandoiu [1], Alexandru Nicolae Mureşan [1,2] and Mugurel Constantin Rusu [1,*]

[1] Division of Anatomy, Faculty of Dental Medicine, "Carol Davila" University of Medicine and Pharmacy, 010825 Bucharest, Romania; andrei.iamandoiu@drd.umfcd.ro (A.V.I.); alexandru-nicolae.muresan@drd.umfcd.ro (A.N.M.)

[2] Research Department, "Dr. Carol Davila" Central Military Emergency Hospital, 010825 Bucharest, Romania

* Correspondence: mugurel.rusu@umfcd.ro

Abstract: (1) Background: The nasopalatine canal (NPC), or the incisive canal (IC) of maxilla, unites the anterior nasal floor with the anterior palatine region. Different morphological variables of the NPC were investigated, indicating it is either anatomically variable or constant. It was therefore decided to perform an additional study of the NPC. (2) Methods: A retrospective cone beam computed tomography (CBCT) study was performed on 89 patient files: 38 males and 51 females. The study documented the presence or absence of a well-defined NPC, the number of openings, and the anatomic variables of the canal (number, course, and secondary canaliculi). (3) Results: the NPC/IC in the coronal plane was classified into five types: (I) NPC present with two superior, nasopalatine foramina (NPFs) (66.29%); (II) NPC absent with two NPFs (2.25%); (III) NPC present and single NPF (17.98%); (IV) NPC present with three NPFs (3.37%); (V) both absent NPC and NPF (10.11%). (4) Conclusions: The anatomical possibilities of the NPC are numerous and diverse; they include the absence of the canal. Therefore, a standardized description of this canal could not be assumed and a radiological assessment is recommended before surgical treatment in the premaxilla area.

Keywords: maxilla; cone beam computed tomography; hard palate; incisive canal; anatomic variation

Citation: Iamandoiu, A.V.; Mureşan, A.N.; Rusu, M.C. Detailed Morphology of the Incisive or Nasopalatine Canal. *Anatomia* **2022**, *1*, 75–85. https://doi.org/10.3390/anatomia1010008

Academic Editors: Gianfranco Natale and Francesco Fornai

Received: 9 June 2022
Accepted: 28 June 2022
Published: 4 July 2022

Publisher's Note: MDPI stays neutral with regard to jurisdictional claims in published maps and institutional affiliations.

1. Introduction

The nasopalatine canal (NPC), or the incisive canal (IC) of maxilla, unites the anterior nasal floor with the anterior palatine region. It is commonly described as consisting of two upper canals (proper NPCs/ICs) that join inferiorly to form the common NPC/IC opening at the incisive foramen (IF) of the hard palate [1]. This results in a "Y" morphology of the canal. The two proper NPCs/ICs are separated by the nasomaxillary, or septo-premaxillary crest [2]. It is commonly assumed that the NPC is traversed by the nasopalatine nerves and arteries. However, histologically, the NPC content is represented by numerous small veins, arteries, and nerve bundles [3].

While the inferior opening of the NPC/IC is indicated as the incisive foramen, the superior openings of that canal are not named in *Terminologia Anatomica* and may be suitably regarded as nasopalatine foramina [4].

Song et al., (2009) invariably found two superior foramina (nasopalatine foramina) and one inferior foramen (incisive foramen) of the NPC/IC while studying 56 anterior maxillae in computed tomography (CT) [3]. We were intrigued by such constant morphology. Abrams et al., (1963) discussed that although the location of the NPCs/ICs is quite constant, their anatomy is not [5]. The large spectrum of anatomical variations of the NPC was confirmed by later studies [6]. Anatomical variations of the NPC remain poorly documented [7].

The topography and morphology of the NPC may impact surgical treatment planning, be it implant restorative treatment, congenital defects repair (cleft palate, congenital syndromes like Down syndrome or cleidocranial dysplasia), traumatic lesions repair, or in orthodontic purpose (orthognathic treatment, palate expanding).

It was therefore decided to perform a cone beam CT (CBCT) retrospective study to check whether or not the morphology of the NPC/IC is constant or variable.

2. Materials and Methods

A retrospective CBCT study on 89 patient files was performed (38 males, 42.69% and 51 females, 57.3%). The patients were all adults who had undergone different dental procedures that required a maxillary CBCT analysis. None of the patients presented maxillary pathological morphology signs (no trauma, congenital defects, or tumor-like lesions) and all the CBCT scans were accurate without signs of distortion.

The subjects were positioned according to the manufacturer instructions and had been scanned using an iCat CBCT machine (Imaging Sciences International (Hatfield, PA, USA)) with the settings at resolution 0.250 pixels/inch, field of view 130 mm, and image matrix size 640 × 640 pixels. The CT data were analyzed using the iCatVision software. The CBCT files were exported as DICOM files, which were additionally analyzed with the Planmeca Romexis Viewer 3.5.0.R software as in other previous studies [8–10]. The two-dimensional multiplanar reconstructions (MPRs) were evaluated in sagittal and axial anatomical planes; coronal slices through the NPCs were used to determine the morphological variables that were aimed. Three-dimensional volume renderings of specific areas were also evaluated. Relevant anatomical features were exported as image files (*.tif).

The patients have given written informed consent for all radiological data to be used for research and teaching purposes, provided the protection of the identity and personal data is maintained. The study was approved (no. 456/04.05.2021) by the responsible authorities (2nd affiliation of the 2nd author).

During the study, the following variables were assessed: (a) the presence or absence of a well-defined NPC/IC; (b) the number of the nasopalatine foramina (NPFs); (c) the number of incisive foramina; (d) the variables of the proper (superior) NPCs/ICs: number, course, parallel or convergent, and intrinsic septa determining secondary canaliculi.

The NPC/IC was classified into 5 different types according to the morphological findings: (I) NPC present with 2 NPFs; (II) NPC absent with 2 NPFs; (III) NPC present with single NPF; (IV) NPC present with 3 NPFs; (V) both absent NPC and NPF.

Subtypes of the NPC/IC classification are presented in Table 1.

Table 1. Characteristics and numbering of the types and subtypes of the nasopalatine/incisive canal.

Type of the NPC/IC	Characteristics of Types	Subtypes	Characteristics of Subtypes
I	NPC/IC present, 2 nasopalatine foramina	Ia	"Y"-shaped NPC/IC, with no secondary canaliculi
		Ib	"Y"-shaped NPC/IC, with secondary canaliculi separated by a sagittal septum
		Ic	"Y"-shaped NPC/IC, with unilateral secondary canaliculi separated by a coronal septum
		Id	"Y"-shaped NPC/IC, with bilateral secondary canaliculi separated by a coronal septum
		Ie	"Y"-shaped NPC/IC, with an added superiorly blind-ended median canal
		If	parallel proper NPCs/ICs separated by septum
		Ig	parallel proper NPCs/ICs unseparated by septum (NPC/IC unique, two nasopalatine foramina)
II	NPC/IC absent, 2 nasopalatine foramina		
III	NPC/IC unique, 1 nasopalatine foramen	IIIa	unique median nasopalatine foramen, inferior to the nasomaxillary crest
		IIIb	unique median nasopalatine foramen, on one side of the nasomaxillary crest
IV	NPC/IC present, 3 nasopalatine foramina, 1 median and 2 lateral		
V	NPC/IC proper absent, absent nasopalatine foramina		

3. Results

Within the study, all types of NPC/IC were found in variable proportions. The incisive foramen was invariably unique in all cases that presented an NPC.

Within the 89 cases that were investigated, Type I of NPC was found in 59 cases (66.29%), Type II in 2 cases (2.25%), Type III in 16 cases (17.98%), Type IV in 3 cases (3.37%), and Type V in 9 cases (10.11%).

In the male lot, the NPC/IC distribution was as follows: Type I in 26 cases (68.42%), Type II in 2 cases (5.26%), Type III in 8 cases (21.05%), and Type V in 2 cases (5.26%). In the female lot, there were no Type II NPC/IC cases found and the other variants were as follows: Type I in 33 cases (64.71%), Type III in 8 cases (15.69%), Type IV in 3 cases (5.88%), and Type V in 7 cases (13.73%).

Type I of NPC/IC had various morphologies (Table 1). In the general lot of 89 cases, we found Subtypes Ia (Figure 1A) in 17 cases (28.81%), Ib (Figure 1B) in 2 cases (3.39%), Ic (Figure 1C) in 7 cases (11.86%), Id (Figure 2) in 1 case (1.69%), Ie (Figure 3A) in 1 case (1.69%), If (Figure 3B) in 19 cases (32.2%), and Ig (Figure 3C) in 12 cases (20.34%). In males, Subtypes I were found as follows: Ia in 10 cases (38.46%), Ib and Ie in one case (3.85%) each, Ic in 5 cases (19.23%), If in 7 cases (26.92%), and Ig in 2 cases (7.69%). We found in females Subtypes: Ia in 7 cases (21.21%), Ib and Id in one case (3.03%) each, Ic in 2 cases (6.06%), If in 12 cases (36.36%), and Ig in 10 cases (30.3%).

In 2 out of 89 cases, both male, Type II (Figure 4) of NPC/IC was found.

Type III of NPC/IC was also divided into Subtypes IIIa (Figure 5A), found in 10 out of the 16 Type III cases, and IIIb (Figure 5B), found in the remaining 6 cases. Seven males and three females presented Subtype IIIa, while Subtype IIIb was found in one male and five female patients. Type IV of NPC (Figure 5C) was found only in female patients. Type V NPC/IC—absent nasopalatine canal and NPFs (Figure 6)—was discovered in two male and seven female patients.

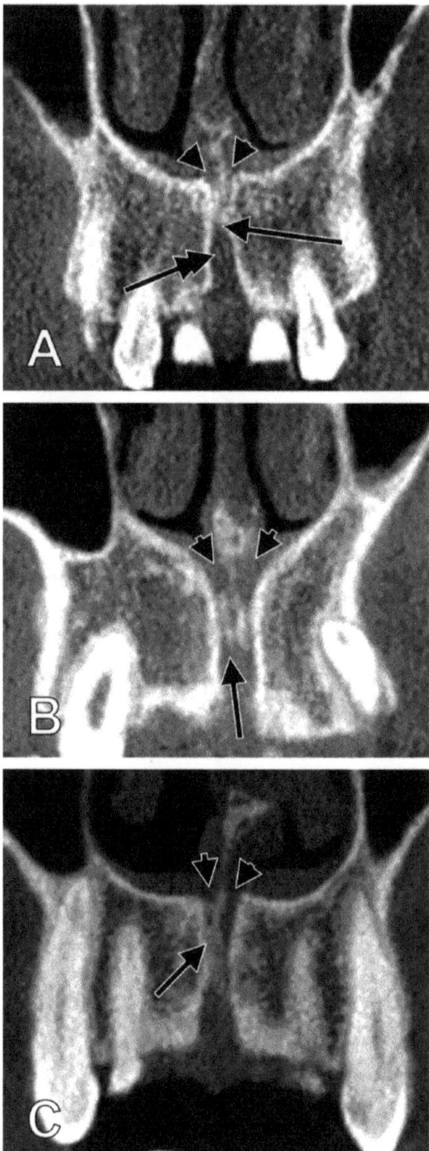

Figure 1. (**A**) Coronal CBCT slice through Type Ia of nasopalatine/incisive canal. The canal is "Y"-shaped, with two superior (nasopalatine) foramina (arrowheads) and one inferior incisive foramen (arrow). The incisive fossa is indicated (double-headed arrow). (**B**) Coronal oblique CTCB slice through Type Ib of nasopalatine/incisive canal. The canal is "Y"-shaped, with two superior (nasopalatine) foramina (arrowheads). A sagittal septum (arrow) separates two secondary canaliculi. (**C**) Coronal CBCT slice through Type Ic of nasopalatine/incisive canal. The canal is "Y"-shaped with two superior (nasopalatine) foramina (arrowheads). A coronal septum (arrow) is found in the right side.

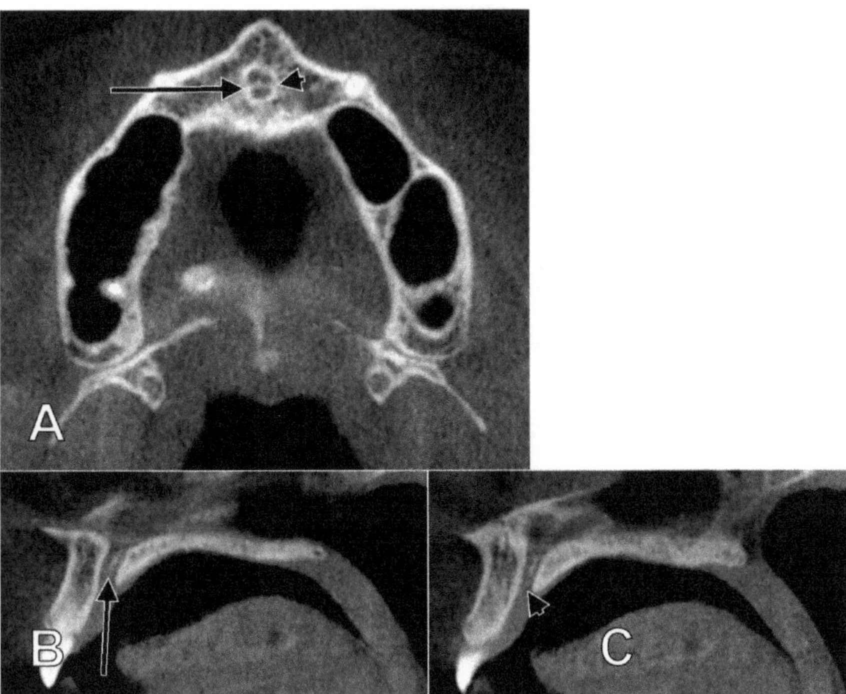

Figure 2. Axial (**A**), right paramedian (**B**), and left paramedian (**C**) CBCT slices through Type Id of nasopalatine/incisive canal. The canal is "Y"-shaped with bilateral secondary canaliculi separated by coronal septa, right (arrow) and left (arrowhead).

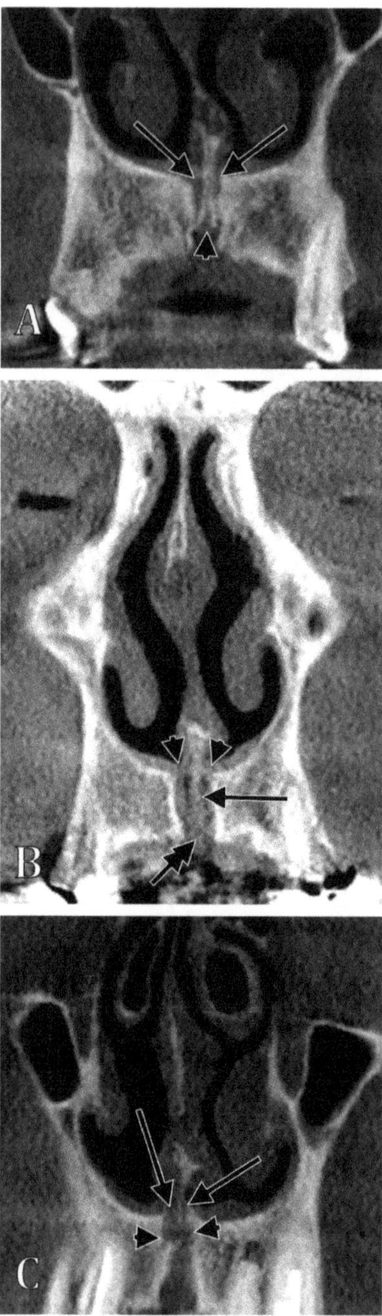

Figure 3. (**A**) Coronal CBCT slice through Type Ie of nasopalatine/incisive canal. The canal is "Y"-shaped with two superior (nasopalatine) foramina (arrows). A superiorly blind-ended median canal is added (arrowhead). (**B**) Coronal CBCT slice through Type If of nasopalatine/incisive canal. The canal has parallel proper NPCs/ICs (arrowheads) separated by a septum (arrow) that open info the incisive fossa (double-headed arrow). (**C**) Coronal CBCT slice through Type Ig of nasopalatine/incisive canal. The upper proper canals (arrowheads) are not separated by septum. Only the nasopalatine foramina (arrows) are separated between by the inferior end of the nasomaxillary (septo-premaxillary) crest.

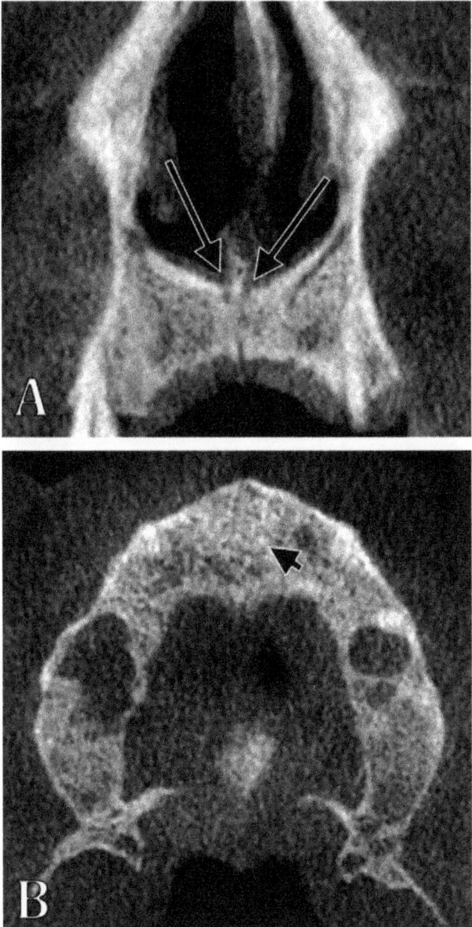

Figure 4. Coronal (**A**) and axial (**B**) CBCT slices through Type II of nasopalatine/incisive canal. Although the nasopalatine foramina can be distinguished ((**A**) arrows), a morphologically configured canal lacks ((**B**) arrowhead).

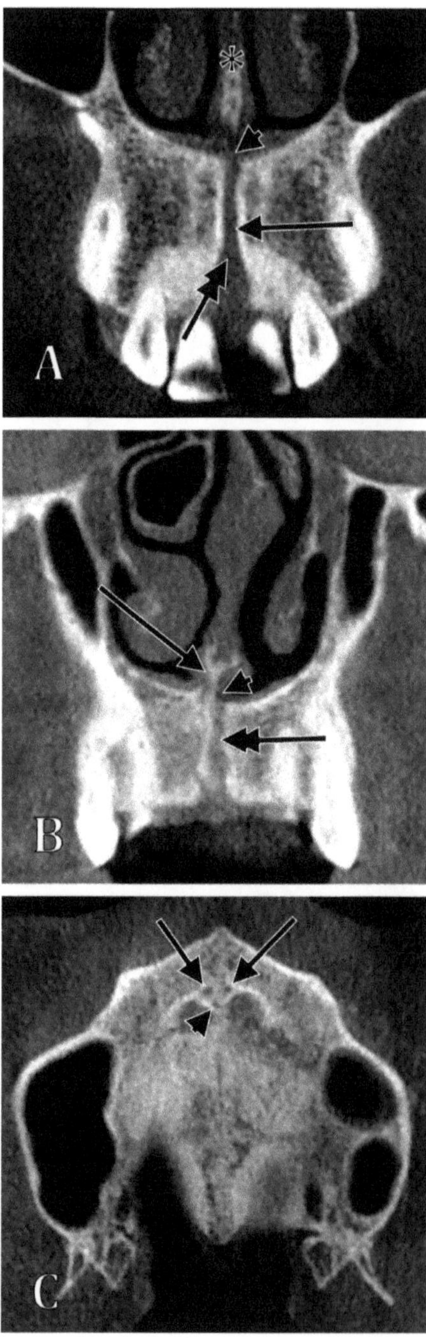

Figure 5. (**A**) Coronal CBCT slice through Type IIIa of unique nasopalatine/incisive canal (arrow) opened superiorly at the nasopalatine foramen (arrowhead) and inferiorly at the incisive foramen (double-headed arrow). (*): nasal septum (**B**) Coronal CBCT slice through Type IIIb of unique nasopalatine/incisive canal (double-headed arrow) opened superiorly at the nasopalatine foramen (arrowhead) located on the left side of the nasomaxillary crest (arrow). (**C**) Axial CBCT slice through Type IV of nasopalatine/incisive canal at the level of the nasal floor. There are three nasopalatine foramina, two lateral (arrows), and one median (arrowhead).

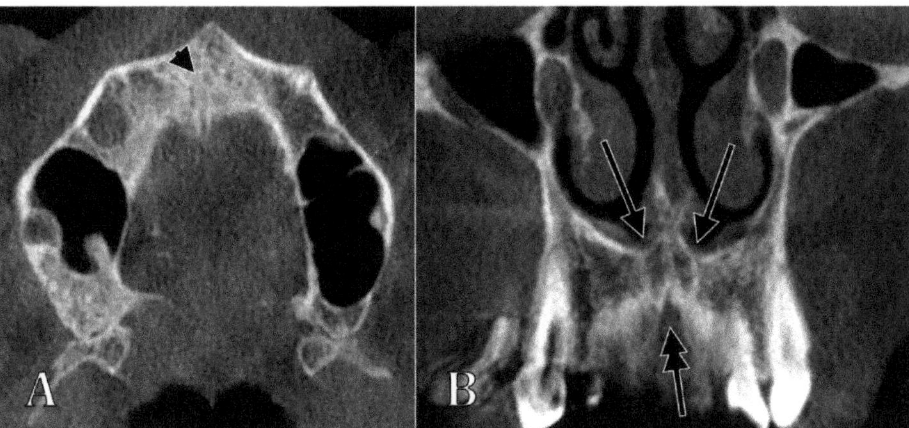

Figure 6. Axial (**A**) and coronal (**B**) CBCT slices through the hard palate in two different cases. Absent nasopalatine/incisive canal proper ((**A**) arrowhead) and nasopalatine foramina ((**B**) arrows) (Type V). The incisive fossa ((**B**) double-headed arrow) is anatomically blind superiorly.

4. Discussion

According to Radlanski, the literature describing the prenatal development of the NPC "is contradictory and partly even bizarre" [11]. This is because different authors indicate that the NPC courses at the border of the primary and secondary palate, which is erroneous [11]. The NPC is formed in the posterior part of the primary palate [12].

The edentulous anterior maxilla is a preferred site for complete maxillary rehabilitation by dental-implant surgery. Therefore, the amount of available bone, as well as the topography and morphology of the NPC, determine the personalized therapeutic plan. These variables are evaluated by CBCT. The resorption of the buccal alveolar plate in the anterior maxilla occurs after tooth extractions, local trauma, periodontal and periradicular pathology, as well as due to cysts or tumors [13]. Dental-implant rehabilitation of the anterior maxilla serves a double purpose: aesthetic and phonetic [14]. Insertion of implants in the vicinity of the NPC may jeopardize a successful procedure [7].

The NPC courses through the premaxilla, or the anterior maxilla. It is traversed by the nasopalatine nerves and arteries. Therefore, injuries to the nasopalatine nerve may result in hypoesthesia, paresthesia, or pain within the premaxilla area (palate buccal and nasal floor mucosa) and the superior frontal teeth [15]. Hemorrhage may result if the nasopalatine artery is damaged, which can lead to local hematoma and compression on the nasopalatine nerves [15]. Such unwanted events could occur during dental-implant rehabilitation of the anterior maxilla or other surgical interventions in the area.

Different previous studies evaluated the NPC by means of CBCT [13,14,16–25]. Some of these did not make the distinction between the proper and common NPCs [7], others did [25]. The authors determined different anatomical variables of the NPC: (a) the diameter of the NPF [14,16–21,23]; (b) the diameter of the IF [14,16–21,23]; (c) the length of the NPC [14,16–24]; (d) the sagittal thickness of the buccal alveolar plate at different height levels (upper, middle, lower) of the NPC [14,16,19,21]; (e) the diameter of the NPC [7,22–24]; (f) the angulation of the NPC [13,18,20]; and (g) the number of NPF and IF [16,17,20]. The course of the NPC in sagittal slices is also variable: (a) vertical–straight; (b) vertical–curved; (c) slanted–straight; and (d) slanted–curved [13,25].

The shape of the NPC was previously determined on axial slices, being classified in six groups: (1) round with separation; (2) round without separation; (3) oval with separation; (4) oval without separation; (5) heart-shaped with separation; (6) heart-shaped without separation [14]. However, as we observed in the figure depicting those shapes (Figure 1 in [14]), there were interpreted axial CBCT slices taken at different heights of those NPCs.

In sagittal slices, the anatomic variants of the NPC shape were classified by Jain into four groups: (1) funnel; (2) cylindrical; (3) hourglass; and (4) banana-like [20]. Sekerci distinguished six different sagittal shapes of the NPC: (1) hourglass; (2) cone; (3) funnel; (4) banana-like; (5) cylindrical; and (6) tree branch-like [26]. Gorurgoz and Ostas created nine groups of sagittal shapes of the NPC: (1) hourglass; (2) spindle; (3) cone; (4) funnel; (5) banana-like; (6) cylindrical; (7) tree branch-like; (8) kink; and (9) other [18]. Other authors found just four of these types of the NPC: cylindrical, funnel, spindle, and hourglass [25,27]. In dry skulls the morphology of the NPC was either conical or cylindrical [6].

Commonly, the superior openings of the NPC are located on each side of the nasal septum [11]. More exactly, they are on each side of the nasomaxillary crest [2]. Their number is individually variable. Numerous authors indicate these NPFs as the foramina of Stensen/Stenson [12,13,17,25–27]. However, as documented by Bahşi et al., Niels Stensen described the IF and not the NPF [4].

The NPC was also evaluated on coronal slices through the premaxilla and was classified in three main types: (1) single canal; (2) two parallel canals; and (3) variations of the Y-type canal in which there was always found a single IF, but the NPF were either two, three, or more than three [13,16,20]. When comparing those three types with our findings, it results that another anatomic possibility should be considered additionally: the absent NPC with or without NPF. An absent NPC does not mean that the minute vessels and nerve bundles are lacking, but that they traverse minute canaliculi of the premaxilla without being gathered within a canal with well-defined boundaries. As the premaxilla is an anatomical situs for implant placement, the absence of the NPC should be of benefit as it increases the amount of bone available for implant placement.

5. Conclusions

The anatomical possibilities of the NPC/IC are numerous and diverse. Therefore, a standardized description of this canal could not be assumed, especially during dental medical anatomical training. The current study adds novelty to the existing scientific literature by including the variant of an absent maxillary incisive, or nasopalatine canal.

Author Contributions: Conceptualization, M.C.R.; methodology, A.V.I. and M.C.R.; validation, M.C.R.; investigation, A.V.I. and A.N.M.; resources, A.N.M. and A.V.I.; writing—original draft preparation, A.V.I. and A.N.M.; writing—review and editing, A.V.I.; supervision, M.C.R. All authors have read and agreed to the published version of the manuscript.

Funding: This research received no external funding.

Institutional Review Board Statement: The study was conducted according to the guidelines of the Declaration of Helsinki and approved by the Ethics Committee of "Dr. Carol Davila" Central Military Emergency Hospital, Bucharest, Romania, (protocol code 456/4 May 2021).

Informed Consent Statement: Informed consent was obtained from all subjects involved in the study.

Data Availability Statement: The data presented in this study are available on request from the corresponding author upon reasonable request.

Conflicts of Interest: The authors declare no conflict of interest.

References

1. Al-Amery, S.M.; Nambiar, P.; Jamaludin, M.; John, J.; Ngeow, W.C. Cone beam computed tomography assessment of the maxillary incisive canal and foramen: Considerations of anatomical variations when placing immediate implants. *PLoS ONE* **2015**, *10*, e0117251. [CrossRef] [PubMed]
2. Iamandoiu, A.V.; Ilie, O.C.; Jianu, A.M.; Rusu, M.C. The nasomaxillary or septo-premaxillary crest. *Med. Evol.* **2021**, *XXVII*, 386–391.
3. Song, W.C.; Jo, D.I.; Lee, J.Y.; Kim, J.N.; Hur, M.S.; Hu, K.S.; Kim, H.J.; Shin, C.; Koh, K.S. Microanatomy of the incisive canal using three-dimensional reconstruction of microCT images: An ex vivo study. *Oral Surg. Oral Med. Oral Pathol. Oral Radiol. Endod.* **2009**, *108*, 583–590. [CrossRef] [PubMed]

4. Bahsi, I.; Orhan, M.; Kervancioglu, P. A sample of morphological eponym confusion: Foramina of Stenson/Stensen. *Surg. Radiol. Anat.* **2017**, *39*, 935–936. [CrossRef]
5. Abrams, A.M.; Howell, F.V.; Bullock, W.K. Nasopalatine cysts. *Oral Surg. Oral Med. Oral Pathol.* **1963**, *16*, 306–332. [CrossRef]
6. Liang, X.; Jacobs, R.; Martens, W.; Hu, Y.; Adriaensens, P.; Quirynen, M.; Lambrichts, I. Macro- and micro-anatomical, histological and computed tomography scan characterization of the nasopalatine canal. *J. Clin. Periodontol.* **2009**, *36*, 598–603. [CrossRef]
7. Mraiwa, N.; Jacobs, R.; Van Cleynenbreugel, J.; Sanderink, G.; Schutyser, F.; Suetens, P.; van Steenberghe, D.; Quirynen, M. The nasopalatine canal revisited using 2D and 3D CT imaging. *Dentomaxillofacial Radiol.* **2004**, *33*, 396–402. [CrossRef]
8. Rusu, M.C.; Sandulescu, M.; Bichir, C. Patterns of pneumatization of the tympanic plate. *Surg. Radiol. Anat.* **2020**, *42*, 347–353. [CrossRef]
9. Bichir, C.; Rusu, M.C.; Vrapciu, A.D.; Maru, N. The temporomandibular joint: Pneumatic temporal cells open into the articular and extradural spaces. *Folia Morphol.* **2019**, *78*, 630–636. [CrossRef]
10. Muresan, A.N.; Mogoanta, C.A.; Stanescu, R.; Rusu, M.C. The sinus septi nasi and other minor pneumatizations of the nasal septum. *Rom. J. Morphol. Embryol.* **2021**, *62*, 227–231. [CrossRef]
11. Radlanski, R.J.; Emmerich, S.; Renz, H. Prenatal morphogenesis of the human incisive canal. *Anat. Embryol.* **2004**, *208*, 265–271. [CrossRef] [PubMed]
12. Von Arx, T.; Lozanoff, S. *Clinical Oral Anatomy: A Comprehensive Review for Dental Practitioners and Researchers*; Springer: Cham, Switzerland, 2016; p. 572.
13. Fernandez-Alonso, A.; Suarez-Quintanilla, J.A.; Muinelo-Lorenzo, J.; Bornstein, M.M.; Blanco-Carrion, A.; Suarez-Cunqueiro, M.M. Three-dimensional study of nasopalatine canal morphology: A descriptive retrospective analysis using cone-beam computed tomography. *Surg. Radiol. Anat.* **2014**, *36*, 895–905. [CrossRef] [PubMed]
14. Acar, B.; Kamburoglu, K. Morphological and volumetric evaluation of the nasopalatinal canal in a Turkish population using cone-beam computed tomography. *Surg. Radiol. Anat.* **2015**, *37*, 259–265. [CrossRef] [PubMed]
15. Neves, F.S.; Oliveira, L.K.; Ramos Mariz, A.C.; Crusoe-Rebello, I.; de Oliveira-Santos, C. Rare anatomical variation related to the nasopalatine canal. *Surg. Radiol. Anat.* **2013**, *35*, 853–855. [CrossRef]
16. Bornstein, M.M.; Balsiger, R.; Sendi, P.; von Arx, T. Morphology of the nasopalatine canal and dental implant surgery: A radiographic analysis of 100 consecutive patients using limited cone-beam computed tomography. *Clin. Oral Implant. Res.* **2011**, *22*, 295–301. [CrossRef]
17. Etoz, M.; Sisman, Y. Evaluation of the nasopalatine canal and variations with cone-beam computed tomography. *Surg. Radiol. Anat.* **2014**, *36*, 805–812. [CrossRef]
18. Gorurgoz, C.; Oztas, B. Anatomic characteristics and dimensions of the nasopalatine canal: A radiographic study using cone-beam computed tomography. *Folia Morphol.* **2021**, *80*, 923–934. [CrossRef]
19. Hakbilen, S.; Magat, G. Evaluation of anatomical and morphological characteristics of the nasopalatine canal in a Turkish population by cone beam computed tomography. *Folia Morphol.* **2018**, *77*, 527–535. [CrossRef]
20. Jain, N.V.; Gharatkar, A.A.; Parekh, B.A.; Musani, S.I.; Shah, U.D. Three-Dimensional Analysis of the Anatomical Characteristics and Dimensions of the Nasopalatine Canal Using Cone Beam Computed Tomography. *J. Maxillofac. Oral Surg.* **2017**, *16*, 197–204. [CrossRef]
21. Milanovic, P.; Selakovic, D.; Vasiljevic, M.; Jovicic, N.U.; Milovanovic, D.; Vasovic, M.; Rosic, G. Morphological Characteristics of the Nasopalatine Canal and the Relationship with the Anterior Maxillary Bone-A Cone Beam Computed Tomography Study. *Diagnostics* **2021**, *11*, 915. [CrossRef]
22. Soumya, P.; Koppolu, P.; Pathakota, K.R.; Chappidi, V. Maxillary Incisive Canal Characteristics: A Radiographic Study Using Cone Beam Computerized Tomography. *Radiol. Res. Pract.* **2019**, *2019*, 6151253. [CrossRef] [PubMed]
23. Suter, V.G.; Jacobs, R.; Brucker, M.R.; Furher, A.; Frank, J.; von Arx, T.; Bornstein, M.M. Evaluation of a possible association between a history of dentoalveolar injury and the shape and size of the nasopalatine canal. *Clin. Oral Investig.* **2016**, *20*, 553–561. [CrossRef] [PubMed]
24. Tozum, T.F.; Guncu, G.N.; Yildirim, Y.D.; Yilmaz, H.G.; Galindo-Moreno, P.; Velasco-Torres, M.; Al-Hezaimi, K.; Al-Sadhan, R.; Karabulut, E.; Wang, H.L. Evaluation of maxillary incisive canal characteristics related to dental implant treatment with computerized tomography: A clinical multicenter study. *J. Periodontol.* **2012**, *83*, 337–343. [CrossRef] [PubMed]
25. Thakur, A.R.; Burde, K.; Guttal, K.; Naikmasur, V.G. Anatomy and morphology of the nasopalatine canal using cone-beam computed tomography. *Imaging Sci. Dent.* **2013**, *43*, 273–281. [CrossRef]
26. Sekerci, A.E.; Buyuk, S.K.; Cantekin, K. Cone-beam computed tomographic analysis of the morphological characterization of the nasopalatine canal in a pediatric population. *Surg. Radiol. Anat.* **2014**, *36*, 925–932. [CrossRef]
27. Nasseh, I.; Aoun, G.; Sokhn, S. Assessment of the Nasopalatine Canal: An Anatomical Study. *Acta Inform. Med.* **2017**, *25*, 34–38. [CrossRef]

Article

Hippocampal Dentation in Children and Adolescents: A Cross-Sectional Analysis from Birth to 18 Years Old

Julia F. Beattie [1,2], Roy C. Martin [3], Edwin W. Cook III [4], Matthew D. Thompson [5], Rajesh K. Kana [6], Ruth Q. Jacobs [7], Tanya A. Correya [4], Anandh K. Ramaniharan [3] and Lawrence W. Ver Hoef [3,8,*]

[1] Behavioral Medicine and Clinical Psychology, Cincinnati Children's Hospital Medical Center, 3333 Burnet Ave., Cincinnati, OH 45229, USA; julia.beattie@cchmc.org

[2] Department of Pediatrics, University of Cincinnati College of Medicine, 3230 Eden Ave, Cincinnati, OH 45267, USA

[3] Department of Neurology, University of Alabama at Birmingham, 1720 University Blvd., Birmingham, AL 35294, USA; rmartin@uabmc.edu (R.C.M.); aramaniharan@uabmc.edu (A.K.R.)

[4] Department of Psychology, University of Alabama at Birmingham, 1300 University Blvd., Birmingham, AL 35233, USA; ecook@uab.edu (E.W.C.III); tanya98@uab.edu (T.A.C.)

[5] Children's Behavioral Health, Children's of Alabama, 1600 7th Ave. S, Birmingham, AL 35233, USA; matthew.thompson@childrensal.org

[6] Center for Innovative Research in Autism, Department of Psychology, University of Alabama, 745 Hackberry Ln, Tuscaloosa, AL 35401, USA; rkkana@ua.edu

[7] Department of Biochemistry and Molecular Genetics, University of Alabama at Birmingham, 705 20th Street S, Birmingham, AL 35233, USA; rqjacobs@uab.edu

[8] Department of Neurology, Birmingham VA Medical Center, 700 University Blvd., Birmingham, AL 35233, USA

* Correspondence: lverhoef@uabmc.edu; Tel.: +1-205-422-8808

Citation: Beattie, J.F.; Martin, R.C.; Cook, E.W., III; Thompson, M.D.; Kana, R.K.; Jacobs, R.Q.; Correya, T.A.; Ramaniharan, A.K.; Ver Hoef, L.W. Hippocampal Dentation in Children and Adolescents: A Cross-Sectional Analysis from Birth to 18 Years Old. *Anatomia* 2022, 1, 41–53. https://doi.org/10.3390/anatomia1010005

Academic Editors: Gianfranco Natale and Francesco Fornai

Received: 10 March 2022
Accepted: 8 April 2022
Published: 15 April 2022

Publisher's Note: MDPI stays neutral with regard to jurisdictional claims in published maps and institutional affiliations.

Abstract: The degree of hippocampal dentation, a morphologic feature of the inferior surface of the human hippocampus, has been positively associated with episodic memory performance in healthy adults. This study examined hippocampal dentation in healthy children and adolescents. The Cincinnati MR Imaging of NeuroDevelopment (C-MIND) dataset was used to examine the relationship between age and hippocampal dentation in 90 healthy children, age < 1 to 18 years old, using T1-weighted MPRAGE scans. Hippocampal dentation was assessed by counting the number of dentes for the left and right hippocampi. Participants had slightly more left than right hippocampal dentes, on average. Dentation did not differ significantly between males and females. Correlational analyses revealed that the numbers of left, right, and total dentes were positively associated with age in this sample. Interestingly, these data highlight the wide variability of dentation in older age groups. While younger children tended to have absent or few dentes, a range of dentation was present in older children and adolescents (ranging from absent to numerous, bilaterally). This is consistent with previous research in a healthy adult cohort, where a range of dentation was also observed. This study is the first to examine hippocampal dentation in children.

Keywords: hippocampus; neurodevelopment; hippocampal anatomy; hippocampal dentation

1. Introduction

Human hippocampal anatomy is particularly complex, given its unique shape and layered cellular structure. A large portion of this structural development occurs during gestation, and by 34 weeks gestational age, hippocampal subfields (i.e., subiculum, CA1, CA2, and CA3) are distinguishable and the dentate gyrus has developed an adult-like appearance [1,2]. Postnatal change includes decreases in cell density [1] and myelination of the hippocampal formation, continuing well into adulthood [3]. Regarding volume, several studies indicate a non-linear relationship between hippocampal volume and age in children [4–6], with increases early in childhood followed by relative stability later in childhood [6] and during adolescence [6]. Despite reported changes in volume, no significant morphological changes are known to occur from childhood into adulthood [1].

Hippocampal dentation is an aspect of hippocampal morphology that has not been previously examined in children. Hippocampal dentation, present on the inferior surface of the hippocampus (primarily CA1/subiculum layer), has been described in few existing studies in adults [7–10], and has been mentioned only briefly in anatomical references [11–13]. This region has only recently been the target of morphological measurement via high-resolution neuroimaging techniques [14]. Hippocampal dentation has been referred to variously as "folds of CA1" [11], and "digitations" or "digitationes hippocampi" [8,12]; however, we differentiate hippocampal dentation, which is a ridged contour on the inferior aspect of the hippocampus that varies considerably across individuals [7], from digitation, which is seen ubiquitously on the anterior superior aspect of the hippocampal head [15]. Research has shown that healthy adults exhibit wide variability in the quantity and prominence of hippocampal dentation, including the complete absence of any observable dentation in 9% of a healthy adult sample. Hippocampal dentation has also recently been shown to be asymmetrically associated with hippocampal sclerosis in temporal lobe epilepsy (TLE), with 70% of the affected hippocampi showing nearly complete loss of dentes [10]. Interestingly, loss of dentation was associated with the epileptic hippocampus, even after correcting for volume loss, suggesting that loss of dentation in TLE is part of the pathologic process and not simply a secondary effect of hippocampal volume loss. Additionally, hippocampal dentation was associated with aspects of both verbal and visual memory performance in a healthy adult cohort, highlighting the functional significance of this structural feature [7]. The present study is the first to examine the development of hippocampal dentation in children and adolescents. Given that (1) postnatal change has been documented within the hippocampus and (2) an increase in cortical gyrification has been observed during childhood and adolescence [16], we hypothesized that hippocampal dentation would increase during this developmental period (i.e., birth to 18 years old).

In addition, given that previous research has shown gender differences in the total volume and developmental trajectory of the hippocampus in males and females [5,17,18], and that males had a greater degree of dentation than females in a healthy adult cohort [7], this study also examined whether hippocampal dentation differs between males and females in a younger age group.

2. Materials and Methods

2.1. Data Acquisition (C-MIND)

The data presented in this work were obtained from the Cincinnati MR Imaging of NeuroDevelopment (C-MIND) database, provided by the Pediatric Functional Neuroimaging Research Network at https://research.cchmc.org/c-mind/ (accessed on 16 March 2017). This Network and the resulting C-MIND database were supported by a contract from the Eunice Kennedy Shriver National Institute of Child Health and Human Development (HHSN275200900018C). Participants were native English speakers with no history of neurological or psychiatric disease in the child or first-degree relatives. Exclusionary criteria included the following: gestation outside 37 to 42 weeks, birth weight less than the 10th percentile, body mass index outside the 5th to 95th percentiles, chronic illness, a school grade average of D+ or below, special education placement based on ability or behavior, previous head trauma, abnormal neurological examination, head circumference outside the 5th to 95th percentiles, and abnormal MRI findings. Please see the C-MIND User Manual, available online at https://research.cchmc.org/c-mind/manual-project-overview (accessed on 16 March 2017), for more information on the study protocol and data collection, which were approved by the Institutional Review Board at Cincinnati Children's Hospital Medical Center (CCHMC). Data were used in the current study with permission from the Pediatric Neuroimaging Research Consortium, as well as the Institutional Review Board at the University of Alabama at Birmingham.

2.2. Sample Demographics

Ninety healthy, right-handed children, between birth and 18 years old were randomly selected from the C-MIND database (age range$_{sample}$: 0 years, 9 months to 18 years, 10 months). Scan quality was rated 0–3 by researchers at the C-MIND Consortium (0 = perfect and 3 = poor). For this study, only scans of quality 0 ("perfect data") or 1 ("good image quality and contrast") were considered for inclusion in order to adequately visualize hippocampal dentation (see Section 8: Quality Assurance Procedures of the C-MIND User Manual at https://research.cchmc.org/c-mind/8-quality-assurance-procedures (accessed on 16 March 2017) for more information).

The sample size was based on feasibility and scan quality, with up to five participants selected at each age, from birth to 18 years old. Scan quantity and/or quality were limited in several age groups, including less than one year of age (2 scans available), 17-year-olds (4 scans), and 18-year-olds (4 scans). For all other age groups, random selection was used to identify five participants for inclusion. The sample included 50 females (56%) and 40 males. Of note, the database uses the term "gender" and does not differentiate gender from biological sex, therefore the data and results are presented using the term gender. Parent race was reported as follows: 66% of mothers in the study were Caucasian, 29% were African American, 1% were Asian, 1% were biracial, and 3% were not reported; fathers' race included 62% Caucasian, 32% African American, 1% Asian, and 4% not reported. Participant ethnicity included 6% Hispanic or Latino. Annual household income was reported in ranges ($0–5000 to over $150,000) and median household income was $50,000–75,000 in this sample. For more information on sample household income as well as parental education level, see Tables A1 and A2.

2.3. Neuroimaging

Neuroimaging data included an anatomical MRI scan (acquisition time: approximately 38 min). The protocol included a desensitization period to acclimate participants to the scanning environment, described previously [19,20] and outlined online at https://research.cchmc.org/c-mind/visitors/preparing (accessed on 16 March 2017). A Philips 3T Achieva scanner with 32-channel head coil was used to acquire 3D T1-weighted anatomical MPRAGE scans ($1 \times 1 \times 1$ mm^3). Scans were visualized using 3D Slicer software, Version 4.5.0.

2.4. Dentation Assessment

Hippocampal dentation, present in the CA1/subiculum of the human hippocampus, was assessed in this study by counting the number of hippocampal dentes for the left and right hippocampus of each participant individually (Figures 1–3). Hippocampal dentation was visualized in the sagittal plane, using all available sagittal slices to determine the total number of dentes for each hippocampus. Hippocampal dentation was described previously in adults using ultra high-resolution structural neuroimaging (HR-MICRA technique with $0.5 \times 0.5 \times 0.75$ mm^3 resolution) [7]. While a previous study described hippocampal dentation in terms of both quantity and prominence, in this study, dentation was measured by quantity only, adapted due to the reduced visibility of dentation and the SRLM layer (stratum radiatum, lacunosum, and moleculare) in this scan resolution. See Beattie and colleagues (2017) for more detailed methodology on the previous study. In the current study, right and left hippocampal dentes were counted for all participants by a researcher trained in hippocampal anatomy (J.F.B.). The main analyses were conducted using these results. Approximately 25% of scans (n = 24) were counted by two additional trained researchers (T.A.C. and R.Q.J.) to assess inter-rater reliability. All raters were blinded to participant variables, including age, while counting dentation.

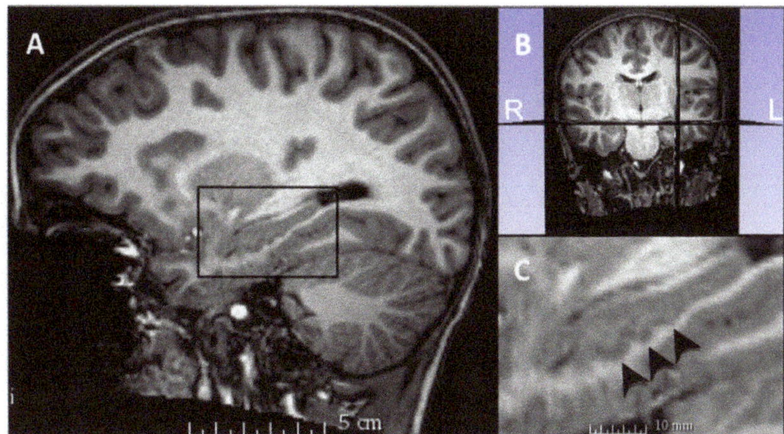

Figure 1. Hippocampal dentation visible in a 7-year-old participant. (**A**) Sagittal view, with the left hippocampus shown in the black box. (**B**) Coronal view, with corresponding sagittal placement of the viewing panes in (**A,C**). A larger image of the hippocampus is visible in (**C**), with black arrowheads indicating three dentes visible in this sagittal slice. All available sagittal slices were used to count the dentes for each hippocampus.

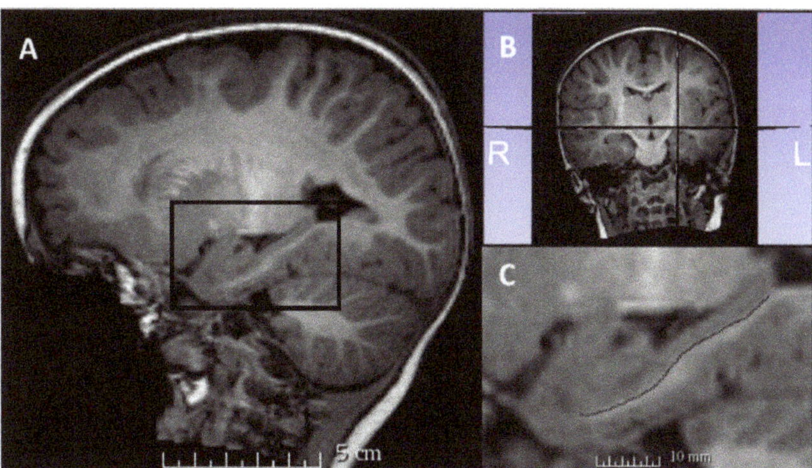

Figure 2. (**A**) Sagittal view of the left hippocampus of a 9-month-old participant. The black box highlights the location of the hippocampus in this view. (**B**) Coronal view indicating the lateral placement of the sagittal plane seen in (**A,C**). (**C**) A more detailed view of the left hippocampus for this participant, with no visible dentes and the smooth contour of the inferior aspect of the hippocampus traced.

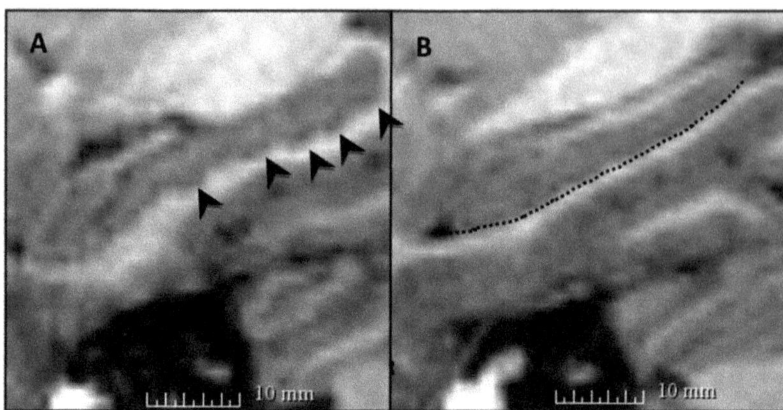

Figure 3. Both (**A**,**B**) show sagittal views of left hippocampi from two different 18-year-old participants. (**A**) A hippocampus with multiple dentes (indicated by black arrowheads), and (**B**) a relatively flat hippocampus, with the inferior surface traced by the black dotted line. This highlights the variability in hippocampal dentation observed among healthy individuals of the same age. Hippocampal dentation is pictured here in a single representative sagittal slice for each participant.

2.5. Experimental Design and Statistical Analysis

Visual inspection of bivariate scatter plots revealed a linear trend between the number of dentes and age. Independent and dependent variables had skewness and kurtosis values less than the absolute value of 2.0. No multivariate outliers were found for the relationships between demographic variables (i.e., age, height, and weight) and left, right, or total number of dentes (Mahalonobis' Distance of less than 15.0 and Cook's Distance less than 1.0). The independence of residuals was observed, based on a Durbin–Watson test statistic close to 2. Normality of residuals was assessed for the relationship between age and number of left, right, and total dentes using the Kolmogorov–Smirnov test for normality. Normality was violated for the relationship between age and number of left dentes, but the null hypothesis of a normal distribution was retained for the relationship between age and number of right dentes and number of total dentes. Visual inspection of bivariate scatter plots of the residual and predicted data points revealed mild heteroscedasticity in the data. Square root and logarithmic transformations were explored; however, homoscedasticity was not achieved with these transformations, therefore, the data were analyzed in their original form to preserve interpretation. Implications were considered. Statistical analysis was performed in SPSS.

There were no missing data for the primary variables of interest, including hippocampal dentation (right, left, and total) and participant age. Birth weight, current height, and current weight were not available for a small number of individuals (see Table 1); however, the number of individuals with missing data points was small and scattered among age groups (<3% of individuals with missing data; see Table 1); therefore, the impact of these missing data was considered minimal [21].

Table 1. Pearson correlations: hippocampal dentation and demographics.

	Number of Hippocampal Dentes		
	Left	**Right**	**Total**
Birth weight (lb) [A]	0.07 ($p = 0.54$)	−0.10 ($p = 0.36$)	−0.02 ($p = 0.89$)
Height (cm) [B]	0.29 ($p = 0.006$) **	0.27 ($p = 0.01$) *	0.31 ($p = 0.004$) **
Weight (kg) [B]	0.21 ($p = 0.052$)	0.13 ($p = 0.23$)	0.19 ($p = 0.08$)

Note: statistics reflect r-value (*p*-value), * denotes significance at $p < 0.05$, ** denotes significance at $p < 0.01$, [A] $n = 89$; 1 unreported birth weight, [B] $n = 87$; 3 unreported height and weight.

Reliability of dentation assessment at this age and scan resolution was measured using an intra-class correlation coefficient ICC (3,1) to assess the absolute agreement for two additional raters (T.A.C. and R.Q.J.) with the dentation counts used in the main analyses (J.F.B.), according to methods reported elsewhere [22]. The analysis included left and right dentes, considered together. Agreement between all three raters was 0.53, with a 95% confidence interval (CI) (0.29, 0.70), and was considered fair. When considered individually, ICCs between rater J.F.B. and raters T.A.C. and R.Q.J. were 0.62 and 0.70, which is considered to be a good reliability [23].

3. Results

A range of dentation was observed for left (range: 0–8, $\bar{x}$= 2.82, s = 1.79), right (range: 0–7, $\bar{x}$ = 2.44, s = 1.69), and total dentes (range: 0–14, $\bar{x}$ = 5.27, s = 3.17). Two-tailed Pearson correlational analyses were conducted to examine the relationship between hippocampal dentation and age in this sample. Analyses revealed a significant positive association between age and number of left (r = 0.34, p = 0.001), right (r = 0.31, p = 0.003), and total dentes (r = 0.36, p = 0.001) (Figures 4 and 5). The means and standard deviations for number of hippocampal dentes by age group are included in Table A3.

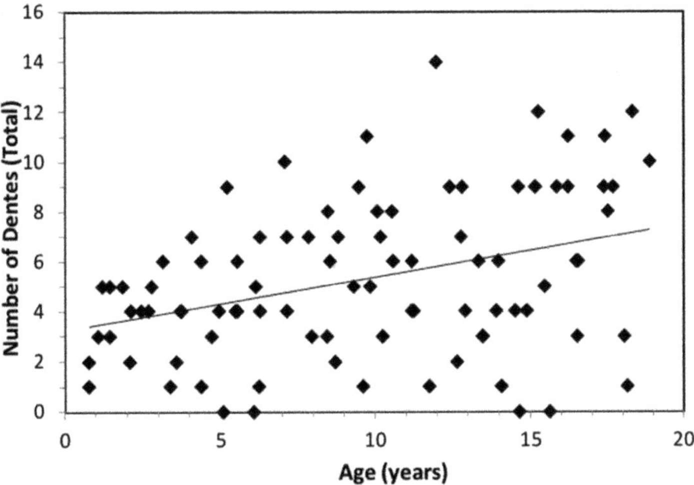

Figure 4. The relationship between total number of hippocampal dentes and age (n = 90), with a linear trend line (p = 0.001). Of note, the variability in number of dentes is high, particularly in older age groups. Individuals with few or no dentes are also seen across the age range.

Visual inspection indicated that variability in dentation appeared to increase with age, such that there was more variability in number of dentes in older children and adolescents when compared with infants and young children. Given the increasing variability in the number of dentes with participant age noted in bivariate scatter plots (Figure 4), an additional analysis was conducted to determine whether there was a significant relationship between variability in hippocampal dentation (i.e., standard deviation) and age group. This analysis, consisting of a two-tailed Pearson correlation, revealed a significant positive association between the age group and standard deviation of the number of total hippocampal dentes (r = 0.56, p = 0.01) (Figure 5).

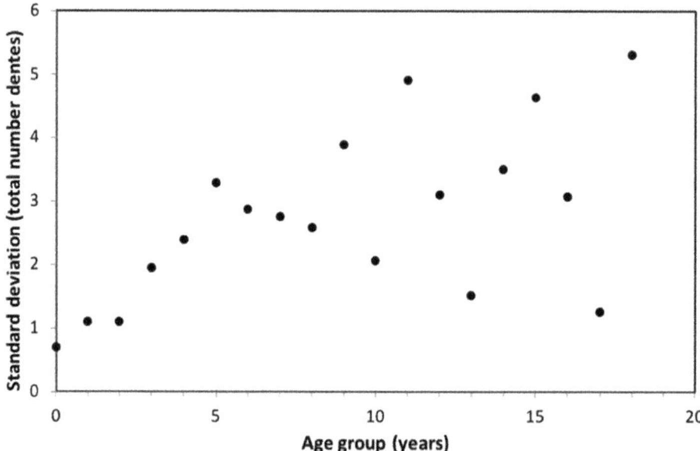

Figure 5. Scatter plot showing the correlation between standard deviation of the number of total hippocampal dentes and age group. The larger standard deviation among older age groups indicates greater variability in the number of hippocampal dentes among older children/adolescents. A significant positive association was found between age group and standard deviation ($p = 0.01$). The range in variability for older age groups may be observed due to the small sample size in each age group.

A paired samples t-test was conducted to determine whether there was a significant difference between the number of left and right hippocampal dentes in the sample. Individuals in this sample tended to have more left than right dentes (difference score (L–R): $\bar{x} = 0.38$, $s = 1.44$), t(89) = 2.49, $p = 0.02$ (Figure 6).

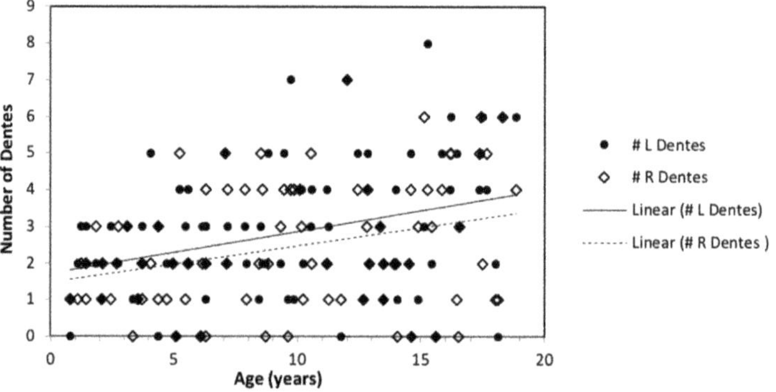

Figure 6. Association between number of left and right dentes individually with age, displayed with linear trend lines. Paired samples t-test revealed a greater number of left dentes for participants, on average.

Two-tailed Pearson correlations were conducted to examine the relationship between the number of hippocampal dentes and demographic variables, including birth weight, current height, and current weight. Analyses revealed significant associations between the number of left, right, and total dentes with height ($p < 0.05$), but not with current weight or birth weight ($p > 0.05$) (Table 1).

Independent samples t-tests were conducted to determine whether hippocampal dentation differed between males and females. No significant differences were observed between groups, including the number of left dentes between males ($\bar{x} = 3.05$, $s = 1.77$) and females ($\bar{x}$ = 2.64, *s* = 1.80); t(88) = 1.08, *p* = 0.28, number of right dentes between males ($\bar{x}$ = 2.75, *s* = 1.79) and females ($\bar{x}$ = 2.20, *s* = 1.58), t(88) = 1.55, *p* = 0.13, or number of total dentes between males ($\bar{x}$ = 5.80, *s* = 3.17) and females ($\bar{x}$ = 4.84, *s* = 3.13) in this sample, t(88) = 1.44, *p* = 0.15 (see Figure 7).

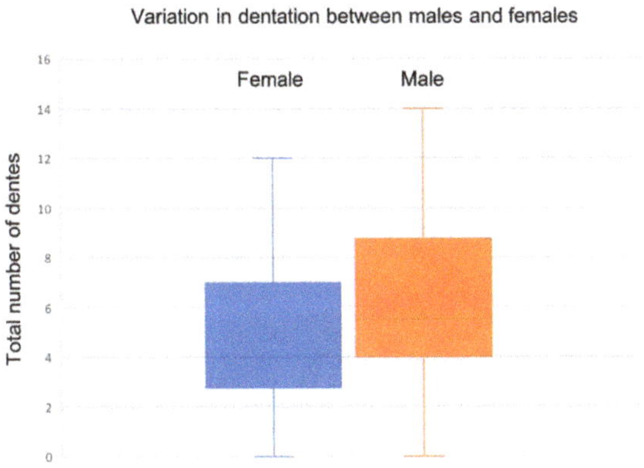

Figure 7. The range of hippocampal dentation among males and females. No significant gender differences were observed (*p* > 0.05).

4. Discussion

4.1. Hippocampal Dentation and Age

Based on a sample of healthy children and adolescents (age 9 months to 18 years old), hippocampal dentation showed a small positive correlation with age, such that older children had a larger number of dentes, on average. This does not appear to directly mirror volumetric changes, which broadly exhibit relative stability following early childhood [17,24]. This differential developmental pattern between hippocampal dentation and volume appears to be generally consistent with past research in adults, which found no association between hippocampal dentation and hippocampal volume [7]. Overall, hippocampal dentation describes shape of the hippocampus rather than the size, and does not appear to be a redundant or surrogate measure of volume. In addition, this finding of a greater number of hippocampal dentes with age introduces a potential form of morphological change that occurs during childhood and adolescence, previously believed to be minimal in the human hippocampus [1]; however, conclusions here are limited by a cross-sectional study design. Given the functional significance of hippocampal dentation in adults, it is important to describe the typical developmental trajectory of dentation—that is, describing when dentation first develops and whether this continues throughout the lifespan. At the same time, there seems to be a significant degree of individual variability, such that not all adults show visible dentation.

4.2. Variability of Hippocampal Dentation by Age

While primary analyses revealed a significant positive association between age and number of hippocampal dentes, visual inspection of the data (see Figure 4) indicated that while younger children tend to have fewer hippocampal dentes, older children and adolescents display wide variability in the number of hippocampal dentes. This was supported by a post-hoc analysis, which revealed a positive association between age group and variability in hippocampal dentation (i.e., standard deviation). The observed degree of variability found in our current study is consistent with a previous finding in a healthy adult cohort, where a wide range of hippocampal dentation was observed [7]. In a study examining volume, substantial variability in hippocampal volume was observed in healthy children aged 4–18 years old; however, the results did not indicate whether the distribution of hippocampal volume varied with age [17].

4.3. Hippocampal Dentation and Gender

Previous research varies in both methodology and reporting when analyzing sex and/or gender differences [5,17,18]. No gender differences were observed between males and females in the right, left, or total number of hippocampal dentes in this sample. Gender differences were previously observed in healthy adults, with males having a greater degree of right and left hippocampal dentation [7]. Observed differences in the two studies could be attributable to age, with gender differences emerging during adulthood. Alternatively, both prominence and quantity were examined in the previous study, thus it is possible that males have a tendency to have more arciform (prominent) dentes, where females have a higher rate of sinusoidal dentes, leading researchers to detect gender differences using one methodology (i.e., the rating system in the previous study) and not the other (i.e., counting dentes used in the present study). More research is needed to elucidate sex and gender differences in hippocampal dentation across the lifespan.

4.4. Hippocampal Dentation and Demographics

In addition to showing a positive relationship with age, the quantity of hippocampal dentes (right, left, and total) was associated with height in this sample. This is not surprising given the strong correlation between age and height in children. No significant correlation was found with weight (current nor birth weight).

Interestingly, prematurity and extremely low birth weight have been associated with memory difficulties during childhood [25,26], which may persist into adulthood [27], justifying an investigation of hippocampal dentation related to birth weight or gestational age. In our study sample, there was no relationship between birth weight and hippocampal dentation; however, due to the study aims of investigating typical development, children with birth weight less than the 10th percentile, as well as those born prematurely (<37 weeks), were excluded from the study. The resulting data were therefore truncated in these parameters. This could lead to limited power to detect an effect related to hippocampal dentation. Future research may examine this relationship in a sample of individuals who were born prematurely and/or at a very low or extremely low birth weight.

4.5. Reliability of Dentation Assessment

Reliability analysis was conducted to determine the consistency of counting dentes at this resolution and in this age range. In a previous study with healthy adults [7], reliability was excellent, ICC (3,1) = 0.94, 95% CI (0.89–0.96). Based on non-overlapping confidence intervals, we can deduce that the ICC in the previous adult study was higher than the current study. While the reliability in this study was good, the ICC values may be relatively lower when compared to the previous study due to the decreased scan resolution and contrast, as well as the younger age range. Given that dentation is associated with age, it is possible that dentation "emerges" during childhood. That is, the prominence of dentes may increase over time (although the current study did not address this hypothesis), making it difficult to distinguish when to "count" a dente (i.e., a threshold). This could contribute to less absolute agreement between raters when counting hippocampal dentation. Additionally, a fixed voxel size appears relatively larger compared to overall hippocampal size in a smaller brain (e.g., infants), making visualization of hippocampal dentation more difficult, assuming the size of the dentes is proportional to the overall hippocampal size. Finally, the methodology differed between the two studies (counting dentes vs. rating quantity and prominence). Reliability in this study was classified as "fair" to "good" and was considered acceptable for measuring this construct; however, the lower reliability in this sample provided support for developing more objective techniques to measure hippocampal dentation.

4.6. Limitations

The limitations of this study included a lower scan resolution compared to previous work examining hippocampal dentation. This presented methodological challenges; however, this scan resolution was determined to be acceptable for viewing the quantity of hippocampal dentes and provided significantly more generalizability for studying hippocampal dentation. Scan acquisition time for an ultra-high resolution neuroimaging session may present substantial barriers, particularly in studying a population of infants and young children. Examining dentation in this image resolution also allows for more readily available clinical use and the interpretation of this construct, in addition to larger sample sizes in research.

Although this study had a relatively large sample size and was well-powered to detect a main effect of age, this sample included a relatively small number of individuals in each age group, particularly in very young children (i.e., less than 12 months of age). Due to rapid brain development during the first years of life, it would be interesting to study additional individuals in this age range so as to better describe characteristics of hippocampal dentation in the early years of life. Sample diversity was generally representative of the region; however, increasing the sample diversity would help detect potential differences in development across groups. The cross-sectional design of this study limits conclusions about developmental trajectory over time.

Participant gender was one of the demographic variables collected in the data set utilized in this study, but was not differentiated from biological sex. As sex (biological descriptive term) and gender (socially-defined construct) are not synonymous, we were unable to explore the nuances of these relationships as part of this study. While we saw no significant gender differences in hippocampal dentation, we cannot definitively rule out sex differences within this sample. This should continue to be explored in future research.

4.7. Future Directions

Future work should examine the development of dentation in longitudinal samples to better assess stability and/or development of this construct over time within individuals. Furthermore, the quantity of hippocampal dentation was related to age in children and adolescents, but not in adulthood, when focusing on early to middle adulthood [7]. It would be interesting to examine this relationship in older adults to determine whether loss of hippocampal dentation occurs during healthy aging, or whether this construct is stable after initial development early in life.

Thus far, hippocampal dentation has been examined primarily in healthy populations to describe the typical development and course of this morphological feature. Important future directions include examining this construct in individuals with neurological conditions, including those where memory impairment is pronounced (e.g., temporal lobe epilepsy, Alzheimer's disease), such as the work recently established by Ramaniharan and colleagues [10]. Additionally, the examination of this feature in individuals with neurodevelopmental differences may shed light on whether this feature plays a role in neurodevelopmental disorders.

Future work should also explore genetic and/or environmental contributors to the development of dentation, including factors such as family socioeconomic status, parent educational attainment, and variables exploring quality of education, as well as parental dentation.

Finally, future directions include the development of a more objective and less time-intensive method to assess hippocampal dentation. Ideally, this measure would incorporate both quantity and prominence, and would be available for use in a relatively accessible scan resolution (e.g., MPRAGE). Recent work towards this effort has been reported [14]. The lower ICCs in this study, while still acceptable, suggests that objective methods would be particularly beneficial for infants and young children, where hippocampal dentation could be more subtle and voxel size appears to be relatively larger.

5. Conclusions

This study provides the first examination of hippocampal dentation in children and adolescents, to the best of our knowledge. The aim was to describe the development of hippocampal dentation in this age range. The results revealed that there is a significant positive association between age and number of hippocampal dentes (left, right, and total). Hippocampal dentation was not related to gender in this sample. Individuals had slightly more left than right dentes. Interestingly, variability in the number of dentes increased significantly with age, such that older children and adolescents displayed a wide range in number of hippocampal dentes, ranging from absent to numerous in both the left and right hippocampi. These results are consistent with a description of hippocampal dentation in a sample of healthy adults, which indicated wide variability in the degree of hippocampal dentation.

Author Contributions: Conceptualization, J.F.B., R.C.M., M.D.T. and R.K.K.; methodology, J.F.B., R.C.M., E.W.C.III, M.D.T. and L.W.V.H.; validation, J.F.B., R.Q.J., T.A.C. and L.W.V.H.; formal analysis, J.F.B., R.C.M., E.W.C.III and L.W.V.H.; investigation, J.F.B., R.C.M. and L.W.V.H.; resources, R.C.M. and L.W.V.H.; data curation, J.F.B., R.Q.J. and T.A.C.; writing—original draft preparation, J.F.B., R.C.M. and L.W.V.H.; writing—review and editing, J.F.B., R.C.M., M.D.T., R.K.K., A.K.R. and L.W.V.H.; visualization, J.F.B., A.K.R. and L.W.V.H.; supervision, R.C.M. and L.W.V.H.; project administration, J.F.B.; funding acquisition, J.F.B., R.C.M. and L.W.V.H. All authors have read and agreed to the published version of the manuscript.

Funding: This research was supported by the Civitan Foundation through a Civitan Emerging Scholars research grant, awarded to the first author.

Institutional Review Board Statement: The study was conducted in accordance with the Declaration of Helsinki, and was approved by the Institutional Review Board of the University of Alabama at Birmingham (protocol number E160205010 (exempt) and date IRB designation issued: 3/24/16).

Informed Consent Statement: Informed consent was obtained from all subjects involved in the study. Informed consent was conducted at the original data collection site.

Data Availability Statement: The CincinnatiMR Imaging of NeuroDevelopment (C-MIND) database is publicly accessible online at http://research.cchmc.org/c-mind (accessed on 16 March 2017).

Acknowledgments: Special thanks to Jerzy Szaflarski, M.D., Ph.D., and Jane Allendorfer, Ph.D., for their consultation and support. We would also like to thank the C-MIND Authorship Consortium at Cincinnati Children's Hospital Medical Center (CCHMC) for access to this data set and support in obtaining this data for our study.

Conflicts of Interest: The authors declare no conflict of interest. The funders had no role in the design of the study; in the collection, analyses, or interpretation of data; in the writing of the manuscript; or in the decision to publish the results.

Appendix A

Table A1. Parents' educational attainment.

	Mothers	Fathers
Less than high school	4% ($n = 4$)	7% ($n = 6$)
High school	17% ($n = 15$)	23% ($n = 21$)
Some college	19% ($n = 17$)	24% ($n = 22$)
College	30% ($n = 27$)	27% ($n = 24$)
Some graduate level	6% ($n = 5$)	($n = 0$)
Graduate level	21% ($n = 19$)	13% ($n = 12$)
Not reported	3% ($n = 3$)	6% ($n = 5$)

Table A2. Annual household income.

	Percent of Sample
$0–$5000	12%
$5000–$10,000	1%
$10,000–$15,000	6%
$15,000–$25,000	8%
$25,000–$35,000	8%
$35,000–$50,000	10%
$50,000–$75,000	14%
$75,000–$100,000	21%
$100,000–$150,000	11%
Over $150,000	4%
Not reported	4%

Table A3. Descriptive statistics: hippocampal dentation by age group.

Age Group (Years)	Frequency	Number of Dentes		
		Left	Right	Total
<1	2	0.50 (0.71)	1.00 (0.00)	1.50 (0.71)
1	5	2.40 (0.55)	1.80 (0.84)	4.20 (1.10)
2	5	2.00 (0.71)	1.80 (0.84)	3.80 (1.10)
3	5	2.00 (1.00)	1.40 (1.14)	3.40 (1.95)
4	5	2.40 (1.82)	1.80 (0.84)	4.20 (2.39)
5	5	2.60 (1.67)	2.00 (1.87)	4.60 (3.29)
6	5	1.80 (1.30)	1.60 (1.67)	3.40 (2.88)
7	5	3.00 (1.22)	3.20 (1.64)	6.20 (2.77)
8	5	2.60 (1.52)	2.60 (1.95)	5.20 (2.59)
9	5	3.20 (2.68)	3.00 (1.73)	6.20 (3.90)
10	5	3.40 (0.89)	3.00 (1.58)	6.40 (2.07)
11	5	3.20 (2.59)	2.60 (2.51)	5.80 (4.92)
12	5	3.40 (1.82)	2.80 (1.30)	6.20 (3.11)
13	5	2.40 (1.14)	2.00 (0.71)	4.40 (1.52)
14	5	1.80 (1.92)	1.80 (1.79)	3.60 (3.51)
15	5	3.60 (3.05)	3.40 (2.19)	7.00 (4.64)
16	5	4.20 (1.30)	2.80 (2.28)	7.00 (3.08)
17	4	4.75 (0.96)	4.50 (1.73)	9.25 (1.26)
18	4	3.50 (3.00)	3.00 (2.45)	6.50 (5.32)

References

1. Insausti, R.; Amaral, D.G. Hippocampal formation. *Hum. Nerv. Syst.* **2004**, *2*, 871–914.
2. Ver Hoef, L.; Deshpande, H.; Cure, J.; Selladurai, G.; Beattie, J.; Kennedy, R.E.; Knowlton, R.C.; Szaflarski, J.P. Clear and Consistent Imaging of Hippocampal Internal Architecture with High Resolution Multiple Image Co-registration and Averaging (HR-MICRA). *Front. Neurosci.* **2021**, *15*, 546312. [CrossRef]
3. Benes, F.M.; Turtle, M.; Khan, Y.; Farol, P. Myelination of a key relay zone in the hippocampal formation occurs in the human brain during childhood, adolescence, and adulthood. *Arch. Gen. Psychiatry* **1994**, *51*, 477–484. [CrossRef]
4. Knickmeyer, R.C.; Gouttard, S.; Kang, C.; Evans, D.; Wilber, K.; Smith, J.K.; Hamer, R.M.; Lin, W.; Gerig, G.; Gilmore, J.H. A structural MRI study of human brain development from birth to 2 years. *J. Neurosci.* **2008**, *28*, 12176–12182. [CrossRef]
5. Uematsu, A.; Matsui, M.; Tanaka, C.; Takahashi, T.; Noguchi, K.; Suzuki, M.; Nishijo, H. Developmental trajectories of amygdala and hippocampus from infancy to early adulthood in healthy individuals. *PLoS ONE* **2012**, *7*, e46970. [CrossRef]
6. Ostby, Y.; Tamnes, C.K.; Fjell, A.M.; Westlye, L.T.; Due-Tonnessen, P.; Walhovd, K.B. Heterogeneity in subcortical brain development: A structural magnetic resonance imaging study of brain maturation from 8 to 30 years. *J. Neurosci.* **2009**, *29*, 11772–11782. [CrossRef]
7. Beattie, J.F.; Martin, R.C.; Kana, R.K.; Deshpande, H.; Lee, S.; Curé, J.; Ver Hoef, L. Hippocampal dentation: Structural variation and its association with episodic memory in healthy adults. *Neuropsychologia* **2017**, *101*, 65–75. [CrossRef]
8. Simic, G.; Kostovic, I.; Winblad, B.; Bogdanovic, N. Volume and number of neurons of the human hippocampal formation in normal aging and Alzheimer's disease. *J. Comp. Neurol.* **1997**, *379*, 482–494. [CrossRef]

9. Ramaniharan, A.K.; Zhang, M.; Martin, R.C.; Parpura, V.; Selladurai, G.; Ver Hoef, L. An objective method to quantify hippocampal dentation and predict the side of seizure onset in temporal lobe epilepsy. In Proceedings of the 2021 IEEE Signal Processing in Medicine and Biology Symposium (SPMB), Philadelphia, PA, USA, 4 December 2021.
10. Kilpattu Ramaniharan, A.; Zhang, M.W.; Selladurai, G.; Martin, R.; Ver Hoef, L. Loss of hippocampal dentation in hippocampal sclerosis and its relationship to memory dysfunction. *Epilepsia* **2022**. [CrossRef]
11. Duvernoy, H.M. *The Human Hippocampus: Functional Anatomy, Vascularization and Serial Sections with MRI*; Springer Science & Business Media: Berlin/Heidelberg, Germany, 2005.
12. Federau, C.; Gallichan, D. Motion-Correction Enabled Ultra-High Resolution In-Vivo 7T-MRI of the Brain. *PLoS ONE* **2016**, *11*, e0154974. [CrossRef]
13. Isaacson, R. *The Limbic System*; Springer Science & Business Media: Berlin/Heidelberg, Germany, 2013.
14. Chang, C.; Huang, C.; Zhou, N.; Li, S.X.; Ver Hoef, L.; Gao, Y. The bumps under the hippocampus. *Hum. Brain Mapp.* **2018**, *39*, 472–490. [CrossRef]
15. Ding, S.L.; Van Hoesen, G.W. Organization and Detailed Parcellation of Human Hippocampal Head and Body Regions Based on a Combined Analysis of Cyto- and Chemoarchitecture. *J. Comp. Neurol.* **2015**, *523*, 2233–2253. [CrossRef]
16. White, T.; Su, S.; Schmidt, M.; Kao, C.-Y.; Sapiro, G. The development of gyrification in childhood and adolescence. *Brain Cogn.* **2010**, *72*, 36–45. [CrossRef]
17. Giedd, J.N.; Snell, J.W.; Lange, N.; Rajapakse, J.C.; Casey, B.J.; Kozuch, P.L.; Vaituzis, A.C.; Vauss, Y.C.; Hamburger, S.D.; Kaysen, D.; et al. Quantitative magnetic resonance imaging of human brain development: Ages 4–18. *Cereb. Cortex* **1996**, *6*, 551–560. [CrossRef]
18. Suzuki, M.; Hagino, H.; Nohara, S.; Zhou, S.Y.; Kawasaki, Y.; Takahashi, T.; Matsui, M.; Seto, H.; Ono, T.; Kurachi, M. Male-specific volume expansion of the human hippocampus during adolescence. *Cereb. Cortex* **2005**, *15*, 187–193. [CrossRef]
19. Byars, A.W.; Holland, S.K.; Strawsburg, R.H.; Bommer, W.; Dunn, R.S.; Schmithorst, V.J.; Plante, E. Practical aspects of conducting large-scale functional magnetic resonance imaging studies in children. *J. Child Neurol.* **2002**, *17*, 885–890. [CrossRef]
20. Vannest, J.; Rajagopal, A.; Cicchino, N.D.; Franks-Henry, J.; Simpson, S.M.; Lee, G.; Altaye, M.; Sroka, C.; Holland, S.K.; the CMIND Authorship Consortium. Factors determining success of awake and asleep magnetic resonance imaging scans in nonsedated children. *Neuropediatrics* **2014**, *45*, 370–377. [CrossRef]
21. Bennett, D.A. How can I deal with missing data in my study? *Aust. N. Z. J. Public Health* **2001**, *25*, 464–469. [CrossRef]
22. Shrout, P.E.; Fleiss, J.L. Intraclass correlations: Uses in assessing rater reliability. *Psychol. Bull.* **1979**, *86*, 420–428. [CrossRef]
23. Cicchetti, D.V.; Sparrow, S.A. Developing criteria for establishing interrater reliability of specific items: Applications to assessment of adaptive behavior. *Am. J. Ment. Defic.* **1981**, *86*, 127–137.
24. Yurgelun-Todd, D.A.; Killgore, W.D.; Cintron, C.B. Cognitive correlates of medial temporal lobe development across adolescence: A magnetic resonance imaging study. *Percept. Mot. Ski.* **2003**, *96*, 3–17. [CrossRef]
25. Isaacs, E.B.; Lucas, A.; Chong, W.K.; Wood, S.J.; Johnson, C.L.; Marshall, C.; Vargha-Khadem, F.; Gadian, D.G. Hippocampal volume and everyday memory in children of very low birth weight. *Pediatr. Res.* **2000**, *47*, 713–720. [CrossRef]
26. Molloy, C.S.; Wilson-Ching, M.; Doyle, L.W.; Anderson, V.A.; Anderson, P.J.; for the Victorian Infant Collaborative Study Group. Visual memory and learning in extremely low-birth-weight/extremely preterm adolescents compared with controls: A geographic study. *J. Pediatr. Psychol.* **2014**, *39*, 316–331. [CrossRef]
27. Aanes, S.; Bjuland, K.J.; Skranes, J.; Lohaugen, G.C. Memory function and hippocampal volumes in preterm born very-low-birth-weight (VLBW) young adults. *Neuroimage* **2015**, *105*, 76–83. [CrossRef]

Article

Diagnostic and Prognostic Value of Quantitative Computed Tomography Parameters of Adrenal Glands in Patients from Internistled ICU with Sepsis and Septic Shock

Moritz Milberg [1,*], Alida Kindt [2], Lisa Luft [1], Ursula Hoffmann [3], Michael Behnes [3], Stefan O. Schoenberg [1] and Sonja Janssen [1]

[1] Clinic of Radiology and Nuclear Medicine, Medical Faculty Mannheim, Heidelberg University, 68167 Mannheim, Germany; lisa.luft@freenet.de (L.L.); stefan.schoenberg@umm.de (S.O.S.); sonja.janssen@medma.uni-heidelberg.de (S.J.)

[2] Division of Analytical Biosciences, Leiden Academic Centre for Drug Research (LACDR), Leiden University, 2333 CC Leiden, The Netherlands; a.s.d.kindt@lacdr.leidenuniv.nl

[3] First Department of Medicine-Cardiology, University Medical Centre Mannheim, 68167 Mannheim, Germany; ursula.hoffmann@umm.de (U.H.); michael.behnes@medma.uni-heidelberg.de (M.B.)

* Correspondence: moritz.milberg@uni-heidelberg.de; Tel.: +49-176-2042-0003

Abstract: The aim was to prospectively evaluate the diagnostic and prognostic value of different quantitative analysis methods assessing adrenal gland parameters on contrast-enhanced CT scans in patients with septic conditions. Seventy–six patients (49 men, 27 women) received CT scans for focus search. Adrenal glands were analyzed by means of three different methods: subjective region of interest (ROI) measurement, organ segmentation and histogram analysis using semi-automated software. Univariate analyses with multiple testing thresholds and receiver operating characteristic curves were performed. Clinical endpoints were 8-days, 28-days and 6-months mortality. Forty-four CT scans were analyzed (ground truth: patients with no sepsis: $n = 6$; patients with sepsis: $n = 15$; patients in septic shock: $n = 21$). Left adrenal gland (LAG) values were analyzed and compared, as data variation was lower than in the right adrenal glands. In patients with septic conditions, the combination of high LAG and Inferior Vena Cava (IVC) density values was highly specific for septic shock with all three methods. Only segmentation values were significantly different between the sepsis and septic shock groups after confounder correction ($p = 0.048$). Total adrenal gland volume was 20% higher in the septic shock patients while a relatively small LAG volume within the septic shock subgroup was associated with higher mortality at day 8 (AUC = 0.8; $p = 0.006$) and at 6 months (AUC = 0.7; $p = 0.035$). However, time-consuming density analysis methods assessing adrenal glands do not provide additional diagnostic value in patients with septic conditions. The combination of high LAG and IVC attenuation values seems to be highly specific for septic shock, regardless of the analysis type. Adrenal gland volume reveals short- and long-term prognostic capacity.

Keywords: sepsis; adrenal glands; computed tomography; ROI; semi-automated segmentation; histogram

Citation: Milberg, M.; Kindt, A.; Luft, L.; Hoffmann, U.; Behnes, M.; Schoenberg, S.O.; Janssen, S. Diagnostic and Prognostic Value of Quantitative Computed Tomography Parameters of Adrenal Glands in Patients from Internistled ICU with Sepsis and Septic Shock. *Anatomia* 2022, 1, 14–32. https://doi.org/10.3390/anatomia1010003

Academic Editor: Eugenio Gaudio

Received: 6 February 2022
Accepted: 4 March 2022
Published: 9 March 2022

Publisher's Note: MDPI stays neutral with regard to jurisdictional claims in published maps and institutional affiliations.

1. Introduction

According to the German Centre for Sepsis Control and Care, approximately 154,000 patients are in-patients due to septic conditions throughout Germany every year, with a mortality as high as 50% and sometimes more [1,2]. Sepsis is—even today—a leading cause of critical illness and hospital mortality and remains a major challenge for modern medicine. Responsible for every third death in Germany, it is the main cause of death in the non-coronary intensive care unit (ICU). In addition, expenses for ICU patients represent about 30% of the overall ICU costs in Germany and amount to approximately 1.77 billion euros. Indirect economical costs are estimated at about 3.5 times as high as direct expenses, caused by work stoppage or early retirement and overall costs are estimated to be around 6.3 billion euros annually [1,3–6].

In 2016, as a result of the release of the *Third International Consensus Definitions for Sepsis and Septic Shock*, new sepsis definitions ("Sepsis 3") were issued by a task force formed by members of the Society of Critical Care Medicine and of the European Society of Intensive Care Medicine [5].

Now, sepsis is defined as a life-threatening organ dysfunction caused by a dysregulated host response to infection. The term also comprises physiological, pathological and biochemical abnormities induced by infection. Organ dysfunction can be represented by an increase in the Sequential (Sepsis-related) Organ Failure Assessment (SOFA) Score of two or more points, which reflects an in-hospital mortality risk of greater than 10%. This score, designed to monitor patients' status and disease development during the stay in ICU, includes six different sub-scores: One for each the respiratory, cardiovascular, hepatic, coagulation, renal and neurological systems.

Septic shock is defined as sepsis in combination with cardiovascular failure resulting in an inadequate tissue perfusion and metabolic dysfunction. Furthermore, patients with septic shock are required to meet the following Sepsis-3 criteria: need of vasopressors to maintain a mean arterial pressure of 65 mmHg or higher and serum lactate level greater than 2 mmol/L (>18 mg/dL) in the absence of hypovolemia [5,7].

Although our understanding of the pathophysiological mechanisms of sepsis has evolved over recent decades, with new therapy regimes being implemented, such as the approach of early goal-directed therapy, timely administration of antibiotics or use of modern ventilation systems, mortality rates remain high [8].

For ICU patients with a suspected septic condition, contrast-enhanced computed tomography (CT) is a fast, cost-efficient, non-invasive and widely available imaging procedure, mostly applied for detecting a possible focus of the infection as well as assessing the patients' clinical course. In this context, the hyperenhancement of adrenal glands is known as a CT sign associated with the "hypoperfusion complex", as described by O' Hara et al. [9] in children with post-traumatic shock, indicating poor prognosis. For adults with intense adrenal enhancement, previous studies indicate poor clinical outcome and high mortality rates in polytraumatized patients and in patients with hypovolemic or septic shock. Additionally, adrenal enhancement may serve as a predictor for organ failure [10,11]. Another recent study by Peng et al. [12] described a new special enhancing pattern (HAGS) of the adrenal glands on dual-phase contrast-enhanced CT in the arterial phase in patients with septic shock, indicating poor prognosis.

Relative adrenal insufficiency (RAI) regularly occurs in patients with septic conditions and describes the body's incapacity to provide the required hormonal response to infection, although the adrenal glands function at their full capacity [12]. The fast response of these organs to stress during septic conditions is vital for survival. The response is mediated by the activation of the hypothalamic-pituitary-adrenal axis (HPA), preventing an over-activation of the immune response [13,14]. The HPA is initiated by cytokines and other inflammatory mediators that trigger the activation of the corticotropin releasing hormone (CRH) from the hypothalamus and leads to the subsequent release of the adrenocorticotropic hormone (ACTH) from the pituitary gland during infection [15]. ACTH then binds to the melanocortin-2 receptor on cells of the adrenal cortex, mediating synthesis of steroids. This feedforward mechanism of cortisol genesis is balanced through a negative feedback interaction by cortisol with ACTH and CRH, inhibiting their release [16,17]. The higher metabolic demand for cortisol and the compensatory mechanism for hypovolemia in patients with septic shock leads to an increased blood flow to the adrenal glands and to their subsequent enlargement, which is seen radiologically as intense adrenal enhancement in contrast-enhanced computed tomography in patients with sepsis and septic shock [9,18,19].

In this study we want to leave behind the approach of qualitative and semiquantitative evaluation of hyperattenuating adrenal glands. Our aim is to assess quantitative data and derive cut-off values in order to distinguish sepsis stages and predict patients' outcomes.

With these endpoints in mind, the authors compared three more or less time-consuming methods of adrenal gland analysis. First, the fast and subjective method of ROI, second, the more elaborate semiautomated quantification of adrenal gland volume and, third, organ segmentation with subsequent histogram analysis. Previous studies have used histogram analysis for adrenal mass characterization, however, to the best of our knowledge, it has never been used for adrenal glands in patients with sepsis and septic shock [20,21].

2. Materials and Methods

2.1. Study Protocol

This prospective single-center study conducted at the University Medical Centre Mannheim (UMM), Germany, was approved by the local ethics committee. A total of seventy-six patients were prospectively enrolled. Informed consent was obtained from all participating patients or their legal representatives. This study was performed according to standards of the Health Insurance Portability and Accountability Act (HIPAA) and the Declaration of Helsinki. Criteria for patients to be included into the study were non-surgical admission to an internist-led ICU and a suspected septic condition as well as performance of a whole-body CT scan for focus search, with i.v. contrast media application via the upper extremities or the internal jugular veins. From all the patients included in the study, venous blood samples for testing parameters known to be associated with a septic condition were collected within 24 h after they had received the contrast-enhanced CT scan. In light of the complete clinical course, the laboratory results and the radiological findings, a baseline could be established, dividing patients into the different study groups at the time-point of CT imaging. Outcomes were determined on day 8, after 28 days and after 6 months.

2.2. CT–Data Acquisition

A clinically indicated contrast-enhanced CT scan was performed on all patients participating in the study and the day of the CT-scan was determined as day 1 regarding outcome analysis. CT data was acquired on a 16-slice single-source CT scanner system (Somatom Emotion, Siemens Healthineers, Erlangen, Germany). In an average-weighted patient, 90 mL of iodinated contrast material was applied with a flow rate of 2.5 mL/s followed by a saline chaser bolus of 20 mL of the same flow rate. The scan protocol consisted of a chest scan acquired during the arterial phase, optionally a neck scan acquired 40 s after the start of contrast material application and an abdominal scan acquired during the portal-venous phase 70 s post contrast material application. The start of the scan was triggered by bolus tracking within the aortic arch. The abdominal portal-venous phase was acquired at a peak tube voltage of 130 kVp, and reference tube current of 110 mAs, pitch of 0.95. CT raw data were reconstructed with a slice thickness of 1.5 mm, slice increment of 0.9 and transferred to our PACS system.

2.3. Clinical Parameters

All collected medical patient data from the internal ICU stay were reviewed by inspection of paper files or by inspection of digital stationary ICU patient files within the hospital's management software SAP. Parameters of interest were basic patient data such as age, previous diseases, cause for admission and days spent on the ICU, as well as sepsis-associated-parameters obtained from the day of the contrast-enhanced CT scan. These included basic vital parameters like breathing rate and systolic blood pressure as well as Glasgow Coma Scale values (GCS), necessary for a SOFA score assessment. To determine sepsis stages according to the "new" Sepsis 3 criteria using the SOFA score, namely paO2, FiO2, flow of catecholamine, creatinine, urinary excretion and lactate were used. If possible, APACHE II-, and SAPS II-scores were assessed as well.

2.4. CT–Image Analysis

"Aycan OsiriX" is an open source Digital Imaging and Communications in Medicine (DICOM) archiving and distribution system that was used to assess CT images and import data for further processing [22]. After the portal-venous abdominal CT scan was imported into the local database, the Aycan-Eclipse-Tool was used to determine a region of interest (ROI) with a target size of 0.1 cm^2 (between 0.08 and 0.34 cm^2).

The following parameters were assessed in duplicate measurements: mean and standard deviation of both adrenal glands and the inferior vena cava and aorta at the level of both adrenal glands within the ROI, as well as the area (cm^2) of the ROI itself.

For the purpose of organ segmentation, "The Medical Imaging Interaction Toolkit" (MITK) was used. This is a free open source software system for the development of interactive medical image processing software, that provides image-guided procedures and image analysis with interaction features to correct results from (semi)automated computation, if necessary [23]. DICOM-portal-venous phase abdominal images of patients' whole-body CT scans were imported into the MITK for segmentation of the adrenal glands. Slice thickness was 1.5 mm and CT scans were either reformatted in the axial or coronal plane depending on the best delineation of the left and right adrenal gland.

3D- or 2D segmentation tools with 3D interpolation were used for the segmentation of adrenal glands. Morphological operations such as "dilation", "closing" or "filling holes" were used, if necessary, for refining automatic segmentation as well as manual segmentation depending on the patients' image. In addition, tumors, blood vessels, fat tissue, calcifications or infarcted areas of the gland were manually excluded from segmentation. The following parameters were assessed separately for both adrenal glands: mean density in Hounsfield Units (HU), median, standard deviation, maximum, minimum, number of voxels (n) and volume of voxel (V in mm^3). Furthermore, histogram data were used by using the "copy-to-clipboard" function of the statistics' tool. Negative pixels were eliminated in order to avoid inclusion of surrounding retroperitoneal fat tissue or small adenomas within the gland and consecutively false-low HU mean values. The histogram data provided pixel attenuation (HU) along the x axis versus the frequency of pixels at each attenuation value along the y axis [20].

2.5. Statistical Analysis

Statistical analysis and graphical representation were performed using dedicated software: R (Version 3.6 for mac); SigmaPlot (Version 14.0); Excel (2019, MSO).

For quantitative data derived from CT scans, the median and interquartile ranges (IQR) were calculated for mean attenuation values. Other quantitative data regarding baseline characteristics are presented as mean ± standard error of the mean (SEM). Categorical variables were reported as numbers and percentages. The Pearson correlation for all variables of interest and analysis of variance wase used (with subsequent Bonferroni correction for multiple testing where appropriate) for comparison of numerical data. To identify combinations of markers predictive of death or sepsis classification at various timepoints, elastic net regression for endpoints and sepsis classification was used, chemical markers were log2 transformed and data was divided into training (80%) and testing (20%) sets for all available data points per outcome. Receiver operating characteristic curves were used to illustrate various cut-offs for mortality risks associated with adrenal gland volume. Attenuation values were corrected for the confounders' age, sex, evidence of presence of germs, APACHE-score and SOFA-score. The sepsis groups were compared with the use of the Mann–Whitney U test for numerical data again with subsequent multiple testing Bonferroni correction. Boxplot diagrams and Pearson correlations were used for all radiological data.

3. Results

3.1. Final Cohort

Baseline characteristics are given in Table 1. From 76 patients included from the internist-led ICU, 54 patients could be assigned according to the new sepsis classification. There were 7 patients (4 male, 3 female) in the non-sepsis group, 23 patients (16 male,

7 female) in the sepsis group and 24 patients (12 male, 12 female) in the septic shock group. Twenty-two patients could not be classified due to early transfer or discharge.

Table 1. Baseline characteristics.

	No Sepsis (*n* = 6)	Sepsis (*n* = 16)	Septic Shock (*n* = 22)
Age, years (mean, range)	74 (33 to 89)	67 (19 to 87)	63 (15 to 82)
Gender, *n* (%)			
Male	3 (50)	11 (69)	10 (46)
Female	3 (50)	5 (31)	12 (54)
Site of infection, *n* (%)			
Lung	4 (67)	10 (63)	15 (68)
Abdominal	1 (33)	2 (13)	2
Urinary tract	-	1 (6)	1
Skin	-	-	1
Heart	-	1 (6)	1
Neck	-	1 (6)	-
Blood	-	-	-
Others	-	1 (6)	-
Laboratory values, mean $\pm$ SEM			
White blood cells, 10^9/L	11.0 ± 1.6	11.5 ± 1.7	18.1 ± 3.2
Platelets, 10^9/L	175.2 ± 39.8	175.6 ± 26.4	199.1 ± 30.2
Creatinine, mg/dL	1.40 ± 0.3	2.0 ± 0.4	8.4 ± 6.6
C-reactive protein, mg/L	152.2 ± 30.2	188.4 ± 28.5	153.8 ± 29.8
pCO_2, mmHg	34.1 ± 6.1	44.3 ± 2.3	43.9 ± 2.8
Lactate, mmol/L	1.4 ± 0.2	1.3 ± 0.2	4.5 ± 0.4
Positive blood cultures, *n* (%)	2 (33)	10 (63)	9 (41)
ICU parameters, mean $\pm$ SEM			
ICU days	15 ± 6.5	20 ± 4	13 ± 3
Ventilation days	13 ± 6.4	10 ± 4	11 ± 3
Catecholamine days	11 ± 5.6	11 ± 3	10 ± 3
Antibiotic treatment days	12 ± 4.3	17 ± 4	12 ± 3
Renal replacement therapy days	5 ± 4.1	2 ± 1	3 ± 1
GCS	12 ± 2	7 ± 1	6 ± 1
APACHE II, mean $\pm$ SEM	18 ± 4	22 ± 2	28 ± 1
SAPS II, mean $\pm$ SEM	36 ± 6	40 ± 5	49 ± 3
SOFA score, mean $\pm$ SEM	7 ± 2	9 ± 1	12 ± 1
All-cause mortality, *n* (%)			
8 days			
Death	0 (0)	4 (25)	10 (46)
Survivor	6 (100)	12 (75)	12 (54)
28 days			
Death	2 (33)	6 (38)	14 (64)
Survivor	4 (66)	10 (62)	8 (36)
6 months			
Death	2 (33)	6 (38)	15 (68)
Survivor	4 (66)	10 (62)	7 (32)

SEM: Standard error of the mean; GCS: Glasgow Coma Scale; APACHE II: Acute physiology and chronic health evaluation II; SAPS: Simplified acute physiology score II; SOFA: Sepsis related organ failure assessment.

In our final patient cohort, we were able to perform image analysis with all three methods in a total of 44 patients assigned to the three sepsis 3-classification groups for both adrenal glands. A total of 10 patients had to be excluded a posteriori. Reasons included poor image quality and organ demarcation, partially due to poor circulatory function, malformed glands due to tumors, respiration-induced artifacts, foreign material artifacts, i.v. contrast material administration via lower extremity veins and individual anatomical variations, e.g., of suprarenal blood vessels. As for the method of segmentation, in five patients only the left adrenal gland (LAG) and in two patients only the right adrenal gland (RAG) was successfully segmented, respectively. Figure 1 presents a flowchart of the formation of the final study cohort.

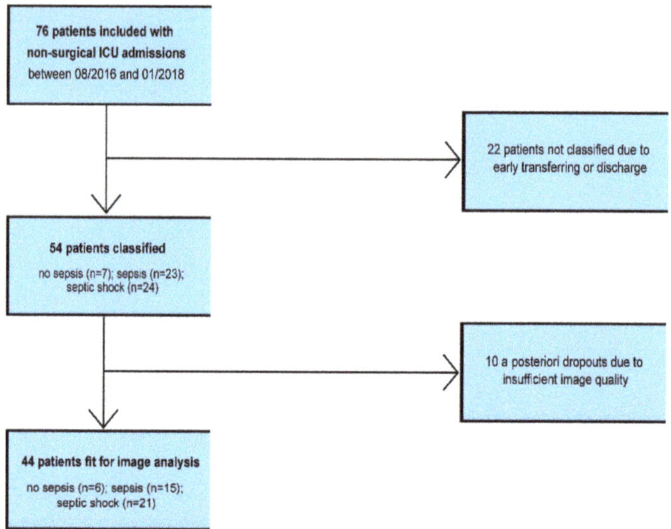

Figure 1. Flowchart of the final study cohort.

3.2. Correlation of HU Attenuation Values for Right and Left Adrenal Gland

Radiological results showed a strong correlation between both adrenal glands for all methods of image analysis. The correlation coefficients (R) for the mean (HU) values for different imaging methods were as follows (that is, for mean (median)): ROI: 0.87; semi-automatic segmentation: 0.88 (0.89); segmentation with histogram analysis: 0.78 (0.77) with all $p < 0.001$. The variation of mean HU values for adrenal glands was higher for RAG (R = 0.59; $p < 0.001$) than for LAG (R = 0.71; $p < 0.001$) for inter-methodical correlation of both ROI and Histogram values, as presented in Figure 2. Due to less data variation, but excellent correlation between both adrenal glands, we then solely focused on the LAG when comparing absolute attenuation values between sepsis groups.

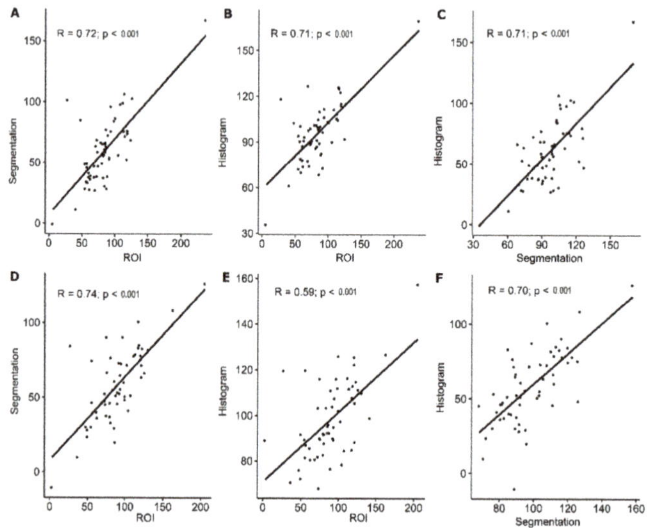

Figure 2. Variation of mean HU values for the left- (**A–C**) and right adrenal gland (**D–F**) comparing inter-methodical correlation of image analysis (HU: Hounsfield unit; ROI: Region of Interest).

3.3. Method of Region of Interest (ROI)

Attenuation values (rounded values) from a ROI (range 0.08 to 0.34 cm^2, average 0.27 cm^2) for the LAG (HU) were as follows (that is, median (IQR)): no sepsis 97 (85 to 111); sepsis 72 (61 to 84); septic shock 91 (77 to 112). The values corrected for the confounders age, sex, evidence of presence of germs, APACHE score and SOFA score were (median (IQR)): no sepsis 24 (10 to 34); sepsis -9 (-17 to 0); septic shock 2 (-15 to 25). Figure 3A illustrates the distribution of absolute mean (HU) values for the LAG according to the different groups of sepsis severity at day 1 of their ICU stay. Between the groups for sepsis (2) and septic shock (3), there was a significant difference with higher attenuation values for patients with septic shock ($p = 0.0020$), which was also significant after multiple testing (ANOVA with subsequent Bonferroni, $p = 0.023$). The differences in between the other groups were not significant ($p > 0.05$).

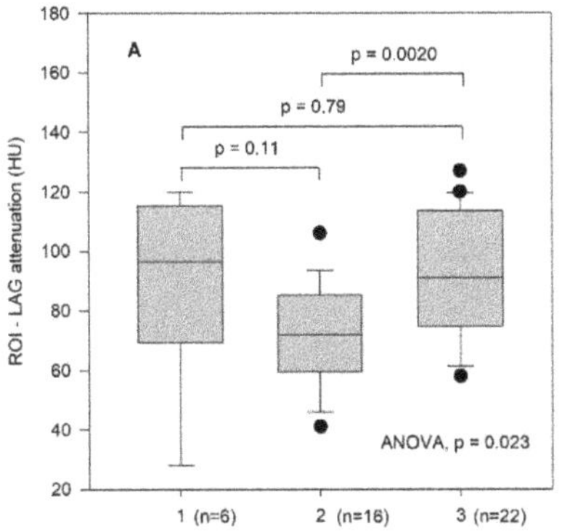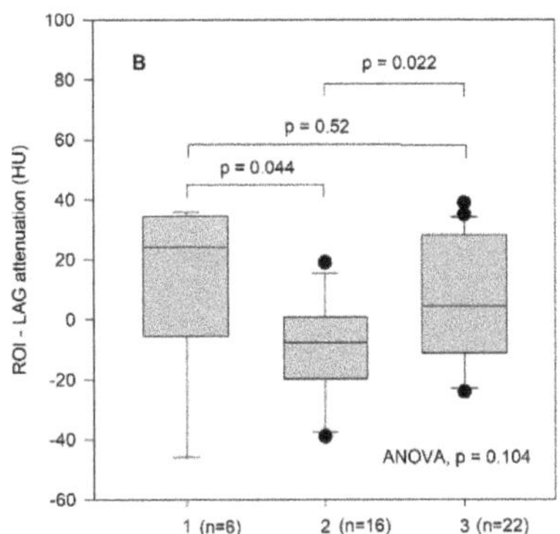

Figure 3. Boxplot diagrams of absolute mean LAG attenuation values (**A**) for the method of ROI for patients not classified with sepsis (1), sepsis (2) and septic shock (3). Mean values were significantly different between sepsis (2) and septic shock (3) ($p = 0.0020$), also significant after multiple testing ($p = 0.023$); Boxplot diagrams for same values corrected for confounders * (**B**), showed differences between groups 2 and 3 ($p = 0.022$) and between groups 1 and 3 ($p = 0.044$), which were no longer significant after multiple testing ($p = 0.104$). (ANOVA: Analysis of Variance; LAG: Left adrenal gland; ROI: Region of Interest; * age, sex, evidence for presence of germs, APACHE and SOFA score).

Figure 3B illustrates the distribution of the corrected ROI (HU) values for the LAG, respectively. There were significant differences between the sepsis group (2) and septic shock group (3) ($p = 0.022$), and between the "no sepsis" and sepsis group (2) ($p = 0.044$), which were no longer significant after multiple testing correction ($p = 0.104$).

3.4. Method of Semi-Automated Organ Segmentation

Attenuation values (rounded) for segmentation with manual adjustments were as follows (that is, median (IQR)): no sepsis 76 (59 to 84); sepsis 49 (32 to 64); septic shock 69 (52 to 81). These values corrected for confounders were (median (IQR)): no sepsis 15 (2 to 26); sepsis -11 (-18 to 3); septic shock 5 (-7 to 16).

Figure 4A illustrates the distribution of absolute segmentation (HU) values for the LAG according to the different groups of sepsis severity at day 1 of their ICU stay. Between the groups for sepsis and septic shock—as demonstrated for the method of ROI—there was a significant difference in attenuation values for patients with septic shock ($p = 0.0035$), which remained significant after multiple testing correction ($p = 0.013$). The significant difference between groups 1 and 2 ($p = 0.036$) was no longer significant after multiple testing correction ($p = 0.074$).

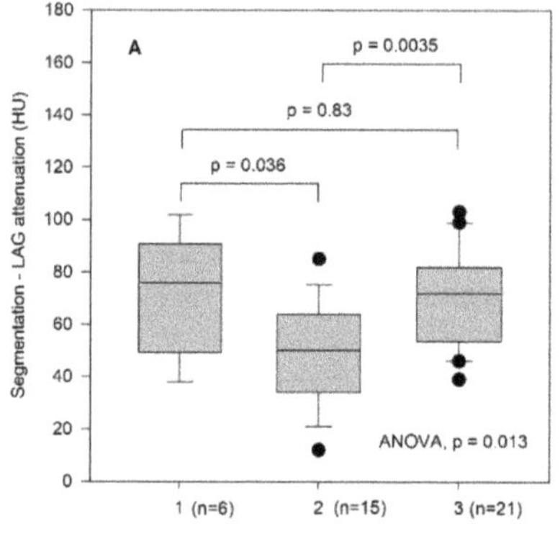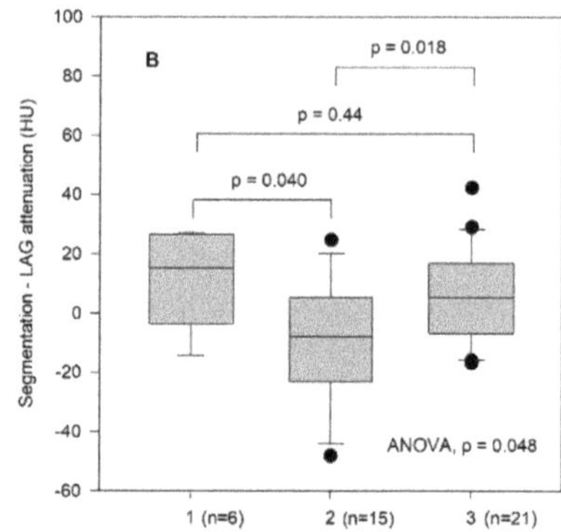

Figure 4. Boxplot diagrams of absolute mean LAG attenuation values (**A**) for the method of segmentation for patients not classified with sepsis (1), sepsis (2) and septic shock (3). Significant difference between sepsis (2) and septic shock (3) ($p = 0.0035$), also significant after multiple testing (ANOVA, $p = 0.013$). Same mean attenuation values corrected for confounders * (**B**). The difference between sepsis (2) and septic shock (3) remained significant ($p = 0.018$) after multiple testing (ANOVA, $p = 0.048$). (ANOVA: Analysis of Variance; LAG: Left adrenal gland; ROI: Region of Interest; * age, sex, evidence for presence of germs, APACHE and SOFA score).

Figure 4B illustrates the distribution of the corrected segmentation HU values, respectively. After correcting for confounders, the difference between mean attenuation values for the sepsis (2) and septic shock group (3) remained significant ($p = 0.018$), also after multiple testing correction ($p = 0.048$).

Adrenal segmentation volumes for both genders are presented in Table 2. Total adrenal gland volume on average increased around 20%, both between the "no sepsis" and septic shock group, and between the sepsis and septic shock group, but not between the no sepsis and sepsis group. The highest increase was around 26% for LAG volume between the "no sepsis" and septic shock group. RAG volume remained unchanged ($\pm 4-6\%$).

Table 2. Adrenal volumes (cm^3).

	All Patients	No Sepsis	Sepsis	Septic Shock
	$n = 42$	$n = 6$	$n = 15$	$n = 21$
LAG		4.3 ± 1.5	4.5 ± 1.9	5.8 ± 2.0
RAG		4.7 ± 1.1	4.2 ± 1.7	4.5 ± 1.9
Total		9.0 ± 1.3	9.1 ± 1.9	10.8 ± 2.1

LAG = Left adrenal gland; RAG = Right adrenal gland.

A small adrenal gland volume of the LAG in the septic shock group was associated with higher mortality, significant for the endpoints day 8 and 6 months. Figure 5A illustrates the receiver operating characteristic curve (ROC) correlating LAG volume and ICU mortality at day 8. A cutoff value of 4.7 cm^3 provided a sensitivity of 70% and a specificity of approx. 82% for the risk of mortality during ICU stay. The likelihood ratio (LR) for death with a LAG volume of less than 4.7 cm^3 was 3.85 (AUC = 0.80; $p = 0.006$). Figure 5B presents the ROC for LAG and mortality at 6 months with a sensitivity of approx. 73% and a specificity of around 83% at a cutoff value of 5.4 cm^3. The LR for death with a LAG volume of less than 5.4 cm^3 was 4.40 (AUC = 0.744; $p = 0.035$).

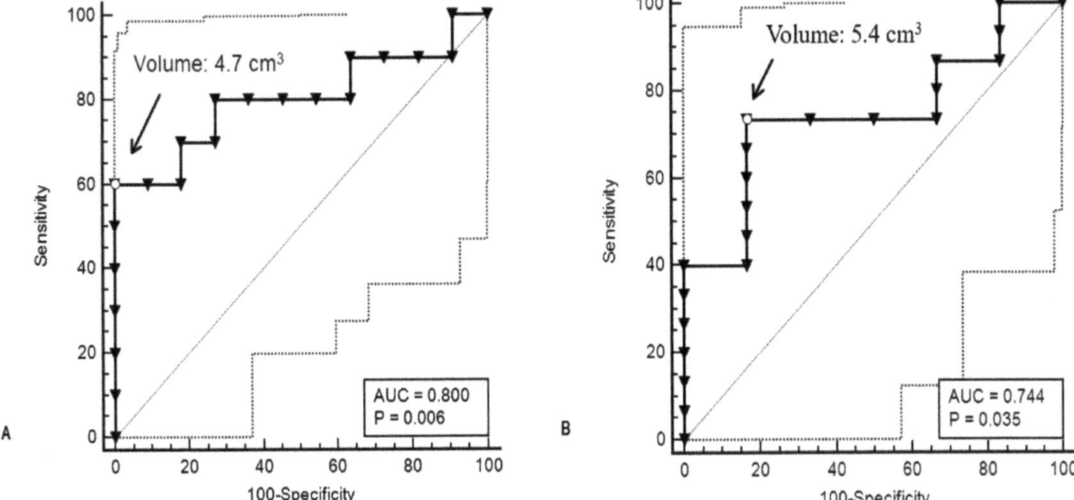

Figure 5. Receiver operating characteristic curve (ROC) between the left adrenal gland volume (NNL) and mortality at day 8 (**A**), showing a sensitivity of 70% and a specificity of around 82% with an AUC = 0.80 (p = 0.06) for a cutoff at 4.7 cm^3. ROC for 6 months (**B**), with a sensitivity of 73% and a specificity of 83% with an AUC = 0.744 for a cutoff at 5.4 cm^3 (p = 0.035).

3.5. Histogram Analysis

Positive attenuation values for LAG (rounded) for histograms subsequent to organ segmentation were as follows (median (IQR)): no sepsis 106 (91 to 113); sepsis 91 (80 to 100); septic shock 102 (91 to 115). These values—corrected for confounders—were (median (IQR)): no sepsis 12 (-6 to 15); sepsis -6 (-11 to 2); septic shock 3 (-7 to 12).

Figure 6A illustrates the distribution of absolute histogram HU values (negative pixels excluded) for the LAG according to the different groups of sepsis severity at day 1 of their ICU stay. As with the methods of ROI and Segmentation, histogram attenuation values likewise showed a significant difference of mean attenuation values between the sepsis (2) and septic shock (3) group with higher attenuation values for patients with septic shock (p = 0.0021), which remained significant after multiple testing correction (ANOVA, p = 0.0010). The difference between the patients not classified with sepsis (1) and sepsis was also significant (p = 0.018) with higher attenuation values for the no sepsis group, which remained significant after multiple testing correction (p = 0.042). There was no difference between group 1 and 3 (p = 0.71).

Figure 6B illustrates the distribution of the corrected histogram HU values, respectively. The significant difference between mean attenuation values for the sepsis- (2) and septic shock (3) group (p = 0.040) was no longer significant after multiple testing (p = 0.079). There was no significant difference between the other groups (p = 0.079, p = 0.88).

The histograms grouped by septic condition are presented in Figure 7A–C. Histograms for the groups were the following: the no sepsis group had a range from -139 to 248 HU, mean attenuation of 72 HU, the sepsis group had a range from -171 to 221 HU, mean attenuation of 49 HU and the septic shock group had a range from -185 to 317 HU, mean attenuation of 71 HU.

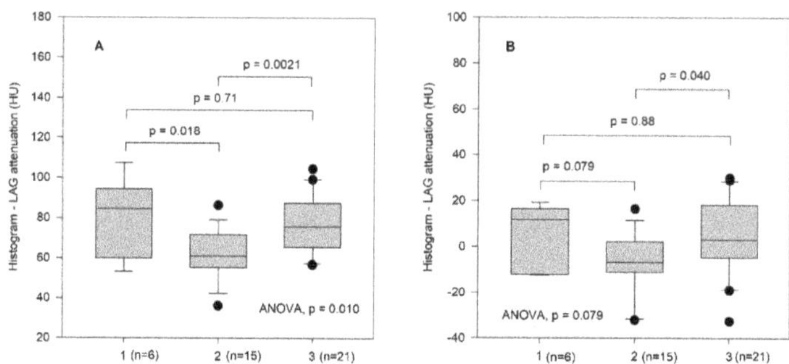

Figure 6. (**A**) illustrates the distribution of absolute histogram HU values (negative pixels excluded) for the LAG according to the different groups of sepsis severity at day 1 of their ICU stay. As with the methods of ROI and Segmentation, histogram attenuation values likewise showed a significant difference of mean attenuation values between the sepsis (2) and septic shock (3) group with higher attenuation values for patients with septic shock ($p = 0.0021$), which was still significant after multiple testing correction (ANOVA, $p = 0.0010$). The difference between the patients not classified with sepsis (1) and sepsis was also significant ($p = 0.018$) with higher attenuation values for the no sepsis group, which remained significant after multiple testing correction ($p = 0.042$). There was no difference between group 1 and 3 ($p = 0.71$); (**B**) illustrates the distribution of the corrected histogram HU values, respectively. The significant difference between mean attenuation values of the sepsis (2) and septic shock (3) group ($p = 0.040$) was no longer significant after multiple testing correction ($p = 0.079$). There was no significant difference between the other groups ($p = 0.079$, $p = 0.88$).

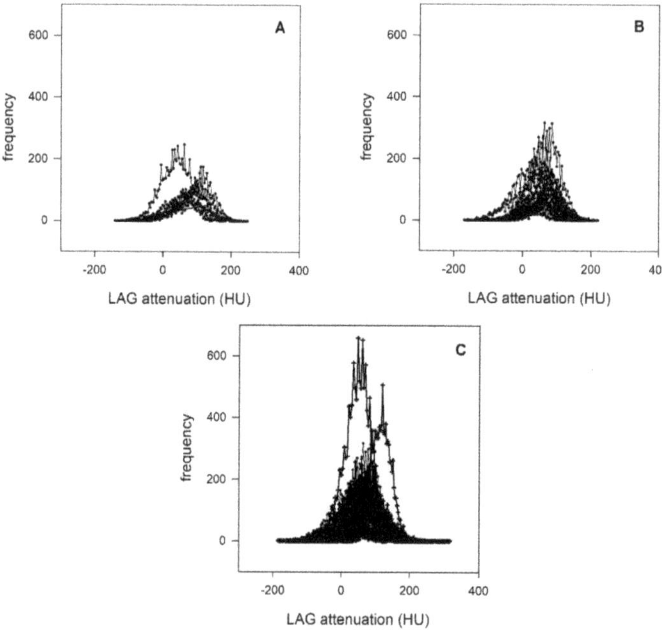

Figure 7. Histograms separated by groups: no sepsis (**A**); sepsis (**B**); septic shock (**C**). Two patients with hyperattenuating glands show high frequencies with high attenuation values.

3.6. Comparison of Image Analysis Methods

Considering the significantly lower range of data variation in LAG analysis, Table A1 (see Appendix A) presents the absolute and corrected mean attenuation values, as well as significant *p*-values for LAG for the three different methods of image analysis and the different groups classified by the Sepsis 3-criteria.

All of the methods showed a significant difference between attenuation values of the sepsis (2) and septic shock group (3), histogram analysis also between the no sepsis (1) and sepsis group (2) ($p < 0.05$). After correction for confounders, only the method of semi-automated segmentation remained significant ($p = 0.048$). None of the three image analysis methods performed in this study was able to differentiate between the groups, no sepsis (1) and septic shock ($p > 0.05$). For the comparison of goodness criteria, scatterplots of mean LAG density values in combination with mean density values of the IVC are presented for all methods of image analysis.

3.6.1. Region of Interest

A scatterplot of LAG uncorrected mean HU values assessed by ROI and the corresponding mean HU values of the inferior vena cava (IVC) for the sepsis and septic shock groups is illustrated in Figure 8A. Patients without septic conditions are excluded from this figure. The first cut-off-value of LAG mean values with a threshold of 90 HU or higher (a) resulted in a sensitivity of 55% with a specificity of 94% for patients with septic shock, with a positive predictive value (PPV) of 92% and a negative predictive value (NPV) of 60%. The second cut-off with a threshold of 112 IVC mean HU values or higher (b) provided a sensitivity of 82% and a specificity of 75%, with a PPV of 82% and a NPV of 75%. The combination of a and b (blue area) resulted in a sensitivity of 50% with a specificity of 100% (PPV = 100%; NPV = 59%). Twelve out of twenty-two (55%) patients with septic shock showed attenuation values of the LAG equal or above 90 HU. Nine of these twelve patients (75%) died (within a range of 1 to 55 days, mean 11 days).

3.6.2. Segmentation

A scatterplot of LAG uncorrected mean segmentation values HU and the corresponding mean values HU of the inferior vena cava (IVC) for the sepsis and septic shock groups is illustrated in Figure 8B. Patients without septic conditions are excluded from this figure. The first cut-off-value of LAG mean values with a threshold of 70 HU or higher (a) produced a sensitivity of 52% with a specificity of 92% for patients with septic shock, with a positive predictive value (PPV) of 92% (PPV = 92%; NPV = 58%). The second cut-off with a threshold of 112 IVC mean values HU or higher (b) provided a sensitivity of 81% and a specificity of 73% (PPV = 82%; NPV = 73%). The combination of a and b (blue area) resulted in a sensitivity of 48% with a specificity of 93% (PPV = 91%; NPV = 56%). Eleven out of twenty-one (52%) patients with septic shock showed attenuation values of the LAG equal or above 70 HU. Seven of these eleven patients (64%) died (within a range of 1 to 55 days, mean 12 days).

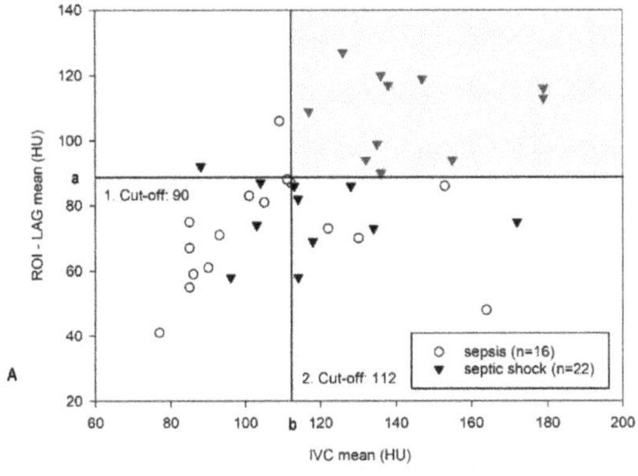

Figure 8. *Cont.*

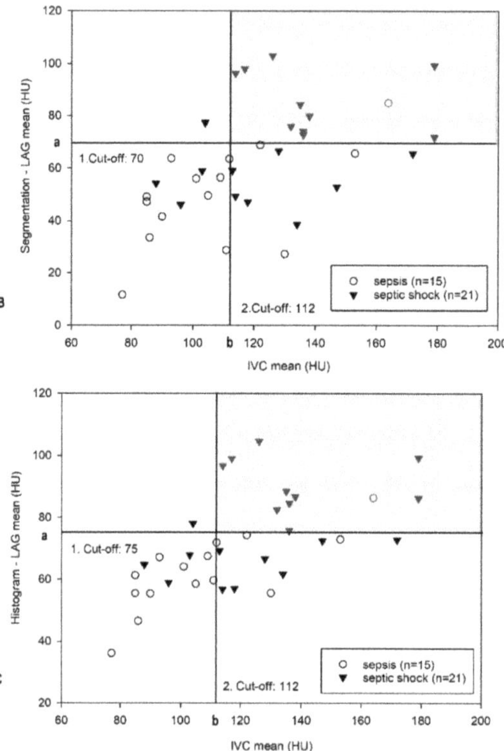

Figure 8. (**A**) A scatterplot of LAG uncorrected mean values HU assessed by ROI (A) and the corresponding mean values HU of the inferior vena cava (IVC) for sepsis and septic shock groups. Patients without septic conditions are excluded in this figure. First cut-off-value of LAG mean values with a threshold of 90 HU or higher (a) with a sensitivity of 55% and a specificity of 94% for patients with septic shock, with a positive predictive value (PPV) of 92% and a negative predictive value (NPV) of 60%. The second cut-off with a threshold of 112 IVC mean values HU or higher (b) with a sensitivity of 82% and a specificity of 75% (PPV = 82%; NPV = 75%. The combination of (a) and (b) (blue area) resulted in a sensitivity of 50% with a specificity of 100% (PPV = 100%; NPV = 59%); (**B**) A scatterplot of LAG uncorrected mean segmentation values (**B**) HU and the corresponding mean values HU of the inferior vena cava (IVC) for sepsis and septic shock. Patients without septic conditions are excluded in this figure. The first cut-off-value of LAG mean values with a threshold of 70 HU or higher (a) produced a sensitivity of 52% with a specificity of 92% for patients with septic shock, with a positive predictive value (PPV) of 92% (PPV = 92%; NPV = 58%). The second cut-off with a threshold of 112 IVC mean values HU or higher (b) provided a sensitivity of 81% and a specificity of 73% (PPV = 82%; NPV = 73%). The combination of (a) and (b) (blue area) resulted in a sensitivity of 48% with a specificity of 93% (PPV = 91%; NPV = 56%); (**C**). A scatterplot of LAG uncorrected positive histogram attenuation values HU and the corresponding mean values HU of the inferior vena cava (IVC) for sepsis and septic shock groups is illustrated in (**C**). Patients without septic conditions are excluded in this figure. The first cut-off-value of LAG mean values with a threshold of 75 HU or higher (a) produced a sensitivity of 52% with a specificity of 93% for patients with septic shock (PPV = 92%; NPV = 58%). The second cut-off with a threshold of 112 IVC mean values HU or higher (b) provided a sensitivity of 81% and a specificity of 67% (PPV = 77%; NPV = 71%). The combination of (a) and (b) (blue area) resulted in a sensitivity of 48% with a specificity of 93% (PPV = 91%; NPV = 56%).

3.6.3. Histogram

A scatterplot of LAG uncorrected positive histogram attenuation values HU and the corresponding mean values HU of the inferior vena cava (IVC) for the sepsis and septic shock groups is illustrated in Figure 8C. Patients without septic conditions are excluded from this figure. The first cut-off-value of LAG mean values with a threshold of 75 HU or higher (a) produced a sensitivity of 52% with a specificity of 93% for patients with septic shock (PPV = 92%; NPV = 58%). The second cut-off with a threshold of 112 IVC mean values HU or higher (b) provided a sensitivity of 81% and a specificity of 67% (PPV = 77%; NPV = 71%). The combination of (a) and (b) (blue area) resulted in a sensitivity of 48% with a specificity of 93% (PPV = 91%; NPV = 56%). Eleven out of twenty-one (52%) patients with septic shock showed attenuation values of the LAG equal or above 75 HU. Nine of these eleven patients (82%) died (within a range of 1 to 55 days, mean 11 days).

4. Discussion

The aim of our study was to evaluate three different quantitative HU density analysis methods as well as the use of volumetry of adrenal glands, which could possibly provide quantitative values with discriminatory power allowing to differentiate between patients either with sepsis, with septic shock or not classified with sepsis and predicting their outcome.

The main findings of our study are three-fold: first, high mean density values of the left adrenal gland (LAG) alone or in combination with high mean density values of the Inferior Vena Cava (IVC) are highly specific for septic shock regardless of the method of image analysis (ROI, Segmentation, Histogram). Second, semi-automated segmentation of the LAG due to least data variation seems to have the highest discriminatory power to differentiate between sepsis and septic shock. It furthermore seems to provide additional short- and long-term prognostic value. Finally, and thirdly, we concluded that none of the three quantitative adrenal gland HU density analysis methods investigated in patients with septic conditions is capable to clearly differentiate between all sepsis stages.

Data variation between ROI and Histogram method was higher for RAG. Therefore, we limited analysis on LAG and suggest doing so when gathering quantitative data. Less HU value variation in LAG may be due to the anatomical proximity of the RAG to the liver, with consequently especially in slim or cachectic patients problems to avoid measuring liver density as partial volume. Another explanation of the differences between LAG and RAG would be due to the fact that the left adrenal veins drain into the left renal vein, while the RAG directly drains into the IVC. As a result, the LAG might be exposed to a higher pressure, thus be more prone to hyperplasia or adenomatous change [24].

4.1. Hyperattenuating Adrenal Glands and Sepsis

According to the definition of hyperattenuating glands by O' Hara et al. [9] (adrenal density HU equal or greater than the inferior Vena Cava), only 2 out of our 44 patients (5%) showed hyperattenuating glands assessed by ROI. They both died after 1 and 4 days of ICU treatment. In our study population due to ROI analysis, 12 out of 22 (55%) patients with septic shock showed attenuation values of the LAG equal or above 90 HU. This cut-off was highly specific for septic shock (94%). Nine out of these twelve patients (75%) died. As for the methods of Segmentation and Histogram, numbers were similar (52% of septic shock patients above cut-offs, with 84%, respectively, 62% mortality). Therefore, we draw the conclusion that adrenal enhancement is an insensitive but highly specific CT sign in patients with septic shock, indicating poor prognosis, even if adrenal HU values are lower than those measured in IVC.

Several previous studies and case reports have described similar phenomena of hyperattenuating glands in severely ill patients, such as the fluctuating occurrence, its prognostic value and the issue of a generally admitted definition.

Rotondo et al. [25] retrospectively reviewed abdominal CT scans from 15 adult patients with clinical hypovolemia suffering from mostly blunt abdominal trauma. In this case study, none of these patients—of whom, all died within 24 h—showed increased adrenal enhancement. However, the definition of abnormal enhancement was subjective and the pathogenesis between their patient cohort with hypovolemic shock and our cohort with septic shock is different. Hence, patient populations and conclusions may not be comparable.

Cheung et al. [26] reported two patients with septic shock and adrenal enhancement with one of these two patients showing no other signs of visceral hypoperfusion. They concluded that persistent adrenal enhancement may only be observable in the early stages of septic shock, due to the initial adrenal stress response, but diminishes within its course due to circulatory failure worsened by vasoconstriction. However, they evaluated abdominal CT scans acquired during the arterial, and not during the portal-venous, phase for hyperenhancement and only had a small series of patients, so data are too limited to draw definitive analogies.

Bollen et al. [11] observed intense adrenal enhancement—defined as enhancement greater than adjacent vascular structures such as the inferior vena cava—in three out of thirty-eight (8%) patients with acute pancreatitis with early organ failure, proposing that hyperenhancing adrenals may be a new prognostic indicator for poor prognosis. The patients in this study represent a subgroup of our patient cohort, hence, study results are comparable.The shortcoming of their study is, again, the small series of patients, making data too limited to draw firm conclusions.

4.2. Adrenal Gland Volume and Septic Shock

Besides the qualitative or semiquantitative CT attenuation assessment of adrenal glands, the diagnostic and prognostic value for quantitative adrenal gland-derived data has been described, namely for adrenal gland volume assessed by segmentation in CT images. In the most severely affected patient group with septic shock, adrenal gland volume seems to be increased significantly and the absence of this enlargement in this subgroup, on the contrary, is associated with an even higher mortality [18,27,28].

One of the mechanisms involved in contributing to better outcomes for patients in septic shock that show adrenal gland enlargement, may be explained by the increased adrenal blood flow in septic conditions in combination with reduced venous drainage, resulting in an increased adrenal volume with a subsequent elevated hormonal response crucial to fight critical illness. However, this effect may only last for a limited time and be dependent on various individual conditions, since other studies have also shown that adrenal gland swelling may be caused by ischemia, edema, microbleeds or necrosis of adrenal glands and that only the early phase of sepsis may be associated with abnormal enlarged adrenals [12,18]. Additionally, Jung et al. [27] showed in their study that the enlarged adrenal glands of some septic shock patients were able to fully reover from their morphologic changes and others did not, underlining the importance of individual factors yet to be assessed in further studies. Although adrenal gland volume, according to our data, seems to be a promising prognostic factor in patients with septic shock, another recently published study by Mongardon et al. [29] suggests that adrenal gland volume is not an adequate surrogate for the outcome of patients with successful cardiopulmonary resuscitation after cardiac arrest, a state sometimes described as a "sepsis-like syndrome" [30].

To inquire about the correlation of hyperattenuating glands and circulatory failure, we plotted patients' absolute mean values of the LAG with the corresponding values of the IVC. Our results showed that any of our image analysis approaches (ROI, Segmentation, Histogram) showed high specificity for the high mean values of the LAG, alone or in combination with the high mean values of the IVC for patients with septic shock. However, this seemingly only holds true when patients with severe courses of diseases other than septic shock are excluded. In our study population, the combination of ROI and IVC values seemed to provide the highest specificity for patients with septic shock with the least amount of time and effort spent on image analysis. However, this may not seem surprising, as patients with septic shock suffer from circulatory backward heart failure, resulting in retrograde accumulation of the i.v. contrast medium in the IVC with subsequent high attenuation values.

The results furthermore show a significant difference between LAG mean values of sepsis and septic shock groups, which remained significant after correction for confounders only for the analysis approach of AdSegmentation. However, we doubt that this promising finding could become relevant in daily clinical routine, as semi-automated segmentation is time-consuming and, at least in our study, no clear cut-offs could be determined. Surprisingly, the group with no septic conditions showed very high attenuation values, comparable to those of patients with septic shock. This, on the one hand, might be explained by the small number ($n = 6$) of patients not classified with sepsis, making it difficult to draw firm conclusions. On the other hand, these patients had severe courses of disease other than septic (e.g., myocardial infarction, central pulmonary embolism) with high adrenal attenuation values underlining once more the low sensitivity of adrenal hyperenhancement for septic conditions.

The lack of clear discrimination between the groups may be explained by the very dynamic nature of septic conditions as the maybe most important limiting factor of all: intense adrenal enhancement in adults may only be apparent in the early stages of shock, as vasoconstriction in subsequent stages of septic shock may not cause abnormal enhancement [26]. This would have ruled out our very severely affected patients with prolonged septic shock from positive correlation with high adrenal attenuation values, as overall adrenal enhancement decreases in late stages of circulatory failure.

Lastly, increased density values in adrenals are also measured during acute adrenal hemorrhage [31].

4.3. Limitations

The study design included a follow-up of six months, so a relatively high number of patients dropped out of this endpoint subanalysis, because the follow-up calls failed. These circumstances led to our final rather small study cohort of 44 patients. A larger number of study patients, in particular for the group without sepsis with six patients in it, would have been required to guarantee sufficient statistical power.

Each of our three different image analysis approaches had its pitfalls and limitations. First, drawing a ROI has the advantage of being a fast technique, easily avoiding partial volume effects or organ areas with artefacts [32]. The drawbacks are its high subjectivity and variability.

Second, semi-automatic whole-organ segmentation and volumetry is less subjective and more representative, but a time-consuming method, because of necessary manual adjustments. Among these were the removal of vascular structures, fatty tissue or the manual accurate demarcation of adjacent anatomical structures.

Third, while the main advantage of histogram analysis might be its objectivity, it is prone to deterioration of CT image quality and increased image noise. Several factors may have impacted image quality in our cohort, e.g., patients' body physique and breathing artifacts, but also the parameters of tube voltage and tube current, collimation, slice thickness, reconstruction kernel, intravenous contrast medium injection flow rate and CT scan delay [20] are known to influence image quality. In our study, however, these limitations only applied to several patients in our study and led to study dropouts or influenced image quality, since not all of these parameters were standardized.

A major finding of our study is that the total adrenal volume computed by semi-automated segmentation seems to be increased by about 20% for patients with septic shock, and LAG volume even increased by around 26% percent. Furthermore, adrenal gland enlargement seems to provide prognostic value for patients with septic shock. We could replicate similar findings of previous studies [18,27,28] and even expand its prognostic value, suggesting that adrenal gland volume could serve as a surrogate for long-term mortality. In our study population, septic shock patients with no increase in LAG volume were approximately four-times more likely to die within eight days and even six months.

Interestingly, the LAG volume of patients not classified with sepsis in comparison to those in septic shock, showed the highest difference in LAG volume (26%), although mean attenuation values were comparably high. Therefore, we assume that adrenal enlargement could be a pathomechanism specific for septic shock, not occurring in circulatory failure of another cause (e.g., coronary failure).

Nougaret et al. [28] made a similar assumption and hypothesized that increased adrenal gland volume could be a surrogate for a rather vital response during sepsis, unlike edema or necrosis of the gland. This phenomenon may be explained by the higher metabolic demand for cortisol and the compensatory mechanism for hypovolemia in patients with septic shock, that led to an increased blood flow to the adrenal glands and to their subsequent enlargement [9,18,19].

Jung et. Al. [27] showed that in septic shock, total adrenal gland volume was an independent prognostic factor for 28-day mortality. Their adrenal gland volume was nearly doubled in the septic shock groups in comparison with the nonseptic ambulatory group, and increased by 35% in comparison to the nonseptic ICU group. Viewing this data, we showed similar enlargement values (20%) for our septic shock group compared to the nonseptic ICU group and extended prognostic value of adrenal gland volume to long-term mortality for our follow-up of six months.

However, some limitations should be acknowledged. First, adrenal gland volume may be affected by factors like gender, weight, body surface area, race or even geographical regions. We did, however, not include any of these factors in our calculations. Furthermore, pre-existing conditions like depression and Cushing's disease are also known to increase adrenal gland volume [12,24,28,33–37]. However, within our study population gender distribution was nearly equal and there were no patients with Cushing's disease and only two patients with depression, one each from the sepsis and septic shock group, so that systematic bias seems unlikely.

5. Conclusions

This study demonstrates that there is no additional diagnostic value in performing time-consuming semi-automated whole-organ adrenal gland segmentation analysis in patients with sepsis or septic shock. High absolute CT density values assessed by simple ROI analysis—alone or in combination with IVC CT density assessemnt—may provide a high specificity for patients with septic shock, which could be used as an additional decision-making support in evaluating their health status and prognosis. A more time-consuming segmentation image analysis may deliver, however, additional prognostic value, as the adrenal gland volume in our cohort was generally increased in patients with septic shock whilst a smaller volume was associated with a higher mortality within the subgroup, even for the long-term survival of six months. For the method of histogram analysis of adrenal glands in patients with septic conditions, we do not see any diagnostic and/or prognostic value justifying time and effort in clinical routine. However, further studies with larger series of patients will be needed, to determine if these encouraging findings will find their way into clinical practice.

Author Contributions: Conceptualization, M.M. and S.J.; methodology, M.M., S.J., A.K., U.H. and M.B.; software, M.M., A.K. and S.J.; validation, M.M., S.J., A.K., M.B. and U.H.; formal analysis, A.K., M.M. and S.J.; investigation, U.H., M.B., S.J., L.L. and M.M; resources, S.J., U.H., M.B. and S.O.S.; data curation, M.M., S.J., A.K. and L.L.; writing—original draft preparation, M.M., S.J. and A.K.; writing—review and editing, S.J., A.K., M.B. and U.H.; visualization, M.M., A.K. and S.J.; supervision, S.J., S.O.S., U.H. and M.B.; project administration, S.J., S.O.S., U.H. and M.B.; funding acquisition, S.J. All authors have read and agreed to the published version of the manuscript.

Funding: This study was part of the research project SU 962/1-1 funded by the DFG (German Research Foundation) within the framework of the DFG Nachwuchsakademie "Kohortenstudien". The funding source was not involved in the study design, collection, analysis and interpretation of data, in the writing of the report or the decision to submit the article for publication.

Institutional Review Board Statement: The study was conducted in accordance with the Declaration of Helsinki, and approved by the Ethics Committee of Medizinische Fakultät Mannheim (protocol code: 2011-384N-MA), date of approval: 28 March 2016.

Informed Consent Statement: This prospective single-center study at the University Medical Centre Mannheim (UMM), Germany, was approved by the local ethics committee of the UMM. A total of seventy-six patients were prospectively included into this study. Informed consent was obtained from all participating patients or their legal representatives.

Data Availability Statement: Not applicable.

Conflicts of Interest: The authors declare no conflict of interest.

Appendix A

Table A1. CT attenuation values HU of the left adrenal gland (LAG) for all methods of image analysis.

Methods	Groups	No Sepsis (1)		Sepsis (2)		Septic Shock (3)	
		$n = 6$		$n = 16$ (15)		$n = 22$ (21)	
Region of Interest (ROI)	mean	90	14 *	71	−10 *	96	5 *
	median	97	24 *	72.0	−9 *	90	2 *
	IQR	26	23 *	23	17 *	39	39 *
	p-value [1] < 0.05				$p = 0.023$		
Segmentation	mean	72	12 *	49	−10 *	71	5 *
	median	76	15 *	49	−11 *	69	5 *
	IQR	25	24 *	32	21 *	29	23 *
	p-value [1] < 0.05				$p = 0.013/0.048$ *		
Histogram	mean	101	6 *	90	−6 *	105	3 *
	median	106	12 *	91	−6 *	102	3 *
	IQR	21	21 *	21	13 *	25	3 *
	p-value [1] < 0.05		$p = 0.042$			$p = 0.010$	

* = Corrected for confounders: age, sex, evidence for presence of germs, creatinine, APACHE-Score, SOFA-Score;
[1] = Bonferroni correction for multiple testing; IQR = Interquartile Range.

References

1. Engel, C.; Brunkhorst, F.M.; Bone, H.-G.; Brunkhorst, R.; Gerlach, H.; Grond, S.; Gründling, M.; Huhle, G.; Jaschinski, U.; John, S.; et al. Epidemiology of sepsis in Germany: Results from a national prospective multicenter study. *Intensive Care Med.* **2007**, *33*, 606–618. [CrossRef] [PubMed]
2. *Sepsis und MODS*; Springer: Berlin/Heidelberg, Germany, 2016.
3. Rhodes, A.; Evans, L.E.; Alhazzani, W.; Levy, M.M.; Antonelli, M.; Ferrer, R.; Kumar, A.; Sevransky, J.E.; Sprung, C.L.; Nunnally, M.E.; et al. Surviving Sepsis Campaign: International Guidelines for Management of Sepsis and Septic Shock: 2016. *Intensive Care Med.* **2017**, *43*, 304–377. [CrossRef] [PubMed]
4. Shankar-Hari, M.; Phillips, G.S.; Levy, M.L.; Seymour, C.W.; Liu, V.X.; Deutschman, C.S.; Angus, D.C.; Rubenfeld, G.D.; Singer, M.; for the Sepsis Definitions Task Force. Developing a New Definition and Assessing New Clinical Criteria for Septic Shock: For the Third International Consensus Definitions for Sepsis and Septic Shock (Sepsis-3). *JAMA* **2016**, *315*, 775–787. [CrossRef] [PubMed]
5. Singer, M.; Deutschman, C.S.; Seymour, C.W.; Shankar-Hari, M.; Annane, D.; Bauer, M.; Bellomo, R.; Bernard, G.R.; Chiche, J.-D.; Coopersmith, C.M.; et al. The Third International Consensus Definitions for Sepsis and Septic Shock (Sepsis-3). *JAMA* **2016**, *315*, 801–810. [CrossRef] [PubMed]
6. Lindner, H.A.; Balaban, Ü.; Sturm, T.; Symbol, C.W.B.; Thiel, M.; Schneider-Lindner, V. An Algorithm for Systemic Inflammatory Response Syndrome Criteria–Based Prediction of Sepsis in a Polytrauma Cohort. *Crit. Care Med.* **2016**, *44*, 2199–2207. [CrossRef]
7. Vincent, J.L.; De Mendonça, A.; Cantraine, F.; Moreno, R.; Takala, J.; Suter, P.; Sprung, C.; FCCM; Colardyn, F.; Blecher, S. Use of the SOFA score to assess the incidence of organ dysfunction/failure in intensive care units: Results of a multicenter, prospective study. *Crit. Care Med.* **1998**, *26*, 1793–1800. [CrossRef]
8. Schlegel, N.; Flemming, S.; Meir, M.; Germer, C.T. Is a different view on the pathophysiology of sepsis the key for novel therapeutic options? *Chirurg* **2014**, *85*, 714–719. [CrossRef]
9. O'Hara, S.M.; Donnelly, L.F. Intense contrast enhancement of the adrenal glands: Another abdominal CT finding associated with hypoperfusion complex in children. *Am. J. Roentgenol.* **1999**, *173*, 995–997. [CrossRef]

10. Schek, J.; Macht, S.; Klasen-Sansone, J.; Heusch, P.; Kröpil, P.; Witte, I.; Antoch, G.; Lanzman, R.S. Clinical impact of hyperattenuation of adrenal glands on contrast-enhanced computed tomography of polytraumatised patients. *Eur. Radiol.* **2014**, *24*, 527–530. [CrossRef]
11. Bollen, T.L.; Dutch Acute Pancreatitis Study Group; Van Santvoort, H.C.; Besselink, M.G.H.; Van Ramshorst, B.; Van Es, H.W.; Gooszen, H.G. Intense adrenal enhancement in patients with acute pancreatitis and early organ failure. *Emerg. Radiol.* **2007**, *14*, 317–322. [CrossRef]
12. Peng, Y.; Xie, Q.; Wang, H.; Lin, Z.; Zhang, F.; Zhou, X.; Guan, J. The hollow adrenal gland sign: A newly described enhancing pattern of the adrenal gland on dual-phase contrast-enhanced CT for predicting the prognosis of patients with septic shock. *Eur. Radiol.* **2019**, *29*, 5378–5385. [CrossRef] [PubMed]
13. Kanczkowski, W.; Sue, M.; Zacharowski, K.; Reincke, M.; Bornstein, S.R. The role of adrenal gland microenvironment in the HPA axis function and dysfunction during sepsis. *Mol. Cell. Endocrinol.* **2015**, *408*, 241–248. [CrossRef] [PubMed]
14. Goodwin, J.E.; Feng, Y.; Velazquez, H.; Sessa, W.C. Endothelial glucocorticoid receptor is required for protection against sepsis. *Proc. Natl. Acad. Sci. USA* **2013**, *110*, 306–311. [CrossRef] [PubMed]
15. Turrin, N.P.; Rivest, S. Unraveling the Molecular Details Involved in the Intimate Link between the Immune and Neuroendocrine Systems. *Exp. Biol. Med.* **2004**, *229*, 996–1006. [CrossRef]
16. Boonen, E.; Bornstein, S.R.; Berghe, G.V.D. New insights into the controversy of adrenal function during critical illness. *Lancet Diabetes Endocrinol.* **2015**, *3*, 805–815. [CrossRef]
17. Boonen, E.; Berghe, G.V.D. Understanding the HPA response to critical illness: Novel insights with clinical implications. *Intensive Care Med.* **2015**, *41*, 131–133. [CrossRef]
18. Chanques, G.; Annane, D.; Jaber, S.; Gallix, B. Enlarged adrenals during septic shock. *Intensive Care Med.* **2007**, *33*, 1671–1672. [CrossRef]
19. Prasad, K.R.; Kumar, A.; Gamanagatti, S.; Chandrashekhara, S.H. CT in post-traumatic hypoperfusion complex—A pictorial review. *Emerg. Radiol.* **2010**, *18*, 139–143. [CrossRef]
20. Bae, K.T.; Fuangtharnthip, P.; Prasad, S.R.; Joe, B.N.; Heiken, J.P. Adrenal Masses: CT Characterization with Histogram Analysis Method. *Radiology* **2003**, *228*, 735–742. [CrossRef]
21. Halefoglu, A.M.; Yasar, A.; Bas, N.; Ozel, A.; Erturk, Ş.M.; Basak, M. Comparison of computed tomography histogram analysis and chemical-shift magnetic resonance imaging for adrenal mass characterization. *Acta Radiol.* **2009**, *50*, 1071–1079. [CrossRef]
22. Spiriev, T.; Nakov, V.; Laleva, L.; Tzekov, C. OsiriX software as a preoperative planning tool in cranial neurosurgery: A step-by-step guide for neurosurgical residents. *Surg. Neurol. Int.* **2017**, *8*, 241. [CrossRef] [PubMed]
23. Wolf, I.; Vetter, M.; Wegner, I.; Nolden, M.; Bottger, T.; Hastenteufel, M.; Schobinger, M.; Kunert, T.; Meinzer, H.P. The medical imaging interaction toolkit (MITK): A toolkit facilitating the creation of interactive software by extending VTK and ITK. In *Medical Imaging 2004: Visualization, Image-Guided Procedures, and Display*; International Society for Optics and Pho-tonics: 2004; SPIE: San Diego, CA, USA, 2014; Volume 5367, pp. 16–27.
24. Wang, X.; Jin, Z.-Y.; Xue, H.-D.; Liu, W.; Sun, H.; Chen, Y.; Xu, K. Evaluation of Normal Adrenal Gland Volume by 64-slice CT. *Chin. Med. Sci. J.* **2012**, *27*, 220–224. [CrossRef]
25. Rotondo, A.; Angelelli, G.; Catalano, O.; Grassi, R.; Scialpi, M.; Stellacci, G.; Derchi, L.E. Abdominal computed tomographic findings in adults with hypovolemic shock. *Emerg. Radiol.* **1997**, *4*, 10–15. [CrossRef]
26. Cheung, S.; Lee, R.; Tung, H.; Chan, F. Persistent Adrenal Enhancement may be the Earliest CT Sign of Significant Hypovolaemic Shock. *Clin. Radiol.* **2003**, *58*, 315–318. [CrossRef]
27. Jung, B.; Nougaret, S.; Chanques, G.; Mercier, G.; Cisse, M.; Aufort, S.; Gallix, B.; Annane, D.; Jaber, S. The absence of adrenal gland enlargement during septic shock predicts mortality: A computed tomography study of 239 patients. *Anesthesiology* **2011**, *115*, 334–343. [CrossRef]
28. Nougaret, S.; Jung, B.; Aufort, S.; Chanques, G.; Jaber, S.; Gallix, B. Adrenal gland volume measurement in septic shock and control patients: A pilot study. *Eur. Radiol.* **2010**, *20*, 2348–2357. [CrossRef]
29. Mongardon, N.; Savary, G.; Geri, G.; El Bejjani, M.-R.; Silvera, S.; Dumas, F.; Charpentier, J.; Pène, F.; Mira, J.-P.; Cariou, A. Prognostic value of adrenal gland volume after cardiac arrest: Association of CT-scan evaluation with shock and mortality. *Resuscitation* **2018**, *129*, 135–140. [CrossRef]
30. Adrie, C.; Adib-Conquy, M.; Laurent, I.; Monchi, M.; Vinsonneau, C.; Fitting, C.; Fraisse, F.; Dinh-Xuan, A.T.; Carli, P.; Spaulding, C.; et al. Successful Cardiopulmonary Resuscitation After Cardiac Arrest as a "Sepsis-Like" Syndrome. *Circulation* **2002**, *106*, 562–568. [CrossRef]
31. Venkatanarasimha, N.; Roobottom, C. Intense Adrenal Enhancement: A Feature of Hypoperfusion Complex. *Am. J. Roentgenol.* **2010**, *195*, W82. [CrossRef]
32. Johnson, P.T.; Horton, K.M.; Fishman, E.K. Adrenal Imaging with Multidetector CT: Evidence-based Protocol Optimization and Interpretative Practice. *Radiographics* **2009**, *29*, 1319–1331. [CrossRef]
33. Taner, A.T.; Schieda, N.; Siegelman, E.S. Pitfalls in Adrenal Imaging. *Semin. Roentgenol.* **2015**, *50*, 260–272. [CrossRef] [PubMed]
34. Freel, E.M.; Nicholas, R.S.; Sudarshan, T.; Priba, L.; Gandy, S.J.; McMillan, N.; Houston, J.G.; Connell, J.M. Assessment of the accuracy and reproducibility of adrenal volume measurements using MRI and its relationship with corticosteroid phenotype: A normal volunteer pilot study. *Clin. Endocrinol.* **2013**, *79*, 484–490. [CrossRef] [PubMed]

35. Grant, L.A.; Napolitano, A.; Miller, S.; Stephens, K.; McHugh, S.M.; Dixon, A.K. A pilot study to assess the feasibility of measurement of adrenal gland volume by magnetic resonance imaging. *Acta Radiol.* **2010**, *51*, 117–120. [CrossRef]
36. Amsterdam, J.D.; Marinelli, D.L.; Arger, P.; Winokur, A. Assessment of adrenal gland volume by computed tomography in depressed patients and healthy volunteers: A pilot study. *Psychiatry Res.* **1987**, *21*, 189–197. [CrossRef]
37. Pojunas, K.; Daniels, D.; Williams, A.; Thorsen, M.; Haughton, V. Pituitary and adrenal CT of Cushing syndrome. *Am. J. Roentgenol.* **1986**, *146*, 1235–1238. [CrossRef] [PubMed]

Review

Juan Valverde de Amusco: Pioneering the Transfer of Post-Vesalian Anatomy

Luis-Alfonso Arráez-Aybar [1,*], **Concepción Reblet** [2] **and José Luis Bueno-López** [2]

[1] Department of Human Anatomy and Embryology, Faculty of Medicine, Complutense University, 28040 Madrid, Spain

[2] Department of Neurosciences, School of Medicine and Nursing, The University of the Basque Country (UPV/EHU), 48940 Leioa, Spain; concha.reblet@ehu.es (C.R.); joseluis.bueno@ehu.es (J.L.B.-L.)

* Correspondence: arraezla@med.ucm.es

Abstract: This article delves into the life and accomplishments of Juan Valverde de Amusco (c. 1525–c. 1587), a Spanish anatomist. Specifically, it focuses on his book titled *HISTORIA de la composición del cuerpo humano*. The book was the first anatomy opus published after Andreas Vesalius' *De humani corporis fabrica libri septem*, written in a Romance language, the Castilian Spanish language, making it the most renowned post-Vesalian anatomy book in Europe and beyond during the 16th and 17th centuries. Compiling complete editions and reproductions of figures, it had 19 editions and several translations. One of its principal contributions was the initial graphical representation of the stapes ossicle. It provided the first accurate description of the pulmonary circulation, vomer bone, and four extraocular rectus muscles. Throughout the book, Valverde corrected numerous of Vesalius' anatomical observations. *HISTORIA de la composición del cuerpo humano* was the first anatomy book to use chalcographic illustrations, which are of superior anatomical quality than those printed from engraved wood in Andreas Vesalius' book. Next, many anatomy textbooks of that time incorporated Valverde's book illustrations. Valverde's book was practical, timely, and well referenced, making it a valuable resource for scholars and non-scholars. The conclusion is that Juan Valverde de Amusco merits a place as a pioneer in scientific knowledge transfer.

Keywords: renaissance anatomy; 16th-century anatomists; history of anatomy; anatomical terminology; carotid circulation; extraocular rectus muscles; oculomotor muscles; pulmonary circulation; stapes ossicle; vomer bone

Citation: Arráez-Aybar, L.-A.; Reblet, C.; Bueno-López, J.L. Juan Valverde de Amusco: Pioneering the Transfer of Post-Vesalian Anatomy. *Anatomia* **2023**, *2*, 450–471. https://doi.org/10.3390/anatomia2040033

Academic Editors: Gianfranco Natale and Francesco Fornai

Received: 5 September 2023
Revised: 23 October 2023
Accepted: 1 December 2023
Published: 11 December 2023

1. Context

Century XVI was the starting point of the Renaissance, the European cultural movement that determined the modern conception of nature and human beings. The Renaissance began in Florence, and its first expansion was in Northern Italy. Gutenberg's printing (c. 1440), universities, and patronage of kings, popes, and prominent families (e.g., such as those of the House of Medici in Florence and the Colonna family in Rome) prompted its development. Humanism was a foundation for Renaissance intellectual thought, derived from significant translations of Greco-Latin Antiquity and Islamic Golden Age authors in the late Middle Ages [1]. New conceptions led to the Protestant Reformation. Then, Pope Paul III (papacy period: 1534–1549) established the Roman Inquisition (1542) and the Council of Trent (1545–1563) as part of the Counter-Reformation, the Catholic answer to Protestantism.

Even though experience and inductive reasoning were gaining impetuses among the learned scholars of Christendom, the early Renaissance was the last and most magnificent phase in the history of Galen medicine [2]. Under Galenism, function, form, and finality are complementary parts. The form turned out to be the basis of the anatomy of the Renaissance at the beginning of the Modern Era. The human body and its parts became a distinctive object of study, not only for medicine and surgery, but also for revealing the implications intertwined with nature, humans, and divinity. Since objective knowledge of the human body was the goal for constructing modern anatomy, the dissection of the human body was preferred [3,4].

An osmosis was established between the fields of art and anatomy. Italian Renaissance artists, including Antonio Polliauolo (1433–1498), Leonardo da Vinci (1452–1519), Michelangelo Buonarroti (1475–1564), and Baccio Bandinelli (1493–shortly after 1560) dissected human bodies. Pope Sixtus IV (papacy period: 1471–1484), who had studied at the Medical School of Bologna, granted permission to carry out human body dissections, subject to the condition that the dissected corpses were buried afterward in a dignified manner. Again, Pope Clement VII (papacy period: 1523–1534) endorsed the teaching of anatomy through dissection in 1531. The practice of dissecting human bodies was also prompted by the custom of stuffing and embalming the remains of individuals who distinguished themselves through their examples and doctrines (since Pope Paul IV (papacy period: 1555–1559) until the present day) [3–6]. On the other hand, the Reformation favored the emergence of modern anatomy by allowing for the dissection of human bodies, at least as early as 1540, in London and other cities [7].

Alessandro Benedetti (1450?–1512), a professor at Padua and a prominent figure in Renaissance humanism, proposed in his book *Anatomice*, published in 1502, the first anatomical theater, which may have been in operation in 1522 [4]. Other prominent humanists who were anatomists as well were Jacques Dubois (latinized as Jacobus Sylvius) (1478–1555), Johann Winter von Andernach (1505–1574), and Miguel Servet (latinized as Michael Servetus) (1511–1553)—who, parenthetically, were teachers of Andreas Vesalius (1514–1564) (Vesalius herein). Among other locations, the private practice of human dissection gained momentum in Paris and Italy. The initial printed work that came into view from that environment is *Anatomica methodus* by Andrés Laguna (1510–1559), which was published in 1535 (Table 1). Andrés Laguna influenced Vesalius' formation [8].

Table 1. Published anatomy books by Spanish anatomists during the XVI century (Spanish Golden Age).

Year	Author	Title (Place of Publication: Publisher)
1535	Andrés Laguna (c. 1510–c. 1559)	*Anatomica methodus seu de sectioni humani corporis contemplatio* (París: Ludouicum Cyaneum)
1542	Luis Lobera de Ávila (c. 1480–c. 1551)	*Libro de Anatomía, es primera parte de "Remedio de cuerpos humanos y silva de experiencias y otras cosas utilísimas"* (Alcalá de Henares: Juan Brocar)
1549	Pedro Jimeno (c. 1515–c. 1551)	*Dialogus de re medica, compendiaria ratione, praeter quaedam alia, universam anatomem humani corporis perstringens* (Valencia: Juan Mey)
1551	Bernardino Montaña de Monserrate (c. 1480–c. 1558)	*Libro de la Anathomía del hombre* (Valladolid: Sebastián Martínez)
1555	Luis Collado (c. 1520–c. 1589)	*Cl. Galeni Pergameni Liber de Ossibus ad tyrones. . . enarrationibus illustratus* (Valencia: Juan Mey)
1556	Juan Valverde de Amusco (c. 1525–c. 1587)	*HISTORIA de la composición del cuerpo humano* (Roma: Antonio Martínez de Salamanca y Antoine Lafréry)
1559	Alfonso Rodríguez de Guevara (c. 1520–c. 1587)	*In pluribus ex iis quibus Galenus impugnatur ab Andrea Vesalio Bruxelensi in de constructione et usu partium corporis humani, defensio: et nonnullorum quae in anatome deficero videbantur supplementum.* (Coimbra: Juan Barreiro)

Two pivotal works of the scientific revolution of the Renaissance appeared in the year 1543: *De revolutionibus orbium coelestium* by Nicolaus Copernicus (1473–1543) and *De humani corporis fabrica libri septem* ("...*Fabrica*..." henceforth) by Vesalius. The latter conveys the perception that, in anatomy, only what can be seen and shown is correct. Vesalius dedicated his monumental work to Charles (reigning period: 1520–1558) (King Carlos I of Spain and Emperor Carolus V of the Holy Roman Empire, among many other titles). Vesalius served as Charles's physician and surgeon. Next, Vesalius likewise served Charles' son Philip (reigning period: 1556–1598) (King Felipe II of Spain and I of Portugal, King of Naples and Sicily, Duke of Milan, Lord of the Seventeen Provinces of the Low Countries, and King *iure oxoris* of England and Ireland, among other titles). Philip designated Vesalius as *Conde Palatino* (Count Palatine) [9].

During the Spanish Golden Age (1492–1659), Spain was the dominant potency in Europe and beyond. The Castilian Spanish language spread. King Charles I of Spain established the initial universities in the Americas: Santo Domingo (Royal and Pontifical University of Saint Thomas Aquinas, 1538); Lima (Royal and Pontifical University of the City of the Kings of Lima, 1551, currently known as National University of San Marcos), and México (Royal and Pontifical University of Mexico, 1551) [10].

However, there was an intense relationship between Spain and Italy regarding artistic, political, and military affairs, human interchanges and humanistic knowledge, not the least that of medicine. The cultural infrastructures of Italy, including the Roman heritage, numerous art collections, universities, printing houses, and libraries, particularly the *Biblioteca Apostolica Vaticana*, served as further incentives for such an exchange. Many Spanish physicians traveled to Italy to find their way into the medical marketplace at the papal, cardinal, and other courts, as well as the numerous hospitals and other medical institutions [11], or to complete their education with the finest anatomists, as the instruction in anatomy and surgery at Spanish universities was somewhat lacking. Sephardi Jewish physicians who had come to Rome from Spain played a significant role in translating medical classical books in the city, often with papal and other ecclesiastical support [12].

Nevertheless, the dissection of human bodies was documented in Spain from an early age. The monks of Guadalupe (Crown of Castile, presently Spain) obtained a papal privilege as early as 1322, which allowed them to open the bodies of deceased pilgrims and investigate their causes of death. The dissection of human bodies received a significant boost owing to King Ferdinand II of Aragon (1452–1516), also known as the Catholic Monarch (who was also King of Sicily, Naples, Navarre, and King *iure oxoris* of Castile). In 1488, King Ferdinand II of Aragon granted a privilege to the physicians of Saragossa (Crown of Aragon, presently Spain), allowing for the dissection of human bodies [13]. Following that, King Charles I of Spain and the fifth emperor of the Holy Roman Empire, as a champion of the Catholic faith, sought the faculty of the Salamanca University (Crown of Castile, presently Spain) for their opinion about human dissection. The faculty responded that it was permissible under the Catholic Church's edicts [6]. The initial autopsy conducted on American soil occurred between 1520 and 1530 by a bachelor of medicine with the surname Barreda [10]. The twins Joana and Melchiora Ballestero were also autopsied in La Hispaniola (at present, this island splits between Haiti and the Dominican Republic) in 1533 to determine whether they shared a heart and, consequently, a soul [14].

Spanish universities were among the first to accept the Vesalius anatomy, thanks to Pedro Jimeno (c. 1515–c. 1551) and Luis Collado (1520–1589). Both were direct disciples of Vesalius at Padua and later held chairs of anatomy and surgery at Valencia University [15]. At Valencia University (Crown of Aragon, presently Spain), a series of structured lectures on surgery (in the year 1501) and anatomy (1549) had been established, marking the inaugural edition of these courses in Spain. These courses served as models for the subsequent courses established at other Spanish universities, including Valladolid in 1550, Salamanca in 1551, and the initial *Complutense* at the town Alcalá de Henares in 1560. In 1559, King Philip II of Spain signed a royal provision so that bodies of unclaimed dead people would be given to the university for the study of anatomy.

In the same year, a Royal Pragmatic (a royal decree) signed by King Philip II and dated 22 November 1559, prohibited the enrollment of subjects from the Spanish Empire in foreign universities—an exception was made to this prohibition with the universities of Bologna, Rome, and Naples [16]. The Royal Pragmatic had two purposes. Initially, the defense against the contamination of the Catholic faith with Reformation ideas. Moreover, it may have served as a shield of contemptuous pride against the Black Legend. The Black Legend is a skewed collection of narratives and literary works that were initially disseminated by the affluent editorial establishment of the Seventeen Provinces (currently the Netherlands, Luxembourg, Belgium, and certain territories of northwestern France) and Italy during and after their rebellion against Spanish rule. The purpose of the Black Legend was to tarnish the image of the Spanish Empire, its people, and their culture [17]. The Royal Pragmatic of 22 November 1559, and the subsequent one signed by King Philip III of Spain on 7 November 1617, were turning points in the isolation of Spanish universities from Europe and the consequent scientific decline of Spain.

Despite the turmoil of that era, Spanish anatomists produced numerous anatomy books. Those published between 1535 and 1559 are listed in Table 1. Of particular importance, among them, is the *HISTORIA de la composición del cuerpo humano* by Juan Valverde de Amusco. No anatomy book written in Spanish was to be translated into another language until 1793, when Antonio de Gimbernat y Arbós published *Nuevo Método de Operar en la Hernia Crural*, translated into English by Thomas Beddoes two years later [18]. Valverde's life and work reflect the cultural upheavals occurring during the Renaissance.

2. Life

Juan Valverde de Amusco (J. Valverde henceforth) (Figure 1) was born in 1525 in the town of Hamusco (now Amusco) in the shire of Tierra de Campos (Crown of Castile, presently Palencia, Spain). There are few precise records of his life. The safest ones come from small comments disseminated throughout his books. The details regarding his childhood and youth remain unconfirmed. His probable Jewish origin has been suggested [19,20].

Figure 1. Line drawings of the two known portraits of Juan Valverde de Amusco (c. 1525–c. 1587). (**A**) The original portrait, painted by Gaspar Becerra [21], is exhibited in the Walters Art Museum of Baltimore, U.S.A. Available at https://bancodeimagenesmedicina.com/imagen/valverde-de-amusc o-hamusco-juan-2/ (last time consulted, 15 October 2023). (**B**) The original portrait is in the book *Anatomia del corpo umano*, Venetia, Giunti (1586), the first Italian edition of Valverde's *HISTORIA de la composición del cuerpo humano.*

In the year 1542, J. Valverde was around 17 years old when he left for Italy. J. Valverde might have resided briefly in Perugia, but this is not supported by clear evidence [21]. Though it is not known where he obtained his medical degree, J. Valverde studied in

Padua, Pisa, and Rome, under the guidance of (Matteo) Realdo Colombo (1516–1559) and Bartolomeo Eustachi (1500 up to 1510–1574). It is assumed that Valverde's move to Padua was due to the renowned *Studi Paduani* (currently known as the University of Padua), where anatomy and surgery were taught by Vesalius. A further highlight was the clinical teaching provided by Giovanni Battista da Monte (1489–1551) at the *Ospedale di San Francesco Grande*. G.B. da Monte introduced clinical medicine into the curriculum to integrate theory and practice. Thanks to him, medical undergraduates and physicians could acquire knowledge at the patient's bedside, perhaps for the first time in Christian Europe [22]. In Padua, G.B. da Monte established the first permanent anatomical theater and botanical garden in 1545 [23].

J. Valverde was interested in Vesalius but met R. Colombo in Padua. At that time—the academic course of 1542–1543 was beginning—Vesalius was involved in the composition of the "...*Fabrica*..." and was ready to move to Basel to prepare the printing with publisher Johannes Oporinus (1507–1568). For the duration of the course, the chancellor of *Studi Paduani* arranged for Pamphilius Montius to be the reader of Mondino de Luzzi's *Anathomia corporis humani* (a book written in 1316), R. Colombo to be the *sector* (i.e the surgeon in charge of dissecting the human body), and Paulus de Crassis to be the *ostensor* responsible for displaying the organs of the human body [24].

Vesalius resigned his chair at Padua after publishing "...*Fabrica*..." to become Emperor Charles V's *archiater*. He followed his father, the Emperor's pharmacist, in such a move [25]. R. Colombo officially assumed the Vesalius' chair of surgery and anatomy at *Studio Paduani* during the academic years 1543–1544 and 1544–1545. For the subsequent courses (1545 up to 1548), the Duke of Tuscany, Cosimo I de Medici, appointed R. Colombo as the chair of surgery and anatomy at Pisa University [23]. J. Valverde, who always acknowledged R. Colombo as his exemplary teacher, was at Colombo's side as a student and, presumably, as an assistant dissector [26]. At Pisa, J. Valverde assisted R. Colombo in investigating the minor (pulmonary) circulation of blood, among other matters [27]. With J. Valverde, R. Colombo dissected not only human bodies but also vivisected animals to study the functioning of the voice; the movement of the lungs, heart, and arteries; the dilation and contraction of the brain; variations in pulse; and other physiological functions [28]. Because of his discoveries, R. Colombo criticized Vesalius, and they engaged in resentful polemics with each other.

On the other hand, R. Colombo became a member of the School of Artists and developed a friendship with Michelangelo [29]. In August 1547, R. Colombo requested a license from the Duke of Tuscany and relocated to Rome, where Michelangelo had to illustrate R. Colombo's book, *De re anatomica libri XV* [30]. The book came out in 1559, but without images. However, R. Colombo and Michelangelo came together to share a great friendship. During R. Colombo's stay in Rome, Pope Paul III appointed R. Colombo to chair the anatomy course at the *Archigimnasio della Sapienza* (also named *Studium Urbis*; presently, *Sapienza-Università di Roma*). In the year 1548, R. Colombo made a lasting move to Rome. Gabriele Falloppio (1523–1562) succeeded him in Pisa. In 1549, Pope Paul III designated R. Colombo as *archiater*, and after the Pope's death, he was appointed surgeon of the conclave that elected Pope Julius III (papacy period: 1550–1555). This year, R. Colombo diagnosed and administered treatment to Michelangelo for nephrolithiasis [31]. R. Colombo became the first Chair of Anatomy at the *Studium Urbis* in 1552 [11]. He probably obtained the Degree of *Philosophia et Medicina* at the *Studium Urbis* around that time [32]. R. Colombo had been initially a sector, but later he became a physician of prestige, protected by Cardinal Girolamo Verallo (1497–1555), prefect of the Supreme Tribunal of the Apostolic Signature and member of the Roman Court of the Holy Office of the Inquisition. R. Colombo then served Juan Álvarez de Toledo (1488–1557), a son of the second Duke of Alba and a Cardinal and General Inquisitor in Rome himself.

While assisting R. Colombo in Rome, J. Valverde actively participated in the cultural and scientific life of the Metropolis. Relevant personalities of the ample Spanish colony and others met in *academia* and *salas*, such as the one at Palazzo Colonna. It included artists such as Michelangelo, Gaspar Becerra, and Pedro Rubiales; cardinals such as G. Verallo and J. Álvarez de Toledo; humanist physicians such as Juan Aguilera (?–1560) (who was at the service of Cardinal J. Álvarez de Toledo and physician of Pope Paul III), and Andrés Laguna and Luis de Lucena (1491–1552), who were both serving Pope Julius III [33]. J. Valverde and J. Aguilera were part of a scientific social group around Cardinal J. Álvarez de Toledo [11]. Thanks to the recommendation of Cardinal J. Álvarez de Toledo, J. Valverde was appointed physician at the *Ospedale di Santo Spirito in Sassia* in Rome in 1555. By that time, J. Valverde was thirty years of age. At the *Ospedale di Santo Spirito in Sassia* (Rome), he honed his clinical skills and devoted himself to teaching and conducting anatomical research. He also embalmed human bodies in Rome, first with R. Colombo, then by himself. Under Colombo's direction, J. Valverde autopsied Cardinal Innocenzo Cybo (1550) and Ignatius of Loyola (1556) [26]. J. Valverde was a highly esteemed physician among the nobility and the affluent. Discrepancies exist regarding whether he was a physician at the court of Pope Paul IV [34]. In the final months of 1557 or early in 1558, J. Valverde acted as a private messenger between King Philip II of Spain, who at that time was residing at Brussels, and Duke Cosimo I de Medici at Firenze, possibly in connection with the invasion of the Pope's states by the King Philip II of Spain [34] (see also the "Dedicatory" subheading below).

In the year 1558, J. Valverde may have returned briefly to his home place in Spain carrying with him a papal bull given by Pope Paul IV for the *Cofradía-Hospital de San Sebastián* ("Brotherhood-Hospital of Saint Sebastian") in Amusco [35]. J. Valverde's date of death is unknown, but he died in Rome, probably circa 1587.

3. Scientific Work

J. Valverde published two books, one in 1552 and another in 1556. He was aged 27 and 31 years, respectively. Both books were intended to disseminate medical and anatomical knowledge to scholars and non-scholar people. One of the books was titled *De animi et corporis sanitate tuenda libellus* and focused on hygiene. The other is *HISTORIA de la composición del cuerpo humano* and targeted on anatomy.

3.1. De animi et Corporis Sanitate Tuenda Libellus ("A Pamphlet on the Preservation of Mental and Physical Health")

This book deals with hygienic and sanitary issues frequently discussed by medical writers during the 16th century. The book is a work in the Latin language, printed in Paris in February 1552. Two editions were published (Figure 2). The initial edition was prepared in *Octavo* by Charles Estienne, a renowned printer and physician also known as Carolus Stephanus (1504–1564). In 164 pages, the book proposes secrets for health conservation and illness evasion. The book's small format, and therefore probable low-cost selling, possibly favored rapid dissemination among scholars but also those affluent curious who paid in exchange for learning the secrets of human nature.

The book's second edition was released in Rome the following year and printed at Domenico Giglio's print house in Venice (also known as Dominicus Lilius). This edition features updates such as a new frontpage, an index, a revised pagination system, and a list of typos. The author dedicated the book to Cardinal Girolamo Verallo (1497–1555), expressing gratitude with the words *"Vale mi Princeps studiosorum Patrone"*, which translates to "Farewell, my Prince, Patron of the studious".

Figure 2. Frontpages of the two editions of *De animi et corporis sanitate tuenda libellus* by Juan Valverde de Amusco.

3.2. *HISTORIA de la Composición del Cuerpo Humano ("HISTORY of the Composition of the Human Body")*

It is an opus written in the Castilian Spanish language. It provides accurate descriptions of human anatomy and explains the functions of the body and its parts. The overall organization of the book responds to a strict functional (Galenic) criterion [36]. The author aimed to not only educate on human anatomy but also to make the information widely available. *HISTORIA de la composición del cuerpo humano* ("*HISTORIA…*" henceforth in the present article) was the most broadly distributed anatomical publication after Vesalius during the *Cinquecento* [37]. It was the culmination of Valverde's research and the knowledge he acquired from R. Colombo and collaborators. The opus includes a cover, dedication, preface to readers, two indexes (one of the chapters and one thematic), seven books (each divided into chapters), and forty-two illustrated plates with figures. Following the princeps edition, many more editions were published (Table 2).

Table 2. Editions and reprinting of Valverde's *HISTORIA de la composición del cuerpo humano*, according to López Piñero [38] and Hernández-Mansilla [34].

Year	Language	Title	Print House/Publisher	Place
1556 *1	Spanish	*HISTORIA de la composición del cuerpo humano*	A. Martínez de Salamanca and A. Lafréry	Rome
1559 *2	Italian	*Anatomia del corpo umano*	Nicolò Bevilacqua	Venice
1560	"	"	Giunta	"
1586	"	"	"	"
1596	"	"	"	"
1606	"	"	"	"
1607	"	"	"	"
1608	"	"	"	"
1657	"	"	"	"
1682	"	"	Giunta/Niccolò Pezzana	"
1589 *3	Latin	*Anatome corporis humani*	Michele Colombo	Venice
1607	"	"	"	"

Table 2. *Cont.*

Year	Language	Title	Print House/Publisher	Place
1566 *4	Latin		Christophe Plantin	Antwerp
1572	"	*Viuæ imágenes partium corporis*	"	"
1579	"	*humaniæreis formis expressæ*	"	"
1568	Dutch	*Anatomie, oft levende beelden vande deelen des menschelicken lichaems: met de verclaringhe van dien, inde*	Christophe Plantin	Antwerp
1583	"	*Neder-duytsche spraecke*	"	"
1583	Dutch	*Bedieninghe der anatomien*	David van Mauden	Antwerp
1646	"	"	"	"

*1 Available at https://books.google.es/books/ucm?vid=UCM5320265722&printsec=frontcover&redir_esc=y#v=onepage&q&f=false (accessed on 4 December 2023); *2 Available at https://collections.nlm.nih.gov/catalog/nlm:nlmuid-9617625-bk (accessed on 4 December 2023); *3 Available at https://books.google.nl/books?id=x_72RuDANkYC&hl=es (accessed on 4 December 2023); *4 Available at https://patrimoniodigital.ucm.es/s/patrimonio/item/578975 (accessed on 4 December 2023). To view the book's contents, please ignore the pop-up and scroll down the websites.

3.2.1. Princeps Edition

The opening publication of "*HISTORIA*..." originated from the presses of the Spaniard Antonio Martínez de Salamanca (c. 1478–1562) and the Frenchman Antoine Lafréry or Lafrerij (1512–1577) in Rome in the year 1556. A. Martínez de Salamanca was instrumental in spreading Spanish culture in Rome. He was concurrently a partner in the printing of "*HISTORIA*..." and Lafréry's primary adversary in the Roman publishing industry [26].

Frontpage and Imprimatur. The first word of the book title is HISTORIA [sic] (Figure 3). Valverde gave the book explicit meaning by emphasizing the parallelism between the description of natural facts of the human body through experience and the story of human events [34]. There is printing permission granted by Pope Paul IV. It also includes a warning of automatic excommunication and a fine of 100 gold ducats for those who copy the book without the author or editors' permission during the next 10 years after that edition.

Dedicatory. After initially considering dedicating "HISTORIA..." to Pope Paul IV [26], J. Valverde decided to dedicate the book to his protector, Cardinal J. Álvarez de Toledo, who was Pope Paul IV's confessor. The precise cause of this shift remains uncertain. Nonetheless, it might be associated with the ongoing conflict between Pope Paul IV and King Philipp II of Spain, a component of the Italian War of 1551–1559. Chiefly, it can be attributed to Pope Paul IV's unsuccessful attempt, with King Henry II of France, to remove King Philip II of Spain as king of Naples in 1556, and to the sub-sequent preparations for the arrival of King Philip II's troops in Rome. This entry fi-nally took place in 1577 and was commanded by Fernando Álvarez de Toledo, third Duke of Alba, and nephew of Cardinal J. Álvarez de Toledo.

The dedicatory provides a clear picture of Valverde's position regarding Vesalius. J. Valverde, acknowledging Vesalius as the master whom he would always follow, "except for certain instances where he exhibited less diligence than necessary (presumably due to fatigue from the arduous task at hand), which I shall note". (Without further comment, this quote and similar ones that follow in this article are translations of Valverde's original Spanish text.)

Preface to Readers. Once again, there is an explicit eulogy for Vesalius' authorship and merit. J. Valverde wrote, "Even though some friends of mine thought I should make new figures, without using the ones from Vesalius, I haven't done it because I want to evade confusion [...] and because his figures are so well-made that it would be, to me, envy or meanness not making use of them".

Next, J. Valverde explained that all Figures belonging to any of the seven "*HISTORIA*..." books are put at the end of the corresponding book because, "being printed from copper

engravings, they could not be mixed with it [the main text] without producing confusion".
J. Valverde also offered a lengthy explanation about the letters and end notes that appear
in each of the legends to the figures [39]. By placing all images at the end of the books, he
made the final format smaller to make it less expensive than other anatomical books like
Vesalius' "…*Fabrica*…" [32].

Figure 3. Frontpage of the princeps edition of *HISTORIA de la composición del cuerpo humano* by
Juan Valverde de Amusco. It is according to the symbolic custom of the time. At the center of the
composition, two telamons hold the coat of arms of Cardinal Juan Álvarez de Toledo, protector of
Juan Valverde de Amusco. The coat of arms features a checked vertical oval surrounded by nine
pennants that represent the possessions and titles of the Alba family, including manors, counties,
marquessates, and duchies. Topping the coat of arms is a cross with trefoil endings and two *putti*
holding the cardinal hat. Below the telamons' feet, the title and author of the book stand within a
horizontal oval surrounded by four shells (two representing birth and life; the third one is the head
of a gargoyle sited there to ward off demons and evil spirits; the fourth item is the skull of a ram,
symbolizing the transience of life).

Books and Chapters. According to Renaissance criteria, "*HISTORIA*…" is divided
into books, each divided into chapters. "*HISTORIA*…" encompasses seven books, akin
to Vesalius' "…*Fabrica*…", but the topics of the books are different in the two opuses
(Table 3). "*HISTORIA*…" and "…*Fabrica*…" do not share either the descriptive order or the
conceptual idea instilling it. "…*Fabrica*…" clearly demonstrates a constructing sequence
comprising supporting structures, union elements, organs, and entrails, according to their
importance categorization. The order in "*HISTORIA*…" varies from book to book. It
is imperative to emphasize again that "*HISTORIA*…" is an opus that served as both an
anatomy book and a popularizing science book, with a significant degree of overlap; in it,
there are chapters intended for general notions, such as Book I, Chapter I; Book II, Chapters

I–II and IV; Book VI, Chapters I–II; Book VII, Chapter I. J. Valverde constructed the general notions produced in "*HISTORIA*…" by abstracting from observation as detailed as possible of anatomical structures, with a rigorous realist criterion. Such a criterion is like that of Vesalius, but distinct from that of Galen and their followers, for whom the form, function, and teleological reasons of the form regarding function are inseparable [40].

Table 3. Book content in Valverde's *HISTORIA de la composición del cuerpo humano* and Vesalius' *De humani corporis fabrica libri septem.*

Book	Valverde's "*HISTORIA*…"	Vesalius' "…*Fabrica*…"
First	*Huesos y ternillas* (bones and cartilages)	Bones and joints
Second	*Ligamentos y músculos* (ligaments and muscles)	Ligaments, muscles, and integumentum
Third	*Miembros de la digestión y la generación* (digestive and reproductive organs)	Veins, arteries, and glands
Fourth	*Miembros de la vida* (life members (lungs and heart))	Nerves and spinal cord
Fifth	*Miembros necesarios al movimiento y sentido* (members needed to sense and movement; encephalon)	Organs of nutrition and generation
Sixth	*Venas y arterias* (veins and arteries)	Heart and associated organs
Seventh	*Nervios* (cranial and spinal nerves)	Encephalon

Books I and II are concerned with body structure, and their descriptive order is morphological. In Books III–V, however, following the Platonic doctrine, the organic cavities are categorized according to their functional order, specifically the rational/cranial, vital/thoracic, and vegetative/abdominal, which would be the respective locations of the three souls: immortal (*tò logistikón*), irascible (*tò thymoeidēs*), and concupiscible (*to epithymētikón*) [41]. In Books VI–VII, ducts, which include veins, arteries, and nerves, are elucidated. Here, J. Valverde disagreed with Galenic theory. Galen's theory correlates veins, arteries, and nerves with the three spirits (pneuma natural, pneuma vital, and pneuma animal) and the spirit generator organs (liver, heart, and encephalon). Galen's theory proposes that pneuma flows mixed with blood through the pulmonary artery and veins. J. Valverde adhered to Colombo's concepts, positing that only blood flows through arteries and veins [42]. Precisely, J. Valverde described the pulmonary circulation of blood (while acknowledging R. Colombo as the author of the discovery). J. Valverde considered that the function of the lungs is to receive air and obtain it for the fabrication of life spirits. Furthermore, the lungs refresh the heart of excessive heat by blasting fresh air. In 1553, three years before the publication of "*HISTORIA*…", the Aragonese Michael Servetus (Villanueva de Sigena, Crown of Aragon, presently Spain; 1511–1553), in his theological treatise *Christianismi restitutio*, had described, for the first time, the pulmonary or minor circulation in the Christian West [43,44].

J. Valverde was not interested in the philosophical discussion of nature, even though Galen's theory was a part of his physiological thinking. He was interested in revealing the body structures that serve as a substrate for the function and its diffusion throughout the body. This view distinguished J. Valverde from other anatomists and medicine theorists —for instance, Thomas Willis (1621–1675), the great experimentalist physician [45]—who often entangled in theoretical disputes about the nature of the spirits, their origin, and the action mechanism of *animae* (movements).

Illustrations. "*HISTORIA*…" was the first anatomy book to use illustrations created with a burin-over-copper technique, which allowed for increased precision and elegance in tracing [46]. The opus contains 42 anatomical plates consisting of 214 numbered images that have caught the attention of historians for their beauty and execution. While some argue that these illustrations tarnish Valverde's reputation for originality, he publicly acknowledged that he derived complete inspiration from "...*Fabrica*..." for their conception. Nonetheless, the opus includes 15 new illustrations, 42 corrections, and 9 clarifications to "...*Fabrica*..." [47]. The artists in charge of drawing and engraving the illustrations in "*HISTORIA*..." are not credited in the opus.

Novel anatomical contributions.

Osteology and Myology. J. Valverde initially described the vomer bone in a written text in "*HISTORIA*...". The discovery probably came out from dissections performed by Colombo's school members, of which J. Valverde was part. J. Valverde states in Book I of "*HISTORIA*...", "The vomer is between the cuneal bone and the palate bones. This bone looks like a plow, Vesalius does not mention it because it lacked, to him, momentum". In the Castilian Spanish language, the term "*Arado*", or "*Reja del Arado*", refers to the plow. In the Latin language, vomer means plow. R. Colombo wrote about this bone in his book *De re anatomica libri XV* (1559) [48].

J. Valverde presented a detailed analysis of the teeth and their supporting structures, vascularization, and innervation in "*HISTORIA*...". He also described techniques for reducing mandibular dislocation [49].

The print of the stapes ossicle appeared in "*HISTORIA*..." for the first time. J. Valverde wrote of it, "...which nobody before me has even mentioned" (Book I, Table V, Figure IIII) (Figure 4). It appears that Giovanni Filippo Ingrassia (1510–1580) was the first to name stapes this ossicle as early as 1546, but this remained unknown until the publication of *In Galeni librum of doctissima ossibus et expertissima commentaria* in 1603. B. Eustachi (c. 1500–1574) also named stapes the ossicle in his magnificent book titled *Tabulae anatomicae* (completed in 1552 yet published in 1714). However, in any case, the first published book in which the term stapes appears for naming the ossicle is the book *Cl. Galeni Pergameni liber de ossibus ad tyrones [...] Medicae doctore* (1555) by L. Collado (Table 1).

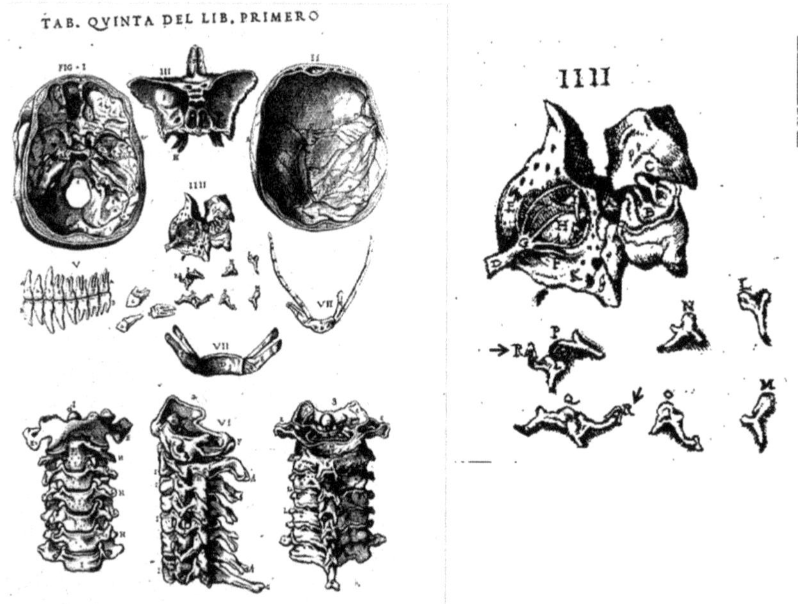

Figure 4. On the left side of this figure is the copy of Table V from Book I of *HISTORIA de la composición del cuerpo humano* by *Juan Valverde de Amusco* (available at: https://books.google.es/books/ucm?vid=UCM5320265722&printsec=frontcover&redir_esc=y#v=onepage&q&f=false; please, ignore the pop-up and scroll down the website;

the last time consulted was 4 December 2023). It displays various bones, including the auditory ossicles. On the right side is a magnification of Figure IIII from Table V of the same book. Letters L–M indicate the first ossicle (the malleus), while N–O represents the second ossicle (the incus). The letter R indicates the third ossicle (the stapes, pointed at by an arrow). Importantly, this is the first instance that the stapes appears illustrated in an anatomy book.

Regarding muscles, there are thirty-two corrections to Vesalius in *"HISTORIA..."*. They concern oculomotor muscles, facial muscles, throat muscles, and muscles of the palm and plant [50]. In Book II, Chapter VII, Valverde wrote, "The eye muscles are four in number, with a fine consistency, and arranged in four directions. This allows the eye to move in four directions.". J. Valverde categorically affirmed the nonexistence of what Vesalius calls the *"coanoid muscle"* or *retractor bulbi*. Galen also mentioned this muscle. Valverde argued that Vesalius found this muscle in dissections of animals, but not human bodies [51].

Neurology and Angiology. In Table I of Book V, the brain is depicted together with the dura mater. The name dura mater originates from the Greco-Latin Antiquity and Islamic Golden Age literatures, but it was first used in Christendom in *"HISTORIA..."*. In addition, J. Valverde elucidated, for the first time, the commencement of the intracranial course of cerebral arteries. He, however, maintained that there is minimal distinction between arterial vessels and veins. G. Falloppio later corrected this error in his *Observationes anatomicae* (1561) [52].

First post-Vesalian anatomical publication in a Romance language. Terminology only has value when used [53]. Besides his anatomical contributions, J. Valverde played a noteworthy role in developing and disseminating anatomical terminologies written in Castilian Spanish, a vernacular Romance language. Hence, he pioneered this practice, which remains associated with the official *Terminologia anatomica* until today [54].

Vesalius' *"...Fabrica..."* was intended for physicians and Latin surgeons and was written in a dark style of the Latin language that is difficult to comprehend even by scholars. J. Valverde penned his *"HISTORIA..."* in the Castilian Spanish language of the era. It was for physicians, Latin surgeons, barber surgeons, midwives, and algebraists (bonesetters) in Spain and its territories, as well as for paramedic practitioners without university education, who had difficulty understanding most anatomy texts written in Latin, but who were "those who most need to understand", as J. Valverde wrote in *"HISTORIA..."*. Authors suggest that this must have been under the direction of Cardinal J. Álvarez de Toledo, as the Cardinal thought the book would be highly convenient for the Spanish nation [11]. Perhaps this is the rationale, or an additional one, behind the dedication of the princeps edition to the Cardinal (but also consider the reasons above, in the "Dedicatory" subheading).

It is worth noting that R. Valverde was not the only anatomist to write in Castilian Spanish in the 16th century. Luis Lobera de Ávila (c. 1480–c. 1551), a physician in Emperor Charles V's House, had already started that venture a little earlier. L. Lobera studied anatomy in France and authored several books in Castilian Spanish covering topics such as medicine, hygiene, and diet and nutrition. One of his works, written in 1542, was the book *Libro de anatomía, es primera parte de "Remedio de cuerpos humanos y silva de experiencias y otras cosas utilísimas"* ("Anatomy book, it is the first part of 'Remedy of human bodies and forest of experiences and other very useful things'"). Another anatomist, Bernardino Montaña de Monserrate (c. 1480–c. 1558), who also served Emperor Charles V, wrote the *Libro de la anathomia del hombre* ("Book of the human anatomy") in 1551 and taught anatomy at the University of Valladolid (Crown of Castile, presently Spain) [55]. However, J. Valverde named more parts of the human anatomy in Castilian Spanish than the other authors and even changed some of the names they used. Many of Valverde's terms have been used in classical literary works, such as Don Quixote by Miguel de Cervantes (published in 1605 and the second part in 1615) [56].

J. Valverde was concerned with form, function, and position to coin the Castilian Spanish term for an anatomical part [40]. In the book *"HISTORIA..."*, Valverde elucidated anatomical terminologies using words derived from vulgar discourse, including many derived from the customary slaughter of pigs [52]. J. Valverde also enriched the anatomical vocabulary with synonyms. When J. Valverde found the common language insufficient or inadequate, he used numbers, such as when he talked about the cartilages of the larynx and wrist bones. Despite losing some rigor, Valverde's terminology gained much for paramedic practitioners and the public to understand [40]. It is enticing to consider why J. Valverde assembled the images of the book from burin-over-copper-engraved plates of such high quality for the non-scholarly and less affluent public. This would surely have a high economic cost. It appears that the success of the book was worth the cost and that the magnificent figures ultimately gave J. Valverde most of his scientific legacy.

3.2.2. Diffusion of the Work

Later editions gave *"HISTORIA…"* and its figures even broader access than that of the works of Vesalius and R. Colombo to readers in Catholic and Reform countries [37].

Nine complete editions in Italian (1559 up to 1682). *"HISTORIA…"*, or its figures, accumulated 18 editions in several languages following the princeps edition (Table 2). The initial nine were in Italian, translated by Antonio Tabo de Albenga, complete, and particularly well known. The title was *Anatomia del Corpo Umano* and the dedication to King Philip II of Spain. The first one of the nine, although printed in Venice, appeared in Rome in 1559 from the same publishers as the princeps edition. The 1586 edition was the most famous of these nine Italian editions. Then, J. Valverde was about 61 years of age, and his demise was imminent. The frontpage of the 1586 edition has an engraved portrait of Valverde (Figure 1) achieved between 1561 and 1565 [21]. It is one of the few portraits of the anatomist that have survived to present day. This 1586 edition has 46 plates with 253 figures, of which 4 are new and interpolated according to the 42 plates of previous editions. All these new plates depict "muscle men" [57].

Two complete editions in Latin (1589 and 1607). In 1589, Michaele Colombo released a whole Latin version of *"HISTORIA…"*, reprinted in 1607 (Table 2). M. Colombo was a son of R. Colombo and physician, philosopher, and translator of some works of Girolamo Mercuriale. The dedicatory is to the Duke of Savoy in this 1589 edition. Such an honoring appears to be very unlikely to have occurred in Valverde's life, which strongly suggests that Valverde died before 1589 [21,26].

Seven editions of *"HISTORIA…"* figures as a part of other books (1566 up to 1646). *Viuæ imagenes partium corporis humani æreis formis expressae* ("Realistic figures of the human body parts from brass molds outputs") is a medley book in Latin published by Christophe Plantin (c. 1520–c. 1589) in Antwerp three times from 1566 up to 1579. The authorship of the book is currently credited to J. Valverde de Amusco, J. Grévin, and A. Vesalius, in that order, as can be seen in https://catalog.nlm.nih.gov/discovery/fulldisplay/alma992329263406676/01NLM_INST:01NLM_INST), or to J. Valverde de Amusco (https://wellcomecollection.org/works/zyhu6y23) (the last time consulted was 4 December 2023). The figures are from *"HISTORIA…"* by J. Valverde. The main text is from Vesalius' book *Andrea Vesalii suorum de humani corporis fabrica librorum epitome* of 1564 (*"…Epitome"* henceforth). *"…Epitome"* is an abridged version of *"…Fabrica…"*. Some short Latin-language texts and one Table are from Jacques Grévin's *Partium omnium corporis differentiae* (c. 1565). Figure legends were translated from the Castilian Spanish language into Latin by the French physician J. Thorius. Yet, the medley lacks commentary on the figure legends that precede every book in whole editions of *"HISTORIA…"*. No text written by J. Valverde appears in the miscellany. With this, any of Valverde's recognition of Vesalius' merit as an anatomist, or the initial Vesalius influence of many of the images in *"HISTORIA…"* disappears. Valverde's authorship is not credited on the frontpage of the Plantin medley. There is also no author list in it. Valverde's authorship of the medley images is only recognized in the dedication that C. Plantin makes to the Senate of Antwerp. Moreover, the medley dedicatory details the steps C. Plantin undertook to obtain the permission of Andreas Wechelus (died 1581) for reproducing the text of the Paris edition of 1564 of the Vesalius' *"…Epitome"* [39]. There is no mention of a printing permit issued by J. Valverde or his editors. To the best of the knowledge of the authors of the present article, it is unknown if this printing had the permission of J. Valverde or his editors. Perhaps it was not necessary because Plantin's medley appeared ten years after the publication of "*HISTORIA…*" princeps edition (see above, princeps frontpage and imprimatur subheading), or because the warning issued by Pope Paul IV was improper at the time of the armed conflict between Flemish Protestants and the Catholic King Philip II of Spain.

Plantin's medley featuring Valverde's "*HISTORIA…*" figures is among the first books printed with copperplate-engraved figures by Plantin Print House. It was a turning point for such a renowned print house [58]. Pieter Huys (c. 1519–c. 1584) engraved the copperplates with his brother Frans. The result was an anatomy synthesis with a practical goal: to satisfy the demand of physicians, Latin surgeons, and medical students. It is undeniable that the understanding of the anatomy of the human body possessed and possesses a large and competitive market. A brief dedicatory from C. Plantin to those who study medicine (*"Artis medicae studiosis"*) underlines the utilitarian approach of the edition. There is also the announcement of a German dictionary, which was in preparation, the *Thesaurus Theotonicae linguae*, published seven years later, in 1573 [59].

Anatomie, oft levende beelden vande deelen des menschelicken lichaems: met de verclaringhe van dien, inde Neder-duytsche spraecke. ("Anatomy, or living images of the parts of the human body: with the explanation thereof, in the Dutch language") is the same miscellany as above, but in Dutch, and published twice (1568 and 1583; Table 2). The favorable reception received by the publication of the medley *Viuæ imagenes partium corporis humani æreis formis expressae* prompted C. Plantin to translate the complete version of the book into the Dutch language. Current references to this book place J. Valverde de Amusco as the author (see https://books.google.es/books?id=tiEttwAACAAJ&printsec=frontcover& hl=es&source=gbs_ge_summary_r&cad=0#v=onepage&q&f=false; moreover, you can browse the book contents, including the illustrations, in this link (the last time consulted was 4 December 2023)).

Bedieninghe der anatomien ("Operation of anatomy") (1583 and 1646; Table 2). Other medleys of anatomy (collected from texts by Galen, Vesalius, Falloppio, and Arantius and accompanied by anatomical plates by Vesalius and Colombo) appeared in Antwerp. David van Mauden (c. 1538–c. 1597) was the author of the book, in Dutch, entitled *Bedieninghe der anatomien*, with explicit reference to J. Valverde as the author of Figures [60]. Plantin Print House published it in 1583 and again in 1646.

Unpublished Greek language complete edition. In the 18th century, Kousis translated the book "*HISTORIA…*" into Greek, though he did not make it to print [61].

4. Reservations

Due to the wide use of the Castilian Spanish and Italian Romance languages and the beauty and anatomical detail of the images, Valverde's "*HISTORIA…*" had a broad impact. Nevertheless, the images of "*HISTORIA…*" have been regarded as a copying of Vesalius' "*…Fabrica…*". The following examines the authorship of these illustrations to discuss whether there is innovation or plagiarism in them.

4.1. Illustrations

A highly influential factor in the success of Valverde's "*HISTORIA…*" was the splendorous 214 numbered figures it contains; furthermore, these images are an essential support for the main text [46]. Vesalius' "*…Fabrica…*" has 379 Figures in its princeps edition (1543) and 4 more in the second edition (1555). The resemblance between the "*HISTORIA…*" figures and the "*…Fabrica…*" figures holds the attention of most historians who have discussed Valverde's book. Some authors have marked the "*HISTORIA…*" figures with the stigma of plagiarism.

Comparative studies of "*…Fabrica…*" and "*HISTORIA…*" have shown that many of the images in "*HISTORIA…*" that are similar to the matching ones in "*…Fabrica…*" contain significant variations that not only improve the quality and clarity [50] but also illustrate more anatomy particulars for the first time [62]. Not least, there are, in *"HISTORIA..."*, 15 wholly original anatomy figures [57]. In other words, *"HISTORIA..."* holds 15 plates (out of 42) that are entirely or partially non-Vesalian [50] (Figure 5).

4.2. Artistic Authorship

Indeed, making artistic illustrations of *"HISTORIA ..."* exhibits technical superiority over those produced by Calcar for Vesalius' *"...Fabrica..."* [46,63]. The artists who made the drawings and copper plate engravings for the images of *"HISTORIA..."* are not mentioned in full name in the book, as was usual then.

Several artists might be the makers of the *"HISTORIA ..."* images. Namely, Gaspar Becerra [64–66], Pedro Rubiales [63], and Nicolas Béatrizet [52]. In Book II, Table III of the princeps edition, Valverde explicitly referred to P. Rubiales and Michelangelo, of whom he said, *"por haberse dado a la Anatomía juntamente con la pintura han venido a ser los mas excelentes y famosos pintores que grandes tiempos han visto"* ("for having given themselves to the anatomy along with painting they have become the most excellent and renowned painters who have seen great times"). However, this reference was a recognition of the artistic and humanistic qualities of P. Rubiales and Michelangelo but not of authorship in Valverde's book. J. Valverde, G. Becerra, P. Rubiales, and Michelangelo were customary of the same circles in Rome at the time. G. Becerra, who had worked in the studio of Michelangelo, was the author of Valverde's portrait at the Walters Art Museum of Baltimore, Maryland, U.S.A [21]. G. Becerra and P. Rubiales presumably collaborated in the preparatory drawings of the *"HISTORIA..."* figures. Probably, P. Rubiales drew the sketches for the entirely original figures or those bearing a more artistic classic theme: for example, those decorating the anatomical preparations with sculptural naked torsos [67].

The engraver on copperplates would have been N. Béatrizet and the craftsman in charge of opening the copper leaves Thomas Barlachi [61]. N. Béatrizet worked between 1540 and 1562 for A. Lafréry and A. Martínez de Salamanca; N. Béatrizet's initials "NB" appear on several of the figure plates in *"HISTORIA..."* princeps and subsequent editions, and as well in Valverde's portrait incorporated in the *"HISTORIA..."* edition of the year 1589.

One of the *"HISTORIA..."* princeps edition's most striking original figures is that of an *écorché* with its skin hanging from its right hand while wielding a dagger with the left, suggesting the *écorché* has inflicted upon itself the skinning (Figure 6A). In addition, the face with the sagging skin has some resemblance to Valverde's face (Figure 6B).

4.3. Innovation or Plagiarism

The initial opus *"HISTORIA..."* came out of the presses in 1556. Vesalius accused *"HISTORIA..."* of plagiarism in a posthumously published book (1564), titled *Anatomicarum Gabrielis Fallopii observationum examen*—where Vesalius criticized *en passant* the scientific ambiance of Charles V's court. Vesalius stated in the book that he did not understand how G. Falloppio could consider J. Valverde a great anatomist. Vesalius added regarding J. Valverde: *"Qui manus sectioni nunquam adhibuit, & medicinae, viti & primarum disciplinarum, est ignarus, & in Hispanam linguam interpretis tantum in nostra hac arte munus, turpis quaestus causa obit"* ("He, who never used his hands for cutting or for medicine, and not for the vines, is ignorant of the main disciplines. And he practices the job of translator into the Hispanic language in this our art only because of filthy lucre"). In the interim, R. Colombo and Vesalius had become bitter rivals by 1555. R. Colombo had censured Vesalius for the same thing that Vesalius did later concerning J. Valverde, that is, avoiding the dissection of human bodies by himself, thus depicting the anatomy of animals instead of humans in his books [28]. Meanwhile, Vesalius severely criticized the findings and merits of R. Colombo in the same way and for the same reason. In the book *"HISTORIA..."*, J. Valverde criticized Vesalius precisely for an equal cause.

All the above suggests that Vesalius' unforgiving criticism of J. Valverde was plausibly directed also towards R. Colombo and those who had dissented from some of Vesalius' findings and techniques. It is reasonable to think too that Vesalius' hostility was venomously aimed at J. Valverde because "*HISTORIA...*" was the first post-Vesalius book—one in which Vesalius' influence is candidly acknowledged besides—but also the first one that explicitly corrected Vesalius, and many times. To make matters worse, "*HISTORIA...*" was a successful book written in a widespread common language and promptly translated into Italian. All this probably irritated Vesalius because he thought it damaged his authority [46]. It looks like a vicious circle of jealousy caused by public reconnaissance of merit and, not least, conceivable prospects of bookselling.

Subsequent high appreciation of Vesalius' works by traditional historiography makes that his critical reference to J. Valverde stays as "the" truth [68]. It might well be that the Black Legend helped do it. Some factors are to be thought of here, such as (A) Valverde's recognition of Vesalius' initial authorship of many images of "*HISTORIA...*" is suppressed in the miscellanies published by C. Plantin in Antwerp and (B) the rebellion against the rule of King Philip II of Spain in the Seven Provinces, and Spain as a global power in many other places, may have prevented J. Valverde from defending his author's morality and copyrights. They were times of war, with all the consequences.

Therefore, Vesalius' derogatory opinion is the one that most frequently has prevailed. It has even led to extreme positions, from absolute ignorance about Valverde's works [69,70] to apologetics defense, even nationalistic [13,71]. Still, consideration of "*HISTORIA...*" just as a mere translated copy of Vesalius' "*...Fabrica...*" [72] or respect for Valverde as a contributor and diffuser of the Vesalius revolution [73,74] always remained. Thus, Johann Gottfried von Berger (1659–1736) declared: "*Fama itaque meritissima Valverdus frueatur, livore etiam frustra obnubilante*" ("Valverde has great fame, despite envy that in vain beclouds") [75]. Albert von Haller (1708–1777) asserted: "*Minorem sanguinis circulationem non ignoravit*" ("He was not ignorant of minor circulation of blood") [76]. François-Joseph-Victor Broussais (1772–1838) acclaimed "*HISTORIA...*" "because it was the only anatomy book published in Spanish-Castilian during the Renaissance" [77]. An untrue fact, since F.-J.-V. Broussais surely was ignorant of Lobera's book *Libro de Anatomía, es primera parte de "Remedio de cuerpos humanos y silva de experiencias y otras cosas utilísimas*" of 1542 and Montaña de Montserrate's book *Libro de la anathomia del hombre* of 1551 (Table 1). Remarkably, figures from "*...Fabrica...*" and "*HISTORIA...*" were precedents of equal merit to Figures in *Teşrih-ül Ebdan ve Tercümânı Kıbale-i Feylesûfan*, the first illustrated anatomy book written in the old Turkish Ottoman language [78]. This book was handwritten in the 17th century and reproduced many times in the 18th century, always during the long-lasting rivalry between the Ottomans and Christians. The authorship of Vesalius and J. Valverde is unacknowledged in the book.

For all the above, plagiarism cannot describe Valverde's work. Plagiarism applies to publishing data that belong to someone else—including applications for grants and a publication submitted in a different language—yet, notably, without precise reference to the genuine author. Other acts of plagiarism are paraphrasing without crediting the source, using "blanket" references, "second-generation" references, and duplicating or repetitive publication of one's own previously published work [79]. A significant historical example was the one by the renowned William Cowper (c. 1666–c. 1709) [80]. W. Cowper, a student in Leiden of Frederick Ruysch (1638–1731), who established the finest anatomical museum in the 17th century, purchased the copperplates of Govert Bidloo's anatomical book (1685), which were prepared by the Belgian painter Gerard de Lairesse (1641–1711) and engraved by Pieter van Gunst (1658 or 1659–1732). W. Cowper added his improved text but did not credit the plates to G. Bidloo. The amalgamation of Bidloo's plates and Cowper's text produced "the most elaborate and beautiful of all 17th century English treatises on anatomy and also one of the most extraordinary plagiarisms in the entire history of medicine" [81].

On the contrary, in the princeps and subsequent full editions of "*HISTORIA...*" (besides those medleys published by C. Plantin in Antwerp), J. Valverde recognized Vesalius' initial authorship and why Valverde wished to take advantage of the images of "*...Fabrica...*". In addition, explicit acknowledgment of the debt of knowledge that the student (J. Valverde) has with the teacher (R. Colombo) appears in the book. R. Valverde assigns R. Colombo authorship in discovering pulmonary circulation and other anatomical and physiological facts. Again, in the dedicatory to King Philip II of Spain in the first Italian edition of "*HISTORIA...*" (1559), J. Valverde states that, "*seruitomi in essa per la maggior parte delle figure del Vessalio, per parermi più degne d'imitatione, che di biasimo: Successe dapoi, che molti non intendendo la lingua Spagnuola, & vedendo le mie figure non-molto diuerse da quelle, cominciarono à dire ch'io hauea tradotta l'historia del Vessalio*" ("making the best use of Vesalius' figures because I find them worthy of imitation more than of censorship, it happened since that moment that many of those who did not understand Spanish, seeing my figures not very different from his, started saying that I had translated my history from Vesalius").

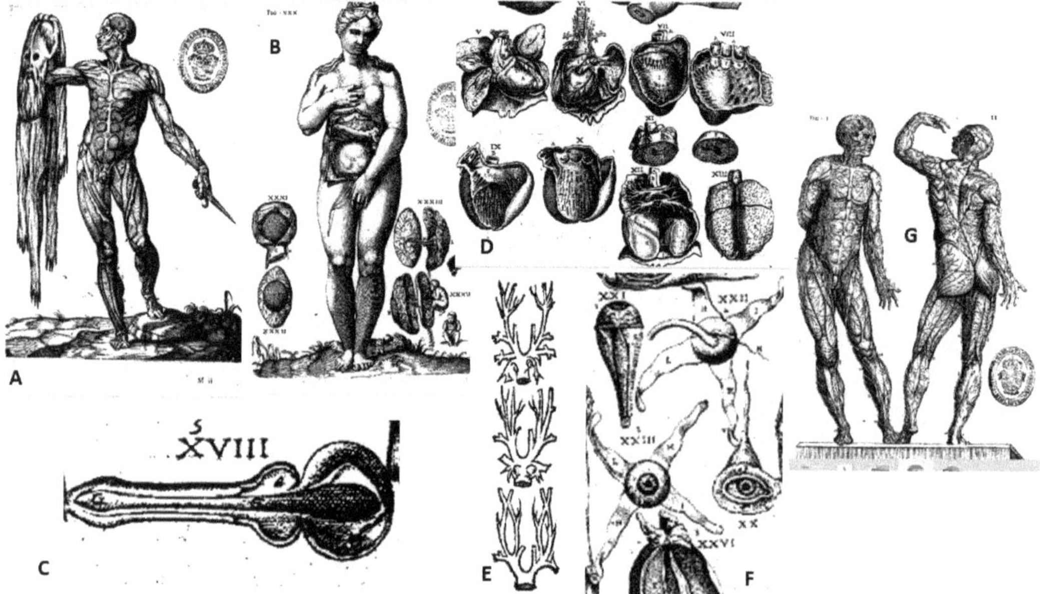

Figure 5. Anatomical illustrations in Valverde's "*HISTORIA de la composición del cuerpo humano*" about which there is consensus on originality. According to Alberti-López [50]; Meyert and Wirt [57]; Huard and Imbault-Huart [62] and Skaarup [82]. (**A**) An écorché with its skin hanging from the right hand and wielding a dagger with the left as if the écorché has inflicted upon itself the skinning (Book II, Plate I, Fig I). (**B**) A Venus with an open abdomen and a detached fetus in the lower right-hand corner of the same plate (Book III, Plate VI, Figure XXX). (**C**) A longitudinal section of the urinary bladder, prostate, urethra, and penis (Book II, Plate XVI, Fig XVIII). (**D**) Images depicting the opened heart (Book II, Plate XV, Figure XX to XXIII). (**E**) A marginal illustration, a diagram, on Book VI, page 87 (reverse side) of the princeps edition and page 123 (obverse side) of the first Italian edition. It depicts the origin of great vessels from the heart. (**F**) An eyeball with its fascial sheath, optic nerve and palpebral fissure, and the extraocular, recti eye muscles (Book II, Plate XV, Figure XX, and Figures XXI to XXIII, respectively). (**G**) Two-skinned standing men (front and back views) with visible subcutaneous veins (Book VI, Plate I, Figures I and II).

Figure 6. (**A**) This image is from *HISTORIA de la composición del cuerpo humano* by Juan Valverde de Amusco. This image has no relation to those in *De humani corporis fabrica libri septem* by Andreas Vesalius. This image shows an écorché with its skin hanging from its right hand while simultaneously wielding a dagger in his left hand, indicating that the écorché is self-skinned. The face in the engraving seems to resemble that of J. Valverde de Amusco. (**B**) There is a similarity between this écorché depicted in *HISTORIA de la composición del cuerpo humano* and the one hanging from Bartholomew's hand, as seen on the ceiling of the Sistine Chapel in the Judgment Section Final, frescoed by Michelangelo between 1536 and 1541. On the other hand, the body posture of the écorché depicted in *HISTORIA de la composición del cuerpo humano* is comparable to that of classical statues, such as (**C**) the Apollo Belvedere in the Vatican Museums, which was a source of inspiration for Michelangelo's David [83] and (**D**) the *Ephebus of Antikythera* in the National Archaeological Museum of Athens.

5. Conclusions

For everything said so far, the first conclusion is that it is unwise and unfair to say Valverde's "*HISTORIA. . .*" is plagiarism. That the book ended up being a selling success must be entirely attributed to the science and judgment of J. Valverde, who published with practical purposes in two widely spoken Romance languages a popularizer anatomical work that was well founded, timely, and with appropriate references to authorship of others. The second conclusion of this article is that, because of his activities as a dissector and author, Juan Valverde de Amusco pioneered the implementation and transfer of post-Vesalian anatomical scientific knowledge.

6. Epilogue

During the 15th and 16th centuries, the study of anatomy made significant advances. The progress was due to observing and interpreting nature directly. From that point forward, the dissection of human bodies, whether macroscopic, microscopic, chemical, or using any other technique, was the primary study method in anatomy. During the Renaissance period, several authors made significant contributions to gross anatomy. These authors include Jacopo Berengario da Carpi, who wrote *Isagoge breves* in 1522, regarded as a turning point by Gabriele Falloppio. Nicolò Massa wrote *Anatomiae libri introductorius* in 1536, while Charles Estienne authored *De dissectione partium corporis humani libri tres* [84] in 1545. Andreas Vesalius contributed highly notably with his work *De humani corporis fabrica libri septem* in 1543. Historians consider Andreas Vesalius the epitome of Renaissance anatomy. At the same time, other notable Renaissance anatomists include Bartolomeo Eustachi (*Tabulae anatomicae*, completed in 1552 yet published in 1714), (Mateo) Realdo Colombo (*De re anatomica libri XV*, 1559), and Gabriele Falloppio (*Observationes anatomicae*, 1561) [68]. (Mateo) Realdo Colombo and Andreas Vesalius were known for their public rivalry. They criticized each other's findings and merits, like how Andreas Vesalius also criticized Juan Valverde de Amusco. *HISTORIA de la composición del cuerpo humano* (1556) incorporates knowledge of anatomy with pragmatism. In this book, Juan Valverde de Amusco used the most advanced techniques of his time, such as the dissection of human bodies; the best possible illustration techniques; and dissemination in prevalent languages. Juan Valverde de Amusco wanted to spread the knowledge of modern anatomy. Today, this multidisciplinary approach is called knowledge management, knowledge transfer, or knowledge translation [85,86].

Author Contributions: Conceptualization and editing, L.-A.A.-A.; writing—original draft preparation, L.-A.A.-A. and J.L.B.-L.; writing—review, J.L.B.-L.; supervision, C.R. All authors have read and agreed to the published version of the manuscript.

Funding: This research did not receive any specific grant from funding agencies in the public, commercial, or not-for-profit sectors.

Institutional Review Board Statement: Not applicable.

Informed Consent Statement: Not applicable.

Data Availability Statement: Not applicable.

Acknowledgments: The authors wish to express gratitude to Ana-Maria Álvarez-Castrosín for her technical assistance with manuscript preparation and Cristina Navarro Collin for her assistance in Figure 1.

Conflicts of Interest: The authors declare no conflict of interest.

List of Abbreviations (of Book Titles)

"Cl. Galeni Pergameni Liber de Ossibus ad tyrones [. . .] Medicae doctore": Cl. Galeni Pergameni Liber de Ossibus ad tyrones; interprete Ferdinando Balamio Siculo; enarrationibus illustratus à Lodouico Collado Valentino publico artis Medicae doctore. *". . .Epitome"*: Andrea Vesalii suorum de humani corporis fabrica librorum epitome. *". . .Fabrica. . ."*: De humani corporis fabrica libri septem. *"HISTORIA. . ."*: HISTORIA de la composición del cuerpo humano.

References

1. Arráez-Aybar, L.; Bueno-López, J.; Raio, N. Toledo School of Translators and their influence on anatomical terminology. *Ann. Anat.* **2015**, *198*, 21–33. [CrossRef] [PubMed]
2. Laín-Entralgo, P. *Historia de la Medicina*; Salvat Editores: Barcelona, Spain, 1978; Volume 4.
3. Ghosh, S.K. Human cadaveric dissection: A historical account from ancient Greece to the modern era. *Anat. Cell Biol.* **2015**, *48*, 153–169. [CrossRef]
4. O'Malley, C.D. Los saberes morfológicos en el renacimiento. La Anatomía. In *Historia Universal de la Medicina*; Laín-Entralgo, P., Ed.; Salvat Editores: Barcelona, Spain, 1972; pp. 43–77.
5. Singer, C.J. Beginnings of Academic Practical Anatomy. In *History and Bibliography of Anatomic Illustration*; Choulant, L., Ed.; Hafner: Lawrence, MA, USA, 1952.
6. Lassek, A.M. *Human Dissection: Its Drama and Struggle*; Ch. Thomas: Springfield, IL, USA, 1958; p. 74.
7. Robinson, J.O. The barber-surgeons of London. *Arch. Surg.* **1984**, *119*, 1171–1175. [CrossRef]
8. O'Malley, C.D. Andrés Laguna and his' Anatomica methodus'. *Physis* **1963**, *5*, 65–69.
9. Romero Reveron, R. Andreas Vesalius (1514–1564): Fundador de la Anatomía Humana Moderna. *Int. J. Morphol.* **2007**, *25*, 847–850. [CrossRef]
10. Laín-Entralgo, P. From galen to magnetic resonance: History of medicine in Latin America. *J. Med. Philos.* **1996**, *21*, 571–591. [CrossRef] [PubMed]
11. Andretta, E. Juan Valverde, or Building a "Spanish Anatomy" in 16th Century Rome. 2009, p. 13. Available online: http://hdl.handle.net/1814/12094 (accessed on 4 September 2023).
12. Andretta, E. Rome Ville anatomique. In *Roma Medica: Anatome d'un systéme médical au XVIe siècle*; Andretta, E., Ed.; École française de Rome: Rome, Italy, 2011; pp. 499–557.
13. Hernández-Morejón, A. *Historia Bibliográfica de la Medicina Española*; Viuda de Jordan e Hijos: Madrid, Spain, 1842–1852.
14. Jimenez, F.A. The first autopsy in the New World. *Bull. N. Y. Acad. Med.* **1978**, *54*, 618–619.
15. López-Piñero, J.M. *Historia de la Medicina Universal*; Ajuntament de València: :Valencia, Spain, 2010; p. 1840.
16. Schulze-Schneider, I. *La leyenda Negra de España: Propaganda en la Guerra de Flandes (1566–1584)*; Editorial Complutense: Madrid, Spain, 2009.
17. Elliott, J.H. *Imperial Spain 1469–1716*; Penguin Books Limited: London, UK, 2002; p. 448.
18. Arráez-Aybar, L.; Bueno-López, J. Antonio Gimbernat y Arbós: An anatomist-surgeon of the Enlightenment (in the 220th anniversary of his "A new method of operating the crural hernia"). *Clin. Anat.* **2013**, *26*, 800–809. [CrossRef]
19. Riera-Palmero, J. Valverde y la anatomia del Renacimiento. In *Historia de la Composición del Cuerp Humano. Juan Valverde de Hamusco (1501–1583)*; Riera-Palmero, J., Ed.; Universidad de Valladolid: Valladolid, Spain, 1981; p. 344.
20. Riera-Climent, C.; Riera-Palmero, J. Biografía de Luis de Lucena. *Llull Rev. Soc. Española Hist. Cienc. Técnicas* **2005**, *28*, 551–562.
21. Guerra, F. Juan Valverde de Amusco: Evidence of the identification of this portrait, claimed to be that of Vesalius, with a reapraisal of this work. *Clio Méd.* **1967**, *2*, 339–363.
22. Ongaro, G. L'insegnamento clinico di Giovan Battista da Monte (1489–1551): Una revisione critica. *Physis Riv. Internazionale Stor. Della Sci.* **1994**, *31*, 357–369.
23. Grendler, P.F. *The Universities of the Italian Renaissance*; The Johns Hopkins University Press: Baltimore, MD, USA, 2002; pp. 341–342.
24. Tosoni, P. *Della Anatomia degli Antichi e Della Scuola Anatomica Padovana, Memoria*; Dalla Tipogr. del Seminario: Padova, Italy, 1844.
25. Singer, C.J. To Vesalius on the fourth centenary of his De Humani Corporis Fabrica. *J. Anat.* **1943**, *77 Pt 4*, 261–265.
26. Riera-Palmero, J. *Juan Valverde de Amusco y la Medicina del Renacimiento*; Universidad de Valladolid: Valladolid, Spain, 1986.
27. Albarracín-Teulón, A. Los orígenes de la fisiología moderna. In *Historia Universal de la Medicina*; Laín-Entralgo, P., Ed.; Salvat Editores: Barcelona, Spain, 1973; pp. 78–85.
28. Cunningham, A. *The Anatomical Renaissance: The Resurrection of the Anatomical Projects of the Ancients*; Scolar Press: London, UK, 1997; p. 292.
29. Gratta, R.D.; Volpi, G.; Ruta, L. *Acta Graduum Academiae Pisanae: 1543–1599*; Gruppo di ricerca dell'Università di Pisa: Pisa, Italy, 1980; p. 6.
30. Premuda, L. *Storia dell' Iconografia Anatomica*; Ciba Edizioni: Milano, Italy, 1993.
31. Ghosh, S.K. Evolution of illustrations in anatomy: A study from the classical period in Europe to modern times. *Anat. Sci. Educ.* **2015**, *8*, 175–188. [CrossRef]
32. Hernández-Mansilla, J.M. Juan Valverde de Amusco 1525–1588 c.a. y la vocación por la anatomía en el Renacimiento hispanoitaliano. *Med. Hist.* **2015**, *4*, 22–34.
33. Amelang, J.S. Exchanges between Italy and Spain: Culture and Religion. In *Spain in Italy Politics, Society and Religion 1500–1700*; Dandelet, T.J., Marino, J.A., Eds.; Brill: Leiden, Netherlands, 2007; p. 422.
34. Hernández-Mansilla, J.M. La idea de Hombre en Juan Valverde Amusco. Ph.D. Thesis, Complutense University of Madrid, Madrid, Spain, 2014.
35. Fernández-Ruiz, C. Historia de la Medicina Palentina. *Publicaciones Inst. Tello Téllez Men.* **1959**, *20*, 1–139.
36. Barona-Vilar, J.L. *Sobre Medicina y Filosofía Natural en el Renacimiento*; Universitat de València: Valencia, Spain, 1993; Volume 3, p. 238.

37. Castiglioni, A. *Historia de la Medicina*; Salvat Editores: Barcelona, Spain, 1941; p. 906.
38. López-Piñero, J.M. *Ciencia y Técnica en la Sociedad Española de los Siglos XVI y XVII*; Editorial Labor: Barcelona, Spain, 1979; p. 511.
39. Cátedra, P.M.; López-Vidriero, M.L.; Andrés-Escapa, P.; Domingo-Malvadi, A.C.L.; Rodríguez-Montederramo, J.L. «Juan de Valverde de Amusco, 'Vivae imagines partium corporis humani. . .'», Fortuna de España. Textos españoles e imprenta europea (siglos XV-XVIII). Centro Virtual Cervantes. Available online: http://cvc.cervantes.es/obref/fortuna/expo/ciencia/cien022.htm (accessed on 4 September 2023).
40. Valle-Inclán, C.d. El léxico anatómico de Bernardino Montaña de Monserrate y de Juan de Valverde. *Asclepio* **1949**, *1*, 121–188.
41. Laín-Entralgo, P. *El Cuerpo Humano: Oriente y Grecia Antigua*; Biblioteca Virtual Miguel de Cervantes: Alicante, Spain, 2012.
42. Hurst, J.W.; Fye, W.B. Realdo Colombo. *Clin. Cardiol.* **2002**, *25*, 135. [CrossRef]
43. Sandoval-Gutiérrez, J.L. Nafis y Servet: Padres de la circulación pulmonar. *Arch. Cardiol. México* **2023**, *93*, 380. [CrossRef] [PubMed]
44. Navarro, F.A. Servet. *Rev. Española Cardiol.* **2023**, *76*, 671. [CrossRef]
45. Arráez-Aybar, L.; Navia-Álvarez, P.; Fuentes-Redondo, T.; Bueno-López, J. Thomas Willis, a pioneer in translational research in anatomy (on the 350th anniversary of Cerebri anatome). *J. Anat.* **2015**, *226*, 289–300. [CrossRef] [PubMed]
46. López-Piñero, J.M. *El Grabado en la Ciencia Hispánica*; Editorial CSIC: Madrid, Spain, 1987; p. 140.
47. Moreno Torres, A. *Aproximación al Léxico de la Anatomía y de la Urología en Romance en el Siglo XVI*; Universidad de Murcia: Murcia, Spain, 2000; p. 1632.
48. García-Jáuregui, C. Aproximación al léxico anatómico del Renacimiento. *Cuad. Inst. Hist. Leng.* **2008**, *1*, 93–110. [CrossRef]
49. López-Valverde, A.; de Diego, R.G.; De Vicente, J. Oral anatomy in the sixteenth century: Juan Valverde de Amusco. *Br. Dent. J.* **2013**, *215*, 141–143. [CrossRef]
50. Alberti-López, L. *La Anatomía y los Anatomistas Españoles del Renacimiento*; CSIC: Madrid, Spain, 1948; pp. 81–130.
51. López de Letona, C. El ojo en la historia de la composición del cuerpo humano (1556)(I). *Arch. Soc. Española Oftalmol.* **2005**, *80*, 117–118.
52. Martín-Araguz, A.; Bustamante-Martínez, C.; Toledo-León, D.; López-Gómez, M.; Moreno-Martínez, J. La neuroanatomía de Juan Valverde de Amusco y la medicina renacentista española. *Rev. Neurol.* **2001**, *32*, 788–797. [CrossRef] [PubMed]
53. Whitmore, I. Terminologia anatomica: New terminology for the new anatomist. *Anat. Rec. (New Anat.)* **1999**, *257*, 50–53. [CrossRef]
54. Kachlik, D.; Baca, V.; Bozdechova, I.; Cech, P.; Musil, V. Anatomical terminology and nomenclature: Past, present and highlights. *Surg. Radiol. Anat.* **2008**, *30*, 459–466. [CrossRef] [PubMed]
55. Romero-Reveròn, R. The first human anatomys book in Spanish: Libro de la anathomía del hombre written by Bernardino Montaña de Monserrate. *Vesalius Acta Int. Hist. Med.* **2019**, *25*, 8–18.
56. Arráez-Aybar, L. Anatomy in the pages of Don Quixote. *Interciencia* **2006**, *31*, 690–694.
57. Meyer, A.W.; Wirt, S.K. The Amuscan Illustrations. *Bull. Hist. Med.* **1943**, *14*, 667–687.
58. Voet, L. *The Golden Compasses*; A history and evaluation of the printing and publishing activities of the Officina Plantiniana at Antwerp; Vangendt: Amsterdam, Netherlands, 1969–1972; Volume 2, p. 273.
59. Voet, L.; Voet-Grisolle, J. *The Plantin Press (1555–1589): A Bibliography of the Works Printed and Published by Christopher Plantin at Antwerp and Leiden*; Van Hoeve: Amsterdam, Neterlands, 1980–1983; p. 3026.
60. Van Hee, R.; Lowis, S. David van mauden (+/−1538 +/− 1597), "sworn medical doctor and surgical prelector of antwerp", and his book on anatomy. *Acta Chir. Belg.* **2006**, *106*, 130–135. [CrossRef]
61. Torres-Pérez, J.M.; González-Martín, R.; San-Julián-Arrupe, T. Anatome Corporis Humani. Universidad de Navarra. Available online: http://www.unav.es/biblioteca/fondoantiguo/hufaexp01/hufaexp01p01.html (accessed on 20 May 2016).
62. Huard, P.; Imbault-Huart, M.-J. *Andrés Vesalio: Iconografía Anatómica (Fabrica, Epitome, Tabula Sex)*; Laboratoros Beecham: Barcelona, Spain, 1983; p. 254.
63. Post, C.R. *A history of Spanish Painting*; Harvard University Press: Cambridge, MA, USA, 1947; Volume 9, p. 931.
64. Carducho, V. *Dialogos de la Pintura sv Defensa, Origen, Esencia, Definicion, Modos y Diferencias*; Francisco Martinez: Madrid, Spain, 1633.
65. Barcia-Pavón, A.M.d. *Catálogo de la Colección de Dibujos Originales de la Biblioteca Nacional de Madrid*; Tipografía de la Revista de Arch., Bibli. y Museos: Madrid, Spain, 1906; p. 962.
66. Martín-González, J. Precisiones sobre Gaspar Becerra. *Arch. Español Arte* **1969**, *42*, 327–356.
67. Kleňhová, K. *Juan Valverde de Hamusco v Kontextu Renesanční Filozofie a Medicíny*; Západočeská Univerzita: Pilsen, Czech Republic, 2011.
68. Barcia-Goyanes, J.J. *El Mito de Vesalio*; Universitat de València: Valencia, Spain, 1994; p. 214.
69. Burggraeve, A. *Histoire de l'Anatomie Physiologique, Pathologique et Philosophique: Avec un Exposé des Principales Découvertes de Cette Science Depuis Son Origine Jusqu'à Nos Jours*; Ch. Chanteaud et Cie: Paris, France, 1880; p. 648.
70. Singer, C.J. *The Evolution of Anatomy: A Short History of Anatomical and Physiological Discovery to Harvey*; Kegan Paul Trench Trubne: New York, NY, USA, 1925.
71. Chinchilla-Piqueras, A. Anales Históricos de la Medicina en General, y Biográfico-Bibliográfico de la Española en Particular: 6 Vols. In *Tribus Relig*; Imprenta de López y Cía.: Valencia, Spain, 1841–1846.

72. Eloy, N.F. *Dictionnaire Historique de la Médecine Ancienne et Moderne. Ou Mémoires Disposés en Ordre Alphabétique Pour Servir à l'Histoire de Cette Science, et à Celle des Médecins, Anatomistes, Botanistes, Chirurgiens et Chymistes de Toutes Nations*; Hoyois: Mons, Belgium, 1778; Volume 4, p. 626.
73. Sprengel, K.P.J. *Versuch Einer Pragmatischen Geschichte der Arzneykunde*; Gebauerschen Buchhandlung: Halle, Germany, 1807; Volume 3, p. 62.
74. O'Malley, C.D. *Andreas Vesalius of Brussels, 1514–1564*; University of California Press: Berkeley, CA, USA, 1964; p. 483.
75. Berger, J.G.v. *Physiologia Medica Sive De Natura Humana Liber Bipartitvs. Iterum in Lucem Prodit Cura Frider. Christiani Cregut....Cujus Dissertatio De Anthropologia Ejusque Praecipuis Tam Antiquis Quam Modernis Scriptoribus Introducionis Loco Praemittitur*; Stock & Schilling: Francofurti (Frankfurt), Germany, 1737.
76. Haller, A.v. *Bibliotheca Anatomica*; Orell, Gessner, Fussli & Co.: Zurich, Switzerland, 1774.
77. Broussais, C. *Atlas Historique et Bibliographique de la Médecine*; ou, Histoire de la Médecine, Composée de Tableaux sur l'Histoire de l'Anatomie, de la Physiologie, de l'Hygiène, de la Chirurgie, de l'Obstétrique, de la Matière Médicale, de la Pharmacie, de la Médecine Légale, et de la Police Médicale, et de la Bibliographie, 2nd ed.; J.B. Baillière: Beauvais, France, 1834.
78. Bahşi, İ.; Bahşi, A. "Teşrih-ül Ebdan ve Tercümânı Kıbale-i Feylesûfan": The first illustrated anatomy handwritten textbook in Ottoman-Turkish medicine. *Surg. Radiol. Anat.* **2019**, *41*, 1135–1146. [CrossRef] [PubMed]
79. Skandalakis, J.E.; Mirilas, P. Plagiarism. *Arch. Surg.* **2004**, *139*, 1022–1024. [CrossRef] [PubMed]
80. Zimmerman, L. Surgery. In *Medicine in Seventeenth Century England: A Symposium Held at UCLA in Honor of CD O'Malley*; Debus, A.G., Ed.; University of California Press: Berkeley, CA, USA, 1974; pp. 46–69.
81. Persaud, T.V. *A history of Anatomy: The Post-Vesalian Era*; Charles C. Thomas Publisher: New York, NY, USA, 1997.
82. Skaarup, B.O. *Anatomy and Anatomists in Early Modern Spain*; Ashgate Publishing, Ltd.: Farnham, UK, 2015; p. 306.
83. Kemp, M. Style and non-style in anatomical illustration: From Renaissance Humanism to Henry Gray. *J. Anat.* **2010**, *216*, 192–208. [CrossRef] [PubMed]
84. Barcia-Goyanes, J.J.; Evans, N.R. Notes on the historical vocabulary of neuroanatomy. *Hist. Psychiatry* **1995**, *6 Pt 4*, 471–482. [CrossRef]
85. Booker, L.D.; Bontis, N.; Serenko, A. The relevance of knowledge management and intellectual capital research. *Knowl. Process Manag.* **2008**, *15*, 235–246. [CrossRef]
86. Parchman, M. Diffusion, Dissemination and Implementation: What Is the Difference. Available online: https://dcricollab.dcri.duke.edu/sites/NIHKR/KR/GR-Slides-10-09-15.pdf (accessed on 4 September 2023).

Review

Pheromone Sensing in Mammals: A Review of the Vomeronasal System

Mateo V. Torres [†], Irene Ortiz-Leal [†] and Pablo Sanchez-Quinteiro *

Department of Anatomy, Animal Production and Clinical Veterinary Sciences, Faculty of Veterinary, University of Santiago de Compostela, Av. Carballo Calero s/n, 27002 Lugo, Spain; mateovazquez.torres@usc.es (M.V.T.); irene.ortiz.leal@usc.es (I.O.-L.)

* Correspondence: pablo.sanchez@usc.es

[†] These authors contributed equally to this work.

Abstract: This review addresses the role of chemical communication in mammals, giving special attention to the vomeronasal system in pheromone-mediated interactions. The vomeronasal system influences many social and sexual behaviors, from reproduction to species recognition. Interestingly, this system shows greater evolutionary variability compared to the olfactory system, emphasizing its complex nature and the need for thorough research. The discussion starts with foundational concepts of chemocommunication, progressing to a detailed exploration of olfactory systems. The neuroanatomy of the vomeronasal system stands in contrast with that of the olfactory system. Further, the sensory part of the vomeronasal system, known as the vomeronasal organ, and the integration center of this information, called the accessory olfactory bulb, receive comprehensive coverage. Secondary projections of both the olfactory and vomeronasal systems receive attention, especially in relation to the dual olfactory hypothesis. The review concludes by examining the organization of the vomeronasal system in four distinct mammalian groups: rodents, marsupials, herpestids, and bovids. The aim is to highlight the unique morphofunctional differences resulting from the adaptive changes each group experienced.

Keywords: chemical communication; olfactory systems; vomeronasal organ; accessory olfactory bulb; dual olfactory hypothesis; mammalian evolution

Citation: Torres, M.V.; Ortiz-Leal, I.; Sanchez-Quinteiro, P. Pheromone Sensing in Mammals: A Review of the Vomeronasal System. *Anatomia* **2023**, *2*, 346–413. https://doi.org/10.3390/anatomia2040031

Academic Editors: Gianfranco Natale and Francesco Fornai

Received: 27 August 2023
Revised: 14 October 2023
Accepted: 7 November 2023
Published: 9 November 2023

1. Introduction

Mammalian chemical communication, a significant and complex domain of study, fundamentally hinges upon the vomeronasal system's ability to sense pheromone-mediated interactions. This system, beyond its basal functions, profoundly influences various social and sexual behaviors such as reproduction, hierarchical dynamics, maternal bonding, and species-specific recognition. What becomes particularly compelling is the pronounced evolutionary variability the vomeronasal system displays; more so when juxtaposed against the variations observed within the olfactory system. Such adaptive diversities present both challenges and opportunities in research, amplifying the call for an in-depth examination of its neuroanatomy and the nuances of its functional morphology.

In the following sections, we will discuss the foundational concepts of chemocommunication and delve into the olfactory systems. The neuroanatomical intricacies of the vomeronasal system will be further elaborated upon, drawing parallels with the olfactory system. Discourse will further encompass the sensory element of the vomeronasal system, focusing on the vomeronasal organ and the accessory olfactory bulb's role. This examination will be augmented by insights into specific models of vomeronasal system organization across evolutionarily distinct mammalian families. While the current literature provides extensive knowledge, there remains a plethora of open questions and untapped areas. The intention of this comprehensive compilation is to serve as both a foundation and a catalyst for future research endeavors in this domain.

In preparing this review, we used a systematic approach. Addressing comprehensive topics such as chemical communication and the neuroanatomy of the vomeronasal system required a degree of subjective selection. We focused our analysis on a qualitative synthesis of articles considered most relevant to the study objectives. While this approach ensures a thorough overview, the review is not exhaustive or strictly historical. Instead, we aimed for clarity and relevance, ensuring readers gain a deep understanding of the subject. The search strategy included standard databases and customary sources in the field. It is pertinent to note that the images incorporated within this review are original contributions from the authors, reinforcing the authenticity of the discourse.

1.1. Chemical Communication in Mammals

Throughout evolution, living organisms have developed intricate systems of chemical communication aimed at interacting with the external environment and enhancing their survival rate [1]. In doing so, they are able to detect a wide array of chemical signals and convert them into sensory information [2]. When these chemical cues are exchanged between different individuals, the substances conveying these messages are termed semiochemicals [3]. These semiochemicals are released through urine, feces, saliva, and secretions from various skin glands, often being deposited in the environment through highly stereotyped behaviors [4]. This form of chemical communication holds a pivotal role, influencing key areas such as social behavior, reproduction, food-seeking, and evasion from potential predators [5]. This is in part because the chemical detection systems project to the limbic system, where processes such as emotions, memory, hunger, sexual instincts, and overall behavior are regulated [6]. Additionally, these semiochemicals operate in tandem with other signals perceived by physical senses (auditory, visual, or tactile), conveying information about species, gender, social status, and physiological and developmental state [7].

Chemical senses emerged in the earliest stages of evolution and are now present in all living organisms, from bacteria to more complex life forms. Hence, all animals are preadapted to perceive a variety of chemical signals [8]. However, primitive organisms detect stimuli through specialized cells, while vertebrates have integrated these receptor cells into highly sophisticated organs. If the perception of a specific chemical stimulus enhances survival or reproductive success, the adaptation aims to reduce the sensitivity threshold to this substance and/or increase the expression of genes responsible for its perception [9]. Yet, identifying such a diverse plethora of molecules (essential lipids, lipoproteins, proteins, steroids, alcohols, etc.), spanning a broad range of volatilities and solubilities, demands a vast array of chemoreceptors. Among these, olfactory receptors stand out as being phylogenetically present in the olfactory rosette of fish [10] and being highly conserved and expanded in mammals [11]. In fact, the largest superfamily of mammalian genes corresponds to the olfactory receptors. They detect a complex web of odor mixtures, allowing the creation of a dynamic three-dimensional image of the surrounding world over time [12]. Depending on the physicochemical characteristics of the detected signals, such as volatility or spatial dispersion, the captured information will multiply [13].

The olfactory system, recognized as one of the most prominent chemical communication systems alongside taste, is one of the first senses to activate, even allowing for prenatal olfactory learning [14]. However, in mammals, odorous chemical signals have the capability to stimulate various chemosensory structures, extending beyond the olfactory organ: such as the vomeronasal organ, Masera's septal organ, Grüneberg's ganglion, and free nerve endings of the trigeminal nerve, among others.

1.1.1. Types of Chemical Signals

In nature, living organisms maintain a constant degree of communication based on the exchange of multiple chemical signals [15] due to a variety of reasons, primarily associated with their need for food, protection, and reproduction. These chemical signals can be classified into two types: hormones, when they act within the same individual that produces them; or semiochemicals, when they act in an individual different from the signal producer. Hormones are chemical signaling molecules produced by tissues or endocrine glands that control various physiological processes within an organism [16]. In parallel, the chemical substances that different organisms use to communicate with each other and perceive their environment are called semiochemicals, a name derived from the Greek root *semeon*, meaning mark or signal [17]. This chemical communication is present in microorganisms, plants, and insects [18] as well as in vertebrates [19] and requires specific secretory mechanisms of semiochemicals, which induce changes in the physiology, metabolism, or behavior of the species that receive the chemosensory signals [20].

Semiochemicals are classified into pheromones, if they act upon individuals of the same species, or allelochemicals, if they affect individuals of a different species [21]. In both cases, these chemical signals may consist of a single chemical compound or a mixture of several. Regarding pheromones, they establish chemical communication among members of the same species, regulating aspects such as development, physiological state, social behavior, reproduction, or territorial marking [22]. On the other hand, allelochemicals are further classified, based on the beneficiary of the interaction, into four types: allomones, when the sender benefits from the message; kairomones, wwhenhere the receiver is the beneficiary; synomones, when both species benefit from the communication, and apneumones, when the chemical signal is emitted by inanimate material and evokes an adaptively favorable behavioral or physiological reaction in the reciever.

Allomones, a term resulting from the combination of the Greek roots *allos* and *hormonas* meaning to excite others, are defined as interspecific chemical signals that give an adaptive advantage to the species that produce them [23]. Allomones from many organisms have been described, from lower plants to higher animals, where those emitted by predators primarily attract prey [24,25], and those produced by prey primarily repel predators [26]. For example, venoms or antibiotics are allomones, as they are chemical compounds produced by microorganisms with the aim of inhibiting the growth of other microorganism species. Conversely, a kairomone is an allelochemical that, when in contact with an individual of another species, evokes an adaptively favorable behavioral or physiological response in the reciever, while being unfavorable for the sender [3]. A notable feature of the main categories of chemical messengers is that they are not mutually exclusive, and an example of this occurs in the predator *Elatophilus hebraicus*, which uses the sex pheromone of its prey *Matsucoccus josephi* as a kairomone, such that, when the predator perceives it, it feels a strong attraction [27]. Another example is the secretion from the mandibular glands of many ants when the nest is disturbed, which acts as a social alarm communication (pheromone), but also has a repelling effect (allomone) against aggressors [28].

Regarding synomones, they can be defined as allelochemicals produced by one organism that, when in contact with an individual of another species, evoke in the receiver a behavior or physiological response that is adaptively favorable for both [21]. Some synomones are repellent, and others are attractive [29], such as repellent molecules that warn of a danger like the toxicity of a plant [30], or chemical signals produced after a parasitic attack aiming to attract the natural predator of the parasite [31].

Lastly, there is a group of allelochemicals originating from inanimate sources, the apneumones, a term formed from the Greek root *a-pne* meaning breathless or lifeless. The receiver benefits from these molecules, but they differ from kairomones in that the producer cannot experience any disadvantage [32]. However, these apneumones are related to the presence of certain prey in specific environments, which is detrimental to organisms of another species associated with these compounds that might be nearby or on the inanimate material. For example, parasites and predators are drawn by apneumones to inanimate substances related to the presence of their hosts or prey [33].

Finally, it is important to emphasize that the exchange of chemosensory stimuli occurs in both terrestrial and aquatic environments. The most significant physicochemical characteristics of these chemical compounds are their size and polarity, as these are the primary factors determining their volatility in the air and solubility in water, respectively. Thus, in terrestrial environments, substances that act at a distance tend to be small and volatile, while in aquatic environments, where solubility is most relevant, even high molecular weight molecules can act at a distance. Therefore, based on their physicochemical properties, which dictate their potential mobility in different mediums, chemical signals can be classified into four types of compounds: volatile-soluble, volatile-insoluble, non-volatile-soluble, and non-volatile-insoluble [13].

Pheromones

Within semiochemicals, pheromones are compounds that allow for chemical signaling between individuals of the same species. In 1959, the entomologist Martin Luscher and the biochemist Peter Karlson introduced the term "pheromone" by combining the Greek roots *pherein* (to carry) and *hormon* (to excite). They defined it as a chemical substance or a mix of chemical substances released by a member of a species that induces specific behavioral responses or physiological changes in other members of the same species upon perception [34]. In fact, the first semiochemical to be isolated and characterized was bombykol, the sexual pheromone of the silkworm moth [35]. Subsequently, two categories of pheromones were defined based on their effect on the benefiting organism: releaser pheromones and primer pheromones. Releaser pheromones trigger immediate behavioral effects in the recipient organisms, whereas primer pheromones induce long-term physiological effects in the recipient individual [36]. Thus, typical releaser pheromones influence aspects related to mating, alarms, trails, and territorial marking. On the other hand, compounds determining the caste of social insects are primer pheromones [37]. Stemming from these discoveries, pheromones have been extensively studied in insects, primarily for their potential in pest control [38]. They have also been the subject of numerous studies in mammals [39].

For decades, the existence of pheromones in mammals has been assumed [40], though some authors have expressed their doubts [41]. Indeed, the concept of a pheromone derives from studies in insects, and the integration of the term in vertebrates has sparked controversy due to the complexity of information received simultaneously through other sensory modalities and the significance of learning in mammals, which can alter behavior, complicating the specific analysis of chemosensory communication [42]. Specifically, in pigs (*Sus scrofa*), male saliva pheromones are essential for courtship and copulation, but physical stimuli produced by applying pressure on the female back are also necessary [43]. As a result, five operational requirements have been introduced for a substance to be considered a pheromone: chemical simplicity of the signal; high stimulus selectivity with its response; an unequivocal behavioral response from the receiver that is morphologically consistent and functionally apparent; reception specificity based on species; and an unconditional linkage between stimulus and response [44]. In their literature review on chemical signals in terrestrial vertebrates, Apps et al. [45] identified up to 63 mammalian pheromones, such as those in mice [46–48], hamsters [49], pigs [50], and elephants [51].

Kairomones

Kairomones, a term derived from the Greek root *kairo* (opportunist), are interspecific allelochemicals that confer an adaptive advantage to the organism receiving the chemical signal, while being disadvantageous to the emitter. Therefore, kairomones are semiochemicals emitted by one species to its own detriment [52]. However, even though kairomones can be detrimental to the emitting organisms, it has been suggested that they could lead to an evolutionary advantage [53]. Although the term kairomone was controversial when introduced, it is now widely accepted [54,55].

Due to the diversity of ecological phenomena associated with kairomones, they can be classified based on various criteria. According to their effect on the benefiting organism, two classes are distinguished: releaser kairomones, which induce an immediate behavioral response, and primer kairomones, which lead to long-term physiological responses in the receiver. They can also be classified based on their function in the receiving individual, resulting in four main groups: foraging kairomones, anti-predator kairomones, sexual kairomones, and aggregation kairomones [56]. Accordingly, foraging kairomones are used to locate food sources [57], anti-predator kairomones to mitigate the negative impact of a natural enemy [58], and sexual kairomones to find mates or for other sexual purposes [59]. Lastly, aggregation kairomones are used by both genders of the receiving species to form aggregations for various purposes, such as optimal exploitation of food resources, mate searching, or defensive reactions. Examples of foraging kairomones include the substances produced by the corn earworm (*Heliothis zea*), used by the parasite *Microplitis croceipes* to locate its host [60], or lactic acid emitted by humans, which attracts the mosquito responsible for transmitting yellow fever, *Aedes aegypti* [61]. In some cases, kairomones used to find a host are produced by other microorganisms that develop within the host [62]. In the aforementioned cases, the chemical signals aid predators in locating their prey or hosts, but the opposite can also occur. Anti-predator kairomones emitted by predators or parasites are used as alarm signals, inducing defensive behavior in receiving organisms [63]. This is the case with certain plants that recognize kairomones produced by herbivores and emit compounds to repel them or to try to attract their natural enemies [64]. Lastly, sexual kairomones indirectly influence sexual communication and can be used to find mates, like the alcohols released by green leaves when female beetles of the *Melolontha* spp. feed on them, attracting males of the same species [65].

1.1.2. Responsible Exocrine Organs

The independent evolution of multiple semiochemicals in mammals has resulted in a vast diversity of produced compounds as well as specialized secretory glands. These signals are commonly emitted externally via feces, urine, or saliva, but they are also found in other secretions like tear ducts, vaginal secretions [66], or are produced in specific glands like perianal ones [7]. Substances excreted through urine or feces can provide information about the age, identity, or gender of the depositor. In fact, there are body postures or movement patterns associated with urination and defecation, like scattering motions, that facilitate this chemical communication [67]. They also provide information about the location of individuals, especially relevant in various contexts such as social structuring or territorial distribution.

In this way, feces play a significant role in marking and intraspecific chemical communication in many mammals. Some odoriferous compounds originate from anal sac secretions [68], but others are a result of bacterial action on food during digestion [69]. In pigs, several semiochemicals present in maternal feces have been detected that attract piglets, resulting in positive behavioral changes in them, acting as a soothing agent and improving growth rates [70].

Both volatile and non-volatile substances can also be identified that act as chemical signals in the urine of most mammals, as it is a fundamental medium used in animal communication [71]. In guinea pigs, urinary chemical substances are involved in gender discrimination [41]. Additionally, the amino acid felinine, responsible for the characteristic odor of urine in this species and implicated in territorial marking, is found in the urine of domestic cats and other members of the Felidae family [72]. It appears in large amounts in male urine from six months of age and is believed to be a precursor to a pheromone attracting females [73].

Multiple semiochemicals are also found in saliva, as in the case of the boar which has androgenic steroids in its saliva that act as sexual attractants, and can even stimulate puberty in juvenile females [74]. These compounds are produced in the testicles and transported by the bloodstream to the submaxillary salivary glands [75] where they are emitted in large quantities during copulation.

Similarly, in mammals, chemical stimuli related to sexual behaviors are produced in or near the genitals. For example, the male pig produces a sexual attractant that is part of the preputial secretions [76], while female hamsters attract sexually experienced males with their vaginal secretions [49]. Before copulation, males sniff and lick these secretions since they contain essential chemicals for adequate sexual behavior [77]. A similar phenomenon occurs in primates, where a chemical compound in female vaginal secretions, copulin, triggers copulatory behavior in male conspecifics. Stimulated by estrogens, this semiochemical enhances sexual responsiveness in rhesus monkeys [78]. In other species, like canids [79] or bovids [80], sexually attractive molecules have also been observed in female vaginal secretions.

On the other hand, various cutaneous glands are involved in the production and emission of chemical signals. These exocrine cutaneous glands release their compounds through a duct system, and depending on the type of secretion, they are divided into sweat-producing (sudoriparous) and sebaceous. The sweat glands are related to sweat production and contain hydrophilic molecules, while the sebaceous glands release sebum and lipophilic substances. Consequently, the various cutaneous glands can produce a complex mixture of chemical signals. Moreover, based on the secretion release mechanism, exocrine glands are also classified as apocrine, holocrine, and merocrine or eccrine. In apocrine glands, a portion of the cytoplasm of the cells becomes part of the secretion. In holocrine glands, cells are almost entirely destroyed and become part of the secretion product. Lastly, in merocrine glands, secretion occurs through an exocytosis mechanism, so there is no injury to the secretory cells. Examples of apocrine glands include mammary glands; holocrine examples are sebaceous glands; and merocrine examples include salivary glands.

Several examples of glands producing semiochemicals in mammals are described below. In rabbits, the apocrine glands of the chin are involved in maintaining social status and are used to mark territory by dominant males [81]. Male gerbils use the secretion of an androgen-dependent sebaceous gland to mark territory [82]. Other apocrine sebaceous glands, the tarsal glands of the male deer (*Odocoileus hemionus columbianus*), secrete a series of molecules, including a semiochemical that causes licking by female congeners [83]. In pronghorn males (*Antilocapra americana*), a compound produced in the subauricular gland was identified that induces licking, marking, or tapping in other males of the same species [84]. Another specific sebaceous gland is the morillo gland, found exclusively on the snout in capybara males, which is related to the production of semiochemicals. These compounds are produced more by dominant males and are used for territorial marking when deposited on vegetation [85]. In the case of guinea pigs (*Cavia porcellus*), the sebaceous secretion of the perineal gland intervenes in dominance [86]. Similarly, in sugar gliders (*Petaurus breviceps*), dominant males have a frontal and a sternal gland, with which they mark territory and other members of their family group. Both male and female sugar gliders emit chemical signals through the paracloacal glands [87]. Merocrine or eccrine glands are abundant in humans and primates, but in the rest of the mammals, their location is restricted to the palms and soles. For instance, the Madagascar tenrec (*Echinops telfairi*) has cutaneous eccrine glands on the footpads, which produce a non-slip secretion containing semiochemicals that inhibit the growth of microorganisms [88]. Lastly, glands producing chemosensory stimuli in minks, that become vestigial after weaning, have been studied. These are apocrine sweat glands located in the neck and in the perineal and inguinal regions related to maternal recognition of the offspring [89].

1.1.3. Ethology and Scent Marking

In mammals, chemical signals can be emitted passively, without being associated with specific behavior, or they can involve a display of behavioral patterns that facilitate the dispersal of such a signal in the environment. For instance, the male pig passively emits a sexual attractant in its preputial secretions [90]. However, it also exhibits specific behaviors, such as chewing movements that increase the production of salivary secretion. This contains semiochemicals like androstenone or androstenol, produced in the testicles and stored in the parotid gland, which stimulate the immobility reflex in female pigs, facilitating the adoption of the mating posture [91]. These specific markings, termed "scent marking", have been described in numerous mammals, with a wide variety of behavioral patterns characterized in different species. This diversity arises from the various functions that chemical communication might entail [92], the multiple deposition sites of the signal [93], or even the location of the exocrine glands on the body.

Regarding function, initially, a relationship was established between marking and territorial defense. However, this behavior can play other roles in communication. In fact, several authors have proposed various functions for chemical marking [94]. Thus, marking can act as a deterrent, avoiding aggression by warning about territory occupation; as a sexual attractant; as an orientation tool; as an indicator of identity, age, or dominance; as an alarm signal for conspecifics; or even as an indicator of population size.

Regarding the territorial function, chemical marks define a specific area and prevent intruders from entering or cause their withdrawal, reducing defensive costs. Therefore, animals deposit these compounds at the boundaries of their territory and at strategic locations as a method of territorial dominance. This behavior is also reflected in experimentally studied mice, as subordinate individuals spend less time in parts of a cage treated with dominant male urine [95]. On the other hand, in rabbits, it has been shown that upon acquiring a dominant status, a male synthesizes 2-phenoxyethanol in its chin gland, which acts as a fixative making its secretions persist in the environment without dissipating [96].

Similarly, marking can have a stimulating or attracting function in the opposite sex, as seen in canids, where females show a higher frequency of urine marking during estrus to attract males [97]. In the case of the striped mongoose (*Mungos mungo*), a social and cooperative herpestid, both males and females increased the marking frequency during the latter estrus [98]. On the other hand, in rodents, the preputial secretion and urine of males can induce or suppress estrus and ovulation in females, in addition to accelerating female sexual maturity [99,100].

Another function of chemical marks is the spatial orientation of the individuals that produce them within their territory, allowing them to have a better familiarity with the environment. In this way, species like the slow loris (*Nycticebus coucang*) use chemical marks to trace paths [101]. Likewise, other mammals, such as coyotes (*Canis latrans*), mark food or hiding places with urine [102].

Marking behaviors are also used to deposit chemical signals that serve as an indicator of individual identity and provide information about dominance, social rank, sex, or age. In this way, beavers (*Castor canadensis*) can discriminate between familiar or unknown individuals [103], just like the sugar gliders (*Petaurus breviceps papuanus*), who are not only able to distinguish both individuals and groups of unknown individuals but also trigger aggressive behaviors in the receiver [87]. This relationship between individual recognition and dominance is reflected in marking patterns. Dominant individuals usually mark more frequently than subdominant or submissive ones and are significantly more likely to adopt specific marking postures, as seen in canids [104]. Likewise, dominant individuals frequently mark in situations where they are intolerant. On the other hand, chemical signals allow recognition between males and females [105] and even induce synchronization of the reproductive state [106].

On occasion, stressed individuals secrete alarm molecules, keeping the members of their group alert. Exposure to these signals produces behavioral changes in the receiver, such as reduced rest, increased heart rate, or stress-induced hyperthermia. This phenomenon has been studied in rats [107]. In cows, the presence of alarm signals in the urine of stressed congeners triggers an increase in cortisol and fearful behaviors [108].

Finally, in certain animals, scent marking has been observed to act as a population self-regulation mechanism, limiting the number of inhabitants before food becomes a limiting factor. This happens in American beavers (*Castor canadensis*), who communicate through marking, delineating occupied territory, and preventing further colonization of a particular area. This prevents the population density from increasing above a limit while maintaining a balance between the amount of resources demanded by the residents and the regeneration capacity their habitat can sustain [109].

Regarding the deposition sites of chemical substances, five contexts have been considered: direct release to the surrounding air, deposition on a specific object in the environment, deposition on the substrate, marking on a social partner, and self-marking or self-anointing. We usually associate the emission of chemical signals with specific marking behavior patterns, but the truth is that mammals also have the ability to release chemical compounds without any visible sign of it. For example, many mammal species emit chemical signals when scared, like the black-tailed deer, which secretes chemical compounds in its metatarsal glands when in situations of fear or stress [83].

On the other hand, marking may be associated with certain objects, so animals deposit feces and urine on them or use rubbing movements with their glandular areas to impregnate them with specific secretions. These marking movements directed at objects often involve adopting unusual postures and allow depositing chemical signals at a certain height, improving communicative efficiency [110]. These objects can be part of a complex communication system between individuals of the same species, as seen in the European brown bear (*Ursus arctos arctos*), which exhibits a rubbing behavior against tree trunks. By standing on their hind legs, they rub their chest, back, and neck while biting and scratching the bark, depositing chemical signals. Males have preferred marking areas, which can be used by a single individual or can be communal and used by several males. Places marked by an unknown adult male are avoided by young bears. On the other hand, females mark in areas separate from the males but can also mark on communal trunks along with other males [111]. In the case of goats, when a male reaches sexual maturity he starts rubbing his head and neck on objects around him as a method to disperse chemical signals with an attractive effect on females [112].

In addition to the deposition of chemical signals through urine or feces in the substrate, mammals show specific marking behaviors on the terrain. In particular, desert-dwelling rodents from the Heteromyidae family perform sand baths [113]. This behavior consists of digging in the substrate and performing a series of specific rubbing movements depending on the species. In this way, the animal grooms itself while spreading chemical secretions in marking areas of the environment, which are sometimes shared. Other mammals have behaviors like scratching the ground, which may involve the deposition of chemical signals produced in the autopodial glands.

Some mammals perform marking on a partner or on conspecific individuals through characteristic behaviors. These behaviors may involve friction movements of specific glandular areas on the receiver [87]. The previously mentioned marking of European rabbits by rubbing their chin on objects or congeners to impregnate them with the secretion of the chin glands allows them to identify members of their social group to establish their territory, thus maintaining the social hierarchy [81,114].

Finally, some mammal species use their own or foreign sources of chemical signals to spread the compounds on their body for various purposes. In this way, the black-tailed deer extends its hind limb to spread the previously described secretions from the tarsal glands over its own head [83]. Likewise, spider monkeys exhibit a self-anointing behavior, rubbing their body with a mixture of saliva and plant material [115]. However, in other cases, individuals use foreign chemical signals, like rats, who impregnate themselves with chemical signals from their main predator, the weasel, to mask their own chemical identity and avoid possible attacks [116].

1.1.4. Chemical Signal Detection Systems

Animals are constantly examining their environment for chemical substances that guide them to food sources or favorable habitats. They also investigate other substances that control social interaction and reproductive behavior [117]. These chemical compounds can be detected through direct contact with salivary or nasal secretions, or by monitoring the respiratory air stream through the nasal cavity. In both cases, the chemical signals are perceived by highly specialized detectors, the chemosensory neurons, which are organized into structurally independent subsystems in the nasal cavity. These olfactory subsystems can be divided based on the anatomical location of their sensory neurons, the type of olfactory receptors they express, the signaling mechanisms they use to transduce chemosensory stimuli, the chemical stimuli to which they respond, and the axonal targets of their sensory neurons in regions of the olfactory forebrain [118].

Initially, two chemical perception systems were identified with sensory neurons located in the nasal cavity: the main or olfactory system and the accessory or vomeronasal system. The olfactory neurons constitute the main olfactory epithelium, which lines the convolute surfaces of both the ethmo- and endoturbinates (Figure 1), as well as the caudal part of the nasal septum (Figure 2) [119]. These neurons send their nerve projections to the main olfactory bulb [120,121]. On the other hand, the vomeronasal system presents a highly specialized peripheral sensory structure where the sensory neurons of the vomeronasal epithelium are located, the vomeronasal organ (VNO). The neurosensory information travels through the vomeronasal nerves to the accessory olfactory bulb, an independent structure adjacent to the main olfactory bulb [122,123].

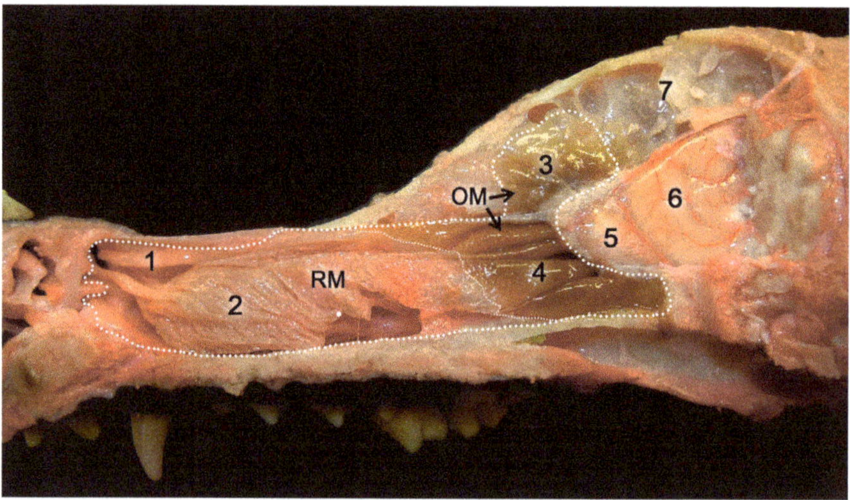

Figure 1. Dissection of the turbinate complex in a dog. The difference between the red respiratory mucosa (RM) and the brown olfactory mucosa (OM) is observed. 1. Dorsal turbinate; 2. Ventral turbinate; 3. Endoturbinate; 4. Ectoturbinates; 5. Olfactory bulb; 6. Telencephalon frontal lobe; 7. Frontal sinus.

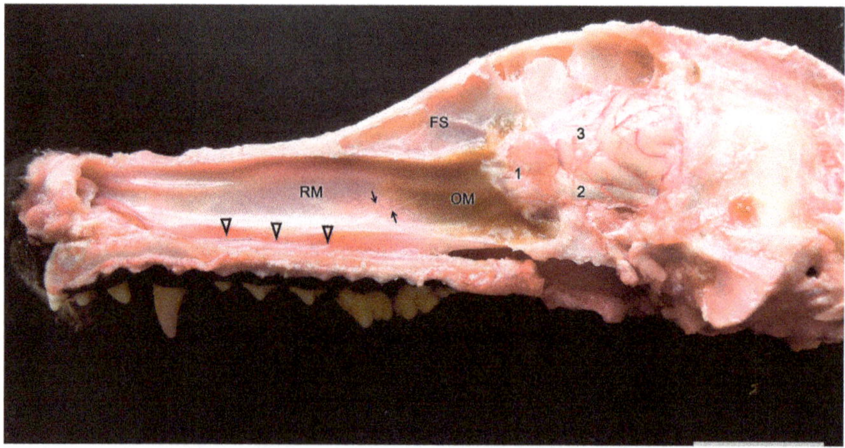

Figure 2. Lateral view of the nasal septum of a dog showing the projection area of the vomeronasal organ (open arrowheads) and the vomeronasal nerves (arrows). FS, frontal sinus; 1. Olfactory bulb, 2. Olfactory peduncle; 3. Telencephalon. Scale bar: 5 cm.

More recently, new sensory systems specialized in the detection of semiochemicals have been discovered in the nasal cavity of mammals, which together are called olfactory subsystems [124]. This is the case of the Grüneberg ganglion, the septal organ, or the solitary chemosensory cells. The Grüneberg ganglion is a structure formed by specific sensory neurons located in the anterior area of the nasal cavity, near the nasal vestibule [125]. On the other hand, the septal organ consists of an isolated area of sensory epithelium located at the base of the nasal septum, ventral to the olfactory epithelium [126]. Finally, the solitary chemosensory cells are distributed in the anterior part of the nasal cavity but are also located in other specific areas such as the entrance of the vomeronasal organ [127]. The first two subsystems mentioned project their neuronal axons to specific areas of the main olfactory bulb, while the axons of the chemosensory cells associated with the VNO are incorporated into the sensitive afferents of the trigeminal nerve [128].

In both the main and the accessory olfactory bulbs, glomeruli are formed in the confluence area of the axonal terminals of the sensory neurons with the dendrites of the second-order neurons with which they establish contact. The vomeronasal glomeruli are part of the accessory olfactory bulb and are clearly differentiated from the glomeruli of the main olfactory bulb by their smaller size and less defined boundaries. However, the glomeruli formed by neurons of the Grüneberg ganglion or the septal organ do not have a clearly differentiated boundary within the main olfactory bulb, although specific areas for each have been described. In fact, a necklace glomerulus complex has been identified in the caudal region of the MOB, consisting of several specific glomerular groups related to the innervation of specific neurosensory cells, such as the cells of the Grüneberg ganglion or cells that express the atypical olfactory receptor guanylyl cyclase type D (GC-D) [129]. Additionally, subbulbar formations have been identified ventral to the accessory olfactory bulb in rats [130], hedgehogs [131], and in lagomorphs—with the latter presenting a particularly complex organization [132]—that could be related to certain specific sensory pathways. The recent characterization of the olfactory limbus—the transition zone between the accessory and main olfactory bulbs—in the fox [133] points to a high morphofunctional complexity in the central integration of chemosensory information.

This wide range of versatile chemodetectors perceives myriads of chemical compounds that vary in their physicochemical properties and function, and that may be involved in controlling multiple behaviors and physiological responses. This multifunctional task is achieved through orchestrated interactions between the various olfactory subsystems, each of which specializes in different functions and uniquely contributes to fulfilling the overwhelming tasks of the sense of smell.

1.1.5. Olfactory Subsystems

In the **main olfactory system**, molecules are detected through direct contact with the olfactory mucosa of the nasal cavity. The fundamental component of the olfactory mucosa is the olfactory neuroepithelium, comprising olfactory sensory neurons. These are bipolar cells whose dendrites reach the epithelial surface. Here, long cilia embedded in nasal mucus provide an extensive surface area for interaction with odorants. The process of neurotransduction occurs within these cilia [134]. The ciliary membrane contains receptor proteins and elements of the olfactory transduction machinery, allowing these cellular compartments to act as chemosensory units, initiating signal detection.

Olfactory receptors consist of various specialized receptor proteins coupled to the G protein (GPCR) [135]. Each olfactory sensory neuron expresses a unique receptor from a family that encompasses about 1000 genes. In fact, olfactory receptors are associated with the largest gene superfamily in vertebrates [136]. Additionally, each olfactory receptor interacts with a broad range of chemical compounds, albeit with different affinities. Therefore, a single olfactory receptor can recognize multiple odorants or olfactory chemical signals, and conversely, multiple olfactory receptors can recognize a single odorant. This combinatorial receptor strategy is utilized to encode odor qualities [137].

The axons of the olfactory neuroepithelial cells converge to form the olfactory nerve (first cranial nerve) [138], whose fibers pass through the cribriform plate of the ethmoid bone to establish their first synapse in the glomeruli, spherical structures found on the periphery of the main olfactory bulbs. Anatomically, the olfactory bulbs are rostral extensions of the cerebral hemispheres, and as such, belong to the rhinencephalon, the olfactory portion of the telencephalon. Thus, they constitute the first synaptic and integrative station of the olfactory system. Notably, from a morphofunctional perspective, all neurons expressing the same type of olfactory receptor send their axons to common glomeruli in the main olfactory bulb [139]. From the main olfactory bulbs, information is relayed through the lateral olfactory tract to deeper brain areas [140].

In contrast to the main olfactory system, molecules stimulating the **vomeronasal system** activate the sensory neurons of the vomeronasal organ, a bilaterally symmetrical tubular structure located dorsally to the floor of the nasal cavity and on both sides of the base of the nasal septum. The characteristics of the vomeronasal epithelium lining the inside of the vomeronasal duct resemble those of the main olfactory epithelium. However, the dendrites of the vomeronasal sensory neurons have microvilli instead of cilia [141]. Both vomeronasal organs are protected by a vomeronasal capsule, which varies in nature depending on the species and is composed of bone, cartilage, or a mixture of both forming specific patterns [142,143]. Caudally, the capsule is typically closed; while rostrally, the vomeronasal organ communicates with the nasal cavity or the nasopalatine canal, depending on the species. In species like rodents, lagomorphs, or some primates, the vomeronasal duct directly opens into the nasal cavity [144], whereas in other species like marsupials, monotremes, or ungulates, the vomeronasal duct opens into the nasopalatine canal, also known as the incisive canal, which connects the oral and nasal cavities [145]. This topographical feature is an initial reflection of the existence among different mammalian species of a wide variability of the vomeronasal organ, at the morphological, topographical, and functional levels. Parallel to the main olfactory system, the axonal terminations of the vomeronasal sensory neurons form the vomeronasal nerve. Subsequently, upon leaving the VNO, they course dorsocaudally to penetrate the cribriform plate of the ethmoid bone and reach the accessory olfactory bulb, constituting its outermost layer, the vomeronasal nerve layer. From the accessory olfactory bulbs, information is transmitted to regions of the central nervous system involved in mediating the pheromonal responses highlighted earlier in this introduction [146].

It has often been hypothesized that the various olfactory subsystems function independently. This idea, in the case of the main olfactory system and the vomeronasal system, is supported by the fact that, while olfactory neurons project to the main olfactory bulb, vomeronasal neurons project to the accessory olfactory bulb. Indeed, highly specific afferent patterns have been described in both cases. In the vomeronasal system of rodents, the apical and basal regions of the vomeronasal neuroepithelium project their axons to the anterior and posterior accessory olfactory bulb, respectively [147]. Similarly, several distinct areas of sensory neuroepithelium in the nasal cavity have been identified that correspond to specific regions of the main olfactory bulb. Initially, four zones were established, with no overlap between them [148], but more recent studies discern up to nine zones with variable overlap [149]. Currently, the information gathered on the olfactory and vomeronasal systems supports the notion that they are distinct entities, each with unique structural characteristics and chemosensory tasks. However, they share a number of morphofunctional aspects. Among these are the presence of vomeronasal receptors in the MOE and vice versa [150,151], and the convergence of secondary projections from both systems in the basal telencephalon [152].

The **Grüneberg ganglion** (Figure 3) consists of a bilateral and compact group of neurons located in the vestibule of the anterior nasal cavity, in a dorsomedial position near the opening of the nostrils [153]. The sensory neurons of the Grüneberg ganglion are embedded in a network of fibroblasts between the nasal septum and the keratinized squamous epithelium that lines the nasal vestibule and is permeable to water [154]. These neurons, which possess various receptors, lack prominent dendrites or microvilli and do not directly access the nasal lumen, though they have numerous cilia that play a role in sensory transduction [117]. Peripherally, glial cells envelop the neurons, trapping the cilia within the ganglion. By detecting the expression of the olfactory marker protein (OMP), it is observed that their axons project along the nasal septum, enter the cribriform plate of the ethmoid bone, and eventually reach the anterior olfactory nucleus of the telencephalon, without synapsing in the main olfactory bulb. Another peculiarity of the neurons of the Grüneberg ganglion is the absence of projections to the vomeronasal organ [155]. Though its function is not yet fully understood, the ganglion could have a role in thermosensation [156].

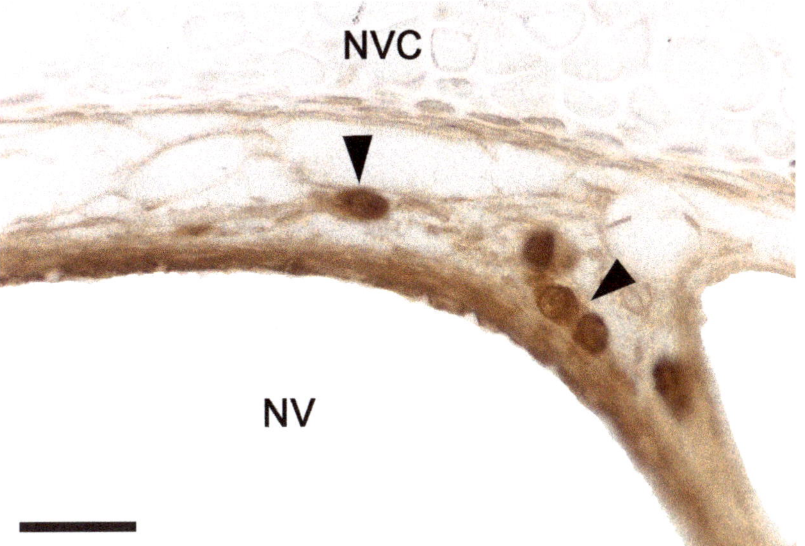

Figure 3. Grüneberg ganglion of a mouse. Transverse section of the dorsal part of the nasal vestibule (NV) immunostained with anti-OMP showing immunopositive ganglion cells (arrowheads). NVC, Cartilage of the nasal vestibule. Scale bar: 100 μm.

The **septal organ** (Figures 4 and 5) is a bilateral area of sensory neuroepithelium located at the base of the nasal septum, ventral to the main olfactory epithelium and rostral to the choanae [157]. Morphologically, it is composed of basal cells, supporting cells, and ciliated sensory neurons with flattened cell bodies and shortened dendrites [124]. The septal organ projects its neuronal axons to certain glomeruli located in the posterior region of the main olfactory bulb, specifically in the ventromedial zone [158].

On the other hand, although the projection pattern of the septal organ is already defined in newborns during the first postnatal days, there is a significant increase in the area of the organ [159] and in the number of septal glomeruli [160]. Based on the location of the septal organ, it has been suggested that it may serve an alert function, detecting odors in the environment when breathing is relaxed, and the air stream does not reach the main olfactory epithelium. In fact, the septal organ may respond to some odors with greater sensitivity than the main olfactory epithelium [161], and both share the main olfactory signal transduction pathways. Another suggested possibility is that it might play a role in detecting low-volatility compounds transferred by licking, which reach both the vomeronasal organ and the septal organ but not the main olfactory epithelium [162]. Thus, the septal organ could detect both odors and chemical signals with sociosexual influence.

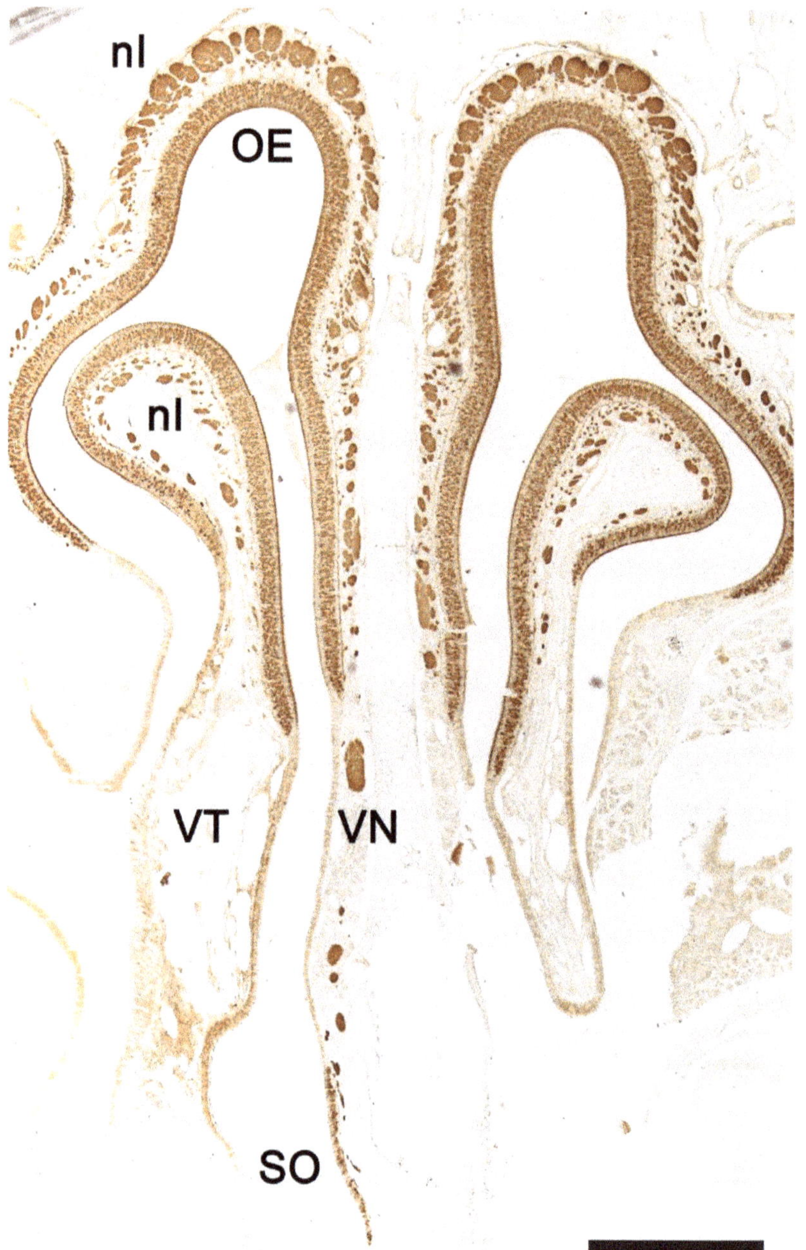

Figure 4. Septal organ of a mouse. Decalcified transverse section of the nasal cavity immunostained with anti-OMP. The immunopositive olfactory epithelium (OE) lines the roof of the nasal cavity, the nasal septum, the dorsal part of the ventral turbinate (VT) and the branches of the olfactory nerve (nI). On the basal part of the nasal septum a patch of OMP-positive epithelium corresponding to the septal organ (SO) can be observed. VN, vomeronasal nerves. Scale bar: 1 mm.

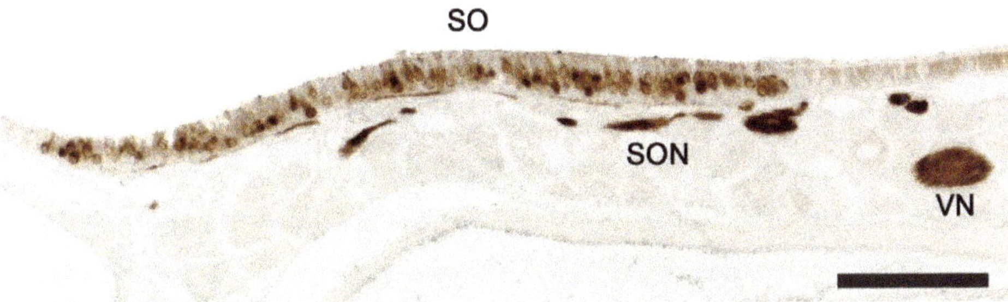

Figure 5. Septal organ of a mouse. Viewed at higher magnification of the septal organ (SO) area than is shown in the Figure 4. Decalcified transverse section of the nasal cavity immunostained with anti-OMP. Immunopositive neuroreceptor cells and the branches of the septal olfactory axons (SON) can be observed. VN, vomeronasal nerves. Scale bar: 100 μm.

Finally, another independent olfactory subsystem has been identified, comprising **solitary chemosensory cells**. The majority of these sensory cells are located in the anterior nasal cavity, but there is also a cluster in the entrance duct of the vomeronasal organ [127], and other populations present in the larynx and in the deeper respiratory tracts [163]. At the same time, fibers from the trigeminal nerve are observed near the solitary chemosensory cells, indicating sensory information transmission to these fibers. The upper respiratory tract is continuously assaulted by harmful substances and xenobiotics carried by the inspiratory air flow that are detected by the trigeminal nerve, which evokes protective reflexes such as sneezing, apnea, and local inflammation of the mucosa. Likewise, certain inhaled pathogens and irritants stimulate the solitary chemosensory cells, which help enhance the chemical response capabilities of the trigeminal nerve. Additionally, solitary chemosensory cells play a significant role in regulating the access of chemical substances to the vomeronasal organ [128], thus providing a mechanism to identify potential environmental irritants early. The fact that anosmic individuals or animals lacking functional olfactory systems still retain the ability to detect a variety of chemical irritants through the trigeminal system, which confirms the crucial role played by this system in nasal chemosensation [164].

2. Vomeronasal System

The vomeronasal system (VNS) or accessory olfactory system is specialized in detecting chemical signals, primarily pheromones, kairomones, and molecules from the major histocompatibility complex. It consists of a set of anatomically and histologically distinguishable structures from the main olfactory system. It is present in most reptiles [165,166] and amphibians [167], but it is particularly developed in mammals, in which this chemosensory system comprises three main components: the vomeronasal organ (VNO) (Figure 6), which acts as the peripheral chemoreceptor organ detecting chemical signals; the vomeronasal nerve, transmitting information to the brain; and the accessory olfactory bulb (AOB), the first neural center where vomeronasal afferent information is processed and integrated before heading to specific areas of the CNS [168].

The significant functional diversity of the vomeronasal system is expressed in the existence of three distinct subpopulations of vomeronasal sensory neurons. Each is associated with a specific family of chemosensory receptors: vomeronasal type 1 receptors (V1R), vomeronasal type 2 receptors (V2R), and formyl peptide receptors (FPR) [169]. However, it is crucial to note that not all mammalian species have functional receptors in all three families, indicating a diversity in chemical signal detection and specialized adaptation based on the biological and ecological needs of each species. This phenomenon underscores the richness and versatility of the VNS in detecting and processing chemical signals, highlighting the importance of chemical communication in mammalian life.

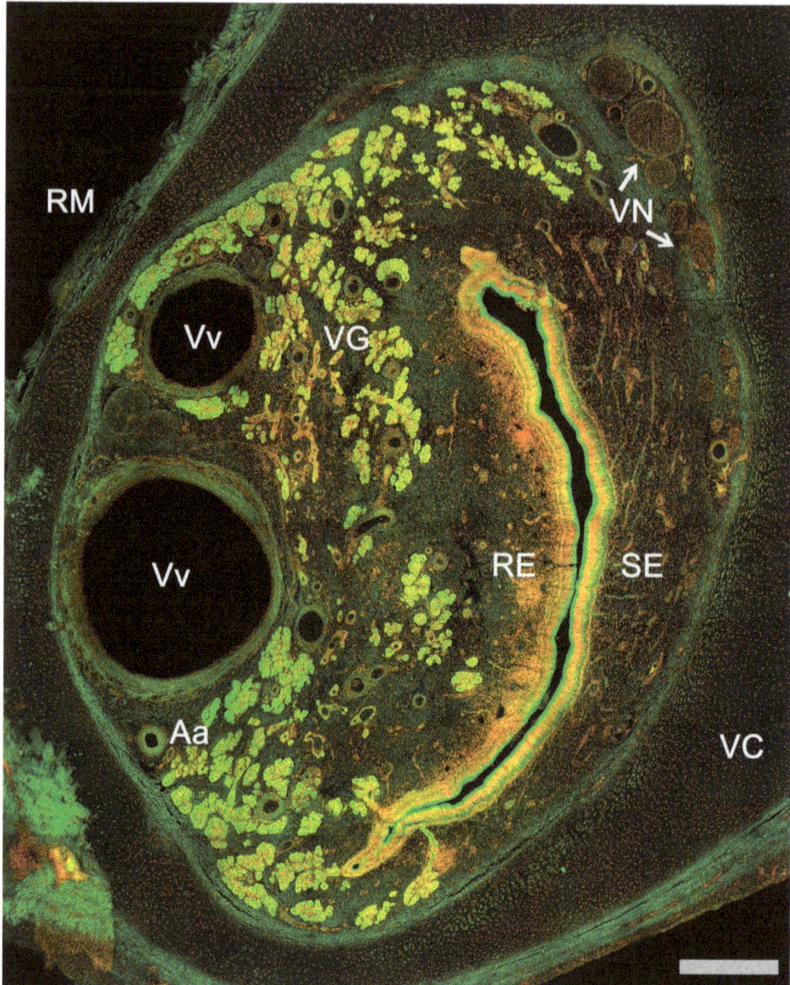

Figure 6. Vomeronasal organ of a horse. Autofluorescence transversal section of the VNO in which all components of the organ are identified. Aa, Artery; RE, Respiratory epithelium; RM, Respiratory mucosa; SE, Sensory epithelium; VC, Vomeronasal cartilage; VG, Vomeronasal glands; VN, Vomeronasal nerves; Vv, Veins. Nuclear contrast: TOPRO-3 iodide. Scale bar: 250 µm.

Identifying these receptor families was essential, and the study of G-proteins was crucial. Initially, by analyzing the expression of Gαo and Gαi2 proteins in rats, it was observed that the nerve endings of the vomeronasal neuroreceptor cells of the AOB were organized into two complementary regions [170]. Subsequently, both G-proteins' involvement in the transduction chain of vomeronasal neurons was determined, and the V1R receptor family in mice was discovered [171]. At that time, the expression pattern of the V1R receptors matched that of the Gαi2 protein. However, the second family of vomeronasal receptors was not identified until two years later, when three separate studies examining the Gαo protein expression in the VNO simultaneously demonstrated the existence of the second vomeronasal receptor family, V2R [172–174]. Finally, the third family of vomeronasal receptors, formylated peptide receptors (FPR), which coexpress with both G-proteins (Gαi2 and Gαo), was identified [175]. On the other hand, each neuronal population maintains a specific projection pattern to the AOB. Despite the fewer types of vomeronasal receptors compared to olfactory receptors, the vomeronasal receptor neurons' projection pattern to multiple AOB glomeruli seems to be more complex than those in the primary olfactory system [176].

There is a general consensus that the VNS is primarily responsible for perceiving pheromones, although it also perceives other non-pheromonal chemical signals, such as kairomones, which mediate defensive behavior [177], and other types of chemical signals vital for tracking prey and attack behavior, as seen in reptiles [178] and urodele amphibians [179]. In certain mammals, such as the gray short-tailed opossum, the VNS also influences food preference [180].

Chemical stimuli found in urine deposits, vaginal secretions, odorous gland secretions, or saliva can be investigated by direct contact. However, specific behaviors are also used to facilitate the entry of non-volatile substances into the VNO, such as facial grooming and "flehmen". The "flehmen" behavior is seen in ungulates and felines and consists of adopting a specific facial posture with the head tilted back, the mouth slightly open, the upper lip everted, and the neck extended for a few seconds [181,182]. It usually occurs after contact with biological secretions from conspecifics, and males exhibit this behavior more frequently [183,184].

Historical interest in the vomeronasal system began with the discovery of the VNO by the Danish anatomist Ludvig Levis Jacobson in the early 19th century, who described its key macroscopic features in a broad range of non-human mammals [185]. While previous illustrations showed the supposed location of the VNO in human nasal septum drawings [186], Jacobson reported this structure absence in *Homo sapiens*. However, the human VNO was later discovered in embryos [187] and a detailed histological description was then carried out both in fetuses and adults [188]. Although Jacobson also contemplated the hypothesis of a possible sensory function of the organ, he mainly suggested a secretory role. The histological contributions of Balogh [189], Klein [190], and Piana [191] revived the hypothesis of the sensory function, but only by the end of the 19th century did the availability of the Golgi technique definitively show the morphological similarity of the neurons of the olfactory and vomeronasal epithelium of the snake, thus establishing the sensory function of the VNO [192]. On the other hand, approximately half a century after the discovery of the VNO, the AOB was identified in sheep by Balogh [189] using traceability and dissection of the vomeronasal nerve. However, the term AOB was coined by Von Gudden [193] following his studies on the vomeronasal nerves in rabbits. Later on, Santiago Ramón y Cajal provided a detailed, accurate, and specific description of the AOB in various mammals, revealing its laminar architecture and the presence of different cell types [194]. It was the North American neuroanatomist Rollo McCotter [122] who established, in a broad range of species, the different nature of the olfactory and vomeronasal nerves and their respective destinations in the MOB and AOB. Finally, in the second half of the 20th century, a clear relationship between the VNS and reproductive behavior was established [195,196] leading to the seminal work of Powers and Winans [197] which convincingly demonstrated the critical role that the VNO plays in rodent reproduction.

Evolutionarily, the vomeronasal system has been linked to the transition of vertebrates to terrestrial environments; however, recent evidence suggests that a precursor VNS exists in teleost fish and its evolutionary origin predates the divergence between teleosts and tetrapods [10,198]. Added to this is the unique case of several species of lungfish that have a vomeronasal system homologous to mammals, showcasing a defined vomeronasal organ and an accessory olfactory bulb [199–202]. Hence, the significance of chemical communication has been a constant throughout the evolutionary history of vertebrates, resulting in significant morphofunctional variations among the chemosensory systems of different species. Specifically, the shift from aquatic to terrestrial life led to changes that significantly impacted pheromonal communication in vertebrates. This arose from the transformation of its key chemical property from solubility to volatility, a process that altered pheromone release mechanisms, accompanied by morphological and physiological changes in the sensory organs [203].

Subsequently, some terrestrial animal species returned to aquatic environments, such as cetaceans, which underwent drastic changes in their olfactory morphology during this migration. While their terrestrial relatives, including hippos, exhibit a defined VNO [204,205], no VNO has been found in any cetacean [206]. In contrast, toothed whales also lost their main olfactory system [207]. However, sea snakes, which evolved from terrestrial tetrapod reptiles, feature a functional and well-developed underwater VNS, while losing their main olfactory system [208]. In snakes, the vomeronasal system is predominantly considered to be the major chemosensory system [209–212]. The failure of their olfactory system to adapt to aquatic life, in contrast to the successful adaptation of their vomeronasal system, underscores the importance and development of the VNS in these reptiles. Other primarily aquatic reptiles, such as sea turtles, have a well-developed VNS [213]. However, while some alligator and crocodile embryos show a VNS, it regresses to be absent in adulthood [214,215]. Yet, most reptiles [216] and amphibians [217] have a functional VNS.

Regarding airborne vertebrates, there is no evidence of pheromonal communication in most birds due to the absence of their VNS [218]. The same is observed in many bats, though certain species possess a particularly well-developed VNS [219]. Finally, the vast majority of terrestrial mammals have a functional VNS, and some, like rodents, lagomorphs, or marsupials, have an especially developed VNS. Semi-aquatic mammals like the capybara [220], hippopotamus [221], beaver [222], and platypus also exhibit a VNS [223].

Among primates, it is believed that the last common ancestor with a functional vomeronasal system might have been small, arboreal, and nocturnal [224,225]. Without adequate light, vision is limited, heightening the reliance on olfactory signals [226]. Presently, primates can be classified into strepsirrhines and haplorhines, based on the presence or absence of a rhinarium: a moist, hairless skin area around the nostrils seen in some mammals. Strepsirrhines, which include lemurs and lorises, possess a rhinarium and are nocturnal, in addition to having a highly developed VNS [227]. On the other hand, haplorhines lack a rhinarium and are mostly diurnal. They include New World monkeys or platyrrhines, and Old World monkeys or catarrhines, among which are the great apes and humans. Regarding the vomeronasal system, platyrrhine monkeys have a well-developed VNO [228]. In parallel, the VNO of catarrhine primates has been generally considered absent; however, a rudimentary VNO in the postnatal stage has been observed in certain chimpanzee and human individuals [229]. Postnatal chimpanzees possess bilateral, ciliated epithelial tubes in the anteroinferior portion of the nasal septum. Both species are similar in possessing a relatively superiorly positioned VNO, which lacks the clear sensory epithelium seen in prosimians and New World primates [230]. It is of utmost importance that further investigations ascertain whether the VNO is retained in other apes. In the meantime, the presence of the VNO in adult Old World primates continues to pose a phylogenetic challenge.

In humans, the VNO undergoes significant development in the early gestational stages, exhibiting a pronounced neural projection from the organ to the olfactory bulb. A salient marker expressed during this period is the luteinizing hormone-releasing hormone (LHRH) [231]. However, a continuous projection from the vomeronasal nerve to the olfactory bulb beyond the 14–28 week period has not been found [232], rendering the function of the human VNO ambiguous. In fact, an accessory olfactory bulb has not been clearly identified since the description by Tryphena Humphrey in 1940 [233], which has not been replicated.

In adult humans, the proper VNO structure is retained in the majority of individuals. The bipolar cells within the human VNO exhibit structural similarities to olfactory receptor cells. Still, there is limited information concerning the histological configuration of its epithelium (VNE). Electron microscopic observations suggest that sections of the duct have a highly specialized epithelium similar to a chemoreceptive organ [234,235]. Yet, the epithelium does not appear to express neuronal features like OMP or PGP 9.5 reactivity, leading to concerns about the functionality of the human VNO. Conversely, the vomeroranasal epithelium displays a distinct arrangement of cell adhesion molecules, distinct from the adjacent nasal epithelia, which could indicate specific chemosensory roles [236]. Consequently, the VNE emerges as a specialized structure with an enigmatic function. Finally, the presence of a nerve connection from the VNO to a presumptive accessory olfactory bulb in adult humans has not been detected, so it remains a major issue in this debate [237–239].

2.1. Anatomy of the Vomeronasal Organ

The vomeronasal organ comprises two tubular structures located bilaterally at the base of the anterior nasal septum. Both organs have a single point of communication with the exterior which, depending on the species, can either be located in the nasopalatine or incisive duct—a conduit that connects the oral and nasal cavities through the palatine fissure—or directly in the nasal cavity [240].

Each organ consists of two clearly differentiated elements: the **vomeronasal duct**, which forms the lumen of the organ and is lined by a pseudostratified columnar epithelium, and the **vomeronasal capsule**, a rigid and protective envelope of either bony or cartilaginous nature, depending on the species [142]. Associated with the vomeronasal duct is the **parenchyma**: a tissue responsible for the organ function, consisting of an accumulation of soft tissue associated with the duct, rich in glands, vessels, nerves, and connective tissue [241].

Upon making a transverse cut in the central part of the organ, the typical crescent shape of the duct lumen at that level is observed. Internally, the duct is lined throughout its surface by two distinguishable epithelia. On its lateral side, it has a pseudostratified and ciliated respiratory epithelium, while on its medial part, the vomeronasal sensory epithelium (Figure 7) is located [242]. This sensory epithelium consists of a thin layer of basal cells, a broad central layer of bipolar neuroreceptor cells, and an outer layer of supporting cells that sustain the dendritic processes of the neuroreceptor cells (Figure 8). These dendritic processes project towards the lumen. At this level, they form microvilli that contain the vomeronasal receptors, which are responsible for recognizing the molecules involved in chemocommunication [243,244].

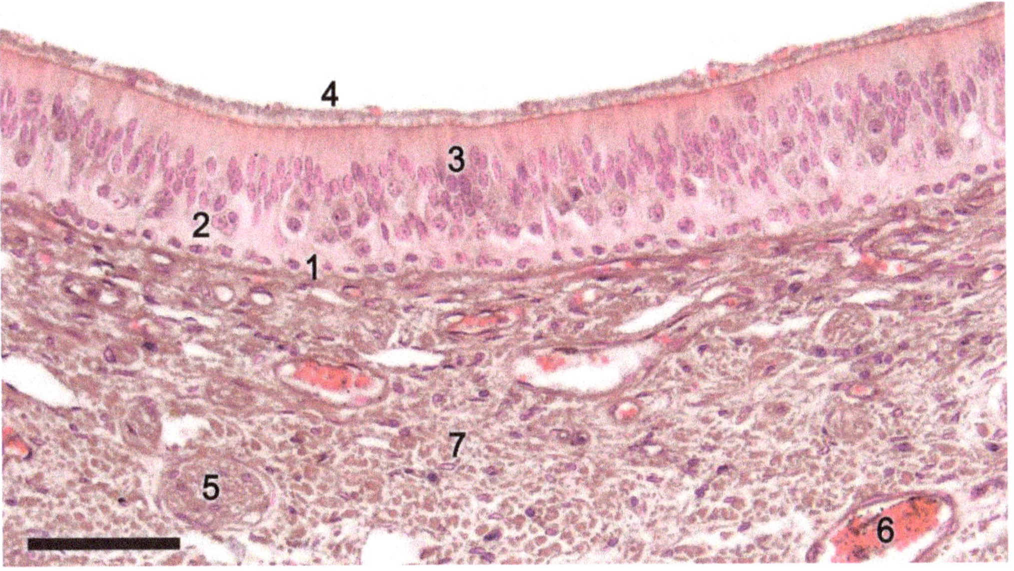

Figure 7. Sensory epithelium of the dama gazelle vomeronasal organ stained with hematoxylin-eosin. 1. Basal cells; 2. Neuroreceptor cells; 3. Sustentacular cells; 4. Mucomicrovillar complex; 5. Vomeronasal nerve axons; 6. Veins; 7. Collagen fibers. Scale bar: 50 µm.

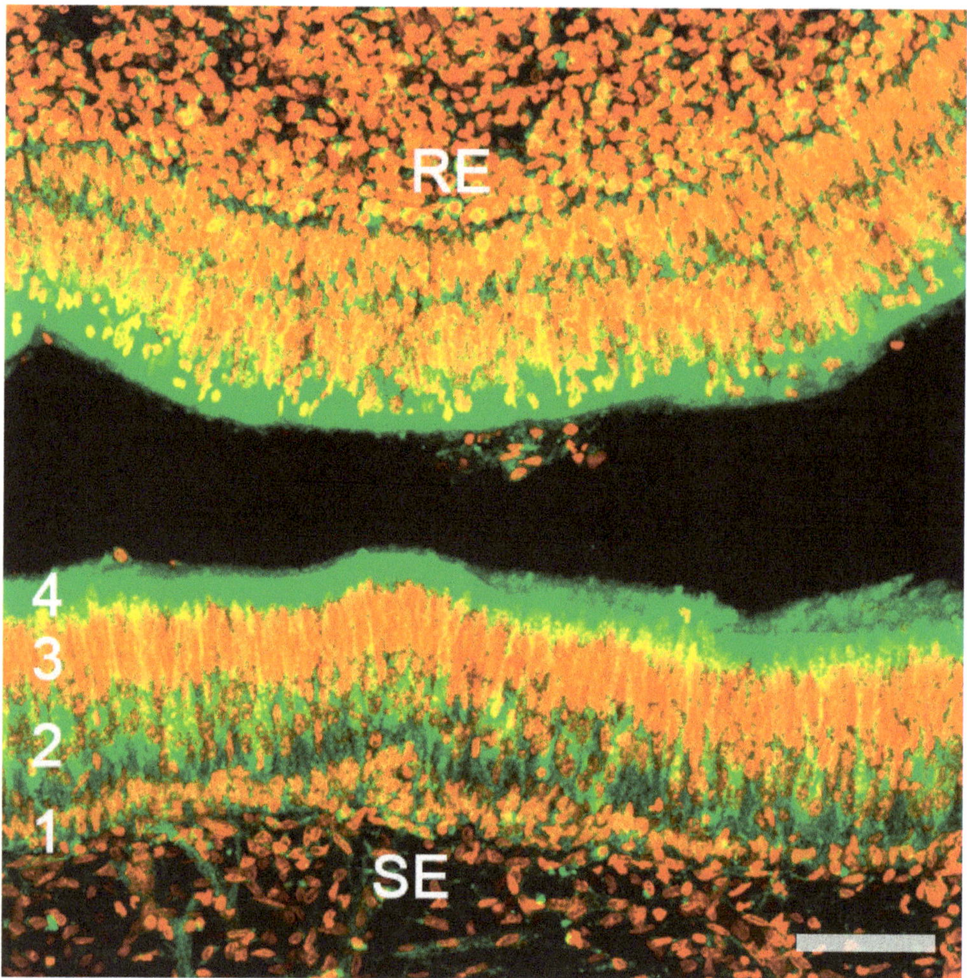

Figure 8. Vomeronasal organ of a horse. Autofluorescence confocal microscopy of the sensory epithelium (SE) allows for the clear differentiation of the zones corresponding to the main strata: Basal (1); Neuroreceptor (2); Sustentacular (3); Mucomicrovillar (4). RE, respiratory epithelium; Red nuclear contrast: TOPRO-3 iodide. Scale bar: 50 μm.

The nerve fascicles of the parenchyma are constituted by the coalescence of the axonal processes of the neuroreceptor cells into bundles, which converge from the vomeronasal epithelium. These nerve bundles are located in the dorsal and medial areas of the VNO, although occasionally they can also run on the ventral or lateral side of the duct. In turn, these bundles come together to form the vomeronasal nerve, which courses dorsocaudally in the submucosa of the nasal septum [245]. After passing through the cribriform plate of the ethmoid on its medial part, the vomeronasal nerve projects towards the anterior area of the telencephalon to synapse in the accessory olfactory bulb [246].

In the central region of the VNO, both the glandular and vascular components of the parenchyma are predominantly situated in the lateral portion of the parenchyma. Meanwhile, the glandular tissue is also concentrated in the caudal part of the VNO. The vomeronasal glands (Figure 9) are responsible for secreting mucus that is discharged into the lumen of the duct, either at its commissures or in the central part of the respiratory epithelium [247]. The mucus facilitates the entry and exit of molecules within the duct, but it also plays a crucial role in vomeronasal perireceptor processes. Histologically, these glands are distinguished by their acinar or tubular morphology, and by their serous or mucoid secretions, respectively [248,249]. Concerning the vascular tissue of the VNO, only arteries of significant caliber can be found in its posterior region; however, it has large veins that run parallel to the vomeronasal duct on the lateral flank [250]. This distinct morphological organization is important as it is involved in the vascular pumping mechanism that is activated to allow the entry of chemical stimuli into the organ. Specifically, when the vessels of the parenchyma contract, the lumen of the duct expands, creating a vacuum effect that draws chemical molecules into the duct. Similarly, when the vessels dilate, a constriction of the duct occurs, expelling the content of the duct outward. This vascular pumping mechanism is known as the vomeronasal pump [251–253].

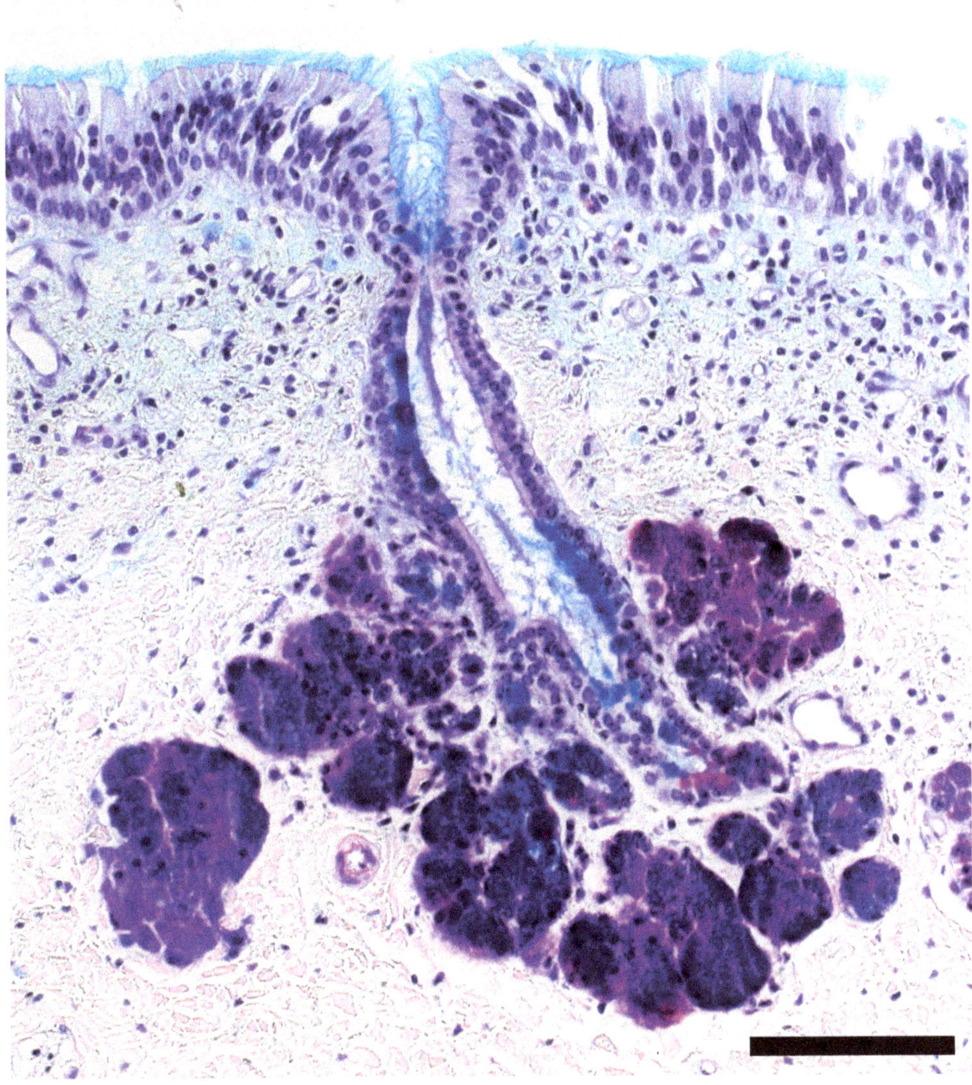

Figure 9. Vomeronasal gland of a cow stained with PAS-Alcian blue-hematoxylin staining. Both the glandular acini and the duct are stained. Scale bar: 150 μm.

The vomeronasal capsule provides structural support for the proper functioning of the vascular pump, counteracting the potential collapse of the parenchyma that the increase in negative pressure might induce, while simultaneously safeguarding the encapsulated structures. Its morphology not only varies along the VNO, but there are also notable differences among various species, both in its nature and its configuration [142]. More typically, towards the front, the capsule is shaped like an incomplete ring with a lateral opening associated with the outlet of the vomeronasal duct. In the central region, the ring becomes complete, but posteriorly the capsule separates again, adopting a "J" shape, with a dorsolateral slit allowing the passage of the vomeronasal nerves. As we progress caudally, on occasion, the ring closes entirely. Its nature can be bony, cartilaginous, or a combination of both in different patterns, making it one of the VNO components with the most diverse design from a comparative anatomy standpoint.

Regarding the sensory transduction of the VNO, in the most studied model, the murine, the neuroreceptor cells are organized into different subpopulations throughout the vomeronasal epithelium, following a clearly defined zoning. In this way, the V1R receptors, linked to the Gαi2 protein, are confined to the neurons whose cell bodies are located in the apical part of the epithelium, while the neurons expressing the V2R receptors, linked to the Gαo protein, locate their cell bodies in the basal layer of the vomeronasal sensory epithelium [254]. In both cases, each neuroreceptor cell exclusively expresses a single receptor among the total population of vomeronasal receptors [173].

Formyl peptide receptors (FPRs) co-express with both G proteins in the two neural layers of the VNO. FPRs react to products from viruses and bacteria, and they are found in various tissues and cell types, mainly associated with immune functions [255]. Within mammals, its expression has only been confirmed in the Muroidea superfamily of rodents and in Lagomorpha [256]. Seven distinct genes encoding FPRs have been identified in mice VNOs [175,257]. Some of these genes, like FPR-rs3, FPR-rs4, FPR-rs6, and FPR-rs7, are specific to the VNO and co-express with Gαi2. However, FPRrs1, recently renamed as FPR3, co-expresses with Gαo in vomeronasal sensory neurons and is also transcribed in immune cells. As there is no evidence that FPRs co-express with V1Rs or V2Rs, it is likely that certain vomeronasal sensory neurons express only FPRs. FPR3 reacts to a limited set of bacterial peptides crucial for managing infection virulence. This links the behavioral and immunological functions of the VNO [258].

On the other hand, it has been demonstrated that the peptides of the major histocompatibility complex (MHC) can perceive non-volatile molecules, both in the olfactory and vomeronasal epithelium of mammals. The MHC was first identified as a primary component in the adaptive immune response, recognizing and presenting antigens to T cells [259]. MHC is encoded in the highly polymorphic H2 and HLA genes for mice and humans, respectively. Current research estimates that over 10^9 distinct MHC antigenic phenotypes exist in outbreeding mammals, which suggests that MHC alleles might be crucial for determining genetic uniqueness [260]. In a lab setting, numerous studies indicate that MHC and its peptide ligands get secreted in animal urine and function as markers for self-identity during individual recognition [261,262]. Both genders of mice display a natural inclination to select mates with varied MHC haplotypes [263]. Additionally, when female mice that have recently mated are introduced to the scent of an unfamiliar male with different MHC peptides, it can lead to a pregnancy block [264]. In human contexts, there are various studies suggesting a connection between MHC differences and preferences in smell and mate selection. Women seem to be drawn to the scent of shirts worn by men who have different MHC genotypes [263].

Mates with specific MHC variations are viewed favorably because of their increased resistance to parasites, potentially leading to enhanced reproductive outcomes for their progeny [265]. Female sticklebacks show a preference for mating with males that exhibit a broad MHC diversity and have a certain MHC haplotype that wards off infections from prevalent parasites [266]. While MHC peptides can be found in both VNO and MOE neurons, they serve different behavioral functions. Specifically, the MOE ability to recognize and prefer urine containing unique MHC peptides contrasts with the VNO-driven Bruce effect, activated with dissimilar MHC peptides. All together suggests a general role of MHC peptides in chemical communication, even in species lacking a functional VNO [267,268].

The transduction cascade of virtually all vomeronasal neurons converges in most species on the activation of a member of the transient receptor potential (TRP) channel family, the TRPC2, which is expressed in both neural layers of the VNO [269]. It was observed that the protein resides in the microvilli of the sensory neurons and coincides with the expression of both Gai2 and Gao [244]. In fact, the genomic sequence of the intact TRPC2 gene is considered to be a predictive marker of the functionality of the vomeronasal organ [218]. When the Trpc2 gene was knocked out in mice, its pivotal role in vomeronasal-mediated behavior became evident. These modified mice exhibited reduced aggressive responses and did not differentiate between males and females, displaying sexual behaviors towards both. This suggests that these mice are unable to discern the gender of their counterparts due to a lack of olfactory signal transduction through VSNs [270,271]. The TRPC2 gene is ubiquitous among mammals, reptiles, amphibians, and fish species. Comprehensive genome studies indicate an absence of TRPC2 genes in avian species [272]. A noted absence of functional TRPC2 in certain primates, such as Old World monkeys, correlates with the lack of VNO [198,273]; although, orangutans and rhesus macaques possess three TRPC2 gene copies, which have been suggested to be involved in mediating the acrosomal response during fertilization [274]. In humans, the loss of a functional TRPC2 gene has been confirmed; however, the presence of functional V1R receptors has been observed in the olfactory mucosa, embryonic cerebral tissues, and non-neuronal cells [275]. It should also be noted that chemical communication in vertebrates originated long before the development of the VNO, and pheromone detection can also be mediated by other olfactory organs [276]. This brings forth questions about the specific impacts of Trpc2$-/-$ on VNO activity. Mombaerts and colleagues integrated the lacZ gene into the Trpc2 site and mapped the pathways of Trpc2-active neurons. Their exploration revealed that two distinct MOE neuron types, which communicate with specific glomeruli on the ventral aspect of the primary olfactory bulb, expressed Trpc2 [277,278]. These observations indicate potential broader roles for Trpc2 within the primary olfactory framework and other cerebral regions. Thus, the behavioral patterns seen in Trpc2$-/-$ mice might not be exclusively attributed to VNO disruption.

While initially, it was thought that all mammals expressed both families of vomeronasal receptors, the absence of the V2R family in the vomeronasal system of various species was subsequently described. Therefore, based on the presence or absence of the two neuronal subpopulations linked to the V1R and V2R vomeronasal receptors, two vomeronasal transduction models were identified: the segregated model and the uniform model. In the segregated model, both types of vomeronasal receptors coexist, while in the uniform model the V2R receptor is absent, and only V1R type neurons are expressed. Recent research on the Tammar wallaby (*Notamacropus eugenii*) concluded that macropods could present a third vomeronasal transduction model, consisting only of V2R neurons, as they did not find expression of the Gαi2 protein in the VNO or in the AOB [279].

A segregated-type vomeronasal system has been characterized in certain species of rodents [280], lagomorphs [143], and marsupials [281], while the rest of the studied mammals, such as ungulates, carnivores, or primates, fall within the uniform model [282–284].

The neurochemical study of the organ allows us to obtain characteristics about its functionality using G proteins, but other markers are also used that provide specific information about this structure. The calcium-binding proteins calbindin (CB) and calretinin (CR) are used to identify neuroactive substances, but also to differentiate cell populations or to define the morphology of vomeronasal neuroreceptor cells [285]. On the other hand, the GAP-43 marker shows neuronal growth and is expressed in the nerve bundles present in the organ [286]. These markers are complemented with routine stains (hematoxylin-eosin) or specific stains (Gallego trichrome), which display the structural characteristics of the VNO. Likewise, using PAS (Periodic acid Schiff) and Alcian blue stains, the nature of the vomeronasal glands can be identified, staining their secretions with neutral and acid mucopolysaccharides, respectively [287].

Likewise, in the study of the VNO, lectins such as *Ulex europaeus* agglutinin (UEA), *Bandeiraea simplicifolia* isolectin B4 (BSI-B4), or *Lycopersicon esculentum* agglutinin (LEA) have been frequently employed, as some are specific to the vomeronasal and/or olfactory system in various mammals [288]. Thus, it is possible to differentiate neuronal populations and study the morphology of the neuroreceptor cells in the sensory epithelium of the VNO. On the other hand, when expressed in the neuronal axons of these cells, it becomes easier to identify the nerve bundles of the VNO and track the vomeronasal nerve by analyzing its topography and anatomical relationships.

Over the past decade, **genomic studies** have paved the way for groundbreaking insights into the VNS. To date, only a limited number of links between chemical ligands, vomeronasal receptors, and behavior have been clarified. This is primarily due to the challenges encountered when dealing with expansive, homologous gene families that have a high degree of sequence similarity. Nonetheless, when examining mouse strains with mutations in genes linked to ligand-VR signal transduction, the role played by the VNO in various social behaviors can be studied in a more specific way. These behaviors include male-to-male and maternal aggression, sexual allure, lordosis, selective pregnancy termination, and interspecies reactions such as aversion and defensive behaviors [289].

One significant discovery has been the broad expansion and diversification of the vomeronasal receptor gene families, V1Rs and V2Rs [273,290]. High-throughput sequencing methodologies, such as RNA-Seq, have enabled researchers to identify and catalog an extensive array of VNRs in various species, showcasing a vast diversity that arguably correlates with species-specific pheromonal communication [11,289,291]. Another groundbreaking revelation is the functional differentiation of V1R and V2R receptors, elucidated through transcriptomic analyses. While V1Rs primarily respond to small volatile molecules, V2Rs are more attuned to larger, peptide-based cues [292].

Comparative genomics indicates that the retention or loss of specific vomeronasal receptors often aligns with the ecological and social structures of the species in question. For instance, species with more intricate social hierarchies or mating systems tend to possess a richer repertoire of functional VNRs [293]. Moreover, the transcriptome analysis of the VNO in mice under different physiological and environmental conditions showed notable variations in the expression of vomeronasal receptors even among individuals of the same species [256,294,295]. For instance, during the final phase of pregnancy, the production of neural progenitors in the VNO of female mice is notably increased. Transcriptome analysis comparing pregnant and control female VNOs reveals differential expression of 101 genes, including 24 vomeronasal receptors and other genes related to cell proliferation and death [294]. Pregnancy or estrogen-driven increases in new chemosensory neurons have significant functional consequences, especially in maternal behaviors. For instance, the ability of the VNO to recognize social odors can induce varied behavioral responses based on the hormonal status, age, gender, or dominance of the individual. In humans, while the VNO appears non-functional, with VRs and transduction molecules being pseudogenized, there is still a sensory target for hormone modulation in the main olfactory epithelium. Further investigations are recommended to discern whether pregnancy also prompts an increase in neurogenesis in the main olfactory epithelium.

The application of transcriptomic techniques has also made it possible to compare the expression of VRs in rabbits of different ages. Some VRs are more expressed in juveniles than in adults [256]. This difference might be attributed to increased exposure to new stimuli during early life stages. The difference in expression between young and adult animals could be crucial for innate or unconditioned responses to specific chemical signals. Likewise, it has been observed that sex-separation induces sex- and stage-specific gene expression differences in the VNO of male and female rabbits, both in adults and juveniles [295].

The presence of sexual dimorphism in the vomeronasal system is an issue of particular interest. Although no qualitative histological and immunohistochemical differences between sexes in both the VNO and the AOB have been identified, morphometric analyses have detected subtle yet significant disparities. In the 1980s, Guillamón and Segovia initiated a series of studies establishing that, in rats, the VNO and the AOB were larger in males than in females [296,297]. Specifically, male rats exhibited a greater numerical density of principal cells in the AOB compared to females [297]. Interestingly, in the AOB of rabbits, the trend was reversed, with females demonstrating higher morphological values [132]. These species-specific manifestations of sexual dimorphism likely reflect the reproductive, behavioral, and physiological characteristics unique to each species. For instance, while female rabbits are reflex ovulators, female rats are spontaneous ovulators. Additionally, there are distinct differences in sexual and maternal behaviors between these species. The contrasting patterns observed in closely related species like rats and rabbits underscore the caution needed when extrapolating findings on brain sexual dimorphism from one species to another.

Sexual dimorphism has also been examined in other mammalian groups. In a study involving wild voles (*Microtus pennsylvanicus* and *M. ochrogaster*), it was found that *M. pennsylvanicus*, which exhibits more pronounced differences in parental behaviors, had larger VNOs in females than in males [298]. A comparison between the VNO and AOB of the monogamous mandarin vole, *Microtus mandarinus*, and the polygamous reed vole, *M. fortis*, revealed significant differences only in the reed voles, suggesting that the degree of sexual dimorphism might be linked to the mating system [299]. Furthermore, a discernable sexual dimorphism in the rate of neuronal generation in the AOB of rats has been identified, with a higher proliferation rate in the anterior AOB in male rats compared to female rats [300].The morphological diversity of the VNO among different orders of mammals is also notable, and indeed, since the advent of Darwinian theory of evolution, the morphological configuration of the VNO has been used as a phylogenetic classification method [301–303]. A significant portion of the studies on the VNS have been conducted in rodents, specifically in myomorphic species like the laboratory rat (*Rattus norvegicus domestica*) or the mouse (*Mus musculus*), which have been used as general VNS models in rodents (Figure 10). Both species display a highly developed vomeronasal organ, with a large neuroreceptor epithelium, of considerable thickness and cell density [304]. The vascular pump consists mainly of a large central vein, which is surrounded by other smaller-diameter veins [305]. The vomeronasal capsule is made up of a thin bone layer in adults; however, in newborn rats, this covering is cartilaginous [241]. A thoroughly investigated aspect of the rodent VNO is the presence of postnatal proliferation in the vomeronasal epithelium. The distribution of these proliferating cells varies with age. In neonatal rats, such cells are distributed fairly evenly across the sensory epithelium. However, starting from P21, the majority of these cells tend to cluster near the boundaries with the non-sensory epithelium [306]. These findings have been more recently corroborated in mice using bromodeoxyuridine (BrdU) immunohistochemistry [307].

Other myomorphic rodents display specific morphological features in their VNO, but in general, they all show a comparable level of development. For instance, the Gambian rat (*Cricetomys gambianus*) displays certain ossification areas in its cartilaginous vomeronasal capsule [308]. Among the sciuromorphic rodents, the Daurian ground squirrel has an exceptionally voluminous vomeronasal vein, which, together with other smaller-caliber veins, constitutes the vascular pump. It also displays a sensory epithelium with significant thickness, and its vomeronasal capsule is exclusively cartilaginous [309]. In hystricomorphic rodents, the long-tailed chinchilla (*Chinchilla lanigera*) showed considerable variation in the primary features of its VNO. Its sensory epithelium shifts from a medial location at the front of the VNO to an entirely dorsal position at the back. Similarly, the vomeronasal capsule in this species consists of cartilaginous tissue at its anterior level, but this shifts dorsally as we move caudally, eventually being completely replaced by a bone layer that encases the VNO at its back [310].

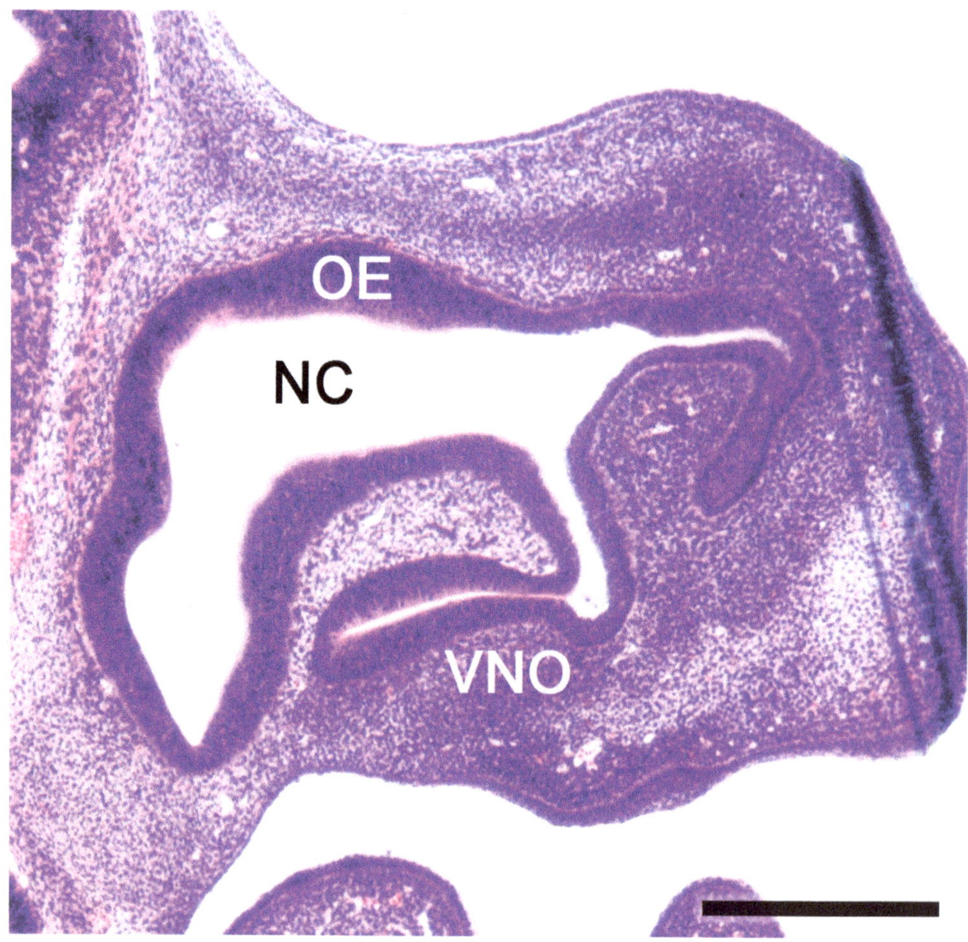

Figure 10. Sagittal section of the nose of a fetal mouse. The vomeronasal organ is located in the base of the nasal septum. A functional opening to the nasal cavity (NC) can be observed. The NC is lined by olfactory epithelium (OE). Hematoxylin-eosin staining. Scale bar: 500 μm.

Concerning the vascular pump, the VNO of the chinchilla has several significant-diameter veins, but features a dominant central vein [311]. Similarly, within the hystricomorphs, African mole rats displayed unique characteristics concerning their vomeronasal organ. In these eusocial rodents, there is hardly any postnatal growth in their vomeronasal neuroepithelium [312]; however, all the components of their VNO exhibit significant and functional development. Thus, it has a sensory epithelium of substantial thickness, and laterally displays large venous sinuses. Transversely, the vomeronasal duct takes on the typical "J" or crescent shape. The parenchyma is protected by a cartilaginous capsule and is also reinforced by an external bone layer that occupies a ventral and/or lateral position depending on the species [313].

As with rodents, lagomorphs have an extensively developed vomeronasal organ (Figure 11). The rabbit (*Oryctolagus cuniculus*) has a double vomeronasal capsule, made up of an external bone covering and an internal cartilaginous lining [314,315]. Its sensory epithelium is thick compared to that of rodents, but its vascular pump comprises multiple large-caliber veins that provide a powerful suction capability to the VNO in this species [143,316].

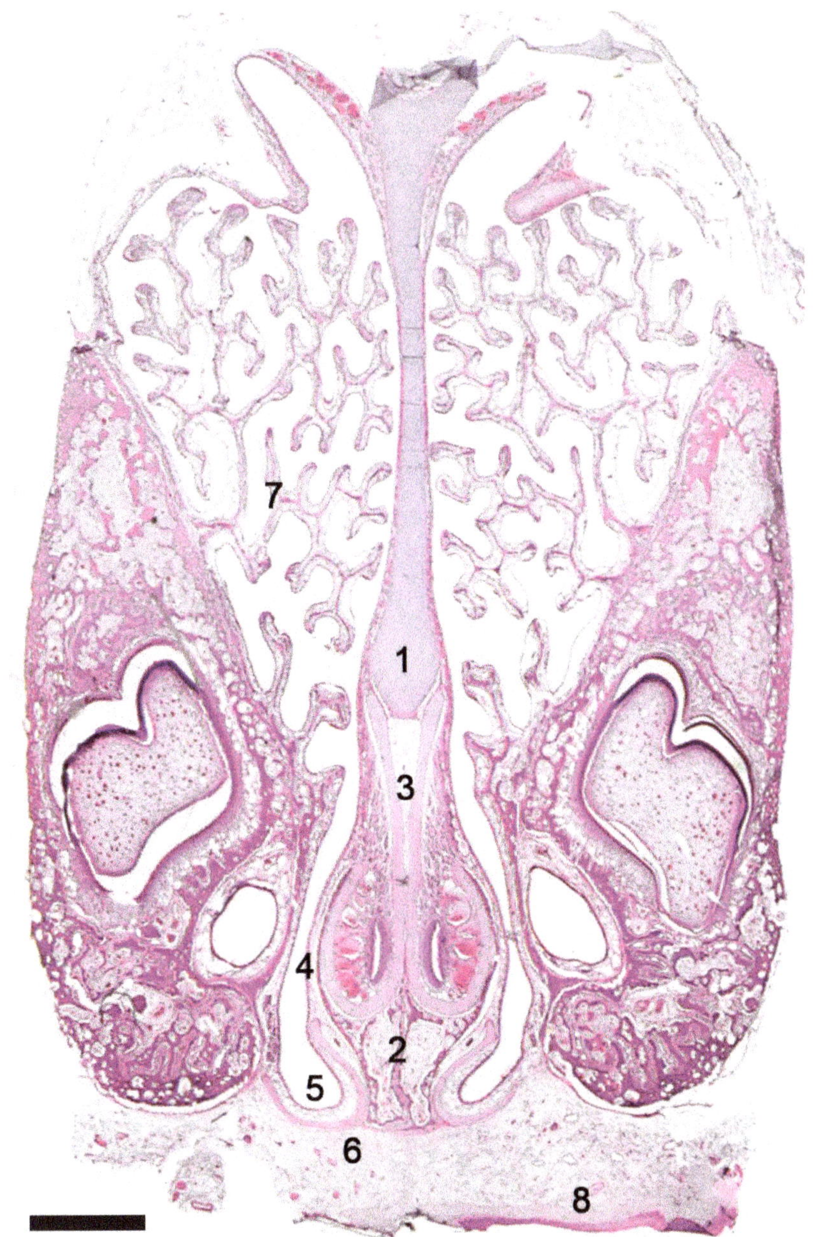

Figure 11. The vomeronasal organ of a rabbit. Hematoxylin-eosin staining. Transverse section of the head showing the nasal septum (1) with the VNOs located over the palatine processes of the incisive bone (2). The J-shaped cartilaginous envelope of the VNO (3) is covered by a thin bony layer (4). The ventral recess of the nasal cavity (5) is shaped by the cartilage of the incisive duct (6). (7). Ventral turbinate; (8). Palate. Scale bar: 500 μm.

Marsupials also display a remarkable development of the vomeronasal organ, suggesting that, as in rodents and lagomorphs, pheromonal communication plays an essential role in these mammals. The Tammar wallaby (*Notamacropus eugenii*), a macropodid of the order Diprotodontia, sports a complete cartilaginous capsule and a densely populated and extensive neuroreceptor epithelium. The vascular tissue enabling the pumping mechanism consists of numerous large-caliber veins, which are mostly distributed on the medial side, but voluminous veins also appear on the medial face of the vomeronasal duct [317]. In the order Didelphimorphia, the gray short-tailed opossum (*Monodelphis domestica*) presents certain unique features. Its vomeronasal capsule is cartilaginous, although at the caudal level, it is replaced by a bony capsule, which encircles a parenchyma very rich in glandular tissue. Its vascular pump is made up of a large main vessel which is located laterally to the vomeronasal duct, and other smaller vessels in the ventromedial and lateral areas. Its sensory epithelium is very thick and is characterized by a rosette-shaped structure that divides said epithelium in half at the caudal part of the organ. This configuration corresponds to the fusion of the main vomeronasal glandular secretion duct with the neuroepithelium, through its opening into the lumen of the vomeronasal duct [318].

In species of the order Monotremata, the VNO shows notable development. Studies on newborn individuals of the platypus and echidna determined that both possess a thick sensory epithelium and a cartilaginous capsule. In contrast, both showed few veins of modest caliber forming their vascular pump [319,320].

Within the order Eulipotyphla, the studied species display a highly developed VNO. The African pygmy hedgehog (*Atelerix albiventris*) has a cartilaginous vomeronasal capsule and a neuroreceptor epithelium of moderate thickness, which extends beyond the usual boundaries, occupying almost the entirety of the lumen in front of the respiratory epithelium. The parenchyma contains mucous and serous glands, and a network of venous sinuses that extend around the entire vomeronasal duct. Through a cross-section, the lumen of its VNO shows a circular shape at the front, which becomes more oval as we move caudally [321]. On the other hand, the VNO of the dasyurid marsupial *Antechinus subtropicus* displayed significant differences compared to the African pygmy hedgehog. In this species, the capsule is also cartilaginous, but it features a primary vein lateral to the vomeronasal duct, along with other secondary veins located both laterally and medially to it, forming its vascular pump. Additionally, its vomeronasal neuroepithelium has a notably greater thickness compared to the African pygmy hedgehog; however, its extension in front of the respiratory epithelium corresponds with the usual pattern, and its lumen shape, when observed in cross-section, is the typical "J" or crescent shape [322].

Mammals of the order Carnivora, both felines and canines, have shown moderate development of their VNO (Figure 12). Studies on the dog [283,323], domestic cat (*Felis silvestris catus*) [324], brown bear (*Ursus arctos*) [325], and European ferret (*Mustela putorius*) [326] revealed that all share the main structural characteristics of the VNO, such as a cartilaginous capsule, the distribution of vessels around the vomeronasal duct forming the vascular pump, and a broad sensory neuroepithelium, albeit thinner than mammals of the orders previously described. However, the red fox (*Vulpes vulpes*) represents an exception among domestic canines. In this species, the vomeronasal epithelium is not only more developed but also displays the expression of G proteins associated with both V1R and V2R receptors [127].

Artiodactyls exhibit a high level of VNO development. Their cartilaginous capsule is complex and developed, displaying numerous morphological differences between species. The length of the sensory vomeronasal epithelium tends to be longer compared to species of the aforementioned orders; however, its thickness is notably inferior, as seen in carnivores. Still, it maintains the three cellular layers that make up the neuroepithelium, which are organized with a clearly defined boundary. Vessels in the vomeronasal parenchyma are numerous, of medium caliber, and are uniformly arranged around the vomeronasal duct [327]. Regarding the morphology that the vomeronasal duct exhibits in cross-section, artiodactyls usually have an oval shape, unlike the usual crescent or "J" shape. Variations also exist between different species like cows [145,328], sheep [329], goats [330,331], deer [332], moose [333], giraffes [334], and duikers [308].

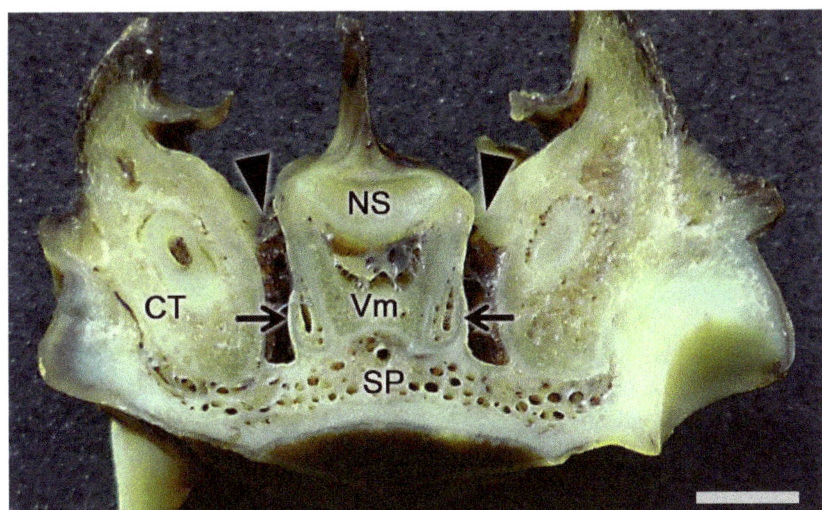

Figure 12. The vomeronasal organ of a fox. Transverse section of the base of the nasal septum (NS) showing the topographic relationships of the vomeronasal organs (arrows) with the ventral recesses of the nasal cavity (arrowheads). CT, canine tooth; SP, soft palate; Vm, vomer bone. Scale bar: 2 mm.

Strepsirrhine primates, which are mostly nocturnal, have a highly developed VNO compared to haplorhine primates, who tend to be diurnal and have a modestly developed VNO. Both have a cartilaginous capsule surrounding the vomeronasal parenchyma; however, the thickness of the vomeronasal epithelium is significantly greater in strepsirrhines. Similarly, the vascular pump in strepsirrhines has numerous large-caliber vessels around the entire vomeronasal duct, in contrast to the few vessels typically displayed by the VNO of haplorhine primates [335].

Among the various bats, the development of the VNO is extremely variable. From species that show a total absence of their VNO even in the embryonic phase, to other species that demonstrate astonishing development of this sensitive structure. Likewise, many other bat groups represent intermediate levels of development [219].

2.2. Neuroanatomy of the Accessory Olfactory Bulb

The accessory olfactory bulb (AOB), the first neuronal integration center of the vomeronasal sensory system, is located in the olfactory areas of the forebrain of certain vertebrate species; what we recognize as the basal rhinencephalon. In mammals, it is usually located in a dorsocaudal position relative to the main olfactory bulb (MOB) [336].

Macroscopically, in most species, the AOB is a relatively difficult structure to discern. Although in most cases it gives rise to a convex prominence in the dorsal transition between the main olfactory bulb and the olfactory tract, its boundaries with these structures are very diffuse. However, histologically, it is easily recognizable due to its laminar morphology. Although this laminar pattern is essentially analogous to that of the MOB, there are significant variations between different groups of mammals. Some species have clearly defined lamination and high cellularity, while others have much less defined layer boundaries (Figures 13 and 14). Thus, from superficial to deep, the AOB layers are:

1. Vomeronasal nerve layer: It consists of bundles of unmyelinated fibers, originating from vomeronasal receptor neurons, surrounded by numerous glial cells, both astrocytic and ensheathing glia [337,338]. Whereas, in the MOB of mammals, each olfactory axon projects to a single glomerulus [139,339], in the AOB, fibers branch out and distribute to more than one glomerulus [340].

2. Glomerular layer: Comprised of spherical structures called glomeruli, these result from synaptic contact between axonal terminations of vomeronasal nerve fibers and the apical dendrites of mitral cells. These glomeruli show a relatively acellular texture rich in neuropil but are bounded by a narrow band of periglomerular cells, especially in their deepest part. Their degree of definition is lower than that observed in the main olfactory bulb.

3. External plexiform/Mitral/Internal plexiform: It is at the level of the three central layers of the accessory olfactory bulb where the greatest differences lie, both compared to the MOB and in the interspecific comparisons that can be made at the level of the AOB itself. The differentiation of an external plexiform layer—formed by the dendrites of the mitral cells and granular cells, as well as by tufted cells and other neuronal types —a mitral layer— formed by a linear band that includes the somas of the mitral cells, the second neuron of the olfactory pathway—and an internal plexiform layer—containing the axons of the mitral cells—which is constant in the organization of the MOB in mammals. However, in the case of the AOB, the existence of plexiform layers has only been proposed in those species where the degree of lamination is highest, mainly rodents, but the issue is controversial. Thus, Cajal [194] concluded that the plexiform layers were absent; however, classic comparative studies of the olfactory system [341,342] included the external plexiform in their description of the AOB but not the internal plexiform. Meisami and Bhatnagar [336] in their exhaustive bibliographic review state that it is inappropriate to extrapolate data from one species to another, indicating that, in mammals with large and well-developed AOBs, both plexiform layers exist; although, it should be recognized that the boundaries of these layers are not recognizable with Nissl staining, requiring more specific stains such as cytochrome oxidase or Galyas staining. However, Salazar et al. [245], in their exhaustive study of the mouse AOB, concluded the impossibility of determining the presence of plexiform layers, as well as differentiating mitral cells from tufted cells, coining the term mitral/tufted layer. In their subsequent study of the cat AOB, Salazar and Sánchez-Quinteiro [343] opted for the term mitral/tufted/plexiform layer which more accurately encompasses the fusion of the three AOB structures equivalent to the three inner layers of the MOB. In later works, this group leaned towards a simpler name: the mitral/plexiform layer, which we have adopted throughout this work. It is important to clarify that the morphology of the main cells of this mitral/plexiform layer rarely corresponds to the typically mitral morphology observed in the mitral layer of the MOB. This has led certain authors to prefer to avoid the term mitral when describing this layer, as is the case with Larriva-Sahd [344] and Villamayor et al. [132] in their respective studies of the AOB in rats and rabbits. In both cases, they opted for the terms "outer cell layer" to designate the set of the mitral/plexiform layer and "principal cells" to designate the projection neurons that constitute this layer. While undoubtedly a more accurate name in terms of morphological reality, the direct functional equivalence between the mitral cells of the MOB and the principal cells of the AOB (projection neurons in both cases) advises maintaining the term mitral cells to designate the principal cells of the AOB. Regarding the use of the term plexiform, Larriva-Sahd [344] considers that, in contrast to the plexus arrangement of the proximal processes of the MOB mitral cells, in the case of the AOB these processes adopt a radial arrangement, resulting in a dense and convoluted dendritic frame, very different from the parallel fibrous texture that forms the plexiform layer of the MOB. The set of radial dendritic arborizations, together with the proximal processes of granular cells and other interneurons, make up a neuropil rather than a plexiform layer. Another interesting aspect that affects the stratification of the AOB is the topography of the lateral olfactory tract (LOT), which can be located either through or below the accessory bulb. Thus, in broad groups of mammals including rodents, insectivores, and primates, the broad axonal bundles that make up the dorsal component of the LOT run through the innermost area of the mitral/plexiform layer or outer cell layer [345].

4. Granule cell layer: It is composed of granular neurons, which lack axons and accumulate compactly. These cells are GABAergic and presumably inhibitory [346]. However, they are not a homogeneous population in terms of the expression of peptidic neuromodulators [347]. Their basal dendrites receive centrifugal impulses from the amygdala, whereas their apical dendrites interact with the basal dendrites of the mitral cells [348]. On the other hand, it has been observed that the number of granular cells is lower in domestic animals and animals kept in zoological facilities [349]. Furthermore, neuronal proliferation has been found in the AOB of adult rats, with most of these new cells being located in the granular layer [350]. Contrary to what one might assume, most of the proliferating neurons present in the OB do not derive from the subventricular zone of the olfactory bulb but from slow-dividing cells that might correspond to the population of resident neural stem cells. These could generate neurons that are incorporated into the OB circuits in vivo [351].

Projection neurons of the olfactory bulb send axons through the lateral olfactory tract (TOL), giving it a laminar structure. Thus, in embryonic stages, the TOL consists of three layers: superficial, middle, and deep. The axons from the BOA are the first to mature, so they are located in the deep layer, whereas the axons from the MOB constitute the middle and superficial strata. The middle layer is composed of mature axons, while the superficial layer consists of newly generated axons that are added, forming a series of stacked axonal laminae at different maturation stages. Initially, the superficial lamina is proportionally larger than the middle lamina, but throughout development, the middle lamina increases, while the superficial lamina drastically decreases [352]. This segregation of both pathways in the TOL, the vomeronasal and olfactory pathways, can be seen in adult animals like rabbits or rats [353,354].

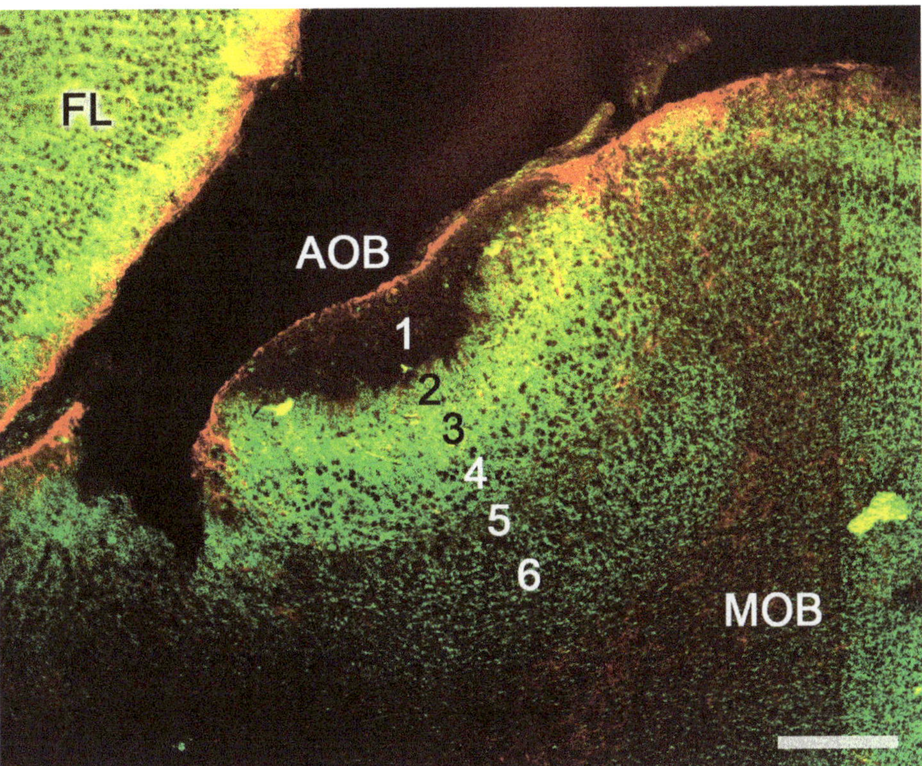

Figure 13. The accessory olfactory bulb of a postnatal mouse. Confocal double immunofluorescence of a sagittal section stained with anti-MAP2 (green) and GFAP (red) showing the lamination of the AOB. 1. Vomeronasal nerve layer; 2. Glomerular layer; 3. Mitral plexiform layer; 4. Somas of mitral cells appear as unstained circular structures; 5. Lateral olfactory tract; 6. Granular cells layer. FL, Frontal lobe. Scale bar: 250 μm.

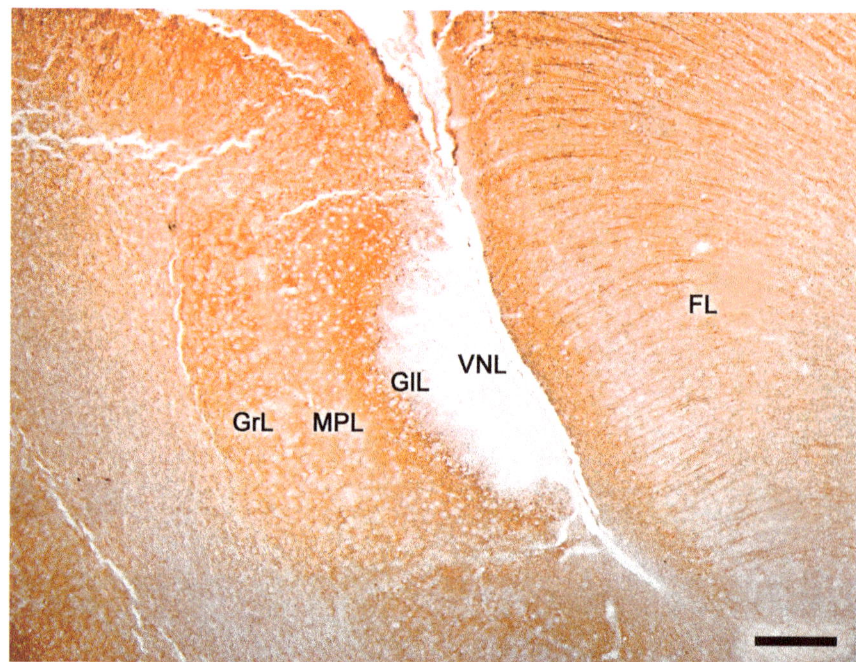

Figure 14. The accessory olfactory bulb of a mouse. Immunohistochemical staining against MAP2. FL. Frontal lobe; GlL, glomerular layer; GrL, granular layer; MPL, mitral-plexiform layer; VNL, vomeronasal nerve layer; 2. Glomerular layer; 3. Mitral plexiform layer; 4. Somas of mitral cells appear as unstained circular structures; 5. Lateral olfactory tract; 6. Granular cells layer. FL, Frontal lobe. Scale bar: 250 μm.

Throughout all the layers that make up the BOA, different cell types coexist. From superficial to deep, we can find periglomerular cells, superficial short-axon neurons, mitral cells, round projection cells, tufted cells, external granular cells, dwarf cells, polygonal neurons, and internal granular cells.

Periglomerular cells (PGs) are small, short-axon neurons found at the base of the glomeruli in both the AOB and the BOP, although they are more numerous in the latter. In the AOB, these cells are primarily GABAergic and have inhibitory functions, possibly playing a modulatory role [355]. There are two classes of PG: amacrine cells, meaning they lack axons, and interneurons. Both have a small fusiform soma, from which they send one or two dendrites to one or several glomeruli. In them, their dendrites branch and originate a dense dendritic plexus. On the other hand, the axon of the interneurons distributes within the neuropil adjacent to the glomeruli.

Superficial short-axon cells are morphologically similar to PG but have some differences. They are larger in size and are located superficially in the external plexiform layer. They have a thick main dendrite that rises to enter a single glomerulus, where it establishes a dendritic network. They also emit other dendrites that distribute ventrally between the somas of PG and other superficial short-axon cells. Their axon interacts with the somas of adjacent PGs [337].

Mitral cells, also called principal cells, are the second neuron in the olfactory and vomeronasal sensory pathways. In the BOP, they are distributed linearly, forming a dense layer of mitral cells. In the AOB, this layer transforms into a diffuse mitral cell band integrated with both plexiform layers, constituting a single mitral-plexiform layer, also called the external cell layer. The cytoarchitecture of the accessory olfactory bulb has been addressed in depth in the rat through the study carried out with the Golgi technique by Larriva-Sahd [344]. This study shows that mitral cells have two types of dendrites: glomerular dendrites (thick) and accessory dendrites (thin). Glomerular dendrites are multiple and extend to the glomeruli located in their respective half of the AOB, according to the anteroposterior axis [356]. The shape of the soma of the mitral cells depends on the number of glomerular dendrites it comprises, so its morphology will be oval, triangular, or polyhedral, presenting two, three, or more primary dendrites, respectively. The axons of these neurons leave the AOB caudally [344].

Another cell type described in rats are the **round projection cells**, which have an oval shape and paired dendrites [344]. Their axon emerges from the base of a proximal dendrite and extends towards the lateral olfactory tract. In addition, from their axon emerge one or two extensions that flow into the soma or dendritic processes of a neighboring cell.

Tufted cells are triangular in shape, although they can have a fusiform morphology. Their size is slightly smaller compared to mitral cells, but they have a similar general organization of their dendrites. However, tufted cells have a single glomerular dendrite that attaches to a single glomerulus. In these cells, the axon usually arises from the base of a non-glomerular dendrite, although it occasionally originates directly from the soma. Then, the axon descends ventrally, traveling a broad, zigzag path to end near the dwarf cells, external granular cells, or mitral cells.

In the granular layer, closer to the surface, we find the **external granular cells**: neurons with round or elliptical somas that have two or three dendrites. The primary dendrites tend to be short, while the secondary branches are longer and branch at their ends. Additionally, their dendrites show the presence of numerous dendritic gemmules. These cells lack an axon, so they are included in the category of amacrine cells. They sometimes integrate into the TOL fibers [344].

In the deeper part of the granular layer are the **internal granular cells**, which have round or triangular somas and also lack an axon [357]. Their dendritic processes consist of a thick ascending dendrite and one or two sets of short, thin branches. Like the external granular cells, they have gemmules on their ascending dendrites. Internal granular cells communicate with the dendrites of tufted cells and mitral cells but also with the somas of certain periglomerular cells and short axon surface neurons [344,358].

In a more internal layer of the granular layer, we find the **dwarf cells** and the **polygonal neurons**. Dwarf cells are interneurons without axons that have a very small, spherical soma, and their dendritic tree lacks gemmules [344]. Polygonal neurons are small fusiform bipolar neurons that have paired dendrites. Their axons branch out, extending through the neuropil, interacting with external granular cells and dwarf cells.

Moreover, in the accessory olfactory bulb, different types of glial cells can be found. The radial glia cells, like the olfactory envelope glial cells, are the most common, but astrocytes and oligodendrocytes are also observed [359].

Regarding the basic circuit that sensory information follows in the AOB, there are functionally three main neuronal components that are activated sequentially through vomeronasal stimulation: the axons of the vomeronasal sensory neurons, the mitral cells, and the granular cells [360]. The axons of the sensory vomeronasal neurons, present in the apical layer of the vomeronasal epithelium (V1R), are distributed in the anterior zone of the AOB [361], while the sensory vomeronasal neurons of the basal layer (V2R) project their axons to the caudal half of it [172]. On the other hand, although mitral cells receive information from several glomeruli, they connect only with axons of sensory vomeronasal neurons of the same type (V1R or V2R). Thus, a convergence of specific sensory inputs occurs in a small population of mitral cells. In the MOB, a much stricter convergence pattern occurs since the information received by each mitral cell comes from a single glomerulus [340].

Morphologically, the AOB shows significant differences among the multiple species of mammals. In those animal groups where this structure is most developed, clear lamination can be identified. However, as we have mentioned, this generally does not allow differentiation of the external plexiform, mitral, and internal plexiform layers, with these forming a broad layer (mitral plexiform/external cellular), consisting of projection cells (mitral/main) distributed within a neuropil. On the other hand, species with a less sophisticated AOB have a very reduced mitral/plexiform layer in which it even becomes difficult to discriminate the projection cells.

The case of the platypus (*Ornithorhynchus anatinus*) is unique among mammals, as in this species the AOB and MOB are of similar size. Likewise, both the platypus and the echidna (*Tachyglossus aculeatus*), both monotremes, have well-differentiated layers forming the laminar structure of the AOB [320].

Rodents also possess an AOB with a high degree of development; however, as has been described, both the rat [344,362] and the mouse [245,363] display poorly differentiated boundaries between the external plexiform, mitral, and internal plexiform layers. Nevertheless, both murine species have a thick mitral-plexiform layer containing a dense network of mitral and tufted cells. In the case of lagomorphs, the AOB has extensive development, and more specifically, the rabbit (*Oryctolagus cuniculus*) exhibits a laminar configuration highlighting a broad mitral-plexiform layer formed by three main types of cells: large, tufted, and rounded cells [132].

Among the different species of bats, there is significant variability in the development of their AOB. Generally, they present only four distinguishable layers; however, in some cases, such as *Glossophaga soricina*, Frahm and Bhatnagar [364] suggested the presence of the plexiform layers (internal and external) and the mitral cell layer. The rest of the studied mammalian orders show the plexiform layers and the mitral layer merged into a single mitral-plexiform layer. Marsupials possess a sophisticated AOB; however, their lamination is restricted to four layers. Despite this, their mitral-plexiform layer has a considerable thickness. These characteristics have been observed in the gray short-tailed opossum (*Monodelphis domestica*) [365] and in the Tammar wallaby (*Notamacropus eugenii*) [279].

Carnivores generally display a surprisingly moderate development of their AOB, with domestic ones being the most studied (Figure 15). There are notable differences between felids and canids in this regard. Although both exhibit four layers in their lamination, felids show a more pronounced development of the mitral-plexiform layer [283,343]. However, in the red fox (*Vulpes vulpes*), a higher degree of AOB differentiation than in dogs has been observed, exhibiting a thick glomerular layer. Additionally, in the latter species, the imaginary lines separating different strata follow irregular trajectories, to the extent that the mitral-plexiform layer shows prominent extensions into the glomerular layer [366]. This raises the hypothesis of a possible involution of the vomeronasal system as a consequence of selection pressure and crossbreeding associated with the domestication of dogs. Differences between domesticated and wild canids are not limited to the VNS but also extend to the olfactory system. In a detailed analysis of the cribriform plate (CP) morphology across 46 dog breeds, a coyote, and a gray wolf using high-tech CT scans, it was observed that all dog breeds, even those known for their olfactory prowess, possess a CP surface area relative to body size that is smaller than in both wild canids [367]. These researchers previously established a correlation between CP size and the number of OR genes in a species, proposing that the CP size might represent evolutionary trends in mammalian olfaction [368].The presence of differences between domesticated animals and their closely related non-domesticated counterparts is not exclusive to canids. Significant differences in the expression of vomeronasal receptors have been identified between *Mus* subspecies and species. This suggests that these receptors could have a role in guiding behavioral adaptations. Furthermore, commonly used, highly inbred laboratory strains exhibit a significantly diminished capacity for differential pheromone-mediated behaviors [369]. From a strictly anatomical perspective, it is noteworthy that the differences in the organization of the olfactory bulb between domestic and wild canids are not so pronounced at the level of the main olfactory bulb. However, recent neuroanatomical observations have identified subtle morphological and neurochemical differences between the MOB of wolves and foxes and the MOB of dogs [370].

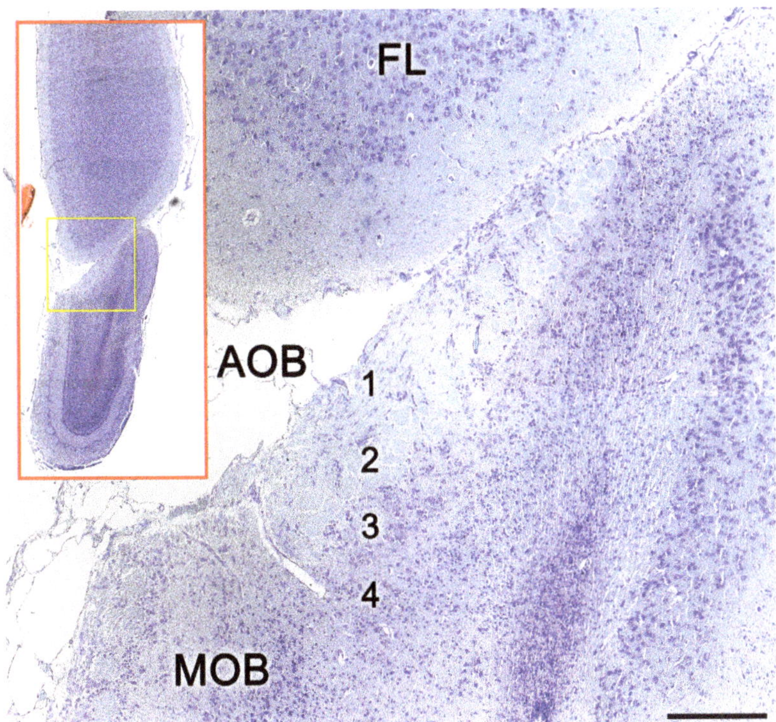

Figure 15. The accessory olfactory bulb of a cat. Nissl staining of a histological transverse section. The red box shows the topographic relationship of the olfactory bulb and the telencephalon frontal lobe. At higher magnification the yellow area containing the AOB allows to differentiate the lamination of the AOB. 1. Vomeronasal nerve layer; 2. Glomerular layer; 3. Mitral-plexiform layer; 4. Granular cell layer. Scale bar: 250 μm.

Despite the presumed regression of the VNS in dogs compared to their wild relatives, the wolves, there is a growing interest in exploring the clinical implications of the VNO in domestic dogs. This has been bolstered by the optimization of an MRI protocol for in vivo visualization of the VNO [371], and the recent identification of a case of canine vomeronasal agenesis, which manifested significant behavioral disorders, such as an inability for sexual discrimination of conspecifics and reduced sexual behavior [372]. Moreover, the use of pheromonotherapy, like the application of dog-appeasing pheromone (DAP) to mitigate symptoms of separation anxiety, including house soiling, vocalizations, and damage [373,374], along with a deepening in the links between VNO inflammation and aggressive behaviors in canids, felids, and livestock [375–378], has further developed this interest in the dog VNS.

Within the carnivores, the Mustelidae family has been the subject of extensive study, as the characterization of its AOB remains unresolved. Whereas the mink (*Mustela vison*) shows a well-developed VNO [379], the characterization of its AOB has proved challenging. Despite efforts, a precise and consistent morphological definition has not been possible using traditional neuroanatomical techniques [380]. However, the AOB has been studied more successfully in the ferret using immunohistochemical markers and lectins, which have revealed a poorly developed AOB [326].

The studied artiodactyls species, such as the sheep (*Ovis orientalis aries*), the Siberian roe deer (*Capreolus pygargus*), or the common hippopotamus (*Hippopotamus amphibius*), have a moderate development of the AOB (Figure 16), higher than carnivores, as they also exhibit four distinct layers forming their laminar organization [325,332,381].

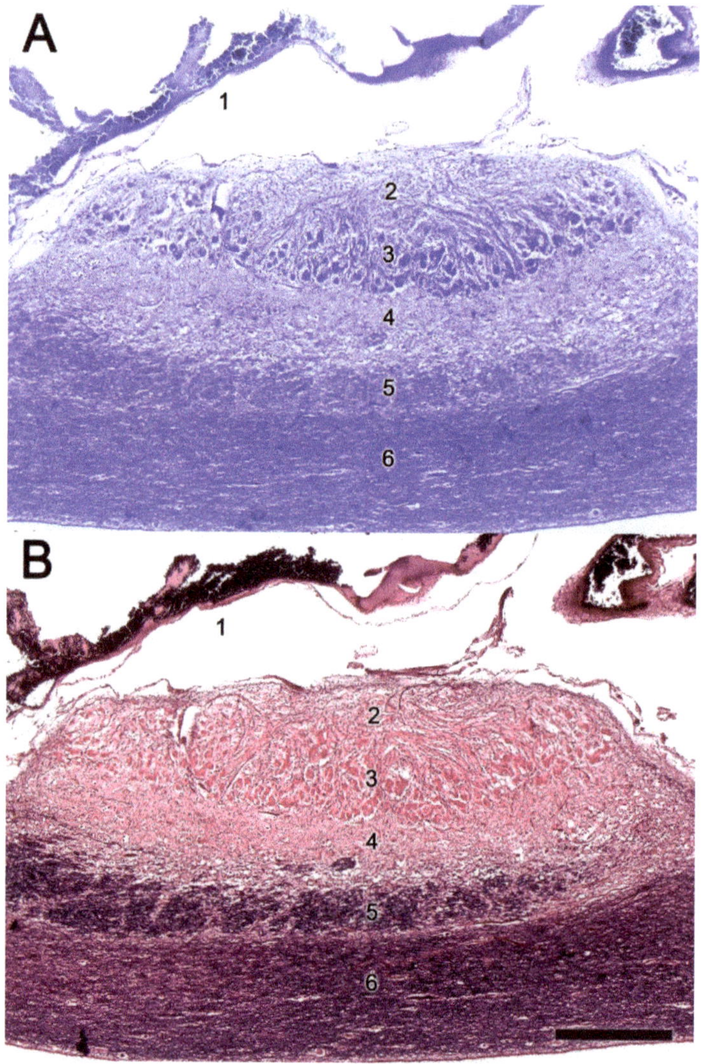

Figure 16. The accessory olfactory bulb of a sheep. Consecutive Nissl (**A**) and Tolivia (**B**) staining of a sagittal section. The red box shows the topographic relationship of the olfactory bulb and the telencephalon frontal lobe. At higher magnification the yellow area containing the AOB allows to differentiate the lamination of the AOB. 1. Duramater; 2. Vomeronasal nerve layer; 3. Glomerular layer; 4. Mitral-plexiform layer; 5. Lateral olfactory tract; 6. Granular cell layer. Scale bar: 500 μm.

Regarding the neurochemical study of the AOB, the use of Gαi2 and Gαo proteins has garnered much attention, as both proteins indicate the expression of vomeronasal receptor families V1R and V2R, respectively, establishing an anteroposterior zonation in species that exhibit both receptors. This delineates the AOB organization into two areas: an anterior region rich in Gαi2 and a posterior region rich in Gαo [169]. Thus, the analysis of both G proteins allows an assessment if the studied species fits within a uniform model, or conversely, exhibits a segregated pattern. For the Gαi2 marker, labeling is restricted to the neural and glomerular layers of the AOB, while the Gαo marker is expressed in both neural and glomerular layers as well as the other strata of the AOB [382].

On the other hand, various immunohistochemical markers have been utilized in AOB studies. Calcium-binding proteins such as calbindin (CB), calretinin (CR), parvalbumin (PV), neurocalcin (NC), and secretagogin are typically expressed in certain layers and cell populations, both in the AOB and BOP [365,383,384]. Additionally, CR labels mitral cells and can be expressed in atypical BOP glomeruli, typically located near the AOB [385]. The glial fibrillary acidic protein (GFAP) is used to generally identify the glial component of the AOB, and specifically astrocytes and ensheathing cells [386]. The microtubule-associated protein 2 (MAP2) is primarily expressed in the dendrites of mitral cells [387], leading to intense labeling in the external plexiform and mitral layers, but can also be expressed in the internal plexiform and granular layers. The growth-associated protein 43 (GAP-43), also termed the plasticity protein, is expressed in neuronal growth cones during development [388]. In the AOB, its immunolabeling is restricted to the neural, glomerular, and granular layers [132]. The olfactory marker protein (OMP) is involved in signal transduction and is used to evaluate neuronal maturity [389] of the olfactory system. In the AOB the expression of this marker is confined to the vomeronasal nerve and glomerular layers. Various markers in the OB are also employed to identify neuronal populations expressing different types of neuronal receptors, such as GABA receptors, dopaminergic receptors [390], cholinergic receptors [391], adrenergic receptors [392], and serotonergic receptors [393]. Additionally, immunohistochemical studies of various neuropeptides in the AOB, such as substance P, cholecystokinin, or neurotensin [394], and neuronal markers like PGP9.5 [395] or doublecortin (DCX) [396], are commonly conducted.

Another technique commonly used in the characterization of the AOB is histochemical labeling with lectins. The labeling pattern produced by each lectin may vary between the different species studied, although the *Ulex europaeus agglutinin* (UEA), *Bandeiraea simplicifolia* isolectin B4 (BSI-B4), and *Lycopersicon esculentum* agglutinin (LEA) are regularly expressed in the nervous and glomerular layers. Specifically, UEA allows differentiation of the anteroposterior zonation of the AOB in species of the segregated model, as it is expressed more intensely in the anterior region (VR1) [147]. In contrast, *Vicia villosa* agglutinin (VVA) shows more intense labeling in the posterior part of the AOB [397]. In this way, different neuronal populations can be identified. Many other lectins are also used in the study of the AOB, such as soybean lectin (SBA), wheat germ lectin (WGA), and *Dolichos biflorus* agglutinin (DBA) [398,399].

2.3. Olfactory Pathways

Regarding the vomeronasal information flow, vomeronasal sensory neurons send the sensory information received from the vomeronasal receptor epithelium to the mitral cells of the AOB [400]. In contrast, olfactory information is captured in the main olfactory epithelium and later travels through the olfactory nerves to the MOB. Then, these stimuli are transmitted through various projections to different brain areas.

The application of neuronal tracer studies to the olfactory pathways allowed the discrimination of the existence of segregated and parallel projections from the main and accessory olfactory bulbs [401,402]. This was known as the **dual olfactory system hypothesis**, a formulation demonstrated for the first time using the Fink–Heimer technique in rabbits [403]. Although this hypothesis is still currently valid, it has been refined over the past decades, especially regarding the study of secondary and tertiary projections, proving the existence of neuronal communication between both pathways.

The efferences of the MOB project ipsilaterally to the anterior olfactory nucleus, tenia tecta, olfactory tubercle, piriform cortex, lateral and medial areas of the entorhinal cortex, and lateral amygdaloid nuclei [404,405]. Among the tertiary projections of the pathway, the communication of the entorhinal cortex with the hippocampus, and the piriform cortex and lateral amygdala with the hypothalamus and mediodorsal thalamic nucleus, stand out [406,407]. It has recently been shown that there is a direct projection of olfactory sensory information to the anterior subdivision of the medial amygdala, a structure traditionally involved in mediating pheromonal information. This supports the hypothesis that both olfactory systems act collaboratively and not differentially in the control of socio-sexual and anti-predator behaviors [408].

Unlike the main olfactory system, efferent projections from the AOB bypass the thalamocortical axis, distributing ipsilaterally to third-order limbic system nuclei, such as the accessory olfactory tract nucleus, the terminal stria nucleus, and the medial and posteromedial cortical amygdaloid nuclei (together forming the vomeronasal amygdala) [402,409–412]. The medial amygdala is strongly interconnected with other structures that receive vomeronasal information from the AOB. This suggests that the information detected by the vomeronasal system undergoes complex intrinsic processing before being transmitted to other structures [413]. The final processing center between vomeronasal information and effectors is the hypothalamus [414]. In this way, the VNS is directly involved in the activity of sex hormones and can facilitate the development of aggressive, defensive, or reproductive behaviors [415]. It is important to note that olfactory and vomeronasal projections show some degree of overlap in the amygdaloid cortex and medial amygdala, suggesting that there are anatomical pathways that allow extensive integration of olfactory and vomeronasal information [152,413,416].

Additionally, the MOB and AOB differ in their centrifugal afferent connections [417]. The MOB receives massive cholinergic and GABAergic projections from the basal forebrain, which mainly originate in the nucleus of the horizontal branch of Broca diagonal band and the magnocellular preoptic nucleus [418]. Cholinergic signaling within the OB modulates olfactory learning and memory, odor discrimination, odor habituation, and social interactions [419]. From a comparative neuroanatomy perspective, the study by Liberia et al. [420] indicates that the synaptic connectivity of the afferent cholinergic circuits is highly conserved in the OB of macrosmatic and microsmatic mammals. Regarding the GABAergic afferents of the basal forebrain, they innervate all the layers of the MOB at least as densely as cholinergic axons, but only a few studies have examined their function in odor processing [421].

The AOB receives significant afferents from a wide range of nerve centers. The bed nucleus of the stria terminalis and the VN amygdala project reciprocally to the AOB, thus forming a feedback circuit [422]. Tracer studies have revealed that both feedback projections to the AOB are topographically organized and use different neurotransmitters [423]. Specifically, GABAergic projections from the bed nucleus terminate in the outer cell layer, while glutamatergic projections from the amygdala are directed to the inner layer of granule cells. A significant number of these feedback neurons in both areas express estrogen receptors ER-α, linking the animal endocrine state with integration in the AOB. The relevance of this pathway was subsequently confirmed with different morphofunctional approaches [424,425].

Other important afferents for the function of the AOB come from noradrenergic structures in the brainstem, which play a crucial role in the formation of olfactory memory [426]. This comes into play, for example, during mating, in which stimulation of the vaginocervical zone leads to sustained increases in the levels of noradrenaline in the AOB, which persist for about 4 h [427]. This time span establishes a crucial phase in which noradrenaline generates plastic changes in the intensity of dendrodendritic synaptic connections [428]. Additionally, the diagonal band of Broca is the origin of numerous cholinergic fibers involved in increasing the excitability of granule cells [429], while from the raphe, abundant serotonergic fibers are sent to both olfactory bulbs that act similarly to the cholinergic ones [393].

In summary, the olfactory and vomeronasal systems represent two separate systems which are functionally interconnected and are in charge of processing olfactory and pheromonal stimuli, respectively. These systems differ in their afferent and efferent connections, morphology, physiology, and functional implications, but both contribute to the complex sensory processing of olfactory and pheromonal information that leads to behavior [430]. The increasing evidence suggests the existence of a more complex and intricate relationship between these systems than previously assumed. Recognizing and understanding the complexities of this interaction will provide a clearer picture of how odors and pheromones shape animal behavior.

3. Specific Study of the Vomeronasal System

The study of the morphological features of the VNS has been utilized as a phylogenetic classification method at both the neuroanatomical [144] and genetic levels [431]. It is also indispensable for fully understanding the physiology, evolution, and functional implications of the VNS in mammals. For this reason, to conclude this review, we have included a final section where, based on our group observations, we exemplify, in four mammal species from evolutionarily distinct families, the vast morphofunctional differences among different mammal groups. This not only pertains to the organization of the vomeronasal organ and the accessory olfactory bulb but also reveals how each species has adapted to various environments, social and reproductive behaviors, feeding habits, and predators, leading to different configurations of the system.

The species selected for this section of the review are the capybara (*Hydrochoerus hydrochaeris*), a hystricomorph rodent; the Bennett's wallaby (*Notamacropus rufogriseus*), a macropod diprotodont; the meerkat (*Suricata suricatta*), a herpestid carnivore; and the dama gazelle (*Nanger dama*), an artiodactyl from the bovid family.

3.1. Capybara (Hydrochoerus hydrochaeris)

There is limited information available on the vomeronasal system (VNS) of rodents, beyond the extensively studied laboratory rodents: rat (*Rattus norvegicus domestica*) and mouse (*Mus musculus*). There is scant data regarding wild rodents. The interest in comparative analysis with wild rodents stems from the fact that most studies on the VNS have been conducted on laboratory rodent strains, exposed to artificial selection that does not reflect the selection pressures in the wild. Consequently, these laboratory strains exhibit significant genetic and behavioral differences compared to wild rodent models [432]. Moreover, these species can have significant disparities with wild rodents due to the domestication process, as seen in canids, where artificial selection in domestic dogs presumably plays a pivotal role in the regression of their VNS [366]. Lastly, the capybara is a semi-aquatic mammal, and studying it helps expand the limited neuroanatomical information on the VNS in such species (Figure 17). This characterization is vital since semi-aquatic habits could influence the organization and functioning of chemosensory systems [433].

A unique feature of the capybara that adds particular interest to its study is the degree of development its VNS reaches in newborn individuals. While studies like the VNS examination of the Asian elephant (Elephas maximus) showed that, at 6 weeks of age, the VNS already has morphological similarities with adults [434], many studies detail the VNS development throughout the embryonic process without clarifying the maturity and functionality level of the VNS at birth [435]—a time that implies numerous transformations involving chemical communication. Hence, the capybara constitutes an excellent model for studying chemical communication development in precocial species.

The expression of G protein subunits in the adult capybara VNS is relevant for understanding the VNS role in this early life stage. The capybara aligns with other studied hystricomorphs like chinchillas, guinea pigs, or degus in having a clear anteroposterior zoning in the AOB [280,310,436]. However, in these species, unlike laboratory rodents, no apical-basal zoning has been found at the VNO level. It is striking that other non-rodent species with a segregated model, like rabbits or wallabies, also lack this apical-basal zoning, suggesting this is a highly specific trait for laboratory rodents [437]. The immunopositivity of both G proteins throughout the neuroepithelium of the VNO, the vomeronasal nerves, and the neural and glomerular layers of the newborn capybara AOB indicates the full maturity of the VNS at birth [438]. A similar early prenatal development pattern in the VNS of other precocial mammals, like pigs [439–441] and sheep [381], was found, but it was determined during the fetal period, and in neither case was the expression of G proteins in the VNS evaluated. Notably, the anteroposterior zoning is also evidenced with the UEA lectin, which stains the anterior area more intensely. Although this pattern aligns with findings in mice, a recent study on UEA-lectin staining in the mouse olfactory system could only identify this zoning in three of the sixteen subjects studied [221]. The authors concluded that these individual differences might be caused by the presence or absence of signals activating the receptor cells or reflect the difference between wild and laboratory rodents. The presence of anteroposterior zoning marked by UEA in capybaras appears to support this theory.

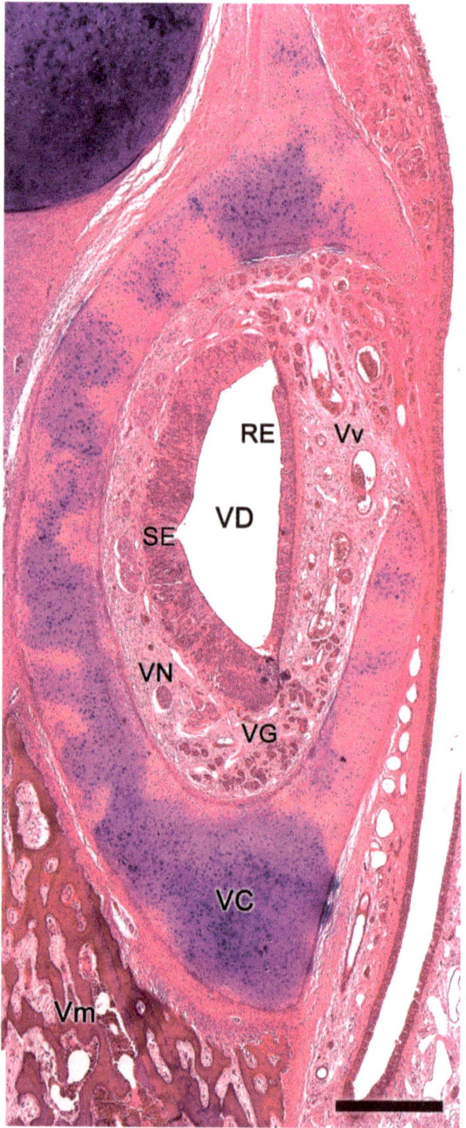

Figure 17. The vomeronasal organ of a capybara. Transverse decalcified histological section stained with hematoxylin-eosin. RE, respiratory epithelium; SE, sensory epithelium; VC, vomeronasal cartilage; VD, vomeronasal duct; VG, vomeronasal glands; VD, vomeronasal duct; Vm, vomer bone; Vv, veins. Scale bar: 500 μm.

Another differential aspect is that the newborn capybara exhibits a bias towards a more prominent caudal AOB. This characteristic was previously described in adult individuals [442]. These results, combined with recent observations on the morphometric parameters of the AOB in two degu species with contrasting social habits, suggest that some AOB structural features are tied to the species lifestyle and arise during an early ontogeny stage [443].

The early functionality of the VNO and AOB in capybara was confirmed in the VNO, where calcium-binding protein markers, calretinin, and calbindin stained the nervous bundles of the parenchyma and almost all sensory neuroepithelial cells, although different cell populations were evidenced. Something similar happens in shrews, where calretinin produces intense staining in almost all receptor cells and vomeronasal nerves of prenatal individuals [444]. Likewise, the robust expression of GFAP in astrocytes from the neural and glomerular layers of the capybara AOB contrasts with the reduced astrocytic development observed at this early stage in other species, like newborn opossums [445]. The expression of olfactory marker protein (OMP) is, however, very reduced in this species' AOB, especially when compared to the MOB. This has been observed in rodents like mice or rats [254] and in marsupials like opossums [389].

In addition to the cited neurochemical aspects, the capybara has exclusive morphological features, such as the nature of the capsule protecting both VNOs, the dorsal location of the VNO in the nasal cavity over the incisive bone palatal process, the high degree of morphological differentiation of the AOB at this early stage, and finally, the presence of a migratory stream from the VNO neuroepithelium towards the vomeronasal nerves.

The dorsal location of both VNOs is a result of them resting on the palatine process of the incisive bone, and not on the vomer bone, as is the case in most species [438]. This topography is only comparable to that described in rabbits [143], where the organs also rest on the palatine process, although in this case, the bony projection is smaller, resulting in a lesser dorsal displacement of both VNOs. While in most species the nature of the vomeronasal capsule is exclusively cartilaginous, there are species such as the mouse where the capsule is formed by a thin bony lamina [446] or like lagomorphs that have a cartilaginous capsule externally reinforced by a bony envelope [143]. In the capybara, the capsule is initially cartilaginous, but from the central area of the organ it is progressively replaced from ventral to dorsal by a bony lamina that ends up completely enveloping the VNO parenchyma in its caudal zone [438].

The secretion from the vomeronasal glands into the vomeronasal duct plays a crucial role in vomeronasal peri-receptor processes [248]. The capybara VNO shows significant development of glandular tissue, especially at the caudal end of the organ [438]. However, in the central part of the duct, the presence of glands is moderate. This pattern is analogous to that found in other rodents such as rats, guinea pigs [447], or mice [247]. Likewise, the glandular nature of the capybara VNO is both PAS+ and AA+, contrasting with most of the studied rodents, which only express positivity for PAS in glandular secretion [310,448,449]. This notable variation in gland characteristics within the same order might be due to the capybara adaptation to its semi-aquatic nature, which could require a specific pheromone-receptor interaction environment.

A unique point, for which, to our knowledge, there is no other example in the literature, is the profuse cellular migration that occurs from the sensory epithelium of the VNO to the nerve bundles of the vomeronasal nerves (Figure 18) [438]. This cellular migration is immunolabelled with anti-GAP-43 and histochemically labelled with UEA and LEA lectins, but immunonegative when labelled with anti-LHRH. This indicates that the migration is not analogous to that described in prenatal rat and mouse fetuses, which occurs from the vomeronasal part of the olfactory placode to the hypothalamus and is mainly composed of LHRH positive cells [450–453]. It is challenging to hypothesize the importance of these cells in the capybara, as this discovery is unprecedented in both olfactory and vomeronasal nerves.

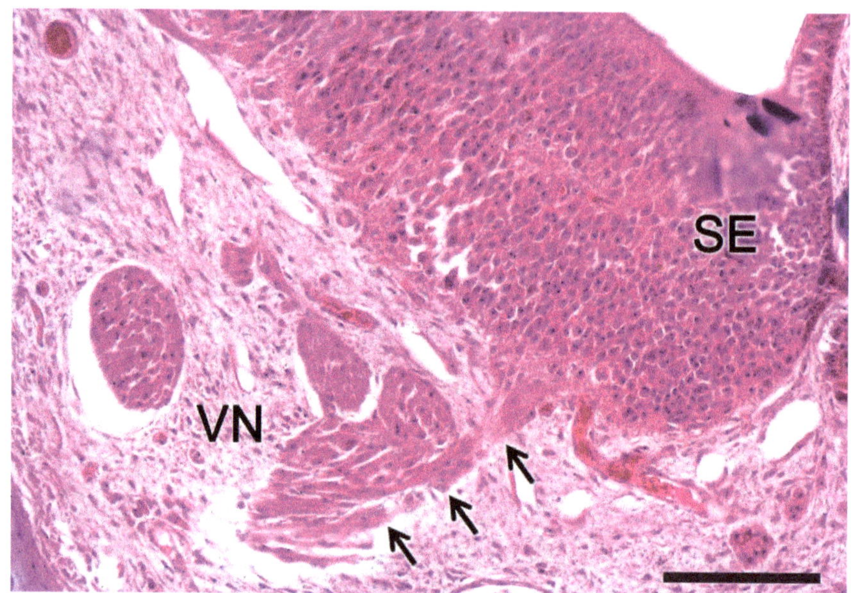

Figure 18. Histological section of a capybara VNO stained with hematoxylin-eosin. A migratory stream (arrows) departing from the sensory epithelium (SE) to the vomeronasal axons (VN) can be observed. Scale bar: 100 μm.

Regarding the morphology of the capybara AOB (Figure 19), it presents a distinctly differentiated laminar pattern [438], similar to that found in other rodents [446], lagomorphs [132], marsupials [454], and prosimians [455]. However, among all these cases, the capybara is the species that has a more distinctly organized mitral layer.

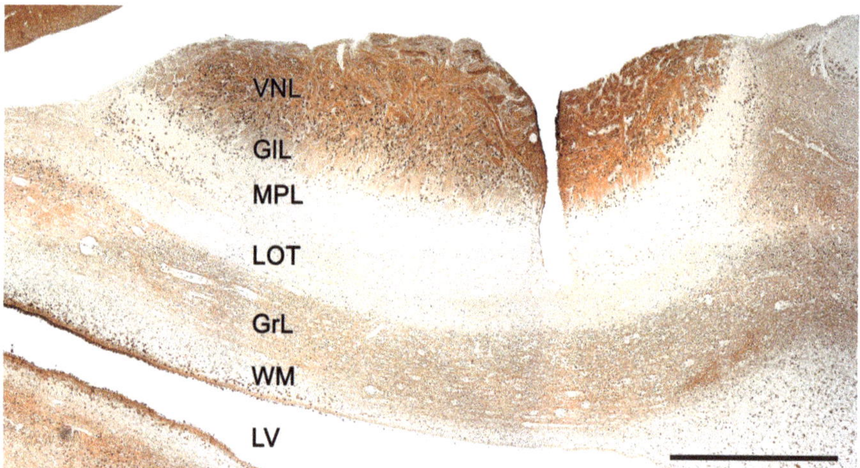

Figure 19. Immunohistochemical staining of a capybara AOB labelled with anti-GAP43 shows strong immunolabelling in the superficial layers: vomeronasal nerve (VNL) and glomerular (GlL) layers. GrL, Granular layer; LOT, Lateral olfactory tract; LV, lateral ventricle; MPL, Mitral-plexiform layer; WM, White matter. Scale bar: 100 μm.

3.2. Bennett's Wallaby (Notamacropus rufogriseus)

The extensive study of the vomeronasal system of the Bennett's wallaby (*Notamacropus eugenii*) conducted by Schneider et al. [279,317,319,456] proposed a new and surprising processing model in the VNS of mammals. Specifically, the investigation into the presence of the G protein subunits, Gαi2 and Gαo, associated with the vomeronasal receptor families V1R and V2R, respectively [457], concluded that the Tammar wallaby only expresses V2R receptors, and therefore, does not fit into either of the two existing models: the uniform model (species that express V1R receptors: Soricidae, Hyracoidea, Perissodactyla, Artiodactyla, Carnivora, or Primates; [458]) and the segregated model (species that express both V1R and V2R receptors: Rodentia, Lagomorpha, Afrosoricida, and Didelphimorphia; [220]).

Aware of the disruption that Schneider et al.'s [279] findings pose to the established knowledge on vomeronasal transduction, the authors of the study warned that their results should be taken with caution. In fact, the observation that the anti-Gαo protein was only expressed in a subpopulation of neuroreceptor cells of the vomeronasal epithelium and not in the entire population of vomeronasal neurons was somewhat atypical, as, in the previously described uniform model, all vomeronasal neurons are marked with Gαi2. To explain this low number of Gαo-positive cells, the authors hypothesized a reduced affinity to the antibody used. However, there are no other examples in the literature of any similar selectivity deficit in response to the use of anti-Gαo antibodies.

More critically, the immunolabeling of the AOB in the Tammar wallaby with the Gαo marker has only been carried out in this study [279], so there are no other references that support or refute the results. The staining protocol followed by these authors resulted in a very weak labeling pattern. It was also restricted to a small area in the superficial layers of the AOB, where projections from the vomeronasal neuroreceptor cells are received. In contrast, in all other studied species, both from the segregated and the uniform models, the Gαo and Gαi2 markers are expressed throughout the thickness of the superficial layers. This includes the neural and glomerular layers. Regrettably, the authors did not specify the antibody used that showed a negative pattern to the Gαi2 marker both in the VNO and the AOB; they did not provide information about the commercial source or the batch number of the antibody, preventing any comparison with other studies on the expression of the Gαi2 protein in the VNS that used the same antibody.

The existence of this third model could only be contrasted with the study of the VNS of the Bennett's wallaby (*Notamacropus rufogriseus*) conducted by Torres et al. [459]. The immunohistochemical study of the G proteins in the VNS of the Bennett's wallaby revealed a canonical labeling pattern for both Gαi2 and Gαo in the VNO and AOB of all the samples studied. The labeling pattern observed in the vomeronasal neuroepithelium of the Bennett's wallaby using the Gαo marker [459] was identical to that described in the Tammar wallaby [279], in which only a small fraction of neuroreceptor cells was marked. However, while Schneider et al. did not obtain a positive Gαi2 immunolabeling in the VNO, Torres et al. [459] observed a large number of αi2 immunopositive cells in the vomeronasal neuroepithelium. Additionally, the strong immunopositivity identified in the vomeronasal axons, both in the proper lamina and in the nasal mucosa, confirmed the neuroreceptor nature of these cells [459]. At the AOB level, a clear anteroposterior zoning pattern, typical of mammalian species belonging to the segregated model, is observed. Vomeronasal axons reaching the anterior zone of the AOB only expressed Gαi2, while those reaching the posterior zone only expressed the Gαo subunit.

The use of the UEA lectin (*Ulex europaeus*) provided further evidence of the segregation of vomeronasal information in the wallaby, revealing selective labeling of the anterior zone of the AOB. The affinity of the UEA lectin for the anterior zone of the AOB has been reported in all species belonging to the segregation model in which this histochemical marker has been investigated, including the hamster [398], the mouse [147,446,460,461], the rat [287], and the capybara [438]. However, studies that employed UEA labeling in species belonging to the uniform model, including the pig [462], the cat [343], the dog [283], and the goat [463], have not shown any evidence of zonation in the AOB in any case.

Beyond the zonation, the main morphological and histological features of the wallaby VNS include the opening of the VNO to the nasopalatine duct, the semilunar shape of the vomeronasal duct, and the stratification and cellularity of the neuroreceptor epithelium, and the presence of numerous PAS-positive vomeronasal glands in the lateral parenchyma (Figures 20–23). Likewise, the arrangement of large blood vessels around the vomeronasal duct along its medial and lateral planes and a profuse, non-myelinated dorsal, medial, ventral, and ventrolateral innervation [317] responsible for transmitting the information collected by the vomeronasal sensory neuroepithelium to the AOB is typical of these species. This arrangement of the vomeronasal axons in the parenchyma is atypical, as it has not been previously described in the VNO of any other non-macropodid mammalian species. Another feature common to the VNO of both macropodids, observed by Schneider et al. [317] and [303] in the Tammar wallaby, is that the caudal vomeronasal capsule does not enclose the VNO parenchyma, so the posterior portion of the VNO is free from the surrounding cartilage. A similar finding has been reported in other marsupials, such as *Notoryctes* [464] and *Caenolestes* [465].

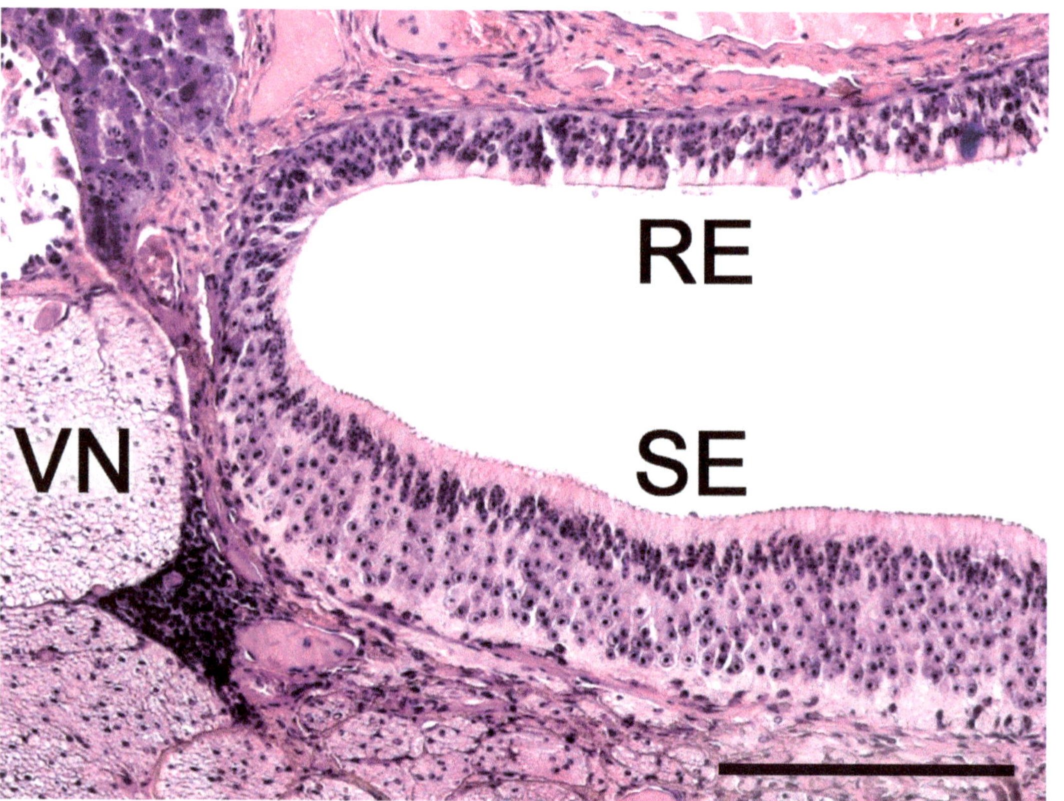

Figure 20. Hematoxylin-eosin histological staining of a wallaby vomeronasal duct. RE. Respiratory epithelium; SE. Sensory epithelium; VN. Vomeronasal nerve. Scale bar: 100 μm.

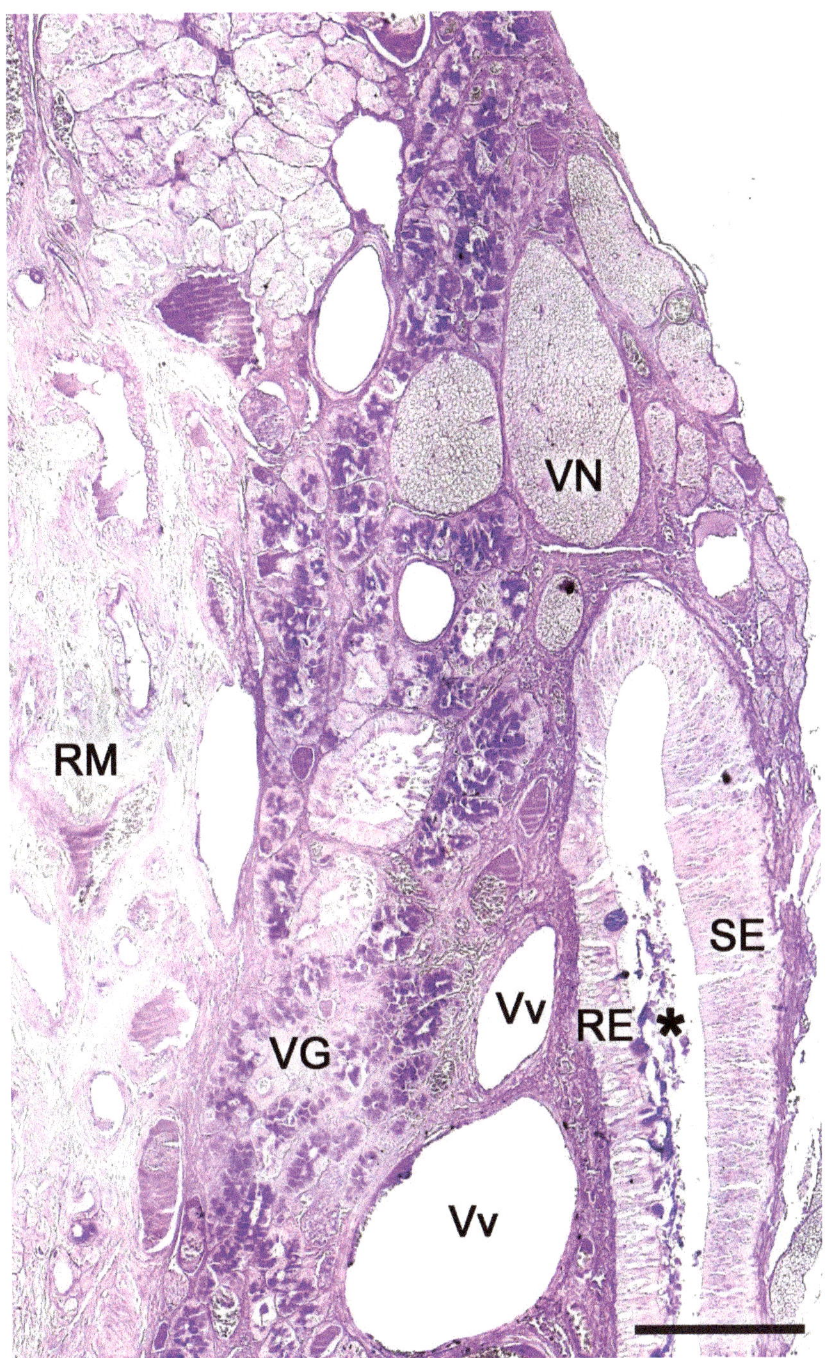

Figure 21. PAS staining of the wallaby VNO showing a high density of PAS+ glandular acini (VG) in the dorsolateral part of the parenchyma. The vomeronasal duct contains PAS+ secretion (*). RE. Respiratory epithelium; RM. Respiratory mucosa; SE. Sensory epithelium; VN. Vomeronasal nerve; Vv. Veins. Scale bar: 200 mm.

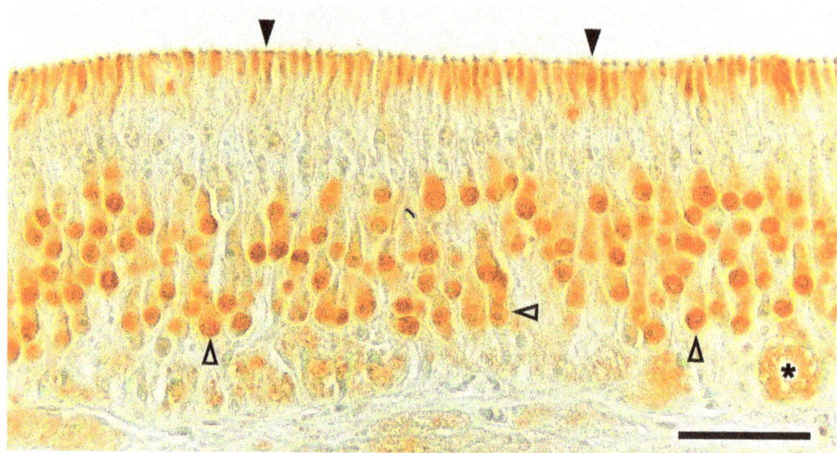

Figure 22. Anti-calretinin immunostaining of a wallaby SE of the VNO. Neuroreceptor cells somata (open arrowheads); Neuroreceptor cells dendritic knobs (arrowheads). Intraepithelial blood vessel (asterisk). Scale bar: 50 μm.

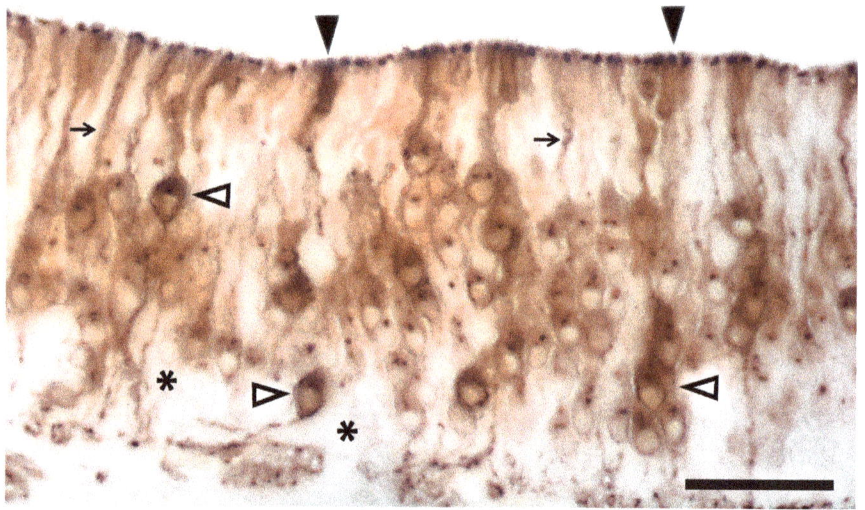

Figure 23. UEA histochemical labelling of a wallaby VNO sensory epithelium. Neuroreceptor cells somata (open arrowheads); Neuroreceptor cells dendritic knobs (arrowheads). Dendritic processes of the neuroreceptor cells (arrows); Intraepithelial blood vessel (asterisk). Scale bar: 50 μm.

Regarding the lamination of the AOB in macropodids, it is characterized by a well-defined lamination, with an organization of mitral cells in a distinct mitral/plexiform layer [459]. Such features appear comparable to those described in species with highly developed AOBs, as seen in rodents [336] and lagomorphs [132]. The lectin LEA, in contrast to UEA, yields a non-zonal staining pattern, similar to that observed for OMP, a marker for mature olfactory and vomeronasal cells [466], and identical to the LEA pattern seen in the rabbit AOBs [132]. Anti-GAP-43 confirms the presence of growing axons in the Bennett's wallaby AOB, with no observed differences between the anterior and posterior zones. However, anti-GFAP displayed more pronounced staining of the glial components in the posterior portion of the AOB compared to the anterior segment. Such a pattern has not been previously reported in other studies of this marker in the mammalian AOB [245,467], necessitating further examination in future research to interpret its implications.

Calbindin (CB) and calretinin (CR) are expressed throughout the VNS of the wallaby, comprising the neuroepithelium of the VNO (Figure 22), vomeronasal nerves, and the AOB. For both markers, VNO immunostaining encompasses the soma and dendrites in a pattern akin to that described in mice [285]. The distribution of CB and CR staining in the Bennett's wallaby AOB is concentrated in the vomeronasal fibers and glomeruli, labelling the neural and glomerular layers in a manner analogous to other species, like the rabbit [132] or capybara [438]. In other marsupials, such as the opossum, significant differences in CB and CR staining patterns in the AOB were observed when compared to the Bennett's wallaby pattern. In the opossum AOB, CB-labeled neurons were present in all layers, excluding the nervous layer and periglomerular cells. CR in the opossum follows a pattern similar to that in the wallaby; however, in the opossum, CR also labels mitral cells and distinguishes an anteroposterior zonation. This is evidenced by the presence of a more intense staining in the posterior zone of the AOB compared to its anterior part [365].

In short, the VNS of macropods displays a degree of differentiation and histochemical and neurochemical diversity comparable to species with a more developed VNS. The existence of the intermediate third type in the processing of vomeronasal information reported in the Tammar wallaby (*Notamacropus eugenii*) is not supported by our histochemical and immunohistochemical findings in the Bennett's wallaby (*Notamacropus rufogriseus*).

3.3. Meerkat (Suricata suricatta) (Figure 24)

To the best of our knowledge, there is no existing neuroanatomical information on the VNO of the Herpestidae family beyond the recent neuroanatomical study of both the AOB and MOB of the meerkat [468]. Within this group, the meerkat holds an emblematic position. It is a gregarious species in which marked dominance is evident, as one male and one female monopolize reproduction in groups of up to forty individuals [469], living primarily in dark burrows they dig themselves. This likely translates into a heightened development of their chemical perception senses. In fact, it has been determined that these animals can recognize kinship through the individual scent of each member [470,471].

The meerkat AOB displays specific inherent features, which is unsurprising considering the vast genomic and morphological diversity described for this structure in other mammals [472,473]. It is intriguing to compare the meerkat AOB with that of other Carnivora species, such as canids and felids, whose VNOs have been deeply studied. The cat (*Felis silvestris catus*) and the dog (*Canis lupus familiaris*) provide striking examples of the vast disparity in the degree of differentiation that can exist in the AOB structural organization. The cat AOB is well developed, macroscopically discernible, and shows clear layer definition [343]. In contrast, the limited differentiation observed in the dog AOB is striking [283,467,474]. In the meerkat, similar to canids, the AOB is hard to identify macroscopically. However, microscopically, the laminar pattern in meerkats is more defined than in dogs, more closely resembling the pattern described in cats. Moreover, the meerkat AOB includes well-defined glomeruli, a broad plexiform mitral layer with relatively numerous principal cells, and granular cells organized in clusters. These similarities are not unexpected since meerkats belong to the Feliformia suborder. However, the layer thickness, the degree of glomerular differentiation, and the number of cells found in the meerkat AOB seem to be inferior to those observed in cats.

These observations gain greater significance in the context of the debate over the taxonomic classification of the meerkat. Its placement within the feliforms has wavered between the Herpestidae and Viverridae families [475]. Currently, the meerkat is considered a herpestid, alongside a wide range of mongoose species that share a broad spectrum of social complexities with it. The extensive morphological diversity observed in the VNO across species has served as a fruitful tool for establishing phylogenetic relationships among various mammalian species and identifying evolutionary pathways [476]. Despite the advent of genomic techniques, which have enabled population-level genetic-variability characterization at a resolution previously unimaginable, the study of morphological features remains a critical element in mammalian taxonomic classification [477]. The VNO continues to be an important reference for these taxonomic studies [303,478]. Future research on the neuroanatomical and neurochemical characteristics of the meerkat VNO and AOB, combined with further observations in other viverrid and herpestid species, may help clarify this phylogenetic debate.

Regarding the expression of the G protein subunits, Gαi2 and Gαo, in the AOB, this aligns the meerkat with the group of mammals that have lost the expression of the V2R receptor family. It also tests positive for OMP, albeit with a slightly weaker expression in the AOB compared to the MOB. This pattern has been similarly reported in leporids [132], didelphids [479], and canids [366]. Anti-MAP-2 (Microtubule Associated Protein 2) and SMI-32 (Neurofilament Protein Marker 32) are routinely used to label neuronal somas in the OB. MAP-2 is mainly expressed in the mitral plexiform layer and in the granular stratum of the AOB, while SMI-32 is not expressed in the AOB, despite inducing a strong reaction in the mitral cells of the MOB.

Among the calcium-binding proteins, ant-CB and anti-CR immunostain the nervous stratum and in the neuropil of all glomeruli. However, periglomerular cells (PGC) are not marked by either anti-CR or anti-CB, which might be related to the low number of PGC detected in the Nissl stain. In contrast, the CR immunostaining pattern in the MOB glomeruli reveals a striking and intense immunostaining of the PGC. However, in the neuropil, the intensity of anti-CR staining is very weak, except for an isolated subpopulation of independent glomeruli located near the AOB, which has strong immunoreactivity. These atypical glomeruli might be involved in processing chemical signals from the vomeronasal nerve, as hypothesized for the olfactory limbus glomeruli in mice [130] and foxes [133].

Although the gorse lectin, UEA, has been employed as a specific marker for the global vomeronasal pathway (VNO, vomeronasal nerves, and AOB) in adult mice [446,480] and dogs, it does not produce any positive labeling in the meerkat AOB. Conversely, the lectin LEA stains the neural and glomerular layers in both the AOB and MOB.

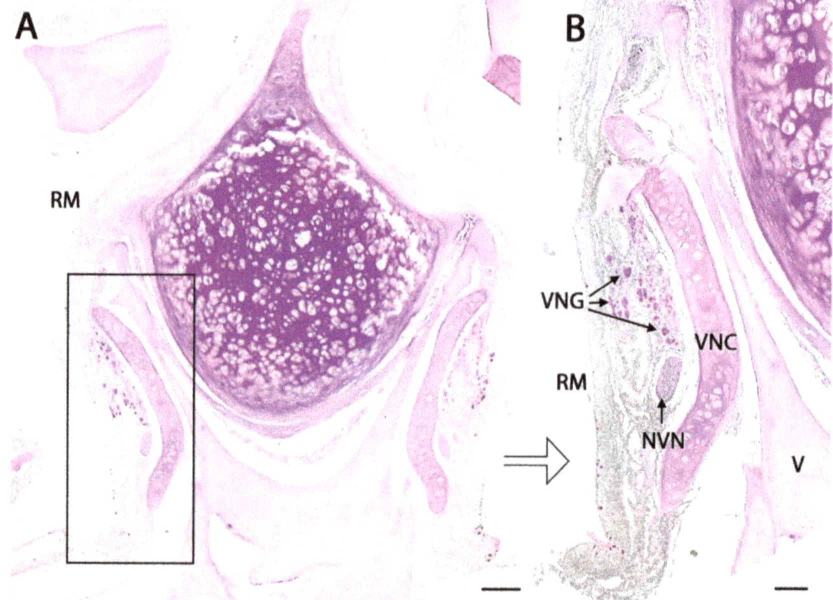

Figure 24. (**A**) The vomeronasal organ of a meerkat. Transverse decalcified histological section stained with PAS. (**B**) Magnification in box A. NVN, vomeronasal nerves; RM, respiratory epithelium; V, vomer; VNC, vomeronasal cartilage; VNG, vomeronasal glands. Scale bars: A, 200 µm; B, 100 µm.

3.4. Dama gazelle (Nanger dama)

The *dama gazelle* is a wild bovid about which only recent information regarding its vomeronasal system has been obtained, and it additionally holds the interest of being an endangered species. In certain species with a well-characterized VNS, methods related to this system are already employed to enhance reproduction, such as the use of pheromones, which help control stress levels and improve fertility [481]. While there currently exists a vast application field for bovine pheromones [482,483], their potential application in the reproduction of the dama gazelle necessitates a prior neuroanatomical and morphofunctional characterization of its vomeronasal organ.

A morphological feature of the VNO in dama gazelles (Figure 25), common to most Ruminantia, is its remarkable length. However, the implications of this trait have only been thoroughly investigated in the dama gazelle, wherein immunohistochemical and lectin-histochemical characteristics of the vomeronasal duct and the vomeronasal nerves were evaluated throughout the VNO entire extent [484]. Beyond this fact, it is significant that in this Antilopinae species, the vomeronasal duct opens into the incisive duct, into which it incorporates its medial side, both sharing the same cartilaginous capsule, which opens into a functional incisive papilla. This fact is especially relevant, as it is not a trait common to all Antilopinae; for instance, the three species of alcelaphine antelopes studied in [485] lack an incisive papilla.

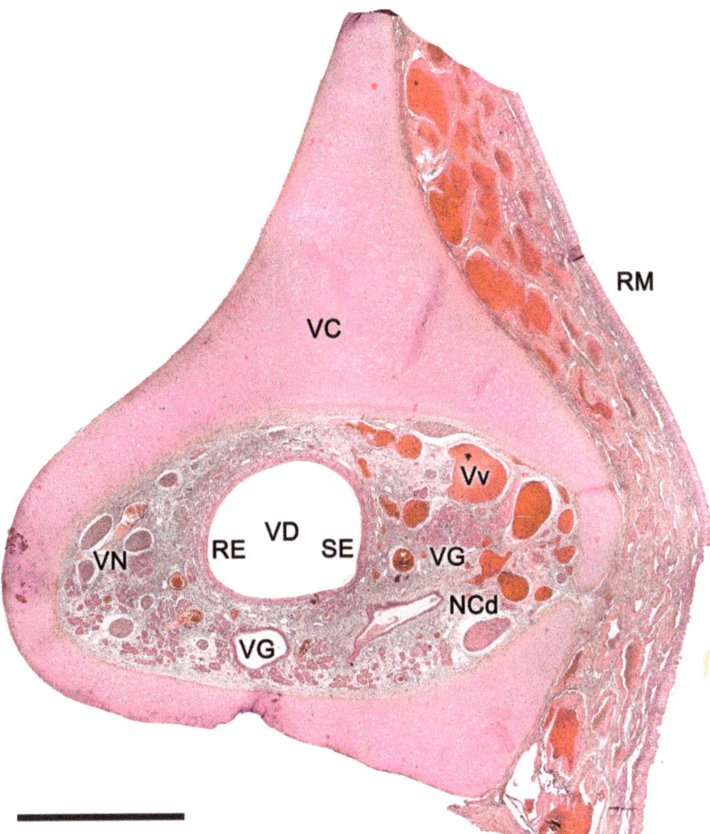

Figure 25. Histological transverse section of the VNO of a dama gazelle showing its main components. NCd, nasal caudal nerve; RE, respiratory epithelium; RM, respiratory mucosa; SE, sensory epithelium; VD, vomeronasal duct; VG, vomeronasal glands; VN, vomeronasal nerve; Vv, veins. Scale bar: 500 µm.

The serial histological study of the VNO (Figure 26) has allowed the determination of the differences along the rostral–caudal axis of the organ concerning the epithelial lining of the vomeronasal duct. Among the most relevant aspects, it is striking that there is a narrowing in its lumen coinciding with the more central levels; however, in the final third of the organ, there is a remarkable increase in the development of the parenchyma and the size of the duct, which retains its medial sensory lining and lateral respiratory lining. The access of pheromones to this extensive caudal chamber requires the presence of a robust vascular pump, responsible for mobilizing pheromones from the external environment to the sensory cells of the VNO. This pump is driven by vasomotor movements, which allow the suction of stimulating substances into the vomeronasal organ and actively expel the content of the vomeronasal duct. These mechanisms are activated by fibers that run through the nasopalatine nerve and cause the constriction of the blood vessels within the VNO capsule [251]. After constriction, the volume of blood in the cavernous tissue is reduced, creating a pressure differential that expands the VNO lumen and extracts fluid from the region surrounding the duct opening [486]. Stimuli enter the vomeronasal organ in solution via the mucus stream that passes to the vomeronasal duct from the incisive duct. Functional vomeronasal stimuli must be molecules soluble in that mucus or substances that become soluble when binding to carrier molecules that may be secreted by the vomeronasal glands [487].

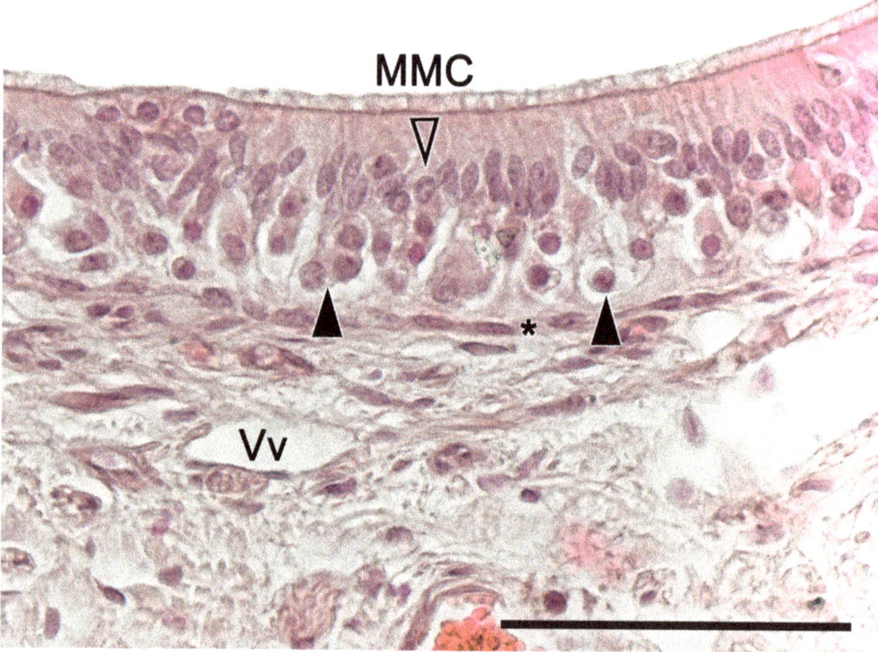

Figure 26. Hematoxilin-eosin staining of a dama gazelle VNO sensory epithelium. Neuroreceptor cells (arrowhead); Sustentacular cells (white arrowheads); Basal cells (asterisk); MMC, mucomicrovillar complex; SE, sensory epithelium; VN, vomeronasal nerve; Vv, veins. Scale bar: 250 μm.

The access of pheromones is determined not only by the action of the pump but also by the physicochemical properties of the molecules. It seems logical to suggest that a diffusion gradient of substances is established along the duct, with only the most soluble substances in the glandular secretion that fills the vomeronasal duct reaching the caudal end, leaving the less soluble ones in the anterior part. It is consistent to hypothesize that the specificity of the vomeronasal receptors along the vomeronasal duct adapts to the differential migration of the substance mixture along the duct, just as compounds are specifically eluted along a chromatography column.

Given the significant variations observed in the structure of the vomeronasal duct and its epithelial lining, the neurochemical study of the VNO has been conducted not only in the central part of the VNO—the most typical—but has been extended throughout the VNO length. The study of G protein expression showed immunopositivity for both markers (Gαi2 and Gαo) in the vomeronasal epithelium. The serial immunohistochemical study throughout the entire vomeronasal duct confirmed that the expression of both markers is not constant in the VNO, as the large dilated caudal segment, despite having a well-characterized neuroepithelium, lacks immunolabeling for both proteins. This suggests that the receptor type expressed in the caudal third of the vomeronasal duct does not correspond to either V1R or V2R, suggesting that the molecules detected by these two receptor types do not reach the caudal end of the organ. This hypothesis implies that only highly soluble compounds in the vomeronasal mucosa, and probably low molecular weight, can be detected in the caudal VNO, and likely by another type of receptors not yet described. The absence of vomeronasal receptor expression in such an extensive segment of the VNO could explain why previous G protein expression studies in the VNO of ruminants have not found positivity for the Gαo marker, as has been the case for goats [488], sheep [329], and sika deer [489].

It is worth noting that, in their study of the cow AOB, the authors of [458] did not find positivity for Gαo, nor was it found in the dog [283] or fox [366] AOB. However, this does not exclude the expression of V2R receptors in the VNO, as it is plausible that the information from these receptors may project to other areas of the olfactory bulb, not necessarily to the AOB. In fact, it has recently been observed that, in the fox olfactory bulb, the expression of Gαo linked to the vomeronasal nerves does not project to the AOB, but to the transition zone located between the main olfactory bulb and the accessory, an area known as the olfactory limb [133].

The use of an additional range of neuronal markers, such as PGP 9.5 and specific neuronal enolase (EN), showed that the expression of these markers extends to the caudal sensory epithelium area, even though neither the analyzed G proteins nor the OMP marker is expressed there. This finding provides evidence that the histologically described neuroepithelial nature of this long caudal stretch of the vomeronasal duct is correct. On the other hand, it suggests that the type of receptors that are expressed differs from those that belong to the V1R and V2R families. Other markers, such as calcium-binding proteins, however, have differential expression. For example, antibodies against CB and CR mark a subpopulation of vomeronasal receptor neurons in the dama gazelle vomeronasal duct. As with G proteins and OMP, the expression of calcium-binding proteins did not occur in the caudal area of the sensory epithelium (Figure 27).

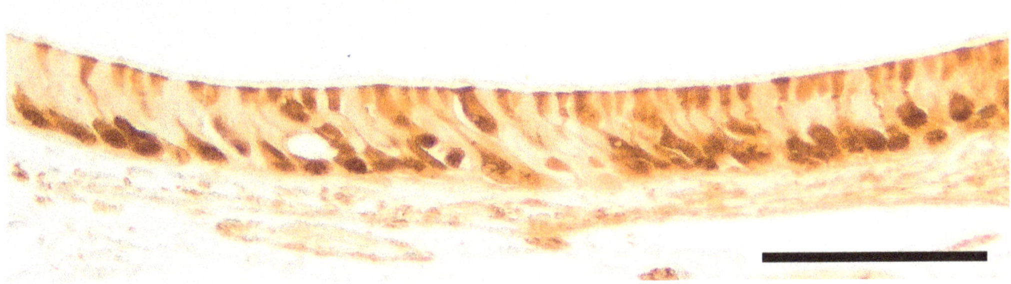

Figure 27. PGP9.5 immunostaining of the SE of a dama gazelle VNO showing the morphology and distribution of the neuroreceptor somata, and their dendritic knobs and processes. Scale bar: 150 μm.

In summary, the research on the vomeronasal organ of the dama gazelle, besides providing valuable information about the configuration and function of this structure in this endangered species, could be used to design and implement programs based on the use of pheromones to enhance reproductive success and increase genetic diversity in captive populations. Such programs can be critical for the species' survival. Furthermore, the discovery in this species of notable differences in the organization of the vomeronasal duct and the expression of neuronal markers along the rostro-caudal axis of the organ underscores the need to consider such differences when conducting future studies on the VNO of other species.

4. Conclusions

Throughout this review, our objective has been to provide a comprehensive and precise portrayal of the crucial role that chemical communication plays in mammals, specifically focusing on the detection of pheromone-mediated communication through the vomeronasal system. Due to comprehensive research and the accumulation of scientific knowledge, we have gained a profound understanding of the significant impact exerted by the vomeronasal system on animal behavior. This intricate sensory system profoundly influences several aspects of social and sexual interactions, encompassing reproductive processes, the formation of hierarchies, maternal bonding, and intraspecies recognition. Furthermore, it is evident that there is a notable range of variability resulting from evolutionary adaptations within this particular system, surpassing the documented variances in the olfactory system. The extensive range of adaptive variability is a challenge in the examination of the vomeronasal system. Therefore, it is vital to direct further research efforts towards a deeper neuroanatomical and morphofunctional understanding of the system. The comprehensive understanding of the impact of pheromones on the physiology and welfare of animals and humans necessitates the pursuit of such an undertaking. Recognizing the intricacies of the vomeronasal system is paramount in the realm of biological research. By delving deeper into its evolutionary origins, functional roles, and neuroanatomical intricacies, we not only gain invaluable insights into its unique position in mammalian communication but also unveil its broader implications in social and reproductive behaviors. The culmination of such basic research has profound potential and it can set the stage for leveraging pheromones as powerful tools. By harnessing this knowledge, we open the doors for innovative approaches in enhancing the well-being and health of both animals and humans.

Author Contributions: Conceptualization, M.V.T., I.O.-L. and P.S.-Q.; writing—original draft preparation, M.V.T., I.O.-L. and P.S.-Q.; writing—review and editing, M.V.T., I.O.-L. and P.S.-Q.; supervision, M.V.T. and P.S.-Q. All authors have read and agreed to the published version of the manuscript.

Funding: This research has received no external funding.

Institutional Review Board Statement: Not applicable.

Informed Consent Statement: Not applicable.

Data Availability Statement: No new data were created.

Conflicts of Interest: The authors declare no conflict of interest.

References

1. Stevens, M. *Sensory Ecology, Behaviour, and Evolution*; Oxford University Press: Oxford, UK, 2013; ISBN 978-0-19-960177-6.
2. Wyatt, T.D. *Pheromones and Animal Behaviour: Communication by Smell and Taste*; Cambridge University Press: Cambridge, UK; New York, NY, USA, 2003; ISBN 978-0-521-48068-0.
3. Wood, W.F. Chemical Ecology: Chemical Communication in Nature. *J. Chem. Educ.* **1983**, *60*, 531. [CrossRef]
4. Castañeda, M.d.L.A.; Martínez-Gómez, M.; Guevara-Guzmán, R.; Hudson, R. Chemical communication in domestic mammals. *Vet. Mex.* **2007**, *38*, 105–123.
5. Kokocińska-Kusiak, A.; Woszczyło, M.; Zybala, M.; Maciocha, J.; Barłowska, K.; Dzięcioł, M. Canine Olfaction: Physiology, Behavior, and Possibilities for Practical Applications. *Animals* **2021**, *11*, 2463. [CrossRef] [PubMed]

6. Iovino, M.; Messana, T.; Iovino, E.; De Pergola, G.; Guastamacchia, E.; Giagulli, V.A.; Triggiani, V. Neuroendocrine Mechanisms Involved in Male Sexual and Emotional Behavior. *Endocr. Metab. Immune Disord. Drug Targets* **2019**, *19*, 472–480. [CrossRef] [PubMed]

7. Beruter, J.; Beauchamp, G.K.; Muetterties, E.L. Complexity of Chemical Communication in Mammals: Urinary Components Mediating Sex Discrimination by Male Guinea Pigs. *Biochem. Biophys. Res. Commun.* **1973**, *53*, 264–271. [CrossRef]

8. Wilson, E.O. Chemical Communication within Animal Species. In *Chemical Ecology*; Academic Press: New York, NY, USA, 1970; pp. 133–155.

9. Surov, A.V.; Maltsev, A.N. Analysis of Chemical Communication in Mammals: Zoological and Ecological Aspects. *Biol. Bull. Russ. Acad. Sci.* **2016**, *43*, 1175–1183. [CrossRef]

10. Villamayor, P.R.; Arana, Á.J.; Coppel, C.; Ortiz-Leal, I.; Torres, M.V.; Sanchez-Quinteiro, P.; Sánchez, L. A Comprehensive Structural, Lectin and Immunohistochemical Characterization of the Zebrafish Olfactory System. *Sci. Rep.* **2021**, *11*, 8865. [CrossRef]

11. Saraiva, L.R.; Ahuja, G.; Ivandic, I.; Syed, A.S.; Marioni, J.C.; Korsching, S.I.; Logan, D.W. Molecular and Neuronal Homology between the Olfactory Systems of Zebrafish and Mouse. *Sci. Rep.* **2015**, *5*, 11487. [CrossRef]

12. Meister, M. On the Dimensionality of Odor Space. *eLife* **2015**, *4*, e07865. [CrossRef]

13. Mollo, E.; Garson, M.J.; Polese, G.; Amodeo, P.; Ghiselin, M.T. Taste and Smell in Aquatic and Terrestrial Environments. *Nat. Prod. Rep.* **2017**, *34*, 496–513. [CrossRef]

14. Hepper, P.G.; Wells, D.L. Perinatal Olfactory Learning in the Domestic Dog. *Chem. Senses* **2006**, *31*, 207–212. [CrossRef] [PubMed]

15. Weissburg, M.J.; Ferner, M.C.; Pisut, D.P.; Smee, D.L. Ecological Consequences of Chemically Mediated Prey Perception. *J. Chem. Ecol.* **2002**, *28*, 1953–1970. [CrossRef] [PubMed]

16. Nordlund, D.A.; Lewis, W.J. Terminology of Chemical Releasing Stimuli in Intraspecific and Interspecific Interactions. *J. Chem. Ecol.* **1976**, *2*, 211–220. [CrossRef]

17. Law, J.H.; Regnier, F.E. Pheromones. *Annu. Rev. Biochem.* **1971**, *40*, 533–548. [CrossRef]

18. Calcagnile, M.; Tredici, S.M.; Talà, A.; Alifano, P. Bacterial Semiochemicals and Transkingdom Interactions with Insects and Plants. *Insects* **2019**, *10*, 441. [CrossRef]

19. Burger, B.V. Mammalian Semiochemicals. In *The Chemistry of Pheromones and Other Semiochemicals II*; Schulz, S., Ed.; Topics in Current Chemistry; Springer: Berlin/Heidelberg, Germany, 2004; Volume 240, pp. 231–278. ISBN 978-3-540-21308-6.

20. Apfelbach, R.; Parsons, M.H.; Soini, H.A.; Novotny, M.V. Are Single Odorous Components of a Predator Sufficient to Elicit Defensive Behaviors in Prey Species? *Front. Neurosci.* **2015**, *9*, 263. [CrossRef]

21. Dicke, M.; Sabelis, M.W. Infochemical Terminology: Based on Cost-Benefit Analysis Rather than Origin of Compounds? *Funct. Ecol.* **1988**, *2*, 131. [CrossRef]

22. Wyatt, T.D. Pheromones and Behavior. In *Chemical Communication in Crustaceans*; Breithaupt, T., Thiel, M., Eds.; Springer: New York, NY, USA, 2010; pp. 23–38. ISBN 978-0-387-77100-7.

23. Sbarbati, A.; Osculati, F. Allelochemical Communication in Vertebrates: Kairomones, Allomones and Synomones. *Cells Tissues Organs* **2006**, *183*, 206–219. [CrossRef]

24. Blum, M.S. Semiochemical Parsimony in the Arthropoda. *Annu. Rev. Entomol.* **1996**, *41*, 353–374. [CrossRef]

25. Gallie, D.R.; Chang, S.C. Signal Transduction in the Carnivorous Plant Sarracenia Purpurea (Regulation of Secretory Hydrolase Expression during Development and in Response to Resources). *Plant Physiol.* **1997**, *115*, 1461–1471. [CrossRef]

26. Apfelbach, R.; Blanchard, C.D.; Blanchard, R.J.; Hayes, R.A.; McGregor, I.S. The Effects of Predator Odors in Mammalian Prey Species: A Review of Field and Laboratory Studies. *Neurosci. Biobehav. Rev.* **2005**, *29*, 1123–1144. [CrossRef] [PubMed]

27. Dunkelblum, E.; Mendel, Z.; Gries, G.; Gries, R.; Zegelman, L.; Hassner, A.; Mori, K. Antennal Response and Field Attraction of the Predator *Elatophilus hebraicus* (Hemiptera: Anthocoridae) to Sex Pheromones and Analogues of Three *Matsucoccus* spp. (Homoptera: Matsucoccidae). *Bioorg. Med. Chem.* **1996**, *4*, 489–494. [CrossRef] [PubMed]

28. Cavill, G.W.K.; Robertson, P.L. Ant Venoms, Attractants, and Repellents: Secretions Are Used by Ants in Attack and Defense and as Chemical Messengers in Their Social Organization. *Science* **1965**, *149*, 1337–1345. [CrossRef] [PubMed]

29. Mattiacci, L.; Dicke, M.; Posthumus, M.A. Induction of Parasitoid Attracting Synomone in Brussels Sprouts Plants by Feeding ofPieris Brassicae Larvae: Role of Mechanical Damage and Herbivore Elicitor. *J. Chem. Ecol.* **1994**, *20*, 2229–2247. [CrossRef] [PubMed]

30. Stowe, M.K.; Turlings, T.C.; Loughrin, J.H.; Lewis, W.J.; Tumlinson, J.H. The Chemistry of Eavesdropping, Alarm, and Deceit. *Proc. Natl. Acad. Sci. USA* **1995**, *92*, 23–28. [CrossRef]

31. Han, B.; Chen, Z. Behavioral and Electrophysiological Responses of Natural Enemies to Synomones from Tea Shoots and Kairomones from Tea Aphids, *Toxoptera aurantii*. *J. Chem. Ecol.* **2002**, *28*, 2203–2219. [CrossRef]

32. Pernaa, J.; Aksela, M. Learning Organic Chemistry through a Study of Semiochemicals. *J. Chem. Educ.* **2011**, *88*, 1644–1647. [CrossRef]

33. Kasinger, H.; Bauer, B.; Denzinger, J. The Meaning of Semiochemicals to the Design of Self-Organizing Systems. In Proceedings of the 2008 Second IEEE International Conference on Self-Adaptive and Self-Organizing Systems, Venezia, Italy, 20–24 October 2008; pp. 139–148.

34. Karlson, P.; Lüscher, M. 'Pheromones': A New Term for a Class of Biologically Active Substances. *Nature* **1959**, *183*, 55–56. [CrossRef]

35. Butenandt, A.; Beckmann, R.; Stamm, D.; Hecker, E. Uber Den Sexsual-Lockstoff Des Seidenspinners Bombyx Mori. Reindarstellung Und Konstitution. *Z. Naturforschg. B* **1959**, *14*, 283–284.

36. Wilson, E.O. Chemical Communication in the Social Insects: Insect Societies Are Organized Principally by Complex Systems of Chemical Signals. *Science* **1965**, *149*, 1064–1071. [CrossRef]

37. Conte, Y.L.; Hefetz, A. Primer Pheromones in Social Hymenoptera. *Annu. Rev. Entomol.* **2008**, *53*, 523–542. [CrossRef] [PubMed]

38. Cork, A.; Kamal, N.; Alam, S.; Choudhury, J.; Talekar, N. Pheromones and Their Applications to Insect Pest Control. *Bangladesh J. Entomol.* **2003**, *13*, 1–13.

39. Liberles, S.D. Mammalian Pheromones. *Annu. Rev. Physiol.* **2014**, *76*, 151–175. [CrossRef]

40. Brennan, P.A.; Zufall, F. Pheromonal Communication in Vertebrates. *Nature* **2006**, *444*, 308–315. [CrossRef] [PubMed]

41. Beauchamp, G.K.; Doty, R.L.; Moulton, D.G.; Mugford, R.A. The Pheromone Concept in Mammalian Chemical Communication: A Critique. In *Mammalian Olfaction, Reproductive Processes, and Behavior*; Elsevier: Amsterdam, The Netherlands, 1976; pp. 143–160. ISBN 978-0-12-221250-5.

42. Brown, R.E. Mammalian Social Odors: A Critical Review. In *Advances in the Study of Behavior*; Elsevier: Amsterdam, The Netherlands, 1979; Volume 10, pp. 103–162. ISBN 978-0-12-004510-5.

43. Booth, W.D.; Signoret, J.P. Olfaction and Reproduction in Ungulates. *Oxf. Rev. Reprod. Biol.* **1992**, *14*, 263–301.

44. Schaal, B.; Coureaud, G.; Langlois, D.; Giniès, C.; Sémon, E.; Perrier, G. Chemical and Behavioural Characterization of the Rabbit Mammary Pheromone. *Nature* **2003**, *424*, 68–72. [CrossRef] [PubMed]

45. Apps, P.J.; Weldon, P.J.; Kramer, M. Chemical Signals in Terrestrial Vertebrates: Search for Design Features. *Nat. Prod. Rep.* **2015**, *32*, 1131–1153. [CrossRef]

46. Ferrero, D.M.; Moeller, L.M.; Osakada, T.; Horio, N.; Li, Q.; Roy, D.S.; Cichy, A.; Spehr, M.; Touhara, K.; Liberles, S.D. A Juvenile Mouse Pheromone Inhibits Sexual Behaviour through the Vomeronasal System. *Nature* **2013**, *502*, 368–371. [CrossRef]

47. Lin, D.Y.; Zhang, S.-Z.; Block, E.; Katz, L.C. Encoding Social Signals in the Mouse Main Olfactory Bulb. *Nature* **2005**, *434*, 470–477. [CrossRef]

48. Novotny, M.; Harvey, S.; Jemiolo, B.; Alberts, J. Synthetic Pheromones That Promote Inter-Male Aggression in Mice. *Proc. Natl. Acad. Sci. USA* **1985**, *82*, 2059–2061. [CrossRef]

49. Singer, A.G.; Agosta, W.C.; O'Connell, R.J.; Pfaffmann, C.; Bowen, D.V.; Field, F.H. Dimethyl Disulfide: An Attractant Pheromone in Hamster Vaginal Secretion. *Science* **1976**, *191*, 948–950. [CrossRef] [PubMed]

50. Dorries, K.M.; Adkins-regan, E.; Halpern, B.P. Olfactory Sensitivity to the Pheromone, Androstenone, Is Sexually Dimorphic in the Pig. *Physiol. Behav.* **1995**, *57*, 255–259. [CrossRef]

51. Rasmussen, L.E.L.; Lazar, J.; Greenwood, D.R. Olfactory Adventures of Elephantine Pheromones. *Biochem. Soc. Trans.* **2003**, *31*, 137–141. [CrossRef]

52. Brown, W.L.; Eisner, T.; Whittaker, R.H. Allomones and Kairomones: Transspecific Chemical Messengers. *BioScience* **1970**, *20*, 21–22. [CrossRef]

53. Lass, S.; Bittner, K. Facing Multiple Enemies: Parasitised Hosts Respond to Predator Kairomones. *Oecologia* **2002**, *132*, 344–349. [CrossRef]

54. Weldon, P.J. In Defense of "Kairomone" as a Class of Chemical Releasing Stimuli. *J. Chem. Ecol.* **1980**, *6*, 719–725. [CrossRef]

55. Pasteels, J.M. Is Kairomone a Valid and Useful Term? *J. Chem. Ecol.* **1982**, *8*, 1079–1081. [CrossRef]

56. Ruther, J.; Meiners, T.; Steidle, J.L.M. Rich in Phenomena-Lacking in Terms. A Classification of Kairomones. *Chemoecology* **2002**, *12*, 161–167. [CrossRef]

57. Ayelo, P.M.; Pirk, C.W.W.; Yusuf, A.A.; Chailleux, A.; Mohamed, S.A.; Deletre, E. Exploring the Kairomone-Based Foraging Behaviour of Natural Enemies to Enhance Biological Control: A Review. *Front. Ecol. Evol.* **2021**, *9*, 641974. [CrossRef]

58. Schonewolf, K.W.; Bell, R.; Rypstra, A.L.; Persons, M.H. Field Evidence of an Airborne Enemy-Avoidance Kairomone in Wolf Spiders. *J. Chem. Ecol.* **2006**, *32*, 1565–1576. [CrossRef] [PubMed]

59. Reinecke, A.; Ruther, J.; Tolasch, T.; Francke, W.; Hilker, M. Alcoholism in Cockchafers: Orientation of Male *Melolontha melolontha* towards Green Leaf Alcohols. *Naturwissenschaften* **2002**, *89*, 265–269. [CrossRef]

60. Jones, R.L.; Lewis, W.J.; Bowman, M.C.; Beroza, M.; Bierl, B.A. Host-Seeking Stimulant for Parasite of Corn Earworm: Isolation, Identification, and Synthesis. *Science* **1971**, *173*, 842–843. [CrossRef]

61. Acree, F.; Turner, R.B.; Gouck, H.K.; Beroza, M.; Smith, N. L-Lactic Acid: A Mosquito Attractant Isolated from Humans. *Science* **1968**, *161*, 1346–1347. [CrossRef]

62. Thibout, E.; Guillot, J.F.; Ferary, S.; Limouzin, P.; Auger, J. Origin and Identification of Bacteria Which Produce Kairomones in the Frass of *Acrolepiopsis assectella* (Lep., Hyponomeutoidea). *Experientia* **1995**, *51*, 1073–1075. [CrossRef]

63. Schoeppner, N.M.; Relyea, R.A. Damage, Digestion, and Defence: The Roles of Alarm Cues and Kairomones for Inducing Prey Defences: Damage, Digestion, and Defence. *Ecol. Lett.* **2005**, *8*, 505–512. [CrossRef] [PubMed]

64. Arimura, G.; Ozawa, R.; Kugimiya, S.; Takabayashi, J.; Bohlmann, J. Herbivore-Induced Defense Response in a Model Legume. Two-Spotted Spider Mites Induce Emission of (E)-β-Ocimene and Transcript Accumulation of (E)-β-Ocimene Synthase in *Lotus japonicus*. *Plant. Physiol.* **2004**, *135*, 1976–1983. [CrossRef]

65. Reinecke, A.; Ruther, J.; Mayer, C.J.; Hilker, M. Optimized Trap Lure for Male *Melolontha* Cockchafers. *J. Appl. Entomol.* **2006**, *130*, 171–176. [CrossRef]

66. Briand, L.; Trotier, D.; Pernollet, J.-C. Aphrodisin, an Aphrodisiac Lipocalin Secreted in Hamster Vaginal Secretions. *Peptides* **2004**, *25*, 1545–1552. [CrossRef]
67. Altmann, D.; Sorensen, L. Zoo Behaviour Science in the Research Centre for Vertebrate Studies of the Academy of Sciences of the German Democratic Republic (in the Animal Park of Berlin). *Appl. Anim. Behav. Sci.* **1987**, *18*, 67–81. [CrossRef]
68. Salazar, I.; Fdez de Troconiz, P.; Prieto, M.D.; Cifuentes, J.M.; Quinteiro, P.S. Anatomy and Cholinergic Innervation of the Sinus Paranalis in Dogs. *Anat. Histol. Embryol.* **1996**, *25*, 49–53. [CrossRef]
69. Martín, J.; Barja, I.; López, P. Chemical Scent Constituents in Feces of Wild Iberian Wolves (*Canis lupus signatus*). *Biochem. Syst. Ecol.* **2010**, *38*, 1096–1102. [CrossRef]
70. Aviles-Rosa, E.O.; Surowiec, K.; McGlone, J. Identification of Faecal Maternal Semiochemicals in Swine (*Sus scrofa*) and Their Effects on Weaned Piglets. *Sci. Rep.* **2020**, *10*, 5349. [CrossRef] [PubMed]
71. Ewer, R.F. *Ethology of Mammals*; Springer: Boston, MA, USA, 1968; ISBN 978-1-4899-4658-4.
72. Bradshaw, J.W.S.; Cameron-Beaumont, C.L. The Signaling Repertoire of the Domestic Cat and Its Undomesticated Relatives. In *The Domestic Cat: The Biology of Its Behaviour*; Cambridge University Press: Cambridge, UK, 2000; pp. 67–93.
73. Miyazaki, M.; Yamashita, T.; Suzuki, Y.; Saito, Y.; Soeta, S.; Taira, H.; Suzuki, A. A Major Urinary Protein of the Domestic Cat Regulates the Production of Felinine, a Putative Pheromone Precursor. *Chem. Biol.* **2006**, *13*, 1071–1079. [CrossRef]
74. Stefańczyk-Krzymowska, S.; Krzymowski, T.; Wasowska, B.; Jana, B.; Słomiński, J. Intramuscular Injections of Male Pheromone 5 Alpha-Androstenol Change the Secretory Ovarian Function in Gilts during Sexual Maturation. *Reprod. Biol.* **2003**, *3*, 241–257.
75. Booth, W.D. Sexual Dimorphism Involving Steroidal Pheromones and Their Binding Protein in the Submaxillary Salivary Gland of the Göttingen Miniature Pig. *J. Endocrinol.* **1984**, *100*, 195–202. [CrossRef]
76. Pietras, R.J. Sex Pheromone Production by Preputial Gland: The Regulatory Role of Estrogen. *Chem. Senses* **1981**, *6*, 391–408. [CrossRef]
77. Murphy, M.R.; Schneider, G.E. Olfactory Bulb Removal Eliminates Mating Behavior in the Male Golden Hamster. *Science* **1970**, *167*, 302–304. [CrossRef]
78. Michael, R.P.; Keverne, E.B. Primate Sex Pheromones of Vaginal Origin. *Nature* **1970**, *225*, 84–85. [CrossRef]
79. Goodwin, M.; Gooding, K.M.; Regnier, F. Sex Pheromone in the Dog. *Science* **1979**, *203*, 559–561. [CrossRef]
80. Rivard, G.; Klemm, W.R. Two Body Fluids Containing Bovine Estrous Pheromone(s). *Chem. Senses* **1989**, *14*, 273–279. [CrossRef]
81. Hayes, R.A.; Richardson, B.J.; Claus, S.C.; Wyllie, S.G. Semiochemicals and Social Signaling in the Wild European Rabbit in Australia: II. Variations in Chemical Composition of Chin Gland Secretion across Sampling Sites. *J. Chem. Ecol.* **2002**, *28*, 2613–2625. [CrossRef]
82. Thiessen, D.D.; Friend, H.C.; Lindzey, G. Androgen Control of Territorial Marking in the Mongolian Gerbil. *Science* **1968**, *160*, 432–434. [CrossRef] [PubMed]
83. Müller-Schwarze, D. Pheromones in Black-Tailed Deer (*Odocoileus hemionus columbianus*). *Anim. Behav.* **1971**, *19*, 141–152. [CrossRef] [PubMed]
84. Müller-Schwarze, D.; Müller-Schwarze, C.; Singer, A.G.; Silverstein, R.M. Mammalian Pheromone: Identification of Active Component in the Subauricular Scent of the Male Pronghorn. *Science* **1974**, *183*, 860–862. [CrossRef]
85. Macdonald, D.W.; Krantz, K.; Aplin, R.T. Behavioural, Anatomical and Chemical Aspects of Scent Marking amongst Capybaras (*Hydrochoerus hydrochaeris*) (Rodentia: Caviomorpha). *J. Zool.* **1984**, *202*, 341–360. [CrossRef]
86. Berüter, J.; Beauchamp, G.K.; Muetterties, E.L. Mammalian Chemical Communication: Perineal Gland Secretion of the Guinea Pig. *Physiol. Zool.* **1974**, *47*, 130–136. [CrossRef]
87. Schultze-Westrum, T. Innerartliche Verständigung durch Düfte beim Gleitbeutler *Petaurus breviceps papuanus* Thomas (Marsupialia, Phalangeridae). *Z. Vergl. Physiol.* **1965**, *50*, 151–220. [CrossRef]
88. Stumpf, P.; Künzle, H.; Welsch, U. Cutaneous Eccrine Glands of the Foot Pads of the Small Madagascan Tenrec (*Echinops telfairi*, Insectivora, Tenrecidae): Skin Glands in a Primitive Mammal. *Cell Tissue Res.* **2004**, *315*, 59–70. [CrossRef]
89. Yager, J.A.; Hunter, D.B.; Wilson, M.R.; Allen, O.B. A Source of Cutaneous Maternal Semiochemicals in the Mink? *Experientia* **1988**, *44*, 79–81. [CrossRef]
90. Izard, M.K. Pheromones and Reproduction in Domestic Animals. In *Pheromones and Reproduction in Mammals*; Academic Press: Cambridge, MA, USA, 1983.
91. Melrose, D.R.; Reed, H.C.B.; Patterson, R.L.S. Androgen Steroids Associated with Boar Odour as an Aid to the Detection of Oestrus in Pig Artificial Insemination. *Br. Vet. J.* **1971**, *127*, 497–502. [CrossRef]
92. Gosling, L.M. A Reassessment of the Function of Scent Marking in Territories. *Z. Tierpsychol.* **1982**, *60*, 89–118. [CrossRef]
93. Eisenberg, J.F.; Kleiman, D.G. Olfactory Communication in Mammals. *Annu. Rev. Ecol. Syst.* **1972**, *3*, 1–32. [CrossRef]
94. Johnson, R.P. Scent Marking in Mammals. *Anim. Behav.* **1973**, *21*, 521–535. [CrossRef]
95. Jones, R.B.; Nowell, N.W. A Comparison of the Aversive and Female Attractant Properties of Urine from Dominant and Subordinate Male Mice. *Anim. Learn. Behav.* **1974**, *2*, 141–144. [CrossRef]
96. Hayes, R.A.; Richardson, B.J.; Wyllie, S.G. To Fix or Not to Fix: The Role of 2-Phenoxyethanol in Rabbit, Oryctolagus Cuniculus, Chin Gland Secretion. *J. Chem. Ecol.* **2003**, *29*, 1051–1064. [CrossRef]
97. Wirant, S.C.; Halvorsen, K.T.; McGuire, B. Preliminary Observations on the Urinary Behaviour of Female Jack Russell Terriers in Relation to Stage of the Oestrous Cycle, Location, and Age. *Appl. Anim. Behav. Sci.* **2007**, *106*, 161–166. [CrossRef]

98. Müller, C.A.; Manser, M.B. Scent-Marking and Intrasexual Competition in a Cooperative Carnivore with Low Reproductive Skew: Scent-Marking and Intrasexual Competition. *Ethology* **2008**, *114*, 174–185. [CrossRef]

99. Bronson, F.H.; Whitten, W.K. Oestrus-Accelerating Pheromone of Mice: Assay, Androgen-Dependency and Presence in Bladder Urine. *Reproduction* **1968**, *15*, 131–134. [CrossRef]

100. Bronson, F.H.; Caroom, D. Preputial Gland of the Male Mouse: Attractant Function. *Reproduction* **1971**, *25*, 279–282. [CrossRef]

101. Seitz, E. Die Bedeutung Geruchlicher Orientierung Beim Plumplori *Nycticebus coucang* Boddaert 1785 (*Prosimii, Lorisidae*). *Z. Tierpsychol.* **1969**, *26*, 73–103. [CrossRef]

102. Harrington, F.H. Urine Marking at Food and Caches in Captive Coyotes. *Can. J. Zool.* **1982**, *60*, 776–782. [CrossRef]

103. Sun, L.; Müller-Schwarze, D. Sibling Recognition in the Beaver: A Field Test for Phenotype Matching. *Anim. Behav.* **1997**, *54*, 493–502. [CrossRef]

104. Jordan, N.R.; Golabek, K.A.; Apps, P.J.; Gilfillan, G.D.; McNutt, J.W. Scent-Mark Identification and Scent-Marking Behaviour in African Wild Dogs (*Lycaon pictus*). *Ethology* **2013**, *119*, 644–652. [CrossRef]

105. Ferkin, M.H.; Johnston, R.E. Meadow Voles, Microtus Pennsylvanicus, Use Multiple Sources of Scent for Sex Recognition. *Anim. Behav.* **1995**, *49*, 37–44. [CrossRef]

106. Hradecký, P. Possible Pheromonal Regulation of Reproduction in Wild Carnivores. *J. Chem. Ecol.* **1985**, *11*, 241–250. [CrossRef]

107. Kiyokawa, Y.; Kikusui, T.; Takeuchi, Y.; Mori, Y. Modulatory Role of Testosterone in Alarm Pheromone Release by Male Rats. *Horm. Behav.* **2004**, *45*, 122–127. [CrossRef]

108. Boissy, A.; Terlouw, C.; Le Neindre, P. Presence of Cues from Stressed Conspecifics Increases Reactivity to Aversive Events in Cattle: Evidence for the Existence of Alarm Substances in Urine. *Physiol. Behav.* **1998**, *63*, 489–495. [CrossRef]

109. Aleksiuk, M. Scent-Mound Communication, Territoriality, and Population Regulation in Beaver (*Castor canadensis* Kuhl). *J. Mammal.* **1968**, *49*, 759–762. [CrossRef]

110. Bartecki, U.; Heymann, E.W. Field Observations on Scent-marking Behaviour in Saddle-back Tamarins, *Saguinus fuscicollis* (Callitrichidae, Primates). *J. Zool.* **1990**, *220*, 87–99. [CrossRef]

111. Tschanz, B.; Meyer-Holzapfel, M.; Bachmann, S. Das Informationssystem Bei Braunbären1. *Z. Tierpsychol.* **2010**, *27*, 47–72. [CrossRef]

112. Wakabayashi, Y.; Iwata, E.; Kikusui, T.; Takeuchi, Y.; Mori, Y. Regional Differences of Pheromone Production in the Sebaceous Glands of Castrated Goats Treated with Testosterone. *J. Vet. Med. Sci.* **2000**, *62*, 1067–1072. [CrossRef] [PubMed]

113. Eisenberg, J.F. A Comparative Study of Sandbathing Behavior in Heteromyid Rodents. *Behavior* **1963**, *22*, 16–23. [CrossRef]

114. Hayes, R.A.; Richardson, B.J.; Wyllie, S.G. Semiochemicals and Social Signaling in the Wild European Rabbit in Australia: I. Scent Profiles of Chin Gland Secretion from the Field. *J. Chem. Ecol.* **2002**, *28*, 363–384. [CrossRef] [PubMed]

115. Laska, M.; Bauer, V.; Salazar, L.T.H. Self-Anointing Behavior in Free-Ranging Spider Monkeys (*Ateles geoffroyi*) in Mexico. *Primates* **2007**, *48*, 160–163. [CrossRef]

116. Xu, Z.; Stoddart, D.M.; Ding, H.; Zhang, J. Self-Anointing Behavior in the Rice-Field Rat, *Rattus rattoides*. *J. Mammal.* **1995**, *76*, 1238–1241. [CrossRef]

117. Breer, H.; Fleischer, J.; Strotmann, J. Signaling in the Chemosensory Systems: The Sense of Smell: Multiple Olfactory Subsystems. *Cell. Mol. Life Sci.* **2006**, *63*, 1465–1475. [CrossRef]

118. Munger, S.D. Noses within Noses. *Nature* **2009**, *459*, 521–522. [CrossRef]

119. Barrios, A.W.; Sanchez Quinteiro, P.; Salazar, I. The Nasal Cavity of the Sheep and Its Olfactory Sensory Epithelium. *Microsc. Res. Tech.* **2014**, *77*, 1052–1059. [CrossRef]

120. Ramón y Cajal, S. Inducciones Fisiológicas de La Morfología y Conexiones de Las Neuronas. *Arch. Pedagog. Cienc. Afines* **1906**, *1*, 216–236.

121. Salazar, I.; Sanchez-Quinteiro, P.; Barrios, A.W.; López Amado, M.; Vega, J.A. Anatomy of the Olfactory Mucosa. In *Handbook of Clinical Neurology*; Elsevier: Amsterdam, The Netherlands, 2019; Volume 164, pp. 47–65. ISBN 978-0-444-63855-7.

122. McCotter, R.E. The Connection of the Vomeronasal Nerves with the Accessory Olfactory Bulb in the Opossum and Other Mammals. *Anat. Rec.* **1912**, *6*, 299–318. [CrossRef]

123. Salazar, I.; Barrios, A.W.; Sánchez-Quinteiro, P. Revisiting the Vomeronasal System From an Integrated Perspective. *Anat. Rec.* **2016**, *299*, 1488–1491. [CrossRef] [PubMed]

124. Barrios, A.W.; Núñez, G.; Sanchez Quinteiro, P.; Salazar, I. Anatomy, Histochemistry, and Immunohistochemistry of the Olfactory Subsystems in Mice. *Front. Neuroanat.* **2014**, *8*, 63. [CrossRef] [PubMed]

125. Fleischer, J.; Breer, H. The Grueneberg Ganglion: A Novel Sensory System in the Nose. *Histol. Histopathol.* **2010**, *25*, 909–915. [CrossRef] [PubMed]

126. Storan, M.J.; Key, B. Septal Organ of Grüneberg Is Part of the Olfactory System. *J. Comp. Neurol.* **2006**, *494*, 834–844. [CrossRef]

127. Ortiz-Leal, I.; Torres, M.V.; Villamayor, P.R.; López-Beceiro, A.; Sanchez-Quinteiro, P. The Vomeronasal Organ of Wild Canids: The Fox (*Vulpes vulpes*) as a Model. *J. Anat.* **2020**, *237*, 890–906. [CrossRef] [PubMed]

128. Ogura, T.; Krosnowski, K.; Zhang, L.; Bekkerman, M.; Lin, W. Chemoreception Regulates Chemical Access to Mouse Vomeronasal Organ: Role of Solitary Chemosensory Cells. *PLoS ONE* **2010**, *5*, e11924. [CrossRef] [PubMed]

129. Juilfs, D.M.; Fülle, H.J.; Zhao, A.Z.; Houslay, M.D.; Garbers, D.L.; Beavo, J.A. A Subset of Olfactory Neurons That Selectively Express cGMP-Stimulated Phosphodiesterase (PDE2) and Guanylyl Cyclase-D Define a Unique Olfactory Signal Transduction Pathway. *Proc. Natl. Acad. Sci. USA* **1997**, *94*, 3388–3395. [CrossRef]

130. Larriva-Sahd, J. Cytological Organization of the Alpha Component of the Anterior Olfactory Nucleus and Olfactory Limbus. *Front. Neuroanat.* **2012**, *6*, 23. [CrossRef]

131. Valverde, F.; López-Mascaraque, L.; De Carlos, J.A. Structure of the Nucleus Olfactorius Anterior of the Hedgehog (*Erinaceus europaeus*). *J. Comp. Neurol.* **1989**, *279*, 581–600. [CrossRef]

132. Villamayor, P.R.; Cifuentes, J.M.; Quintela, L.; Barcia, R.; Sanchez-Quinteiro, P. Structural, Morphometric and Immunohistochemical Study of the Rabbit Accessory Olfactory Bulb. *Brain Struct. Funct.* **2020**, *225*, 203–226. [CrossRef]

133. Ortiz-Leal, I.; Torres, M.V.; Vargas-Barroso, V.; Fidalgo, L.E.; López-Beceiro, A.M.; Larriva-Sahd, J.A.; Sánchez-Quinteiro, P. The Olfactory Limbus of the Red Fox (*Vulpes vulpes*). New Insights Regarding a Noncanonical Olfactory Bulb Pathway. *Front. Neuroanat.* **2023**, *16*, 1097467. [CrossRef] [PubMed]

134. Menco, B. Ultrastructural Studies on Membrane, Cytoskeletal, Mucous, and Protective Compartments in Olfaction. *Microsc. Res. Tech.* **1992**, *22*, 215–224. [CrossRef] [PubMed]

135. Dryer, L.; Berghard, A. Odorant Receptors: A Plethora of G-Protein-Coupled Receptors. *Trends Pharmacol. Sci.* **1999**, *20*, 413–417. [CrossRef]

136. Zhang, X.; Firestein, S. The Olfactory Receptor Gene Superfamily of the Mouse. *Nat. Neurosci.* **2002**, *5*, 124–133. [CrossRef]

137. Malnic, B.; Hirono, J.; Sato, T.; Buck, L.B. Combinatorial Receptor Codes for Odors. *Cell* **1999**, *96*, 713–723. [CrossRef] [PubMed]

138. Crespo, C.; Liberia, T.; Blasco-Ibáñez, J.M.; Nácher, J.; Varea, E. Cranial Pair I: The Olfactory Nerve: The olfactory nerve. *Anat. Rec.* **2019**, *302*, 405–427. [CrossRef]

139. Mombaerts, P.; Wang, F.; Dulac, C.; Chao, S.K.; Nemes, A.; Mendelsohn, M.; Edmondson, J.; Axel, R. Visualizing an Olfactory Sensory Map. *Cell* **1996**, *87*, 675–686. [CrossRef]

140. Price, J.L.; Powell, T.P. Certain Observations on the Olfactory Pathway. *J. Anat.* **1971**, *110*, 105–126.

141. Höfer, D.; Shin, D.; Drenckhahn, D. Identification of Cytoskeletal Markers for the Different Microvilli and Cell Types of the Rat Vomeronasal Sensory Epithelium. *J. Neurocytol.* **2000**, *29*, 147–156. [CrossRef]

142. Salazar, I.; Quinteiro, P.S.; Cifuentes, J.M. Comparative Anatomy of the Vomeronasal Cartilage in Mammals: Mink, Cat, Dog, Pig, Cow and Horse. *Ann. Anat.* **1995**, *177*, 475–481. [CrossRef]

143. Villamayor, P.R.; Cifuentes, J.M.; Fdz-de-Troconiz, P.; Sanchez-Quinteiro, P. Morphological and Immunohistochemical Study of the Rabbit Vomeronasal Organ. *J. Anat.* **2018**, *233*, 814–827. [CrossRef] [PubMed]

144. Wöhrmann-Repenning, A. Comparative anatomical studies of the vomeronasal complex and the rostral palate of various mammals. *Gegenbaurs Morphol. Jahrb.* **1984**, *130*, 501–530. [PubMed]

145. Salazar, I.; Sánchez-Quinteiro, P.; Alemañ, N.; Prieto, D. Anatomical, Immunohistochemical and Physiological Characteristics of the Vomeronasal Vessels in Cows and Their Possible Role in Vomeronasal Reception. *J. Anat.* **2008**, *212*, 686–696. [CrossRef] [PubMed]

146. Wysocki, C.J. Neurobehavioral Evidence for the Involvement of the Vomeronasal System in Mammalian Reproduction. *Neurosci. Biobehav. Rev.* **1979**, *3*, 301–341. [CrossRef] [PubMed]

147. Salazar, I.; Sánchez Quinteiro, P. Differential Development of Binding Sites for Four Lectins in the Vomeronasal System of Juvenile Mouse: From the Sensory Transduction Site to the First Relay Stage. *Brain Res.* **2003**, *979*, 15–26. [CrossRef]

148. Ressler, K.J.; Sullivan, S.L.; Buck, L.B. A Zonal Organization of Odorant Receptor Gene Expression in the Olfactory Epithelium. *Cell* **1993**, *73*, 597–609. [CrossRef]

149. Zapiec, B.; Mombaerts, P. The Zonal Organization of Odorant Receptor Gene Choice in the Main Olfactory Epithelium of the Mouse. *Cell Rep.* **2020**, *30*, 4220–4234.e5. [CrossRef]

150. Trinh, K.; Storm, D.R. Vomeronasal Organ Detects Odorants in Absence of Signaling through Main Olfactory Epithelium. *Nat. Neurosci.* **2003**, *6*, 519–525. [CrossRef]

151. Hohenbrink, P.; Mundy, N.I.; Zimmermann, E.; Radespiel, U. First Evidence for Functional Vomeronasal 2 Receptor Genes in Primates. *Biol. Lett.* **2013**, *9*, 20121006. [CrossRef]

152. Pro-Sistiaga, P.; Mohedano-Moriano, A.; Ubeda-Bañon, I.; Del Mar Arroyo-Jimenez, M.; Marcos, P.; Artacho-Pérula, E.; Crespo, C.; Insausti, R.; Martinez-Marcos, A. Convergence of Olfactory and Vomeronasal Projections in the Rat Basal Telencephalon. *J. Comp. Neurol.* **2007**, *504*, 346–362. [CrossRef]

153. Grüneberg, H. A Ganglion Probably Belonging to the N. Terminalis System in the Nasal Mucosa of the Mouse. *Z. Anat. Entwickl. Gesch.* **1973**, *140*, 39–52. [CrossRef]

154. Fleischer, J.; Schwarzenbacher, K.; Breer, H. Expression of Trace Amine-Associated Receptors in the Grueneberg Ganglion. *Chem. Senses* **2007**, *32*, 623–631. [CrossRef] [PubMed]

155. Fleischer, J.; Schwarzenbacher, K.; Besser, S.; Hass, N.; Breer, H. Olfactory Receptors and Signalling Elements in the Grueneberg Ganglion. *J. Neurochem.* **2006**, *98*, 543–554. [CrossRef] [PubMed]

156. Schmid, A.; Pyrski, M.; Biel, M.; Leinders-Zufall, T.; Zufall, F. Grueneberg Ganglion Neurons Are Finely Tuned Cold Sensors. *J. Neurosci.* **2010**, *30*, 7563–7568. [CrossRef]

157. Rodolfo-Masera, T. Su l'esistenza di un particolare organo olfattivo nel setto nasale della cavia e di altri roditori. *Arch. Ital. Anat. Embriol.* **1943**, *48*, 157–213.

158. Tian, H.; Ma, M. Molecular Organization of the Olfactory Septal Organ. *J. Neurosci.* **2004**, *24*, 8383–8390. [CrossRef]

159. Weiler, E.; Farbman, A.I. The Septal Organ of the Rat During Postnatal Development. *Chem. Senses* **2003**, *28*, 581–593. [CrossRef]

160. Ma, M.; Grosmaitre, X.; Iwema, C.L.; Baker, H.; Greer, C.A.; Shepherd, G.M. Olfactory Signal Transduction in the Mouse Septal Organ. *J. Neurosci.* **2003**, *23*, 317–324. [CrossRef]

161. Marshall, D.A.; Maruniak, J.A. Masera's Organ Responds to Odorants. *Brain Res.* **1986**, *366*, 329–332. [CrossRef]

162. Wysocki, C.J.; Wellington, J.L.; Beauchamp, G.K. Access of Urinary Nonvolatiles to the Mammalian Vomeronasal Organ. *Science* **1980**, *207*, 781–783. [CrossRef]

163. Tizzano, M.; Cristofoletti, M.; Sbarbati, A.; Finger, T.E. Expression of Taste Receptors in Solitary Chemosensory Cells of Rodent Airways. *BMC Pulm. Med.* **2011**, *11*, 3. [CrossRef] [PubMed]

164. Silver, W.L.; Moulton, D.G. Chemosensitivity of Rat Nasal Trigeminal Receptors. *Physiol. Behav.* **1982**, *28*, 927–931. [CrossRef] [PubMed]

165. Filoramo, N.I.; Schwenk, K. The Mechanism of Chemical Delivery to the Vomeronasal Organs in Squamate Reptiles: A Comparative Morphological Approach. *J. Exp. Zool.* **2009**, *311*, 20–34. [CrossRef] [PubMed]

166. Kondoh, D.; Kaneoya, Y.; Tonomori, W.; Kitayama, C. Histological Features and $G\alpha_{olf}$ Expression Patterns in the Nasal Cavity of Sea Turtles. *J. Anat.* **2023**, *243*, 486–503. [CrossRef] [PubMed]

167. Reiss, J.O.; Eisthen, H.L. Comparative Anatomy and Physiology of Chemical Senses in Amphibians. In *Sensory Evolution on the Threshold: Adaptations in Secondarily Aquatic Vertebrates*; University of California Press: Berkeley, CA, USA, 2008; ISBN 978-0-520-25278-3.

168. Monti-Bloch, L.; Jennings-White, C.; Berliner, D.L. The Human Vomeronasal System: A Review. *Ann. N. Y. Acad. Sci.* **1998**, *855*, 373–389. [CrossRef]

169. Tirindelli, R. Coding of Pheromones by Vomeronasal Receptors. *Cell Tissue Res.* **2021**, *383*, 367–386. [CrossRef]

170. Shinohara, H.; Asano, T.; Kato, K. Differential Localization of G-Proteins Gi and Go in the Accessory Olfactory Bulb of the Rat. *J. Neurosci.* **1992**, *12*, 1275–1279. [CrossRef]

171. Dulac, C.; Axel, R. A Novel Family of Genes Encoding Putative Pheromone Receptors in Mammals. *Cell* **1995**, *83*, 195–206. [CrossRef]

172. Herrada, G.; Dulac, C. A Novel Family of Putative Pheromone Receptors in Mammals with a Topographically Organized and Sexually Dimorphic Distribution. *Cell* **1997**, *90*, 763–773. [CrossRef]

173. Matsunami, H.; Buck, L.B. A Multigene Family Encoding a Diverse Array of Putative Pheromone Receptors in Mammals. *Cell* **1997**, *90*, 775–784. [CrossRef]

174. Ryba, N.J.P.; Tirindelli, R. A New Multigene Family of Putative Pheromone Receptors. *Neuron* **1997**, *19*, 371–379. [CrossRef] [PubMed]

175. Rivière, S.; Challet, L.; Fluegge, D.; Spehr, M.; Rodriguez, I. Formyl Peptide Receptor-like Proteins Are a Novel Family of Vomeronasal Chemosensors. *Nature* **2009**, *459*, 574–577. [CrossRef] [PubMed]

176. Brennan, P.A. The Vomeronasal System. *Cell. Mol. Life Sci.* **2001**, *58*, 546–555. [CrossRef] [PubMed]

177. Papes, F.; Logan, D.W.; Stowers, L. The Vomeronasal Organ Mediates Interspecies Defensive Behaviors through Detection of Protein Pheromone Homologs. *Cell* **2010**, *141*, 692–703. [CrossRef] [PubMed]

178. Halpern, M.; Kubie, J.L. The Role of the Ophidian Vomeronasal System in Species-Typical Behavior. *Trends Neurosci.* **1984**, *7*, 472–477. [CrossRef]

179. Placyk, J.S., Jr.; Graves, B.M. Prey Detection by Vomeronasal Chemoreception in a Plethodontid Salamander. *J. Chem. Ecol.* **2002**, *28*, 1017–1036. [CrossRef]

180. Halpern, M.; Daniels, Y.; Zuri, I. The Role of the Vomeronasal System in Food Preferences of the Gray Short-Tailed Opossum, Monodelphis Domestica. *Nutr. Metab.* **2005**, *2*, 6. [CrossRef]

181. Dagg, A.I.; Taub, A. Flehmen. *Mammalia* **1970**, *34*, 686–695. [CrossRef]

182. Estes, R.D. The Role of the Vomeronasal Organ in Mammalian Reproduction. *Mammalia* **1972**, *36*, 315–341. [CrossRef]

183. Hart, B.L. Flehmen Behavior and Vomeronasal Organ Function. In *Chemical Signals in Vertebrates 3*; Müller-Schwarze, D., Silverstein, R.M., Eds.; Springer: Boston, MA, USA, 1983; pp. 87–103. ISBN 978-1-4757-9652-0.

184. Hart, L.A.; Hart, B.L. Flehmen, Osteophagia, and Other Behaviors of Giraffes (*Giraffa giraffa angolensis*): Vomeronasal Organ Adaptation. *Animals* **2023**, *13*, 354. [CrossRef]

185. Jacobson, L. Anatomisk Beskrivelse over et nyt Organ i Huusdyrenes Naese. *Vet. Selsk. Skr.* **1813**, *2*, 209–246.

186. Ruysch, F. *Thesaurus Anatomicus Tertius*; J. Wolters: Amsterdam, Netherlands, 1703; pp. 48–49.

187. Dursy, E. *Zur Entwicklungsgeschichte des Kopfes des Menschen und der höheren Wirbelthiere: Mit Holzschnitten und einem Atlas von neun Kupfertafeln mit erklärendem Texte*; H. Laupp: Tübingen, Germany, 1869.

188. Kölliker, A. Über Die Jacobson'schen Organe Des Menschen. In *Festschrift zu dem 40 jährign Professoren-Jubiläum des Herrn Franz von Rinecker 31 März 1877*; Wilhelm Engelmann: Leipzig, Germany, 1877; pp. 3–11.

189. Balogh, C. Das Jacobson'sche Organ des Schafes, *Sitzungsberichte Der Kais. Akad. Der Wiss. Wien.* **1860**, *42*, 449–476.

190. Klein, E. Memoirs: Contributions to the Minute Anatomy of the Nasal Mucous Membrane. *J. Cell Sci.* **1881**, *2*, 98–113. [CrossRef]

191. Piana, G. Contribuzioni alla conoscenza della strutture e della funzione dell' organo di jacobson. *Deusch. Zeitsch. f. Tiermedizin* **1882**, *7*, 325.

192. Retzius, G. Die riechzellen der ophidier in der riechschleimhaut und im jacobson'schen organ. *Biol. Untersuch. Neue Folge* **1894**, *6*, 48–51.

193. Gudden, B.V. Experimentaluntersuchungen Über Das Peripherische Und Centrale Nervensystem. *Arch. Psychiatr. Nervenkr.* **1870**, *2*, 693–723. [CrossRef]
194. Ramón y Cajal, S.R. Textura Del Lobulo Olfativo Accesorio. *Rev. Micros.* **1902**, *1*, 141–150.
195. Vandenbergh, J.G. Male Odor Accelerates Female Sexual Maturation in Mice. *Endocrinology* **1969**, *84*, 658–660. [CrossRef]
196. Whitten, W.K. Modification of the Oestrous Cycle of the Mouse by External Stimuli Associated with the Male. *J. Endocrinol.* **1956**, *13*, 399–404. [CrossRef]
197. Powers, J.B.; Winans, S.S. Vomeronasal Organ: Critical Role in Mediating Sexual Behavior of the Male Hamster. *Science* **1975**, *187*, 961–963. [CrossRef]
198. Grus, W.E.; Zhang, J. Origin and Evolution of the Vertebrate Vomeronasal System Viewed through System-Specific Genes. *Bioessays* **2006**, *28*, 709–718. [CrossRef] [PubMed]
199. González, A. Lungfishes, like Tetrapods, Possess a Vomeronasal System. *Front. Neuroanat.* **2010**, *4*, 130. [CrossRef] [PubMed]
200. Nakamuta, S.; Nakamuta, N.; Taniguchi, K.; Taniguchi, K. Histological and Ultrastructural Characteristics of the Primordial Vomeronasal Organ in Lungfish. *Anat. Rec.* **2012**, *295*, 481–491. [CrossRef] [PubMed]
201. Nakamuta, S.; Yamamoto, Y.; Miyazaki, M.; Sakuma, A.; Nikaido, M.; Nakamuta, N. Type 1 Vomeronasal Receptors Expressed in the Olfactory Organs of Two African Lungfish, *Protopterus annectens* and *Protopterus amphibius*. *J. Comp. Neurol.* **2023**, *531*, 116–131. [CrossRef]
202. Wittmer, C.; Nowack, C. Epithelial Crypts: A Complex and Enigmatic Olfactory Organ in African and South American Lungfish (Lepidosireniformes, Dipnoi). *J. Morphol.* **2017**, *278*, 791–800. [CrossRef]
203. Swaney, W.T.; Keverne, E.B. The Evolution of Pheromonal Communication. *Behav. Brain Res.* **2009**, *200*, 239–247. [CrossRef]
204. Pihlström, H. Comparative Anatomy and Physiology of Chemical Senses in Aquatic Mammals. In *Sensory Evolution on the Threshold: Adaptations in Secondarily Aquatic Vertebrates*; Thewissen, J.G.M., Ed.; University of California Press: Berkeley, CA, USA, 2008; ISBN 978-0-520-25278-3.
205. Suzuki, H.; Nishida, H.; Kondo, H.; Yoda, R.; Iwata, T.; Nakayama, K.; Enomoto, T.; Wu, J.; Moriya-Ito, K.; Miyazaki, M.; et al. A Single Pheromone Receptor Gene Conserved across 400 Million Years of Vertebrate Evolution. *Mol. Biol. Evol.* **2018**, *35*, 2928–2939. [CrossRef]
206. Breathnach, A.S. The Cetacean Central Nervous System. *Biol. Rev.* **1960**, *35*, 187–230. [CrossRef]
207. McGowen, M.R.; Clark, C.; Gatesy, J. The Vestigial Olfactory Receptor Subgenome of Odontocete Whales: Phylogenetic Congruence between Gene-Tree Reconciliation and Supermatrix Methods. *Syst. Biol.* **2008**, *57*, 574–590. [CrossRef]
208. Kishida, T. Olfaction of Aquatic Amniotes. *Cell Tissue Res.* **2021**, *383*, 353–365. [CrossRef]
209. Halpern, M. Nasal Chemical Senses in Reptiles: Structure and Function. In *Hormones, Brain, and Behaviour. Biology of the Reptilia*; University of Chicago Press: Chicago, IL, USA, 1992; Volume 18, pp. 423–523.
210. Schwenk, K. Of Tongues and Noses: Chemoreception in Lizards and Snakes. *Trends Ecol. Evol.* **1995**, *10*, 7–12. [CrossRef] [PubMed]
211. Kondoh, D.; Yamamoto, Y.; Nakamuta, N.; Taniguchi, K.; Taniguchi, K. Lectin Histochemical Studies on the Olfactory Epithelium and Vomeronasal Organ in the Japanese Striped Snake, Elaphe Quadrivirgata. *J. Morphol.* **2010**, *271*, 1197–1203. [CrossRef] [PubMed]
212. Kondoh, D.; Yamamoto, Y.; Nakamuta, N.; Taniguchi, K.; Taniguchi, K. Seasonal Changes in the Histochemical Properties of the Olfactory Epithelium and Vomeronasal Organ in the Japanese Striped Snake, Elaphe Quadrivirgata: Seasonal Changes in Snake Olfactory Organs. *Anat. Histol. Embryol.* **2012**, *41*, 41–53. [CrossRef] [PubMed]
213. Hatanaka, T.; Matsuzaki, O. Odor Responses of the Vomeronasal System in Reeve's Turtle, *Geoclemys reevesii*. *Brain Behav. Evol.* **1993**, *41*, 183–186. [CrossRef] [PubMed]
214. Parsons, T.S. Studies on the Comparative Embryology of the Reptilian Nose. *Bull. Mus. Comp. Zool. Harv. Coll.* **1959**, *120*, 261–275.
215. Parsons, T.S. Evolution of the Nasal Structure in the Lower Tetrapods. *Am. Zool.* **1967**, *7*, 397–413. [CrossRef]
216. Houck, L.D. Pheromone Communication in Amphibians and Reptiles. *Annu. Rev. Physiol.* **2009**, *71*, 161–176. [CrossRef]
217. Eisthen, H.L. Presence of the Vomeronasal System in Aquatic Salamanders. *Philos. Trans. R. Soc. Lond. B* **2000**, *355*, 1209–1213. [CrossRef]
218. Silva, L.; Antunes, A. Vomeronasal Receptors in Vertebrates and the Evolution of Pheromone Detection. *Annu. Rev. Anim. Biosci.* **2017**, *5*, 353–370. [CrossRef]
219. Bhatnagar, K.P.; Meisami, E. Vomeronasal Organ in Bats and Primates: Extremes of Structural Variability and Its Phylogenetic Implications. *Microsc. Res. Tech.* **1998**, *43*, 465–475. [CrossRef]
220. Suárez, R.; Fernández-Aburto, P.; Manger, P.R.; Mpodozis, J. Deterioration of the Gαo Vomeronasal Pathway in Sexually Dimorphic Mammals. *PLoS ONE* **2011**, *6*, e26436. [CrossRef]
221. Kondoh, D.; Watanabe, K.; Nishihara, K.; Ono, Y.S.; Nakamura, K.G.; Yuhara, K.; Tomikawa, S.; Sugimoto, M.; Kobayashi, S.; Horiuchi, N.; et al. Histological Properties of Main and Accessory Olfactory Bulbs in the Common Hippopotamus. *Brain Behav. Evol.* **2017**, *90*, 224–231. [CrossRef] [PubMed]
222. Tomiyasu, J.; Korzekwa, A.; Kawai, Y.K.; Robstad, C.A.; Rosell, F.; Kondoh, D. The Vomeronasal System in Semiaquatic Beavers. *J. Anat.* **2022**, *241*, 809–819. [CrossRef] [PubMed]
223. Grus, W.E.; Shi, P.; Zhang, J. Largest Vertebrate Vomeronasal Type 1 Receptor Gene Repertoire in the Semiaquatic Platypus. *Mol. Biol. Evol.* **2007**, *24*, 2153–2157. [CrossRef]

224. Cartmill, M. Rethinking Primate Origins: The Characteristic Primate Traits Cannot Be Explained Simply as Adaptations to Arboreal Life. *Science* **1974**, *184*, 436–443. [CrossRef] [PubMed]
225. Heesy, C.P.; Ross, C.F. Evolution of Activity Patterns and Chromatic Vision in Primates: Morphometrics, Genetics and Cladistics. *J. Hum. Evol.* **2001**, *40*, 111–149. [CrossRef]
226. Wang, G.; Shi, P.; Zhu, Z.; Zhang, Y.-P. More Functional V1R Genes Occur in Nest-Living and Nocturnal Terricolous Mammals. *Genom. Biol. Evol.* **2010**, *2*, 277–283. [CrossRef]
227. Garrett, E.C.; Dennis, J.C.; Bhatnagar, K.P.; Durham, E.L.; Burrows, A.M.; Bonar, C.J.; Steckler, N.K.; Morrison, E.E.; Smith, T.D. The Vomeronasal Complex of Nocturnal Strepsirhines and Implications for the Ancestral Condition in Primates: Vomeronasal Complex of Nocturnal Strepsirhines. *Anat. Rec.* **2013**, *296*, 1881–1894. [CrossRef]
228. Smith, T.D.; Garrett, E.C.; Bhatnagar, K.P.; Bonar, C.J.; Bruening, A.E.; Dennis, J.C.; Kinznger, J.H.; Johnson, E.W.; Morrison, E.E. The Vomeronasal Organ of New World Monkeys (*Platyrrhini*). *Anat. Rec.* **2011**, *294*, 2158–2178. [CrossRef]
229. Smith, T.D.; Siegel, M.I.; Bonar, C.J.; Bhatnagar, K.P.; Mooney, M.P.; Burrows, A.M.; Smith, M.A.; Maico, L.M. The Existence of the Vomeronasal Organ in Postnatal Chimpanzees and Evidence for Its Homology with That of Humans. *J. Anat.* **2001**, *198*, 77–82. [CrossRef]
230. Smith, T.D.; Siegel, M.I.; Bhatnagar, K.P. Reappraisal of the Vomeronasal System of Catarrhine Primates: Ontogeny, Morphology, Functionality, and Persisting Questions. *Anat. Rec.* **2001**, *265*, 176–192. [CrossRef] [PubMed]
231. Boehm, N.; Roos, J.; Gasser, B. Luteinizing Hormone-Releasing Hormone (LHRH)-Expressing Cells in the Nasal Septum of Human Fetuses. *Dev. Brain Res.* **1994**, *82*, 175–180. [CrossRef] [PubMed]
232. Kjær, I.; Hansen, B.F. The Human Vomeronasal Organ: Prenatal Developmental Stages and Distribution of Luteinizing Hormone-Releasing Hormone. *Eur. J. Oral Sci.* **1996**, *104*, 34–40. [CrossRef] [PubMed]
233. Humphrey, T. The Development of the Olfactory and the Accessory Olfactory Formations in Human Embryos and Fetuses. *J. Comp. Neurol.* **1940**, *73*, 431–468. [CrossRef]
234. Moran, D.T.; Jafek, B.W.; Rowley, J.C. The Vomeronasal (Jacobson's) Organ in Man: Ultrastructure and Frequency of Occurence. *J. Steroid Biochem. Mol. Biol.* **1991**, *39*, 545–552. [CrossRef]
235. Hazem, A.G.; Ahmed, A.T.; Tantawy, A.M.; Diaa, M.H.; Hesham, M.S. The Vomeronasal (Jacobson's) Organ in Adult Humans: Frequency of Occurrence and Enzymatic Study. *Acta Otolaryngol.* **1998**, *118*, 409–412. [CrossRef]
236. Witt, M.; Hummel, T. Vomeronasal Versus Olfactory Epithelium: Is There a Cellular Basis for Human Vomeronasal Perception? *Int. Rev. Cytol.* **2006**, *248*, 209–259. [CrossRef]
237. Meredith, M. Human Vomeronasal Organ Function: A Critical Review of Best and Worst Cases. *Chem. Senses* **2001**, *26*, 433–445. [CrossRef]
238. Zhang, J.; Webb, D.M. Evolutionary Deterioration of the Vomeronasal Pheromone Transduction Pathway in Catarrhine Primates. *Proc. Natl. Acad. Sci. USA* **2003**, *100*, 8337–8341. [CrossRef]
239. Smith, T.D.; Laitman, J.T.; Bhatnagar, K.P. The Shrinking Anthropoid Nose, the Human Vomeronasal Organ, and the Language of Anatomical Reduction: The Shrinking Anthropoid Nose. *Anat. Rec.* **2014**, *297*, 2196–2204. [CrossRef]
240. Negus, V.E. The Organ of Jacobson. *J. Anat.* **1956**, *90*, 515–519. [PubMed]
241. Salazar, I.; Sanchez-Quinteiro, P. Supporting Tissue and Vasculature of the Mammalian Vomeronasal Organ: The Rat as a Model. *Microsc. Res. Tech.* **1998**, *41*, 492–505. [CrossRef]
242. Døving, K.B.; Trotier, D. Structure and Function of the Vomeronasal Organ. *J. Exp. Biol.* **1998**, *201*, 2913–2925. [CrossRef] [PubMed]
243. Menco, B.P.M. Ultrastructural Aspects of Olfactory Signaling. *Chem. Senses* **1997**, *22*, 295–311. [CrossRef]
244. Menco, B.P.M.; Carr, V.M.; Ezeh, P.I.; Liman, E.R.; Yankova, M.P. Ultrastructural Localization of G-Proteins and the Channel Protein TRP2 to Microvilli of Rat Vomeronasal Receptor Cells. *J. Comp. Neurol.* **2001**, *438*, 468–489. [CrossRef]
245. Salazar, I.; Sanchez-Quinteiro, P.; Cifuentes, J.M.; De Troconiz, P.F. General Organization of the Perinatal and Adult Accessory Olfactory Bulb in Mice. *Anat. Rec.* **2006**, *288*, 1009–1025. [CrossRef]
246. Jungblut, L.D.; Reiss, J.O.; Pozzi, A.G. Olfactory Subsystems in the Peripheral Olfactory Organ of Anuran Amphibians. *Cell Tissue Res.* **2021**, *383*, 289–299. [CrossRef]
247. Mendoza, A.S.; Kühnel, W. Morphological Evidence for a Direct Innervation of the Mouse Vomeronasal Glands. *Cell Tissue Res.* **1987**, *247*, 457–459. [CrossRef]
248. Takami, S.; Getchell, M.L.; Getchell, T.V. Resolution of Sensory and Mucoid Glycoconjugates with Terminal α-Galactose Residues in the Mucomicrovillar Complex of the Vomeronasal Sensory Epithelium by Dual Confocal Laser Scanning Microscopy. *Cell Tissue Res.* **1995**, *280*, 211–216. [CrossRef]
249. Kondoh, D.; Tomiyasu, J.; Itakura, R.; Sugahara, M.; Yanagawa, M.; Watanabe, K.; Alviola, P.A.; Yap, S.A.; Cosico, E.A.; Cruz, F.A.; et al. Comparative Histological Studies on Properties of Polysaccharides Secreted by Vomeronasal Glands of Eight Laurasiatheria Species. *Acta Histochem.* **2020**, *122*, 151515. [CrossRef]
250. Salazar, I.; Sánchez Quinteiro, P.; Cifuentes, J.M.; Fernández, P.; Lombardero, M. Distribution of the Arterial Supply to the Vomeronasal Organ in the Cat. *Anat. Rec.* **1997**, *247*, 129–136. [CrossRef]
251. Meredith, M.; O'Connell, R.J. Efferent Control of Stimulus Access to the Hamster Vomeronasal Organ. *J. Physiol.* **1979**, *286*, 301–316. [CrossRef] [PubMed]
252. Eccles, R. Autonomic Innervation of the Vomeronasal Organ of the Cat. *Physiol. Behav.* **1982**, *28*, 1011–1015. [CrossRef] [PubMed]

253. Meredith, M. Chronic Recording of Vomeronasal Pump Activation in Awake Behaving Hamsters. *Physiol. Behav.* **1994**, *56*, 345–354. [CrossRef]
254. Halpern, M.; Shapiro, L.S.; Jia, C. Heterogeneity in the Accessory Olfactory System. *Chem. Senses* **1998**, *23*, 477–481. [CrossRef]
255. Weiß, E.; Kretschmer, D. Formyl-Peptide Receptors in Infection, Inflammation, and Cancer. *Trends Immunol.* **2018**, *39*, 815–829. [CrossRef]
256. Villamayor, P.R.; Robledo, D.; Fernández, C.; Gullón, J.; Quintela, L.; Sánchez-Quinteiro, P.; Martínez, P. Analysis of the Vomeronasal Organ Transcriptome Reveals Variable Gene Expression Depending on Age and Function in Rabbits. *Genomics* **2021**, *113*, 2240–2252. [CrossRef]
257. Liberles, S.D.; Horowitz, L.F.; Kuang, D.; Contos, J.J.; Wilson, K.L.; Siltberg-Liberles, J.; Liberles, D.A.; Buck, L.B. Formyl Peptide Receptors Are Candidate Chemosensory Receptors in the Vomeronasal Organ. *Proc. Natl. Acad. Sci. USA* **2009**, *106*, 9842–9847. [CrossRef]
258. Bufe, B.; Teuchert, Y.; Schmid, A.; Pyrski, M.; Pérez-Gómez, A.; Eisenbeis, J.; Timm, T.; Ishii, T.; Lochnit, G.; Bischoff, M.; et al. Bacterial MgrB Peptide Activates Chemoreceptor Fpr3 in Mouse Accessory Olfactory System and Drives Avoidance Behaviour. *Nat. Commun.* **2019**, *10*, 4889. [CrossRef]
259. Swain, S.L. T Cell Subsets and the Recognition of MHC Class. *Immunol. Rev.* **1983**, *74*, 129–142. [CrossRef]
260. Singh, P. Chemosensation and Genetic Individuality. *Reproduction* **2001**, *121*, 529–539. [CrossRef] [PubMed]
261. Brennan, P.A.; Kendrick, K.M. Mammalian Social Odours: Attraction and Individual Recognition. *Philos. Trans. R. Soc. B* **2006**, *361*, 2061–2078. [CrossRef] [PubMed]
262. Ruff, J.S.; Nelson, A.C.; Kubinak, J.L.; Potts, W.K. MHC Signaling during Social Communication. In *Self and Nonself*; Advances in Experimental Medicine and Biology; López-Larrea, C., Ed.; Springer: New York, NY, USA, 2012; Volume 738, pp. 290–313. ISBN 978-1-4614-1679-1.
263. Yamazaki, K.; Beauchamp, G.K. Genetic Basis for MHC-Dependent Mate Choice. In *Advances in Genetics*; Elsevier: Amsterdam, The Netherlands, 2007; Volume 59, pp. 129–145. ISBN 978-0-12-017660-1.
264. Leinders-Zufall, T.; Brennan, P.; Widmayer, P.; Shivalingappa, P.C.; Maul-Pavicic, A.; Jäger, M.; Li, X.-H.; Breer, H.; Zufall, F.; Boehm, T. MHC Class I Peptides as Chemosensory Signals in the Vomeronasal Organ. *Science* **2004**, *306*, 1033–1037. [CrossRef] [PubMed]
265. Konjević, D.; Erman, V.; Bujanić, M.; Svetličić, I.; Arbanasić, H.; Lubura Strunjak, S.; Galov, A. Wild Boar (*Sus scrofa*)—Fascioloides Magna Interaction from the Perspective of the MHC Genes. *Pathogens* **2022**, *11*, 1359. [CrossRef]
266. Milinski, M.; Griffiths, S.; Wegner, K.M.; Reusch, T.B.H.; Haas-Assenbaum, A.; Boehm, T. Mate Choice Decisions of Stickleback Females Predictably Modified by MHC Peptide Ligands. *Proc. Natl. Acad. Sci. USA* **2005**, *102*, 4414–4418. [CrossRef]
267. Leinders-Zufall, T.; Lane, A.P.; Puche, A.C.; Ma, W.; Novotny, M.V.; Shipley, M.T.; Zufall, F. Ultrasensitive Pheromone Detection by Mammalian Vomeronasal Neurons. *Nature* **2000**, *405*, 792–796. [CrossRef]
268. Spehr, M.; Kelliher, K.R.; Li, X.-H.; Boehm, T.; Leinders-Zufall, T.; Zufall, F. Essential Role of the Main Olfactory System in Social Recognition of Major Histocompatibility Complex Peptide Ligands. *J. Neurosci.* **2006**, *26*, 1961–1970. [CrossRef]
269. Liman, E.R.; Corey, D.P.; Dulac, C. TRP2: A Candidate Transduction Channel for Mammalian Pheromone Sensory Signaling. *Proc. Natl. Acad. Sci. USA* **1999**, *96*, 5791–5796. [CrossRef]
270. Leypold, B.G.; Yu, C.R.; Leinders-Zufall, T.; Kim, M.M.; Zufall, F.; Axel, R. Altered Sexual and Social Behaviors in Trp2 Mutant Mice. *Proc. Natl. Acad. Sci. USA* **2002**, *99*, 6376–6381. [CrossRef]
271. Stowers, L.; Holy, T.E.; Meister, M.; Dulac, C.; Koentges, G. Loss of Sex Discrimination and Male-Male Aggression in Mice Deficient for TRP2. *Science* **2002**, *295*, 1493–1500. [CrossRef]
272. Zhang, G.; Li, C.; Li, Q.; Li, B.; Larkin, D.M.; Lee, C.; Storz, J.F.; Antunes, A.; Greenwold, M.J.; Meredith, R.W.; et al. Comparative Genomics Reveals Insights into Avian Genome Evolution and Adaptation. *Science* **2014**, *346*, 1311–1320. [CrossRef] [PubMed]
273. Shi, P.; Zhang, J. Comparative Genomic Analysis Identifies an Evolutionary Shift of Vomeronasal Receptor Gene Repertoires in the Vertebrate Transition from Water to Land. *Genome Res.* **2007**, *17*, 166–174. [CrossRef] [PubMed]
274. Young, J.M.; Massa, H.F.; Hsu, L.; Trask, B.J. Extreme Variability among Mammalian V1R Gene Families. *Genome Res.* **2010**, *20*, 10–18. [CrossRef] [PubMed]
275. Rodriguez, I.; Greer, C.A.; Mok, M.Y.; Mombaerts, P. A Putative Pheromone Receptor Gene Expressed in Human Olfactory Mucosa. *Nat. Genet.* **2000**, *26*, 18–19. [CrossRef] [PubMed]
276. Baum, M.J.; Kelliher, K.R. Complementary Roles of the Main and Accessory Olfactory Systems in Mammalian Mate Recognition. *Annu. Rev. Physiol.* **2009**, *71*, 141–160. [CrossRef] [PubMed]
277. Omura, M.; Mombaerts, P. Trpc2-Expressing Sensory Neurons in the Main Olfactory Epithelium of the Mouse. *Cell Rep.* **2014**, *8*, 583–595. [CrossRef]
278. Omura, M.; Mombaerts, P. Trpc2-Expressing Sensory Neurons in the Mouse Main Olfactory Epithelium of Type B Express the Soluble Guanylate Cyclase Gucy1b2. *Mol. Cell. Neurosci.* **2015**, *65*, 114–124. [CrossRef]
279. Schneider, N.Y.; Fletcher, T.P.; Shaw, G.; Renfree, M.B. Goα Expression in the Vomeronasal Organ and Olfactory Bulb of the Tammar Wallaby. *Chem. Senses* **2012**, *37*, 567–577. [CrossRef]
280. Suárez, R.; Mpodozis, J. Heterogeneities of Size and Sexual Dimorphism between the Subdomains of the Lateral-Innervated Accessory Olfactory Bulb (AOB) of *Octodon degus* (Rodentia: Hystricognathi). *Behav. Brain Res.* **2009**, *198*, 306–312. [CrossRef]

281. Halpern, M.; Shapiro, L.S.; Jia, C. Differential Localization of G Proteins in the Opossum Vomeronasal System. *Brain Res.* **1995**, *677*, 157–161. [CrossRef]
282. Young, J.M.; Trask, B.J. V2R Gene Families Degenerated in Primates, Dog and Cow, but Expanded in Opossum. *Trends Genet.* **2007**, *23*, 212–215. [CrossRef] [PubMed]
283. Salazar, I.; Cifuentes, J.M.; Sánchez-Quinteiro, P. Morphological and Immunohistochemical Features of the Vomeronasal System in Dogs. *Anat. Rec.* **2013**, *296*, 146–155. [CrossRef] [PubMed]
284. Moriya-Ito, K.; Hayakawa, T.; Suzuki, H.; Hagino-Yamagishi, K.; Nikaido, M. Evolution of Vomeronasal Receptor 1 (V1R) Genes in the Common Marmoset (*Callithrix jacchus*). *Gene* **2018**, *642*, 343–353. [CrossRef]
285. Kishimoto, J.; Keverne, E.B.; Emson, P.C. Calretinin, Calbindin-D28k and Parvalbumin-like Immunoreactivity in Mouse Chemoreceptor Neurons. *Brain Res.* **1993**, *610*, 325–329. [CrossRef] [PubMed]
286. Verhaagen, J.; Oestreicher, A.; Gispen, W.; Margolis, F. The Expression of the Growth Associated Protein B50/GAP43 in the Olfactory System of Neonatal and Adult Rats. *J. Neurosci.* **1989**, *9*, 683–691. [CrossRef]
287. Salazar, I.; Sánchez Quinteiro, P. Lectin Binding Patterns in the Vomeronasal Organ and Accessory Olfactory Bulb of the Rat. *Anat. Embryol.* **1998**, *198*, 331–339. [CrossRef]
288. Ichikawa, M.; Osada, T.; Ikai, A. Bandeiraea Simplicifolia Lectin I and Vicia Villosa Agglutinin Bind Specifically to the Vomeronasal Axons in the Accessory Olfactory Bulb of the Rat. *Neurosci. Res.* **1992**, *13*, 73–79. [CrossRef]
289. Ibarra-Soria, X.; Levitin, M.O.; Saraiva, L.R.; Logan, D.W. The Olfactory Transcriptomes of Mice. *PLoS Genet.* **2014**, *10*, e1004593. [CrossRef]
290. Francia, S.; Silvotti, L.; Ghirardi, F.; Catzeflis, F.; Percudani, R.; Tirindelli, R. Evolution of Spatially Coexpressed Families of Type-2 Vomeronasal Receptors in Rodents. *Genom. Biol. Evol.* **2015**, *7*, 272–285. [CrossRef]
291. Yohe, L.R.; Davies, K.T.J.; Rossiter, S.J.; Dávalos, L.M. Expressed Vomeronasal Type-1 Receptors (V1rs) in Bats Uncover Conserved Sequences Underlying Social Chemical Signaling. *Genom. Biol. Evol.* **2019**, *11*, 2741–2749. [CrossRef]
292. Haga, S.; Hattori, T.; Sato, T.; Sato, K.; Matsuda, S.; Kobayakawa, R.; Sakano, H.; Yoshihara, Y.; Kikusui, T.; Touhara, K. The Male Mouse Pheromone ESP1 Enhances Female Sexual Receptive Behaviour through a Specific Vomeronasal Receptor. *Nature* **2010**, *466*, 118–122. [CrossRef] [PubMed]
293. Silva, L.; Mendes, T.; Antunes, A. Acquisition of Social Behavior in Mammalian Lineages Is Related with Duplication Events of FPR Genes. *Genomics* **2020**, *112*, 2778–2783. [CrossRef] [PubMed]
294. Oboti, L.; Ibarra-Soria, X.; Pérez-Gómez, A.; Schmid, A.; Pyrski, M.; Paschek, N.; Kircher, S.; Logan, D.W.; Leinders-Zufall, T.; Zufall, F.; et al. Pregnancy and Estrogen Enhance Neural Progenitor-Cell Proliferation in the Vomeronasal Sensory Epithelium. *BMC Biol.* **2015**, *13*, 104. [CrossRef]
295. Villamayor, P.R.; Gullón, J.; Quintela, L.; Sánchez-Quinteiro, P.; Martínez, P.; Robledo, D. Sex Separation Unveils the Functional Plasticity of the Vomeronasal Organ in Rabbits. *Front. Mol. Neurosci.* **2022**, *15*, 1034254. [CrossRef]
296. Segovia, S.; Guillamón, A. Effects of Sex Steroids on the Development of the Vomeronasal Organ in the Rat. *Dev. Brain Res.* **1982**, *5*, 209–212. [CrossRef] [PubMed]
297. Guillamón, A.; Segovia, S. Sex Differences in the Vomeronasal System. *Brain Res. Bull.* **1997**, *44*, 377–382. [CrossRef]
298. Maico, L.M.; Burrows, A.M.; Mooney, M.P.; Siegel, M.I.; Bhatnagar, K.P.; Smith, T.D. Size of the Vomeronasal Organ in Wild Microtus with Different Mating Strategies. *Acta Biol. Hung.* **2003**, *54*, 263–274. [CrossRef]
299. Tai, F.D.; Wang, T.Z.; Zhang, Y.H.; Sun, R.Y. Sexual Dimorphism of the Vomeronasal Organ and the Accessory Olfactory Bulb of the Mandarin Vole *Microtus mandarinus* and the Reed Vole *M. fortis*. *Acta Theriol.* **2004**, *49*, 33–42. [CrossRef]
300. Peretto, P.; Giachino, C.; Panzica, G.; Fasolo, A. Sexually Dimorphic Neurogenesis Is Topographically Matched with the Anterior Accessory Olfactory Bulb of the Adult Rat. *Cell Tissue Res.* **2001**, *306*, 385–389. [CrossRef]
301. Broom, R. A Contribution to the Comparative Anatomy of the Mammalian Organ of Jacobson. *Trans. R. Soc. Edinb.* **1900**, *39*, 231–255. [CrossRef]
302. Wöhrmann-Repenning, A. Phylogenetic aspects of the Jacobson's organ and nasopalatine duct topography in insectivores, primates, Tupaia and Didelphis. *Anat. Anz.* **1984**, *157*, 137–149. [PubMed]
303. Sánchez-Villagra, M.R. Ontogenetic and Phylogenetic Transformations of the Vomeronasal Complex and Nasal Floor Elements in Marsupial Mammals. *Zool. J. Linn. Soc.* **2001**, *131*, 459–479. [CrossRef]
304. Vaccarezza, O.L.; Sepich, L.N.; Tramezzani, J.H. The Vomeronasal Organ of the Rat. *J. Anat.* **1981**, *132*, 167–185. [PubMed]
305. Mechin, V.; Pageat, P.; Teruel, E.; Asproni, P. Histological and Immunohistochemical Characterization of Vomeronasal Organ Aging in Mice. *Animals* **2021**, *11*, 1211. [CrossRef] [PubMed]
306. Weiler, E.; McCulloch, M.A.; Farbman, A.I. Proliferation in the Vomeronasal Organ of the Rat during Postnatal Development. *Eur. J. Neurosci.* **1999**, *11*, 700–711. [CrossRef] [PubMed]
307. Giacobini, P.; Benedetto, A.; Tirindelli, R.; Fasolo, A. Proliferation and Migration of Receptor Neurons in the Vomeronasal Organ of the Adult Mouse. *Dev. Brain Res.* **2000**, *123*, 33–40. [CrossRef]
308. Ibokwe, C.; Okpe, G.C. Morphological Studies Of Vomeronasal Organ In The Wild Juvenile Red-Flanked Duiker Cephalophus Rufilatus (GRAY 1864). *Anim. Res. Int.* **2009**, *6*, 932–937. [CrossRef]
309. Wang, H.; Wang, J.; Yang, C.; He, Y. Histological Structure of the Vomeronasal Organ and Accessory Olfactory Bulb and the Seasonal Changes of Olfactory Bulb C-Fos Expression in *Spermophilus dauricus*. *Acta Theriol. Sin.* **2021**, *41*, 685. [CrossRef]

310. Oikawa, T.; Shimamura, K.; Saito, T.R.; Taniguchi, K. Fine Structure of the Vomeronasal Organ in the Chinchilla (*Chinchilla laniger*). *Exp. Anim.* **1994**, *43*, 487–497. [CrossRef]

311. Jurcisek, J.A.; Durbin, J.E.; Kusewitt, D.F.; Bakaletz, L.O. Anatomy of the Nasal Cavity in the Chinchilla. *Cells Tissues Organs* **2003**, *174*, 136–152. [CrossRef]

312. Smith, T.D.; Alport, L.J.; Burrows, A.M.; Bhatnagar, K.P.; Dennis, J.C.; Tuladhar, P.; Morrison, E.E. Perinatal Size and Maturation of the Olfactory and Vomeronasal Neuroepithelia in Lorisoids and Lemuroids. *Am. J. Primatol.* **2007**, *69*, 74–85. [CrossRef] [PubMed]

313. Dennis, J.C.; Stilwell, N.K.; Smith, T.D.; Park, T.J.; Bhatnagar, K.P.; Morrison, E.E. Is the Mole Rat Vomeronasal Organ Functional? *Anat. Rec.* **2020**, *303*, 318–329. [CrossRef] [PubMed]

314. Luckhaus, G. Light and electron microscopic findings in the epithelial lamina of the vomeronasal organ of the rabbit. *Anat. Anz.* **1969**, *124*, 477–489. [PubMed]

315. Mahdy, E.; El Behery, E.; Mohamed, S. Comparative Morpho-Histological Analysis on the Vomeronasal Organ and the Accessory Olfactory Bulb in Balady Dogs (*Canis familiaris*) and New Zealand Rabbits (*Oryctolagus cuniculus*). *J. Adv. Vet. Res.* **2019**, *6*, 506. [CrossRef] [PubMed]

316. Elgayar, S.A.M.; Eltony, S.A.; Othman, M.A. Morphology of Non-Sensory Epithelium during Post-Natal Development of the Rabbit Vomeronasal Organ. *Anat. Histol. Embryol.* **2014**, *43*, 282–293. [CrossRef]

317. Schneider, N.Y.; Fletcher, T.P.; Shaw, G.; Renfree, M.B. The Vomeronasal Organ of the Tammar Wallaby. *J. Anat.* **2008**, *213*, 93–105. [CrossRef]

318. Poran, N.S. Vomeronasal Organ and Its Associated Structures in the opossum *Monodelphis domestica*. *Microsc. Res. Tech.* **1998**, *43*, 500–510. [CrossRef]

319. Schneider, N.Y. The Development of the Olfactory Organs in Newly Hatched Monotremes and Neonate Marsupials: Olfaction in Monotremes and Marsupials. *J. Anat.* **2011**, *219*, 229–242. [CrossRef]

320. Ashwell, K.W.S. Development of the Olfactory Pathways in Platypus and Echidna. *Brain Behav. Evol.* **2012**, *79*, 45–56. [CrossRef]

321. Kondoh, D.; Tanaka, Y.; Kawai, Y.K.; Mineshige, T.; Watanabe, K.; Kobayashi, Y. Morphological and Histological Features of the Vomeronasal Organ in African Pygmy Hedgehog (*Atelerix albiventris*). *Animals* **2021**, *11*, 1462. [CrossRef]

322. Aland, R.C.; Gosden, E.; Bradley, A.J. Seasonal Morphometry of the Vomeronasal Organ in the Marsupial Mouse, *Antechinus subtropicus*. *J. Morphol.* **2016**, *277*, 1517–1530. [CrossRef] [PubMed]

323. Dennis, J.C.; Allgier, J.G.; Desouza, L.S.; Eward, W.C.; Morrison, E.E. Immunohistochemistry of the Canine Vomeronasal Organ. *J. Anat.* **2003**, *203*, 329–338. [CrossRef] [PubMed]

324. Salazar, I.; Sanchez Quinteiro, P.; Cifuentes, J.M.; Garcia Caballero, T. The Vomeronasal Organ of the Cat. *J. Anat.* **1996**, *188 Pt 2*, 445–454.

325. Tomiyasu, J.; Kondoh, D.; Sakamoto, H.; Matsumoto, N.; Sasaki, M.; Kitamura, N.; Haneda, S.; Matsui, M. Morphological and Histological Features of the Vomeronasal Organ in the Brown Bear. *J. Anat.* **2017**, *231*, 749–757. [CrossRef] [PubMed]

326. Kelliher, K.R.; Baum, M.J.; Meredith, M. The Ferret's Vomeronasal Organ and Accessory Olfactory Bulb: Effect of Hormone Manipulation in Adult Males and Females. *Anat. Rec.* **2001**, *263*, 280–288. [CrossRef] [PubMed]

327. Salazar, I.; Lombardero, M.; Sánchez-Quinteiro, P.; Roel, P.; Cifuentes, J.M. Origin and Regional Distribution of the Arterial Vessels of the Vomeronasal Organ in the Sheep. A Methodological Investigation with Scanning Electron Microscopy and Cutting-Grinding Technique. *Ann. Anat.* **1998**, *180*, 181–187. [CrossRef]

328. Salazar, I.; Quinteiro, P.S.; Cifuentes, J.M. The Soft-Tissue Components of the Vomeronasal Organ in Pigs, Cows and Horses. *Anat. Histol. Embryol.* **1997**, *26*, 179–186. [CrossRef]

329. Salazar, I.; Quinteiro, P.S.; Alemañ, N.; Cifuentes, J.M.; Troconiz, P.F. Diversity of the Vomeronasal System in Mammals: The Singularities of the Sheep Model. *Microsc. Res. Tech.* **2007**, *70*, 752–762. [CrossRef]

330. Ichikawa, M.; Shin, T.; Kang, M.S. Fine Structure of the Vomeronasal Sensory Epithelium of Korean Goats (Capra Hircus). *J. Rep. Dev.* **1999**, *45*, 81–89. [CrossRef]

331. Yang, W.; Choi, Y.; Park, C.; Lee, K.-H.; Ahn, M.; Kang, W.; Heo, S.-D.; Kim, J.; Shin, T. Histological and Lectin Histochemical Studies in the Vomeronasal Organ of the Korean Black Goat, *Capra hircus coreanae*. *Acta Histochem.* **2021**, *123*, 151684. [CrossRef]

332. Park, C.; Ahn, M.; Lee, J.-Y.; Lee, S.; Yun, Y.; Lim, Y.-K.; Taniguchi, K.; Shin, T. A Morphological Study of the Vomeronasal Organ and the Accessory Olfactory Bulb in the Korean Roe Deer, *Capreolus pygargus*. *Acta Histochem.* **2014**, *116*, 258–264. [CrossRef] [PubMed]

333. Vedin, V.; Eriksson, B.; Berghard, A. Organization of the Chemosensory Neuroepithelium of the Vomeronasal Organ of the Scandinavian Moose *Alces alces*. *Brain Res.* **2010**, *1306*, 53–61. [CrossRef] [PubMed]

334. Kondoh, D.; Nakamura, K.G.; Ono, Y.S.; Yuhara, K.; Bando, G.; Watanabe, K.; Horiuchi, N.; Kobayashi, Y.; Sasaki, M.; Kitamura, N. Histological Features of the Vomeronasal Organ in the Giraffe, *Giraffa camelopardalis*. *Microsc. Res. Tech.* **2017**, *80*, 652–656. [CrossRef] [PubMed]

335. Smith, T.D.; Bhatnagar, K.P.; Shimp, K.L.; Kinzinger, J.H.; Bonar, C.J.; Burrows, A.M.; Mooney, M.P.; Siegel, M.I. Histological Definition of the Vomeronasal Organ in Humans and Chimpanzees, with a Comparison to Other Primates. *Anat. Rec.* **2002**, *267*, 166–176. [CrossRef] [PubMed]

336. Meisami, E.; Bhatnagar, K.P. Structure and Diversity in Mammalian Accessory Olfactory Bulb. *Microsc. Res. Tech.* **1998**, *43*, 476–499. [CrossRef]

337. Takami, S.; Graziadei, P.P.C. Morphological Complexity of the Glomerulus in the Rat Accessory Olfactory Bulb—A Golgi Study. *Brain Res.* **1990**, *510*, 339–342. [CrossRef]

338. Nakajima, M.; Tsuruta, M.; Mori, H.; Nishikawa, C.; Okuyama, S.; Furukawa, Y. A Comparative Study of Axon-Surrounding Cells in the Two Nasal Nerve Tracts from Mouse Olfactory Epithelium and Vomeronasal Organ. *Brain Res.* **2013**, *1503*, 16–23. [CrossRef]

339. Vassar, R.; Chao, S.K.; Sitcheran, R.; Nuñez, J.M.; Vosshall, L.B.; Axel, R. Topographic Organization of Sensory Projections to the Olfactory Bulb. *Cell* **1994**, *79*, 981–991. [CrossRef]

340. Del Punta, K.; Puche, A.; Adams, N.C.; Rodriguez, I.; Mombaerts, P. A Divergent Pattern of Sensory Axonal Projections Is Rendered Convergent by Second-Order Neurons in the Accessory Olfactory Bulb. *Neuron* **2002**, *35*, 1057–1066. [CrossRef]

341. Crosby, E.C.; Humphrey, T. Studies of the Vertebrate Telencephalon. II. The Nuclear Pattern of the Anterior Olfactory Nucleus, Tuberculum Olfactorium and the Amygdaloid Complex in Adult Man. *J. Comp. Neurol.* **1941**, *74*, 309–352. [CrossRef]

342. Allison, A.C. The Structure of the Olfactory Bulb and Its Relationship to the Olfactory Pathways in the Rabbit and the Rat. *J. Comp. Neurol.* **1953**, *98*, 309–353. [CrossRef] [PubMed]

343. Salazar, I.; Sánchez-Quinteiro, P. A Detailed Morphological Study of the Vomeronasal Organ and the Accessory Olfactory Bulb of Cats. *Microsc. Res. Tech.* **2011**, *74*, 1109–1120. [CrossRef] [PubMed]

344. Larriva-Sahd, J. The Accessory Olfactory Bulb in the Adult Rat: A Cytological Study of Its Cell Types, Neuropil, Neuronal Modules, and Interactions with the Main Olfactory System. *J. Comp. Neurol.* **2008**, *510*, 309–350. [CrossRef] [PubMed]

345. Switzer, R.C., III; Johnson, J.I.; Kirsch, J.A.W. Phylogeny Through Brain Traits. *Brain Behav. Evol.* **1980**, *17*, 339–363. [CrossRef] [PubMed]

346. Baker, H.; Towle, A.C.; Margolis, F.L. Differential Afferent Regulation of Dopaminergic and GABAergic Neurons in the Mouse Main Olfactory Bulb. *Brain Res.* **1988**, *450*, 69–80. [CrossRef]

347. Gouda, M.; Matsutani, S.; Senba, E.; Tohyama, M. Peptidergic Granule Cell Populations in the Rat Main and Accessory Olfactory Bulb. *Brain Res.* **1990**, *512*, 339–342. [CrossRef]

348. McLean, J.H.; Shipley, M.T. Neuroanatomical Substrates of Olfaction. In *Science of Olfaction*; Serby, M.J., Chobor, K.L., Eds.; Springer: New York, NY, USA, 1992; pp. 126–171. ISBN 978-1-4612-7690-6.

349. Frahm, H.D.; Stephan, H.; Baron, G. Comparison of Brain Structure Volumes in Insectivora and Primates. V. Area Striata (AS). *J. Hirnforsch.* **1984**, *25*, 537–557.

350. Bonfanti, L.; Peretto, P.; Merighi, A.; Fasolo, A. Newly-Generated Cells from the Rostral Migratory Stream in the Accessory Olfactory Bulb of the Adult Rat. *Neuroscience* **1997**, *81*, 489–502. [CrossRef]

351. Defteralı, Ç.; Moreno-Estellés, M.; Crespo, C.; Díaz-Guerra, E.; Díaz-Moreno, M.; Vergaño-Vera, E.; Nieto-Estévez, V.; Hurtado-Chong, A.; Consiglio, A.; Mira, H.; et al. Neural Stem Cells in the Adult Olfactory Bulb Core Generate Mature Neurons In Vivo. *Stem Cells* **2021**, *39*, 1253–1269. [CrossRef]

352. Inaki, K.; Nishimura, S.; Nakashiba, T.; Itohara, S.; Yoshihara, Y. Laminar Organization of the Developing Lateral Olfactory Tract Revealed by Differential Expression of Cell Recognition Molecules. *J. Comp. Neurol.* **2004**, *479*, 243–256. [CrossRef]

353. Broadwell, R.D. Olfactory Relationships of the Telencephlaon and Diencephalon in the Rabbit. 1. An Autoradiographic Study of the Efferent Connections of the Main and Accessory Olfactory Bulbs. *J. Comp. Neurol.* **1975**, *163*, 329–345. [CrossRef] [PubMed]

354. Schwob, J.E.; Price, J.L. The Development of Axonal Connections in the Central Olfactory System of Rats. *J. Comp. Neurol.* **1984**, *223*, 177–202. [CrossRef] [PubMed]

355. Takami, S.; Fernandez, G.D.; Graziadei, P.P.C. The Morphology of GABA-Immunoreactive Neurons in the Accessory Olfactory Bulb of Rats. *Brain Res.* **1992**, *588*, 317–323. [CrossRef] [PubMed]

356. Takami, S.; Graziadei, P.P.C. Light Microscopic Golgi Study of Mitral/Tufted Cells in the Accessory Olfactory Bulb of the Adult Rat. *J. Comp. Neurol.* **1991**, *311*, 65–83. [CrossRef]

357. Hinds, J.W. Autoradiographic Study of Histogenesis in the Mouse Olfactory Bulb I. Time of Origin of Neurons and Neuroglia. *J. Comp. Neurol.* **1968**, *134*, 287–304. [CrossRef]

358. Rall, W.; Shepherd, G.M.; Reese, T.S.; Brightman, M.W. Dendrodendritic Synaptic Pathway for Inhibition in the Olfactory Bulb. *Exp. Neurol.* **1966**, *14*, 44–56. [CrossRef]

359. Martín-López, E.; Corona, R.; López-Mascaraque, L. Postnatal Characterization of Cells in the Accessory Olfactory Bulb of Wild Type and Reeler Mice. *Front. Neuroanat.* **2012**, *6*, 15. [CrossRef]

360. Jia, C.; Chen, W.R.; Shepherd, G.M. Synaptic Organization and Neurotransmitters in the Rat Accessory Olfactory Bulb. *J. Neurophysiol.* **1999**, *81*, 345–355. [CrossRef]

361. Del Punta, K. Sequence Diversity and Genomic Organization of Vomeronasal Receptor Genes in the Mouse. *Genom. Res.* **2000**, *10*, 1958–1967. [CrossRef]

362. Segovia, S.; Orensanz, L.M.; Valencia, A.; Guillamón, A. Effects of Sex Steroids on the Development of the Accessory Olfactory Bulb in the Rat: A Volumetric Study. *Dev. Brain Res.* **1984**, *16*, 312–314. [CrossRef]

363. Yokosuka, M. Histological Properties of the Glomerular Layer in the Mouse Accessory Olfactory Bulb. *Exp. Anim.* **2012**, *61*, 13–24. [CrossRef] [PubMed]

364. Frahm, H.D.; Bhatnagar, K.P. Comparative Morphology of the Accessory Olfactory Bulb in Bats. *J. Anat.* **1980**, *130*, 349–365. [PubMed]

365. Jia, C.; Halpern, M. Calbindin D28k, Parvalbumin, and Calretinin Immunoreactivity in the Main and Accessory Olfactory Bulbs of the Gray Short-Tailed Opossum, *Monodelphis domestica*. *J. Morphol.* **2004**, *259*, 271–280. [CrossRef] [PubMed]

366. Ortiz-Leal, I.; Torres, M.V.; Villamayor, P.R.; Fidalgo, L.E.; López-Beceiro, A.; Sanchez-Quinteiro, P. Can Domestication Shape Canidae Brain Morphology? The Accessory Olfactory Bulb of the Red Fox as a Case in Point. *Ann. Anat.* **2022**, *240*, 151881. [CrossRef]

367. Bird, D.J.; Jacquemetton, C.; Buelow, S.A.; Evans, A.W.; Van Valkenburgh, B. Domesticating Olfaction: Dog Breeds, Including Scent Hounds, Have Reduced Cribriform Plate Morphology Relative to Wolves. *Anat. Rec.* **2021**, *304*, 139–153. [CrossRef]

368. Bird, D.J.; Murphy, W.J.; Fox-Rosales, L.; Hamid, I.; Eagle, R.A.; Van Valkenburgh, B. Olfaction Written in Bone: Cribriform Plate Size Parallels Olfactory Receptor Gene Repertoires in Mammalia. *Proc. R. Soc. B* **2018**, *285*, 20180100. [CrossRef]

369. Wynn, E.H.; Sánchez-Andrade, G.; Carss, K.J.; Logan, D.W. Genomic Variation in the Vomeronasal Receptor Gene Repertoires of Inbred Mice. *BMC Genom.* **2012**, *13*, 415. [CrossRef]

370. Ortiz-Leal, I.; Torres, M.V.; López-Callejo, L.N.; Fidalgo, L.E.; López-Beceiro, A.; Sanchez-Quinteiro, P. Comparative Neuroanatomical Study of the Main Olfactory Bulb in Domestic and Wild Canids: Dog, Wolf and Red Fox. *Animals* **2022**, *12*, 1079. [CrossRef]

371. Dzięcioł, M.; Podgórski, P.; Stańczyk, E.; Szumny, A.; Woszczyło, M.; Pieczewska, B.; Niżański, W.; Nicpoń, J.; Wrzosek, M.A. MRI Features of the Vomeronasal Organ in Dogs (*Canis familiaris*). *Front. Vet. Sci.* **2020**, *7*, 159. [CrossRef]

372. Muñiz-de Miguel, S.; Barreiro-Vázquez, J.D.; Sánchez-Quinteiro, P.; Ortiz-Leal, I.; González-Martínez, Á. Behavioural Disorder in a Dog with Congenital Agenesis of the Vomeronasal Organ and the Septum Pellucidum. *Vet. Rec. Case Rep.* **2023**, *11*, e571. [CrossRef]

373. Pageat, P.; Gaultier, E. Current Research in Canine and Feline Pheromones. *Vet. Clin. N. Am. Small Anim. Pract.* **2003**, *33*, 187–211. [CrossRef] [PubMed]

374. Puglisi, I.; Masucci, M.; Cozzi, A.; Teruel, E.; Navarra, M.; Cirmi, S.; Pennisi, M.G.; Siracusa, C. Effects of a Novel Gel Formulation of Dog Appeasing Pheromone (DAP) on Behavioral and Physiological Stress Responses in Dogs Undergoing Clinical Examination. *Animals* **2022**, *12*, 2472. [CrossRef] [PubMed]

375. Asproni, P.; Cozzi, A.; Verin, R.; Lafont-Lecuelle, C.; Bienboire-Frosini, C.; Poli, A.; Pageat, P. Pathology and Behaviour in Feline Medicine: Investigating the Link between Vomeronasalitis and Aggression. *J. Feline Med. Surg.* **2016**, *18*, 997–1002. [CrossRef] [PubMed]

376. Asproni, P.; Mainau, E.; Cozzi, A.; Carreras, R.; Bienboire-Frosini, C.; Teruel, E.; Pageat, P. Is There a Link between Vomeronasalitis and Aggression in Stable Social Groups of Female Pigs? *Animals* **2022**, *12*, 303. [CrossRef]

377. Mechin, V.; Asproni, P.; Bienboire-Frosini, C.; Cozzi, A.; Chabaud, C.; Arroub, S.; Mainau, E.; Nagnan-Le Meillour, P.; Pageat, P. Inflammation Interferes with Chemoreception in Pigs by Altering the Neuronal Layout of the Vomeronasal Sensory Epithelium. *Front. Vet. Sci.* **2022**, *9*, 936838. [CrossRef]

378. Mechin, V.; Pageat, P.; Boutry, M.; Teruel, E.; Portalier, C.; Asproni, P. Does the Environmental Air Impact the Condition of the Vomeronasal Organ? A Mouse Model for Intensive Farming. *Animals* **2023**, *13*, 1902. [CrossRef]

379. Salazar, I.; Cifuentes, J.M.; Quinteiro, P.S.; Caballero, G. The Vomeronasal System of the Mink, Mustela Vison. I. The Vomeronasal Organ. *Funct. Dev. Morphol.* **1994**, *4*, 113–117.

380. Salazar, I.; Quinteiro, P.S.; Cifuentes, J.M.; Lombardero, M. The Accessory Olfactory Bulb of the Mink, Mustela Vison: A Morphological and Lectin Histochemical Study. *Anat. Histol. Embryol.* **1998**, *27*, 297–300. [CrossRef]

381. Salazar, I.; Lombardero, M.; Alemañ, N.; Sánchez Quinteiro, P. Development of the Vomeronasal Receptor Epithelium and the Accessory Olfactory Bulb in Sheep. *Microsc. Res. Tech.* **2003**, *61*, 438–447. [CrossRef]

382. Tirindelli, R.; Mucignat-Caretta, C.; Ryba, N.J.P. Molecular Aspects of Pheromonal Communication via the Vomeronasal Organ of Mammals. *Trends Neurosci.* **1998**, *21*, 482–486. [CrossRef]

383. Alonso, J.R.; Briñón, J.G.; Crespo, C.; Bravo, I.G.; Arévalo, R.; Aijón, J. Chemical Organization of the Macaque Monkey Olfactory Bulb: II. Calretinin, Calbindin D-28k, Parvalbumin, and Neurocalcin Immunoreactivity: Calcium-Binding Proteins in the Monkey OB. *J. Comp. Neurol.* **2001**, *432*, 389–407. [CrossRef] [PubMed]

384. Briñón, J.G.; Weruaga, E.; Crespo, C.; Porteros, A.; Arévalo, R.; Aijón, J.; Alonso, J.R. Calretinin-, Neurocalcin-, and Parvalbumin-Immunoreactive Elements in the Olfactory Bulb of the Hedgehog (*Erinaceus europaeus*). *J. Comp. Neurol.* **2001**, *429*, 554–570. [CrossRef] [PubMed]

385. Weruaga, E. A Sexually Dimorphic Group of Atypical Glomeruli in the Mouse Olfactory Bulb. *Chem. Senses* **2001**, *26*, 7–15. [CrossRef]

386. Smithson, L.J.; Kawaja, M.D. A Comparative Examination of Biomarkers for Olfactory Ensheathing Cells in Cats and Guinea Pigs. *Brain Res.* **2009**, *1284*, 41–53. [CrossRef] [PubMed]

387. Dehmelt, L.; Halpain, S. The MAP2/Tau Family of Microtubule-Associated Proteins. *Genome Biol.* **2005**, *6*, 204. [CrossRef]

388. Benowitz, L.I.; Perrone-Bizzozero, N.I.; Neve, R.L.; Rodriguez, W. Chapter 26 GAP-43 as a Marker for Structural Plasticity in the Mature CNS. In *Progress in Brain Research*; Elsevier: Amsterdam, The Netherlands, 1990; Volume 86, pp. 309–320. ISBN 978-0-444-81121-9.

389. Shapiro, L.S.; Roland, R.M.; Halpern, M. Development of Olfactory Marker Protein and N-CAM Expression in Chemosensory Systems of the Opossum, *Monodelphis domestica*. *J. Morphol.* **1997**, *234*, 109–129. [CrossRef]

390. Mugnaini, E.; Oertel, W.H.; Wouterlood, F.F. Immunocytochemical Localization of GABA Neurons and Dopamine Neurons in the Rat Main and Accessory Olfactory Bulbs. *Neurosci. Lett.* **1984**, *47*, 221–226. [CrossRef]

391. Le Jeune, H.; Aubert, I.; Jourdan, F.; Quirion, R. Comparative Laminar Distribution of Various Autoradiographic Cholinergic Markers in Adult Rat Main Olfactory Bulb. *J. Chem. Neuroanat.* **1995**, *9*, 99–112. [CrossRef]

392. Woo, C.C.; Leon, M. Distribution and Development of β-Adrenergic Receptors in the Rat Olfactory Bulb. *J. Comp. Neurol.* **1995**, *352*, 1–10. [CrossRef]

393. Huang, Z.; Thiebaud, N.; Fadool, D.A. Differential Serotonergic Modulation across the Main and Accessory Olfactory Bulbs: Differential Serotonin Modulation in Olfactory System. *J. Physiol.* **2017**, *595*, 3515–3533. [CrossRef]

394. Matsutani, S.; Senba, E.; Tohyama, M. Neuropeptide- and Neurotransmitter-Related Immunoreactivities in the Developing Rat Olfactory Bulb. *J. Comp. Neurol.* **1988**, *272*, 331–342. [CrossRef] [PubMed]

395. Lundberg, L.-M.; Alm, P.; Wharton, J.; Polak, J.M. Protein Gene Product 9.5 (PGP 9.5): A New Neuronal Marker Visualizing the Whole Uterine Innervation and Pregnancy-Induced and Developmental Changes in the *Guinea pig*. *Histochemistry* **1988**, *90*, 9–17. [CrossRef] [PubMed]

396. Nacher, J.; Crespo, C.; McEwen, B.S. Doublecortin Expression in the Adult Rat Telencephalon: Doublecortin Expression in the Adult Rat. *Eur. J. Neurosci.* **2001**, *14*, 629–644. [CrossRef] [PubMed]

397. Takami, S.; Graziadei, P.P.C.; Ichikawa, M. The Differential Staining Patterns of Two Lectins in the Accessory Olfactory Bulb of the Rat. *Brain Res.* **1992**, *598*, 337–342. [CrossRef] [PubMed]

398. Taniguchi, K.; Nii, Y.; Ogawa, K. Subdivisions of the Accessory Olfactory Bulb, as Demonstrated by Lectin-Histochemistry in the Golden Hamster. *Neurosci. Lett.* **1993**, *158*, 185–188. [CrossRef]

399. Shapiro, L.S.; Halpern, M.; Ee, P.-L. Lectin Histochemical Identification of Carbohydrate Moieties in Opossum Chemosensory Systems during Development, with Special Emphasis on VVA-Identified Subdivisions in the Accessory Olfactory Bulb. *J. Morphol.* **1995**, *224*, 331–349. [CrossRef]

400. Hovis, K.R.; Ramnath, R.; Dahlen, J.E.; Romanova, A.L.; LaRocca, G.; Bier, M.E.; Urban, N.N. Activity Regulates Functional Connectivity from the Vomeronasal Organ to the Accessory Olfactory Bulb. *J. Neurosci.* **2012**, *32*, 7907–7916. [CrossRef]

401. Raisman, G. An Experimental Study of the Projection of the Amygdala to the Accessory Olfactory Bulb and Its Relationship to the Concept of a Dual Olfactory System. *Exp. Brain Res.* **1972**, *14*, 395–408. [CrossRef]

402. Scalia, F.; Winans, S.S. The Differential Projections of the Olfactory Bulb and Accessory Olfactory Bulb in Mammals. *J. Comp. Neurol.* **1975**, *161*, 31–55. [CrossRef]

403. Winans, S.S.; Scalia, F. Amygdaloid Nucleus: New Afferent Input from the Vomeronasal Organ. *Science* **1970**, *170*, 330–332. [CrossRef]

404. Shipley, M.T.; Adamek, G.D. The Connections of the Mouse Olfactory Bulb: A Study Using Orthograde and Retrograde Transport of Wheat Germ Agglutinin Conjugated to Horseradish Peroxidase. *Brain Res. Bull.* **1984**, *12*, 669–688. [CrossRef] [PubMed]

405. Dulac, C.; Wagner, S. Genetic Analysis of Brain Circuits Underlying Pheromone Signaling. *Annu. Rev. Genet.* **2006**, *40*, 449–467. [CrossRef] [PubMed]

406. Price, J.L. Beyond the Primary Olfactory Cortex: Olfactory-Related Areas in the Neocortex, Thalamus and Hypothalamus. *Chem. Senses* **1985**, *10*, 239–258. [CrossRef]

407. Courtiol, E.; Wilson, D.A. The Olfactory Thalamus: Unanswered Questions about the Role of the Mediodorsal Thalamic Nucleus in Olfaction. *Front. Neural Circuits* **2015**, *9*, 49. [CrossRef]

408. Cádiz-Moretti, B.; Otero-García, M.; Martínez-García, F.; Lanuza, E. Afferent Projections to the Different Medial Amygdala Subdivisions: A Retrograde Tracing Study in the Mouse. *Brain Struct. Funct.* **2016**, *221*, 1033–1065. [CrossRef]

409. Devor, M. Fiber Trajectories of Olfactory Bulb Efferents in the Hamster. *J. Comp. Neurol.* **1976**, *166*, 31–47. [CrossRef]

410. Martinez-Marcos, A. On the Organization of Olfactory and Vomeronasal Cortices. *Prog. Neurobiol.* **2009**, *87*, 21–30. [CrossRef]

411. Gutiérrez-Castellanos, N.; Pardo-Bellver, C.; Martínez-García, F.; Lanuza, E. The Vomeronasal Cortex—Afferent and Efferent Projections of the Posteromedial Cortical Nucleus of the Amygdala in Mice. *Eur. J. Neurosci.* **2014**, *39*, 141–158. [CrossRef]

412. Stowers, L.; Liberles, S.D. State-Dependent Responses to Sex Pheromones in Mouse. *Curr. Opin. Neurobiol.* **2016**, *38*, 74–79. [CrossRef]

413. Pardo-Bellver, C.; Cádiz-Moretti, B.; Novejarque, A.; Martínez-García, F.; Lanuza, E. Differential Efferent Projections of the Anterior, Posteroventral, and Posterodorsal Subdivisions of the Medial Amygdala in Mice. *Front. Neuroanat.* **2012**, *6*, 33. [CrossRef]

414. Lo, L.; Anderson, D.J. A Cre-Dependent, Anterograde Transsynaptic Viral Tracer for Mapping Output Pathways of Genetically Marked Neurons. *Neuron* **2011**, *72*, 938–950. [CrossRef] [PubMed]

415. Trotier, D. Vomeronasal Organ and Human Pheromones. *Eur. Ann. Otorhinolaryngol. Head Neck Dis.* **2011**, *128*, 184–190. [CrossRef] [PubMed]

416. Cadiz, B.; Martinez-Garcia, F.; Lanuza, E. Neural Substrate to Associate Odorants and Pheromones: Convergence of Projections from the Main and Accessory Olfactory Bulbs in Mice. In *Chemical Signals in Vertebrates 12*; Springer: New York, NY, USA, 2013; pp. 3–16. ISBN 978-1-4614-5926-2.

417. Mohrhardt, J.; Nagel, M.; Fleck, D.; Ben-Shaul, Y.; Spehr, M. Signal Detection and Coding in the Accessory Olfactory System. *Chem. Senses* **2018**, *43*, 667–695. [CrossRef] [PubMed]

418. Záborszky, L.; Carlsen, J.; Brashear, H.R.; Heimer, L. Cholinergic and GABAergic Afferents to the Olfactory Bulb in the Rat with Special Emphasis on the Projection Neurons in the Nucleus of the Horizontal Limb of the Diagonal Band. *J. Comp. Neurol.* **1986**, *243*, 488–509. [CrossRef]
419. De Saint Jan, D. Target-Specific Control of Olfactory Bulb Periglomerular Cells by GABAergic and Cholinergic Basal Forebrain Inputs. *eLife* **2022**, *11*, e71965. [CrossRef]
420. Liberia, T.; Blasco Ibáñez, J.M.; Nájcher, J.; Varea, E.; Lanciego, J.L.; Crespo, C. Synaptic Connectivity of the Cholinergic Axons in the Olfactory Bulb of the Cynomolgus Monkey. *Front. Neuroanat.* **2015**, *9*, 28. [CrossRef]
421. Böhm, E.; Brunert, D.; Rothermel, M. Input Dependent Modulation of Olfactory Bulb Activity by HDB GABAergic Projections. *Sci. Rep.* **2020**, *10*, 10696. [CrossRef]
422. Gomez, D.M.; Newman, S.W. Differential Projections of the Anterior and Posterior Regions of the Medial Amygdaloid Nucleus in the Syrian Hamster. *J. Comp. Neurol.* **1992**, *317*, 195–218. [CrossRef]
423. Fan, S.; Luo, M. The Organization of Feedback Projections in a Pathway Important for Processing Pheromonal Signals. *Neuroscience* **2009**, *161*, 489–500. [CrossRef]
424. Inbar, T.; Davis, R.; Bergan, J.F. A Sex-specific Feedback Projection from Aromatase-expressing Neurons in the Medial Amygdala to the Accessory Olfactory Bulb. *J. Comp. Neurol.* **2022**, *530*, 648–655. [CrossRef]
425. Pardo-Bellver, C.; Vila-Martin, M.E.; Martínez-Bellver, S.; Villafranca-Faus, M.; Teruel-Sanchis, A.; Savarelli-Balsamo, C.A.; Drabik, S.M.; Martínez-Ricós, J.; Cervera-Ferri, A.; Martínez-García, F.; et al. Neural Activity Patterns in the Chemosensory Network Encoding Vomeronasal and Olfactory Information in Mice. *Front. Neuroanat.* **2022**, *16*, 988015. [CrossRef] [PubMed]
426. Keverne, E.B.; Brennan, P.A. Olfactory Recognition Memory. *J. Physiol.* **1996**, *90*, 399–401. [CrossRef] [PubMed]
427. Brennan, P.A.; Kendrick, K.M.; Keverne, E.B. Neurotransmitter Release in the Accessory Olfactory Bulb during and after the Formation of an Olfactory Memory in Mice. *Neuroscience* **1995**, *69*, 1075–1086. [CrossRef] [PubMed]
428. Brennan, P.A. The Nose Knows Who's Who: Chemosensory Individuality and Mate Recognition in Mice. *Horm. Behav.* **2004**, *46*, 231–240. [CrossRef] [PubMed]
429. Takahashi, Y.; Kaba, H. Muscarinic Receptor Type 1 (M1) Stimulation, Probably through KCNQ/Kv7 Channel Closure, Increases Spontaneous GABA Release at the Dendrodendritic Synapse in the Mouse Accessory Olfactory Bulb. *Brain Res.* **2010**, *1339*, 26–40. [CrossRef] [PubMed]
430. Wilson, R.I.; Mainen, Z.F. Early Events in Olfactory Processing. *Annu. Rev. Neurosci.* **2006**, *29*, 163–201. [CrossRef]
431. Grus, W.E.; Zhang, J. Origin of the Genetic Components of the Vomeronasal System in the Common Ancestor of All Extant Vertebrates. *Mol. Biol. Evol.* **2009**, *26*, 407–419. [CrossRef]
432. Smith, J.; Hurst, J.L.; Barnard, C.J. Comparing Behaviour in Wild and Laboratory Strains of the House Mouse: Levels of Comparison and Functional Inference. *Behav. Process.* **1994**, *32*, 79–86. [CrossRef]
433. Catania, K.C. Underwater "sniffing" by Semi-Aquatic Mammals. *Nature* **2006**, *444*, 1024–1025. [CrossRef]
434. Johnson, E.W.; Rasmussen, L. Morphological Characteristics of the Vomeronasal Organ of the Newborn Asian Elephant (*Elephas maximus*). *Anat. Rec.* **2002**, *267*, 252–259. [CrossRef]
435. Garrosa, M.; Gayoso, M.J.; Esteban, F.J. Prenatal Development of the Mammalian Vomeronasal Organ. *Microsc. Res. Tech.* **1998**, *41*, 456–470. [CrossRef]
436. Sugai, T.; Sugitani, M.; Onoda, N. Subdivisions of the Guinea-Pig Accessory Olfactory Bulb Revealed by the Combined Method with Immunohistochemistry, Electrophysiological, and Optical Recordings. *Neuroscience* **1997**, *79*, 871–885. [CrossRef] [PubMed]
437. Fieni, F. Apical and Basal Neurones Isolated from the Mouse Vomeronasal Organ Differ for Voltage-Dependent Currents. *J. Physiol.* **2003**, *552*, 425–436. [CrossRef] [PubMed]
438. Torres, M.V.; Ortiz-Leal, I.; Villamayor, P.R.; Ferreiro, A.; Rois, J.L.; Sanchez-Quinteiro, P. The Vomeronasal System of the Newborn Capybara: A Morphological and Immunohistochemical Study. *Sci. Rep.* **2020**, *10*, 13304. [CrossRef]
439. Salazar, I.; Lombardero, M.; Cifuentes, J.M.; Quinteiro, P.S.; Aleman, N. Morphogenesis and Growth of the Soft Tissue and Cartilage of the Vomeronasal Organ in Pigs. *J. Anat.* **2003**, *202*, 503–514. [CrossRef]
440. Salazar, I.; Sánchez Quinteiro, P.; Lombardero, M.; Aleman, N.; Fernández de Trocóniz, P. The Prenatal Maturity of the Accessory Olfactory Bulb in Pigs. *Chem. Senses* **2004**, *29*, 3–11. [CrossRef]
441. Park, J.; Lee, W.; Jeong, C.; Kim, H.; Taniguchi, K.; Shin, T. Developmental Changes Affecting Lectin Binding in the Vomeronasal Organ of Domestic Pigs, *Sus scrofa. Acta Histochem.* **2012**, *114*, 24–30. [CrossRef]
442. Suárez, R.; Santibáñez, R.; Parra, D.; Coppi, A.A.; Abrahão, L.M.B.; Sasahara, T.H.C.; Mpodozis, J. Shared and Differential Traits in the Accessory Olfactory Bulb of Caviomorph Rodents with Particular Reference to the Semiaquatic Capybara: The AOB of Capybaras and Other Caviomorphs. *J. Anat.* **2011**, *218*, 558–565. [CrossRef]
443. Fernández-Aburto, P.; Delgado, S.E.; Sobrero, R.; Mpodozis, J. Can Social Behaviour Drive Accessory Olfactory Bulb Asymmetries? Sister Species of Caviomorph Rodents as a Case in Point. *J. Anat.* **2020**, *236*, 612–621. [CrossRef]
444. Malz, C.R.; Knabe, W.; Kuhn, H.-J. Calretinin Immunoreactivity in the Prenatally Developing Olfactory Systems of the Tree Shrew *Tupaia belangeri. Anat. Embryol.* **2002**, *205*, 83–97. [CrossRef]
445. Brunjes, P.C.; Jazaeri, A.; Sutherland, M.J. Olfactory Bulb Organization and Development in *Monodelphis domestica* (Grey Short-Tailed Opossum). *J. Comp. Neurol.* **1992**, *320*, 544–554. [CrossRef] [PubMed]
446. Salazar, I.; Sanchez-Quinteiro, P.; Lombardero, M.; Cifuentes, J.M. Histochemical Identification of Carbohydrate Moieties in the Accessory Olfactory Bulb of the Mouse Using a Panel of Lectins. *Chem. Senses* **2001**, *26*, 645–652. [CrossRef] [PubMed]

447. Bojsen-Møller, F. Topography of the Nasal Glands in Rats and Some Other Mammals. *Anat. Rec.* **1964**, *150*, 11–24. [CrossRef]

448. Pastor, L.M.; Frutos, M.J.; Graña, L.; Ramos, D.; Gallego-Huidobro, J.; Calvo, A. Histochemical Study of Glycoconjugates in the Nasal Mucosa of the Rat and *Guinea pig. Histochem. J.* **1992**, *24*, 727–736. [CrossRef] [PubMed]

449. Roslinski, D.L.; Bhatnagar, K.P.; Burrows, A.M.; Smith, T.D. Comparative Morphology and Histochemistry of Glands Associated with the Vomeronasal Organ in Humans, Mouse Lemurs, and Voles. *Anat. Rec.* **2000**, *260*, 92–101. [CrossRef]

450. Schwanzel-Fukuda, M.; Pfaff, D.W. Origin of Luteinizing Hormone-Releasing Hormone Neurons. *Nature* **1989**, *338*, 161–164. [CrossRef]

451. Schwanzel-Fukuda, M.; Pfaff, D.W. The Migration of Luteinizing Hormone-Releasing Hormone (LHRH) Neurons from the Medial Alfactory Placode into the Medial Basal Forebrain. *Experientia* **1990**, *46*, 956–962. [CrossRef]

452. Schwanzel-Fukuda, M.; Pfaff, D.W. Migration of LHRH-Immunoreactive Neurons from the Olfactory Placode Rationalizes Olfacto-Hormonal Relationships. *J. Steroid Biochem. Mol. Biol.* **1991**, *39*, 565–572. [CrossRef]

453. Schwanzel-Fukuda, M.; Pfaff, D.W. Luteinizing Hormone-Releasing Hormone (LHRH) and Neural Cell Adhesion Molecule (NCAM)-Immunoreactivity in Development of the Forebrain and Reproductive System. *Ann. Endocrinol.* **1994**, *55*, 235–241.

454. Skeen, L.C.; Hall, W.C. Efferent Projections of the Main and the Accessory Olfactory Bulb in the Tree Shrew (*Tupaia glis*). *J. Comp. Neurol.* **1977**, *172*, 1–35. [CrossRef]

455. Stephan, H. Der Bulbus Olfatorius Accessorius Bei Insektivoren Und Primaten. *Cells Tissues Organs* **1965**, *62*, 215–253. [CrossRef]

456. Schneider, N.Y.; Fletcher, T.P.; Shaw, G.; Renfree, M.B. The Olfactory System of the Tammar Wallaby Is Developed at Birth and Directs the Neonate to Its Mother's Pouch Odours. *Reproduction* **2009**, *138*, 849–857. [CrossRef] [PubMed]

457. Wettschureck, N.; Offermanns, S. Mammalian G Proteins and Their Cell Type Specific Functions. *Physiol. Rev.* **2005**, *85*, 1159–1204. [CrossRef] [PubMed]

458. Kondoh, D.; Kawai, Y.K.; Watanabe, K.; Muranishi, Y. Artiodactyl Livestock Species Have a Uniform Vomeronasal System with a Vomeronasal Type 1 Receptor (V1R) Pathway. *Tissue Cell* **2022**, *77*, 101863. [CrossRef] [PubMed]

459. Torres, M.V.; Ortiz-Leal, I.; Villamayor, P.R.; Ferreiro, A.; Rois, J.L.; Sanchez-Quinteiro, P. Does a Third Intermediate Model for the Vomeronasal Processing of Information Exist? Insights from the Macropodid Neuroanatomy. *Brain. Struct. Funct.* **2022**, *227*, 881–899. [CrossRef]

460. Kondoh, D.; Kamikawa, A.; Sasaki, M.; Kitamura, N. Localization of A1-2 Fucose Glycan in the Mouse Olfactory Pathway. *Cells Tissues Organs* **2017**, *203*, 20–28. [CrossRef]

461. Keller, L.-A.; Niedermeier, S.; Claassen, L.; Popp, A. Comparative Lectin Histochemistry on the Murine Respiratory Tract and Primary Olfactory Pathway Using a Fully Automated Staining Procedure. *Acta Histochem.* **2022**, *124*, 151877. [CrossRef]

462. Salazar, I.; Sanchez-Quinteiro, P.; Lombardero, M.; Cifuentes, J.M. A Descriptive and Comparative Lectin Histochemical Study of the Vomeronasal System in Pigs and Sheep. *J. Anat.* **2000**, *196*, 15–22. [CrossRef]

463. Mogi, K.; Sakurai, K.; Ichimaru, T.; Ohkura, S.; Mori, Y.; Okamura, H. Structure and Chemical Organization of the Accessory Olfactory Bulb in the Goat. *Anat. Rec.* **2007**, *290*, 301–310. [CrossRef]

464. Sweet, G. Memoirs: Contributions to Our Knowledge of the Anatomy of Notoryctes Typhlops, Stirling. *J. Cell Sci.* **1906**, *2*, 547–572. [CrossRef]

465. Broom, R. On the Organ of Jacobson and Some Other Structures in the Nose of *Cænolestes*. By R. BROOM, D.Sc., P.R.S., C.M.Z.S. *Proc. Zool. Soc. Lond.* **1926**, *96*, 419–424. [CrossRef]

466. Bock, P.; Rohn, K.; Beineke, A.; Baumgärtner, W.; Wewetzer, K. Site-Specific Population Dynamics and Variable Olfactory Marker Protein Expression in the Postnatal Canine Olfactory Epithelium. *J. Anat.* **2009**, *215*, 522–535. [CrossRef] [PubMed]

467. Salazar, I.; Cifuentes, J.M.; Quinteiro, P.S.; Caballero, T.G. Structural, Morphometric, and Immunohistological Study of the Accessory Olfactory Bulb in the Dog. *Anat. Rec.* **1994**, *240*, 277–285. [CrossRef] [PubMed]

468. Torres, M.V.; Ortiz-Leal, I.; Ferreiro, A.; Rois, J.L.; Sanchez-Quinteiro, P. Neuroanatomical and Immunohistological Study of the Main and Accessory Olfactory Bulbs of the Meerkat (*Suricata suricatta*). *Animals* **2021**, *12*, 91. [CrossRef]

469. O'Riain, M.J.; Bennett, N.C.; Brotherton, P.N.M.; McIlrath, G.; Clutton-Brock, T.H. Reproductive Suppression and Inbreeding Avoidance in Wild Populations of Co-Operatively Breeding Meerkats (*Suricata suricatta*). *Behav. Ecol. Sociobiol.* **2000**, *48*, 471–477. [CrossRef]

470. Leclaire, S.; Nielsen, J.F.; Thavarajah, N.K.; Manser, M.; Clutton-Brock, T.H. Odour-Based Kin Discrimination in the Cooperatively Breeding Meerkat. *Biol. Lett.* **2013**, *9*, 20121054. [CrossRef]

471. Leclaire, S.; Jacob, S.; Greene, L.K.; Dubay, G.R.; Drea, C.M. Social Odours Covary with Bacterial Community in the Anal Secretions of Wild Meerkats. *Sci. Rep.* **2017**, *7*, 3240. [CrossRef]

472. Grus, W.E.; Zhang, J. Rapid Turnover and Species-Specificity of Vomeronasal Pheromone Receptor Genes in Mice and Rats. *Gene* **2004**, *340*, 303–312. [CrossRef]

473. Salazar, I.; Sánchez-Quinteiro, P. The Risk of Extrapolation in Neuroanatomy: The Case of the Mammalian Accessory Olfactory Bulb. *Front. Neuroanat.* **2009**, *3*, 22. [CrossRef]

474. Nakajima, T.; Sakaue, M.; Kato, M.; Saito, S.; Ogawa, K.; Taniguchi, K. Immunohistochemical and Enzyme-Histochemical Study on the Accessory Olfactory Bulb of the Dog. *Anat. Rec.* **1998**, *252*, 393–402. [CrossRef]

475. Hughes, L. Meerkats: Essential Wildlife; Character-19: 2020. Available online: https://www.amazon.com/Meerkats-Essential-Wildlife-Lisa-Hughes-ebook/dp/B08GY42Z76 (accessed on 13 October 2023).

476. Broom, R. On the Organ of Jacobson in the Hyrax. *J. Anat. Physiol.* **1898**, *32*, 709–713. [PubMed]

477. Yohe, L.R.; Krell, N.T. An Updated Synthesis of and Outstanding Questions in the Olfactory and Vomeronasal Systems in Bats: Genetics Asks Questions Only Anatomy Can Answer. *Anat. Rec.* **2023**, *306*, 2765–2780. [CrossRef] [PubMed]
478. Wöhrmann-Repenning, A. Functional Aspects of the Vomeronasal Complex in Mammals. *Zool. Jb. Anat.* **1991**, *121*, 71–80.
479. Shnayder, L.; Schwanzel-Fukuda, M.; Halpern, M. Differential OMP Expression in Opossum Accessory Olfactory Bulb. *NeuroReport* **1993**, *5*, 193–196. [CrossRef]
480. Kondoh, D.; Sasaki, M.; Kitamura, N. Age-Dependent Decrease in Glomeruli and Receptor Cells Containing A1–2 Fucose Glycan in the Mouse Main Olfactory System but Not in the Vomeronasal System. *Cell Tissue Res.* **2018**, *373*, 361–366. [CrossRef]
481. Tiwari, R.P.; Ahmed, A.; Mishra, G.K. Role of Pheromones and Biostimulation in Animal Reproduction—An Overview. *J. Vet. Sci. Tech.* **2014**, *3*, 15–20.
482. Rekwot, P.I.; Ogwu, D.; Oyedipe, E.O.; Sekoni, V.O. The Role of Pheromones and Biostimulation in Animal Reproduction. *Anim. Reprod. Sci.* **2001**, *65*, 157–170. [CrossRef]
483. Landaeta-Hernández, A.J.; Ungerfeld, R.; Chenoweth, P.J. Biostimulation and Pheromones in Livestock: A Review. *Anim. Reprod. Sci.* **2023**, *248*, 107154. [CrossRef]
484. Torres, M.V.; Ortiz-Leal, I.; Ferreiro, A.; Rois, J.L.; Sanchez-Quinteiro, P. Immunohistological Study of the Unexplored Vomeronasal Organ of an Endangered Mammal, the Dama Gazelle (*Nanger dama*). *Microsc. Res. Tech.* **2023**, *86*, 1206–1233. [CrossRef]
485. Hart, B.L.; Hart, L.A.; Maina, J.N. Alteration in Vomeronasal System Anatomy in Alcelaphine Antelopes: Correlation with Alteration in Chemosensory Investigation. *Physiol. Behav.* **1988**, *42*, 155–162. [CrossRef]
486. Meredith, M.; Marques, D.M.; O'Connell, R.J.; Stern, F.L. Vomeronasal Pump: Significance for Male Hamster Sexual Behavior. *Science* **1980**, *207*, 1224–1226. [CrossRef] [PubMed]
487. Krishna, N.S.R.; Getchell, M.L.; Margolis, F.L.; Getchell, T.V. Differential Expression of Vomeromodulin and Odorant-Binding Protein, Putative Pheromone and Odorant Transporters, in the Developing Rat Nasal Chemosensory Mucosae. *J. Neurosci. Res.* **1995**, *40*, 54–71. [CrossRef] [PubMed]
488. Takigami, S.; Mori, Y.; Ichikawa, M. Projection Pattern of Vomeronasal Neurons to the Accessory Olfactory Bulb in Goats. *Chem. Senses* **2000**, *25*, 387–393. [CrossRef] [PubMed]
489. Matsubara, K.; Akaogi, S.; Nakamuta, S.; Tsujimoto, T.; Nakamuta, N. Characteristics of Olfactory Organs in Sika Deer (*Cervus nippon*). *Jpn. J. Zoo Wildl. Med.* **2019**, *24*, 115–122. [CrossRef]

Review

Anatomist and Co-Founder of Polish Veterinary Education—Ludwik Henryk Bojanus (1776–1827)

Jarosław Sobolewski [1] and Maciej Zdun [2,*]

[1] Department of Public Health Protection and Animal Welfare, Institute of Veterinary Medicine, Faculty of Biological and Veterinary Sciences, Nicolaus Copernicus University, ul. Lwowska 1, 87-100 Torun, Poland; jsobolewski@umk.pl

[2] Department of Basic and Preclinical Sciences, Institute of Veterinary Medicine, Faculty of Biological and Veterinary Sciences, Nicolaus Copernicus University, ul. Lwowska 1, 87-100 Torun, Poland

* Correspondence: maciejzdun@umk.pl

Abstract: Ludwig Henry Bojanus was born on 16 July 1776 in Buchsweiler, Alsace. After studying in Jena and Vienna, L. H. Bojanus enrolled at the University of Jena for his doctoral studies. Bojanus's scientific activities are closely associated with Vilnius, where he was a professor of veterinary medicine from 1806 (he was elected to this position in 1804). In 1815, he became a professor of comparative anatomy. These were the times of the greatest flourishing of Vilnius University, where the foundations of modern Polish science were being laid. At Vilnius University, he established a technical and anatomical-pathological office for the zoo, a veterinary clinic and a model forge for shoeing horses in 1823. Bojanus founded a veterinary school in Vilnius and drew up a plan for a veterinary institute, which was not opened until 1832, simultaneously with the opening of the medico-surgical academy. He became known as one of Europe's most prominent anatomists and zoologists. A lasting memorial to the scientist is the monograph "Anatomy of the Tortoise", which many scholars still point to today as a model of accurate and precise anatomical research. He was the first to identify the anatomical differences between the European bison (*Bos bonasus*) and the aurochs (*Bos primigenius*). In his lectures on comparative anatomy, Bojanus presented the principle of uninterrupted development. He can be described as one of the most decisive and consistent evolutionists before Darwin. He died in 1827.

Keywords: Louis Bojanus; history of anatomy; history of veterinary medicine; origins of veterinary education

Citation: Sobolewski, J.; Zdun, M. Anatomist and Co-Founder of Polish Veterinary Education—Ludwik Henryk Bojanus (1776–1827). *Anatomia* **2023**, 2, 261–270. https://doi.org/10.3390/anatomia2030024

Academic Editors: Gianfranco Natale and Francesco Fornai

Received: 27 July 2023
Revised: 17 August 2023
Accepted: 31 August 2023
Published: 4 September 2023

1. Introduction

Ludwik Henryk Bojanus (Figure 1) is known first and foremost as the founder of the Polish veterinary school; only secondly is he spoken of as an outstanding anatomist and representative of medical science. In this work, we would like to present the figure of Prof. Bojanus concerning his anatomical and organisational achievements related to the creation of the veterinary school in Vilnius. We also must remember the contribution of our hero to the education of his successors, who developed the veterinary education system in the Polish lands. The end of the 18th century and the beginning of the 19th century were a period of revival of the sciences in Poland. Thanks to the efforts of the Commission of National Education, established in 1773, the Polish education system was reformed, and two academies were reorganised and named Main Schools. The Crown Main School was established in Kraków, while the Main School of the Grand Duchy of Lithuania was established in Vilnius. As Poland did not have sufficient scientists at the time, importing scholars from abroad became necessary. Among them was Ludwik Henryk Bojanus. His name became permanently associated with the history of anatomy, veterinary science and zoology in Poland, particularly with the history of the Imperial University of Vilnius (the name given to the Main School of Vilnius in 1803).

Figure 1. Ludwik H. Bojanus (lithograph by Elichwald from the first half of the 19th century. Archive of the Polish Academy of Sciences in Warsaw).

2. Materials and Methods

In the preparation of this article, heuristic methods and a search of source materials collected in the State Archive in Warsaw, the Lithuanian Central State Archive in Vilnius, the Veterinary History Room at the Voivodeship Veterinary Inspectorate in Bydgoszcz and the Veterinary Museum at the Ks. Krzysztof Kluk Museum of Agriculture in Ciechanowiec were used. While writing the thesis, scientific literature was also used, particularly on works published in Polish scientific and veterinary journals. The search included full yearbooks of "Przeglad Weterynaryjny", "Medycyna Weterynaryjna", "Życie Weterynaryjne".

3. Youth and Scientific Career

He was born on 16 July 1776 (Table 1) in France at Buschweiler (Buschwiller) near Strasbourg and, by the standards of the time, as a subject of Louis XVI, was French. He spent his youth and early school years in French Alsace. He began his education at a French-speaking secondary school in Buschweiler and finished at a German-speaking one in Darmstadt near Frankfurt am Main, where his German family moved when he was thirteen. After the incorporation of Alsace into France, the Bojanus family moved to Darmstadt in Hesse, due to the fact that Ludwig's father, Jacob, was an official of the ducal forests. He was Protestant and treated both languages equally. Furthermore, he studied medicine at the University of Jena, receiving his doctorate in 1797 [1,2]. After defending his doctoral dissertation, he travelled to Vienna to further his medical knowledge. He presumably came into contact with scholars of the then-highly regarded Viennese veterinary school. After a year, he returned to Darmstadt, where he practised as a doctor for two years. In 1801, he travelled to France and studied at a veterinary school—the Ecole Veterinaire d'Alfortville in Alfort, near Paris, and at the National Museum of Natural History in Paris. He became a veterinarian after studying at the veterinary school [3]. These journeys were funded by the Hessian landgrave Wilhelm IX. In Paris, he began his first anatomical studies of animals under the famous professor of comparative anatomy, Georges Cuvier [4]. He then went on to study at veterinary schools in Lyon, London, Hanover, Vienna, Berlin and Copenhagen, as well as at major animal breeding centres. In doing so, he acquired extensive veterinary knowledge and the organisational knowledge for running an animal doctor's training. This is important because there was no coherent veterinary education system at the beginning of the nineteenth century. At that time, schools had different views and methods of treating animals, often conflicting. In 1803, he returned to Darmstadt, where a new veterinary school was planned to open, which Bojanus was to head [5]. However, the facility failed to get off the ground, and due to a competition announced by Vilnius University, Bojanus decided to use his knowledge to create a veterinary school at that very institution [6,7].

Table 1. Key dates in the life of L. H. Bojanus [8,9].

Year	Events
1776	born in Bousville (Buchsweiler) in Alsace
1797	obtains a doctorate in medicine and surgery in Jena
1801–1803	scientific journey to become acquainted with the veterinary schools in Alfort, Paris, Lyon, London, Hanover, Vienna, Berlin and Copenhagen
1802	becomes a member of the Society for the Study of Man in Paris
1804	wins the competition to become professor at Vilnius University as the chair of "bovine treatment"
1810	becomes a member of La Societe Imperiale des Naturalistes de Moscou
1810	becomes an honorary member of the Mediko-Chirurgiczeskoj Akadiemii in St. Petersburg
1814	becomes a member of the Imperial Academy of Sciences in St. Petersburg
1815	begins lectures on comparative anatomy
1818	becomes a member of the Kaiserlich Leopoldinisch-Carolinisch Deutsche Akademie der Naturforscher in Bonn
1818	becomes a member of the Royal Society for the Improvement of Veterinary Medicine in Copenhagen
1818	becomes a member of the Wernerian Natural History Society in Edinburgh
1821	becomes a member of the Königliga Svenska Vetenskapsakademien in Stockholm
1820–1822	takes part in the committee appointed by the University Council to create a project for the reform of studies at Vilnius University
1824	becomes a member of the Medical and Surgical Society in Berlin
1824	becomes a member of the Impieratorskogo Moskovskogo Obszczestva Sielskogo Choziajstwa
1827	dies in Darmstadt

4. Founder of Veterinary Education on Polish Soil

When considering Ludwik Bojanus as the founder of the Polish school of veterinary medicine, it should be remembered that he was active during the Partitions of the Polish–Lithuanian Commonwealth, when this country did not exist on the maps of Europe. The fact is that the school he created functioned on Polish soil, and the language of instruction was Polish. The students were overwhelmingly Polish. This entitles us to conclude that Ludwik Bojanus was the real founder of Polish veterinary education. Having been informed of a competition held by the University of Vilnius, he put forward his candidature for the chair of "cattle treatment", a unit of veterinary medicine functioning in the Faculty of Medicine [8,10]. He presented a work on the organisation of veterinary schools known to him, "Über den Zweck und die Organization Thierarznei Schulen" (later published in print in 1805 in Frankfurt am Main). In it, he wrote that veterinary medicine was a worthy profession and that by engaging in it, one could become an excellent scientist. He heralded the strictly scientific direction of veterinary training. He also presented the opinions about veterinary medicine prevailing in various European countries [11]. He did not arrive in Vilnius until 20 May 1806, bringing his extensive library of veterinary works as a gift to the University [9]. He was appointed professor and taught veterinary knowledge to fourth-year medical students six hours a week.

Furthermore, he lectured in excellent Latin, so interestingly that crowds of people came to his lectures, including professors from various faculties of the University and from outside its walls. From 1815 he also taught classes in comparative anatomy. At the same time, he devoted himself to intensive research work in zoology and veterinary medicine, assisted by Steven Drew, a medical doctor who arrived from England [9]. He began his scientific work by organising the first animal anatomy clinic in post-partition Poland and Lithuania. In a short time, he collected several hundred specimens [12]. He published many works in Latin, German, French and Polish. Before arriving in Vilnius, he was the author of several veterinary treatises, including an article in 1805 on the progress of veterinary medicine in the last three centuries [5], "Kritische Übersicht der Fortschritte der Thierarzneykunde in den leztverflossen drey Jahrunderten...". Printed in Marburg, in this publication he recommended a thorough knowledge of animal anatomy and discussed the state of veterinary knowledge of the last three centuries. One of Bojanus's ideas was to separate scientific veterinary medicine, which was to train veterinary surgeons, from the practical part of the profession, which was to be carried out by veterinary fieldmen. The same year, Bojanus's article on the education of veterinary surgeons 'Über den Zweck und Organization der Thierarzneyschulen' was published in Frankfurt am Main [9].

It should be mentioned that Ludwik Bojanus's experience in organising veterinary schools contributed to the fact that the Government Commission for Religious Denominations and Public Enlightenment of the Confederate Kingdom of Poland asked Bojanus (in addition to Prof. Jerzy Sick and Dr. Adam Rudnicki) to prepare plans for a practical veterinary school in Warsaw (based on the Imperial Decree of 25 September 1816). Despite his involvement in establishing the school in Vilnius, Ludwik Bojanus submitted a project, which he presented at a meeting of professors at Vilnius University on 15 February 1821. It proposed the division of the Institute into three separate establishments: A chair at the University of Warsaw for the training of medical students in veterinary science; A Practical School of Veterinary Medicine for the training of competent animal doctors; A separate scientific and veterinary establishment for scientific experiments. The Government Commission, in a letter of 21 March 1822, expressed appreciation and thanks for the plans prepared and awarded Bojanus the medal of the University of Warsaw but chose Rudnicki's project as the more practical option [13].

5. Epizootiologist

At the time of the great foot-and-mouth epizootic that occurred in 1808, when half of the cattle in Lithuania died, Bojanus became intensively involved in epizootiology. This was reflected in a book published in 1810 in Riga in German and translated into Polish, "O ważniejszych zarazach bydła rogatego i koni". The Polish edition of this publication was, for many years, the basic textbook on epizootiology and contributed significantly to the communication of the principles of epizootic management to animal physicians in Poland. The work was reprinted in 1832 and 1846 and sent to district medical doctors as a manual for action in cases of infectious animal diseases [14]. The book was so important for the agricultural economy and public health that Bojanus received the Order of St. Vladimir from the Tsarist authorities for it. In this publication, he discussed the more important infectious diseases of animals, such as horned cattle plague (rinderpest), splenitis, carbuncle, anthrax, muzzle plague (foot-and-mouth disease) and many other diseases. When describing cattle blight, he surmised that it was caused by contagion, which was confirmed in later studies by other scholars. At the time, rinderpest caused huge losses on cattle farms. To combat it, he recommended the use of vaccination: "*A four- or five-ounce cotton string, half a foot long, is soaked in the moisture of the eyes or nostrils, or in the blood or bile of cattle suffering from the plague. For this, the infectious matter is chosen between the 8th and 10th day of the plague, that is, from the second to the ninth day of the manifestation of the disease: for earlier and later, the matter is ineffective. This cord, soaked in fresh material, is pulled with a coarse needle or a suture through the folds of the skin of the hind legs, in the loincloth, between the forelegs, or in any other place on the body of the cattle to be vaccinated. After pulling the string from the top to the bottom, move it back and forth so that the material rubs off, and then tie the ends of the string together. To be more secure, you can inoculate in two places together. On the third to fifth day after inoculation, the wound is usually inflamed; on the sixth and seventh day, the disease appears, of course, and the cord is removed as unnecessary"*. [14] (p. 28). Bojanus recommended a method of vaccinating animals that was well-known in the West but had found no practical use in Lithuania. This method of vaccinating animals was called 'putting on the veil'. He believed that contagious diseases were caused by infection. He was probably familiar with the views of Bourgelat, who believed that glanders in horses was caused by infection, as well as a pamphlet by the English physician Edward Jenner published at the end of the 18th century. Jenner proved that inoculation with cowpox gave humans immunity to smallpox. This was called variolysis (variola vera—smallpox) and later vaccinia (vacca—cow). The concept of contagion was known as early as the 16th century and was introduced by the father of epidemiology, Girolamo Fracastoro (1483–1553), the papal court physician. In 1810, Bojanus published a treatise in Polish entitled "O kuciu koni podług zasad Kolemana" (On the Forging of Horses According to the Rules of Kolemans) and published a work on epizootiology entitled "Über die Ausrottung der Rindvieh—Pest (in Polen und Lithuanian)...", in which he recommends that in order to combat cattle infestation, cattle from neighbouring countries should not be allowed into Poland and Lithuania for ten years [15].

6. Anatomy

Bojanus became interested in anatomy questions already during his scientific tour of European veterinary colleges. While in Hanover, he described the horse's muscular system, comparing it with that of the human, thus giving rise to comparative myology. Bojanus's pupil, Karol Muyschel, wrote about this [16]. On his return to Vilnius, he began his lectures on comparative anatomy in 1815 with an introductory lecture entitled. "Introductio in anatomen comparatam" [17]. In this lecture, he compared the organs and functions of various groups of organisms, beginning with plants and ending with man, demonstrating at every step the relationships and gradual complication and refinement of structure.

This introductory lesson made Bojanus one of the forerunners of evolution, already formulated in his time in the writings of Lamarck, Geoffroy St. Hilaire and other scholars and so brilliantly developed later by Charles Darwin [18]. The lectures on comparative anatomy by Bojanus were the first lectures on this subject on Polish soil. However, their beginning should be assigned not to 1815 but to 1806, as Bojanus had lectured on the comparative anatomy of domestic animals since his arrival in Vilnius. In his veterinary works on anatomy, he discussed horse tendons and the forging teeth in horses and sheep. In embryology, he described the foetal membranes of sheep, horses, hares and humans and the structure of the omohypophysis and villi in dog foetuses. In osteological work, he elaborated on the cranial structure of sheep, goats, fish, birds and other vertebrates. He studied the fossil bones of the mammoth and the aurochs, finally extinct as a species in the first half of the 17th century in Poland. He gave it the Latin name *'Bos primigenius'*.

Furthermore, he was the first to prove that the bison and the aurochs were separate animal species [19]. He described the anatomical structure of the leech, clam, spider and crayfish. He also published parasitological works. He became world-famous for his monograph on the turtle's anatomy, now known as the mud turtle, 'Anatome testudinis europaeae'. Furthermore, he compiled it after ten years of studying 500 specimens of this reptile [20,21]. It was published in Vilnius in 1819–1821. Two hundred and one drawings were made by the author's hand for this publication, from which Ferdinand Lehmann, an engraver (engraver) brought from Darmstat, produced copperplate engravings [9]. Eighty copies of this book were published at the author's expense. In the Library of the Lithuanian

Academy of Sciences (former Wróblewski Library) in Vilnius, a copy with one colour engraving and the autograph of the future professor of the Medical and Surgical Academy Adam Adamowicz, a pupil of Bojanus, has been preserved. This copy was donated to the library in 1930 by the then-professor of the Stefan Batory University in Vilnius and, after the war, of the Nicolaus Copernicus University in Toruń, Professor Stefan Narębski. Two copies with a set of colour engravings are also known to exist [8]. Bojanus also compiled an anatomical monograph of the sheep, for which he made several hundred drawings. Due to technical difficulties and the voluminous nature of this work, it was not published but remained in manuscript and was lost in Darmstadt years later [20]. His views on evolution were ahead of the times, as he believed that certain species of plants and animals could evolve into others and that gradual evolution occurs over time. In his view, there are transitional forms between groups of plants and animals, the so-called 'Zoophytia' [9].

In addition to vertebrate anatomy, Bojanus presented many papers on the anatomy of invertebrate animals (including parasites). In his work 'Kurze Nachricht über Cerkarien' [22], he stated that cercariae, commonly considered at the time to be adult independent animals, are the developmental stages of flukes, the intermediate host being the snail Lymnaea stagnalis. Bojanus also worked out the anatomy of the leech and described it in the works: Die Anatomie des Blutegels [23], Was wis sen wir denn eigentlich vom Bau des Blutegels and Observations nouvelles sur la sangsue [24]. A separate line of research was the study of the anatomy and biology of Anodonta cygnea (the great rats). He published many papers [25], the most important of which is Über die Athem—und Kreislaufwerkzeuge der zweischaligen Muscheln. He undertook studies on the reproduction of Anodonta cygnea, but also described the respiratory organs, the excretory and vascular systems, and established many unknown facts, including describing the excretory organ, later called the 'Bojanus organ' (he erroneously classified it as a respiratory organ). It is important to emphasise the wide range of Bojanus's research interests, which included both vertebrates and invertebrates, adult forms and developmental forms. Bojanus's laboratory in Vilnius was the first zoological research workshop in the Polish lands.

Bojanus belonged to the 'natural philosophers' because, like Goethe, Oken, Spix and Carus, he propounded the idea that the skull is the transformed anterior end of the vertebral column and consists of deformed and fused vertebrae. There were many arguments in favour of the theory: the skull, like the vertebral column, initially passes through a cartilaginous stage and later ossifies; the skull includes the anterior end of the dorsal cords; there is a cavity in the skull filled with the brain, which can be considered an extension of the medulla oblongata. Goethe, who was the first to publish the "vertebral theory of the skull" as early as 1790, had a similar understanding. Oken came up with a similar idea in 1807, independently of Goethe. In the pages of "Isis", a discussion began on this topic, in which, in addition to Oken, Spix and Frank, Carus and also Bojanus took part [9].

In addition to anatomy, medicine and veterinary science, he was interested in print-making and painting and was himself a gifted draughtsman. This helped him to document his scientific work. In 1817, he lectured at the University of Vilnius on a novel graphic technique, lithography, which began to be used in printing book engravings. This lecture was translated from French into Polish and published in print as the 'Lecture on Lithographic Art'.

7. Anatomical Theatre

Even before the opening of the first veterinary school in Vilnius, after he had already taken up the "chair of cattle treatment", Bojanus set about creating an anatomical theatre. In 1808, the University acquired the ruins of the Uniate Spasskaya Orthodox Church (Figure 2) and the larger and smaller palaces on the opposite side of Spasskaya (Metropolitan) Street, which belonged to the Uniate Metropolitan. These buildings were utterly devastated during the siege of Vilnius by the Russian army in 1794 [26]. In their place, the Anatomical Theatre and two buildings designed by the architect Szulc were built to house studios and flats for University staff. It was probably not realised that buried in this church were the parents of the Polish King Wladyslaw Jagiello, Prince Olgierd and Juliana, and later in 1513 the wife of Alexander Jagiellon and daughter of Ivan III the Terrible, Helena, as well.

Figure 2. Spasskaya church engraved in the first half of the 19th century.

The Anatomical Theatre housed the comparative anatomy department and veterinary surgeries and laboratories. A large, oval, amphitheatre-shaped lecture theatre was created on the ground floor and an animal prosectorium on the ground floor. At the formal opening of the prosectorium on 13 December 1815, Professor Bojanus delivered a lecture in French, "Des principales causes de la dēgēnēration des chevaaux.", which translated into Polish as "On the causes of the disappearance of good horse species". In this lecture, he predicted the extinction of the pure breed of Polish horses due to the importation of breeding horses from England [27].

Lectures on veterinary medicine and zoology were held in the amphitheatre hall and, from 1815, on comparative anatomy as well. Gradually, exhibits of stuffed domestic animals and Lithuanian fauna and their skeletons began to be placed there. Many species and breeds of stuffed birds were exhibited in the neighbouring halls known as the zoological museum.

On the opposite side of Spasskaya Street, a larger building housed studios for macerating bones and stuffing animal exhibits, and on the floors there were flats for the staff. The attic, had a "blich" (boiler) for bleaching the bones. An animal clinic was built next door. In the neighbouring smaller building, a blacksmith's shop, carriage house and storerooms were set up on the ground floor. On the first floor, Professor Bojanus took up residence. After the Anatomicum was built, the staff of the medical and natural sciences faculties had very good conditions for scientific research and lectures. Bojanus performed most of his scientific work in the Anatomical Theatre (Figure 3) [28,29].

Figure 3. Anatomical theatre engraving 1st half of the 19th century.

8. Interests

In addition to comparative anatomy and veterinary medicine issues, Bojanus approached various aspects of everyday life with curiosity, which he usually considered from the point of view of science. In 1807, during the war in which the Russian army and the Prussian army fought against Napoleon, Bojanus worked in the hospitals of Vilnius, caring for wounded soldiers [9,30]. He noticed recurring lung problems, which he attributed to the faulty design of the army satchel and the poorly designed thongs from which it was suspended. The observation led him to devise a different solution for attaching the military satchel. Bojanus left for St. Petersburg when Napoleon Bonaparte's Grand Army entered Russia in 1812 and did not return to Vilnius after the withdrawal of the French [9]. In 1821, during Tsar Alexander I's stay in Vilnius, Ludwig Bojanus presented a new type of satchel that came into use in the Tsar's army. As a reward for this modification, Tsar Alexander I awarded him a special ring. The hero of this article also became famous as a talented draughtsman. By attaching drawings to his works, he to the time to reproduce details accurately and produce artistic effects. In 1817, he presented a lecture on the art of lithography at a session of the University, which he later published in print: Lecture on the Art of Lithography [9]. In 1818, when he proceeded to publish his Anatomy of a Tortoise, having found no suitable equipment locally, he built the press for printing the engravings himself. Also, he made the paints himself [31]. He brought an engraver from Germany, Ferdinand Lehmann, to whom he entrusted the engraving of his prepared drawings on copper and their reflection on the newly equipped press. Bojanus also painted portraits of himself, his wife and his friend Gregory Langsdorf. A lithograph depicting his self-portrait is in the collection of the Vilnius Medical Association [7].

9. Conclusions

Ludwig H. Bojanus was, for his time, an active scientist, as evidenced by the publication of 70 scientific papers and his membership in many scientific societies in Europe. There was a break in his work in 1812 when the French army was approaching Vilnius. His organisational activities are also admirable. He expanded the zootomic cabinet, collecting 1653 preparations and skeletons of various animal species and 144 preparations of parasites called 'visceral worms'. In 1820, he proposed a reform of the organisation and curriculum of the University. He was widely liked, respected and admired. In 1818, he was offered the position of professor and director of the College of Veterinary Medicine in Berlin. However, he refused, as he also did when the University of Vilnius asked him in 1822 to accept the position of rector. His aim was to organise a veterinary faculty at the University in the longer term. He established a clinic and a school but failed to establish an institute because he lacked sufficiently trained scientific staff. In 1824, he fell ill, probably with tuberculosis. The University granted him indefinite medical leave.

The bibliography of Bojanus's works demonstrates his extraordinary diligence and diversity of interests. However, he was not only the founder and propagator of a new direction of zoological research in Poland but was also able to gather around him many disciples who later worked in the direction indicated by the master of what we can rightly call the Bojanus school. The most prominent among them are Adam Ferdynand Adamowicz, initially a professor at the veterinary school for fieldpersons and later at the Medical and Surgical Academy in Vilnius, author of many works on anatomy and veterinary medicine; Konstanty Balbiani, author of the anatomy of the medical leech; Adam Bielkiewicz, professor at Vilnius University and the Medical and Surgical Academy in Vilnius, anatomist and physiologist; Karol Muyschel, professor at the Veterinary School, later professor of pet anatomy at the Vilnius Medical and Surgical Academy; and finally Fortunat Jurewicz, prematurely deceased but promising deputy professor of zoology and comparative anatomy at Vilnius University.

Bojanus left Vilnius and went to Germany for treatment, where he died on 2 April 1827 in Darmstadt. A self-portrait and portrait of his wife by Bojanus have been preserved in the collection of the Vilnius Medical Association. Among the most important achievements of Ludwig Henry Bojanus we can include:

1. the organisation and launch of the first veterinary school in the Polish lands in 1823
2. the compilation of a monograph on the anatomy of the European pond turtle "Anatome testudinis europaeae" between 1819 and 1821
3. the development of methods for dealing with infectious diseases of animals and publishing them in the book "O ważniejszych zarazach bydła rogatego i koni".

Author Contributions: Conceptualization, J.S.; formal analysis, J.S.; investigation, J.S.; writing—original draft preparation, J.S. and M.Z.; writing—review and editing, J.S. and M.Z. All authors have read and agreed to the published version of the manuscript.

Funding: This research received no external funding.

Institutional Review Board Statement: Not applicable.

Informed Consent Statement: Not applicable.

Data Availability Statement: The data presented in this study are available on request from the corresponding author.

References

1. Chodynicki, I. *Dykcjonarz Uczonych Polaków, Zawierający Krótkie Rysy Ich Życia*; Kuhn i Millikowski: Lviv, Poland, 1833; pp. 1–397.
2. Kośmiński, S. *Słownik Lekarzów Polskich*; Author's Edition: Warsaw, Poland, 1888; pp. 1–665.
3. Adamowicz, A.F. Wiadomość o Życiu I Pismach Ludwika Bojanusa. *Tyg Petersburski* **1835**, 80–84.
4. Daszkiewicz, P. Poszukiwanie Śladów Ludwika Bojanusa w Paryżu. *Kwart. Hist. Nauk. Tech.* **2002**, *47*, 63.
5. Sobieszczański, F.M. *Wiadomość Bibliograliczna O Życiu I Pismach L. H. Bojanusa, vol. 1*; Bibl Warsz: Warsaw, Poland, 1849.
6. Jundziłł, S. Cudzoziemcy w Uniwersytecie. Urywek pamiętnika. In *W Promieniach Wilna i Krzemieńca*; Janowski, L., Ed.; Księg Józefa Zawadzkiego: Vilnius, Poland, 1923; pp. 1–271.
7. Rostafiński, J. *Księga Pamiątkowa Uniwersytetu Stefana Batorego*; Univ St Batory: Wilno, Poland, 1929; pp. 1–438.
8. Rostafiński, J. *Polski Słownik Biograficzny, vol. 2*; Polska Akademia Umiejętności: Kraków, Poland, 1936; pp. 218–347.
9. Fedorowicz, Z. *Ludwik Henryk Bojanus*; Polska Akademia Nauk, Instytut Zoologiczny, Memorabilia Wrocław-Warszawa: Kraków, Poland, 1958; pp. 1–46.

10. Królikowski, C. *O Polskich Zakładach Weterynaryjnych*; Author's Edition: Lwów, Poland, 1889.
11. Bojanus, L.H. *Über Den Zweck Und Die Organization Thierarznei Schulen*; Andreaeischen Buchandlimg: Frankfurt am Main, Germany, 1805; pp. 1–206.
12. Adamowicz, A.F. *Krótki Rys Początku I Postępu Anatomii W Polsce I Litwie*; Księg Józefa Zawadzkiego: Vilnius, Russia, 1855.
13. Sobolewski, J. Ludwik Bojanus współtwórca polskiego szkolnictwa weterynaryjnego. *Weter. Zesz. Hist.* **2004**, *1*, 12–13.
14. Bojanus, L.H. *O ważniejszych Zarazach Bydła Rogatego I Koni*; Księg Józefa Zawadzkiego: Vilnius, Russia, 1810; pp. 1–151.
15. Bieliński, J. *Stan Nauk Lekarskich Za Czasów Akademii Medyko-Chirurgicznej Wileńskiej Biograficznie Przedstawiony*; Towarzystwo Lekarskie Warszawskie: Warsaw, Poland, 1888; pp. 1–907.
16. Muyschel, K. *Systema Myologica Equi*; Księg Józefa Zawadzkiego: Vilnius, Russia, 1829.
17. Bojanus, L.H. *Introductio in Anatomen Comparatam*; Oratio Academica, Księg Józefa Zawadzkiego: Vilnius, Poland, 1815.
18. Adamowicz, A.F. *Ludwik Bojanus*; Wizerunki i Roztrząsania Naukowe: Vilnius, Poland, 1836.
19. Ślósarski, A. *O Głowach Tura (Bos Primigenius. Bojanus) Znalezionych W Królestwie Polskiem*; Orgelbranda Synowie: Warszawa, Poland, 1881.
20. Daszkiewicz, P. Egzemplarz Anatome Testudinis Europae ze zbiorów biblioteki Wróblewskich w Wilnie—Interesujący dokument historii nauki. *Kwart. Hist. Nauk. Tech.* **2018**, *3*, 15.
21. Daszkiewicz, P.; Edel, P. Poszukiwanie egzemplarzy Anatome testudinis europaeae w europejskich i amerykańskich bibliotekach. *Z Badań Książką Księgozbiorami Hist.* **2016**, *10*, 119–130.
22. Bojanus, L.H. Kurze Nachricht über Cerkarien und ihren Fundort. *Isis V. Oken* **1818**, *4*, 729–730.
23. Bojanus, L.H. Die Anatomie des Blutegels. *Isis V. Oken* **1817**, *7*, 873–884.
24. Bojanus, L.H. Observations nouvelles sur l'organisation de la Sangsue (*Hirudo medicinalis*). *J. Phys. Paris* **1818**, *88*, 468.
25. Bojanus, L.H. Sendschreiben an den Her rn Chevalier G. de Cuvier über Athem- und Kreislaufwerkzeuge der zweischaligen Muscheln, insbesondere des Anodon cygneum. *Isis V. Oken* **1819**, *1*, 41–100.
26. Magowska, A. Veterinary science in the nascent state: The animal hospital in Vilnius 1834–1842. *Med. Weter* **2017**, *73*, 252–256. [CrossRef]
27. Królikowski, S. *Bibliografia Polska Weterynarii I Hodowli Zwierząt*; Przegląd Weterynarski: Lviv, Poland, 1891.
28. Brzęk, G. *Historia Zoologii W Polsce Do Roku 1918*; Wyd Uniw Marii Skłodowskiej Curie: Lublin, Poland, 1947; pp. 1–251.
29. Hoyer, H. *Zarys Dziejów Zoologii W Polsce*; Polska Akademia Umiejętności: Kraków, Poland, 1948; pp. 1–25.
30. Szumowski, W. *Fryderyk Hechell O Swoich Studiach W Uniwersytecie Wileńskim*; University St Batory: Vilnius, Poland, 1929; pp. 1–438.
31. Hoyer, H. *Anatomia Porównawcza Zwierząt Kręgowych*; Wyd Kasy im Mianowskiego: Warsaw, Poland, 1931; pp. 223–228.

Review

Jean Cruveilhier (1791–1874), a Predecessor of Evidence-Based Medicine

Luis-Alfonso Arráez-Aybar [1,*], Talía Fuentes-Redondo [2], José-Luis Bueno-López [3] and Rafael Romero-Reverón [4]

1   Department of Human Anatomy and Embryology, Faculty of Medicine, Complutense University, 28040 Madrid, Spain
2   Department of Pediatrics, Complejo Hospitalario de Toledo, 45007 Toledo, Spain
3   Department of Neurosciences, School of Medicine and Nursing, The University of the Basque Country (UPV/EHU), 48940 Leioa, Spain
4   Department of Human Anatomy, José María Vargas Medical School, Faculty of Medicine, Central University of Venezuela, Caracas 1040, Venezuela
*   Correspondence: arraezla@med.ucm.es

Abstract: This article focuses on Jean Cruveilhier and particularly on his book *Anatomie descriptive*, which was a great success during the author's lifetime. (Notwithstanding this, it is pertinent to point out that the five editions of *Anatomie descriptive* were surpassed in number by others of the Cruveilhier's creations, such as *Anatomie pathologique* and *Traité d'Anatomie pathologique*.) Unlike other texts of the time and later, *Anatomie descriptive* presents the anatomy of the human body in a way that can be applied both by students and medical professionals. The objectives of *Anatomie descriptive* were to make understand how the functions of an organ can be inferred from its structure, and to encourage students and professionals to investigate the anatomical origin of health and disease phenomena. Depending on which sections of the book, the parts of the body were described with morphological, topographic or functional criteria. Many of Cruveilhier's contributions influenced anatomical eponymy and keep today's *Terminologia Anatomica* alive. All of this has made consider Jean Cruveilhier the most outstanding anatomist in France of the first half of the nineteenth century. Due to the scientific rigor Cruveilhier always applied and asked to be applied in the investigation of the anatomic changes linked to pathological processes, he could certainly be considered a predecessor of the objectivity sought by evidence-based medicine.

Keywords: 19th century anatomy; 19th century anatomists; anatomical terminology; anatomo-clinical method; clinical-pathological method; Cruveilhier

Citation: Arráez-Aybar, L.-A.; Fuentes-Redondo, T.; Bueno-López, J.-L.; Romero-Reverón, R. Jean Cruveilhier (1791–1874), a Predecessor of Evidence-Based Medicine. *Anatomia* **2023**, *2*, 206–221. https://doi.org/10.3390/anatomia2030019

Academic Editors: Gianfranco Natale and Francesco Fornai

Received: 5 May 2023
Revised: 5 June 2023
Accepted: 4 July 2023
Published: 7 July 2023

1. Anatomy in the Time of Cruveilhier

The late eighteenth and early nineteenth centuries, so revolutionary in socio-political aspects, marked the end of the Vesalius revolution and the influence of the *naturphilosophie* in the world of anatomy. New exploratory techniques and conceptual changes emerged [1]. We highlight two aspects on the technical side: the thorough study of anatomopathological lesions that, when done systematically, showed specific morpho-functional "systems", and the experimental induction of lesions to study the morpho-functional consequences. On the conceptual side, both the anatomical–comparative paradigm and the anatomical–tissue paradigm spread. The first one, derived from the comparative zoological morphology of the eighteenth century, had a non-evolutionist orientation during the first half of the nineteenth century which turned into two versions, the speculative and the positive. The speculative version, initiated by J. W. von Goethe (1749–1832) and pursued by some *naturphilosophen*, put embryology at the service of comparative anatomy. The positive version of the anatomical–comparative paradigm—which was supported by zoologists, such as G.

Cuvier (1769–1832), E.-G. Saint-Hilaire (1772–1844), and J.-B. de Lamark (1744–1829) in the post-Revolution France—became one of the basic disciplines of modern Biology [1].

Under M.-F.-X. Bichat (1771–1802), the anatomical–tissue paradigm was initially sensualist and vitalist—that is, following the doctrine that sensations and perception are the basic and most important form of true cognition, on the one hand, but also the belief based on the premise that living organisms are fundamentally different from non-living entities because they are governed by different principles, on the other -and afterwards it became cellular under F.-G.-J. Henle (1809–1885). M.-F.-X. Bichat studied wide homogeneous components of the organism, the so-called *uniform parts ('ta homoiomere')* by Aristotle (384 B.C.–322 B.C.) or *'tissu muqueux'* by T. de Bordeu (1722–1776). From M.-F.-X. Bichat's sensualist and vitalist mentality arose the concept of *'tissue'* considered a major anatomical unit in the explanation of the physiological properties and pathological changes of the organism. F.-G.J. Henle, in his *Allgemeine Anatomie* (1841), integrated the concept of tissue into M.-J. Scheiden's (1804–1881) and T. Schwann's (1810–1882) cellular theory [1].

2. L'École de Santé de Paris

The *Révolution française* proclaimed academic freedom and destroyed the whole medical organization of the *Ancien Régime*, in all its healthcare, professional, educational, and institutional aspects. Alleging that disease had to disappear in a well organized society, it also abolished hospitals, faculties (18 August 1792), and academies (1793), including various local *sociétés de médecine* and the unifying *Société royale de médecine* (recently driven by F. Viq d'Azyr, in 1778), as well as the slightly older *Académie royale de chirurgie* (1731). The latter would awake and merge with the five Academies working in all fields of knowledge and the arts into the *Institut de France* (25 October 1795). After this revolutionary storm, reality imposed. Hospitals were reorganized and became places where physicians co-operated with surgeons in professional equality, opening the way to merge both into a unified profession and an equally unified teaching not under state control, but as civil societies run by municipal administrations.

Three *Écoles de santé* were created in Paris, Montpellier, and Strasbourg (12 April 1794), which would eventually become their respective *Facultés de médecine* (17 March 1808). The initial purpose of the *Écoles* was to train surgeons for the armies of the Republic, but then this training was extended to civil assistance physicians. The new medical education was based on four principles: the fusion of medicine and surgery, as they are two branches of the same science; the setting up of a clinical teaching practice in hospitals; the competitive selection of students and teachers through examinations; and the obtaining of a doctorate of universal value [2–4].

The pedagogy of medical teaching was rebuilt, based on the sensualist paradigm of E. Bonnot de Condillac (*l'Ábbé de Condillac*; 1714–1780). In the words of P.-J.-G. Cabanis (1757–1808), friend and follower of Condillac, *"the true instruction of young doctors is not received from books, but at the bedside"*. Thus, the medical student should be trained in chemistry experiments, anatomical dissection, and surgical interventions. *"Read little, see much, and do much"* [5,6].

The *Révolution française* also promoted the foundation of independent higher education institutions. The *Société philomatique* (founded in 1788) served as a meeting point to F. Vicq d'Azyr (1748–1794), J.-B. de Lamark (1744–1829), M.-F.-X. Bichat (1771–1802), G. Cuvier (1769–1832), E.-G. Saint-Hilaire (1722–1844), F. Chaussier (1746–1828), G. Dupuytren (1777–1835), J.-G. Cloquet (1787–1840), F. Magendie (1783–1855), E.-R.-A. Serres (1786–1868), and A.-A.-L. Velpeau (1795–1867), among others. P.-A. Beclard (1785–1825) was an independent professor before he occupied a chair at *l'École de santé* of Paris in 1818. Private anatomical amphitheaters proliferated. There were up to fifteen in Paris before their abolition in 1813 [3].

In close concord with the revolutionary political changes, but also with the changes in the way of practicing and teaching medicine that occurred not only in France, but also in other countries (in particular, in the union of medicine with surgery and in the need to perform dissections to better know the human body, a need that emerged from warships and battlefields) it was founded *l'École de santé de Paris* in 1794. In operation since 1750, *l'École pratique de dissection de Paris* is one of the few higher education establishments that has not changed its name from Louis XV to the present day. *l'École pratique de dissection de Paris* was highly appreciated when the events of this historical review occurred because it opposed the practical learning and free discussion that took place in it to the corseted activities of the other educational institutions [7]. These facilities were widened in 1813 by the annexation of the former *Collège de chirurgie* and the purchasing of contiguous buildings [4]. In *La Charité* and the *Hôtel Dieu* hospitals, teaching began focusing on the study of the patient from the bedside [5]. In these hospitals, the coexistence between physicians and surgeons enabled to tie clinical observations together with the anatomical lesions found postmortem. This fact served to develop the anatomo-clinical mentality of the first half of the 19th century and led to the birth of a new discipline: pathology (in France and other countries, but only developments in France are discussed in this review). M.-F.-X. Bichat was the cornerstone of this discipline. To him, the clinical symptoms and their nosographic arrangement should be related to the anatomical lesion that originated those [8]. This idea was initially developed in clinical practice by J.-N. Corvisart (1755–1821) and his disciples R.- T. Laënnec (1781–1826) and G. L. Bayle (1774–1816) [6].

In 1800, the *Société de l'école de médecine* was founded, which was based on the *l'École de santé of Paris*. The *Société*, apart from being a centre of scientific research, served as an advisory body to the French government until the foundation of the *Académie nationale de médecine* in December 1820 [4].

The Revolution also polarized the relationship between professionals and professors [2]. During the *Ancien Régime*, professionals were educated either in the *Facultés de médecine* or in the *Collège de chirurgie*. The rivalry then was between physicians and surgeons. The Revolution brought about another conflict: both the independent professors and the new hierarchies of the hospital (whether they were physicians or surgeons) accused the school/faculty professors of not taking courses or publishing, and therefore questioned whether the new academic hierarchy was qualified enough to teach medicine and surgery [3]. In 1810, the designation of school professors was established by public competition, which was removed by the Restoration (1814–1830) and then re-established again in 1823. Mateo Orfila (born in Mahón, Spain, 1787; died in Paris, 1853), often called the 'father of toxicology', was dean of *l'École de Paris* between 1831 and 1848. He reorganized the *l'École de santé de Paris*, raised educational requirements for admission, and instituted more rigorous examination procedures. He also helped to establish hospitals and museums, specialty clinics, botanical gardens, a center for dissection in Clamart, France, and a new medical school in Tours, France (https://www.nlm.nih.gov/exhibition/visibleproofs/galleries/biographies/orfila.html, accessed on 5 July 2023). He abolished the *titre d'Officier de santé* and made mandatory the obtain the *Baccalauréat dès sciences* to achieve the title of *Docteur en médecine*. As a result of Orfila's reform, the access to the membership and chairs of the *Écoles de médecine* was invigorated. Student selection was also competitive [2]. *L'École de Paris* admitted about 300 students each year (a not always respected number), and among them was Jean Cruveilhier (quoted as J. Cruveilhier henceforth) (Figure 1), the most outstanding anatomist in France in the first half of the nineteenth century.

Figure 1. A portrait of Jean Cruveilhier (1791–1874).

3. His Life and Work

J. Cruveilhier was born on 9 February 1791 in Limoges, France [9]. For two centuries, the members of the Cruveilhier family were born and buried in Limoges. His grandfather Joseph (1726–1762) was a master surgeon and his father Léonard (1760–1836) was an important military surgeon, an attending surgeon at *l'Hôpital Saint-Alexis* in Limoges, and also a revolutionary Jacobin fanatic [10]. His mother, Anna Reix was a devout Catholic and extremely pious woman from whom J. Cruveilhier inherited an indelible sense of the catholic mysticism that he maintained throughout his life. His uncle, Jean Reix, was a priest expelled to Spain in 1792 for refusing to take the oath to the Civil Constitution of the Clergy. After his return to France, Jean Reix was the Vicar of the *Cathédrale Saint-Étienne* in Limoges [11].

J. Cruveilhier studied in *l'École Centrale de Limoges* (formerly, the Chapel of the Visitation) and after that, at the *École impériale* where he received an award of excellence and various prizes of honour (Latin, Literature, Mathematics, and Chemistry).

Until 1809, J. Cruveilhier, together with G.-L. Bayle and R.-T.-H. Laënnec, used to frequent the *Congrégation de la Sainte Vierge*, founded in 1801 by the Jesuit J.-B. Bourdier. Against his religious vocation, his paternal insistence led him to begin medical studies in Paris, although he abandoned them because of the disgust and horror he felt at the dissecting rooms. These events exacerbated his religious vocation so that he took refuge at the *Séminaire Saint-Sulpice* to study Theology. There he met D.-A.-L. de Frayssinous (1765–1841), who later became Bishop of Hermopolis *in partibus* and Minister of Public Instruction [12–14]. Again, paternal intervention forced him to resume his medical studies in Paris. As a student, he obtained several prizes (Table 1). He spent his whole intern period at the hôspital *Hôtel-Dieu* along with G. Dupuytren (1777–1835), for whom he developed a great empathy and high admiration, despite their ideological differences. Dupuytren was a Freemason, a member of the *Sainte-Catherine du Grand Orient Loge* in Paris; however, he protected the career of his pupil J.Cruveilhier, despite his independent nature and Catholic devotion [14].

On 24 January 1816, J. Cruveilhier defended his doctoral thesis (Table 2), dedicated to his father Leonard and his master G. Dupuytren. Influenced by the works of M.-F.-X. Bichat and his friend R.-T. Laënnec, J. Cruveilhier showed his special interest in the study of the anatomical lesion [12]. Once he obtained his doctoral degree, J. Cruveilhier returned to

Limoges to succeed his father. He married Jenny Grellet des Prades de Fleurelle (1801–1849), daughter of a notable banker from Limoges and the manager of *l'hôpital Saint-Alexis*. In October of that year, the prefect of the Haute-Vienne (France) asked him for a report about the major epidemic of typhoid fever, a disease that J. Cruveilhier called *'enteromesenteric fever'* [15] while masterfully describing the anatomical lesions of the ileum. Between 1818 and 1821, he applied for the direction of the Childbirth Course of the Limoges Hospital and then for the charge of chief surgeon, without any success [11].

In 1823, he took the restored *'Concours d'agrégation'* (competition for professorship) (Table 1) again under the guidance of his father and supported by G. Dupuytren, J. Cruveilhier was the first of the five promoted *agrégés* over the twenty-six candidates presented, among them being A. Velpeau. Following G. Dupuytren's recommendation, J. Cruveilhier chose to be *agrégé de médecine opératoire* (professor of Surgical Medicine) in Montpellier, where he went in July 1824. Nevertheless, this place did not please him and J. Cruveilhier returned to his native town resolved to dedicate himself to the most unfortunate sick people. When everything was ready for this return to Limoges, P.A. Béclard (1785–1825) died. Thereafter, J. Cruveilhier received this message from G. Dupuytren: *'Béclard passed away, come to Paris, you have a chance'* [11,13]. D.-A.-L. de Frayssinous, who was already *Grand Maître de l'Université*, also encouraged him to go to Paris to seat in the Chair of Anatomy left by P.A. Béclard. However, G. Breschet (1784–1845) and J. Cloquet also applied for this position. The designation of J. Cruveilhier was considered as an act of ministerial authority. J. Cruveilhier was received with hostility by the students when he presented at the *grand amphithéâtre* of the Faculty of Paris on 10 November 1825. His personality and sincere modesty quickly conquered his audience [12]. Thinking of his students, J. Cruveilhier began to compose his work *Anatomie descriptive* (see below). Among his partners in *l'École pratique de dissection de Paris* were E.-P.-M. Chassaignac (1804–1879), C.-L. Bonamy (1812–1887), P. Broca (1824–1880) [16–18] and the artist Emile Beau (1810–1872), who also assisted in drawing anatomical atlases by authors such as A.-L. Foville (1799–1878) and his *Traité complet de l'anatomie, de la physiologie et de la pathologie du système nerveux cérébro-spinal* (1844), which is regarded as one of the best works on the subject before the invention of the microscope [19].

In 1826, J. Cruveilhier was appointed chief physician of the hospitals of Paris, which was followed by his success in a series of clinical appointments until 1849. (Table 1). On 12 October 1826, he restored the *'Société d'anatomie'*, founded in 1803 by G. Dupuytren and dissolved in 1808 under the chairmanship of R.-T. Laënnec. J. Cruveilhier was its president until 1866 when G. Breschet was appointed [20].

In 1835 G. Dupuytren died. His last volition was to endow funding for the creation of the Chair of Pathological Anatomy of Paris, and designed his disciple J. Cruveilhier as its first holder. Appointment that he accepted. In 1836, G. Breschet succeeded J. Cruveilhier to the Chair of Anatomy at Paris University.

J. Cruveilhier's professional life shared two culminations: a religious devotion to the sick and the development of a rigorously scientific career with many honours [12]. Along with his work at the *Charité* and *Salpêtrière* hospitals, he was involved in a very active clinical practice in Paris under the rules of a very strict ethic that he summarized in his speech *Des devoirs et de la moralité du médecin* (Table 2). J. Cruveilhier founded a charity to help humble people. He had significant customers, national and foreign, and from all social classes to whom he gave equal treatment. When he was invited to be the physician of Napoleon III (1807–1873), J. Cruveilhier answered ' ... *qu'je le soignerait comme mes maladies d'hôpital'* /' ... *I would take after him as I do with my patients at the hospital'*. On a separate occasion, he was advised to make a courtesy visit to the emperor, and so he said: ' ... *s'il n'est pas malade, ma visite est inutile'* /' ... *if he is not sick, my visit is useless'*. Cruveilhier's attitude upset Napoleon III, who vetoed his election to the *Institut de France* [14].

In 1866, at the age of 75 and at the insistence of his family, he retired. He left Paris on 18 September 1870, a day before the siege of Paris by the Prussian forces and went to

Sussac (Haute-Vienne, France) where he died of pneumonia on 7 March 1874. The funeral was held in the church where he had been baptized.

The scientific authority of J. Cruveilhier was widely recognized both in France and abroad. J. Cruveilhier's academic career, distinctions, and prizes are shown in Table 1 (as collected from the *Académie nationale de médecine* [8]; Orcel and Vetter [11]; Androutsos and Vladimiros [12]; Vayre, [13] and Huard, [19]). The only lack in his brilliant career was not to become a member of the *Institut de France* due to the personal veto of Napoleon III [14].

Table 1. Jean Cruveilhier's academic career, distinctions and prizes.

Year	Academic Career, Distinctions and Prizes
1811	*Major du concours de l'internat des hôpitaux de Paris*
1812	*Prix des hospices civils de Paris*
1813	*Prix de l'école pratique* *Prix de médecine opératoire*
1816	*Dissertation*
1823	*Major du concours d'agrégation en médecine* *Professeur agrégé de médecine opératoire à Montpellier*
1824	*Membre associé non-résidant de l'Académie royale de médecine (Académie nationale de médecine* nowadays*)*
1825	*Professeur titulaire del la chaire d'anatomie à Paris*
1826	*Président de la Société anatomique* (until 1866) *Chef du department de médecine à l'hôpital de la Salpêtrière*
1828	*Médecin suppléant à la Maison royale de la santé*
1830	*Médecin chef de la maternité de Paris* *Chirurgien chef du service des hôpitaux de Paris*
1833	*Medaille recompense epidemie cholera*
1836	*Premier professeur titulaire de la chaire d'anatomopathologie de Paris* *Membre élu de l'Académie royale de médecine*
1843	*Membre de la Société de chirurgie de Paris*
1847	*Chef du service de chirurgie de l'hôpital de la Charité*
1848	*Membre du comité consultatif d'hygiène*
1849	*Medaille recompense epidemie cholera* *Chef de service de chirurgie de l'hôpital de la Salpêtrière*
1855	*Medaille recompense epidemie cholera*
1856	*Docteur émérite des hôpitaux*
1859	*Président de l' Académie impériale de médecine (Académie nationale de médecine* nowadays*)*
1863	*Commandeur de la Légion d'Honneur*
1866	*Professeur émérite*

Table 2. The Key publications of Jean Cruveilhier (according to Orcel and Vetter [11]; Androutsos and Vladimiros [12]; Vayre [13]; Huard [19]).

Year	Name of Publication
1816	*Essai sur l'anatomie pathologique en général et sur les transformations et productions organiques en particular (2 voumes.). Doctorate thesis*, Paris
1821	*Médicine pratique éclairée par l'anatomie et la physiologie pathologiques.* J.B. Baillière et fils, Paris
1824	*An omnis pulmonum exulceratio vel etiam excavatio insanabilis ? Concours d'agrégation en médecine thesis*, Montpellier
1825	*Discours sur l'histoire de l'anatomie.* Opening lecture of his anatomy course as professor of descriptive anatomy in Paris
1828	*Anatomie pathologique du corps humaine, ou descriptions, avec figures lithographiées et coloriées, des diverses alterations morbides dont le corps humain est susceptible (2 volumes).*
1842	Illustrated Atlas. T1: 118 plates [82 hand-coloured]; T.II: 115 plates [2 double, 85 hand-coloured]. J.B. Baillière et fils, Paris
1829–1836	In the *Dictionnaire de médecine et de chirurgie pratique* (15 volumes) the following articles: *"Abdomen"*, *"Acéphalocystes"*, *"Adhérences"*, *"Adhésions"* *"Anatomie chirurgicale médicale"*, *"Anatomie pathologique"*, *"Artères (maladies des)"*, *" Articulations (maladies)"*, *"Entozoaires"*, *"Estomac"*, *"Muscles"* and *"Phlébites"*. Ed: Gabon, Méquignon-Marvis, Paris.
1830	*Cours d'etudes anatomiques (2 volumes) chez Béchet Jeune*, Paris.
1833	*Traité de médecine pratique éclairé par l'anatomie et la physiologie*, Paris
1834–1836	*Anatomie descriptive (4 volumes).* Béchet jeune, Paris,
1835	*Deux cas d'anomalie dans la distribution de l'artère brachiale, Bulletins et memoires de la Société Anatomique de Paris, 1835: 2*
1836	*Académie royale de médecine: Trois rapports sur un mémoire de M. Jules Guérin, relatifs aux déviations simulées de la colonne vertébrale: faits à l'Académie royale de médecine, au nom d'une commission.*
1837	*Des devoirs et de la moralité du médecin, Discours prononcé dans la séance publique de la Faculté de médecine -de Paris-, le 3 Novembre 1836*
1838	*Académie royale de médecine: Mémoire sur les déviations simulées de la colonne vertébrale, et les moyens de les distinguer des déviations pathologiques, présenté à, le 31 mai 1836 / Précédé de trois rapports faits à l'Académie royale de médecine [par J.Cruveilhier] et suivi des comptes rendus des discussions soulevées à l'Académie à l'occasion de ce mémoire; 2ᵉ mémoire sur les difformités.* Auteurs: Jules Guérin; Jean Cruveilhier
1838	*Anatomie du système nerveux de l'Homme*
1839	*Académie royale de médecine –Rapport fait à cette académie dans la séance du 22 octobre 1839 sur les pièces pathologiques modelées en relief et publiées par le docteur Félix Thibert, auteur d'un nouveau procédé*
1841	*Vie de Dupuytren. Ed. Béchet jeune et Labé, Paris*
1844	Atlas the anatomy of the human body with Constantin L.Bonami y Emile Beau. H. Bailliere, London.
1846	*Histoire de l'anatomie pathologique in Annales de l'anatomie et de la physiologie pathologiques (1846), 9–18, 37–46, 75–88.*
1849–1864	*Traité d'anatomie pathologique générale (5 volumes) J.B. Baillière et fils, Paris.*
1853	*Sur la paralysie progressive atrophique in Bulletin de l'Académie de médecine,* **18** (8 March 1853), 490–502 (29 March 1853), 546–584.
1858	*Communication a l'Académie impériale de médecine: De la fièvre puerpérale, de sa nature et de son traitement (30 March 1858), 127–155.*

4. Medical Work

J. Cruveilhier was at the same time an anatomist, a pathologist, and an experimenter [21]. He was a man with a wide humanist culture [17] who knew about the art of observation and expression; according to [22], this is visible in all his papers, but mainly in two booklets: *Vie de Dupuytren* and the so mentioned above *Des devoires et de la moralité du médicin*. The latter is considered as a profession of deontological faith in which J. Cruveilhier's sense of duty appears in all its rigour [12].

His observational ability led him to realize the importance of isolation to prevent contagion. Thereby, from his stay in the Maternity and the large number of deaths due to puerperal fever, he encouraged the creation of small clinical units instead of large departments to limit the progression of infections, preceding this way to Semmelweis' concepts on nosocomial infections (1850) [23].

His injections of mercury into the blood vessels and the bronchial system of cadavers led him to support the theory of phlebitis, which, he said, *'dominates the whole of pathology'*. It made possible the concepts of embolism and infarction, which were developed by R. Virchow, beginning in 1846. Yet, while R. Virchow considered the vascular thrombosis to be the primary lesion and the lesion in the venous wall to be secondary, J. Cruveilhier thought that alteration of the venous wall generated the thrombosis [21].

In his doctoral thesis (1816) J. Cruveilhier worked on a methodology based on the correlation of medical history with anatomical causes [23]. He expressed this shift towards a 'scientific medicine', which was taking place in France and other countries, with the phrase, *'Les systèmes passent, les faits restent'* / *'The systems pass and only the facts remain'* [13]. In 1833, in his *Traité de médicine pratique* (Table 1) he justified his particular interest in detailed anatomical studies because of the necessity to simplify this knowledge to develop proper treatments. When J. Cruveilhier was appointed Chair of Anatomy in Paris, there was no laboratory of experimental surgery, so he turned the pavilions of the *l'École pratique de dissection de Paris* into a meeting point for young surgeons and a training centre for new techniques. For more than thirty years, J. Cruveilhier spent his days in those pavilions of *l'École pratique de dissection de Paris,* collecting all the information that allowed him to develop his scientific and anatomical work.

In 1829, J. Cruveilhier began the composition of his atlas *Anatomie pathologique*, with 233 lithographed plates (Table 2), before holding the Chair of Pathological Anatomy in Paris (1836) (Figure 2). These plates were made by Antoine Chazal (1793–1854), French painter, engraver, art teacher at *Muséum National d'histoire naturelle,* and great-uncle of Paul Gauguin (1848–1903). The singularity of *Anatomie pathologique* consisted in showing the information about the clinical cases collected while the patient was still alive and relating them with the pathological findings observed systematically after the postmortem dissection of the same patients' body. This was the importance of the anatomical–clinical method developed in the nineteenth century [6]. In the same way as the comparative anatomy collections existing in the anatomical cabinets of the time, the first collections of clinical-anatomy specimens arose. In *Anatomie pathologique,* J. Cruveilhier made multiple contributions—either by description, by illustration, or both—such as the hypertrophic pyloric stenosis, the ulceration of the stomach due to hyperacidity and the dilation of the veins of the abdominal wall giving the appearance of the head of Medusa [24]; the latter condition was called Cruveilhier-Baumgarten cirrhosis. J. Cruveilhier was particularly innovative in associating the location of intracranial tumors with observed symptoms. Some examples are acoustic neuroma and the intercranial epidermoid, intracranial, and spinal meningiomas. J. Cruveilhier also provided an adequate description of disseminated sclerosis and progressive muscle atrophy [25–29].

J. Cruveilhier's histological knowledge was very limited. He only made a few histological allusions in the fifth volume of *Traité d'anatomie pathologique générale* which was edited by his students. As said above in Subheading 1, the anatomical work of J. Cruveilhier is sensualist, like M.-F.-X. Bichat's. However, J. Cruveilhier's anatomo-pathological work has aged better than some recent authors' work, despite they were able to take advantage of the use of the microscope [21].

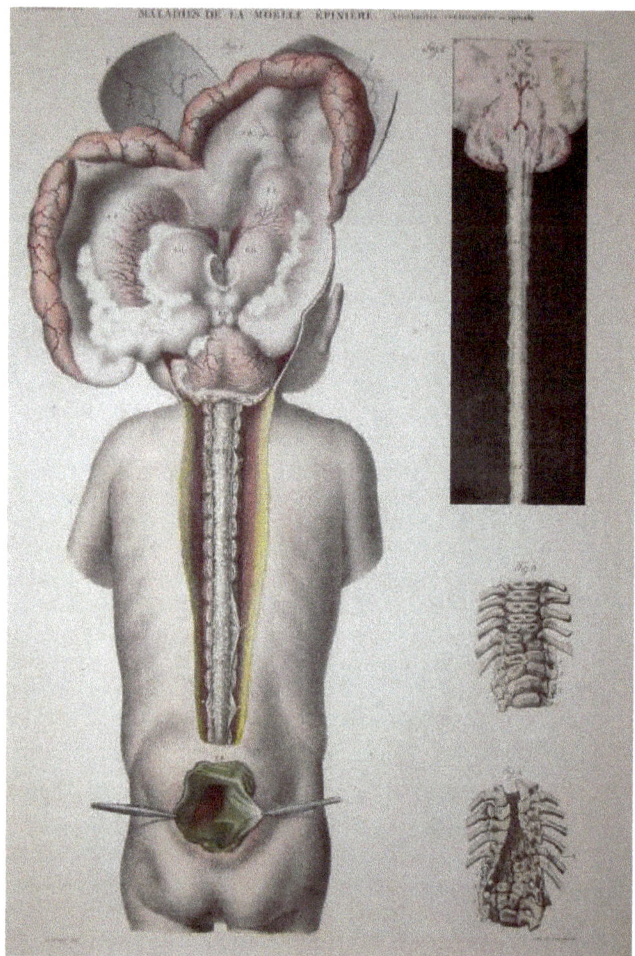

Figure 2. Diseases of the spinal cord: spina bifida and spinal and ventricular subarachnoid meningitis. Coloured lithograph by illustrator Antoine Chazal from Anatomie Pathologique Du Corps Humain (Pathological Anatomy of the Human Body) by Jean Cruveilhier, 1842.

Anatomie Descriptive

Anatomie pathologique du corps humain (1828–1842) and *Traité d'Anatomie pathologique générale* (1849–1864) (Table 2) are the best examples of J. Cruvelhier's work nowadays. This is well deserved, as seen above; however, it has obscured the other authoritative book of his authorship that merited multiple editions in his time. It is the case for Cruveilhier's *Anatomie descriptive*, which now is considered one of the best summaries of anatomy of the time [30].

The first edition of *Anatomie descriptive* was published between 1834 and 1836 (Figure 3). It was the result of an enlargement of the 24 lessons of the *Cours d'Etudes anatomiques*, which played an important role in the progress of anatomical studies at *l'École de médecine de Paris*. *Anatomie descriptive* had five editions, with successive variations in title and number of volumes (Table 3). The first two editions were entitled *Anatomie descriptive* and consisted of four volumes each. Édouard Chassaignac contributed to them [16]. The third edition, published in 1851, also consisted of four volumes, but the title changed into *Traité d'anatomie descriptive*. It was revised by J. Cruveilhier himself and is considered the best and most complete edition of this work [21]. The fourth and fifth editions were reduced to three volumes with figures (Table 3); they involved Marc D. Sée (1827–1912), *professeur agrégé* to the Faculty of Medicine of Paris and his son Edward, who appears in the fourth edition as *aide d'anatomie* and in the fifth edition as *professeur agrégé* at the Faculty of Medicine of Paris.

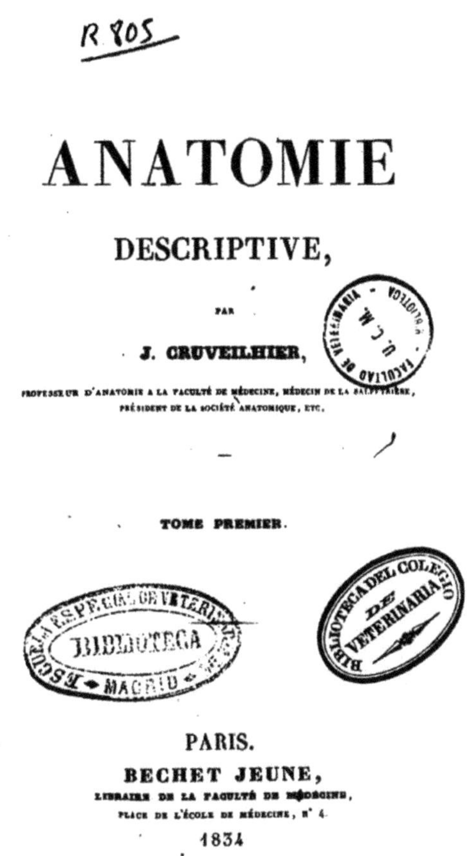

Figure 3. The cover of the first tome of J. Cruveilhier's masterwork, *Anatomie descriptive*, Paris (1834).

Unlike the anatomy treatises of the time which showed an abstract, dry, and fastidious image of the anatomy, *Anatomie descriptive* was written not only *"to exhibit the current state of the science of anatomy"*, but also with the aim of teaching an applied anatomy, as it is said on its preface. For this purpose, J. Cruveilhier always added an applicative indication, whether it was functional, surgical, or medical, after the corresponding anatomical description. He tried to impress the student so that from the beginning of its medical studies he understood the immediate applications of the anatomy and dedicated himself fervently to its study. J. Cruveilhier also wanted the reader to understand that the functions of an organ are inferred essentially from its structure: *'physiology is nothing more than the interpretation of anatomy'* (Author's preface, page XI, first English edition).

Table 3. The main editions of Cruveilhier's *Anatomie descriptive*.

Year and Place of Publication	Title (Edition, Website, Volume, Contents, Year of Publication, Pages)
1834–1836 Paris First ed.	**Anatomie descriptive (First edition)** Vol. 1. *Osteologie, Arthrologie and Dens* (1834, 543 pages) Vol. 2. *Myologie, Aponeurologie, Splanchnologie (organes de la digestion, respiration, genito-urinaires, de la génération)* (1834, 830 pages) Vol. 3. *Angéiologie, organes des sens* (1834, 539 pages ***) Vol. 4. *Névrologie,* (1836, 528 pages ***)
1841–1842 London	**Descriptive anatomy (First edition)** Vol. 1. *Osteology, Arthrology, Odontology, Myology, Aponeurology, Splanchnologie* (1841, pp: 1–638, 190 figures) Vol. 2. *Angeiology, Neurology* (1842, pp: 639–1217, 112 figures)
1843–1845 Paris	**Anatomie descriptive (Second edition)** Vol. 1. *Osteologie, Arthrologie and Dens* (1843, 615 pages) Vol. 2. *Myologie, splanchnologie (organes de la digestion, respiration, genito-urinaires, de la génération)* (1843, 764 pages) Vol. 3. *Angéiologie, organes des sens* (1843, 730 pages) Vol. 4. *Névrologie,* (1845, 839 pages)
1844 New York	**The Anatomy of the Human Body (First edition)** 1 Vol. (1844, 907 pp, 302 figures)
1851–1852 Paris	**Traité d'anatomie descriptive (Third edition)** Vol. 1. *Osteologie, Arthrologie and Dens* (1851, 620 pages) Vol. 2. *Myologie, Description du Cœur et l'Artériologie* (1852, 7844 pages) Vol. 3. *Veines, Vaisseaux lymphatiques, Splanchnologie (organes de la digestion, respiration, genito-urinaires, de la génération)* (1852, 768 pages) Vol. 4. *Appareil des sensations, Névrologie and Ovologie ou Embryogénie,* (1852, 852 pages)
1862–1867 Paris	*Traité d'anatomie descriptive* **(Fourth edition)** Vol. 1. *Ostéologie. Arthrologie. Myologie* (1862, 860 pages) Vol. 2. *Splanchnologie (organes de la digestion, respiration, genito-urinaires), organes de sens* (1865, 728 pages) Vol. 3. *Angéiologie, névrologie* (1867, 712 pages)
1871–1877 Paris	*Traité d'anatomie descriptive* **(Fifth edition)** Vol. 1. *Ostéologie. Arthrologie. Myologie* (1871, 851 pages) Vol. 2. *Splanchnologie, organes de sens* (1876, 758 pages) Vol. 3. *Angéiologie, névrologie* (1877, 736 pages)

*** Page numbering is continous along T3 and T4, T3 pages being 1–526 (index: 527–539) and T4 pages being 527–1034 (index: 1035–1046).

He also wanted the medical professional to be always aware that without anatomy, medicine would always revolve around the same circle of conceptual mistakes (from mechanism, chemism and vitalism) and that *"the study of the physiological or healthy state of organisation and of life should precede that of their pathological or diseased conditions"*(Author's preface, page VII, first English edition). J. Cruveilhier wanted students and professionals to get used to investigate eagerly the anatomical reasons for the phenomena in healthy or pathological states, and to comprehend the difference between the anatomical findings and the prior concepts of the philosophical anatomy. It could be said that J. Cruveilhier wrote *Anatomie descriptive* from the scalpel, *'to exhibit the actual state of the science of anatomy"*(Author's Introductory page, first English edition). In J. Cruveilhier's words, *'All the descriptions have been made from actual dissections. It was only after having completed from nature the account of each organ that I consulted writers, whose imposing authority could then no longer confine my thoughts, but always excited me to renewed investigations wherever any discrepancy existed '*(Author's preface page XI, first English edition). Thus, J. Cruveilhier rectified the errors transmitted from work to work and year to year by 'cabinet' anatomists. His descriptions are usually of a great accuracy, and neither the purity nor the elegance of language are excluded. His first three French editions do not contain figures, just text. Although the concept established centuries ago that illustrations were convenient and necessary on the anatomy treatises for a better comprehension of the findings reported [31], J. Cruveilhier claimed scientific descriptions fix more deeply in memory when they are well written [22].

Numerous J. Cruveilhier's contributions were collected by the anatomical eponymy of the time, and some of them still remain in the *Terminologia Anatomica* of our days [32]. Some examples are shown on Table 4, as said by various authors [9,16,32–34].

Table 4. Cruveilhier's eponyms.

Terminologia Anatomica (Latin Language) FIPAT	Cruveilhier's (C.) Eponimy
A02.1.05.047. Fossa scaphoidea	Cruveilhier fossa [1]; fossa navicularis **C.** [3]
A03.2.04.001. Articulatio atlantoaxialis mediana	Cruveilhier joint [1]; articulation de **C.** [3]
A03.3.04.010. foramen costotransversarium	trou de conjugaison postérieur de **C.** [2]
A04.1.03.011. Pars alaris Musculi nasalis	muscle pinnal transverse de **C.** [2,3]
A04.1.03.012. M. depressor septi nasi	músculo pinnal radiado de **C.** [3]
A04.3.01.005. M. transversus nuchae	músculo cutáneo suboccipital de **C.** [3]
A04.5.04.016. Corpus anococcygeum; Ligamentum anococcygeum	ligne blanche de Cruveilhier [2]
A04.7.02.026. M. adductor longus	m. segundo adductor superficial de **C.** [3]
A04.7.02.027. M. adductor brevis	m. pequeño adductor profundo de **C.** [3]
A04.7.02.028. M. adductor magnus	m. gran adductor profundo de **C.** [3]
A05.3.01.027: Fascia pharyngobasilaris	Aponévrose céphalo-pharyngienne du **C.** [2,3] Aponeurosis pterigo-faringea de **C.** [3]
A05.4.01.014. Tela submucosa	túnica fibrosa de **C.** [3]
A05.7.03.013. Taeniae coli	brides musculeuses de Cruveilhier [2]
A08.3.01.021. M. vesicoprostaticus	músculo vésico-prostático, de **C.** [3]
A09.5.02.002. Fascia perinei; Fascia investiens perinei superficialis	Cruveilhier fascia[1]; fascia de **C.** [3]
A09.5.03.004. M. transversus perinei profundus	músculo transverso anal de **C.** [3]
A09.5.02.005. M bulbospongiosus	músculo contractor de la vagina de **C.** [3]
A12.1.00.028. Tendo infundibuli	muscle compresseur de la valvule tricuspide de Sappey et **C.** [2]
A12.1.04.006. Cuspides commissurales	Cruveilhier nodules [1]
A12.2.05.011. Arteria meningea posterior (arteria pharyngea ascendens)	artère méningée de **C.** [2]
A12.2.05.036. ramus descendens arteriae occipitalis	artère cervicale postérieure de **C.** [2]
A12.2.08.043. arteria thyroidea inferior	Cruveilhier artery [4]
A12.2.08.053. arteria transversa colli; arteria transversa cervicis	artère trapézienne de **C.** [2]
A12.3.05.103. confluens sinuum (sinus durae matris)	confluent occipital de **C.** [2]

Table 4. *Cont.*

Terminologia Anatomica (Latin Language) FIPAT	Cruveilhier's (C.) Eponimy
A12.3.06.012. V. anastomotica superior (V. media superficialis cerebri)	vena de **C**. [3]
A12.3.11.007. V. saphena accessoria (V. saphena magna)	veine saphène accessoire de **C**. [2]
A12.3.11.020. V. marginalis lateralis (V. saphena parva)	veine dorsale externe de **C**. [2]
A12.3.11.021. V. marginalis medialis (V. saphena parva)	veine dorsale interne de **C**. [2]
A14.2.02.010. Plexus cervicalis posterior	plexus de **C**. [2] Cruveilhier plexus [1]
A14.3.03.047. Nervus hypogastricus (N. presacralis)	cordon plexiforme de **C**. [2]
A15.2.07.058. Pars orbitalis (Glandula lacrimalis)	glande lacrymale orbitaire de **C**. [2]
A15.2.07.071. Plica lacrimalis (Ductus nasolacrimalis)	valvule de **C**. [2]

As said by [1] Bartolucci et al. [31]; [2] Académie Nationale de Médecine [8]; [3] Rodríguez –Rivero [14]; [4] Tubbs et al. [32] and Terminologia Anatomica (FIPAT) [30].

Conceptually, J. Cruveilhier's anatomy has Vesalius foundations. J. Cruveilhier skillfully combined what he called *'rapport physiologiques'* with the description of anatomical structures (*Avant-propos*, first French edition). J. Cruveilhier claimed in his work, *'sous le rapport de l'organization, l'homme est du ressort de l'anatomie'* / *'the organization or structure of a man is the object of anatomy'* and *'sous le rapport des fonctions, l'homme est l'objet de la physiologie'* / *'the vital functions of a man are the object of physiology'*. In line with his eclecticism, he defended the necessity to resort to objects and geometric figures to facilitate the description of bones. Although he was aware of the inaccuracy of this procedure, so often used by the ancients, J. Cruveilhier claimed not to proscribe it entirely from science (*'si familier aux anciens, ne saurait être proscrit entièrement de la science'*) [30]. J. Cruveilhier followed the of-the-time typical way of describing the body, which outlined the anatomical parts and their morphological accidents, sometimes with a topographic order and sometimes with a physiological order. Thus, he preferred the topographical order for myology but the physiological order for splanchnology. Whatever the structure employed, Cruveilhier always graduated from less to more difficulty in studying the structures described, *'for the great aim in a work of instruction should be to conduct the mind gradully, from simple and easy objects to those which are more complicated'* (Page 6, English edition). Besides, to facilitate dissections, J. Cruveilhier provided in his work summaries of the best way of preparing the organ before explaining it in depth. In many respects, J. Cruveilhier was a great follower of S.-T. Sömmerring. For this reason, many of the lines of work pointed to in S.-T. Sömmerring's work were further developed in *Anatomie descriptive* [30].

J. Cruveilhier's interest in the nervous system shows to what extent he was concerned about the problems of the anatomy of his time: *'de tous les organes, il n'en est aucun dont la structure excite davantage notre curiosité, et malheureusement, il n'en est aucun don't la structure soit enveloppée de plus épaisses ténébres'* / *'the structure of no other organ in the body excites so much curiosity, and, unfortunately, there is none whose structure is involved in greater obscurity'*. J. Cruveilhier's neuroanatomical observations, being of great interest, surely deserve a review apart from the brief one made in this article.

J. Cruveilhier separated *'de l'appareil locomoteur les muscles de la face; ces muscles, aujourd'hui mieux connus, constituent an appareil musculaire spécial lequel nous plaçons en tête des organes des sens'* / *'of the musculoskeletal system the muscles of the face; these muscles, now better known, constitute a special muscular apparatus, which we place at the head of the sensory organs'*. Muscles to which J. Cruveilhier specifically intended to dedicate the second volume of his *Anatomie descriptive du système nerveux de l'Homme..*

Meanwhile, J. Cruveilhier needed to contribute to the then current lines of work about the explanation of the anatomy of the adult human body by both *l'anatomie du foetus* and the comparative anatomy (*'anatomie de l'évolution'*). He rarely dealt with *'that species of induction and analogical reasoning which, in a great measure, constitutes philosophical anatomy'*, which he considered made of *'views almost always ingenious, but usually bold and speculative'* (Author's preface page XI, first English edition).

The English edition of Cruveilhier's *Anatomy* (Table 3) was published in London and had two volumes, and the first came to light in 1841, while the second did in the following year. It was translated by Dr. W. Herries Madden (?–1883) and reviewed by Prof. W. Sharpey of University College of London. This English edition encompassed 1232 closely printed pages and upwards of 300 illustrations. It is striking that this edition had illustrations while the French edition did not. As stated in the prelude to the English edition: *'The illustrations have been selected with great care from the best source, which will be duly acknowledged. For the selection of these illustrations, and for superintending their execution, as well as for much valuable assistance in preparing the work, the Editor begs to express his obligations to Mr. John Marshall'*.

The American edition (Table 3) was published in New York in 1844, and it was a partial translation of the second French edition. It was edited by G. S. Pattison, Professor of Anatomy at the University of New York, and Member of the *Societé philomatique de Paris*. In the Editor's Preface, G.S. Pattison wrote *'Since the English edition of J. Cruveilhier has been published in London, the first and second volumes of a second edition of the work have been published by the author in Paris. The editor has carefully compared the second edition with the first, so far it has been published, and has incorporated in the American edition whatever he thought could increase its value. He has, however, only followed the second edition when he thought that the changes introduced were improvements'*. Pattison did not like the modifications that the second French edition incorporated in the myology section, the reason he kept the description of the muscles of the first French edition on his American edition. G.S. Pattison also noted *'*
... in the original work there are no engravings; this is a great desideratum, which has been removed in the English edition by the introduction of numerous woodcuts, selected with care from the best anatomical engravings, and marked with letters of reference. This greatly enhances the value of the work'. And finally, he pointed out *'Systems of Anatomy generally offer little interest except to the anatomical student. This cannot be said of the system of Anatomy of Cruveilhier. It imbodies a fund of information, in connexion with Physiology and Pathology ... '*.

5. Conclusions

By the early 19th century, the key gross parts of the human body, considered within the Enlightenment man–machine paradigm, had been described and classified primarily through the dissection of human cadavers. As a result, anatomy was the first of the so-called 'basic' sciences throughout the 19th century, and there was no medical school in which anatomy was not thoroughly taught. During that century, France was one of the main actors in a particularly brilliant period in terms of the teaching and organization of medicine, which has been attributed to the intellectual freedom directly inherited not only from the Enlightenment but also from the Revolution. Several events allowed the development of the anatomic–clinical method and led to the birth of a new discipline, pathology. The first holder of the Chair of Pathological Anatomy in Paris was Jean Cruveilhier.

Jean Cruveilhier's anatomical work, similar to M.-F.-X. Bichat's, was sensualist. His six volumes on anatomical pathology were a perfect testimony to the use of the anatomo-clinical method during the 19th century, on the one hand, and laid the foundations for the birth of pathology as a medical specialty, on the other [33]. Jean Cruveilhier's pathological work eclipsed his contributions to descriptive, gross anatomy. He published these contributions to gross anatomy in his work *Anatomie descriptive*. Many of Jean Cruveilhier's contributions to descriptive anatomy were initially collected by the anatomical eponymy of the time, and some of them remain in the *Terminologia Anatomica* of today.

Anatomie descriptive was written with two targets: firstly, *' ... to exhibit the actual state of the science of anatomy'*, and, secondly, to teach an applied anatomy. For Jean Cruveilhier, *' ... anatomy forms the first link in the chain of medical science'*. As highlighted in Jean Cruveilhier's preface to *Anatomie descriptive*, *' ... anatomy being the basis of medical science, we would greatly misapprehend its nature, did we not consider it the chief of the accessory sciences of medicine'*.

Besides, *Anatomie descriptive* offers all the clues of what descriptive anatomy will experience later. All this means that Jean Cruveilhier is recognized today as one of the most outstanding French anatomists in the first half of the 19th century. His work *Anatomie descriptive* is considered one of the best anatomical treatises of the time.

Finally, —and not only because of Jean Cruvehilier's scientific contribution but also because of the rigor that he always applied and that asked to be applied in the investigation of anatomo-clinical events linked to individual and systematic pathological processes—it is our contention that Jean Cruveilhier was a precursor of objectivity and applicability sought by evidence-based medicine today. Jean Cruveilhier, in parallel with other authors of the time, found a way to apply the scientific method in its practicality to cure patients and reform the teaching of medicine. Not using an abstract and theoretical deduction, but from direct observations beneath the bed, or in the operating room, or in the dissection room. This is the substance of Jean Cruveilhier's excellence, in the opinion of the authors of the present article.

Author Contributions: Conceptualization and editing, L.-A.A.-A.; writing—original draft preparation, L.-A.A.-A., T.F.-R. and J.-L.B.-L.; writing—review, J.-L.B.-L.; supervision, R.R.-R. All authors have read and agreed to the published version of the manuscript.

Funding: This research did not receive any specific grant from funding agencies in the public, commercial, or not-for-profit sectors.

Institutional Review Board Statement: Not applicable.

Informed Consent Statement: Not applicable.

Data Availability Statement: Not applicable.

Acknowledgments: The authors wish to express gratitude to Ana-Maria Álvarez-Castrosín for her technical assistance with manuscript preparation and Christina Navarro Collin for her assistance in Figure 1.

Conflicts of Interest: The authors declare no conflict of interest.

References

1. Laín-Entralgo, P. *Historia de la Medicina*; Salvat Editores, S.A.: Barcelona, Spain, 1978; Volume 4.
2. Huard, P.; Imbault-Huart, M.-J. Concepts et réalités de l'éducation et de la profession médico-chirurgicales pendant la Révolution. *J. Des Savants* **1973**, *2*, 126–150. [CrossRef]
3. Huard, P.; Imbault-Huart, M.-J. L'enseignement libre de la médecine à Paris au XIX e siècle. *Rev. D'histoire Des Sci.* **1974**, *27*, 45–62. [CrossRef]
4. Huard, P.; Imbault-Huart, M.-J. Structure et fonctionnement de la Faculté de Médecine de Paris en 1813. *Rev. D'histoire Des Sci.* **1975**, *28*, 139–168. [CrossRef]
5. Foucault, M. *El nacimiento de la Clínica: Una Arqueología de la Mirada Médica*, 20th ed.; Siglo XXI Editores: Mexico City, Mexico, 2001; p. 293.
6. de Saint-Maur, P.P. The birth of the clinicopathological method in France: The rise of morbid anatomy in France during the first half of the nineteenth century. *Virchows Arch.* **2012**, *460*, 109–117. [CrossRef]
7. Imbault-Huard, M.-J. L'École pratique de dissection de Paris de 1750 à 1822. *Annu. De L'école Prat. Des Hautes Études* **1971**, *103*, 841–850. [CrossRef]
8. Shoja, M.M.; Tubbs, R.S.; Loukas, M.; Shokouhi, G.; Ardalan, M.R. Marie-François Xavier Bichat (1771–1802) and his contributions to the foundations of pathological anatomy and modern medicine. *Ann. Anat.* **2008**, *190*, 413–420. [CrossRef]
9. Médecine, A.N.d. Cruveilhier. In *Dictionnaire Médical de l'Académie de Médecine*; CILF, 11 rue de Navarin: Paris, Italy, 2016.
10. Vayre, P. Heurs et malheurs de trois chirurgiens limousins de la Révolution française au Second Empire. *Hist. Des Sci. Medicales* **2010**, *44*, 179–187.
11. Delhoume, L. *L'école de Dupuytren; Jean Cruveilhier*; J.-B. Bailliére: Paris, Italy, 1937.
12. Orcel, L.; Vetter, T. Dupuytren, cruveilhier and the anatomical society (author's transl). *Arch. D'anatomie Et De Cytol. Pathol.* **1976**, *24*, 167–179.
13. Androutsos, G.; Vladimiros, L. The eminent French pathologist Jean Cruveilhier (1791–1874) and his works on cancer. *J. BUON Off. J. Balk. Union Oncol.* **2006**, *11*, 369–376.
14. Vayre, P. Jean Cruveilhier (1791–1874) Chirurgien promoteur de la preuve par les faits à la médecine fondée sur la preuve. *e-Mémoires De L'académie Natl. De Chir.* **2008**, *7*, 1–12.
15. Petit, M.A. *Traité de la Fièvre Entéromesentérique: Observée . . . a L'hôtel-Dieu de Paris, Dans le Années 1811, 1812 et 1813*; Hacquart: Paris, Italy, 1813.
16. Rodriguez-Rivero, P.D. *Eponimias Anatomicas*; Sociedad Venezolana de Historia de la Medicina: Caracas, Venezuela, 1939.

17. Waring, J.I. William Middleton Michel in Paris, 1842–1846 A Vignette of Cruveilhier. *J. Hist. Med. Allied Sci.* **1968**, *23*, 349–355. [CrossRef]
18. Tricoire, J.-L. L'anatomie et les anatomistes toulousains de 1789 à 1940. *Morphologie* **2016**, *100*, 112–113. [CrossRef]
19. Brogna, C.; Fiengo, L.; Türe, U. Achille Louis Foville's atlas of brain anatomy and the Defoville syndrome. *Neurosurgery* **2012**, *70*, 1265–1273. [CrossRef] [PubMed]
20. Roussy, G. Eloge de Jean Cruveilhier. *Assoc. D'anatomie Pathol. Et D'anatomie Norm. Chir.* **1926**, *9*, 1–19.
21. Huard, P. Jean Cruveilhier. In *Dictionary of Scientific Biography*; Gillispe, C.C., Ed.; Charles Scribner's Sons: New York, NY, USA, 1980; Volume 3, pp. 489–490.
22. Richelot, H. Cruveilhier (Jean). In *Dictionnaire de la Conversation et de la Lecture*; Garnier Frères, Libraires: Paris, Italy, 1845; Volume 58, pp. 63–65.
23. Berhouma, M.; Dubourg, J.; Messerer, M. Cruveilhier's legacy to skull base surgery: Premise of an evidence-based neuropathology in the 19th century. *Clin. Neurol. Neurosurg.* **2013**, *115*, 702–707. [CrossRef] [PubMed]
24. Park, R.; Park, M.P. Caput Medusae in medicine and art. *BMJ Br. Med. J.* **1988**, *297*, 1677–1679. [CrossRef] [PubMed]
25. Flamm, E.S. The neurology of Jean Cruveilhier. *Med. Hist.* **1973**, *17*, 343–355. [CrossRef]
26. Compston, A. The 150th anniversary of the first depiction of the lesions of multiple sclerosis. *J. Neurol. Neurosurg. Psychiatry* **1988**, *51*, 1249–1252. [CrossRef]
27. Pearce, J.M. Cruveilhier and acoustic neuroma. *J. Neurol. Neurosurg. Psychiatry* **2003**, *74*, 1015. [CrossRef]
28. Pearce, J.M. Historical descriptions of multiple sclerosis. *Eur. Neurol.* **2005**, *54*, 49–53. [CrossRef]
29. Drouin, E.; Drouin, A.-S.; Péréon, Y. Cruveilhier versus Charcot. *Lancet Neurol.* **2016**, *15*, 362. [CrossRef] [PubMed]
30. Balaguer-Perigüell, E.; Ballester-Añón, R. Los saberes morfológicos durante el romanticismo. La Anatomía. In *Historia Universal de la Medicina. Barcelona: Salvat*; Laín-Entralgo, P., Ed.; Salvat, S.A.: Barcelona, Spain, 1973; Volume 5, pp. 179–180.
31. Kemp, M. Style and non-style in anatomical illustration: From Renaissance Humanism to Henry Gray. *J. Anat.* **2010**, *216*, 192–208. [CrossRef] [PubMed]
32. Federative Committee on Anatomical Terminology (FCAT). *Terminologia Anatomica*, 2nd ed.; Georg Thieme Verlag: Stuttgart, Germany, 2011; p. 292.
33. Bartolucci, S.L.; Stedman, T.L.; Forbis, P. *Stedman's Medical Eponyms*, 2nd ed.; Lippincott Williams & Wilkins: Baltimore, MD, USA, 2005; pp. 163–164.
34. Tubbs, R.S.; Shoja, M.M.; Loukas, M. *Bergman's Comprehensive Encyclopedia of Human Anatomic Variation*; Wiley: Hoboken, NJ, USA, 2016; p. 1432.

Review

René-Édouard Claparède (1832–1871), Pioneer Protozoologist and Comparative Anatomist

Penelope A. Kollarou and Lazaros C. Triarhou *

Department of Psychology, Aristotelian University Faculty of Philosophy, 54124 Thessaloniki, Greece
* Correspondence: triarhou@psy.auth.gr

Abstract: The pioneer Swiss naturalist René-Édouard Claparède (1832–1871), professor at the University of Geneva, left important contributions to diverse areas of natural science, biology, and comparative anatomy, including the structure of infusoria, annelids, and earthworms, the evolution of arthropods, and the embryology of spiders. He also published observations on marine invertebrates. This essay presents a brief overview of his academic life and work and makes the distinction from his nephew with the same name, the neurologist and educational psychologist Édouard Claparède (1873–1840).

Keywords: comparative anatomy; history of biology; neuroscience; zoology

Citation: Kollarou, P.A.; Triarhou, L.C. René-Édouard Claparède (1832–1871), Pioneer Protozoologist and Comparative Anatomist. *Anatomia* **2023**, *2*, 165–175. https://doi.org/10.3390/anatomia2020015

Academic Editors: Gianfranco Natale and Francesco Fornai

Received: 14 March 2023
Revised: 6 May 2023
Accepted: 2 June 2023
Published: 6 June 2023

1. Introduction

The Swiss naturalist René-Édouard Claparède (Figure 1A) was one of the most skilful, laborious, and honoured European zoologists [1]. He left important contributions to various areas of biology and natural science, including the structure of infusoria, annelids, and earthworms, the evolution of arthropods, and the embryology of spiders. He also published observations on diverse, mostly marine, invertebrates, including protozoa. His zoological publications left their mark on the evolution of comparative anatomy. There is little information about his life and work in the modern biomedical literature. In some instances, his work is confused with that of his nephew, the neurologist Édouard Claparède (1873–1940). The aim of the present study is to provide a brief outline of the academic life and research work of René-Édouard Claparède, 'senior'.

Figure 1. (A) René-Édouard Claparède. Undated, oil on canvas by unknown artist. Credit: Bibliothèque de Genève, Switzerland. https://notrehistoire.ch/entries/2PDBm5mPBbk (accessed on 22 February 2023). Signature from a manuscript dated 11 April 1865. Credit: The Waller Manuscript Collection, Uppsala Universitetsbibliotek, Sweden. http://waller.ub.uu.se/23553.html (accessed on

22 February 2023). (**B**) Élie Metchnikoff in the early 1910s. Credit: George Grantham Bain Collection, Library of Congress Prints and Photographs, Washington, DC. https://www.loc.gov/pictures/item/2014694632 (accessed on 22 February 2023). Signature from a manuscript dated 28 March 1914. Credit: The Waller Manuscript Collection, Uppsala Universitetsbibliotek, Sweden. http://waller.ub.uu.se/36382.html (accessed on 22 February 2023).

2. Early Life and Academic Career

René-Édouard Claparède lived and worked in an era that witnessed some landmark events in the biological sciences and technology (Table 1). He was born on 24 April 1832 in Chancey, a Swiss village near the French border, not far from Geneva. Of French ancestry, the family had taken refuge after the revocation of the Edict of Nantes, a religious persecution. Claparède's father was a pastor. Although the family was highly cultivated, René-Édouard was the first to be involved in scientific research. Driven by an interest in nature and an inquisitive mind, the meticulous analyst sought precision in the details. Through rational reasoning, René-Édouard delved into the organisation of animals, as well as their similarities and differences [1,2].

Table 1. Chronology of select events in science and technology during Claparède's lifetime.

Year	Event
1832	René-Édouard Claparède born on 24 April; French naturalist Georges Cuvier dies on 13 May
1833	Swedish inventor Alfred Nobel born
1834	German zoologist Ernst Haeckel born
1837	Louis Daguerre invents photography
1838	Theodor Schwann formulates cell theory
1843	Italian pathologist Camillo Golgi born; German physician Robert Koch born
1845	Russian-French microbiologist Élie Metchnikoff born
1849	Hippolyte Fizeau and Léon Foucault measure speed of light; Russian physiologist Ivan P. Pavlov born
1852	Spanish histologist Santiago Ramón y Cajal born
1855	Alfred R. Wallace publishes *Law which has regulated the introduction of new species*
1856	Austrian psychoanalyst Sigmund Freud born; Serbian-American inventor Nikola Tesla born
1857	Russian physiologist Vladimir M. Bekhterev born; English physiologist Charles S. Sherrington born
1859	Charles R. Darwin publishes *Origin of species*; Rudolf Virchow publishes *Die Cellularpathologie*
1863	Thomas H. Huxley publishes *Man's place in nature*
1865	Claude Bernard publishes *Introduction à l'étude de médecine expérimentale*
1866	Gregor Mendel proposes basic principles of heredity
1869	Friedrich Miescher discovers 'nuclein' (now known as DNA)
1871	Charles R. Darwin publishes *Descent of man* on 24 February; Claparède dies on 31 May

In 1852, Claparède began to study natural science, medicine, and northern languages in Berlin, where he became a pupil of the eminent zoologist Johannes Müller (1801–1858) a year later. With Müller, they journeyed to Norway in 1855 on a desolate reef. The young student spent time drawing details of marine animals through the microscope [2].

Claparède was also influenced to a great degree by Christian Gottfried Ehrenberg (1795–1876), the German naturalist, zoologist, comparative anatomist, geologist, and microscopist, whose research concerned the anatomy and metamorphoses of echinoderms. Due to the influence of Ehrenberg, Claparède developed a strong interest in micrography.

Between 1851 and 1852, Claparède attended courses taught by renowned experts in paleontology, including Georges Cuvier (1769–1832), Étienne Geoffroy Saint-Hilaire (1772–1844), and Augustin Pyramus de Candolle (1778–1841) [1].

Claparède obtained his M.D. degree in 1857, but he never practised clinical medicine. In 1862, he was appointed professor of comparative anatomy at the University of Geneva. He became a member of the Physical Society, the Medical Society, and the Geneva National Institute. At the Geneva Academy, he was trained by François-Jules Pictet de La Rive (1809–1872), a Swiss zoologist, palaeontologist, and advocate of progressive creationism. As a professor of zoology, next to Pictet who always recognised the academic potential in his apprentice, Claparède, in his lectures, defended opposing views to those of earlier authors, a fact that created great displeasure in the church circles. Moreover, he became involved in editing the series titled 'Universal Library of Sciences, Literature and Arts' (*Bibliothèque Universelle des Sciences, Belles-Lettres et Arts*), and served as a reviewer for the 'Swiss Archives' (*Archives Suisses*), with "his admirable style of writing, especially the reviews and criticisms" [3]. Claparède was one of the most laborious editors of the 'Bulletin of the Archives of the Universal Library' (*Bulletin des Archives de la Bibliothèque Universelle*), publishing numerous scholarly analyses and reviews over a period of 15 years [2]. He had already authored several important papers during his student years, most of them appearing in 'Müller's Archive for Anatomy, Physiology and Scientific Medicine' (*Archiv für Anatomie, Physiologie und Wissenschaftliche Medicin von Dr. Müller*), such as the 'Anatomy and developmental history of *Neritina fluviatilis*' (1857), 'Supplement to a memoir by G.-R. Wagener: On *Dicyema* et cetera' (1857), and 'Contribution to the anatomy of *Cyclostoma elegans*' (1858), which were works that secured him an important place among zoologists of that era. For instance, in a review on *Actinophrys eichhornii*, Claparède documented a large contractile vesicle considered to be a heart-shaped organ [2,4]. He further described the mode of digestion in these animals, capable of enveloping and digesting vegetable and animal matter through any part of their body through an orifice that served either as a mouth or an anus, a feature that would classify them among the rhizopods [2].

René-Édouard could write in French and German equally well; therefore, some of his works are found in German periodicals, others in French, such as in the 'Transactions of the Academy of Geneva'. Being bilingual, he had access to both the French and German scientific literature and, consequently, appeared critical to naturalists who might exhibit superficiality or dishonesty in their field of study. For instance, the research of the French embryologist Édouard-Gérard Balbiani (1823–1899) on the development of *Aphides* became the subject of a three-year investigation by Claparède. After meticulously studying the publications of Balbiani on the embryology of *Aphis* while in Naples, Claparède came to the conclusion that they were unfounded and inconsistent with the work of Balbiani's predecessors. Félix Dujardin (1801–1860) was also treated harshly in *Recherches sur les infusoires*, as was Charles Marie Benjamin Rouget (1824–1904), who appeared to have personally resented Claparède's rectification [3]. In 1860, Claparède married his cousin, Eveline Claparède (1840–1910), and became financially independent. His house became a centre of scientific exchange [2].

Although famous in the scientific world, Claparède was not widely known to the general public. It was in 1860 that he gained renown through a popular course that he offered in Geneva. Without a doubt, he was one of the individuals who contributed the most to overthrowing the old conservative prejudices against modern science [2].

Upon publication of Darwin's *Origin of species*, Claparède became one of the first scientists in Switzerland and the French-speaking world to promote Darwinian natural selection and to further support it with his own observations, especially the detention organs of certain species of mites. At the end of his article on the development of mites, Claparède devotes a chapter in support of Darwin's theory, by showing that the apparatus which serves as a holdfast in parasitic mites escaped the law of homology. In effect, it was not a fixed organ, fulfilling such functions, but on the contrary, an organ modified in various species by adaptation to similar functions. In some species, it became the forelimb, in others the hindlimb; in *Listophorus larisi*, it is the lower lip that transformed into a fixed organ [2].

He epitomized Darwin's theories in the quote: "It is better to be a perfect ape than a degenerate Adam" [5]. Along with his colleague Carl Vogt (1817–1895), a prominent naturalist in Geneva, and Thomas Henry Huxley (1825–1895) in England [1], Claparède made efforts to popularise Darwinism [6]. Claparède adopted Darwin's theory of natural selection and published a series of pertinent articles on the subject [7]. He further discussed the insight of the leading evolutionary thinker Alfred Wallace on natural selection [8]. Although Darwin hardly considered the various reviews noteworthy, he singled out Claparède with the remark: "There is a favourable and long review in *Revue Germanique*, which is important from coming from so good observer, as Claparède" [4,9].

In 1862, the Society of Sciences in Utrecht awarded Claparède a gold medal for his research on the embryonic development of spiders (dorsal position of the embryo during the first development period, instead of turning over to roll up in the belly as in other arthropods). Claparède further investigated their blood circulation and observed that the blood escaping from the heart circulated not from back to the front, but in the opposite direction, as Franz von Leydig (1821–1908) had already proposed [2,10].

3. Contributions to Zoology and Comparative Anatomy

In Berlin, Claparède, in collaboration with Johannes Lachmann (1832–1860), made a substantial effort to study infusoria and rhizopoda (*Études sur les infusoires et les rhizopodes*, 3 parts, 1859–1868), which was one of the earlier extensive works, and a most impressive monograph, comprising some 200 protist species, most described for the first time [11,12]. In that study, the authors presented parasitic forms (the ciliate *Balantidium*), a variety of invertebrate groups, tintinnid ciliate species, and dinoflagellate species. The observations on dinoflagellates, known as *peridiniens*, led them to state that many forms appeared to contain ingested food items, suggesting that dinoflagellates were more animal-like than plant-like [10]; Claparède and Lachmann submitted their tome *Sur la réproduction des infusoires* in 1855. It consisted of 302 pages and 11 plates containing 192 figures. This work, along with another authored by Nicolaus Lieberkühn, earned the grand prize in the physical sciences from the Academy of Sciences of Paris for the year 1856. Claparède and Lachmann were quite young when they won, both 26 years old, while Lieberkühn was 36, and later became an expert on sponges [3,10,12,13]. Lachmann died before the work was completed, in 1860, having not quite reached his 28th birthday. Additionally, this study immediately placed its authors among the experts in zoology and formed the foundation of the modern views on infusoria. Furthermore, Claparède and Lachmann showed that infusoria are neither as complicated as Ehrenberg had argued, nor as simple as claimed by Félix Dujardin (1801–1860), whose theory had long dominated the field [10]. Although Claparède and Lachmann initially supported the 'polygastric theory' of Ehrenberg, they subsequently made taxonomic refinements and advances, firmly establishing the suctoria as a separate and definitive group. Lachmann [11] proved wrong the view of Friedrich Stein (1818–1885), wherein suctoria were larval stages of peritrichs. They established the affinities of infusoria with the worms and the coelenterates on one hand, and with rhizopods on the other, satisfactorily classifying them for the first time. They also distinguished 10 families, described several species, and extended the knowledge on their organisation [2,9,11].

Although their work was overshadowed by subsequent publications by Georg August Zenker (1855–1922), Ronald Cohn (1943–2022), and other investigators who had better instruments at their disposal, the studies of Claparède and Lachmann are the ones that set the foundations of modern research on infusoria [9].

In the summer of 1859, Claparède visited the English physician William Benjamin Carpenter (1813–1885), an invertebrate zoologist, physiologist, and one of the founders of the modern theory of the adaptive unconscious. They worked together on the microscope in the Hebrides, chiefly on worms and annelids. Their expedition resulted in an account of new marine species related to earthworms, richly illustrated with plates and many observations on turbellarian worms. Papers were published in the *Proceedings of the Royal Physical Society of Edinburgh*, 'Reichert's Archive', and the 'Reviews of the Physical Society and Natural History of Geneva' (*Mémoires de la Société de Physique et d'Histoire Naturelle de Genève*). Jointly, with Carpenter in 1860, Claparède also published certain observations on *Tomopteris onisciformis* in the *Transactions of the Linnean Society of London* [2,3].

In Geneva, Claparède continued to observe the limicolous annelids. In his 'Anatomical research on oligochaetes', published by the Geneva Academy, he described the structure of the worm nervous system and the three large tubular fibres for the first time. This research appeared in print in the 'Reviews of the Physical Society and Natural History of Geneva' (*Mémoires de la Société de Physique et d'Histoire Naturelle de Genève*). With this work, he enriched zoological knowledge with a complete account of anatomical and systematic differences among many worms, such as the demonstration of the homology of the segmental organ with the reproductive ducts. Until then, such details were neglected and misunderstood [2,3].

From 1854 to 1857, Claparède and Lachmann were members of a study group in Berlin under Müller's guidance. Another member of that group, who later became a lifelong friend, was the zoologist and naturalist Ernst Haeckel (1834–1919), who promoted Darwin's work in Germany. Claparède and Lachmann had a great influence on Haeckel's work on protists, as well as Radiolaria, in particular [10].

In the winter of 1866, despite his severe health problems, Claparède devoted himself to his immense studies on the annelids of the Gulf on Naples and other seacoasts (Normandy, Hebrides), a "most striking discovery" [3] that largely filled volumes 19 and 20 of the *Mémoires de la Société de Physique et d'Histoire Naturelle de Genève*. This work presented many new forms and a wealth of anatomical and physiological facts depicted in more than 50 heavy plates. Claparède found that *Nereis dumerilii* lays sexually-fertilised ova, which produce a worm that belongs to a distinct genus (*Heteronereis*) after hatching. This worm lays similar ova, which sometimes produce a second kind of *Heteronereis*, or in other seasons, the original form of *Nereis dumerilii* again. This is a real case of alternation of 'sexual generations' on record [3]. His research on the structure of sedentary annelids (*Recherches sur la structure des Annélides sédentaires*) was posthumously published in 1873 [4,12].

In 1867, Claparède published a study on the nervous system of the earthworm (*Lumbricidae*), where he also observed the arrangement of the dorsal intromittent organ (*cirrus*) and announced its developmental history at the Congress of the Swiss Natural Research Society, held in Einsiedeln [9]. Moreover, between 1867 and 1869, Claparède collaborated with Élie Metchnikoff (1845–1916), who was working in Odessa at the time, on the developmental history of the chaetopods, publishing an early and rare work with six plates [13] (Figure 2).

Metchnikoff (Figure 1B), a Russian-French zoologist, cytologist, embryologist, and immunologist, is considered the 'father of natural immunity'. In 1882, he discovered phagocytes or macrophages, the differentiated form of mononuclear leukocytes, when they migrate from the blood into other tissues. In 1908, he was jointly awarded the 1908 Nobel prize in physiology or medicine with the German physician Paul Ehrlich (1854–1915) for their discoveries concerning immunity. Metchnikoff established the concept of cellular immunity, while Ehrlich established the concept of humoral immunity. Having graduated from the University of Kharkov in 1864, Metchnikoff studied invertebrate and fish embryology at several European centres and earned his doctorate from the University of Saint Petersburg in 1867. After teaching and conducting research in Saint Petersburg and Odessa for a year, he occupied a post in Messina, Italy. It was there that he began his immunological studies. He hypothesized that there were cells that become actively mobilised against foreign bodies to either kill and assimilate them or to destroy themselves in the process. On his way back to Odessa, he stopped in Vienna, where Carl Claus (1835–1899), the professor of zoology, suggested to him the terms 'phagocyte' and 'phagocytosis'. Metchnikoff confirmed the defence role of phagocytes by conducting experiments with the freshwater crustacean *Daphnia* after administering spores of the parasitic fungus *Monospora bicuspidata*, which were attacked by the host's phagocytes.

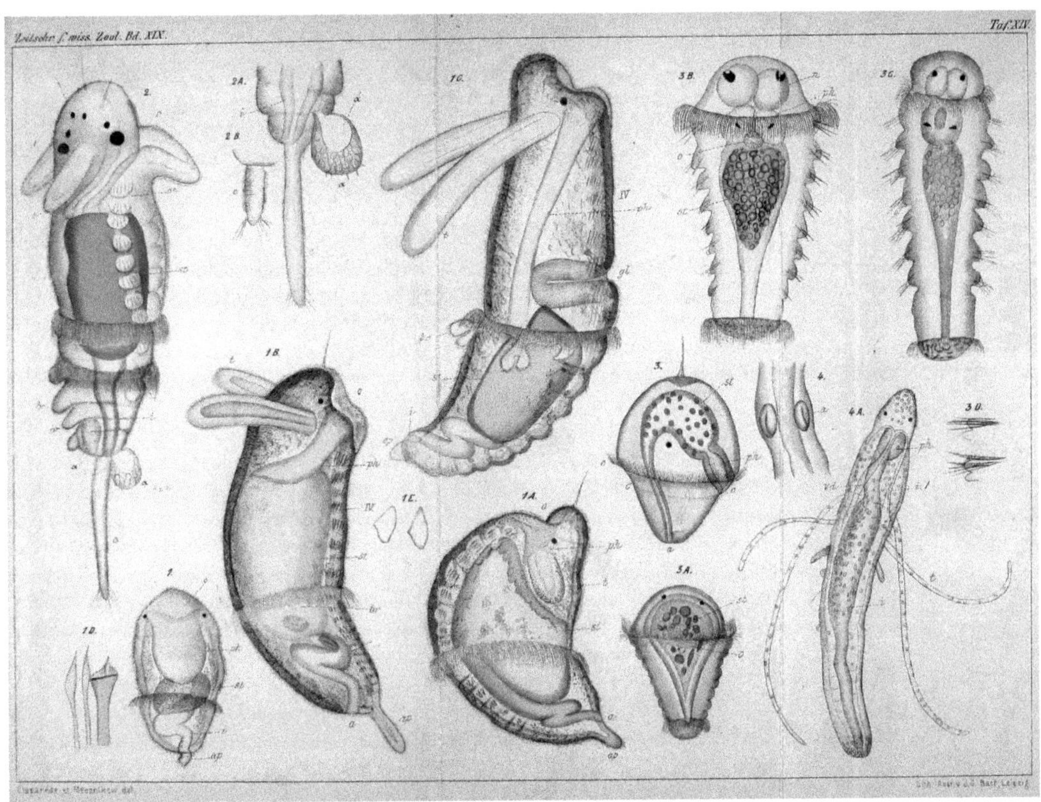

Figure 2. A plate from the study by Claparède and Metchnikoff [14] depicting the larval and later developmental stages in several indigenous marine worms from the Gulf of Naples. Drawings 1–1E: *Telepsavus Costarum Claparède*. ×75 (1), ×40 (1A, 1B, 1E), ×35 (1C), ×150 (1D). Drawing 2: A larva probably belonging to *Phyllochaetopterus socialis Claparède*. ×50. Drawings 3–3D: Development of *Nephthys scolopendroides Delle Chiaje*. ×130 (3), ×150 (3A–3D). Drawings 4–4A: Developmental stages of *Cirratulus spectabilis Kinberg*. ×20. Source: The Biodiversity Heritage Library. https://www.biodiversitylibrary.org/item/159258 (accessed on 22 February 2023).

Metchnikoff and Claparède were both influenced by Darwin's theory of natural selection and by Haeckel's biogenetic principle. In the last phase of his career, Metchnikoff delved into the study of the ageing of organisms and the extension of life. He is the investigator who introduced in medicine the term 'Gerontology' for the emerging science of ageing and longevity. Moreover, he found that certain bacteria (today's 'probiotics') may protect against pathogenic microbes in the intestinal flora.

Although Claparède had a predilection for lower species, he occupied himself with the most varied subjects and often wrote extensive notices aimed at offering a summary of recent work on diverse scientific topics. One finds many reviews of this kind related to topics in physiology, zoology, geology, and even archaeology, while in other articles, he addressed advanced problems of natural philosophy [2].

In 1858, Claparède made certain comments on the theory of binocular vision and the horopter, leading him to study the visual physiology of the compound eye and publish various reviews. By completing numerous experiments, he confirmed the demonstrations of Jean-Louis Prévost (1838–1927) regarding the visual area covered by the two eyes through the aiming point and the optical centres. These studies led him to the study of the compound eye of arthropods, the evolution of which he followed in various nymphs. The resulting paper was published in 1863 in Leipzig [2,9,15]. He concluded that the theory of vision in insects, as formulated by Müller, was not tenable, as the animal would be so nearsighted that it could barely see a few feet away.

Additionally, there was a collection of miscellaneous observations, finely illustrated and published on zootomic gleanings (*Glanures zootomiques*) and observations on invertebrates (*Beobachtungen über wirbelloser Thiere*) about annelids, free-living worms, and forms of paradoxical marine larvae. His works covered the circulation of spiders (transparent young of the genus *Lycosia*), new modes of reproduction, new anatomical details and physiological observations of rare forms, and the development of the marine gastropod *Neritina fluviatilis*. Claparède showed that this gastropod was not a hermaphrodite, and its testaceous operculum had a different structure from that of the shell. Such a claim went against the view of John Edward Gray (1800–1875), who argued that the operculum was a second atrophied valve [9].

Claparède's doctoral dissertation centred on *Cyclostoma elegans*, a freshwater bivalve, a gastropod, in which he described a calcareous organ composed of concentric layers, an organ that had remained unknown until then in gastropods; it appeared in Müller's *Archiv für Anatomie und Physiologie* [2,12] (Figure 3A). Moreover, Claparède produced an elaborate work on the development of nematode worms, with many new and detailed facts (including the significance of the parts of their ova), in German. His studies on the anatomy of acarids provided new details, such as the modification of dissimilar parts in different genera to form identical organs, which Claparède explained satisfactorily through Darwinian evolutionary theory [2,3]. He also clarified the special conditions of the blood circulation in *Acarina* mites (the existence of lateral openings in the heart that gape during diastole) [9].

Another study, abundant with curious facts, concerned the particular discovery of a double and even triple interlocking of the egg, a phenomenon that the author designated with the terms *deutovum* and *tritovum*. However, such a singular phase of development is not found in all species; it is lacking in *Tetranychus*, which lives on plants; the *deutovum* is observed in *Unionicola bonzi* (*Atax bonzi*), common parasites of freshwater mussels living on the gills or mantle of their hosts, and the *tritovum* appears in *Myobia*, in particular *Tyrophagus muris*, a parasite that infects mice [2].

His last published work was in the 'Journal of Scientific Zoology' (*Zeitchrift für Wissenschaftliche Zoologie*) and consisted of observations on the anatomy and reproduction of marine polyzoa, accompanied by three colour plates. The minutest and most careful piece of work that he ever produced was a review on the histology of the earthworm, illustrated with six colour plates [3,16]. The biologist Émile Yung (1854–1918) compiled, in 1904, some of Claparède's works. Henri de Saussure (1829–1905), a Swiss mineralogist and entomologist and a prolific taxonomist, in his 'Notice sur Édouard Claparède' (1871), listed 58 published and unpublished works by Claparède, including analyses of works of other authors by Claparède for the 'Bulletin of the Archives of the Universal Library' [1,2]. Due to Claparède's premature death, a large work on the embryology of insects and microscopic preparations on the histology of annelids remained unpublished [2,3].

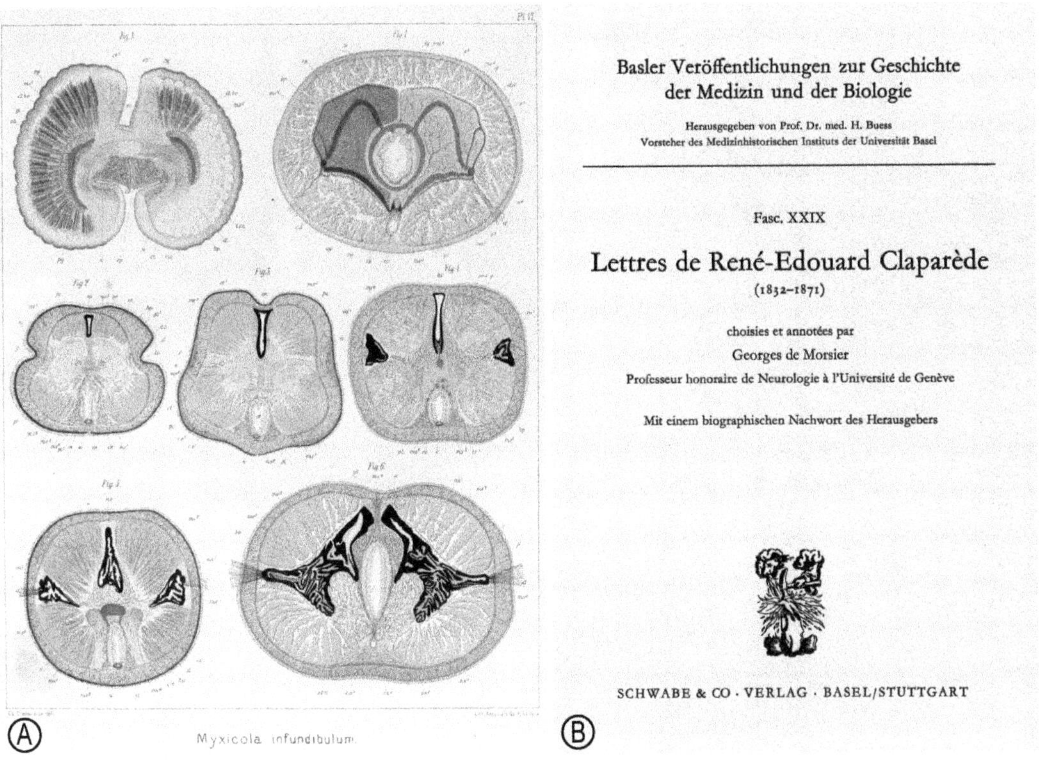

Figure 3. (**A**) Histological structure of the polychaete worm *Myxicola infundibulum*, from Claparède's posthumous work on sedentary annelids [12]. Source: The Internet Archive. (**B**) The book of Claparède's correspondence to his family, compiled and edited by the French-Swiss neurologist Georges de Morsier (1894–1982), former director of the Geneva University Neurological Clinic, including a biographical note by the Swiss gynaecologist and medical historian Heinrich Buess (1911–1984), former director of the Department of Medical History at the University of Basel [9]. (Author's private library).

4. Personal Life

Table 2 summarises the chronicle of main events in Claparède's life and scientific work. To commemorate the centenary of the death of Claparède, Georges de Morsier (1894–1982), a French-Swiss neurologist and honorary professor at the University of Geneva, produced a booklet with extracts from 231 of Claparède's letters, written between 1855 and 1871 and addressed to family members (Figure 3B). In those letters there were ironic allusions to Samuel Hahnemann (1755–1843), a German physician, known as the founder of homeopathy, to Müller and to Louis-Napoleon Bonaparte (1808–1873), the future Napoleon III and first President of France. Additionally, in some of these correspondence items, Claparède, similar to many Swiss citizens, made satirical remarks about Prussian politics, when relations between Switzerland and Prussia were quite tense because of the Neuchâtel Affair [9,15]. There is an afterword by Heinrich Buess (1911–1984), a Swiss gynaecologist and head of the Medical History Institute of the University of Basel. It is written in German, while the rest of the book is written in French [15].

Table 2. Chronology of events in the life and scientific career of René-Édouard Claparède.

Year	Events
1832	Born in Chancey, Switzerland
1852	Enters Friedrich Wilhelm University in Berlin
1853	Studies under Johannes Müller
1855	Journeys to Norway; completes study on infusoria with Johannes Lachmann
1856	Awarded *Grand Prix* from the French Academy of Sciences
1857	Earns M.D. degree from Berlin University
1858	Comments on theory of binocular vision
1860	Marries Eveline Claparède
1862	Appointed professor of comparative anatomy in Geneva; receives Gold Medal from Utrecht Society of Sciences
1866	Investigates coastal annelids from Naples, Normandy, and Hebrides
1867	Publishes on the nervous system of the earthworm; begins collaboration with Élie Metchnikoff
1871	Dies in Siena, Italy

Claparède was a resolute advocate of the philosophy of Immanuel Kant (1724–1804) and a convinced subjectivist. He seemed to lean toward dynamic pantheism, in other words, the argument that repulsive force is required for matter to fill space. Such a trend was common among naturalists, and led, early in his career, to frequent disagreement with persons adhering to dogma. Nevertheless, a few years later, as he gained fame, Claparède became appreciated by his critics [2].

Based on a series of unpublished letters, Claparède reportedly suffered from various illnesses over two decades, beginning in his student years, when he was stricken by arthritis, which also affected his heart and led to further complications in 1854. Despite the lack of effective remedies at the time, Claparède responded to his treatment by internist Heinrich Quincke (1842–1922), who is known for the introduction of lumbar puncture as a diagnostic and therapeutic method. Claparède also underwent squint surgery by Albrecht von Graefe (1828–1870), the Prussian pioneer in ophthalmology [9]. The strict diet that Claparède followed in order to bring about some relief of symptoms only worsened his weakness and led to further complications that even his physicians could not predict. A frequent neuralgia caused excruciating pain and, to cope with it, Claparède resorted to extreme means [2]. Shortly before his final exams in 1857, he suffered a severe episode of heart failure. In 1865–1866, he survived typhoid fever, and his work was interrupted. He also suffered from rheumatic pain in the right knee, which worsened as a result of incorrect medical treatment, and from haemorrhagic cough, in addition to hearing problems. Yet, the researcher in him presented one treatise after another and gained international reputation. He died on 31 May 1871 in Siena, one month after his 39th birthday, most likely of tuberculosis. He was survived by his wife and their two young children [2,4,9,10].

Claparède bequeathed his scientific library of books and documents to the City of Geneva by a deed of will. The rich collection filled an important gap, which had long occurred in the Geneva Public Library. His wish was that, at the conclusion of his own career, others would continue to benefit from the records that he had accumulated during a lifetime of research work. Moreover, he also wished to promote scientific development of his native city. Thus, through this act of beneficence, future generations would have a chance to become aware of his personal efforts [2].

In Geveva, the Claparède Plaza was dedicated in his memory. Furthermore, the Lacustrine Zoological Station, created by Yung, was also named after Édouard Claparède senior. In October 1911, a fund was raised by Claparède's daughter, which allowed the purchase of a motorboat for lake searches. There is an old postcard, from the Julien Brothers personal collection, which depicts that boat.

Claparède's nephew, the pioneer neurologist, psychologist, and educator Édouard Claparède (Figure 4B) had also a distinguished career [17,18]. Details on his life and work from a historical and a modern perspective will form the subject of a separate study. In brief, he earned his M.D. degree in 1897 from the University of Geneva, and he left important contributions to diverse areas, including hemiplegia, the association of ideas, the nature of sleep, stereopsis, synaesthesia, memory systems in the brain, psychoanalysis, animal behaviour, comparative and evolutionary psychology, and the developmental human psychology. In 1912, he co-founded, with Pierre Bovet (1878–1965), the *Jean-Jacques Rousseau Institute*, devoted to experimental research on child development and educational methods. While René-Édouard Claparède 'senior' (Figure 4A) promoted Darwin's ideas in the French-speaking world [7], Édouard Claparède 'junior' played a key role as the mentor of developmental psychologist Jean Piaget (1896–1980) [19].

Figure 4. (**A**) Portrait of the zoologist René-Édouard Claparède (1832–1871), circa 1870. Photograph by Attilio Runcaldier (1801–1884) and François Artus (1823–1909). Credit: Bibliothèque de Genève, Switzerland. https://bge-geneve.ch/iconographie/oeuvre.icon-o-1924-47 (accessed 22 February 2023). (**B**) Portrait of the neurologist Édouard Claparède (1873–1940), sitting in his office. Photo contributed by Elphège Gobet. Galeries de l'Institut J.-J.Rousseau. Credit: Bibliothèque de Genève, Switzerland. https://notrehistoire.ch/entries/bl3Wxnd0B9m (accessed 21 March 2023).

5. Conclusions

Claparède was an extraordinary researcher and one of the greatest Swiss zoologists [1,15]. With his "indomitable spirit" [11], he enriched science. One can only speculate about the work that he might have produced had he lived longer. An ardent explorer, he served zoology and comparative anatomy with passion and unravelled scientific facts hitherto unknown. His scientific exploits set the tone for further research by his contemporaries and by future scholars.

Author Contributions: Conceptualization, P.A.K. and L.C.T.; methodology, P.A.K. and L.C.T.; investigation, P.A.K. and L.C.T.; writing—original draft preparation, P.A.K. and L.C.T.; writing—review and editing, P.A.K. and L.C.T. All authors have read and agreed to the published version of the manuscript.

Funding: This research received no external funding.

Institutional Review Board Statement: Not applicable (theoretical review paper not involving humans or animals).

Informed Consent Statement: Not applicable.

Data Availability Statement: No new data were created or analyzed in this study. Data sharing is not applicable to this article.

Conflicts of Interest: The authors declare no conflict of interest.

References

1. Yung, E. Discours sur Édouard Claparède (rapporté par J. Briquet, secrétaire général). *Bull. Inst. Natl. Genev.* **1907**, *37*, 23–34. Available online: https://archive.org/details/bulletindeling01genegoog/page/n26 (accessed on 5 May 2023).
2. de Saussure, H. *Notice sur Édouard Claparède*; Imprimerie Ramboz et Schuchardt: Genève, Switzerland, 1871; Available online: https://archive.org/details/noticesuredouard00saus (accessed on 5 May 2023).
3. Lancaster, E.R. Édouard René Claparède. *Nature* **1871**, *4*, 224–225. [CrossRef]
4. Anonymous. Claparède, Jean Louis René Antoine Édouard (1832–1870). *Encycl. Br.* **1911**, *6*, 422–423. Available online: https://en.wikisource.org/wiki/1911_Encyclop%C3%A6dia_Britannica/Clapar%C3%A8de,_Jean_Louis (accessed on 5 May 2023).
5. Monnier, M. Édouard Claparède. *J. Débats Polit. Litt.* 13 September 1987. Available online: https://gallica.bnf.fr/ark:/12148/bpt6k458028x/f3 (accessed on 5 May 2023).
6. Buscaglia, M.; Duboule, D. Developmental biology in Geneva: A three century-long tradition. *Int. J. Dev. Biol.* **2002**, *46*, 5–13.
7. Claparède, É. Monsieur Darwin et sa théorie de la formation des espèces. *Rev. Ger. Fr. Étrang.* **1861**, *16*, 523–559. Available online: https://gallica.bnf.fr/ark:/12148/bpt6k9657839f/f529 (accessed on 5 May 2023).
8. Claparède, É. Remarques a propos de l' ouvrage de Monsieur Alfred Russel Wallace sur la théorie de la selection naturelle. *Arch. Sci. Phys. Nat.* **1870**, *38*, 160–189. Available online: https://www.biodiversitylibrary.org/item/92829#page/164 (accessed on 5 May 2023).
9. Buess, H. Nachwort. In *Lettres de René-Édouard Claparède (1832–1871)*; de Morsier, G., Ed.; Schwabe & Co. Verlag: Basel, Switzerland; Stuttgart, Germany, 1971; pp. 71–74.
10. Dolan, J.R. On Édouard Claparède and Johannes Lachmann (Clap & Lach) and their "Études sur les infusoires et les rhizo podes". *Eur. J. Protistol.* **2021**, *81*, 125822. [CrossRef] [PubMed]
11. Corliss, J.O. A salute to fifty-four great microscopists of the past: A pictorial footnote to the history of protozoology. Part I. *Trans. Am. Microsc. Soc.* **1978**, *97*, 419–458. [CrossRef] [PubMed]
12. Claparède, É. *Recherches sur la Structure des Annélides Sédentaires*; H. Georg Librairie-Éditeur: Genève, Switzerland, 1873; Available online: https://archive.org/details/recherchessurlas00clap (accessed on 5 May 2023).
13. Anonymous. Sciences physiques concours pour l'année 1857: Rapport sur le concours pour le Grand Prix des Sciences Physiques, proposé en 1854 pour 1856 et rémis au concours pour 1857. *C. R. Hébd. Séances Acad. Sci.* **1858**, *46*, 274–279. Available online: https://gallica.bnf.fr/ark:/12148/bpt6k3003h/f274 (accessed on 5 May 2023).
14. Claparède, É.; Mecznikow, É. Beiträge zur Kenntnis der Entwicklungsgeschichte der Chaetopoden. *Z. Wiss. Zool.* **1869**, *19*, 163–205. Available online: https://www.biodiversitylibrary.org/page/45006728#page/193 (accessed on 5 May 2023).
15. Théodoridès, J. Lettres de René-Édouard Claparède (1832–1871) by G. de Morsier (book review). *Isis* **1974**, *65*, 285–286. [CrossRef]
16. Claparède, É. Histologische Untersuchungen über den Regenwurm (Lumbricus terrestris Linné). *Z. Wiss. Zool.* **1869**, *19*, 563–624. Available online: https://www.biodiversitylibrary.org/item/159258#page/641 (accessed on 5 May 2023).
17. Lerner, E. Édouard Claparède, 1873–1940. *Am. J. Psychol.* **1941**, *54*, 296–299.
18. Lovell, H.T. Édouard Claparède (1873–1940). *Australas. J. Psychol. Philos.* **1942**, *20*, 81–85. [CrossRef]
19. Kollarou, P.A.; Triarhou, L.C. Édouard Claparède (1873–1940). *J. Neurol.* **2023**, *270*, 3280–3282. [CrossRef] [PubMed]

Review

The Microbiota Is Not an Organ: Introducing the Muco-Microbiotic Layer as a Novel Morphofunctional Structure

Alberto Fucarino [1,2], Stefano Burgio [1], Letizia Paladino [1,3,*], Celeste Caruso Bavisotto [1,3], Alessandro Pitruzzella [1,4], Fabio Bucchieri [1,5] and Francesco Cappello [1,3]

[1] Department of Biomedicine, Neuroscience and Advanced Diagnostics-University of Palermo, 90127 Palermo, Italy
[2] SMART Engineering Solutions & Technologies Research Center, eCampus University, 90100 Palermo, Italy
[3] Euro-Mediterranean Institute of Science and Technology (IEMEST), 90136 Palermo, Italy
[4] Consorzio Universitario Caltanissetta, University of Palermo, 93100 Caltanissetta, Italy
[5] Institute of Translational Pharmacology (IFT), National Research Council, 90146 Palermo, Italy
* Correspondence: letizia.paladino@unipa.it

Abstract: In this paper, we want to refute the notion that the microbiota should be considered an organ, given that an organ comprises tissue of similar or different embryological origin, while the microbiota is a pool of different microbial species originating individually from single replications and not from a common ancestral cellular element. Hence, we would like to propose a new morphological interpretation of its nature, based on the comprehensive context in which these microbes live: a muco-microbiotic layer of hollow organs, such as the airways and the bowel. The above concept should represent not only a new terminological annotation but also a more accurate portrayal of the physiology and pathophysiology of these organs. Indeed, a better understanding of the biological nature of this part of the human body can help scientists develop more specific experimental protocols, potentially leading to the establishment of better therapeutic strategies.

Keywords: airways; bowel; respiratory system; digestive system; microbiota; muco-microbiotic layer; cell differentiation; tissue homeostasis; organ remodeling; nanovesicles

Citation: Fucarino, A.; Burgio, S.; Paladino, L.; Caruso Bavisotto, C.; Pitruzzella, A.; Bucchieri, F.; Cappello, F. The Microbiota Is Not an Organ: Introducing the Muco-Microbiotic Layer as a Novel Morphofunctional Structure. *Anatomia* **2022**, *1*, 186–203. https://doi.org/10.3390/anatomia1020019

Academic Editors: Gianfranco Natale and Francesco Fornai

Received: 20 July 2022
Accepted: 8 November 2022
Published: 14 November 2022

Publisher's Note: MDPI stays neutral with regard to jurisdictional claims in published maps and institutional affiliations.

1. Brief Introduction to the History and Terminology Related to the Term "Microbiota" and about Its Definition in Scientific Literature

Although it is often claimed that the terms "microbiota" and "microbiome" were coined by the Nobel laureate microbiologist Joshua Lederberg in 2001, Susan L. Prescott [1] clarified that this is not correct. Lederberg did not come up with these terms, simply because they are both basic microbiology terms that had already been in common use for many decades before becoming widely used. The terms are often used synonymously by the scientific community; however, they are two separate entities. In fact, "microbiota" describes specific microbe populations that are found within a specific environment (i.e., a hollow organ) comprising bacteria, fungi, archaea, protists, helminths and viruses that symbiotically inhabit the host [2], while "microbiome" refers to the sum of all microbes and their genes. In this paper, we discuss only the microbiota.

Furthermore, a number of researchers would describe the microbiota as an "additional organ" of our body [3], using such terms as "organ-like collection of microbes," "microbial organ," "microbial system," or "metabolic organ" [4], all of which are devoid of scientific significance. At the same time, novel conceptions have been formulated to address its biological and ontological significance, such as "holobiont" and "hologenome" [5,6], although these proposals are still under debate.

In this brief paper, we present our vision of the microbiota from a strictly morphofunctional point of view. In particular, we illustrate the microbiota in a histological context, focusing our attention on its significance in the physiology and pathophysiology of two anatomical districts—the airways and the intestine.

2. Microscopic Anatomy of Human Airways and Bowel: Similarities and Differences

We would like to include just a few lines to briefly recapitulate some basic anatomic and histologic knowledge that can be useful for those readers of this manuscript who may not be experts in this field. For further details, any good anatomy and histology textbook can be consulted.

In human anatomy, organs are subdivided into two categories: hollow and parenchymatous. Hollow organs, in structural terms, have a wall composed of several layers. These layers are formed by tissue, which in turn are formed by cells, each one having a functional specialization deriving from differentiation.

Human anatomy aims to describe cell differentiation, tissue homeostasis and organ remodeling during the life span of an individual (from zygote formation to the death of the organism). Any alteration in cell differentiation (for any reason, intrinsic or extrinsic) leads to tissue disruption and organ dysfunction, in turn determining the onset of a disease. Therefore, diagnostic approaches, as well as the development and perfection of treatment approaches, are reliant on sufficiently understanding these processes.

Airways are part of the respiratory system, together with the lungs. They are divided into upper and lower ones. The former includes the nasal and paranasal cavities and pharynx; the latter includes the larynx, trachea, and bronchi. Both the upper and the lower airways are hollow structures. The walls of the lower airways are made up of three layers: mucosa, submucosa (or fibromuscular cartilaginous layer) and adventitia (Figure 1A). The mucosa is the most internal one, comprising a lining of pseudostratified ciliated mucous epithelium lying on a connective lamina propria hosting mucous glands, as well as hematic and lymphatic vessels, and populated by fibroblasts and immune cells. More recently, the concept of an epithelial–mesenchymal trophic unit (EMTU) has been introduced to better explain the close relationship between the epithelium and the lamina propria in all homeostatic processes in these tissue types [7,8].

Evans et al. theorized the idea of the EMTU at the beginning of the 2000s. According to their hypothesis, the epithelial layer and underlying mesenchymal cells communicate to control and regulate responses to external environmental stimuli [9]. This communication network plays a key role in many different aspects of airway homeostasis. Both the internal and external communication of the EMTU hinges on cytokines and signaling molecules released by both cell populations [10,11]. Processes like embryogenesis and airway differentiation use EMTU signaling [12,13], and a dysregulation of these pathways can lead to the development of various pathologies, such as chronic obstructive pulmonary disease (COPD), asthma, bronchiolitis obliterans syndrome, and idiopathic pulmonary fibrosis [14–17].

The bowel is part of the digestive system and comprises the small and large bowel. Both are hollow structures and have a wall divided into four layers (Figure 1B): mucosa, submucosa, muscularis propria, and adventitia (replaced by the serosa, i.e., the peritoneum, in some areas). The bowel mucosa is made of three overlapped tissue types: the lining of monolayered columnar epithelium composed mainly of adsorbent and mucous cells (but containing also other cytotypes; see specialized textbooks for details); the lamina propria, containing, among others, glands, hematic and lymphatic vessels, fibroblasts, and immune cells; and the muscularis mucosae, representing the border between the connective tissue of the lamina propria and that of the submucosa.

The bowel and the airways can be considered two remarkably similar structures from the following points of view. Both are in fact covered with a layer of mucus that is produced by the mucous cells and glands of the epithelium and houses the microbiota, constituted of about 100 trillion bacterial, archaeal, viral, and eukaryotic domains, encompassing more than 1000 different species [18–23]. This mucus also houses extracellular vesicles that are very useful tools of trafficking between human cells and the microbiota, actively involved in maintaining their homeostasis.

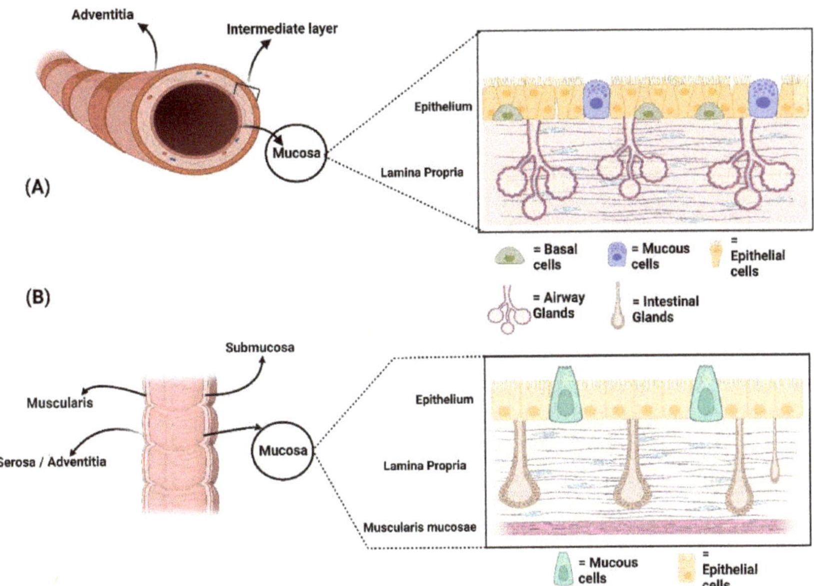

Figure 1. Components of the airway (**A**) and bowel (**B**) mucosa, as classically described in anatomical and histological literature. In brief, both organs have a monolayer of epithelial cells with an underlying lamina propria, in which glands can be found. We can appreciate morphological differences between airway and intestinal glands: airway glands have a tubuloacinar structure, single terminal ducts and branched secretory tubules that end in serous acini; intestinal glands generally present unbranched tubules that end in a single acinus. The muscularis mucosa is present only in the bowel, separating the connective tissue of the lamina propria from that of the submucosa. See the text for further details.

The microbiota presents great differences in relation to age, ethnicity, eating habits, seasonal changes, immune status, and other physiological variables, both in the airways [21–24] and in the bowel [25–30]. These findings have prompted several studies on its role during the pathogenesis of various diseases of the respiratory and the digestive systems (including inflammatory and tumoral ones), the systemic complications caused by such disorders, and last but not least, the therapeutic potential of altering the microbiota with the use of prebiotics and probiotic or microbiotic transplants [31–36]. Here we want to focus on and compare the microbiota of the airways and the bowel, in the context of microenvironment.

3. The "Ghost" Layer

The classic description of airway and bowel structure made in the previous paragraph does not take into account the fact that in living subjects, the innermost lining of the intestinal lumen is not the mucous layer, but mainly consists of a mix of mucus, microbiota, and myriad soluble factors in which nanovesicles are dispersed. For this reason, we propose using the term "muco-microbiotic (MuMi) layer" to describe the innermost lining of both airway and bowel walls in living subjects (Figure 2).

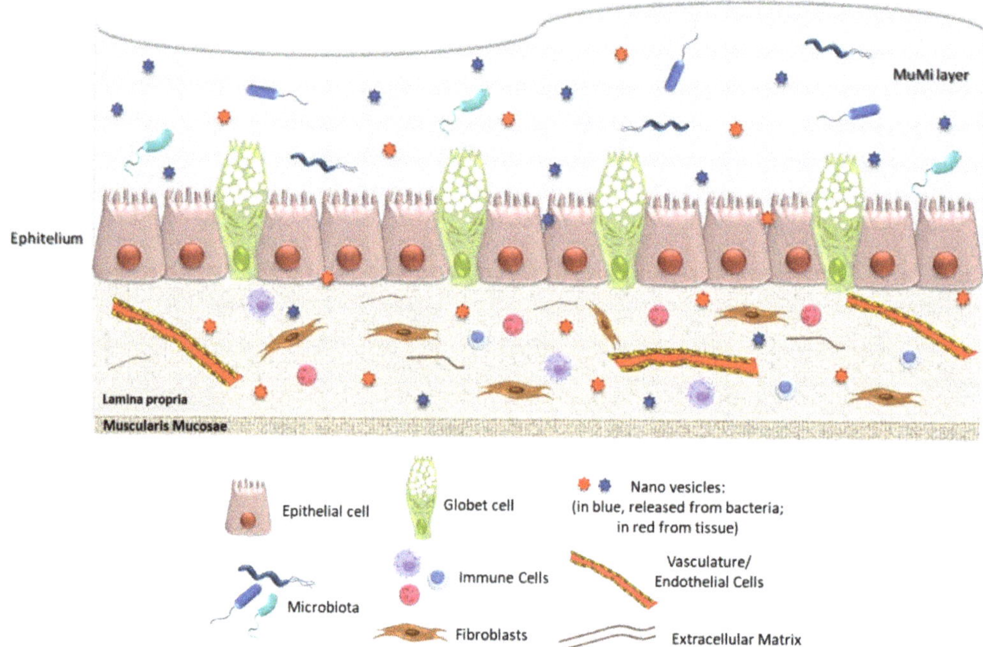

Figure 2. Main components of the muco-microbiotic layer (i.e., mucus, microbiota, and nanovesicles) in the bowel mucosae and its relationships with epithelial elements and connective cells of lamina propria. See the text for further details.

Enterocytes, the cells responsible for the absorption of nutrients, and goblet cells, the elements secreting the mucus that covers the intestinal wall, are the most common cytotypes that make up the bowel epithelium. The MuMi layer represents the first line of defense against various types of stress, such as ingested material, hydrochloric acid, proteolytic enzymes, etc., that the bowel is exposed to. The mucus, synthesized and secreted by goblet cells, is colorless, since it is mainly composed of water (up to 95%), lipids and small proteins. Its mechanical and viscoelastic properties are guaranteed by the presence of high-molecular-weight glycoproteins, identified as mucins [37,38]. The mucus is denser at the epithelial level and less so closer to the lumen, being also exposed to bacterial proteolytic activity. In fact, different microbial species of the normal microbiota are present in the MuMi layer. This allows the formation of a layer innermost to the intestinal lumen, that we call for the first time the "MuMi layer", produced by the union of the mucus layer, microorganisms, and soluble factors, including extracellular vesicles (EVs) [39]. EVs represent the main dialogue tool between the microbiota and human cells of the mucosal layer, including not only epithelial cells but also fibroblasts and immune elements distributed in the lamina propria.

The MuMi layer is not visible under routine histologic examination because it is lost during the processing of biopsies for microscopic observation, due to the solubility of the mucus in alcoholic solutions. Consequently, this internal lining has been systematically missed by histological studies and generally ignored for a long time, despite the fact that in our opinion, it has a key role in airway and intestinal physiology and pathophysiology (as discussed below).

It is also important to remember that both the airway and intestinal tracts establish multiple relationships with other anatomical districts through nanovesicular trafficking. In fact, exosomes and outer membrane vesicles—produced, respectively, by human and bacterial cells—are able to reach, through the bloodstream, virtually any anatomical district, including some "protected sanctuaries," such as the brain, the testicles, and the thymus [35,36]. Due to its wide extension all along the hollow organs, this layer releases a great quantity of nanovesicles of both bacterial and human origin, in addition to other soluble factors with autocrine- and paracrine-like effects capable of reaching the general circulation and producing systemic effects.

In this context, the MuMi layer should be seen as a morphofunctional layer more than merely an anatomical one, because it is not a derivative of any embryonic structure, unlike the remaining layers of the GI tract or the airway. However, since it represents the symbiotic relationship between the microbiota and the host cells, we cannot ignore the existence of this layer, for reasons better clarified in the following paragraphs.

4. Extracellular Vesicles: Physiology and Pathophysiology

Extracellular vesicles (EVs) are nanosized lipid membrane vesicles released by virtually all human cells involved in many cellular processes, both in physiological and pathological conditions, such as cell differentiation and tissue homeostasis [40–42].

EVs are classified into different subtypes based on their biophysical features, such as size, density, and biochemical composition. According to this classification, it is possible to describe "small EVs" with ranges between 50 nm and 200 nm and "medium/large EVs" sized >200 nm. However, this classification frequently refers to the traditional nomenclature based on EV dimension and mechanism of biogenesis: (a) exosomes of 20–200 nm of diameter that are formed within the endosomal network and released upon the fusion of multivesicular bodies with the plasma membrane; (b) microvesicles with a diameter between 200 nm 1000 nm (also called microparticles and ectosomes) that are produced by outward budding and fission of the plasma membrane; and (c) apoptotic bodies sized >1000 nm that are released as blebs of cells during apoptosis [43,44].

EVs are present in many different biofluids, such as blood, lymph, breast milk, saliva, urine, and cerebrospinal fluid, carrying into the extracellular compartment as well as into distant districts, lipids, nucleic acids and proteins, which represents a way of long-distance communication and horizontal transfer of genetic material. EV content varies according to the type of cell that released it, thus providing information on its status [40,45]. Furthermore, it has been shown that EV-carried molecules modulate and reprogram recipient cells, mediating signal transduction between cells and influencing their behavior [46].

In addition to participating in some fundamental physiological processes, EVs have also been implicated in disease development, favoring, for instance, tumor cell proliferation, angiogenesis, extracellular matrix degradation, and immune response deregulation [47]. Therefore, investigation of EVs is useful to highlight mechanisms that may determine health or disease. Furthermore, as EVs carry genetic material and proteins that may affect different signaling pathways in the target cells and organs, and since they are commonly found in biological fluids and tissue, they can be used as noninvasive and easily accessible disease biomarkers [48–53]. As stated above, in different diseases, cells can produce EVs with alterations in number and molecular content compared to physiological conditions, and accumulating evidence demonstrates their role in cancer, in which the application of EVs may be helpful for early diagnostics and the identification of new therapeutic targets [52].

In the airways, secreted EVs are involved in lung homeostasis and pathologies, such as inflammatory disease and lung cancer [54], and have been reported in plasma and secretions from airways, such as saliva [55], induced sputum [56] or bronchoalveolar lavage fluid (BALF) [57]. Among the cells of the pulmonary microenvironment that secrete EVs, bronchial epithelial cells and alveolar macrophages are the major sources, and it has been demonstrated that their molecular content has some immunomodulatory effects and defense functions of the airway in lung inflammatory and chronic diseases, such as asthma, chronic obstructive pulmonary disease, pulmonary fibrosis, and lung cancer [54,58].

Furthermore, in the bowel tract, secreted EVs act as communicators between intestinal cells and as long-distance mediators involved in maintaining immune homeostasis. EVs produced by intestinal layers, including the MuMi one, have different effects, such as playing a role in the repair mechanisms of epithelium integrity [59] and in the activation of immunological pathways [60,61]. At the same time, they can become a disease biomarker when isolated from circulation or other biological fluids [53]. Moreover, EVs released by intestinal layers seem to be involved in the defense mechanism against pathogenic microbes [60], and consequently have important roles in the cross talk between the host body and resident microorganisms [62].

As a matter of fact, EVs are naturally produced by both Gram-negative and Gram-positive bacteria harbored in the airways and the bowel. They are called outer membrane vesicles (OMVs) or membrane vesicles (MVs), respectively [63–65]. These particles are ubiquitously produced by blebbing of the outer membrane or following bacterial lysis, and contain a variety of molecules, such as proteins, RNAs, DNAs, lipids and enzymes, whose loading mechanism is not yet fully understood [63–65].

EVs produced by the airway microbiota have important roles in activating immune responses, such as *Staphylococcus aureus*- and *Pseudomonas aeruginosa*-derived EVs, which may induce neutrophilic pulmonary inflammation [66,67]. Initially, it was believed that intestinal bacterial EVs may have a pathogenic role, carrying toxins and virulence factors favorable to establishment of infection [68]. Subsequently, it was observed that microbiota-derived EVs can also initiate host immune responses and cross the intact intestinal epithelial barrier to reach distant tissue [69]. In fact, like the exosomes, OMVs and MVs play an important role in communication with neighboring bacteria and the environment, being able to travel long distances, transporting their contents throughout the body. This could lead us to hypothesize that OMVs also act elsewhere, other than the airway and bowel lumen, amplifying the role of the microbiota [70].

The following paragraphs analyze the composition of both the mucus and the microbiota in the airways and the bowel in better detail to add important information for a better understanding of the role of the MuMi layer in human physiology and pathophysiology.

5. Airway Composition

The respiratory tract represents a large surface area that interacts with the external environment. As a physical barrier, the airway epithelium and the bronchial mucus are classically considered essential for the protection of both the airways and the lungs from any external etiological agents, including microbes and allergens [71]. Airway mucus, produced by the submucosal glands, is a complex dilute aqueous solution resulting from the sum of proteins, glycoconjugates, and lipids, containing electrolytes, enzymes, anti-enzymes, oxidants and antioxidants [72]. Under normal conditions, mucus is scarce in quantity but still present, and forms a thin layer that homogeneously covers the surface of the airways, from the upper airways to the bronchi. In fact, the first, most fundamental function of this thin mucous layer is to ensure the maintenance of appropriate humidity and temperature levels of the incoming air flowing through the respiratory tree. This is also important to ensure a favorable environment for the colonization of bacteria, viruses and fungi that compose the pulmonary microbiota.

In fact, the respiratory tract is inhabited by niche-specific communities of bacteria acting as gatekeepers that provide resistance to colonization by respiratory pathogens. A variety of different microbial species colonize the airways right after birth, depending on the type of delivery. Significant changes occur during the first year of life, driven by both immune system maturation and diet [73]. Thus, the microbial community in infants and children gradually transforms into the adult respiratory tract microbiome, becoming less dense but more diverse [74].

The lower respiratory tract is one of the least-populated surfaces of the human body in terms of the microbiome, with approximately 10–100 bacteria per 1000 human cells [75]. In particular, airways exhibit a gradient of bacteria burden that is relatively high in the upper tract and substantially diminished in the lower one (Table 1) [76]. The mucous surfaces of the upper respiratory system are colonized by a wide range of bacteria belonging to the genera Firmicutes, Actinobacteria, Bacteroidetes and Proteobacteria. While it was previously assumed that the lower respiratory tract was sterile, except during infection, recent studies have shown the presence of bacteria, such as Firmicutes, in the lungs of healthy individuals [16,74,77–79]. The different distribution of bacteria genera is probably ascribable to the different characteristics of the airway's tracts. In the upper section of the respiratory tract, the inhaled air is heated, humidified, and filtered. These processes affect both bacteria from the external environment and those that reside permanently in our body. The selection of genera leads to the formation of a specific symbiotic microbiome for the different portions of the airways [76].

Generally, differences in temperature, pH, mucus secretion, and relative oxygen concentration regulate bacterial colonization in the respiratory tract. Several environmental factors, such as age, weight, and tobacco smoke exposure, can damage the epithelium and facilitate microbial entry into the host, contributing to dysbiosis [80]. Similarly to the gut, airway dysbiosis is associated with local inflammation and exacerbation of many pulmonary disorders, such as COPD or asthma [81]. Clinical features of inflammation include increased vascular permeability, facilitating the release of immune cells and mucus secretion [82,83].

Moreover, the airway microbiota, along with the mucus, plays a fundamental role in the barrier function of the airway epithelium, maintaining a stable homeostasis [71], and a dialogue is likely to develop between specific bacterial communities and the host in a dynamic way. In particular, the loss of barrier function can affect the composition of the respiratory microbiota, allowing the entry of pathogens and other particles. As mentioned before, the mucus maintains an efficient airway environment, acting as a physical and biological barrier and ensuring the humidification of the air that passes through the trachea and the large bronchi.

Numerous experimental studies have shown that in the absence of mucus, the respiratory tract changes profoundly, often causing substantial damage or alterations to the anatomical structures of the lung [84]. It is well known, for example, that the absence of mucus in the airways can cause an irritative inflammatory reaction [84]. It also constitutes an important filter and dilution tool: many types of gaseous substances that are irritating or toxic to the respiratory tract are subjected to a process of filtration and dilution by the mucous surface of the airways. It is also used to remove particles, dust and biological agents.

On the other hand, emerging experiments have highlighted a crucial cross talk between the intestinal microbiota and the respiratory one, called the "gut–lung axis," highlighting the possible involvement of microaspiration of intestinal microbes in the development of the airway microbiota [75,84]. In fact, a change in the constituents of the gut microbiome could alter immune responses and homeostasis in the airways. Various factors, such as diet, have been shown to not only impact the composition of the intestinal microbiota but also the respiratory one, indicating that the intestinal and respiratory compartments are closely interconnected, and that changes at one of the two sites could impact the other [71,75,84].

Short-chain fatty acids (SCFAs), the metabolites produced by the gut microbiome in the intestinal microenvironment, have been shown to reach other organs and to play a role in respiratory diseases, such as asthma [75]. In fact, several studies have shown that a diet rich in fiber causes an increase in circulating SCFA levels produced by the fermentation of anaerobic bacteria. SCFAs promote leukocyte recruitment and immune regulation within the inflammatory process, preventing allergy- and asthma-related lower airway inflammation [16,85,86].

However, despite several studies demonstrating a crucial and beneficial role of the microbiota, the usefulness of probiotics as a therapeutic strategy for respiratory diseases has still not been conclusively proven [87]. Consequently, it may be necessary to further broaden our horizons to fully understand the actions and mechanisms of probiotics in the prevention and treatment of respiratory tract diseases [87–89].

6. Bowel Mucus Composition

The mucus of the gastrointestinal tract, coating the intestinal epithelium, acts as a physical barrier, representing the first line of defense against ingested material, digestive enzymes and microbial by-products [90].

The mucus is synthesized and secreted by goblet cells and is mainly composed of water, lipids and mucins, a complex structure of glycoproteins with specific O-linked glycans (O-glycans) capable of guaranteeing the appropriate mechanical and viscoelastic properties [37,38,91,92]. Among the many mucins, Muc2 is the main mucin component of the intestinal mucus layer [93]. In addition to the mucins, mucus contains immune mediators that target the gut microbiota, providing a diffusion barrier. In fact, IgA, secreted into the lumen of the duodenum and the ileum at a high concentration, is able to bind bacteria or viruses, leading to slower diffusion and reducing bacterial mobility in the mucus [92].

The composition and thickness of the mucus vary along the length of the intestine. While in the small intestine, the mucus forms a single layer, in the large intestine, due to the higher density of microorganisms, the mucus is organized in two different layers: the inner one separates the commensal bacteria from the host epithelium and the outer one represents the real natural habitat for the commensal bacteria [94]. Moreover, in the small intestine, the mucus is penetrable, but the bacteria are kept away from the epithelium by antibacterial mediators, whereas in the large intestine, the inner mucus layer is impenetrable to bacteria and only the outer mucus layer hosts the bacteria [95]. Therefore, the inner mucus layer is denser and the outer is more dissolved, as it is exposed to the proteolytic activity of bacteria [91,96]. The mucus composition has an important role in creating a favorable microenvironment to host the gut microbiota, also providing nutrients and other physiological substances for its homeostasis [97]. For example, some bacteria, using MUC2 O-linked glycans, are capable of colonizing tissue and releasing a mucous biofilm, a complex self-produced polymer matrix in which microorganisms can attach to each other and be attached to the mucosal surface [98]. In addition, some bacteria can also use pili, fimbriae, or flagella to bind mucus [99,100]. Indeed, the main flagellar subunit protein has also been shown to be involved in bacteria–mucus adhesion. Finally, it is likely that the components of the mucus layer, such as secretory IgA and mucins, play a role in the interaction between bacteria and mucus [91,101].

7. Bowel Microbiota Composition

The gastrointestinal tract is one of the microbial ecosystems with the highest population density. It is estimated that about 10^{14} microorganisms are present in the intestine of an adult, exceeding the total number of cells in the human body (10^{13}), and including 100 times more genes than the human genome [102–104]. In general, the development of an individual's intestinal microflora depends on the number and type of microorganisms with which the individual comes into contact in the early stages of growth, and on the genetic composition of the individual [105]. The phenotype composition can therefore be regulated by such factors as diet, environment and stress. Infection, disease and antibiotics are among the many factors that may alter microbial composition [106].

The human intestinal microbiota contains bacteria, fungi, archaea, protists, helminths and viruses that symbiotically inhabit the human digestive system, with five bacterium phyla—Bacillota, Bacteroidota, Actinomycetota, Pseudomonadota and Verrucomicrobiota—representing the predominant microorganisms in the gut (Table 2) [2]. Along the intestinal tract, both the quantity and the quality of the microbiota vary. It is notable that bacterial concentration tends to increase along the distal ileum, reaching very high concentrations at the colon level. In addition, the horizontal stratification of the microbiota is also remarkable: different commensal microorganisms are located in the intestinal lumen, the mucus layer and the epithelial crypts [107].

Due to poor intestinal motility, the highest number of bacteria with the greatest microbial diversity are located in the large intestine (10^{11}–10^{12} CFU/mL of luminal content). This colonization allows the formation of a symbiotic and commensal relationship between the host and the microbiota, dependent on the production and the release of multiple bioactive metabolites [108,109]. The intestinal microbiota thus influences normal intestinal homeostasis, acting on the proliferation and apoptosis of intestinal epithelial cells (IECs).

Among the main metabolic functions performed by the microbiota, there are the synthesis of vitamins (B12 and K), absorption of Ca, Fe and Mg ions and the production of short-chain fatty acids (SCFAs) [110,111]. In general, SCFAs, such as butyrate, acetate and propionate, derived from the fermentation of indigestible fibers, play an important role in human physiology, maintaining intestinal barrier integrity by preventing microbial translocation, which is known to be associated with local intestinal inflammation, systemic inflammation and neuroinflammation [108,112].

The intestinal microbiota also plays an important role in the development and maintenance of the immune system [113], as confirmed by studies conducted on germ-free mice that were more sensitive to infections and had low IgA concentration [114–116]. In general, the immune system is a complex network of chemical and cellular mediators that can recognize any form of insult in order to keep the body healthy. However, it is important for the intestinal immune system to be able to recognize pathogenic microorganisms from commensal ones. In particular, during the development of the immune system, the Toll-like receptors (TLRs)—expressed on IECs and lymphoid cells—promote immunological tolerance to the normal components of the microbiota, recognizing several general molecules associated with microbes (i.e., microbe- or pathogen-associated molecular patterns—MAMPs or PAMPs, respectively), such as peptidoglycans and capsular components [117]. In this way, the intestinal microbiota not only manages to evade the host's immune system but also contributes to the development, maturation and regulation of the immune system [108].

Furthermore, although it is classically thought that the mucus produced by the goblet cells of the intestinal mucosa represents the first defense barrier, even the microbiota does not allow the colonization of exogenous/pathogenic microbes. In fact, there is competition between commensal and pathogenic bacteria. The colonization of exogenous bacteria can be prevented due to the limited availability of nutrients or to the production, during sugar metabolism, of organic acids that determine a reduction in intestinal pH with consequent inhibition of the growth of acid-sensitive bacterial species. Other than these mechanisms, one should also consider the production of antimicrobial substances by commensal bacteria, such as bacteriocins, allowing the microflora to control the growth of exogenous microorganisms [118].

It has been shown that the microbiota of the gastrointestinal tract is able to modulate the migration of the neutrophils resident in the lamina propria, to favor the differentiation of TCD8+ cells toward TCD4+, and finally to release IgA through the activation of B cells [119]. Hence, the complex interaction between nonpathogenic bacteria, IECs and immune cells of the mucous layer is a fundamental prerequisite for the development of immune functions and defense mechanisms in the adult intestine.

However, dysbiosis of the gut microbiota, associated with an increased abundance of potentially detrimental bacteria, can compromise gut barrier integrity through bacterial production of endotoxins (e.g., lipopolysaccharide (LPS)) capable of altering immune response, initiating proinflammatory pathways and directly damaging intestinal epithelial cells (known as leaky gut), with changes in the distribution and localization of its tight junction proteins. This promotes the translocation of bacterial components from the intestinal lumen to the systemic circulation and other organs, including the central nervous system (CNS). In fact, gut dysbiosis has also been observed in various neurological and psychiatric conditions, including severe depression and Parkinson's and Alzheimer's diseases [120]. For instance, several studies have reported the potential connection between α-syn-related pathology and dysbiosis [120,121].

It is thus widely understood that the microbiota communicates with the host under both physiological and pathological conditions. This type of communication takes place not only by direct contact but also through humoral signaling molecules and hormonal components. As previously described, intercellular communication with the host also occurs through the release of micro- and nanovesicles, which can enter the systemic circulation. In particular, the maintenance of tissue homeostasis is also mediated by the release of EVs, such as exosomes and OMVs/MVs, contributing significantly to coordinated signaling events and communication between microbiota, IECs, endothelial cells and immune cells [122]. Among proteins transported within the EVs, there are metalloproteinases, growth factors and chemokines, used as secondary messengers for the coordination of cellular responses [123]. Several studies have shown that the release of EVs could play a role in response to vaccination and therapeutic applications, as a vehicle in the delivery of drugs and targeted therapy [50,124–126]. In general, IEC-derived EVs are able to regulate the integrity of the epithelial barrier thanks to the transport of desmosomal cadherins that stabilize cell–cell epithelial adhesions, as well as being able to protect against pathogenic infections thanks to the transport of some antimicrobial peptides, such as beta-defensin [127,128].

In recent years, several studies have focused on the role of OMVs produced by the intestinal microbiota, highlighting their importance as immunological mediators. In fact, it has been shown that the large capsular polysaccharide A (PSA) is selectively packaged within OMVs. The OMVs are then internalized into dendritic cells, which induce the differentiation of T-regulatory cells to produce IL-10. This represents an example of how host immunotolerance towards the symbiote is determined [129,130].

Other studies show the entire intestinal microbial population benefits from OMV production A mutual support is established by the bacterial species that release OMVs and others that receive them. In fact, some bacteroides are able to pack hydrolases inside OMVs, making them usable to other bacteria that are privy to them. Since hydrolases are used for the digestion of polysaccharides, this mechanism promotes the growth of other bacterial species that are unable to hydrolyze polysaccharides. This again supports the role of OMVs in creating and maintaining the balance of the gut microbiota [131].

Table 1. The main bacteria that can be found in the airway microbiota. Percentages can change depending on age, lifestyle, epigenetic stimuli, and diseases [129].

Anatomy	Density	Kingdoms	Phylum	Genus	(%)	References
URT (upper respiratory tract)		Bacteria				[16,73,74,76–78,107,132–134]
Nasal cavity	10^3		Actinomycetota	*Corynebacterium* spp.	20–25%	
			Bacillota	*Staphylococcus* spp.	7–12%	
				Streptococcus spp.	8.5–13%	
			Pseudomonadota	*Haemophilus* spp.	2–7%	
				Moraxella spp.	13–18%	
Nasopharynx	10^3		Actinomycetota	*Corynebacterium* spp.	6–11%	
			Bacillota	*Staphylococcus* spp.	12–17%	
				Streptococcus spp.	10.5–15%	
			Pseudomonadota	*Moraxella* spp.	14–19%	
Oropharynx	10^6		Actinomycetota	*Rothia* spp.	5.5–10%	
			Bacteroidota	*Prevotella* spp.	12–17%	
			Bacillota	*Streptococcus* spp.	23–28,5%	
				Lactobacillus spp.	1–5%	
				Veillonella spp.	4–9%	
			Fusobacteriota	*Leptotrichia* spp.	4–9%	
LRT (lower respiratory tract	Lungs	10^2	Bacteria	Bacteroidota	Prevotella spp.	75–80%
				Bacillota	*Lactobacillus* spp. / *Streptococcus* spp. / *Veillonella* spp.	78–83%
				Pseudomonadota	*Halomonas* spp.	41–46%

Table 2. The main bacteria that can be found in the bowel microbiota. Percentages can change depending on age, lifestyle, epigenetic stimuli, and diseases [135].

Domain	Kingdoms	Phylum	Genus	(%)	References
Procaryota	Bacteria	Bacteroidota	Bacteroides Prevotella	73.13 ± 22.16%	[136–139]
		Bacillota	Clostridium Lactobacillus	22.2 ± 18.66%	
		Actinomycetota	Bifidobacterium	1.67 ± 2.94	
		Fusobacteriota	Fusobacterium	19 ± 21.4%	
		Pseudomonadota	Escherichia Shigella	2.15 ± 10.39%	
		Verrucomicrobiota	Akkermansia	1 ± 2%	
		Spirochaetota	Brachyspira	5 ± 7.8%	
	Archea	Euryarchaeota	Halobacterium	0.1 ± 21.3%	[140,141]

Domain	Kingdoms	Phylum	Family	(%)	References
Eukaryota	Fungi	Ascomycota	Aspergillaceae Debaryomycetaceae Dipodascaceae Saccharomycetaceae	47.8 ± 99.5%	[142–144]
		Basidiomycota	Tremellaceae Malasseziaceae	47.8 ± 99.5%	
Virii	Bacteriophage	Uroviricota	Demerecviridae Ackermannviridae Autographiviridae	80 ± 84%	[139,145–147]

8. Conclusions

The MuMi layer of the airways and bowel should not be misinterpreted any longer: its identification and study can enable a better understanding of the physiology and pathophysiology of these organs and, in turn, of the entire organism. Further studies are essential to better gauge not only the microbe population of the MuMi layer (both in qualitative and quantitative terms) but also the mucous component, i.e., the "matrix" in which these microbes live, proliferate, and traffic soluble molecules and nanovesicles during the entire life span of an individual, both in physiological and pathophysiological conditions. A better understanding of this important constituent of the human body could lead to the enhancement of diagnostics and treatment strategies for airway and bowel diseases, as well as their extraorgan complications.

Author Contributions: Conceptualization, F.B. and F.C.; software, S.B.; writing—original draft preparation, A.F., S.B., L.P., C.C.B., A.P., F.B. and F.C. All authors have read and agreed to the published version of the manuscript.

Funding: This research received no external funding.

Institutional Review Board Statement: Not applicable

Informed Consent Statement: Not applicable

Data Availability Statement: Not applicable

Acknowledgments: We thank Gabriele Chiaramonte (University of Palermo) for microbiological advice.

Conflicts of Interest: The authors declare no conflict of interest.

References

1. Prescott, S.L. History of medicine: Origin of the term microbiome and why it matters. *Hum. Microbiome J.* **2017**, *4*, 24–25. [CrossRef]
2. Aya, V.; Flórez, A.; Perez, L.; Ramírez, J.D. Association between physical activity and changes in intestinal microbiota composition: A systematic review. *PLoS ONE* **2021**, *16*, e0247039. [CrossRef] [PubMed]
3. Riccio, P.; Rossano, R. The human gut microbiota is neither an organ nor a commensal. *FEBS Lett.* **2020**, *594*, 3262–3271. [CrossRef] [PubMed]
4. Bäckhed, F.; Ding, H.; Wang, T.; Hooper, L.V.; Koh, G.Y.; Nagy, A.; Semenkovich, C.F.; Gordon, J.I. The gut microbiota as an environmental factor that regulates fat storage. *Proc. Natl. Acad. Sci. USA* **2004**, *101*, 15718–15723. [CrossRef] [PubMed]
5. Rosenberg, E.; Zilber-Rosenberg, I. The hologenome concept of evolution after 10 years. *Microbiome* **2018**, *6*, 78. [CrossRef]
6. Triviño, V.; Suárez, J. Holobionts: Ecological communities, hybrids, or biological individuals? A metaphysical perspective on multispecies systems. *Stud. Hist. Philos. Biol. Biomed. Sci.* **2020**, *84*, 101323. [CrossRef]
7. Richter, A.; Puddicombe, S.M.; Lordan, J.L.; Bucchieri, F.; Wilson, S.J.; Djukanovic, R.; Dent, G.; Holgate, S.T.; Davies, D.E. The contribution of interleukin (IL)-4 and IL-13 to the epithelial-mesenchymal trophic unit in asthma. *Am. J. Respir. Cell Mol. Biol.* **2001**, *25*, 385–391. [CrossRef]
8. Bucchieri, F.; Pitruzzella, A.; Fucarino, A.; Gammazza, A.M.; Bavisotto, C.C.; Marcianò, V.; Cajozzo, M.; Lo Iacono, G.; Marchese, R.; Zummo, G.; et al. Functional characterization of a novel 3D model of the epithelial-mesenchymal trophic unit. *Exp. Lung Res.* **2017**, *43*, 82–92. [CrossRef]
9. Evans, M.J.; Winkle, L.S.V.; Fanucchi, M.V.; Plopper, C.G. The Attenuated Fibroblast Sheath of the Respiratory Tract Epithelial-Mesenchymal Trophic Unit. *Am. J. Respir. Cell Mol. Biol.* **1999**, *21*, 655–657. [CrossRef]
10. Sangiorgi, C.; Vallese, D.; Gnemmi, I.; Bucchieri, F.; Balbi, B.; Brun, P.; Leone, A.; Giordano, A.; Conway de Macario, E.; Macario, A.J.; et al. HSP60 activity on human bronchial epithelial cells. *Int. J. Immunopathol. Pharm.* **2017**, *30*, 333–340. [CrossRef]
11. Sunadome, H.; Matsumoto, H.; Petrova, G.; Kanemitsu, Y.; Tohda, Y.; Horiguchi, T.; Kita, H.; Kuwabara, K.; Tomii, K.; Otsuka, K.; et al. IL4Rα and ADAM33 as genetic markers in asthma exacerbations and type-2 inflammatory endotype. *Clin. Exp. Allergy* **2017**, *47*, 998–1006. [CrossRef] [PubMed]
12. Minoo, P.; King, R.J. Epithelial-Mesenchymal Interactions in Lung Development. *Annu. Rev. Physiol.* **1994**, *56*, 13–45. [CrossRef] [PubMed]
13. Bartis, D.; Mise, N.; Mahida, R.Y.; Eickelberg, O.; Thickett, D.R. Epithelial–mesenchymal transition in lung development and disease: Does it exist and is it important? *Thorax* **2014**, *69*, 760–765. [CrossRef] [PubMed]
14. Jolly, M.K.; Ward, C.; Eapen, M.S.; Myers, S.; Hallgren, O.; Levine, H.; Sohal, S.S. Epithelial–mesenchymal transition, a spectrum of states: Role in lung development, homeostasis, and disease. *Dev. Dyn.* **2018**, *247*, 346–358. [CrossRef]

15. Hansel, N.N.; Paré, P.D.; Rafaels, N.; Sin, D.D.; Sandford, A.; Daley, D.; Vergara, C.; Huang, L.; Elliott, W.M.; Pascoe, C.D.; et al. Genome-Wide Association Study Identification of Novel Loci Associated with Airway Responsiveness in Chronic Obstructive Pulmonary Disease. *Am. J. Respir. Cell Mol. Biol.* **2015**, *53*, 226–234. [CrossRef] [PubMed]

16. Al-Muhsen, S.; Johnson, J.R.; Hamid, Q. Remodeling in asthma. *J. Allergy Clin. Immunol.* **2011**, *128*, 451–462. [CrossRef] [PubMed]

17. Pulvirenti, G.; Parisi, G.F.; Giallongo, A.; Papale, M.; Manti, S.; Savasta, S.; Licari, A.; Marseglia, G.L.; Leonardi, S. Lower Airway Microbiota. *Front. Pediatrics* **2019**, *7*, 393. [CrossRef] [PubMed]

18. Rinninella, E.; Raoul, P.; Cintoni, M.; Franceschi, F.; Miggiano, G.A.D.; Gasbarrini, A.; Mele, M.C. What is the Healthy Gut Microbiota Composition? A Changing Ecosystem across Age, Environment, Diet, and Diseases. *Microorganisms* **2019**, *7*, 14. [CrossRef]

19. Rajilić-Stojanović, M.; de Vos, W.M. The first 1000 cultured species of the human gastrointestinal microbiota. *FEMS Microbiol. Rev.* **2014**, *38*, 996–1047. [CrossRef]

20. D'Argenio, V.; Salvatore, F. The role of the gut microbiome in the healthy adult status. *Clin. Chim. Acta* **2015**, *451*, 97–102. [CrossRef]

21. Wagner Mackenzie, B.; Chang, K.; Zoing, M.; Jain, R.; Hoggard, M.; Biswas, K.; Douglas, R.G.; Taylor, M.W. Longitudinal study of the bacterial and fungal microbiota in the human sinuses reveals seasonal and annual changes in diversity. *Sci. Rep.* **2019**, *9*, 17416. [CrossRef] [PubMed]

22. Lee, S.Y.; Mac Aogáin, M.; Fam, K.D.; Chia, K.L.; Binte Mohamed Ali, N.A.; Yap, M.M.C.; Yap, E.P.H.; Chotirmall, S.H.; Lim, C.L. Airway microbiome composition correlates with lung function and arterial stiffness in an age-dependent manner. *PLoS ONE* **2019**, *14*, e0225636. [CrossRef] [PubMed]

23. Yamanishi, S.; Pawankar, R. Current advances on the microbiome and role of probiotics in upper airways disease. *Curr. Opin. Allergy Clin. Immunol.* **2020**, *20*, 30–35. [CrossRef]

24. Man, W.H.; Scheltema, N.M.; Clerc, M.; van Houten, M.A.; Nibbelke, E.E.; Achten, N.B.; Arp, K.; Sanders, E.A.M.; Bont, L.J.; Bogaert, D. Infant respiratory syncytial virus prophylaxis and nasopharyngeal microbiota until 6 years of life: A subanalysis of the MAKI randomised controlled trial. *Lancet Respir. Med.* **2020**, *8*, 1022–1031. [CrossRef]

25. Simrén, M.; Barbara, G.; Flint, H.J.; Spiegel, B.M.; Spiller, R.C.; Vanner, S.; Verdu, E.F.; Whorwell, P.J.; Zoetendal, E.G. Intestinal microbiota in functional bowel disorders: A Rome foundation report. *Gut* **2013**, *62*, 159–176. [CrossRef] [PubMed]

26. Hillman, E.T.; Lu, H.; Yao, T.; Nakatsu, C.H. Microbial Ecology along the Gastrointestinal Tract. *Microbes Environ.* **2017**, *32*, 300–313. [CrossRef] [PubMed]

27. Eun, C.S.; Kwak, M.J.; Han, D.S.; Lee, A.R.; Park, D.I.; Yang, S.K.; Kim, Y.S.; Kim, J.F. Does the intestinal microbial community of Korean Crohn's disease patients differ from that of western patients? *BMC Gastroenterol.* **2016**, *16*, 28. [CrossRef]

28. Fulde, M.; Sommer, F.; Chassaing, B.; van Vorst, K.; Dupont, A.; Hensel, M.; Basic, M.; Klopfleisch, R.; Rosenstiel, P.; Bleich, A.; et al. Neonatal selection by Toll-like receptor 5 influences long-term gut microbiota composition. *Nature* **2018**, *560*, 489–493. [CrossRef]

29. Maynard, C.; Weinkove, D. The Gut Microbiota and Ageing. *Sub-Cell. Biochem.* **2018**, *90*, 351–371. [CrossRef]

30. Cebula, A.; Seweryn, M.; Rempala, G.A.; Pabla, S.S.; McIndoe, R.A.; Denning, T.L.; Bry, L.; Kraj, P.; Kisielow, P.; Ignatowicz, L. Thymus-derived regulatory T cells contribute to tolerance to commensal microbiota. *Nature* **2013**, *497*, 258–262. [CrossRef]

31. Martens, K.; Pugin, B.; De Boeck, I.; Spacova, I.; Steelant, B.; Seys, S.F.; Lebeer, S.; Hellings, P.W. Probiotics for the airways: Potential to improve epithelial and immune homeostasis. *Allergy* **2018**, *73*, 1954–1963. [CrossRef] [PubMed]

32. Caverly, L.J.; Huang, Y.J.; Sze, M.A. Past, Present, and Future Research on the Lung Microbiome in Inflammatory Airway Disease. *Chest* **2019**, *156*, 376–382. [CrossRef] [PubMed]

33. Caruso, R.; Lo, B.C.; Núñez, G. Host-microbiota interactions in inflammatory bowel disease. *Nat. Rev. Immunol.* **2020**, *20*, 411–426. [CrossRef] [PubMed]

34. Lavelle, A.; Sokol, H. Gut microbiota-derived metabolites as key actors in inflammatory bowel disease. *Nat. Rev. Gastroenterol. Hepatol.* **2020**, *17*, 223–237. [CrossRef] [PubMed]

35. Fais, S.; O'Driscoll, L.; Borras, F.E.; Buzas, E.; Camussi, G.; Cappello, F.; Carvalho, J.; Cordeiro da Silva, A.; Del Portillo, H.; El Andaloussi, S.; et al. Evidence-Based Clinical Use of Nanoscale Extracellular Vesicles in Nanomedicine. *ACS Nano* **2016**, *10*, 3886–3899. [CrossRef]

36. Yáñez-Mó, M.; Siljander, P.R.; Andreu, Z.; Zavec, A.B.; Borràs, F.E.; Buzas, E.I.; Buzas, K.; Casal, E.; Cappello, F.; Carvalho, J.; et al. Biological properties of extracellular vesicles and their physiological functions. *J. Extracell. Vesicles* **2015**, *4*, 27066. [CrossRef]

37. Rondelli, V.; Di Cola, E.; Koutsioubas, A.; Alongi, J.; Ferruti, P.; Ranucci, E.; Brocca, P. Mucin Thin Layers: A Model for Mucus-Covered Tissues. *Int. J. Mol. Sci.* **2019**, *20*, 3712. [CrossRef]

38. Johansson, M.E.V.; Larsson, J.M.H.; Hansson, G.C. The two mucus layers of colon are organized by the MUC2 mucin, whereas the outer layer is a legislator of host—microbial interactions. *Proc. Natl. Acad. Sci. USA* **2011**, *108*, 4659–4665. [CrossRef]

39. Cappello, F.; Rappa, F.; Canepa, F.; Carini, F.; Mazzola, M.; Tomasello, G.; Bonaventura, G.; Giuliana, G.; Leone, A.; Saguto, D.; et al. Probiotics Can Cure Oral Aphthous-Like Ulcers in Inflammatory Bowel Disease Patients: A Review of the Literature and a Working Hypothesis. *Int. J. Mol. Sci.* **2019**, *20*, 5026. [CrossRef]

40. Simons, M.; Raposo, G. Exosomes—Vesicular carriers for intercellular communication. *Curr. Opin. Cell Biol.* **2009**, *21*, 575–581. [CrossRef]

41. Sabrina David, F.R.; Antonella Marino Gammazza, Alberto Giuseppe Fucarino, Celeste Caruso Bavisotto, Alessandro Pitruzzella, Claudia Campanella. Exosomes: Can doctors still ignore their existence? *EuroMediterranean Biomed. J.* **2013**, *8*, 4. [CrossRef]

42. Campanella, C.; Caruso Bavisotto, C.; Logozzi, M.; Marino Gammazza, A.; Mizzoni, D.; Cappello, F.; Fais, S. On the Choice of the Extracellular Vesicles for Therapeutic Purposes. *Int. J. Mol. Sci.* **2019**, *20*, 236. [CrossRef] [PubMed]

43. Théry, C.; Witwer, K.W.; Aikawa, E.; Alcaraz, M.J.; Anderson, J.D.; Andriantsitohaina, R.; Antoniou, A.; Arab, T.; Archer, F.; Atkin-Smith, G.K.; et al. Minimal information for studies of extracellular vesicles 2018 (MISEV2018): A position statement of the International Society for Extracellular Vesicles and update of the MISEV2014 guidelines. *J. Extracell. Vesicles* **2018**, *7*, 1535750. [CrossRef] [PubMed]

44. Caruso Bavisotto, C.; Marino Gammazza, A.; Campanella, C.; Bucchieri, F.; Cappello, F. Extracellular heat shock proteins in cancer: From early diagnosis to new therapeutic approach. In *Seminars in Cancer Biology*; Academic Press: Cambridge, MA, USA, 2021. [CrossRef]

45. Van Niel, G.; D'Angelo, G.; Raposo, G. Shedding light on the cell biology of extracellular vesicles. *Nat. Rev. Mol. Cell Biol.* **2018**, *19*, 213–228. [CrossRef]

46. Kooijmans, S.A.A.; Schiffelers, R.M.; Zarovni, N.; Vago, R. Modulation of tissue tropism and biological activity of exosomes and other extracellular vesicles: New nanotools for cancer treatment. *Pharmacol. Res.* **2016**, *111*, 487–500. [CrossRef]

47. Tickner, J.A.; Urquhart, A.J.; Stephenson, S.A.; Richard, D.J.; O'Byrne, K.J. Functions and therapeutic roles of exosomes in cancer. *Front. Oncol.* **2014**, *4*, 127. [CrossRef]

48. Graziano, F.; Iacopino, D.G.; Cammarata, G.; Scalia, G.; Campanella, C.; Giannone, A.G.; Porcasi, R.; Florena, A.M.; Conway de Macario, E.; Macario, A.J.L.; et al. The Triad Hsp60-miRNAs-Extracellular Vesicles in Brain Tumors: Assessing Its Components for Understanding Tumorigenesis and Monitoring Patients. *Appl. Sci.* **2021**, *11*, 2867. [CrossRef]

49. Vitale, A.M.; Santonocito, R.; Vergilio, G.; Marino Gammazza, A.; Campanella, C.; Conway de Macario, E.; Bucchieri, F.; Macario, A.J.L.; Caruso Bavisotto, C. Brain Tumor-Derived Extracellular Vesicles as Carriers of Disease Markers: Molecular Chaperones and MicroRNAs. *Appl. Sci.* **2020**, *10*, 6961. [CrossRef]

50. Caruso Bavisotto, C.; Cipolla, C.; Graceffa, G.; Barone, R.; Bucchieri, F.; Bulone, D.; Cabibi, D.; Campanella, C.; Marino Gammazza, A.; Pitruzzella, A.; et al. Immunomorphological Pattern of Molecular Chaperones in Normal and Pathological Thyroid Tissues and Circulating Exosomes: Potential Use in Clinics. *Int. J. Mol. Sci.* **2019**, *20*, 4496. [CrossRef]

51. Caruso Bavisotto, C.; Cappello, F.; Macario, A.J.L.; Conway de Macario, E.; Logozzi, M.; Fais, S.; Campanella, C. Exosomal HSP60: A potentially useful biomarker for diagnosis, assessing prognosis, and monitoring response to treatment. *Expert Rev. Mol. Diagn.* **2017**, *17*, 815–822. [CrossRef]

52. Cappello, F.; Logozzi, M.; Campanella, C.; Bavisotto, C.C.; Marcilla, A.; Properzi, F.; Fais, S. Exosome levels in human body fluids: A tumor marker by themselves? *Eur. J. Pharm. Sci. Off. J. Eur. Fed. Pharm. Sci.* **2017**, *96*, 93–98. [CrossRef]

53. Campanella, C.; Rappa, F.; Sciumè, C.; Marino Gammazza, A.; Barone, R.; Bucchieri, F.; David, S.; Curcurù, G.; Caruso Bavisotto, C.; Pitruzzella, A.; et al. Heat shock protein 60 levels in tissue and circulating exosomes in human large bowel cancer before and after ablative surgery. *Cancer* **2015**, *121*, 3230–3239. [CrossRef] [PubMed]

54. Campanella, C.; D'Anneo, A.; Marino Gammazza, A.; Caruso Bavisotto, C.; Barone, R.; Emanuele, S.; Lo Cascio, F.; Mocciaro, E.; Fais, S.; Conway De Macario, E.; et al. The histone deacetylase inhibitor SAHA induces HSP60 nitration and its extracellular release by exosomal vesicles in human lung-derived carcinoma cells. *Oncotarget* **2016**, *7*, 28849–28867. [CrossRef] [PubMed]

55. Lässer, C.; Alikhani, V.S.; Ekström, K.; Eldh, M.; Paredes, P.T.; Bossios, A.; Sjöstrand, M.; Gabrielsson, S.; Lötvall, J.; Valadi, H. Human saliva, plasma and breast milk exosomes contain RNA: Uptake by macrophages. *J. Transl. Med.* **2011**, *9*, 9. [CrossRef]

56. Sánchez-Vidaurre, S.; Eldh, M.; Larssen, P.; Daham, K.; Martinez-Bravo, M.J.; Dahlén, S.E.; Dahlén, B.; van Hage, M.; Gabrielsson, S. RNA-containing exosomes in induced sputum of asthmatic patients. *J. Allergy Clin. Immunol.* **2017**, *140*, 1459.e1452–1461.e1452. [CrossRef]

57. Admyre, C.; Grunewald, J.; Thyberg, J.; Gripenbäck, S.; Tornling, G.; Eklund, A.; Scheynius, A.; Gabrielsson, S. Exosomes with major histocompatibility complex class II and co-stimulatory molecules are present in human BAL fluid. *Eur. Respir. J.* **2003**, *22*, 578–583. [CrossRef]

58. Pastor, L.; Vera, E.; Marin, J.M.; Sanz-Rubio, D. Extracellular Vesicles from Airway Secretions: New Insights in Lung Diseases. *Int. J. Mol. Sci.* **2021**, *22*, 583. [CrossRef]

59. Leoni, G.; Neumann, P.A.; Kamaly, N.; Quiros, M.; Nishio, H.; Jones, H.R.; Sumagin, R.; Hilgarth, R.S.; Alam, A.; Fredman, G.; et al. Annexin A1-containing extracellular vesicles and polymeric nanoparticles promote epithelial wound repair. *J. Clin. Investig.* **2015**, *125*, 1215–1227. [CrossRef]

60. Zhang, X.; Deeke, S.A.; Ning, Z.; Starr, A.E.; Butcher, J.; Li, J.; Mayne, J.; Cheng, K.; Liao, B.; Li, L.; et al. Metaproteomics reveals associations between microbiome and intestinal extracellular vesicle proteins in pediatric inflammatory bowel disease. *Nat. Commun.* **2018**, *9*, 2873. [CrossRef]

61. Mitsuhashi, S.; Feldbrügge, L.; Csizmadia, E.; Mitsuhashi, M.; Robson, S.C.; Moss, A.C. Luminal Extracellular Vesicles (EVs) in Inflammatory Bowel Disease (IBD) Exhibit Proinflammatory Effects on Epithelial Cells and Macrophages. *Inflamm. Bowel Dis.* **2016**, *22*, 1587–1595. [CrossRef]

62. Sommer, F.; Anderson, J.M.; Bharti, R.; Raes, J.; Rosenstiel, P. The resilience of the intestinal microbiota influences health and disease. *Nat. Rev. Microbiol.* **2017**, *15*, 630–638. [CrossRef] [PubMed]

63. Avila-Calderón, E.D.; Araiza-Villanueva, M.G.; Cancino-Diaz, J.C.; López-Villegas, E.O.; Sriranganathan, N.; Boyle, S.M.; Contreras-Rodríguez, A. Roles of bacterial membrane vesicles. *Arch. Microbiol.* **2015**, *197*, 1–10. [CrossRef] [PubMed]

64. Olaya-Abril, A.; Prados-Rosales, R.; McConnell, M.J.; Martín-Peña, R.; González-Reyes, J.A.; Jiménez-Munguía, I.; Gómez-Gascón, L.; Fernández, J.; Luque-García, J.L.; García-Lidón, C.; et al. Characterization of protective extracellular membrane-derived vesicles produced by Streptococcus pneumoniae. *J. Proteom.* **2014**, *106*, 46–60. [CrossRef] [PubMed]
65. Roier, S.; Zingl, F.G.; Cakar, F.; Durakovic, S.; Kohl, P.; Eichmann, T.O.; Klug, L.; Gadermaier, B.; Weinzerl, K.; Prassl, R.; et al. A novel mechanism for the biogenesis of outer membrane vesicles in Gram-negative bacteria. *Nat Commun* **2016**, *7*, 10515. [CrossRef] [PubMed]
66. Kim, M.R.; Hong, S.W.; Choi, E.B.; Lee, W.H.; Kim, Y.S.; Jeon, S.G.; Jang, M.H.; Gho, Y.S.; Kim, Y.K. *Staphylococcus aureus*-derived extracellular vesicles induce neutrophilic pulmonary inflammation via both Th1 and Th17 cell responses. *Allergy* **2012**, *67*, 1271–1281. [CrossRef]
67. Park, K.S.; Lee, J.; Jang, S.C.; Kim, S.R.; Jang, M.H.; Lötvall, J.; Kim, Y.K.; Gho, Y.S. Pulmonary inflammation induced by bacteria-free outer membrane vesicles from *Pseudomonas aeruginosa*. *Am. J. Respir. Cell Mol. Biol.* **2013**, *49*, 637–645. [CrossRef]
68. Stentz, R.; Carvalho, A.L.; Jones, E.J.; Carding, S.R. Fantastic voyage: The journey of intestinal microbiota-derived microvesicles through the body. *Biochem. Soc. Trans.* **2018**, *46*, 1021–1027. [CrossRef]
69. Durant, L.; Stentz, R.; Noble, A.; Brooks, J.; Gicheva, N.; Reddi, D.; O'Connor, M.J.; Hoyles, L.; McCartney, A.L.; Man, R.; et al. Bacteroides thetaiotaomicron-derived outer membrane vesicles promote regulatory dendritic cell responses in health but not in inflammatory bowel disease. *Microbiome* **2020**, *8*, 88. [CrossRef]
70. Kim, J.H.; Lee, J.; Park, J.; Gho, Y.S. Gram-negative and Gram-positive bacterial extracellular vesicles. *Semin. Cell Dev. Biol.* **2015**, *40*, 97–104. [CrossRef]
71. Invernizzi, R.; Lloyd, C.M.; Molyneaux, P.L. Respiratory microbiome and epithelial interactions shape immunity in the lungs. *Immunology* **2020**, *160*, 171–182. [CrossRef]
72. Knowles, M.R.; Boucher, R.C. Mucus clearance as a primary innate defense mechanism for mammalian airways. *J. Clin. Investig.* **2002**, *109*, 571–577. [CrossRef] [PubMed]
73. Bassis, C.M.; Erb-Downward, J.R.; Dickson, R.P.; Freeman, C.M.; Schmidt, T.M.; Young, V.B.; Beck, J.M.; Curtis, J.L.; Huffnagle, G.B. Analysis of the upper respiratory tract microbiotas as the source of the lung and gastric microbiotas in healthy individuals. *mBio* **2015**, *6*, e00037. [CrossRef] [PubMed]
74. Rao, R.; Dsouza, J.M.; Mathew, J.L. Comparison of microbiota in the upper versus lower respiratory tract in children during health and respiratory disease: Protocol for a systematic review. *Syst. Rev.* **2021**, *10*, 253. [CrossRef] [PubMed]
75. Marsland, B.J.; Trompette, A.; Gollwitzer, E.S. The Gut-Lung Axis in Respiratory Disease. *Ann. Am. Thorac. Soc.* **2015**, *12* (Suppl. 2), S150–S156. [CrossRef]
76. Lynch, S.V. The Lung Microbiome and Airway Disease. *Ann. Am. Thorac. Soc.* **2016**, *13* (Suppl. 2), S462–S465. [CrossRef]
77. Dickson, R.P.; Erb-Downward, J.R.; Freeman, C.M.; McCloskey, L.; Falkowski, N.R.; Huffnagle, G.B.; Curtis, J.L. Bacterial Topography of the Healthy Human Lower Respiratory Tract. *mBio* **2017**, *8*, e02287-16. [CrossRef]
78. Sahin-Yilmaz, A.; Naclerio, R.M. Anatomy and physiology of the upper airway. *Proc. Am. Thorac. Soc.* **2011**, *8*, 31–39. [CrossRef]
79. Man, W.H.; de Steenhuijsen Piters, W.A.; Bogaert, D. The microbiota of the respiratory tract: Gatekeeper to respiratory health. *Nat. Rev. Microbiol.* **2017**, *15*, 259–270. [CrossRef]
80. Goto, T. Airway Microbiota as a Modulator of Lung Cancer. *Int. J. Mol. Sci.* **2020**, *21*, 3044. [CrossRef]
81. Elgamal, Z.; Singh, P.; Geraghty, P. The Upper Airway Microbiota, Environmental Exposures, Inflammation, and Disease. *Medicina* **2021**, *57*, 823. [CrossRef]
82. Larsen, G.L.; Holt, P.G. The concept of airway inflammation. *Am. J. Respir. Crit. Care Med.* **2000**, *162*, S2–S6. [CrossRef]
83. Zhou-Suckow, Z.; Duerr, J.; Hagner, M.; Agrawal, R.; Mall, M.A. Airway mucus, inflammation and remodeling: Emerging links in the pathogenesis of chronic lung diseases. *Cell Tissue Res.* **2017**, *367*, 537–550. [CrossRef] [PubMed]
84. Dang, A.T.; Marsland, B.J. Microbes, metabolites, and the gut-lung axis. *Mucosal Immunol.* **2019**, *12*, 843–850. [CrossRef] [PubMed]
85. Fujimura, K.E.; Lynch, S.V. Microbiota in allergy and asthma and the emerging relationship with the gut microbiome. *Cell Host Microbe* **2015**, *17*, 592–602. [CrossRef] [PubMed]
86. Vinolo, M.A.; Rodrigues, H.G.; Nachbar, R.T.; Curi, R. Regulation of inflammation by short chain fatty acids. *Nutrients* **2011**, *3*, 858–876. [CrossRef]
87. Chiu, C.J.; Huang, M.T. Asthma in the Precision Medicine Era: Biologics and Probiotics. *Int. J. Mol. Sci.* **2021**, *22*, 4528. [CrossRef]
88. Yang, K.; Dong, W. Perspectives on Probiotics and Bronchopulmonary Dysplasia. *Front. Pediatrics* **2020**, *8*, 570247. [CrossRef]
89. Sestito, S.; D'Auria, E.; Baldassarre, M.E.; Salvatore, S.; Tallarico, V.; Stefanelli, E.; Tarsitano, F.; Concolino, D.; Pensabene, L. The Role of Prebiotics and Probiotics in Prevention of Allergic Diseases in Infants. *Front. Pediatrics* **2020**, *8*, 583946. [CrossRef]
90. Herath, M.; Hosie, S.; Bornstein, J.C.; Franks, A.E.; Hill-Yardin, E.L. The Role of the Gastrointestinal Mucus System in Intestinal Homeostasis: Implications for Neurological Disorders. *Front. Cell. Infect. Microbiol.* **2020**, *10*, 248. [CrossRef]
91. Sicard, J.F.; Le Bihan, G.; Vogeleer, P.; Jacques, M.; Harel, J. Interactions of Intestinal Bacteria with Components of the Intestinal Mucus. *Front. Cell. Infect. Microbiol.* **2017**, *7*, 387. [CrossRef]
92. Jakobsson, H.E.; Rodríguez-Piñeiro, A.M.; Schütte, A.; Ermund, A.; Boysen, P.; Bemark, M.; Sommer, F.; Bäckhed, F.; Hansson, G.C.; Johansson, M.E. The composition of the gut microbiota shapes the colon mucus barrier. *EMBO Rep.* **2015**, *16*, 164–177. [CrossRef] [PubMed]

93. Allaire, J.M.; Morampudi, V.; Crowley, S.M.; Stahl, M.; Yu, H.; Bhullar, K.; Knodler, L.A.; Bressler, B.; Jacobson, K.; Vallance, B.A. Frontline defenders: Goblet cell mediators dictate host-microbe interactions in the intestinal tract during health and disease. *Am. J. Physiol. Gastrointest. Liver Physiol.* **2018**, *314*, G360–G377. [CrossRef] [PubMed]

94. Pelaseyed, T.; Bergström, J.H.; Gustafsson, J.K.; Ermund, A.; Birchenough, G.M.; Schütte, A.; van der Post, S.; Svensson, F.; Rodríguez-Piñeiro, A.M.; Nyström, E.E.; et al. The mucus and mucins of the goblet cells and enterocytes provide the first defense line of the gastrointestinal tract and interact with the immune system. *Immunol. Rev.* **2014**, *260*, 8–20. [CrossRef] [PubMed]

95. Johansson, M.E.; Hansson, G.C. Immunological aspects of intestinal mucus and mucins. *Nat. Rev. Immunol.* **2016**, *16*, 639–649. [CrossRef]

96. Etzold, S.; Juge, N. Structural insights into bacterial recognition of intestinal mucins. *Curr. Opin. Struct. Biol.* **2014**, *28*, 23–31. [CrossRef]

97. Arike, L.; Hansson, G.C. The Densely O-Glycosylated MUC2 Mucin Protects the Intestine and Provides Food for the Commensal Bacteria. *J. Mol. Biol.* **2016**, *428*, 3221–3229. [CrossRef]

98. Juge, N. Microbial adhesins to gastrointestinal mucus. *Trends Microbiol.* **2012**, *20*, 30–39. [CrossRef]

99. Douillard, F.o.P.; Ribbera, A.; Jï¿½rvinen, H.M.; Kant, R.; Pietilï¿½, T.E.; Randazzo, C.; Paulin, L.; Laine, P.K.; Caggia, C.; Ossowski, I.v.; et al. Comparative Genomic and Functional Analysis of *Lactobacillus casei* and *Lactobacillus rhamnosus* Strains Marketed as Probiotics. *Appl. Environ. Microbiol.* **2013**, *79*, 1923–1933. [CrossRef]

100. Kankainen, M.; Paulin, L.; Tynkkynen, S.; Ossowski, I.v.; Reunanen, J.; Partanen, P.; Satokari, R.; Vesterlund, S.; Hendrickx, A.P.A.; Lebeer, S.; et al. Comparative genomic analysis of *Lactobacillus rhamnosus* GG reveals pili containing a human—Mucus binding protein. *Proc. Natl. Acad. Sci. USA* **2009**, *106*, 17193–17198. [CrossRef]

101. Slížová, M.; Nemcová, R.; Mad'ar, M.; Hadryová, J.; Gancarčíková, S.; Popper, M.; Pistl, J. Analysis of biofilm formation by intestinal lactobacilli. *Can. J. Microbiol.* **2015**, *61*, 437–446. [CrossRef]

102. Belkaid, Y.; Hand, T.W. Role of the microbiota in immunity and inflammation. *Cell* **2014**, *157*, 121–141. [CrossRef] [PubMed]

103. Sender, R.; Fuchs, S.; Milo, R. Revised Estimates for the Number of Human and Bacteria Cells in the Body. *PLoS Biol.* **2016**, *14*, e1002533. [CrossRef] [PubMed]

104. Karlsson, F.; Tremaroli, V.; Nielsen, J.; Bäckhed, F. Assessing the human gut microbiota in metabolic diseases. *Diabetes* **2013**, *62*, 3341–3349. [CrossRef] [PubMed]

105. Milani, C.; Duranti, S.; Bottacini, F.; Casey, E.; Turroni, F.; Mahony, J.; Belzer, C.; Palacio, S.D.; Montes, S.A.; Mancabelli, L.; et al. The First Microbial Colonizers of the Human Gut: Composition, Activities, and Health Implications of the Infant Gut Microbiota. *Microbiol. Mol. Biol. Rev.* **2017**, *81*, e00036-17. [CrossRef] [PubMed]

106. Kinross, J.M.; Darzi, A.W.; Nicholson, J.K. Gut microbiome-host interactions in health and disease. *Genome Med.* **2011**, *3*, 14. [CrossRef]

107. Wang, H.; Dai, W.; Feng, X.; Zhou, Q.; Wang, H.; Yang, Y.; Li, S.; Zheng, Y. Microbiota Composition in Upper Respiratory Tracts of Healthy Children in Shenzhen, China, Differed with Respiratory Sites and Ages. *BioMed Res. Int.* **2018**, *2018*, 6515670. [CrossRef]

108. Wang, G.; Huang, S.; Wang, Y.; Cai, S.; Yu, H.; Liu, H.; Zeng, X.; Zhang, G.; Qiao, S. Bridging intestinal immunity and gut microbiota by metabolites. *Cell. Mol. Life Sci.* **2019**, *76*, 3917–3937. [CrossRef]

109. Sánchez, B.; Delgado, S.; Blanco-Míguez, A.; Lourenço, A.; Gueimonde, M.; Margolles, A. Probiotics, gut microbiota, and their influence on host health and disease. *Mol. Nutr. Food Res.* **2017**, *61*, 1600240. [CrossRef]

110. Hill, M.J. Intestinal flora and endogenous vitamin synthesis. *Eur. J. Cancer Prev. Off. J. Eur. Cancer Prev. Organ.* **1997**, *6* (Suppl. 1), S43–S45. [CrossRef]

111. Morrison, D.J.; Preston, T. Formation of short chain fatty acids by the gut microbiota and their impact on human metabolism. *Gut Microbes* **2016**, *7*, 189–200. [CrossRef]

112. Scheppach, W. Effects of short chain fatty acids on gut morphology and function. *Gut* **1994**, *35*, S35–S38. [CrossRef] [PubMed]

113. Jiao, Y.; Wu, L.; Huntington, N.D.; Zhang, X. Crosstalk Between Gut Microbiota and Innate Immunity and Its Implication in Autoimmune Diseases. *Front. Immunol.* **2020**, *11*, 282. [CrossRef] [PubMed]

114. Hernández-Chirlaque, C.; Aranda, C.J.; Ocón, B.; Capitán-Cañadas, F.; Ortega-González, M.; Carrero, J.J.; Suárez, M.D.; Zarzuelo, A.; Sánchez de Medina, F.; Martínez-Augustin, O. Germ-free and Antibiotic-treated Mice are Highly Susceptible to Epithelial Injury in DSS Colitis. *J. Crohns Colitis* **2016**, *10*, 1324–1335. [CrossRef] [PubMed]

115. Bhattarai, Y.; Kashyap, P.C. Germ-Free Mice Model for Studying Host–Microbial Interactions. In *Mouse Models for Drug Discovery: Methods and Protocols*; Proetzel, G., Wiles, M.V., Eds.; Springer: New York, NY, USA, 2016; pp. 123–135.

116. Grover, M.; Kashyap, P.C. Germ-free mice as a model to study effect of gut microbiota on host physiology. *Neurogastroenterol. Motil.* **2014**, *26*, 745–748. [CrossRef]

117. Lazar, V.; Ditu, L.-M.; Pircalabioru, G.G.; Gheorghe, I.; Curutiu, C.; Holban, A.M.; Picu, A.; Petcu, L.; Chifiriuc, M.C. Aspects of Gut Microbiota and Immune System Interactions in Infectious Diseases, Immunopathology, and Cancer. *Front. Immunol.* **2018**, *9*, 1830. [CrossRef] [PubMed]

118. Liévin, V.; Peiffer, I.; Hudault, S.; Rochat, F.; Brassart, D.; Neeser, J.-R.; Servin, A.L. Bifidobacterium strains from resident infant human gastrointestinal microflora exert antimicrobial activity. *Gut* **2000**, *47*, 646–652. [CrossRef]

119. Wang, X.; Zhang, P.; Zhang, X. Probiotics Regulate Gut Microbiota: An Effective Method to Improve Immunity. *Molecules* **2021**, *26*, 6076. [CrossRef]

120. Klann, E.M.; Dissanayake, U.; Gurrala, A.; Farrer, M.; Shukla, A.W.; Ramirez-Zamora, A.; Mai, V.; Vedam-Mai, V. The Gut-Brain Axis and Its Relation to Parkinson's Disease: A Review. *Front Aging Neurosci.* **2022**, *13*, 782082. [CrossRef]
121. Gustafsson, G.; Lööv, C.; Persson, E.; Lázaro, D.F.; Takeda, S.; Bergström, J.; Erlandsson, A.; Sehlin, D.; Balaj, L.; György, B.; et al. Secretion and Uptake of α-Synuclein Via Extracellular Vesicles in Cultured Cells. *Cell Mol. Neurobiol.* **2018**, *38*, 1539–1550. [CrossRef]
122. Bui, T.M.; Mascarenhas, L.A.; Sumagin, R. Extracellular vesicles regulate immune responses and cellular function in intestinal inflammation and repair. *Tissue Barriers* **2018**, *6*, e1431038. [CrossRef]
123. Lo Cicero, A.; Stahl, P.D.; Raposo, G. Extracellular vesicles shuffling intercellular messages: For good or for bad. *Curr. Opin. Cell Biol.* **2015**, *35*, 69–77. [CrossRef] [PubMed]
124. Ahmadi Badi, S.; Moshiri, A.; Fateh, A.; Rahimi Jamnani, F.; Sarshar, M.; Vaziri, F.; Siadat, S.D. Microbiota-Derived Extracellular Vesicles as New Systemic Regulators. *Front. Microbiol.* **2017**, *8*, 1610. [CrossRef] [PubMed]
125. Lagos, L.; Tandberg, J.; Kashulin-Bekkelund, A.; Colquhoun, D.J.; Sørum, H.; Winther-Larsen, H.C. Isolation and Characterization of Serum Extracellular Vesicles (EVs) from Atlantic Salmon Infected with Piscirickettsia Salmonis. *Proteomes* **2017**, *5*, 34. [CrossRef] [PubMed]
126. Moshiri, A.; Dashtbani-Roozbehani, A.; Najar Peerayeh, S.; Siadat, S.D. Outer membrane vesicle: A macromolecule with multifunctional activity. *Hum. Vaccines Immunother.* **2012**, *8*, 953–955. [CrossRef] [PubMed]
127. Campbell, H.K.; Maiers, J.L.; DeMali, K.A. Interplay between tight junctions & adherens junctions. *Exp. Cell Res.* **2017**, *358*, 39–44. [CrossRef]
128. Hu, G.; Gong, A.-Y.; Roth, A.L.; Huang, B.Q.; Ward, H.D.; Zhu, G.; LaRusso, N.F.; Hanson, N.D.; Chen, X.-M. Release of Luminal Exosomes Contributes to TLR4-Mediated Epithelial Antimicrobial Defense. *PLOS Pathog.* **2013**, *9*, e1003261. [CrossRef]
129. Muraca, M.; Putignani, L.; Fierabracci, A.; Teti, A.; Perilongo, G. Gut microbiota-derived outer membrane vesicles: Under-recognized major players in health and disease? *Discov. Med.* **2015**, *19*, 343–348.
130. Shen, Y.; Giardino Torchia, M.L.; Lawson, G.W.; Karp, C.L.; Ashwell, J.D.; Mazmanian, S.K. Outer membrane vesicles of a human commensal mediate immune regulation and disease protection. *Cell Host Microbe* **2012**, *12*, 509–520. [CrossRef]
131. Haurat, M.F.; Elhenawy, W.; Feldman, M.F. Prokaryotic membrane vesicles: New insights on biogenesis and biological roles. *Biol. Chem.* **2015**, *396*, 95–109. [CrossRef]
132. Li, K.J.; Chen, Z.L.; Huang, Y.; Zhang, R.; Luan, X.Q.; Lei, T.T.; Chen, L. Dysbiosis of lower respiratory tract microbiome are associated with inflammation and microbial function variety. *Respir. Res.* **2019**, *20*, 272. [CrossRef]
133. Hernández-Terán, A.; Vega-Sánchez, A.E.; Mejía-Nepomuceno, F.; Serna-Muñoz, R.; Rodríguez-Llamazares, S.; Salido-Guadarrama, I.; Romero-Espinoza, J.A.; Guadarrama-Pérez, C.; Sandoval, J.; Campos, F.; et al. Microbiota composition in the lower respiratory tract is associated with severity in patients with acute respiratory distress by influenza. *medRxiv* **2021**. Available online: https://www.medrxiv.org/content/10.1101/2021.12.07.21267419v1 (accessed on 8 October 2022).
134. Wang, H.; Gu, X.; Weng, Y.; Xu, T.; Fu, Z.; Peng, W.; Yu, W. Quantitative analysis of pathogens in the lower respiratory tract of patients with chronic obstructive pulmonary disease. *BMC Pulm. Med.* **2015**, *15*, 94. [CrossRef] [PubMed]
135. Oren, A.; Garrity, G.M. Valid publication of the names of forty-two phyla of prokaryotes. *Int. J. Syst. Evol. Microbiol.* **2021**, *71*, 005056. [CrossRef] [PubMed]
136. Eckburg, P.B.; Bik, E.M.; Bernstein, C.N.; Purdom, E.; Dethlefsen, L.; Sargent, M.; Gill, S.R.; Nelson, K.E.; Relman, D.A. Diversity of the Human Intestinal Microbial Flora. *Science* **2005**, *308*, 1635–1638. [CrossRef]
137. Bäckhed, F.; Ley, R.E.; Sonnenburg, J.L.; Peterson, D.A.; Gordon, J.I. Host-bacterial mutualism in the human intestine. *Science* **2005**, *307*, 1915–1920. [CrossRef]
138. King, C.H.; Desai, H.; Sylvetsky, A.C.; LoTempio, J.; Ayanyan, S.; Carrie, J.; Crandall, K.A.; Fochtman, B.C.; Gasparyan, L.; Gulzar, N.; et al. Baseline human gut microbiota profile in healthy people and standard reporting template. *PLoS ONE* **2019**, *14*, e0206484. [CrossRef]
139. Nayfach, S.; Páez-Espino, D.; Call, L.; Low, S.J.; Sberro, H.; Ivanova, N.N.; Proal, A.D.; Fischbach, M.A.; Bhatt, A.S.; Hugenholtz, P.; et al. Metagenomic compendium of 189,680 DNA viruses from the human gut microbiome. *Nat. Microbiol.* **2021**, *6*, 960–970. [CrossRef]
140. Miller, T.L.; Wolin, M.J. Methanogens in human and animal intestinal Tracts. *Syst. Appl. Microbiol.* **1986**, *7*, 223–229. [CrossRef]
141. Kim, J.Y.; Whon, T.W.; Lim, M.Y.; Kim, Y.B.; Kim, N.; Kwon, M.S.; Kim, J.; Lee, S.H.; Choi, H.J.; Nam, I.H.; et al. The human gut archaeome: Identification of diverse haloarchaea in Korean subjects. *Microbiome* **2020**, *8*, 114. [CrossRef]
142. Kühbacher, T.; Ott, S.J.; Helwig, U.; Mimura, T.; Rizzello, F.; Kleessen, B.; Gionchetti, P.; Blaut, M.; Campieri, M.; Fölsch, U.R.; et al. Bacterial and fungal microbiota in relation to probiotic therapy (VSL#3) in pouchitis. *Gut* **2006**, *55*, 833–841. [CrossRef]
143. Raimondi, S.; Amaretti, A.; Gozzoli, C.; Simone, M.; Righini, L.; Candeliere, F.; Brun, P.; Ardizzoni, A.; Colombari, B.; Paulone, S.; et al. Longitudinal Survey of Fungi in the Human Gut: ITS Profiling, Phenotyping, and Colonization. *Front. Microbiol.* **2019**, *10*, 1575. [CrossRef] [PubMed]
144. Li, J.; Chen, D.; Yu, B.; He, J.; Zheng, P.; Mao, X.; Yu, J.; Luo, J.; Tian, G.; Huang, Z.; et al. Fungi in Gastrointestinal Tracts of Human and Mice: From Community to Functions. *Microb. Ecol.* **2018**, *75*, 821–829. [CrossRef]

145. Zhang, T.; Breitbart, M.; Lee, W.H.; Run, J.-Q.; Wei, C.L.; Soh, S.W.L.; Hibberd, M.L.; Liu, E.T.; Rohwer, F.; Ruan, Y. RNA Viral Community in Human Feces: Prevalence of Plant Pathogenic Viruses. *PLoS Biol.* **2005**, *4*, e3. [CrossRef] [PubMed]
146. Manrique, P.; Dills, M.; Young, M.J. The Human Gut Phage Community and Its Implications for Health and Disease. *Viruses* **2017**, *9*, 141. [CrossRef] [PubMed]
147. Mayneris-Perxachs, J.; Castells-Nobau, A.; Arnoriaga-Rodríguez, M.; Garre-Olmo, J.; Puig, J.; Ramos, R.; Martínez-Hernández, F.; Burokas, A.; Coll, C.; Moreno-Navarrete, J.M.; et al. Caudovirales bacteriophages are associated with improved executive function and memory in flies, mice, and humans. *Cell Host Microbe* **2022**, *30*, 340–356.e348. [CrossRef] [PubMed]

Review

Anatomy of Cerebral Arteries with Clinical Aspects in Patients with Ischemic Stroke

Francesco Barbato [1],*, Roberto Allocca [1], Giorgio Bosso [1,2] and Fabio Giuliano Numis [1,2]

[1] Department of Emergency and Urgent Medicine, Stroke Unit, Santa Maria Delle Grazie Hospital, 80078 Naples, Italy
[2] Department of Emergency and Urgent Medicine, Emergency Medicine, Santa Maria Delle Grazie Hospital, 80078 Naples, Italy
* Correspondence: francesco.barbato@aslnapoli2nord.it

Abstract: Computed tomography (CT) angiography is the main method for the initial evaluation of cerebral circulation in acute stroke. A comprehensive CT examination that includes a review of the three-dimensional and maximum-intensity projection images of the main intra and extracranial arteries allows the identification of most abnormalities and normal variants. Anatomical knowledge of the presence of any normal variants, such as fenestration, duplications, and persistent fetal arteries, plays a crucial role in the diagnosis and therapeutic management of acute stroke. However, the opposite is also true. In fact, sometimes it is the clinical picture that allows weighing how relevant or not the alteration found is. Therefore, in this review, a concise representation of the clinical picture attributable to a given arterial vessel will be included.

Keywords: cerebral vessels; ischemic stroke; neurovascular anatomy; cerebral arteries

Citation: Barbato, F.; Allocca, R.; Bosso, G.; Numis, F.G. Anatomy of Cerebral Arteries with Clinical Aspects in Patients with Ischemic Stroke. *Anatomia* **2022**, *1*, 152–169. https://doi.org/10.3390/anatomia1020016

Academic Editors: Gianfranco Natale and Francesco Fornai

Received: 15 September 2022
Accepted: 19 October 2022
Published: 21 October 2022

1. Introduction

Ischemic stroke is the second most common cause of death worldwide and is a leading cause of disability, with an increasing incidence in developing countries [1]. It occurs when a blood clot blocks or narrows an artery that carries blood to the brain. The clinical management of a stroke is complex and multidisciplinary [2]. Neurological evaluation is a fundamental moment as it allows one to carry out a differential diagnosis, a prognostic, and finally, a therapeutic evaluation [2]. Therefore, knowing exactly the anatomy of the cerebral vessels is a necessary assumption to avoid gross medical errors [2,3]. Although digital subtraction angiography (ASD) remains the gold standard, computed tomography (CT) angiography is the main method for the initial assessment of cerebral circulation in acute stroke. Changes in cerebral circulation, particularly in the Willis polygon, are common [1,4]. It is important to know the appearance of these normal variants, their prevalence, and their clinical relevance. In this review paper, we will address both the anatomy of the main extra and intracranial vessels associated with acute cerebrovascular events and the clinical picture associated with them.

2. Description of Vascular Anatomy

We conducted a search of reviews referring to the anatomy of the cerebral vessels and their variants in the PubMed database and in the book "Anatomia Umana" by Anastasi et al. In the PubMed database, we searched through the terms [anatomy of the cerebral vessel] AND [variants of cerebral supply] AND [normal variants of cerebral vessels] with no filter. The result was 119 papers, 35 containing direct information relevant to the question. For the clinical point, a further review of the literature was carried out using PubMed and a book, Harrison's neurology in clinical medicine (2016). The literature review produced 11 papers, of which 8 were selected for this work. In both searches, case reports were excluded.

2.1. Extracranial Circulation

2.1.1. Arch of the Aorta

The aortic arch follows the ascending aorta and ideally begins at the level of the upper margin of the second sternocostal joint on the right. Concave downwards, its path follows an oblique plane bringing itself posteriorly and to the left on the anterior surface of the trachea, reaching the height of the body of the fourth thoracic vertebra, where it continues in the descending aorta. This level corresponds to a narrowing of the vessel, called the aortic isthmus. From the convex face of the arch of the aorta originate the right brachiocephalic trunk (TABC) or anonymous trunk, the left common carotid artery, and the left subclavian artery [5].

Clinical point: The arch of the aorta is involved in atherothrombotic disease, aneurysmal pathology, dissection, and inflammatory diseases such as aortitis. The clinical picture, often impressive, is determined by the degree and number of vessels involved. In symptomatic cases with distal embolization or aneurysm thrombosis, the clinical presentation depends on the size of embolic debris and the location of arterial occlusions. This is added to retrosternal pain, often lacerating, and neurovegetative symptomatology [6–10].

2.1.2. Supra-Aortic Vessels (SAT)

They consist of the unnamed artery, or brachiocephalic artery (BCT) on the right, common carotid artery (CCA), and subclavian artery (SA) bilaterally. These arterial vessels are most frequently arranged in seven types [11]. The most common is type 1, present in 80.9% of the population. In these cases, BCT is divided into CCA and SA on the right, while CCA and SA on the left are formed directly from the aorta. The other variants in the origin of SAT are found in 19.1% of cases. The second most common variation is type 2, the bovine arc, with an incidence of 13.6% [11,12]. Here the left CCA originates directly from the BCT. Type 3 instead has a frequency of 2.9%, in which the left vertebral artery originates directly from the aortic arch. The other 4 variants have an overall frequency of 1% [11,12]. We also include another non-exceptional variant, the formation of the right SA, not from the BCT, but directly from the arch of the aorta, completely to the left, downstream of the left SA, where it takes the name of artery "lusoria" or ARSA (aberrant right subclavian artery) (Figure 1). It passes behind the esophagus and then finds its usual path. It can be a source of dysphagia, which results from a defect in the resorption of the fourth right aortic arch during fetal life. Finally, the direct origin of the vertebral artery from the aortic arch (4%) between the CCA and the SA on the left may occur [11–13].

2.1.3. Common Carotid Artery (CCA)

It originates from the BCT on the right and the aortic arch on the left. It runs within the carotid sulcus, between the internal jugular vein that interposes laterally and the vagus nerve posteriorly, wrapped in a connective sheath called the vascular-nervous bundle of the neck. Its terminal portion is projected at the height of the body of the C4 vertebra and gives rise to the internal carotid artery (with intracranial destination) and the external carotid artery (with cervicofacial destination). The projection of the carotid ending can occur at a variable level; the most frequently occurs at the level of the upper margin of the thyroid cartilage of the larynx, between the C6 plane and the angle of the jaw [5,14].

Clinical point: The common carotid artery can be the site of atherothrombotic disease and sometimes dissection. The carotid artery most frequently affected by these processes is the left since it originates directly from the aortic arch, while the site is preferably located at the level of the bifurcations and at the origin of the collateral branches of the carotids [6,7]. Carotid dolicoarteriopathies (CDA) are even a common finding. The global prevalence of CDA is 12.9%, and carotid kinking is more frequent in females and in the left carotid axis [15]. CDA is not associated with a major occurrence of cerebrovascular events. Bilateral common carotid artery occlusions at their origin may occur in Takayasu's arteritis [6,16]. The neurological symptomatology is variable, as, in some patients, it may go unnoticed, while in other patients presents with hemiparesis and contralateral hemianesthesia associated with dysarthria and amaurosis. Mandibular claudication may be present after chewing [6–8].

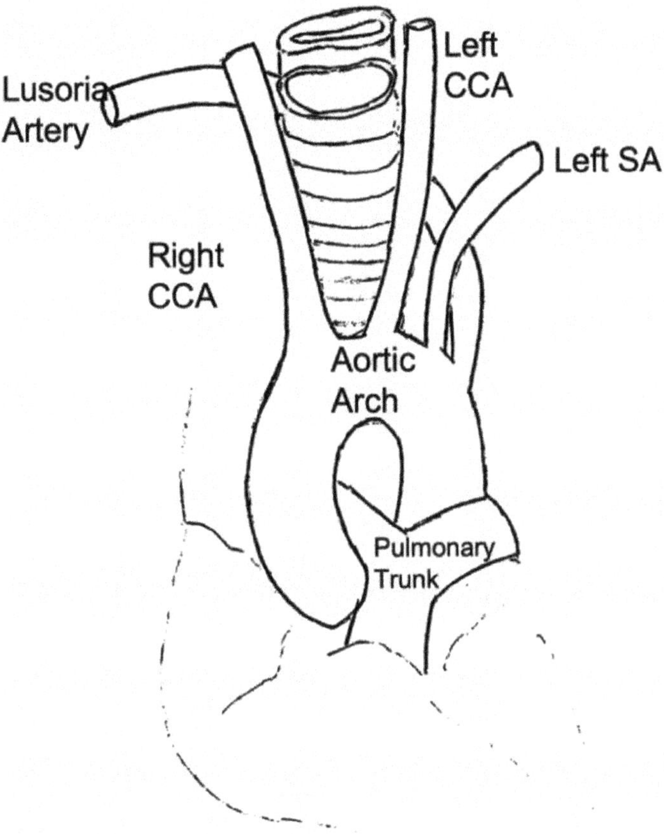

Figure 1. The image represents the picture typically associated with the lusoria artery. Instead of being the first branch (with the right common carotid as the brachiocephalic artery), it arises on its own as the fourth branch, distal to the left subclavian artery. It then hooks back to reach the right side with a strong relationship to the esophagus.

2.1.4. Internal Carotid Artery (ICA)—Extracranial Tract

On an anterior projection, it presents, in most cases, an outermost path of the external carotid artery at its origin. Four segments are distinguished: the cervical segment, the petrous segment, the intracavernous segment, and the supracavernous segment (Figure 2) [5]. It runs most frequently in its first segment within the maxillopharyngeal space, accompanied by the internal jugular vein. It is retropharyngeal in 10% of cases, running posterior to the wall of the oropharynx [5,17]. In any case, it then penetrates the petrous rock and runs in the carotid canal, which first has a vertical and then horizontal orientation. It returns to the intracranial plane through the torn foramen, then runs into the cavernous loggia; then, it enters the subarachnoid spaces in its supraclinoid portion. It emits on its anterior face the ophthalmic artery, then on the posterior face, spaced a few millimeters, the posterior communicating arteries, and anterior choroid. Finally, it divides to give the anterior and middle cerebral arteries [5,14]. In 20% of cases, a fetal origin of the posterior cerebral artery is recognized, which therefore originates from the internal carotid artery [18]. There are some anatomical variations. An aberrant internal carotid artery (AbICA) is a rare variation. It is considered a result of agenesis of the first cervical ICA segment. Abnormal vessel develops from the fusion of the inferior tympanic branch of the ascending pharyngeal artery with the caroticotympanic artery. Agenesis of ICA occurs in less than 0.01% of the population, while the bilateral absence of the artery is seen in less than 10% of cases of agenesis [13,14,19]. Prevalence of hypoplasia is 0.079% [19]. Other variants of the internal carotid artery include duplication, ICA fenestration, and high or low branching of the carotid artery (from T2 to C1 level). Rarely, ICA and ECA may originate directly from the aorta [20].

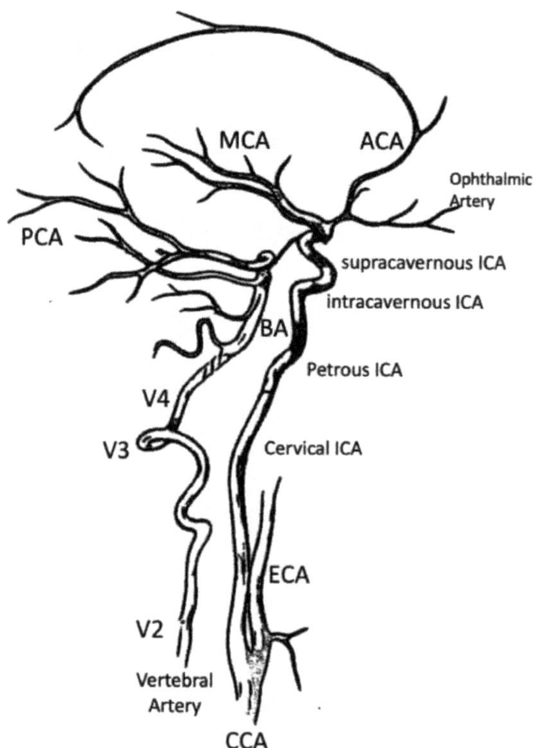

Figure 2. In this image are represented the main segments of the arteries afferent to the brain. On the right, it is possible to see the four segments of the ICA (cervical, petrous, intracavernous, supracavernous) with the carotid siphon and the main terminal vessels (ACA, MCA) and collateral (ophthalmic artery). On the left, we observe the vertebral artery in its intraspinal tract (V2), the loop at the level of C2 (V3), and its intracranial portion (V4) before joining with the contralateral to form the basilar artery. It is also possible to observe the origin and course of PCA.

Clinical point: The clinical picture of the occlusion of the internal carotid artery is extremely variable [6]; in fact, as in the case of a common carotid artery involvement, it could go unnoticed if the compensations determined by the Willis polygon and the carotid circles are efficient [7]. The cerebral cortex pertaining to the middle cerebral artery (MCA) is affected more frequently [6,7]. In this case, the symptoms are identical to the occlusion of proximal MCA (see the clinical point on the middle cerebral artery). When both the origins of the anterior cerebral artery (ACA) and middle (MCA) are occluded, the disorder of consciousness is accompanied by hemiplegia and complete hemianesthesia, aphasia or dysarthria, and anosognosia depending on whether the dominant hemisphere is affected or not [6]. When the posterior cerebral artery (PCA) is fetal (therefore, in 20% of cases), it can also become occluded and give rise to symptoms referable to its peripheral territory [7]. In addition to supplying the ipsilateral brain, the internal carotid artery perfuses the optic nerve and retina through the ophthalmic artery. In 25% of symptomatic internal carotid disease, recurrent transient monocular blindness (amaurosis fugax) precedes cerebral ischemia. When retinal ischemia has been established, eye blindness can be recognized because the photomotor reflex is not present in the affected eye [6–8,16].

2.1.5. Vertebral Artery (VA)

The vertebral arteries are the first ascending branches of the division of the subclavian arteries and originate from the posterior and upper wall of the pre-scalenic tract. Four portions [5,21] are distinguished (Figure 2):

- V1 segment, the segment of the vertebral artery before its entry into the transverse canal (C6);
- V2 segment, part of the vertebral artery that runs in the transverse canal of C6 up to C2;
- V3 segment, which bypasses the lateral masses of C1 before redirecting medially to penetrate the atlanto-occipital membrane and through the occipital hole, makes its entry into the cranial cavity;
- V4 segment of intracranial localization, just before the meeting of the two vertebral arteries. It is noted that this V4 segment is of subarachnoid collocation.

Anatomical variants of the origin and course of the vertebral arteries are frequent. The most frequent (found in up to 2.9% of cases) is the origin of the left vertebral artery directly from the aortic arch, between the left common carotid artery and the left subclavian artery. The vertebral arteries also frequently have an asymmetrical caliber. The term "dominant vertebral artery" refers to the most voluminous (generally the left vertebral artery). It is also possible to find hypoplasia of a vertebral artery in 2–6%, characterized by a reduction in the diameter of the lumen < 2.2 mm starting from its origin [22]. Intracranial vertebral fenestrations were found in 1.1% of patients and are the most common fenestration of cerebral vessels [23,24].

Clinical point: The pathologies that can give an acute cerebrovascular syndrome starting from an extracranial vertebral artery are atherothrombosis with artery-to-artery embolism and dissection [9,10]. The formation of atherothrombotic plaques is most common in the V1 and V4 segments [10]. If the subclavian artery is occluded proximal to the origin of the vertebral artery, there is a reversal in the direction of blood flow in the ipsilateral vertebral artery. Exercise of the ipsilateral arm may increase demand on vertebral flow, producing posterior circulation TIAs, or "subclavian steal syndrome" [6,9,10].

In the case of artery-to-artery embolism, the intraarterial emboli travel to reach the ipsilateral intracranial vertebral artery and sometimes travel on to block the rostral basilar artery and/or its branches [6,10]. Dissections are mostly located in the pars transversaria segment (V2, 35% of cases) or in the atlas loop segment (V3, 34% of cases) [10,25]. Patients present with a variety of signs and symptoms, most frequently with neck pain and headache (typically occipital) as well as posterior fossa ischemic events manifesting as nausea, ataxia, dysarthria, lateral medullary syndrome, or even collapse and coma. Other presentations include spinal cord infarction and even cervical nerve root impairment [25].

2.2. Intracranial Circulation

2.2.1. Willis Polygon

The Willis polygon represents a compensatory network of the intracranial arteries. It has the shape of a heptagon whose sides are: the precommunicating segments of the anterior cerebral arteries, the anterior communicating artery, the posterior communicating arteries, and the precommunicating segments of the posterior cerebral arteries [5]. It is characterized by great variability [5,12,13]. About 22 different forms of Willis polygons have been described (Figure 3). It is complete only in 13–21% of cases [12,13].

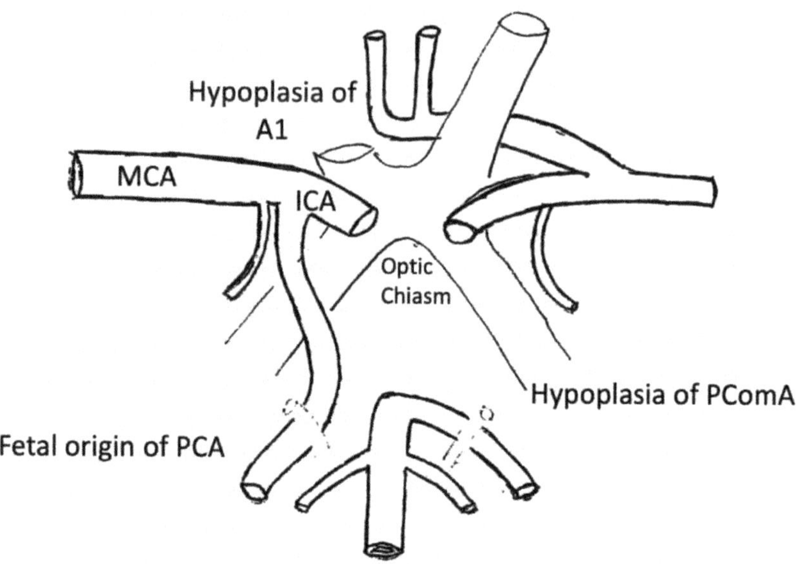

Figure 3. In this figure is arranged the Willis polygon (heptagon), emphasizing the possible hypoplasia/agenesis of the posterior communicating artery on the right, the fetal origin of the PCA directly of the ICA, and the hypoplasia of the left anterior cerebral artery. About 22 different types of Willis polygon are described.

Fenestration may occur in all cerebral arteries, but it most commonly involves the Willis polygon, basilar artery, and anterior cerebral artery. Fenestration is the presence of two arterial channels corresponding to a single path with partial or segmental arterial duplication. It involves the segmental division of a vessel to form two separate lumens with endothelial walls. These lumens may be surrounded by the same adventitia or have separate external laminae [26].

Among the anatomical variants, we recognize the carotid-basilar anastomoses. These anastomoses correspond to the persistence of vestigial arteries, which should normally regress during the development of the embryo, between the internal carotid artery and the vertebrobasilar system [12,13,26].

The most frequent is the persistence of the trigeminal artery (0.02–0.6%). It connects the C4 or C5 segment of the internal carotid artery with the anterior face of the middle third/upper third junction of the basilar artery. It passes through the Meckel cord after bypassing the *clivus* (parasellar course), then runs near the trigeminal nerve in its cisternal portion before anastomizing with the basilar artery. It can also pass through an orifice right into the back of the *sella turcica* (trans-sellar course). It is extremely important to recognize it as it can be crucial to plan a neurosurgical intervention aimed at the pituitary gland. Hypoplasia or absence of the homolateral posterior communicating artery and the P1 segment of the PCA is often associated with this variant. An absence/hypoplasia of the homolateral vertebral artery can also be observed, and, in this case, the basilar artery is fed by the persistent trigeminal artery. Another anastomotic persistence is the presence of the acoustic artery, which is extremely rare. It connects, crossing the petrous rock, the internal carotid artery in the carotid canal, and the caudal part of the basilar trunk. The presence of the hypoglossal artery is the second in order of frequency among carotid-basilar anastomoses (from 0.027% to 0.26%). It originates on the posterior face of the upper portion of the internal carotid artery in the cervical segment, at the height of the C1–C2 vertebral bodies. It then runs posteriorly and upwards in the hypoglossal foramen (or condylic canal), which appears enlarged. It has a rear concavity curve, then curves to reach the midline. It is often associated with hypoplasia of the vertebral and posterior communicating arteries [27,28].

The last one we mention is the presence of the proatlantal artery. This carotid-vertebrobasilar anastomosis is rare. There are two types: the first coming from the internal carotid artery and the second from the external carotid artery. It originates from the posterior face of the internal carotid artery at the C2–C3 vertebral segments (more rarely C4) or from the proximal portion of the external carotid artery. It has a posterior and ascending course; it surrounds the lateral masses of C1 and accompanies the homolateral vertebral artery in its entry into the foramen magnum [21,27,28].

2.2.2. Internal Carotid Artery (ICA)—Intracranial Tract

The intracranial internal carotid artery presents, in its intracavernous portion, a series of curves that give it an Italic "S" appearance open upwards. This curved portion bears the name of the carotid siphon. Using the classification of Fischer, five segments for the carotid siphon are distinguished, named from C5 proximal to C1 distally, in the opposite direction to blood flow [29]. Segments C4 and C3 have an intracavernous localization, and segments C2 and C1 have a distribution above cavernosa (Figure 4).

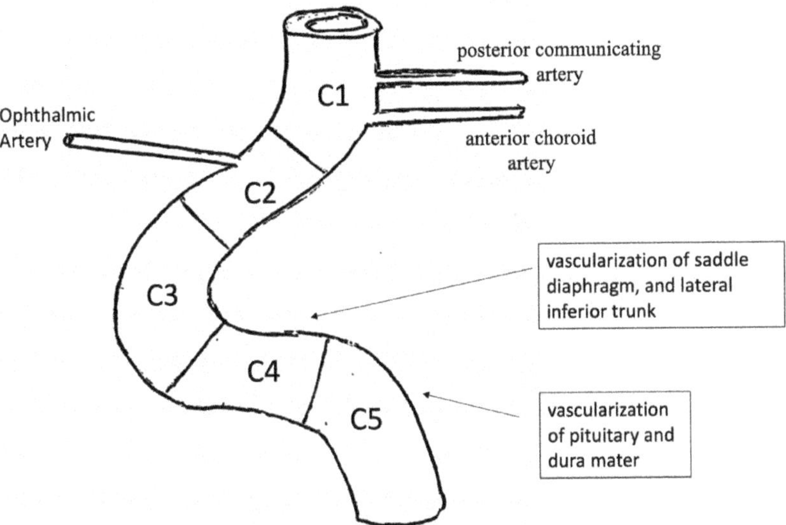

Figure 4. It is possible to observe Fisher's subdivision of the carotid siphon. Five segments for the carotid siphon are distinguished, named from C5 proximal to C1 distally. Segments C4 and C3 have an intracavernous localization, and segments C2 and C1 have a distribution above cavernosa.

- C5 segment gives rise to the meningo-pituitary trunk, from which emerges above all some branches that participate in the vascularization of the pituitary and adjacent dura mater;
- C4 segment has a landscape orientation and is directed forward. Its main collaterals are the capsular arteries, which vascularize the saddle diaphragm and the lateral inferior trunk. The lateral inferior trunk is divided into three branches: upper, anterior, and posterior. They participate, respectively, in the vascularization of the roof of the cavernous sinus and oculomotor nerves;
- C3 segment is curved and has a front convexity;
- C2 segment, above cavernous and infra-clinoid, is short and has a posterior, horizontal, or slightly ascending orientation. It gives rise to the superior pituitary arteries, which vascularize the anterior loggia of the pituitary gland. From the anterior part of the upper portion of the C2 segment originates the ophthalmic artery. The ophthalmic

artery is of subarachnoid localization, more rarely intradural. It has an anterior orientation to reach the optic canal, where it runs against the lateral inferior face of the optic nerve. From this segment, the intradural arteries do not possess an elastic lamina and are represented by a small tunica adventitia and poor elastic fibers in the tunica media;

- Finally, segment C1 is above cavernous and supraclinoid.

In its supra-clinoid portion (anatomically subarachnoid), the ICA is divided into four branches: posterior communicating artery, anterior choroid artery, anterior cerebral artery, and middle cerebral artery.

Clinical point: it is the same as the extracranial counterpart. The atherothrombotic disease is the most frequently encountered disease, particularly at the level of the carotid siphon. Intracranial stenosis may be present in the C1 and C2 tracts with clinical transient amaurosis and contralateral hemiparesis. At the entrance to the carotid canal, it is sometimes possible that the internal carotid artery undergoes dissection with artery-to-artery embolism. In this case, due to the low number of elastic fibers in the middle tunic and the almost absent elastic lamina in the case of dissection, it is preferable to use a double antiplate aggregation rather than an anticoagulant.

2.2.3. Anterior Cerebral Artery (ACA)

The anterior cerebral artery (ACA) originates from the carotid ending on its anteromedial face. It is divided, according to Fischer [30], into five segments:

- The A1 segment, known as 'pre-communicating';
- The A2 segment, which has an ascending course up to the corpus callosum;
- The A3 segment, which surrounds the knee of the corpus callosum;
- A4 and A5, which continue their course around the corpus callosum.

The ACA presents, in its proximal portion (segment A1), an anteromedial course. It bypasses the optic chiasm to reach the interhemispheric cleavage. The A1 segment gives rise to an artery that has a recurrent posterior course: Heubner's recurrent artery, which vascularizes the head of the caudate nucleus, the anteroinferior part of the inner capsule, and the anterior portion of the putamen. The recurrent artery of Heubner (RAH) is also called a median striatal artery [5,30]. The A1 segment of the left ACA is characterized by its great variability in caliber; an asymmetry is present in 55% of cases. Hypoplasia of segment A1 is present in 10% of cases [30–33]. It is defined by a diameter of less than 1 mm. It is frequently associated with aneurysms of the anterior communicating artery [32]. The ACA then bypasses the knee of the corpus callosum. It is called the "pericallosal artery". From the pericallosal artery, some cortical branches are born, among which are identified: the orbito-frontal artery, the fronto-polar artery, the medial, anterior, and posterior frontal arteries, the paracentral artery, the superior (or pre-cuneal), and inferior medial parietal arteries. It ends with the posterior peri-callous artery and gives rise in more than 80% of cases to a callose-marginal artery, which runs in the callose-marginal sulcus [5,30].

There are several anatomical variants [30–33]. The median artery of the corpus callosum, which participates in the vascularization of the corpus callosum, the bihemispheric arteries, which correspond to an ACA that vascularizes the contralateral hemisphere (Figure 5) and, finally, the origin of a single common trunk that divides at the knee of the corpus callosum, are distinguished. This is also referred to as ACA Azygos (2%). Here, the A2 segments are fused, and the anterior communicating artery is absent (Figure 5) [33].

Clinical point: Proximal ACA occlusion is usually asymptomatic due to collateral flow through the anterior communicating artery and collaterals through MCA and PCA [6,7]. The occlusion of a single A2 segment causes contralateral motor and/or sensory deficits (leg > arm and face). If both A2 segments are derived from a single anterior brain stem, the occlusion can affect both hemispheres. [7] Deep abulia (a delay in verbal and motor response) and bilateral pyramidal signs with paraparesis and urinary incontinence. Apraxia of the march is also possible [7].

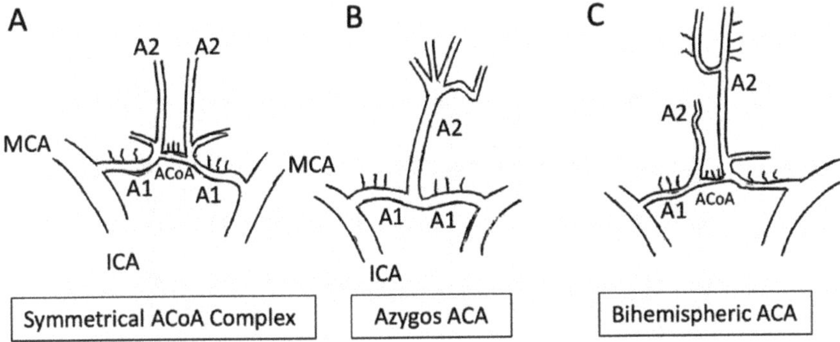

Figure 5. In Panel (**A**), it is possible to observe the modal arrangement of the ACA. In Panel (**B**), the origin of a single common trunk that divides at the knee of the corpus callosum, named ACA Azygos (2%), with A2 segments fused, and the anterior communicating artery is absent. The bihemispheric arteries, which correspond to an ACA that vascularizes the contralateral hemisphere, are on Panel (**C**).

2.2.4. Middle Cerebral Artery (MCA)

Also called the "Silvian artery", the middle cerebral artery (MCA) represents the most voluminous subdivision branch of the ICA [5]. It measures at its origin about 3 mm in diameter. Four segments are generally distinguished: the M1 segment or basal segment, the M2 segment or insular segment, the M3 segment or opercular segment, and, finally, the M4 cortical segment (Figure 6). The M1 segment extends from the origin of the artery to its entry into the Silvian cavity. It has a lateral course under the anterior perforated space. The M1 segment gives rise to the lenticolostriate branches, which arise perpendicularly and penetrate the anterior perforated space. In variable numbers (from 6 to 20), these arteries participate in the vascularization of the lenticular nucleus, the inner capsule, and the head of the caudate nucleus. They have an Italic 'S' course in frontal projection [5,19,30].

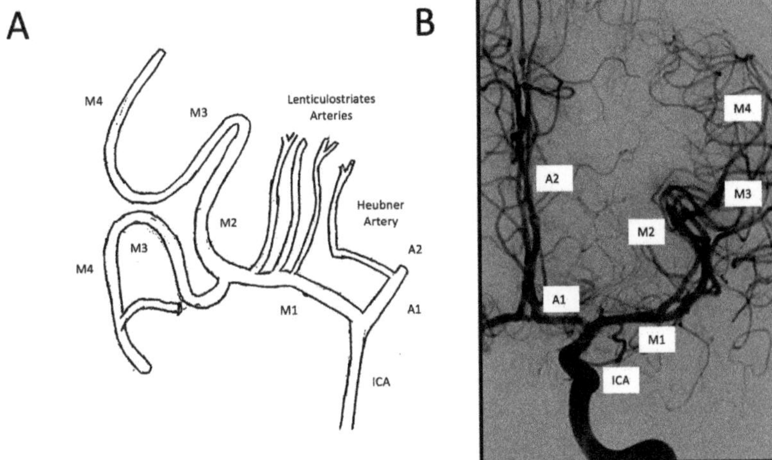

Figure 6. Arrangement of the middle cerebral artery in a schematization of the coronal section (panel **A**) and in the images obtained by angiography (panel **B**). It represents a terminal division branch of the ICA. Its course can be divided horizontally (MI), whose collateral branches are represented by lenticulostriate arteries. It ends in a knee where the insular segment (M2) begins. From here, two trunks are formed: upper and lower, both called M3. From the endings of both trunks of M3 cortical branches, M4 is formed.

We distinguish insular branches and cortical branches [19,30]. The insular branches, subdivision branches of the Silvian artery, curve to the lower pole of the insula to run on its outer face. They have a postero-superior oblique orientation and are divided into terminal arteries. The ascending terminal branches are distinguished, which have a first circuit with lower concavity; they are reflected on the upper line of the insula (which has a slight upper convexity), then form an upper concavity circuit before branching to the outer face of the brain. The descending branches have only an upper concavity and line the outer face of the temporal lobe. In lateral projection, the line joining the reflection points of the upper terminal branches draws a line called an "insular line", which corresponds to the upper margin of the insula and has a straight or slightly convex course upwards.

The cortical branches of the MCA are divided into three groups: ascending, posterior, and descending. From the front to the back are distinguished:

- The frontal orbital artery, which vascularizes the lower face of the frontal lobe;
- The anterior frontal artery, which vascularizes the external faces of the F2 and F3 convolutions and the frontal operculum;
- The ascending frontal artery, which runs in the central and post-central grooves. It ensures the vascularization of the ascending frontal and parietal convolutions;
- The anterior parietal artery;
- The posterior parietal artery;
- The artery of the curved fold or angular artery, which prolongs the axis of the Silvian artery. It ensures the vascularization of the posterior frontal region.

The descending branches are three:

- The anterior temporal artery, which vascularizes the temporal pole;
- The middle temporal artery, which has an oblique course at the bottom and backward in the middle part of the temporal lobe. It ensures the vascularization of the middle and posterior parts of T1eT2;
- The posterior temporal artery or temporo-occipital, which has a course roughly parallel to the previous one. It vascularizes the outer face of the occipital lobe and the back of the outer face of the temporal lobe.

We recognize as anatomical variants that the relatively rare accessory media cerebral artery (0.3–1%) corresponds to an MCA that emerges both directly from the internal carotid artery and from the homolateral ACA [34,35]. In other rare variants, MCA fenestrations typically affect the first millimeters of the M1 segment. Finally, variations in the distribution of cortical branches are very frequently found [26].

Clinical point: If the entire MCA is occluded at its origin (blocking both its penetrating and cortical branches) and the distal collaterals are limited, the clinical picture is characterized by complete contralateral hemiplegia (due to the involvement of the penetrating vessels and, therefore, of the inner capsule) (face and arm > thigh > foot), hemianesthesia, and lastly homonymous lateral hemianopsia [6,7]. When the dominant hemisphere is involved, global aphasia is also present, while when the non-dominant hemisphere is affected, anosognosia, constructive apraxia, and neglect are found. Cortical collateral blood flow and different arterial configurations are likely responsible for the development of many partial syndromes [6]. Partial syndromes due to embolic occlusion of a single branch include brachial syndrome or facial weakness with or without non-flowing Broca aphasia (frontal opercular syndrome). A combination of sensory disturbances, motor weakness, and non-fluent aphasia suggests that an embolus occluded the proximal superior division and infarcted large portions of the frontal and parietal cortexes [7]. If a fluent Wernicke's aphasia occurs without weakness, the lower division of the MCA that supplies the posterior part (temporal cortex) of the dominant hemisphere is likely involved [6,8]. Sludge language and the inability to understand written and spoken language are prominent features, often accompanied by a contralateral upper quadrantanopia. Hemineglect or spatial agnosia without weakness indicates that the lower division of MCA in the non-dominant hemisphere is involved [8]. The occlusion of a lenticulostriate vessel may produce a lacunar infarct. This produces a purely motor, purely sensory, or contralateral motor sensory picture of the injury. Ischemia inside the knee of the inner capsule mainly causes facial weakness followed by weakness of the arm and then of the legs. Alternatively, the contralateral hand may become ataxic, and the dysarthria will be prominent (clumsy hand, lacunar syndrome of dysarthria). Lacunar infarction affecting globus pallidus and putamen often has few clinical signs, but parkinsonism and hemiballism have been reported [6–8].

2.2.5. Anterior Choroid Artery

It originates from the posterior face of the terminal part of the ICA, a few millimeters above the posterior communicating artery. It has a fine caliber (0.6–2 mm). Its path takes place on the inner face of the temporal lobe [5]. Its course is divided into three segments: the proximal, cisternal, and distal (intraventricular) segments. After its cisternal course, where it emits branches intended for the globus pallidus, the posterior arm of the inner capsule, the tail of the caudate nucleus, the lateral ventral nucleus of the thalamus, and the red nucleus, the anterior choroid artery then follows the choroid fissure, then penetrates the temporal horn changing the orientation of its course and its caliber. It ends by branching at the level of the choroid plexuses of the lateral ventricles. It has, in its cisternal path, a posteroinferior direction, therefore, an ascending course. It has a bayonet course at the time of its entry into the ventricular horn, so it bypasses the pulvinar in its terminal portion. In frontal projection, it takes the form of an inverse "S", with the inferior medial concavity corresponding to its cisternal segment and the superior lateral concavity corresponding to its ventricular segment. The cortical territory of the anterior choroid artery is reduced to temporal branches that ensure vascularization of the anterior and medial parts of the temporal lobe [5,36].

Clinical point: Complete syndrome of anterior choroid artery occlusion consists of contralateral hemiplegia, hemianesthesia, and homonymous lateral hemianopia. However, since this territory is also provided by penetrating vessels of the proximal MCA and posterior choroid arteries, minimal deficits may occur, and patients often recover substantially [6–8]. Strokes of the anterior choroid artery are usually the result of in situ thrombosis of the vessel [7].

2.2.6. Anterior Communicating Artery (ACoA)

The anterior communicating artery is short, arranged transversely in front and slightly below the portion of the optic chiasm. It closes the Willis polygon [5] in front (Figure 5). This artery has numerous variations in caliber [19,24,26] (ranging up to hypoplasia that makes the two ACA independent) and in number (split in 30% of cases, tripled in 10%, exceptionally vascular network). The absence of the anterior communicating artery has been found in 5% of surgical dissections. Fenestrations in the ACoA were found in 5.3% of the population [37]. Most fenestrations were associated with 1 or more aneurysms of the AcoA [26,37].

2.2.7. Posterior Communicating Artery (PCoA)

It does not belong to the posterior circulation (vertebrobasilar) but to the anterior circle (carotid) due to its embryological origin [13]. It originates from the posterior face of the supracavernous ICA (C1–C2). Most often modest in size (1–2 mm on average), it measures about 10–15 mm in length. It connects the terminal portion of the ACI and the homolateral posterior cerebral artery (Figure 5). It gives rise to some thin collateral branches that participate in the vascularization of the thalamus, hypothalamus, hippocampus, optic tract, and posterior arm of the inner capsule [5,30].

2.2.8. Intracranial Vertebral Artery (VA)

It pierces the atlantoaxial membrane to penetrate the posterior cerebral fossa. It anastomizes with its contralateral homolog at the bulbo-pontine sulcus after bypassing the lateral face of the bulb to form the basilar artery (Figure 2). The V4 segment emits a thin accessory branch that runs downwards and frequently anastomizes with the contralateral branch to form the anterior spinal arterial axis of the high cervical medulla [5,12,30].

Clinical point: An atherothrombotic disease of the fourth distal segment (V4) of the vertebral artery can promote the formation of a thrombus and manifest as an embolism or with the propagation of thrombosis up to the basilar artery. Proximal stenosis at the origin of PICA can cause lateral ischemia of the medulla and of the lower posterior surface of the cerebellum [9,10].

It can therefore manifest itself with a constellation of symptoms such as headache, dizziness, numbness of the ipsilateral face and contralateral limbs, diplopia, hoarseness, dysarthria, dysphagia, and ipsilateral Horner syndrome. This is called lateral (or Wallenberg's) medullary syndrome [6,9]. The occlusion of the penetrating medullary branches of the vertebral artery or PICA causes partial syndromes. Hemiparesis is not a feature of vertebral artery occlusion; however, tetra paresis can result from occlusion of the anterior spinal artery. Cerebellar infarction can present with ataxia, headache, and nausea. Distinguishing these symptoms from those of viral labyrinthitis can be a challenge, but headaches, neck stiffness, and unilateral dysmetria favor stroke [10]. It should be noted that a medial medullary syndrome rarely occurs. The cause is ischemia of the bulbar pyramid with contralateral hemiparesis of the arm and leg and sparing of the face. If the medial lemniscus and the nucleus of the hypoglossal nerve (XII c.n.) are involved, contralateral loss of the sense of joint position and weakness of the ipsilateral tongue occur [9,10]. In the intracranial dissection of the V4 segment of the spine, as opposed to its extracranial portion, there is a high risk of subarachnoid hemorrhage (up to 50% for vertebrobasilar dissections) due to the microscopic anatomy of the intracranial arteries as the elastic fibers are in the subendothelial elastic lamina, which overall is less thick [10].

2.2.9. Posteroinferior Cerebellar Artery (PICA)

It generally originates from the vertebral artery in its V4 portion, 15–20 mm from its end. It then runs between the tonsil and the roof of the fourth ventricle. At its peak, it forms a circuit with superior convexity. In the proximal portion of its course, it emits some bulbar branches, tonsils, and the posterior spinal arteries. In the cranial circuit of its distal portion, it gives rise to the choroid branches directed to the fourth ventricle. In its terminal portion, it generates the lower vermian branches and the hemispherical branches [5,30]. The origin of the inferior cerebellar artery is subject to many variations. It can form directly from the basilar artery (7–10%), from the internal carotid artery in its C5 portion (through a persistent trigeminal artery), or from the ascending pharyngeal artery (starting from its hypoglossal branch), but it can also arise from a common trunk that also gives rise to the middle cerebellar artery [12–14].

Clinical point: Cerebellar infarction in the PICA distribution may involve only the vermis, the lateral surface, or the entire territory of the PICA. Infarctions of the complete PICA territory are often accompanied by the formation of edema and the mass effect of increased intracranial pressure (ICP) with the headache associated with cerebellar symptoms [6]. About 15% of cerebellar infarctions of the PICA territory are accompanied by ischemia in the dorsolateral medulla (Wallenberg syndrome). The combination of lateral medullary infarction and cerebellar PICA occurs when the intracranial vertebral artery is occluded and blocks the orifice of both the PICA and the lateral medullary penetrating vessels [6,9,10]. Medial worm-limited ischemia in the medial territory of PICA usually causes a vertiginous labyrinthine syndrome that mimics peripheral vestibulopathy. Severe dizziness, ipsilateral later pulsion, and nystagmus are the main symptoms [10].

Infarcts of the PICA territory of the lateral cerebellar hemisphere are characterized by ataxia and ipsilateral lateral pulsion but without dizziness or dysarthria. Sometimes adiadochokinesia is present. When the entire cerebellar territory PICA is involved, the neurological symptomatology described above is associated with headache, usually present in the occiput on the ipsilateral side [10]. The head can also be tilted, with the occiput tending to tilt toward the ipsilateral side [6,10].

2.2.10. Basilar Artery

Single and median, it is formed by the conjunction of the two vertebral arteries at the height of the bulbopontine sulcus (Figure 2). It results from the fusion of the posterior longitudinal arteries. Usually of a length of 2.5–3.5 cm, it has an anterosuperior orientation. Its diameter is 3–4 mm. It runs in front of the brainstem to end at the level of the cerebral peduncles at the height of the ponto-mesencephalic sulcus, emitting the posterior cerebral arteries (PCA). It is generally rectilinear in the young subject and may have curvatures in the elderly subject. In addition to PCA, it gives rise to anteroinferior cerebellar and anterosuperior cerebellar arteries [5,12,30]. It can rarely give rise to posteroinferior cerebellar arteries. On its posterolateral face are born the circumferential branches (from four to six pairs) and the perforating branches, numerous in its distal portion. Among the anatomical variants of the basilar artery, fenestrations and hypoplasia are mentioned. Basilar artery fenestration has been found in 5% of autopsies [26,33]. Basilar artery fenestrations are mostly located in the proximal basilar trunk, close to the vertebrobasilar junction. The reported frequency of aneurysm formation in cases of basilar artery fenestration is 7%. In the latter case, the PCA shall become dependent on the carotid territory [12,26,30].

Clinical point: It is possible to distinguish the clinical picture determined in proximal or distal occlusion of the basilar artery [6,9,10,30]. *Proximal Basilar artery occlusion* most often presents as ischemia in the pons. The major burden of ischemia is in the middle of the pons, mostly in the paramedian base and often also in the paramedian tegmentum. This is because the lateral tegmentum is also supplied collateral circles coming to the PICA from the AICA and the SCA. The paramedian pontine base contains descending long motor tract and crossing cerebellar fibers. The paramedian tegmentum contains mostly oculomotor fibers. As a result, the predominant symptoms and signs in patients with basilar artery occlusive disease are motor (various combinations of hemiparesis or tetra paresis with palatal myoclonus until locked in syndrome) and oculomotor signs such as pinpoint miosis, horizontal conjugate gaze palsy or internuclear ophthalmoplegia (INO). Skew deviation of the eyes and ocular bobbing may also be present. Bulbar symptoms include facial weakness, dysphonia, dysarthria, dysphagia, and limited jaw movements. The face, pharynx, larynx, and tongue are most often involved. Alteration in the level of consciousness is an important sign in patients with basilar artery occlusion [6,10]. They may present with coma when the bilateral medial pontine tegmentum is ischemic. *Occlusion of the rostral portion of the basilar artery* (the "top of the basilar") can cause ischemia of the midbrain, thalami, and temporal and occipital lobe hemispheric territories supplied by the posterior cerebral artery branches of the basilar artery. It presents with tetraparesis, paralysis of the cranial nerves, and coma. The pupils may not be miotic, and sometimes it is possible to recognize spontaneous eye movements such as reverse ocular bobbing. This clinical picture is sometimes very difficult to distinguish from diffuse cerebral suffering of metabolic origin [10,38].

2.2.11. Posterior Cerebral Artery (PCA)

The posterior cerebral arteries arise from the terminal subdivision of the basilar artery at the height of the ponto-mesencephalic sulcus. The PCA has a short course in the interpeduncular cistern, so it bypasses the cerebral peduncles. About 1 cm from its origin, it receives the posterior communicating artery that anastomizes it to the internal carotid artery (Figures 2 and 3). It ensures vascularization of the internal and lower faces of the temporal lobe, the inner face of the occipital lobe and thalamus, the third ventricle, and the lateral ventricles [5,30].

The PCA divides to give rise to the internal occipital artery, ascending and the inferior temporal descending branch. The PCA is divided into four segments:

- The P1 or precommunicating segment;
- The P2 segment, which bypasses the cerebral peduncle through III c.n;
- The P3 segment, which runs along the underside of the temporal lobe;
- The P4 segment that arrive at the calcarine fissure.

The most frequent anatomical variant is the persistence of fetal organization (20%). In this case, the PCA arises directly from the back of the ICA, and the P1 segment has agenesis or hypoplasia [18]. This form corresponds to a failure to regression of modal vascularization in the fetus. In fact, during fetal life, the PCA is fed by ICA through the segment that corresponds to the future posterior communicating artery. This segment regresses in adulthood, and the P1 portion then takes charge of the vascularization of the PCA [18]. In any case, *diencephalic branches* are present. These correspond to thin branches that originate from the PCA or from the posterior communicating artery [5,12,30]. They are mainly distinguished [5,30]:

- The inferior dorsomedial arteries, which vascularize the lower portion of the thalamus and originate from the posterior communicating artery;
- The posteromedian choroid artery, which originates from the P2 segment of the ACP. It runs along the upper margin of the posterior communicating artery, then on the upper edge of the thalamus, emitting branches that vascularize the anterior nuclei of the thalamus. Finally, it penetrates the choroid canvas of the third ventricle, which it sprays;
- The posterolateral choroid artery. Ensures the vascularization of the choroid plexuses. It crosses the inner face of the pulvinar and penetrates the choroid fissure of the lateral ventricle;
- The posterior perichallose artery, which runs on the posterior face of the splenium of the corpus callosum.

Clinical point: PCA syndromes usually result from atheroma formation or emboli that lodge at the top of the basilar artery; posterior circulation disease may also be caused by dissection of either the vertebral artery or fibromuscular dysplasia. Two clinical syndromes are commonly observed with occlusion of the PCA:

- P1 syndrome: midbrain, subthalamic, and thalamic signs, which are due to disease of the proximal P1 segment of the PCA or its penetrating branches (thalamogeniculate, Percheron, and posterior choroidal arteries);
- P2 syndrome: cortical temporal and occipital lobe signs due to occlusion of the segment distal to the junction of the PCA with the posterior communicating artery.

In P1 syndromes, infarction usually occurs in the ipsilateral subthalamus and medial thalamus and in the ipsilateral cerebral peduncle and midbrain. A third nerve palsy with contralateral ataxia (Claude's syndrome) or with contralateral hemiplegia (Weber's syndrome) may result [6,9]. The ataxia indicates involvement of the red nucleus or dentatorubrothalamic tract; the hemiplegia is localized to the cerebral peduncle. If the subthalamic nucleus is involved, contralateral hemiballismus may occur [6,9]. Occlusion of the artery of Percheron produces paresis of upward gaze and drowsiness, and often abulia [39]. Extensive infarction in the midbrain and subthalamus occurring with bilateral proximal PCA occlusion presents as coma, unreactive pupils, bilateral pyramidal signs, and decerebrate rigidity [6,9]. Occlusion of the penetrating branches of thalamic and thalamogeniculate arteries produces less extensive thalamic and thalamocapsular lacunar syndromes [39]. The thalamic Déjerine-Roussy syndrome consists of contralateral hemisensory loss followed later by an agonizing, searing, or burning pain in the affected areas. It is persistent and responds poorly to analgesics [9].

In P2 syndromes, occlusion of the distal PCA causes infarction of the medial temporal and occipital lobes occurs. Contralateral homonymous hemianopia with macula sparing is the usual manifestation. Occasionally, only the upper quadrant of the visual field is involved. If the visual association areas are spared, and only the calcarine cortex is involved, the patient may be aware of visual defects. Medial temporal lobe and hippocampal involvement may cause an acute disturbance in memory, particularly if it occurs in the dominant hemisphere. The defect usually clears because memory has a bilateral representation. If the dominant hemisphere is affected and the infarct extends to involve the splenium of the corpus callosum, the patient may demonstrate alexia without agraphia. Visual agnosia for faces, objects, mathematical symbols, and colors and anomia with paraphasic errors (amnestic aphasia) may also occur in this setting, even without callosal involvement. Occlusion of the posterior cerebral artery can produce peduncular hallucinosis (visual hallucinations of brightly colored scenes and objects) [6–10,30].

Bilateral infarction in the distal PCAs produces cortical blindness (blindness with preserved pupillary light reaction). The patient is often unaware of the blindness or may even deny it (Anton's syndrome). Bilateral visual association area lesions may result in Balint's syndrome, a disorder of the orderly visual scanning of the environment, usually resulting from infarctions secondary to low flow in the "watershed" between the distal PCA and MCA territories, as occurs after cardiac arrest. Patients may experience the persistence of a visual image for several minutes despite gazing at another scene (palinopsia) or an inability to synthesize the whole of an image (simultagnosia) [6,30].

2.2.12. Anteroinferior Cerebellar Artery (or Middle Cerebellar Artery, AICA)

It originates in the lower third of the basilar artery in more than 3/4 of cases and in the vertebrobasilar junction only in about 10% of cases [40]. From its origin, the artery runs laterally and inferiorly on the front face of the bridge, giving rise to multiple perforating branches that vascularize the lower 2/3 of the bridge and the upper part of the medulla oblongata. The artery then enters the cistern at the angle of the cerebellum and near the facial nerve and vestibulocochlear nerve. It then heads laterally, in a higher position than the cerebellar flocculus, then folds medially, crossing the lower face of the cerebellar hemispheres, where it sends some collateral branches that vascularize these structures, as well as portions of the lower part of the worm and, in depth, the homolateral toothed nucleus [5,40].

Clinical point: The symptomatology is like that of lateral medullary syndrome but with peripheral paralysis of the facial nerve (VII c.n) and of the eighth cranial nerve (impairment of homolateral hearing to the lesion) [6,10].

2.2.13. Anterosuperior Cerebellar Artery (or Superior Cerebellar Artery, SCA)

It originates from the terminal portion of the basilar artery. Immediately it moves laterally, just below the oculomotor nerve, which separates it from the posterior cerebral artery, and turns around the brain stem, near the trochlear nerve, to reach the cerebellomesencephalic fissure reaching the upper surface of the cerebellum. Here it is divided into branches into the pia mater that anastomize with those of the anterior inferior cerebellar artery (AICA) and posterior (PICA). In a percentage of subjects between 12% and 14%, the artery, once originated, can go to duplicate [5,13,40].

Clinical point: The clinic is characterized by dynamic ataxia, dysmetria, dysarthria, dizziness, and headache [6,10].

3. Discussion

The WHO (World Health Organization) defines stroke as "a sudden appearance of signs and/or symptoms referable to deficits in brain functions, localized or global lasting more than 24 h or to an unfortunate outcome not attributable to any other apparent cause than cerebral vasculopathy" [41].

As a prerequisite for ischemic stroke, therefore, there is the occlusion of a cerebral artery, resulting in a reduction in blood supply and, therefore, oxygen and nutrients to the brain tissue. This causes alterations in the functionality of brain cells that quickly lead to necrosis of the portion of brain tissue most affected by ischemia (the ischemic core), around which there is a portion of tissue in ischemic suffering but still recoverable (the ischemic penumbra). The mechanisms that induce cell death (homeostasis of Ca^{2+}, Na^+/Cl^-, K^+, excitotoxicity, peri-infarction depolarization, oxidative stress, inflammation, etc.) are not limited to neurons. They involve the whole of the neurovascular unit, oligodendrocytes, and microglia [30,42].

In addition, systemic immunodepression caused by ischemic stroke is responsible for infections that worsen the functional prognosis. [43]. Therefore, it becomes essential to recognize and distinguish neurological symptoms attributable to the involvement of a brain area dependent on a vessel. This is not so much for a mere anecdotal purpose but for its clinical and prognostic implication since, in this way, it is possible to choose the best possible therapeutic option. Historically, studies of the clinical consequences of strokes, and their relation to vascular territories in the brain, provided information about the location of various brain functions [43]. Now, non-invasive functional imaging techniques have largely supplanted the correlation of clinical signs and symptoms with the location of tissue damage observed at autopsy [42,43]. This also helps in understanding the clinical significance of possible anatomic variants found in patients.

4. Conclusions

The anatomy of the cerebral circulation and the main extracranial vessels involved in cerebral ischemia is subject to significant variability. Knowledge of normal anatomy and its variants is the essential premise in the clinical approach to the stroke patient. In fact, the neurological clinical picture is strongly dependent on the vessels affected by the pathological process.

The methodical use of all patients suspected of having an ischemic stroke, of radiological investigations such as CT angiography, perfusional CT, and sometimes even cerebral angiography will allow the deepening of the anatomy of the vessels of the cerebral district and, at the same time, relating it with the clinical picture.

5. Future Directions

The preliminary anatomical knowledge, the mode of representation, and the incidence of vascular variations of the brain help us as a future perspective in understanding the vascular model. If, on the one side, thanks to the study of the neurovascular unit, we understand more and more the pathophysiological mechanisms underlying ischemic stroke, with the aim of reaching new therapeutic targets. On the other hand, emergency imaging methods allow us to expand our knowledge of the macroscopic anatomy of the cerebral vessels and to relate it to the clinical picture of a patient in urgency. In addition, anatomical variations such as fenestrations, agenesis, hypoplasia, or the aberrant nature of vessels provide the essential prerequisite for surgeons before planning neurovascular surgeries.

Author Contributions: Conceptualization: F.B. Methodology: F.B. and G.B. Software: R.A. Validation: F.G.N. Formal analysis: F.G.N. Investigation: F.B. and R.A. Resources: F.G.N. Data curation: G.B. Writing—original draft preparation: F.B. Writing—review and editing: F.B. Visualization: All authors. Supervision: F.G.N. Project administration: F.G.N. All authors have read and agreed to the published version of the manuscript.

Funding: This research received no external funding.

Institutional Review Board Statement: Not applicable.

Informed Consent Statement: Not applicable.

Conflicts of Interest: The authors declare no conflict of interest.

References

1. Campbell, B.C.V.; De Silva, D.A.; Macleod, M.R.; Coutts, S.B.; Schwamm, L.H.; Davis, S.M.; Donnan, G.A. Ischaemic stroke. *Nat. Rev. Dis. Primers* **2019**, *5*, 70. [CrossRef] [PubMed]
2. De Falco, F.A.; Sterzi, R.; Toso, V.; Consoli, D.; Guidetti, D.; Provinciali, L.; Leone, M.; Beghi, E. The neurologist in the emergency department. An Italian nationwide epidemiological survey. *Neurol. Sci.* **2008**, *29*, 67–75. [CrossRef] [PubMed]
3. Sañudo, J.; Vázquez, R.; Puerta, J. Meaning and clinical interest of the anatomical variations in the 21st century. *Eur. J. Anat.* **2003**, *7*, 1–3.
4. Harrigan, M.R.; Deveikis, J.P. Essential Neurovascular Anatomy. In *Handbook of Cerebrovascular Disease and Neurointerventional Technique Series: Contemporary Medical Imaging*; Harrigan, M.R., Deveikis, J.P., Eds.; Humana Press: New York, NY, USA, 2009; pp. 1–87.

5. Anastasi, G.; Gaudio, E.; Tacchetti, C. *Anatomia Umana—Atlante*; Edi.Ermes: Milano, Italy, 2013; Volume 1, ISBN 9788870513486.
6. Hauser, S. *Harrison's Neurology in Clinical Medicine*; Mcgraw-Hill Education: New York, NY, USA, 2016.
7. Caplan, L.R. Diagnosis and the Clinical Encounter. In *Caplan's Stroke: A Clinical Approach*, 4th ed.; Saunders: Philadelphia, PA, USA, 2009; p. 64.
8. Perry, J.M.; McCabe, K.K. Recognition and initial management of acute ischemic stroke. *Emerg. Med. Clin. N. Am.* **2012**, *30*, 637–657. [CrossRef]
9. Caplan, L.R. *Posterior Circulation Disease: Clinical Findings, Diagnosis, and Management*; Blackwell Science: Boston, MA, USA, 1996.
10. Savitz, S.I.; Caplan, L.R. Vertebrobasilar disease. *N. Engl. J. Med.* **2005**, *352*, 2618. [CrossRef]
11. Popieluszko, P.; Henry, B.M.; Sanna, B.; Hsieh, W.C.; Saganiak, K.; Pękala, P.A.; Walocha, J.A.; Tomaszewski, K.A. A systematic review and meta-analysis of variations in branching patterns of the adult aortic arch. *J. Vasc. Surg.* **2018**, *68*, 298–306.e10. [CrossRef]
12. Dimmick, S.J.; Faulder, K.C. Normal variants of the cerebral circulation at multidetector CT angiography. *Radiographics* **2009**, *29*, 1027–1043. [CrossRef] [PubMed]
13. Okahara, M.; Kiyosue, H.; Mori, H.; Tanoue, S.; Sainou, M.; Nagatomi, H. Anatomic variations of the cerebral arteries and their embryology: A pictorial review. *Eur. Radiol.* **2002**, *12*, 2548–2561. [CrossRef] [PubMed]
14. Caldemeyer, K.S.; Carrico, J.B.; Mathews, V.P. The radiology and embryology of anomalous arteries of the head and neck. *Am. J. Roentgenol.* **1998**, *170*, 197–203. [CrossRef] [PubMed]
15. Valvano, A.; Bosso, G.; Apuzzi, V.; Mercurio, V.; Di Simone, V.; Panicara, V.; De Luca, M.; Tomas, C.; Cammarota, F.; Cittadini, A.; et al. Long-term follow-up in high-risk hypertensive patients with carotid dolicoarteriopathies. *Int. Angiol.* **2020**, *39*, 24–28. [CrossRef] [PubMed]
16. Petty, G.W.; Brown, R.D., Jr.; Whisnant, J.P.; Sicks, J.D.; O'Fallon, W.M.; Wiebers, D.O. Ischemic stroke subtypes: A population-based study of incidence and risk factors. *Stroke* **1999**, *30*, 2513. [CrossRef]
17. Ekici, F.; Tekbas, G.; Onder, H.; Gumus, H.; Cetincakmak, M.G.; Palanci, Y.; Bakir, S.; Bilici, A. Course anomalies of extracranial internal carotid artery and their relationship with pharyngeal wall: An evaluation with multislice CT. *Surg. Radiol. Anat.* **2012**, *34*, 625–631. [CrossRef] [PubMed]
18. Shaban, A.; Albright, K.C.; Boehme, A.K.; Martin-Schild, S. Circle of Willis variants: Fetal PCA. *Stroke Res. Treat.* **2013**, *2013*, 105937. [CrossRef]
19. Hakim, A.; Gralla, J.; Rozeik, C.; Mordasini, P.; Leidolt, L.; Piechowiak, E.; Ozdoba, C.; El-Koussy, M. Anomalies and Normal Variants of the Cerebral Arterial Supply: A Comprehensive Pictorial Review with a Proposed Workflow for Classification and Significance. *J. Neuroimaging* **2018**, *28*, 14–35. [CrossRef]
20. Horowitz, M.; Bansal, S.; Dastur, K. Aortic Arch Origin of the Left External Carotid Artery and Type II Proatlantal Fetal Anastomosis. *Am. J. Neuroradiol. Mar.* **2003**, *24*, 323–325.
21. Menshawi, K.; Mohr, J.P.; Gutierrez, J. A functional perspective on the embryology and anatomy of the cerebral blood supply. *J. Stroke* **2015**, *17*, 144–158. [CrossRef] [PubMed]
22. Thierfelder, K.M.; Baumann, A.B.; Sommer, W.H. Vertebral artery hypoplasia: Frequency and effect on cerebellar blood flow characteristics. *Stroke* **2014**, *45*, 1363–1368. [CrossRef]
23. Saade, C.; Bourne, R.; Wilkinson, M.; Brennan, P.C. MDCT angiography of the major congenital anomalies of the extracranial arteries: Pictorial review. *J. Med. Imaging Radiat. Oncol.* **2013**, *57*, 321–328. [CrossRef] [PubMed]
24. Kovač, J.D.; Stanković, A.; Stanković, D.; Kovač, B.; Šaranović, D. Intracranial arterial variations: A comprehensive evaluation using CT angiography. *Med. Sci. Monit.* **2014**, *20*, 420–427. [CrossRef] [PubMed]
25. Debette, S.; Compter, A.; Labeyrie, M.A.; Uyttenboogaart, M.; Metso, T.M.; Majersik, J.J.; Goeggel-Simonetti, B.; Engelter, S.T.; Pezzini, A.; Bijlenga, P.; et al. Epidemiology, pathophysiology, diagnosis, and management of intracranial artery dissection. *Lancet Neurol.* **2015**, *14*, 640–654. [CrossRef] [PubMed]
26. Hudák, I.; Lenzsér, G.; Lunenkova, V.; Dóczi, T. Cerebral arterial fenestrations: A common phenomenon in unexplained subarachnoid haemorrhage. *Acta Neurochir.* **2013**, *155*, 217–222. [CrossRef] [PubMed]
27. Hoksbergen, A.W.J.; Fulesdi, B.; Legemate, D.A.; Csiba, L. Collateral Configuration of the Circle of Willis. Transcranial Color-Coded Duplex Ultra- sonography and Comparison with Postmortem Anatomy. *Stroke* **2000**, *31*, 1346–1351. [CrossRef]
28. Iqbal, S. A comprehensive study of the anatomical variations of the circle of willis in adult human brains. *J. Clin. Diagn. Res.* **2013**, *7*, 2423–2427. [CrossRef] [PubMed]
29. Chmielewski, R.; Ciszek, B. Internal Carotid Artery Classification Systems. *Pol. J. Aviat. Med. Bioeng. Psychol.* **2018**, *24*, 27–35. [CrossRef]
30. Chandra, A.; Li, W.A.; Stone, C.R.; Geng, X.; Ding, Y. The cerebral circulation and cerebrovascular disease I: Anatomy. *Brain Circ.* **2017**, *3*, 45–56. [PubMed]
31. Shapiro, M. Anterior Cerebral Artery. Available online: http://neuroangio.org/anatomy-and-variants/anterior-cerebral-artery/ (accessed on 17 February 2017).
32. Chuang, Y.M.; Liu, C.Y.; Pan, P.J.; Lin, C.-P. Anterior cerebral artery A1 segment hypoplasia may contribute to A1 hypoplasia syndrome. *Eur. Neurol.* **2007**, *57*, 208–211. [CrossRef] [PubMed]
33. Uchino, A.; Nomiyama, K.; Takase, Y.; Kudo, S. Anterior cerebral artery variations detected by MR angiography. *Neuroradiology* **2006**, *48*, 647–652. [CrossRef]
34. Komiyama, M.; Nakajima, H.; Nishikawa, M.; Yasui, T. Middle cerebral artery variations: Duplicated and accessory arteries. *Am. J. Neuroradiol.* **1998**, *19*, 45–49.

35. Takahashi, S.; Hoshino, F.; Uemura, K.; Takahashi, A.; Sakamoto, K. Accessory middle cerebral artery: Is it a variant form of the recurrent artery of Heubner? *Am. J. Neuroradiol.* **1989**, *10*, 563–568.
36. Shapiro, M. Anterior Choroidal Artery. Available online: http://neuroangio.org/anatomy-and-variants/anterior-choroidal-artery/ (accessed on 2 February 2017).
37. De Gast, A.N.; van Rooij, W.J.; Sluzewski, M. Fenestrations of the Anterior Communicating Artery: Incidence on 3D Angiography and Relationship to Aneurysms. *Am. J. Neuroradiol.* **2008**, *29*, 296–298. [CrossRef]
38. Barbato, F.; Allocca, R.; Serra, C.; Bosso, G.; Numis, F.G. Spontaneous eye movements in myxedematous coma. *Intern. Emerg. Med.* **2022**, *17*, 2063–2064. [CrossRef] [PubMed]
39. Percheron, G. Arteries of the human thalamus. *Rev. Neurol.* **1976**, *132*, 297–307. [PubMed]
40. Amarenco, P.; Hauw, J.J. Anatomy of the cerebellar arteries. *Rev. Neurol.* **1989**, *145*, 267. [PubMed]
41. Coupland, A.P.; Thapar, A.; Qureshi, M.I.; Jenkins, H.; Davies, A.H. The definition of stroke. *J. R. Soc. Med.* **2017**, *110*, 9–12. [CrossRef]
42. Kluytmans, M.; van der Grond, J.; van Everdingen, K.J.; Klijn, C.J.M.; Kappelle, L.J.; Viergever, M.A. Cerebral hemodynamics in relation to patterns of collateral flow. *Stroke* **1999**, *30*, 1432–1439. [CrossRef] [PubMed]
43. Kuriakose, D.; Xiao, Z. Pathophysiology and Treatment of Stroke: Present Status and Future Perspectives. *Int. J. Mol. Sci.* **2020**, *21*, 7609. [CrossRef]

anatomia

Review

Common Anatomical Variations of Neurovascular Canals and Foramina Relevant to Oral Surgeons: A Review

Laura Sferlazza [1,*]**, Fabrizio Zaccheo** [2]**, Maria Elisabetta Campogrande** [3]**, Giulia Petroni** [1] **and Andrea Cicconetti** [1,*]

[1] Department of Oral and Maxillofacial Sciences, Sapienza University of Rome, 00161 Rome, Italy
[2] Independent Researcher, 00199 Rome, Italy
[3] Department of Odontostomatologic Science, University of Rome Tor Vergata, 00133 Rome, Italy
* Correspondence: laura.sferlazza@uniroma1.it (L.S.); andrea.cicconetti@uniroma1.it (A.C.)

Abstract: (1) Background: The anatomical variations of neurovascular canals that are encountered in oral surgery are often overlooked by anatomy textbooks or provided with insufficient information. The aim of this study is to analyze the most common variations, describing their morphology, prevalence and clinical implications. (2) Methods: A review of published literature from the 20th century onwards was performed using the PubMed electronic database as well as anatomical textbooks. The variations being investigated were: retromolar canal (RMC) and foramen (RMF), accessory mental foramen (AMF), midline (MLF) and lateral (MLF) lingual foramina and canalis sinuosus (CS). (3) Results: Anatomical variants of neurovascular canals and foramina have a significant incidence and important clinical implications in the most common oral surgery procedures such as third molar extraction, bone harvesting and implantology. (4) Conclusions: Knowledge of these variables is highly important both for students to have a more accurate anatomical awareness and for professional surgeons to be able to provide better diagnoses and prevent complications during oral surgery techniques.

Keywords: mandible; anatomical variations; mandibular canal variations; retromolar canal; retromolar foramen; accessory mental foramen; mandibular lingual foramina; lingual foramina; canalis sinuosus; cone-beam computed tomography

Citation: Sferlazza, L.; Zaccheo, F.; Campogrande, M.E.; Petroni, G.; Cicconetti, A. Common Anatomical Variations of Neurovascular Canals and Foramina Relevant to Oral Surgeons: A Review. *Anatomia* **2022**, *1*, 91–106. https://doi.org/10.3390/anatomia1010010

Academic Editors: Gianfranco Natale and Francesco Fornai

Received: 12 June 2022
Accepted: 1 August 2022
Published: 8 August 2022

Publisher's Note: MDPI stays neutral with regard to jurisdictional claims in published maps and institutional affiliations.

1. Introduction

Anatomy represents the fundament of surgery and forms the basis for diagnosis, treatment and care. Anatomical knowledge is crucial for meeting diagnostic challenges but also for developing surgical procedures. Therefore, anatomy is a major requirement for all surgical specialities, including oral and maxillofacial surgery.

The concept of *normality* in anatomy is called relative: this is inferred from a series of repeated observations. Organisms vary within the same population but also the same species. Anatomical variations may contribute to unusual manifestations of clinical symptoms, influence the course of clinical examination or the interpretation of imaging and prove of primary importance during surgical procedures [1].

Insufficient anatomical awareness can lead to serious complications and poor postoperative results both aesthetically and functionally. Therefore, to perform safe and effective surgeries, it is essential to have a thorough understanding of clinical anatomy and potential anatomical differences that may be found between individuals.

Morphological changes caused by atrophy of alveolar processes following tooth loss may sometimes prevent the surgeon from locating anatomical landmarks.

This condition is complicated by anatomical variations, representing potential risk factors in oral and maxillofacial surgery. Variations in nerves and blood vessels can lead to serious complications such as uncontrollable haemorrhage with potentially fatal outcomes and permanent nerve trunk injuries which negatively impact patients' quality of life. To avoid such complications, surgeons must be aware of anatomical changes and their clinical management.

We decided to analyse the most common anatomical variations of neurovascular canals and foramina relevant in the treatment planning and outcome of the most frequent dental surgery: third molar surgery, bone harvesting and implantology.

The anatomical variations we decided to investigate were: retromolar canal (RMC) and foramina (RMF), accessory mental foramen (AMF), midline (MLF) and lateral (MLF) lingual foramina and canalis sinuosus (CS).

The aim of this study was to assess the clinical anatomy and surgical significance of the main variations of neurovascular canals and foramina by reviewing the literature. These results may contribute to a more accurate awareness of variations in oral anatomy and may assist the oral surgeon in better diagnosis, preventing complications during oral surgical techniques.

2. Materials and Methods

The literature selected for this review was limited to work published in English from the 20th century onwards. Standard anatomical textbooks as well as keyword searches using the online PubMed and Google Scholar databases were used. Key terms used were: "Anatomical Variations", "Anatomical Variants", "Mandibular Canal Variations", "Retromolar Canal", "Retromolar Foramen", "Retromolar Foramina", "Mental Foramen", "Mental Foramina", "Mental Foramen Variation", "Accessory Mental Foramen, "Accessory Mental Foramina", "Mandibular Lingual Foramina", "Lingual Foramen", "Lingual Foramina", "Canalis Sinuosus", and "Cone-beam Computed Tomography". Further relevant papers were identified by examination of the reference lists of the useful articles found.

3. Results

The most common anatomical variants that this review selected were: Retromolar Canal and Foramen, Accessory Mental Foramina, Lingual Foramina, and Canalis Sinuosus.

3.1. Retromolar Canal and Foramen

The mandibular canal (MC) is now known as the main trunk from which multiple minor branches originate, running approximately parallel to it [2]. These accessory canals result from failure of primitive canal fusion during the prenatal period and are usually detected as incidental radiographic findings since they have no clinical landmarks [3]. However, a better understanding of the openings and location of mandibular accessory canals is important to ensure the safety of surgical procedures in the posterior part of the mandible.

3.1.1. Retromolar Canal

The retromolar canal (RMC) is a bifid variant of the MC that involves branches of the inferior alveolar nerve (IAN). These branches divide from the top of the canal and travel antero-superiorly within the bone to emerge from the retromolar foramen (RMF) into the retromolar fossa or retromolar triangle [4].

Classification of RMC

The retromolar canal was first classified by Ossenberg (1987) [5] and later by Von Arx et al. (2011) [6] into three main types. Later Patil et al. (2013) [7] introduced an additional subclassification in two subtypes.

The resulting classification is as follows: (Figure 1)

- **Type A1:** RMC with a vertical course that branches off from the MC and courses postero-superiorly to open into the retromolar fossa.
- **Type A2:** RMC with a vertical course that forms an additional horizontal anterior branch before opening posterosuperiorly into the retromolar fossa
- **Type B1:** RMC with a posteriorly curved course

- **Type B2:** RMC with posteriorly curved course forming an additional anterior horizontal branch before opening into the retromolar fossa
- **Type C:** RMC with a horizontal or transverse course with a posterior opening behind the temporal crest and an anterior opening in front of the temporal crest in the retromolar fossa.

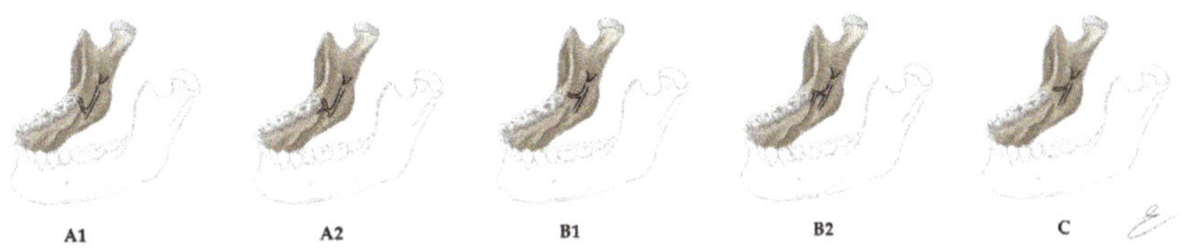

Figure 1. Retromolar Canal's Classification according to Patil et al. (2013). (**A1**) RMC with a vertical course; (**A2**) RMC with a vertical course forming an additional horizontal anterior branch; (**B1**) RMC with a posteriorly curved course; (**B2**) RMC with a posteriorly curved course forming an additional horizontal anterior branch; (**C**) RMC with a horizontal or transverse course with a posterior opening behind the temporal crest and an anterior opening in the retromolar fossa. Copyright © 2022 Maria Elisabetta Campogrande.

Zhang et al. [8] also described this classification system into three subtypes:

- Subtype 1, which runs directly over the surface of the bone;
- Subtype 2, which reaches the retromolar region with a single curve, giving the impression of a "V";
- Subtype 3, with three segments and two main curves before ending in the retromolar region, resembling a "U".

Another classification was proposed by Luangchana et al. (2018) [9], which distinguished five types of RMC by further modifying the classification of Patil et al. (2013) [7]:

- Type A: superior type;
- Type B: radicular retromolar type;
- Type C: dental type;
- Type D: plexus type;
- Type E: forward type. Further subclassified in: E1, where RMC branches off the MC and runs forward without fusing to the MC; E2, where RMC branches off the MC, runs forward for some distance and then fuses with the MC.

In spite of the numerous classification systems proposed to describe RMC in its various locations and configurations, most studies conducted over the years have focused on the most common type of RMC (type 1) proposed by Ossenberg (1987) [10,11].

Prevalence of RMC

According to the literature, RMC has an extremely variable prevalence ranging from 0% to 75.4%, depending on the method applied. Studies based on orthopantomography reported a low prevalence of RMC [5,12–15] while studies in which OPT and CBCT were associated found significantly higher results. These outcomes indicate the importance of three-dimensional investigations and their great advantage in detecting these anatomical structures [5,13,14,16]. The prevalence of RMC reported by studies carried out on CBCT ranges from 24.65% to 75.4% [7,14].

Other methods to evaluate RMCs have included anatomical dissection on cadavers [16], micro-CT studies, CBCT [11] and, more recently, endoscope studies [17].

Content of the RMC

Based on radiologic studies, cadaver studies, and biopsies, it has been reported that the retromolar canal contains a neurovascular bundle that is predominantly endowed by thin myelinated nerve fibres, venules and arterioles covered by collagen fibres, and adipose tissue in small amounts.

A recent study has mentioned that the diameter of nerve bundles varied between 40 to 60 microns, and the larger ones ranged from 80 to 180 microns. The largest arteriole had a diameter of a maximum of 600 microns [12].

Morphometric Measurement of RMC

The distance from the mandibular foramen to the origin of the RMC is 21.5 (11.2) mm [11]. The diameter of the RMC has been reported to range from a mean of 0.75 mm to 2.28 mm [8,18]. The narrowest RMC reported was 0.27 mm, while the widest was 3.29 mm in diameter [19]. The reported mean length of the RMC ranges from 6.9 mm to 16.2 mm [19,20]. The longest RMC ever reported was 33.2 [19]. The mean height of the canal has been reported to range from 6.66 (2.18) mm to 15.3 (4.6) mm from the MC [5,19].

3.1.2. Retromolar Foramen

RMF is the accessory foramen that allows RMC outflow and is often located on the medial [21,22] or lingual [23] surface of the retromolar trigone.

It has been suggested that the foramen on the medial aspect of the retromolar trigone may carry fibres of the mylohyoid nerve; this may be responsible in part for the dolorific sensibility of the third molar. The RMF often occurs as a single unilateral or bilateral foramen, although up to three foramina may be present at a single site [24].

Prevalence of RMF

The prevalence of the RMF as reported in studies of dry mandibles ranges from 3.2% to 92.7% [25]. Studies using orthopantomography reported a prevalence of only 5.3% [26], while CBCT studies reported a range between 7.33% and 12.4% [25,26]. This wide range has been attributed to ethnic differences, environmental and genetic factors, techniques of study, and differences in the number of samples studied [25].

Morphometric Measurement of RMF

RMF was first analysed in its size and location by Lofgren in 1957 [27].

The average diameter of the RMF ranged from 1.2 mm to 2.97 mm [28,29]. The smallest reported diameter was 0.1 mm [28]. Studies by Haas et al. (2016) [30] reported that RMF on the right side of the mandible was located more distally than on the contralateral side in the retromolar area. RMF is approximately 4.23 mm to 9.71 mm distant from the posterior edge of the third molar's socket [29,30]. However, it can be located as close as 1 mm from the tooth [29]. It lies 8.02 mm from the anterior aspect of the branch [4,31]. It is closer to the buccal cortex than to the lingual plate [4].

Clinical Significance

Currently, in clinical practice, a large number of procedures are performed in the retromolar area, and this is evidence of how RMC is a structure of considerable importance that deserves more attention in daily surgical practices.

The main complications that can be caused by accessory canals relate to failure to achieve complete local anaesthesia and intraoperative injury to neurovascular bundles if not detected before surgery [32].

The presence of an RMC is often associated with the so-called "escape pain phenomenon", which is the persistence of painful symptomatology in procedures such as third molar extraction despite proper local anaesthesia.

This escape pain is related to the presence of nerve fibres in the RMC responsible for innervation of the retromolar triangle mucosa, buccal mucosa and gingiva of the mandibular posterior region [33].

Local Anesthetic Failure

Accessory channels are often blamed for the failure to achieve a complete local anaesthetic block. Some studies, however, have suggested that this can be achieved provided the clinician has awareness of the anatomy of the region.

Suazo et al. (2008) [34] proposed a technique accounting for RMC, which allows the achievement of 72.5% anaesthesia with a latency of ten minutes. They argued that the technique allows for easy tracking of the puncture site, which can be observed during the entire deposition of the anaesthetic agent, without bleeding. Owing to the limited vascularization and strongly adherent periosteum of this area, this technique was proposed by the authors as the first-choice approach for anaesthesia in patients with blood dyscrasia because of its low risk of vascular injury.

Haemorrhage

The presence of a retromolar canal, given its neurovascular content, is one of the causes of excessive bleeding during surgical procedures in the retromolar trigone. There may be an initial haemorrhage from blood vessels that emerge from the retromolar canal which are torn when the flap is raised, or spontaneous haemorrhage may occur later during healing and lead to a hematoma.

These complications can be managed locally by crushing the bone in the area occupied by the canal or filling the opening with bone wax [35].

Neurosensory Disturbance

According to numerous case reports, such as that reported by Sigh in 1981 [36], the retromolar canal may contain an aberrant branch of the buccal nerve.

This anatomical variation is of considerable clinical significance since surgical procedures in the retromolar area could result in complications such as unilateral paresthesia or hypoesthesia subsequent to its damage.

Normally, the buccal nerve emerges from the mandibular nerve in the infratemporal fossa before heading low to the inner portion of the branch and crossing the external oblique line up to 3 mm from the deepest concavity of the anterior aspect of the branch. In the event that an aberrant long buccal nerve emerges from the retromolar foramen, it would be at a closer distance than expected from the surgical site. Thus, an injury to the RMC nerve bundle potentially results in neurosensory disturbances such as temporary or permanent sensory impairment or traumatic neurinoma formation, negatively impacting the patient's quality of life.

3.2. Accessory Mental Foramina

The mental foramen (MF) is the opening through which the mental nerve emerges from the mandible in its anterolateral aspect. It is usually located either between the roots of the first and second mandibular premolars or apical to the second premolar. The mental nerve represents one of the terminal branches of the mandibular nerve and divides into three branches supplying the lower lip, cheeks, chin, and the vestibular gingival of mandibular incisors [37]. Although the anatomy of the mandibular nerve is well established, some anatomical variations have been reported that must be taken into consideration to avoid clinical complications. One of these is the accessory mental nerve, which runs through small foramina in the area surrounding the MF, known as accessory mental foramina (AMF) [38].

The accessory mental nerve is a relevant anatomic structure in dental practice with special relevance to local anaesthesia and surgical procedures involving this region.

AMF are smaller than mental foramina and outflow from the mandibular canal. Naitoh et al. (2009) distinguished AMFs from nutrient foramina, which may also border the MF but are not connected to MC [39].

3.2.1. Prevalence and Morphometric Measurement of AMF

The AMF has been first reported by Toh et al.'s study in 1992 [38]. The study was on a cadaver case report of three samples only. Once CBCT became a common practice in the dental field, many researchers reported a higher prevalence of AMF [40–42].

Studies show a prevalence of AMF that range between 2% and 13% [43–45], with an average reported prevalence of 8.3%, according to the cumulative results [46].

AMF has been reported in multiple configurations for instant, separated, diffused or continuous in relation to MF (Figure 2). Muinelo-Lorenzo et al. (2014) observed that the presence of AMF influences the size of MF, finding that on the sides where AMF occurred, MF was significantly smaller [47]. Naitoh et al. (2009) also found MF size to be smaller on sides with AMF but did not find statistical significance [39]. The presence of the anterior loop of mandibular nerve (ALM) is an added complexity to the mental foraminal area.

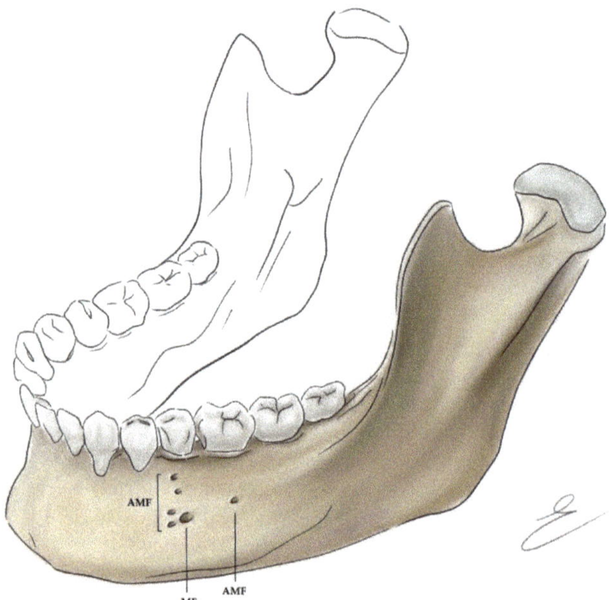

Figure 2. Schematic illustration of Mental Foramen (MF) and Accessory Mental Foramen (AMF) in its possible localization. Copyright © 2022 Maria Elisabetta Campogrande.

The average size of the reported measurement of the AMF is 1.55 mm from all the studies that calculated the AMF opening size. Iwanaga et al. (2017) detailed the bundle that exits from AMF, he and his colleagues in a dissection anatomical investigation showed nerve and blood vessels are passing through this foramen [42].

3.2.2. Clinical Significance

The position of the Mental Foramen (MF) and the Accessory Mental Foramen (AMF) is of utmost importance to modern dentistry and oral surgeons in general. The location and prevalence of any particular variation of the inferior alveolar nerve (IAN) and MN are vital for dental practitioners. Injuries to the mandibular nerve bundle can cause permanent lifelong disability and psychological damage. These include numbness of the lower lip and chin, tingling sensation of the same area and possibly dysesthesia. Many reported injuries in the dental field to the MN due to endodontic therapy [48], implant placement [49] and dental extraction [50].

Modern developments in radiology facilitate the detection of such landmarks by CBCT and panoramic imaging compared to periapical radiograph [46]. Panoramic imaging has a great advantage of ease of use and availability in many dental practices. Using panoramic imaging is beneficial to localize the inferior alveolar never and the MF location in relation to dentitions. However, this modality has been reported to be inefficient to show the location of AMF due to AMF's relatively small size [47]. In addition, the panoramic image provides only a two-dimensional image that masks other important anatomical structures, such as the ALM [51].

On the other hand, the CBCT exams present several important features related to its feasibility in dentistry, such as low dose of radiations, high level of accuracy (an average of 95.5%), high resolution, and three-dimensional orientation [46]. Importantly, AMF is visible on CBCT and has been showing a relatively high incidence: Lam et al. (2019) reported a 6.4% presence of AMF in a 4000 CBCT study [52]. Interestingly, the AMF was known to the dental community since 1992 [38] through cadaver studies. However, the prevalence of AMF in radiology exams increased after the CBCT was introduced to dentistry. Other important anatomical structures in the mandible such as the anterior loop and the bifid mandibular canal [53] were also not delineated before the CBCT introduction to dentistry.

3.3. Mandibular Lingual Foramina

Lingual foramina are accessory foramina on the lingual surface of the mandible. They are typically located in the interforaminal area [54] but can also be commonly observed in the area of the second premolar [54–56]. Vascular and nerve anastomoses arising from branches of the submental artery, lingual artery, and mylohyoid nerve are found in these structures.

The lingual foramen is also known as medial lingual canal, lingual vascular canal, lateral lingual canal, and genial spinal foramen, which is itself subdivided into superior, when it is at or above the mental or genial spines, or inferior, when it is below the mental or genial spines. Other names by which it is described include supraspinous foramen (when it is located above the mental or genial spines); interspinous foramen (when it is located at the level of the mental or genial spines); infraspinous foramen (when it is located below the mental or genial spines); mental spinal foramen, lingual accessory foramen, mandibular accessory foramen, and mandibular lingual foramen [54,55].

The interforaminal zone, delimited by first premolars, is a mandibular area that is traditionally considered safe from a surgical point of view [57]. This area, on its vestibular side, is routinely used as a donor site for bone grafts of the mental block and implant placements [58]. However, these practices are often not supported by clinical protocols that take into account possible interference with lingual foramina. In fact, these structures are not described by current dental anatomy textbooks [59], and only radiographic anatomy textbooks have occasionally reported their presence [59,60].

3.3.1. Classification of Lingual Foramina

Lingual foramina are currently classified according to their location on the inner surface of the mandible into (Figure 3):

- Median or midline lingual foramina (MLF), located on the midline of the lingual aspect of the mandible, close to the genial tubercles;
- Paramedian lingual foramina, located up to the posterior margin of the canine;
- Posterior lingual foramina, located distal to the canine [61].

Many authors define all lingual foramina located beyond the midline as lateral lingual foramina (LLF). MLF is found very frequently regardless of the study method applied, while the finding of LLF in comparison is very minor.

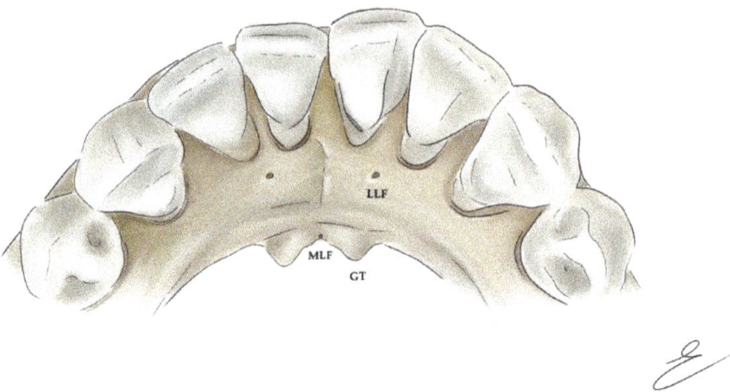

Figure 3. Schematic illustration of inner aspect of the jaw; LLF: Lateral Lingual Foramina; MLF: Midline Lingual Foramina; GT: Genial Tubercles. Copyright © 2022 Maria Elisabetta Campogrande.

Midline Lingual Foramina

MLF are the mandibular lingual foramina that are most often encountered and documented, with a prevalence ranging from 96% to 100% [62].

MLFs can often be divided into superior, middle, and inferior lingual foramina depending on their vertical position in relation to genial tubercles (or bilateral mental spines). As a result, these foramina are called supraspinous, interspinous, and infraspinous foramina [62]. Generally, one to three foramina are found, although some authors have described up to four midline lingual foramina [56]. Most studies describe two MLFs as the most common occurrence [62].

Studies have reported that if only one MLF is present, it is usually located above the mental spine. In mandibles with two foramina, the largest was located above the genial tubercles. Unlike, most mandibles with three MLF showed two of the three foramina located below the genial tubercles [59].

Lateral Lingual Foramina

LLF has been less commonly evaluated than MLF and has been observed in a lower number of cadavers and patient mandibles. The frequency of LLF generally reported in the literature ranges from 6 to 80% [61]. According to a study published by Katakami et al. (2009) [56], LLF is encountered with a relatively high frequency in the area corresponding to the first premolar (36%) but also in central incisor and canine areas (each 19.4%).

Von Arx et al. (2011) described frequencies of 22.6% and 20.3% for LLF in the first and second premolar areas, respectively [62]. The high occurrence of LLF under the first premolar was also confirmed in a cadaver study where 45.7% of the sides of dry cadaveric skulls had a notable lingual foramen in that region [63].

Concerning bilateral recurrence of LLF, in a CT study of 70 patients, bilateral lingual canals were observed in 29.7% of the cases [64]. Another CT study carried out on 200 patients found a 44% frequency of bilateral LLF [54]. However, no information on the symmetry of LLFs was provided in both studies.

3.3.2. Morphometric Measurement of Lingual Foramina

The average size of MLF reported in the literature ranges from 0.6 to 1.17 mm. The smallest size was 0.1 mm in diameter and the largest was 2.29 mm. Higher mean values were found for the superior MLF (0.75–1.12 mm) than for the inferior MLF (0.58–0.9 mm) [61]. CBCT studies found a greater height than the width in the MLF, while an inverted ratio in the LLF [62]. Regarding ratios from the MLF or LLF to adjacent anatomical structures, distances from the lower edge of the mandible have been observed by several studies.

The average distance to the lower edge of the mandible ranged from 12 to 18.4 mm for the superior MLF and from 2.2 to 7 mm the inferior MLF [61]. Only two studies [65,66] evaluated the distance from the MLF to the alveolar crest because measurements and comparisons are hampered by periodontal diseases or atrophic bone. The average distances from the superior MLF to the alveolar crest were 14.2 mm [65] and 14.4 mm [66]. For LLF, the average distances from the lower edge of the mandible ranged from 5.3 to 11.5 mm. The reported distance of LLF of the incisor/canine area from the inferior edge of the mandible was 11.5 mm, while the distance of LLF of the premolar/molar area was 7.1 mm.

3.3.3. Canals of Lingual Foramina

MLF is always associated with a canal, often described in radiology textbooks as a radiopaque area surrounding the lingual foramen.

The medial lingual canal may be located superiorly, at the same level, or inferior to the genial tubercles. The MLC is almost always (99%) perpendicular to the mandibular lingual borders, from an axial view [62]. From the sagittal view, the superior MLC usually runs in an anteroinferior direction, whereas the inferior MLC runs in an anterosuperior direction [54,62]. In addition, the s-MLC can rarely run horizontally or upward (3% and 1%), while the i-MLC sometimes runs horizontally or downward (3% and 21%) [66]. As reported in some studies, the MLC may join the lingual and labial plaques [60].

It was reported that a lingual canal associated with LLF occurs in 95% of individuals. From the axial view, LLCs usually course anterior to the foramen (43%), even if 21% also course perpendicular to the lingual bone surface of the mandible [61,62].

Despite the fact that the most common type of lingual vascular canal is the single canal (77%), studies observed that 20% have bifurcations and 3% triple canals [67].

3.3.4. Content of Lingual Foramina and Canals

The LLF contains a branch of the submental artery, which may form an anastomosis with the lingual branch of the inferior alveolar artery or be independent and give branches supplying the incisional area and the lower lip.

In cases where anastomosis occurs, the inferior alveolar artery also supplies the incisive area [68,69].

LLC arising from the premolar region anastomoses with the incisor canal about 87% of the time and with the mandibular canal, 38% of the time [62]. Anastomoses with the mandibular canal can form in the anterior loop, in the area adjacent to the mental foramen, or posterior to the mental foramen [56]. In the incisor–canine region, all communication occurs with the incisor canal [62], thus connecting the CLL with the area concerning the mental nerve. For that reason, the occurrence of the lateral lingual foramen is a strong predictor of communication with the incisor canal [61].

The superior MLC contains branches of the lingual artery and vein. The superior MLC and anterior MLC are crossed by an anastomosing branch of the sublingual artery [59,60]. The inferior MLC includes submental or sublingual branches [60]. The MLCs do not connect in 92% of cases [62]. When they anastomose, the artery in the MLC may anastomose with the incisive artery in 8–40% of cases but less frequently than in MLC anastomoses with the incisive canal. That is because this canal is not typically present in the middle of the symphysis [56,62,67].

The superior MLC contains a branch of the lingual nerve, while the inferior MLC contains a branch of the mylohyoid nerve [60].

3.3.5. Clinical Significance

In light of the significant neurovascular content of both MLC and LLC and given the involvement of the interforaminal region in many surgical procedures, a thorough awareness of the location, size, and content of these structures is of utmost importance in daily clinical practice.

Dental Implantation

Implant rehabilitation represents one of the most common rehabilitative procedures that are performed in the interforaminal area in edentulous patients with bone atrophy. Implant placement allows recovery of function and aesthetics by taking advantage of an anatomical area that is considered surgically safe.

Nevertheless, after tooth loss, horizontal bone resorption occurs, drastically reducing the distance between the lingual foramen and the bone crest: this results in an increased risk of intra- and post-operative complications. Thus, the presence of at least one lingual foramen in almost all patients has prompted increased interest in identifying the exact position of MLF and LLF and the possible management of vessel injury therein.

Haemorrhage

Hemorrhages occurring in association with surgery can be late or early. The latter is most often caused by perforation of the lingual cortical. Intraoperative injury to the interforaminal lingual cortical may result in damage to the sublingual artery, submental artery as well as the inferior alveolar artery branch, which leads to an oral floor haematoma [70].

Since the interforaminal lingual cortical area is characterized by a rich blood supply, if the haemorrhage persists, it could cause an upward dislocation of the oral floor and tongue, leading to a potentially fatal upper airway obstruction [59,66].

In addition, in some individuals with mylohyoid muscle defects, haemorrhage from the sublingual space may extend to the submandibular space, causing airway obstruction as well [68]. Airway patency can be regained by intubation and tracheostomy. A better assessment of mandibular anatomy is therefore essential to avoid complications related to lingual cortical perforation [70].

Knowledge of the distance between the lingual canal and the alveolar ridge is particularly important in surgical procedures such as implant placement, as immediate implants placed deeper than the natural alveolus and implant platform preparation can damage the lingual canals [67]. Long dental implants should not be placed in canine regions and atrophic jaws. In all cases in which implant placement caused life-threatening haemorrhage, a drilling depth greater than 15 mm was observed to cause perforation of the lingual cortical [54,71]. Edema is the result of surgical bleeding and, if observed, should be immediately controlled. Additional prescription of antibiotics and steroids might be helpful when the bleeding resolves [72].

3.4. Canalis Sinuosus

The Canalis Sinuosus (CS) is a tortuous intraosseous canal that originates from the infraorbital nerve. It emerges posterior to the infraorbital foramen and runs in an anterolateral direction up to the anterior wall of the nasal cavity below the orbital margin, at this point, it turns sharply downwards along the pyriform opening describing an S-shaped course and then heading towards the low and emerge into the palatine mucosa through an accessory foramen (Figure 4). The term Canalis Sinuosus (CS) describes the double curvature of the latter, which runs for about 55 mm along the maxilla and is characterized by a thin overlying bone that makes it more susceptible to injury in case of trauma. For this reason, it is considered an important branch of the orbital nerve (ION). The anterosuperior alveolar nerve (ASAN) and corresponding veins and arteries run within it. The CS commonly presents anatomic variations anterior to the incisive canals in the anterior palate, called accessory channels (ACs). The neurovascular branches in the CS innervate incisor and canine region and adjacent soft tissues through the dental plexus [73].

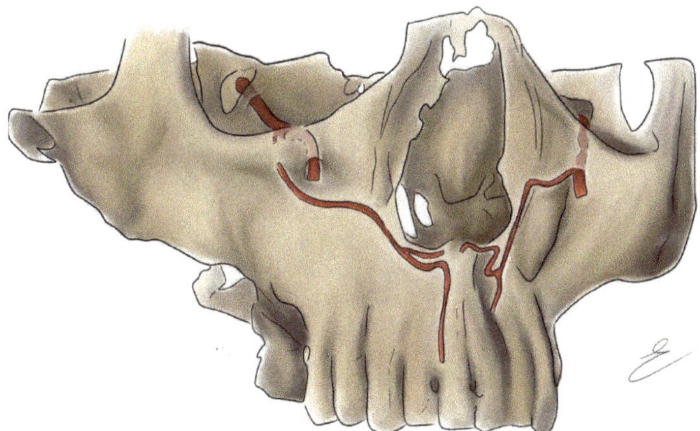

Figure 4. Schematic illustration of Canalis Sinuosus (CS) originating from the infraorbital canal. Copyright © 2022 Maria Elisabetta Campogrande.

3.4.1. Prevalence of CS

According to Wanzeler et al. (2015) and Gurler et al. (2017), the frequency of CS in the patients of these studies was 88% and 100% respectively [74,75].

In a study conducted by Oliveria-Santos et al. (2013), of a total of 178 patients, 28 presented ACs that were at least 1 mm in diameter, of which fourteen had a direct extension with the CS [73]. In Von Arx et al.'s (2015) study, ACs larger than 1.00 mm were found in 49 patients, and 56.7% of the ACs had a direct extension with the CS [76]. Machado et al. (2016), Ghandourah et al. (2017), and Orhan et al.'s (2018) studies provided similar results, reporting 52.1%, 67.6%, and 70.8% of ACs being interconnected with CSs, respectively [77–79]. The studies by Wanzeler et al. (2015) and Gurler et al. (2017) characterized the frequency of CS as 88% and 100%, respectively [74,75]. Oliveira-Santos et al. (2013) analyzed the frequency of the ACs, discovering that out of a total of 178 patients, 28 presented ACs that were at least 1 mm in diameter and that fourteen of their samples had a direct extension with the CS [73]. In the investigation of Von Arx et al. [76], ACs larger than 1.00 mm were found in 49 patients, and 38 (56.7%) of the ACs had a direct extension with the CS. Similar findings were observed in the studies by Machado et al. (2016), Ghandourah et al. (2017), and Orhan et al. (2018), who reported 52.1%, 67.6%, and 70.8% of the ACs, respectively, being interconnected with the CS [77–79].

3.4.2. Morphometric Measurement of CS

Most studies have focused on the terminal portion of the CS, showing that the final portions of the CS and the ACs variations are more frequent in the incisor and canine region near the palate [75,80].

Most of the studies found no significant statistical difference in frequency [74,75,80]; only one study showed the left side as more commonly affected [81].

The average diameter of the CS was found to be approximately 1 mm.

Regarding the diameter of ACs: Oliveira Santos et al. found a mean value of 1.4 mm [73]; Von Arx et al. of 1.3 mm [76]; Machado et al. of 1.0 mm [77].

3.4.3. Clinical Significance

There are many surgical procedures involving the anterior maxilla from all branches of dentistry: endodontics, periodontology, implantology, oral surgery and orthognathic [76].

Because of its anatomic course, the CS should be always considered in the treatment planning and outcome of dental surgery.

In implantology, invasion of the neurovascular bundle of the CS and Acs can compromise osseointegration and cause temporary or permanent paresthesia as well as blood loss and neurovascular disturbance in the region [82,83].

CS injuries may result in a clinical condition known as post-traumatic midface pain, characterized by paroxysmal neuropathic pain localized to the central and lateral incisors, canines, and maxilla. These symptoms are related to ASAN involvement. Injuries to ASAN can occur after midface fractures but can also be iatrogenic, following Lefort I-type osteotomies or dental surgery [84].

For all the reasons just listed, it is essential to know the trajectory and calibre of CS during the surgical programming phase. At present, some authors report essential the individuation of the CS in order to avoid the risk of haemorrhage and paraesthesia [83,85]. Jacobs et al. (2000) analysed the anatomical structures of the maxillary bones and any associated risks. The authors claim that there is great individual variability regarding the presence and the course of the CS and Acs. The identification of these anatomical structures in any surgical intervention in the maxillary region is essential. To avoid complications, it is very important surgical programming through radiographic examinations of the second level [86].

4. Discussion

Anatomically, the structures of significance in oral surgery contain neurovascular bundles and related foramina that, both in their classical anatomy and in their anatomical variations, are of crucial importance in the vascularization and innervation of dental structures and their surrounding tissues. Their knowledge allows professionals to select an appropriate approach to the patient's requirements.

This study shows that the reported prevalence of RMC is 24.6–75.4%, RMF is 7.33–12.4%, AMF is 2–13%, and MLF is 96–100%, while that of CS is 88–100%. Thus, these are anatomical variants that can be termed anything but rare and may be encountered with relative frequency in surgical practice.

The diagnosis of these variations has been changing as technology has evolved. In the first reported studies, the findings were presented on the basis of cadaveric mandibles and in the best of cases PANs [87], where it was often difficult to find them due to the ghost images that are generated in relation to the superposition of adjacent structures which is expected within a 2D image.

Currently, the range of tools available to diagnose anatomical variations of any neurovascular channel in the oro-maxillofacial territory has made the CBCT technique the first preference or the one with the best clinical acceptance. This is thanks to the high quality of 3D images that allow better investigation forms the presence of these variations within the mandibular and maxillary structure.

This review describes the clinical implications of encountering each of the listed variables. The main complications that can be found arise from the neurovascular content of these structures and are represented by marked bleeding, temporary or permanent sensitivity changes of the supplied areas, as well as local anaesthesia failure. Further, these accessory foramina may present an entry point and the related canals may act as a route of infection or, most rarely, tumour metastases [32,88].

The clinician's awareness of such eventualities should always be kept in mind during surgical procedures in order to be able to manage them properly.

5. Conclusions

The anatomical variations of the oromaxillofacial district occur with a relatively high incidence. Knowledge of them is of crucial importance both for professors, who provide morphological training, and for professional surgeons, who work daily in the oromaxillofacial territory. In addition, it is very important to have new lines of research that focus their efforts on describing and diagnosing these anatomical variations and being able to associate in which ethnic groups or genders it occurs in greater frequency.

In conclusion, all health professionals must continually challenge themselves in knowing the different anatomical variations that the human body presents and how these may affect clinical practice in order to perform a surgery that is as safe and minimally invasive as possible.

Author Contributions: Conceptualization, L.S. and F.Z.; methodology, F.Z.; software, L.S.; validation, A.C.; formal analysis, A.C.; investigation, L.S., F.Z. and G.P.; resources, A.C.; data curation, L.S., M.E.C. and G.P.; writing—original draft preparation, L.S. and F.Z.; writing—review and editing, L.S.; visualization, A.C.; supervision, A.C.; project administration, A.C. All authors have read and agreed to the published version of the manuscript.

Funding: This research received no external funding.

Institutional Review Board Statement: Not applicable.

Informed Consent Statement: Not applicable.

Conflicts of Interest: The authors declare no conflict of interest.

References

1. Żytkowski, A.; Tubbs, R.S.; Iwanaga, J.; Clarke, E.; Polguj, M.; Wysiadecki, G. Anatomical normality and variability: Historical perspective and methodological considerations. *Transl. Res. Anat.* **2021**, *23*, 100105. [CrossRef]
2. Kieser, J.; Kieser, D.; Hauman, T. The course and distribution of the inferior alveolar nerve in the edentulous mandible. *J. Craniofac. Surg.* **2005**, *16*, 6–9. [CrossRef] [PubMed]
3. White, S.C.; Pharoah, M.J. *Oral Radiology-E-Book: Principles and Interpretation*; Elsevier Health Sciences: St. Louis, MI, USA, 2014; ISBN 0-323-09634-4.
4. Motamedi, M.H.K.; Gharedaghi, J.; Mehralizadeh, S.; Navi, F.; Badkoobeh, A.; Valaei, N.; Azizi, T. Anthropomorphic assessment of the retromolar foramen and retromolar nerve: Anomaly or variation of normal anatomy? *Int. J. Oral Maxillofac. Surg.* **2016**, *45*, 241–244. [CrossRef]
5. Ossenberg, N.S. Retromolar foramen of the human mandible. *Am. J. Phys. Anthropol.* **1987**, *73*, 119–128. [CrossRef]
6. Von Arx, T.; Hänni, A.; Sendi, P.; Buser, D.; Bornstein, M.M. Radiographic study of the mandibular retromolar canal: An anatomic structure with clinical importance. *J. Endod.* **2011**, *37*, 1630–1635. [CrossRef]
7. Patil, S.; Matsuda, Y.; Nakajima, K.; Araki, K.; Okano, T. Retromolar canals as observed on cone-beam computed tomography: Their incidence, course, and characteristics. *Oral Surg. Oral Med. Oral Pathol. Oral Radiol.* **2013**, *115*, 692–699. [CrossRef]
8. Zhang, Y.-Q.; Zhao, Y.-N.; Liu, D.-G.; Meng, Y.; Ma, X.-C. Bifid variations of the mandibular canal: Cone beam computed tomography evaluation of 1000 Northern Chinese patients. *Oral Surg. Oral Med. Oral Pathol. Oral Radiol.* **2018**, *126*, e271–e278. [CrossRef]
9. Luangchana, P.; Pornprasertsuk-Damrongsri, S.; Kitisubkanchana, J.; Wongchuensoontorn, C. The retromolar canal and its variations: Classification using cone beam computed tomography. *Quintessence Int.* **2018**, *49*, 61–67.
10. Lizio, G.; Pelliccioni, G.A.; Ghigi, G.; Fanelli, A.; Marchetti, C. Radiographic assessment of the mandibular retromolar canal using cone-beam computed tomography. *Acta Odontol. Scand.* **2013**, *71*, 650–655. [CrossRef]
11. Park, M.-K.; Jung, W.; Bae, J.-H.; Kwak, H.-H. Anatomical and radiographic study of the mandibular retromolar canal. *J. Dent. Sci.* **2016**, *11*, 370–376. [CrossRef]
12. Capote, T.S.d.O.; de Almeida Gonçalves, M.; Campos, J.Á.D.B. Retromolar canal associated with age, side, sex, bifid mandibular canal, and accessory mental foramen in panoramic radiographs of Brazilians. *Anat. Res. Int.* **2015**, *2015*, 434083. [CrossRef] [PubMed]
13. Sisman, Y.; Ercan-Sekerci, A.; Payveren-Arıkan, M.; Sahman, H. Diagnostic accuracy of cone-beam CT compared with panoramic images in predicting retromolar canal during extraction of impacted mandibular third molars. *Med. Oral Patol. Oral Cir. Bucal* **2015**, *20*, e74. [CrossRef] [PubMed]
14. Palma, L.F.; Buck, A.F.; Kfouri, F.d.Á.; Blachman, I.T.; Lombardi, L.A.; Cavalli, M.A. Evaluation of retromolar canals on cone beam computerized tomography scans and digital panoramic radiographs. *Oral Maxillofac. Surg.* **2017**, *21*, 307–312. [CrossRef] [PubMed]
15. Fuentes, R.; Arias, A.; Farfán, C.; Astete, N.; Garay, I.; Navarro, P.; Dias, F.J. Morphological variations of the mandibular canal in digital panoramic radiographs: A retrospective study in a Chilean population. *Folia Morphol.* **2019**, *78*, 163–170. [CrossRef] [PubMed]
16. Kikuta, S.; Iwanaga, J.; Nakamura, K.; Hino, K.; Nakamura, M.; Kusukawa, J. The retromolar canals and foramina: Radiographic observation and application to oral surgery. *Surg. Radiol. Anat.* **2018**, *40*, 647–652. [CrossRef] [PubMed]
17. Iwanaga, J.; Watanabe, K.; Saga, T.; Tubbs, R.S.; Tanaka, K.; Kikuta, S.; Tabira, Y.; Fisahn, C.; Kamura, Y.; Kusukawa, J. A novel method for observation of the mandibular foramen: Application to a better understanding of dental anatomy. *Anat. Rec.* **2017**, *300*, 1875–1880. [CrossRef] [PubMed]

18. Li, Y.; Yafei, C.; Jun, P.; Yuanyuan, L.; Shuqun, Q.; Jian, P. Cone beam computed tomography evaluation of bifid mandibular canals in the adult population in Sichuan Province. *Hua Xi Kou Qiang Yi Xue Za Zhi Huaxi Kouqiang Yixue Zazhi West China J. Stomatol.* **2017**, *35*, 82–88.

19. Kang, J.-H.; Lee, K.-S.; Oh, M.-G.; Choi, H.-Y.; Lee, S.-R.; Oh, S.-H.; Choi, Y.-J.; Kim, G.-T.; Choi, Y.-S.; Hwang, E.-H. The incidence and configuration of the bifid mandibular canal in Koreans by using cone-beam computed tomography. *Imaging Sci. Dent.* **2014**, *44*, 53–60. [CrossRef]

20. Alves, N.; Deana, N.F. Anatomical and radiographical study of the retromolar canal and retromolar foramen in macerated mandibles. *Int. J. Clin. Exp. Med.* **2015**, *8*, 4292.

21. Jeyaseelan, N.; Sharma, J.K. Morphological study of unnamed foramina in north Indian human mandibles and its possible role in neurovascular transmission. *Int. J. Oral Surg.* **1984**, *13*, 239–242. [CrossRef]

22. Potu, B.K.; Kumar, V.; Salem, A.-H.; Abu-Hijleh, M. Occurrence of the retromolar foramen in dry mandibles of South-eastern part of India: A morphological study with review of the literature. *Anat. Res. Int.* **2014**, *2014*, 296717. [CrossRef] [PubMed]

23. Kawai, T.; Asaumi, R.; Sato, I.; Kumazawa, Y.; Yosue, T. Observation of the retromolar foramen and canal of the mandible: A CBCT and macroscopic study. *Oral Radiol.* **2012**, *28*, 10–14. [CrossRef]

24. Rossi, A.C.; Freire, A.R.; Prado, G.B.; Prado, F.B.; Botacin, P.R.; Ferreira Caria, P.H. Incidence of retromolar foramen in human mandibles: Ethnic and clinical aspects. *Int. J. Morphol.* **2012**, *30*, 1074–1078. [CrossRef]

25. Ngeow, W.C.; Chai, W.L. The clinical significance of the retromolar canal and foramen in dentistry. *Clin. Anat.* **2021**, *34*, 512–521. [CrossRef] [PubMed]

26. Muinelo-Lorenzo, J.; Suárez-Quintanilla, J.A.; Fernández-Alonso, A.; Marsillas-Rascado, S.; Suárez-Cunqueiro, M.M. Descriptive study of the bifid mandibular canals and retromolar foramina: Cone beam CT vs. panoramic radiography. *Dentomaxillofac. Radiol.* **2014**, *43*, 20140090. [CrossRef] [PubMed]

27. Lofgren, A.B. Foramina retromolaria mandibulae. *Odont Tidskr* **1957**, *65*, 552–570.

28. Bilodi, A.K.S.; Singh, S.; Ebenezer, D.A.; Suman, P.; Kumar, K. A study on retromolar foramen and other accessory foramina in human mandibles of Tamil Nadu region. *Int. J. Health Sci. Res.* **2013**, *3*, 61–65.

29. Shantharam, V.; Manjunath, K.Y.; Aruna, N.; Shastri, D. Retromolar foramen in South Indian mandibles. *Anat. Karnataka* **2013**, *7*, 34–37.

30. Haas, L.F.; Dutra, K.; Porporatti, A.L.; Mezzomo, L.A.; De Luca Canto, G.; Flores-Mir, C.; Corrêa, M. Anatomical variations of mandibular canal detected by panoramic radiography and CT: A systematic review and meta-analysis. *Dentomaxillofac. Radiol.* **2016**, *45*, 20150310. [CrossRef]

31. Tiwari, S.; Ramakrishna, R.; Sangeeta, M. A study on the incidence of retromolar foramen in South Indian adult dried human mandibles and its clinical relevance. *Int. J. Res. Med. Sci.* **2015**, *3*, 1383–1387. [CrossRef]

32. Bilecenoglu, B.; Tuncer, N. Clinical and anatomical study of retromolar foramen and canal. *J. Oral Maxillofac. Surg.* **2006**, *64*, 1493–1497. [CrossRef] [PubMed]

33. Arakeri, G.; Sagoo, M.G.; Brennan, P.A. Neurovascular plexus theory for "escape pain phenomenon" in lower third molar surgery. *Plast Aesthet Res.* **2015**, *2*, 107–110. [CrossRef]

34. Suazo Galdames, I.C.; Cantín López, M.G.; Zavando Matamala, D.A. Inferior alveolar nerve block anesthesia via the retromolar triangle, an alternative for patients with blood dyscrasias. *Med. Oral Patol. Oral Cir. Bucal* **2008**, *13*, E43–E47. [PubMed]

35. Azaz, B.; Lustmann, J. Anatomical configurations in dry mandibles. *Br. J. Oral Surg.* **1973**, *11*, 1–9. [CrossRef]

36. Singh, S. Aberrant buccal nerve encountered at third molar surgery. *Oral Surg. Oral Med. Oral Pathol.* **1981**, *52*, 142. [CrossRef]

37. Greenstein, G.; Tarnow, D. The mental foramen and nerve: Clinical and anatomical factors related to dental implant placement: A literature review. *J. Periodontol.* **2006**, *77*, 1933–1943. [CrossRef] [PubMed]

38. Toh, H.; Kodama, J.; Yanagisako, M.; Ohmori, T. Anatomical study of the accessory mental foramen and the distribution of its nerve. *Okajimas Folia Anat. Jpn.* **1992**, *69*, 85–88. [CrossRef] [PubMed]

39. Naitoh, M.; Hiraiwa, Y.; Aimiya, H.; Gotoh, K.; Ariji, E. Accessory mental foramen assessment using cone-beam computed tomography. *Oral Surg. Oral Med. Oral Pathol. Oral Radiol. Endodontol.* **2009**, *107*, 289–294. [CrossRef]

40. Iwanaga, J.; Saga, T.; Tabira, Y.; Nakamura, M.; Kitashima, S.; Watanabe, K.; Kusukawa, J.; Yamaki, K.-I. The clinical anatomy of accessory mental nerves and foramina. *Clin. Anat.* **2015**, *28*, 848–856. [CrossRef]

41. Iwanaga, J.; Saga, T.; Tabira, Y.; Watanabe, K.; Yamaki, K. A novel method for visualization of the inferior alveolar nerve for clinical and educational purposes. *J. Oral Biosci.* **2016**, *58*, 66–68. [CrossRef]

42. Iwanaga, J.; Watanabe, K.; Saga, T.; Kikuta, S.; Tabira, Y.; Kitashima, S.; Fisahn, C.; Alonso, F.; Tubbs, R.S.; Kusukawa, J. Undetected small accessory mental foramina using cone-beam computed tomography. *Cureus* **2017**, *9*, e1210. [CrossRef] [PubMed]

43. Aytugar, E.; Özeren, C.; Lacin, N.; Veli, I.; Çene, E. Cone-beam computed tomographic evaluation of accessory mental foramen in a Turkish population. *Anat. Sci. Int.* **2019**, *94*, 257–265. [CrossRef]

44. Rawlinson, J.E.; Bass, L.; Campoy, L.; Lesser, C.; Prytherch, B. Evaluation of the equine mental foramen block: Cadaveric and in vivo injectate diffusion. *Vet. Anaesth. Analg.* **2018**, *45*, 839–848. [CrossRef]

45. Velasco-Torres, M.; Padial-Molina, M.; Avila-Ortiz, G.; García-Delgado, R.; Catena, A.; Galindo-Moreno, P. Inferior alveolar nerve trajectory, mental foramen location and incidence of mental nerve anterior loop. *Med. Oral Patol. Oral Cir. Bucal* **2017**, *22*, e630. [CrossRef] [PubMed]

46. Zainy, M.A.A.M. *Mental Foramen and Accessory Mental Foramen Variations: A Systematic Review*; Boston University: Boston, MA, USA, 2020.
47. Muinelo-Lorenzo, J.; Suárez-Quintanilla, J.-A.; Fernández-Alonso, A.; Varela-Mallou, J.; Suárez-Cunqueiro, M.-M. Anatomical characteristics and visibility of mental foramen and accessory mental foramen: Panoramic radiography vs. cone beam CT. *Med. Oral Patol. Oral Cir. Bucal* **2015**, *20*, e707. [CrossRef] [PubMed]
48. Scarano, A.; Di Carlo, F.; Quaranta, A.; Piattelli, A. Injury of the inferior alveolar nerve after overfilling of the root canal with endodontic cement: A case report. *Oral Surg. Oral Med. Oral Pathol. Oral Radiol. Endodontol.* **2007**, *104*, e56–e59. [CrossRef] [PubMed]
49. Wismeijer, D.; Van Waas, M.A.J.; Vermeeren, J.I.J.F.; Kalk, W. Patients' perception of sensory disturbances of the mental nerve before and after implant surgery: A prospective study of 110 patients. *Br. J. Oral Maxillofac. Surg.* **1997**, *35*, 254–259. [CrossRef]
50. Elahi, F.; Manolitsis, N.; Ranganath, Y.S.; Reddy, C.G. Mental nerve neuropathy following dental extraction. *Pain Physician* **2014**, *17*, E375. [CrossRef]
51. Raju, N.; Zhang, W.; Jadhav, A.; Ioannou, A.; Eswaran, S.; Weltman, R. Cone-beam computed tomography analysis of the prevalence, length, and passage of the anterior loop of the mandibular canal. *J. Oral Implantol.* **2019**, *45*, 463–468. [CrossRef]
52. Lam, M.; Koong, C.; Kruger, E.; Tennant, M. Prevalence of accessory mental foramina: A study of 4000 CBCT scans. *Clin. Anat.* **2019**, *32*, 1048–1052. [CrossRef]
53. Fuentes, R.; Farfán, C.; Astete, N.; Garay, I.; Dias, F.; Arias, A. Bilateral bifid mandibular canal: A case report using cone beam computed tomography. *Folia Morphol.* **2018**, *77*, 780–784. [CrossRef] [PubMed]
54. Tagaya, A.; Matsuda, Y.; Nakajima, K.; Seki, K.; Okano, T. Assessment of the blood supply to the lingual surface of the mandible for reduction of bleeding during implant surgery. *Clin. Oral Implants Res.* **2009**, *20*, 351–355. [CrossRef] [PubMed]
55. Ikuta, C.R.S.; da Silva Ramos, L.M.P.; Poleti, M.L.; Capelozza, A.L.A.; Rubira-Bullen, I.R.F. Anatomical study of the posterior mandible: Lateral lingual foramina in cone beam computed tomography. *Implant Dent.* **2016**, *25*, 247–251. [CrossRef]
56. Katakami, K.; Mishima, A.; Kuribayashi, A.; Shimoda, S.; Hamada, Y.; Kobayashi, K. Anatomical characteristics of the mandibular lingual foramina observed on limited cone-beam CT images. *Clin. Oral Implants Res.* **2009**, *20*, 386–390. [CrossRef]
57. Di Bari, R.; Coronelli, R.; Cicconetti, A. Intraosseous vascularization of anterior mandible: A radiographic analysis. *J. Craniofac. Surg.* **2014**, *25*, 872–879. [CrossRef] [PubMed]
58. Tolstunov, L. Implant zones of the jaws: Implant location and related success rate. *J. Oral Implantol.* **2007**, *33*, 211–220. [CrossRef]
59. Rosano, G.; Taschieri, S.; Gaudy, J.F.; Testori, T.; Del Fabbro, M. Anatomic assessment of the anterior mandible and relative hemorrhage risk in implant dentistry: A cadaveric study. *Clin. Oral Implants Res.* **2009**, *20*, 791–795. [CrossRef]
60. Liang, X.; Jacobs, R.; Lambrichts, I.; Vandewalle, G. Lingual foramina on the mandibular midline revisited: A macroanatomical study. *Clin. Anat. Off. J. Am. Assoc. Clin. Anat. Br. Assoc. Clin. Anat.* **2007**, *20*, 246–251. [CrossRef]
61. von Arx, T.; Lozanoff, S. Lingual foramina and canals. In *Clinical Oral Anatomy*; Springer: Berlin/Heidelberg, Germany, 2017; pp. 463–487.
62. Von Arx, T.; Matter, D.; Buser, D.; Bornstein, M.M. Evaluation of location and dimensions of lingual foramina using limited cone-beam computed tomography. *J. Oral Maxillofac. Surg.* **2011**, *69*, 2777–2785. [CrossRef]
63. Yoshida, S.; Kawai, T.; Okutsu, K.; Yosue, T.; Takamori, H.; Sunohara, M.; Sato, I. The appearance of foramen in the internal aspect of the mental region of mandible from Japanese cadavers and dry skulls under macroscopic observation and three-dimensional CT images. *Okajimas Folia Anat. Jpn.* **2005**, *82*, 83–88. [CrossRef]
64. Tepper, G.; Hofschneider, U.B.; Gahleitner, A.; Ulm, C. Computed tomographic diagnosis and localization of bone canals in the mandibular interforaminal region for prevention of bleeding complications during implant surgery. *Int. J. Oral Maxillofac. Implants* **2001**, *16*, 68–72. [PubMed]
65. Babiuc, I.; Tarlungeanu, I.; Pauna, M. Cone beam computed tomography observations of the lingual foramina and their bony canals in the median region of the mandible. *Rom. J. Morphol. Embryol.* **2011**, *52*, 827–829.
66. Sheikhi, M.; Mosavat, F.; Ahmadi, A. Assessing the anatomical variations of lingual foramen and its bony canals with CBCT taken from 102 patients in Isfahan. *Dent. Res. J.* **2012**, *9*, S45.
67. Oettlé, A.C.; Fourie, J.; Human-Baron, R.; van Zyl, A.W. The midline mandibular lingual canal: Importance in implant surgery. *Clin. Implant Dent. Relat. Res.* **2015**, *17*, 93–101. [CrossRef] [PubMed]
68. Nakajima, K.; Tagaya, A.; Otonari-Yamamoto, M.; Seki, K.; Araki, K.; Sano, T.; Okano, T.; Nakamura, M. Composition of the blood supply in the sublingual and submandibular spaces and its relationship to the lateral lingual foramen of the mandible. *Oral Surg. Oral Med. Oral Pathol. Oral Radiol.* **2014**, *117*, e32–e38. [CrossRef]
69. Kawai, T.; Sato, I.; Yosue, T.; Takamori, H.; Sunohara, M. Anastomosis between the inferior alveolar artery branches and submental artery in human mandible. *Surg. Radiol. Anat.* **2006**, *28*, 308–310. [CrossRef]
70. He, P.; Truong, M.K.; Adeeb, N.; Tubbs, R.S.; Iwanaga, J. Clinical anatomy and surgical significance of the lingual foramina and their canals. *Clin. Anat.* **2017**, *30*, 194–204. [CrossRef] [PubMed]
71. Scaravilli, M.S.; Mariniello, M.; Sammartino, G. Mandibular lingual vascular canals (MLVC): Evaluation on dental CTs of a case series. *Eur. J. Radiol.* **2010**, *76*, 173–176. [CrossRef]
72. Kim, D.H.; Kim, M.Y.; Kim, C.-H. Distribution of the lingual foramina in mandibular cortical bone in Koreans. *J. Korean Assoc. Oral Maxillofac. Surg.* **2013**, *39*, 263. [CrossRef]

73. de Oliveira-Santos, C.; Rubira-Bullen, I.R.; Monteiro, S.A.; León, J.E.; Jacobs, R. Neurovascular anatomical variations in the anterior palate observed on CBCT images. *Clin. Oral Implants Res.* **2013**, *24*, 1044–1048. [CrossRef]

74. Wanzeler, A.M.V.; Marinho, C.G.; Junior, S.M.A.; Manzi, F.R.; Tuji, F.M. Anatomical study of the canalis sinuosus in 100 cone beam computed tomography examinations. *Oral Maxillofac. Surg.* **2015**, *19*, 49–53. [CrossRef] [PubMed]

75. Gurler, G.; Delilbasi, C.; Ogut, E.E.; Aydin, K.; Sakul, U. Evaluation of the morphology of the canalis sinuosus using cone-beam computed tomography in patients with maxillary impacted canines. *Imaging Sci. Dent.* **2017**, *47*, 69–74. [CrossRef] [PubMed]

76. von Arx, T.; Lozanoff, S. Anterior superior alveolar nerve (ASAN). *Swiss Dent. J.* **2015**, *125*, 1202–1209. [PubMed]

77. Machado, V.d.C.; Chrcanovic, B.R.; Felippe, M.B.; Júnior, L.M.; De Carvalho, P.S.P. Assessment of accessory canals of the canalis sinuosus: A study of 1000 cone beam computed tomography examinations. *Int. J. Oral Maxillofac. Surg.* **2016**, *45*, 1586–1591. [CrossRef] [PubMed]

78. Ghandourah, A.O.; Rashad, A.; Heiland, M.; Hamzi, B.M.; Friedrich, R.E. Cone-beam tomographic analysis of canalis sinuosus accessory intraosseous canals in the maxilla. *GMS Ger. Med. Sci.* **2017**, *15*, Doc20.

79. Orhan, K.; Gorurgoz, C.; Akyol, M.; Ozarslanturk, S.; Avsever, H. An anatomical variant: Evaluation of accessory canals of the canalis sinuosus using cone beam computed tomography. *Folia Morphol.* **2018**, *77*, 551–557. [CrossRef]

80. Von Arx, T.; Lozanoff, S.; Sendi, P.; Bornstein, M.M. Assessment of bone channels other than the nasopalatine canal in the anterior maxilla using limited cone beam computed tomography. *Surg. Radiol. Anat.* **2013**, *35*, 783–790. [CrossRef]

81. Manhães Júnior, L.R.C.; Villaça-Carvalho, M.F.L.; Moraes, M.E.L.; Lopes, S.L.P.d.C.; Silva, M.B.F.; Junqueira, J.L.C. Location and classification of Canalis sinuosus for cone beam computed tomography: Avoiding misdiagnosis. *Braz. Oral Res.* **2016**, *30*, e49. [CrossRef]

82. McCrea, S.J. Aberrations causing neurovascular damage in the anterior maxilla during dental implant placement. *Case Rep. Dent.* **2017**, *2017*, 5969643. [CrossRef]

83. Torres, M.G.G.; de Faro Valverde, L.; Vidal, M.T.A.; Crusoé-Rebello, I.M. Branch of the canalis sinuosus: A rare anatomical variation—A case report. *Surg. Radiol. Anat.* **2015**, *37*, 879–881. [CrossRef]

84. Olenczak, J.B.; Hui-Chou, H.G.; Aguila III, D.J.; Shaeffer, C.A.; Dellon, A.L.; Manson, P.N. Posttraumatic midface pain: Clinical significance of the anterior superior alveolar nerve and canalis sinuosus. *Ann. Plast. Surg.* **2015**, *75*, 543–547. [CrossRef] [PubMed]

85. Arruda, J.A.; Silva, P.; Silva, L.; Álvares, P.; Silva, L.; Zavanelli, R.; Rodrigues, C.; Gerbi, M.; Sobral, A.P.; Silveira, M. Dental implant in the canalis sinuosus: A case report and review of the literature. *Case Rep. Dent.* **2017**, *2017*, 4810123. [CrossRef] [PubMed]

86. Jacobs, R.; Quirynen, M.; Bornstein, M.M. Neurovascular disturbances after implant surgery. *Periodontol. 2000* **2014**, *66*, 188–202. [CrossRef] [PubMed]

87. Valenzuela-Fuenzalida, J.J.; Cariseo, C.; Gold, M.; Díaz, D.; Orellana, M.; Iwanaga, J. Anatomical variations of the mandibular canal and their clinical implications in dental practice: A literature review. *Surg. Radiol. Anat.* **2021**, *43*, 1259–1272. [CrossRef]

88. Fanibunda, K.; Matthews, J.N.S. The relationship between accessory foramina and tumour spread on the medial mandibular surface. *J. Anat.* **2000**, *196*, 23–29. [CrossRef]

Case Report

A Rare Bilateral Variation in the Branches of the Internal Thoracic Artery: A Case Report

Jihad S. Hawi [1], Rosalyn A. Jurjus [2], Hisham S. Daouk [2], Maya N. Ghazi [2], Charbel A. Basset [3], Francesco Cappello [3], Inaya Hajj Hussein [4], Angelo Leone [3] and Abdo R. Jurjus [2],*

[1] Faculty of Medicine and Medical Sciences, University of Balamand, Tripoli P.O. Box 100, Lebanon; jihad.hawi@balamand.edu.lb

[2] Cell Biology and Physiological Sciences, Department of Anatomy, Faculty of Medicine, American University of Beirut, Beirut P.O. Box 11-0236, Lebanon; rosalynej@gmail.com (R.A.J.); hd09@aub.edu.lb (H.S.D.); mayanghazi99@gmail.com (M.N.G.)

[3] Department of Biomedicine, Neuroscience and Advanced Diagnostics, Institute of Human Anatomy and Histology, University of Palermo, 90128 Palermo, Italy; cab19@mail.aub.edu (C.A.B.); francapp@hotmail.com (F.C.); angelo.leone@unipa.it (A.L.)

[4] Department of Foundational Medical Studies, William Beaumont School of Medicine, Oakland University, Rochester, MI 48309, USA; hajjhuss@oakland.edu

* Correspondence: aj00@aub.edu.lb

Abstract: Background: Anatomical variations and, in particular, arterial variations constitute an important chapter in the learning of Clinical Anatomy. **Purpose:** The purpose of this report is to describe a rare bilateral anatomical variation in the internal thoracic artery (ITA) in a 60-year-old corpse and to depict its extreme clinical importance in coronary artery bypass surgery. **Methods:** The rare bilateral aberrant branches of the internal thoracic artery and their course in the thorax were incidentally discovered during routine anatomy dissection of the thorax at the Faculty of Medicine and Medical Sciences of the University of Balamand. The findings were thoroughly documented using digital photography, and the dissection followed the instructions from the "16th Edition of Grant's Dissector". **Results:** In the observed case, the left aberrant branch of ITA descends laterally and gives medial and lateral anterior intercostal branches at the first six intercostal spaces. Conversely, the right aberrant artery, which branches from the ITA, descends laterally and gives medial and lateral branches to every intercostal space in the first five intercostal spaces. **Conclusion:** This report emphasizes that any unexpected variations in the lateral aberrant branches of the internal thoracic artery may complicate the surgical procedure. Bilateral aberrant lateral branches of the internal thoracic artery constitute rare anatomical variations of the internal thoracic artery and have been rarely reported in the literature. Such a course for aberrant lateral branches in the thorax poses possible lethal complications during several procedures involving the thorax, including basic coronary artery bypass graft, thoracocentesis and intercostal paracentesis and breast reconstruction.

Keywords: internal thoracic artery; aberrant branches; arterial variation; thoracic wall; coronary artery bypass grafting; thoracocentesis; breast surgery

Citation: Hawi, J.S.; Jurjus, R.A.; Daouk, H.S.; Ghazi, M.N.; Basset, C.A.; Cappello, F.; Hajj Hussein, I.; Leone, A.; Jurjus, A.R. A Rare Bilateral Variation in the Branches of the Internal Thoracic Artery: A Case Report. *Anatomia* **2023**, *2*, 320–327. https://doi.org/10.3390/anatomia2040028

Academic Editors: Gianfranco Natale and Francesco Fornai

Received: 21 July 2023
Revised: 9 October 2023
Accepted: 11 October 2023
Published: 12 October 2023

1. Introduction

The internal thoracic artery (ITA), also referred to as the internal mammary artery (IMA), arises from the antero-inferior branch of the first part of the subclavian artery [1,2]. As the largest artery in the thoracic wall, the ITA descends into the thorax region, coursing anterior to the pleura and separated from the posterior surface of the ribs by the transversus thoracis muscle [3,4]. Along its course, the ITA gives rise to branches that supply various structures, including the thymus, breast, mediastinum and sternum [3,5,6]. Typically, it gives rise to several branches, including the anterior intercostal arteries, perforating branches, pericardiophrenic, sternal and mediastinal, and terminates as the superior epigastric artery and musculophrenic artery [1,7–10]. Additionally, various cases of accessory internal thoracic vessels have been reported, which may give rise to superficial and deep cervical branches [11,12].

Furthermore, prior anatomical studies of the ITA employing ink injection have reflected the territories and have elucidated the links with the adjacent vessels [13]. These studies also unveiled the presence of connecting vessels between the ITA and tissue perfusion [14]. Recent data emanating from plastic and reconstructive surgery of the breast, as well as head and neck procedures, highlighted the importance of ITA and its perforating medial intercostal branches for adipocutaneous flaps [15]. Moreover, the ITA is increasingly becoming the preferred choice for microsurgical breast reconstruction and cardiothoracic surgery, with an excellent prognostic outcome [16]. Understanding the variations concerning the ITA is crucial for better outcomes. The selection process relies on clinical observations that reveal distinct structural and behavioral differences between the internal thoracic artery (ITA) and other nearby pectoral vessels, such as the thoracodorsal artery and inferior epigastric artery [17]. These distinctions are imperative and play a crucial role in guiding the appropriate choice of the ITA for specific procedures. Microscopic examinations have demonstrated that for the ITA, once clamped and severed with scissors, its wall separates into sleeve-like layers [18]. The middle layer, known as the tunica media, becomes separated from the adventitia and extrudes like sleeves. Light microscopy studies of formalin-fixed ITA and multiple stainings, with routine hematoxylin and eosin or the Weigert van Geison and Verhoeff for elastic laminae, showed a relatively thick tunica media with several well-formed elastic lamellae more than the other nearby arteries [19]. In addition, there was no significant difference in the thickness of the tunica intima or adventitia. These data were confirmed via transmission electron microscopy, whereby the media of the ITA had less packed smooth muscle cells, sparsely distributed with a low number of fenestrations in the elastic lamellae [20]. Such histological characteristics provide protective effects against intimal thickening and smooth muscle penetrations, an important element for the prevention of atherosclerosis [21]

Aberrant lateral branches of the ITA are rare anatomical and vascular variations that have received limited attention in the medical literature [22]. These variations have an incidence between 10% and 40% [23]. As such, aberrant right ITA can arise directly from the aorta and not from the subclavian artery [24]. Table 1 summarizes some of the rare aberrant branches of the ITA (Table 1). This case report sheds light on the discovery of bilateral aberrant lateral branches of the ITA, detailing their course and exploring their clinical significance. By documenting this uncommon anatomical finding, the report contributes to our understanding of vascular variations and their potential implications in clinical practice. Such variation can involve the origin, number of branches or the path they take during their progression, making the vascular system different from one person to another.

Table 1. Some of the rare variations in the aberrant branches of the ITA.

Variation	Description
Origination from extrascalenic (third part) of the SCA [25]	Incidence rate: 0.5–1% in anatomical studies. Descends inferomedially, anterior to the scalenus anterior muscle.
Origination from the third part of the subclavian artery [26]	Medial and lateral branches of ITA with high bifurcation.
Bilateral hypoplasia of the ITAs [27]	Link up between ventral somatic anastomosis fails
Hypoplasia of the left ITA [27]	Hypoplasia from the first part of SCA
Bilateral aberrant branches of the ITA [28]	Rare. Present within the upper thorax. Can be used during CABG for better myocardium perfusion.

2. Case Report

This is a case of a remarkable new discovery during anatomical dissections at The Faculty of Medicine, University of Balamand, following the examination of over 400 corpses. The discovery was made in a 60-year-old male corpse, where both left and right aberrant lateral branches were identified originating from the left and right ITAs, respectively. These aberrant lateral branches exhibited an unusual pattern, as they appeared to supply the upper six anterior intercostal spaces on both the left and right sides, deviating from the typical course of the ITA. Specifically, the left aberrant branch of the ITA was situated 2.5 cm away from the manubrium. Descending laterally, it gave rise to medial and lateral anterior intercostal branches at the first four intercostal spaces. The branch terminated at the fifth intercostal space, bifurcating 14 cm away from the body of the sternum, into medial and lateral anterior intercostal branches. Similarly, the right aberrant artery, after branching off the internal thoracic, 2.5 cm away from the manubrium, descends laterally and gives rise to medial and lateral branches at every intercostal space. It terminates at the sixth intercostal space, 16.5 cm away from the body of the sternum, and bifurcates into medial and lateral collateral branches (Figures 1 and 2).

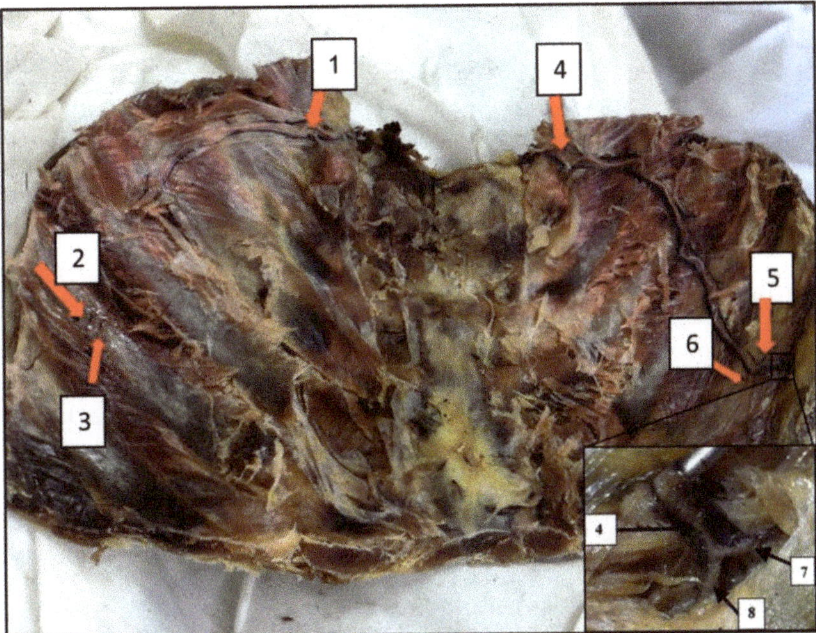

Figure 1. This photograph displays the bilateral aberrant lateral branches of the internal thoracic artery, with the following labeled structures: (1) left aberrant lateral artery; (2) left anterior lateral intercostal artery; (3) left anterior medial intercostal artery; (4) right aberrant lateral artery; (5) right anterior lateral intercostal artery; (6) right anterior medial intercostal artery; (7) right lateral collateral artery; (8) right medial collateral artery.

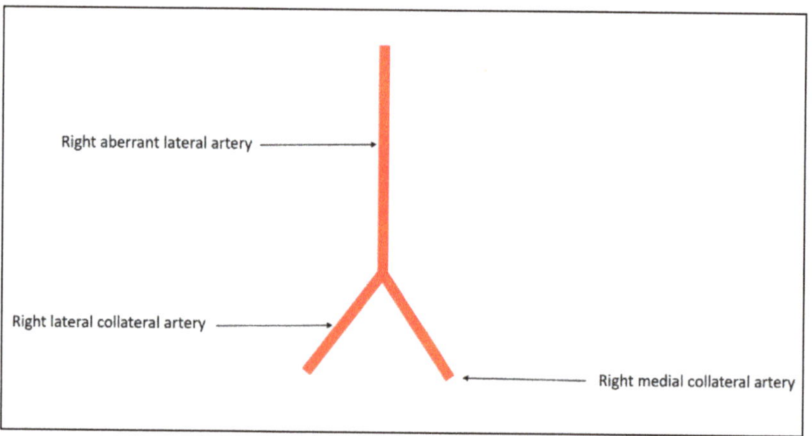

Figure 2. Simple scheme showing the right aberrant lateral artery with its lateral branches and medial collateral arteries.

3. Discussion

The discovery of such bilateral variation in the ITA has not been previously reported in the existing literature. Other case reports and studies have identified different ITA variations, including variation in the ITA on the right side, in a female dead body [26]. In this corpse, the ITA descends to the right second costal cartilage where it splits into medial and lateral branches that give rise to the sternal and anterior intercostal arteries, respectively. The medial and lateral branches anastomose in a horizontal pattern. Puri and colleagues studied 100 adult corpses, of which 24 were females and 76 were males [29]. According to their study, the ITA on the left in 96 specimens took origin from the first part of the subclavian artery, while in 4 other dead bodies, it was a branch of a common trunk. The ITA on the right took origin from the first part of the subclavian in 88 corpses, and in 12 cases, it was a branch of a common trunk. The artery was of similar length on both sides in most specimens. On the left side, the mean length of the ITA was 1.925 ± 0.23 cm, while on the right, the mean length was 1.954 ± 0.224 cm [18]. In most of the corpses studied, the ITA terminated on both sides at the sixth intercostal space [30].

According to the literature, the lateral costal artery (LCA), the lateral branch of the ITA, is known to be the largest artery in the thoracic wall, arising from the subclavian artery and located 1.5 cm lateral to the sternum [31,32]. Because of its critical location, understanding the development of this artery can give insight into the development of these unexpected bilateral aberrant lateral branches and their subsequent clinical significance [31]. Due to the clinical importance of this area, particularly in thoracic surgery, knowledge of possible anatomic variations is essential in preventing clinical errors.

As cited in the "Compendium of Human Anatomic Variations" by Ronald A. Bergman and co-workers, the ITA may develop from different embryologic origins [8,33]. It can develop from the subclavian artery and passes either in front or behind the scalenus anterior muscle. Additionally, the ITA can ascend from a common trunk with the inferior thyroid artery. It can also develop from the brachiocephalic artery that is branching from the ascending aorta [34]. As the ITA descends into the thorax, it reaches the sixth intercostal space, where it anastomoses with the intercostal arteries. In fact, the lateral ITA can originate from different arteries. It can branch out from the thyrocervical trunk or the costocervical trunk. It can even arise from the ascending cervical artery [35]. These diverse embryologic origins contribute to the anatomical complexity of the ITA and offer insights into potential variations in its course and branching patterns. Hence, in any surgical procedure performed in this area, the possible variations in the ITA must be taken into consideration and must be of great concern to physicians [6].

The ITA is the origin of several branches, including the pericardial, pericardiophrenic, sternal, mediastinal, anterior intercostal, terminal and perforating branches [18,24]. It is considered to be the main blood supply to the sternum, and any mistake during procedures involving the thorax can limit the sternum supply and result in sternal wound complication [36]. Therefore, care must be taken during procedures on the thoracic walls, including percutaneous subclavian vein catheterization and introducing pacemakers [37]. This area must be avoided during intercostal paracentesis of the pericardial sac, and the needle must be distanced from the sternal margin [38]. The ITA is most popular for its use in coronary artery bypass grafts (CABGs) [39]. During CABG procedures, the ITA is mobilized from its location and surgically anastomosed with the coronary artery to revascularize the myocardium [40]. Consequent to the critical location and passage of the ITA and its involvement in several critical procedures, any missed anatomic variation, such as the one reported here, may lead to complications involving the CABG procedure or any other thoracic wall procedure.

The long-term clinical benefits of ITA in the field of cardiothoracic surgery have been established, and several studies investigated the detailed histologic characteristics behind such benefits. The consensus is that the ITA is a transitional-type artery with an elastic nature in its upper part (second intercostal space), then elasto-muscular later and finally a muscular-type artery in the rest of the chest [41]. Such histological differences could impact on its successful long-term use in cardiac bypass surgery. For instance, studies have found that the histology of ITA plays a mandatory role in its function and duration [42]. Furthermore, the ITA and its perforators have been reported in the flap reconstruction of the tracheostoma and anterior neck to replace the deltopectoral flap with less flap necrosis [43,44].

Recently, post-mastectomy reconstruction of the breast using ITA branches for the flaps has become an appealing procedure. Such a procedure utilizes the tissue flaps along with ITA perforators to attain the appropriate symmetry and correct deformities. Such a technique makes good use of a dermoglandular perforating branch of the IMA present in each of the 5–6 intercostal spaces, laterodorsal to the lateral border of the sternum [45]. Such branches run superficially in a laterocaudal direction to supply skin of the medial two-thirds of the pectoral region in a sequential order. In addition, perforators of the fourth and fifth intercostal spaces have been proven to contribute to the blood supply of the areola and the direct inframammary fold and area [46]. Information on the vascular anatomy of the region, in particular, the IMA preoperatively to detect possible variation patterns, is essential.

4. Conclusions

Bilateral aberrant branches of the ITA were found during anatomy laboratory dissection, which had an unusual vascular pattern. They introduced a new vascular pattern that challenges surgical procedures. Such findings are exceptionally uncommon and have received limited attention in the existing medical literature. Hence, such a discovery emphasizes the importance of further investigations concerning genetic development and the evolution aspects of these variations. However, these findings highlight the diversity of the anatomical variations that might be present in the human vascular system. Perhaps understanding these variations is crucial for the impact on multiple medical procedures. Given the critical role of the ITA in various thoracic procedures, particularly the coronary artery bypass graft (CABG) procedure, where surgeons rely on this artery, any undetected anatomical variation can lead to potentially life-threatening complications. Such an unexpected anatomical variation can force surgeons to go for further investigations, implicating imaging studies for better vascular architecture comprehension. In addition, the presence of these aberrant branches can have a significant impact on breast plastic and reconstructive surgeries. Understanding the variations is mandatory for a good surgical plan that might require the adjustment of conventional surgical procedures.

While the presence of bilateral aberrant branches holds clinical importance and potential applicability, their extremely rare occurrence necessitates further exploration to fully comprehend their clinical implications in thoracic procedures. The understanding of such anatomical variations can significantly contribute to surgical precision and patient safety in cardiothoracic and reconstructive interventions. The ITA does not have a unique structure, and it can vary among the population. This case report sheds light on the need for continued research exploring the complexity of the human body and highlighting the importance of the variation, especially in heart and chest surgeries. Thus, careful investigation and documentation of these unique findings may provide adjustments to surgical techniques accordingly and may broaden our knowledge of vascular anatomy and enhance medical practices to better serve patients and advance the medical field.

Author Contributions: J.S.H.: The author identified the anatomical variation, wrote the case report, collected data, and provided supervision throughout the development of the manuscript. H.S.D.: The author collected data and contributed to the manuscript writing. C.A.B.: The author contributed to the manuscript writing, and formatting. F.C.: The author contributed to the manuscript writing, manuscript format editing. I.H.H.: The author contributed to the manuscript writing, formatting, and editing. M.N.G.: The author contributed to manuscript writing, editing, formatting, and reviewing. A.L. and R.A.J.: The authors contributed to manuscript writing, formatting, and editing. A.R.J.: The author contributed to manuscript writing, formatting, editing, and provided supervision throughout the development of the manuscript. All authors have read and agreed to the published version of the manuscript.

Funding: This research received no external funding.

Institutional Review Board Statement: The IRB at the Department of Health and Human Services (DHHS) at the American University of Beirut reviewed the manuscript and declare that no living individual was concerned in this paper, hence the proposed activity is not "Human Subject Research". Code of Federal Regulations for the Protection of Human Subjects ("The Common Rule") 45CFR46, subparts A, B, C, and D, with 21CFR56; and operate in a manner consistent with the Belmont report, FDA guidance, Good Clinical Practices under the ICH guidelines, and applicable national/local regulations.

Informed Consent Statement: This work received an exemption from IRB.

Data Availability Statement: Data is available with first and corresponding authors.

Conflicts of Interest: The authors declare that they have no conflict of interest. The dead body was treated ethically, and the dissection was conducted in accordance with the "16th Edition of Grant's Dissector" instructions.

Abbreviations

ITA	Internal thoracic artery
IMA	Internal mammary artery
CABG	Coronary artery bypass surgery
SCA	Subclavian artery
LCA	Lateral costal artery

References

1. Hadley, G. Essential Clinical Anatomy. *J. Anat.* **2007**, *211*, 413. Available online: https://www.ncbi.nlm.nih.gov/pmc/articles/PMC2375806/ (accessed on 29 August 2023). [CrossRef]
2. Internal Thoracic Artery—An Overview | ScienceDirect Topics. Available online: https://www.sciencedirect.com/topics/medicine-and-dentistry/internal-thoracic-artery (accessed on 9 October 2023).
3. Shahoud, J.S.; Kerndt, C.C.; Burns, B. Anatomy, Thorax, Internal Mammary (Internal Thoracic) Arteries. In *StatPearls*; StatPearls Publishing: Treasure Island, FL, USA, 2023. Available online: http://www.ncbi.nlm.nih.gov/books/NBK537337/ (accessed on 29 August 2023).
4. Otaki, M.; Lust, R.M.; Sun, Y.S.; Norton, T.O.; Spence, P.A.; Zeri, R.S.; Hopson, S.B.; Chitwood, W.R. Bilateral vs Single Internal Thoracic Artery Grafting for Left Main Coronary Artery Occlusion. *Chest* **1994**, *106*, 1260–1263. Available online: https://www.sciencedirect.com/science/article/pii/S0012369216333839 (accessed on 29 August 2023). [CrossRef]

5.   Stoddard, N.; Heil, J.R.; Lowery, D.R. Anatomy, Thorax, Mediastinum. In *StatPearls*; StatPearls Publishing: Treasure Island, FL, USA, 2023. Available online: http://www.ncbi.nlm.nih.gov/books/NBK539819/ (accessed on 29 August 2023).

6.   Sajja, L.R.; Mannam, G. Internal thoracic artery: Anatomical and biological characteristics revisited. *Asian Cardiovasc. Thorac. Ann.* **2015**, *23*, 88–99. [CrossRef]

7.   Palanisamy, V.; Mohandoss, B.K.; Ravikumar, M.S.; Raman, R.K.; Rajakumar, A.P.; Kurian, V.M. Uncommon Anatomic Variation of Left Internal Mammary Artery: God-Created Y-Graft Conduit. *Ann. Thorac. Surg.* **2020**, *109*, e113–e114. Available online: https://www.sciencedirect.com/science/article/pii/S000349751930877X (accessed on 29 August 2023). [CrossRef]

8.   Internal Thoracic Artery. Available online: https://www.kenhub.com/en/library/anatomy/internal-thoracic-artery (accessed on 29 August 2023).

9.   Moore, K.L.; Dalley, A.F.; Agur, A.M.R. *Clinically Oriented Anatomy*; Lippincott Williams & Wilkins: Philadelphia, PA, USA, 2013; 1171p, ISBN 978-1-4511-1945-9.

10.   Vorster, W.; du Plooy, P.T.; Meiring, J.H. Abnormal origin of internal thoracic and vertebral arteries. *Clin. Anat.* **1998**, *11*, 33–37. [CrossRef]

11.   Akgun, V.; Hamcan, S.; Bozkurt, Y.; Battal, B. Thoracic Aorta. In *Bergman's Comprehensive Encyclopedia of Human Anatomic Variation*; John Wiley & Sons, Ltd.: New York, NY, USA, 2016; pp. 501–529. ISBN 978-1-118-43030-9. Available online: https://onlinelibrary.wiley.com/doi/abs/10.1002/9781118430309.ch49 (accessed on 29 August 2023).

12.   Tunali, S. Subclavian Artery. In *Bergman's Comprehensive Encyclopedia of Human Anatomic Variation*; John Wiley & Sons, Ltd.: New York, NY, USA, 2016; pp. 575–582. ISBN 978-1-118-43030-9. Available online: https://onlinelibrary.wiley.com/doi/abs/10.1002/9781118430309.ch52 (accessed on 29 August 2023).

13.   Palmer, J.H.; Ian Taylor, G. The vascular territories of the anterior chest wall. *Br. J. Plast. Surg.* **1986**, *39*, 287–299. Available online: https://www.sciencedirect.com/science/article/pii/0007122686900378 (accessed on 29 August 2023). [CrossRef]

14.   Tafner, P.F.d.A.; Chen, F.K.; Rabello Filho, R.; Corrêa, T.D.; de Freitas Chaves, R.C.; Serpa Neto, A. Recent advances in bedside microcirculation assessment in critically ill patients. *Rev. Bras. Ter. Intensiv.* **2017**, *29*, 238–247. Available online: https://www.ncbi.nlm.nih.gov/pmc/articles/PMC5496759/ (accessed on 29 August 2023). [CrossRef]

15.   Deramo, P.; Rose, J. Flaps: Muscle and Musculocutaneous. In *StatPearls*; StatPearls Publishing: Treasure Island, FL, USA, 2023. Available online: http://www.ncbi.nlm.nih.gov/books/NBK546581/ (accessed on 29 August 2023).

16.   Fischer, S.; Diehm, Y.F.; Kotsougiani-Fischer, D.; Gazyakan, E.; Radu, C.A.; Kremer, T.; Hirche, C.; Kneser, U. Teaching Microsurgical Breast Reconstruction—A Retrospective Cohort Study. *J. Clin. Med.* **2021**, *10*, 5875. Available online: https://www.ncbi.nlm.nih.gov/pmc/articles/PMC8707719/ (accessed on 29 August 2023). [CrossRef]

17.   Conduits for Coronary Bypass: Arteries Other Than the Internal Thoracic Artery's—PMC. Available online: https://www.ncbi.nlm.nih.gov/pmc/articles/PMC3680601/ (accessed on 29 August 2023).

18.   Paliouras, D.; Rallis, T.; Gogakos, A.; Asteriou, C.; Chatzinikolaou, F.; Georgios, T.; Tsirgogianni, K.; Tsakiridis, K.; Mpakas, A.; Sachpekidis, N.; et al. Surgical anatomy of the internal thoracic arteries and their branching pattern: A cadaveric study. *Ann. Transl. Med.* **2015**, *3*, 212. Available online: https://www.ncbi.nlm.nih.gov/pmc/articles/PMC4583587/ (accessed on 29 August 2023).

19.   Piccinin, M.A.; Schwartz, J. Histology, Verhoeff Stain. In *StatPearls*; StatPearls Publishing: Treasure Island, FL, USA, 2023. Available online: http://www.ncbi.nlm.nih.gov/books/NBK519050/ (accessed on 29 August 2023).

20.   Nag, A.C. Study of non-muscle cells of the adult mammalian heart: A fine structural analysis and distribution. *Cytobios* **1980**, *28*, 41–61.

21.   Camaré, C.; Pucelle, M.; Nègre-Salvayre, A.; Salvayre, R. Angiogenesis in the atherosclerotic plaque. *Redox Biol.* **2017**, *12*, 18–34. Available online: https://www.ncbi.nlm.nih.gov/pmc/articles/PMC5312547/ (accessed on 29 August 2023). [CrossRef]

22.   Anagiotos, A.; Kazantzi, M.; Tapis, M. Aberrant internal carotid artery in the middle ear: The duplication variant. *BMJ Case Rep.* **2019**, *12*, e228865. [CrossRef]

23.   Bernardes, M.N.D.; Cascudo, N.C.M.; El Cheikh, M.R.; Gonçalves, V.F.; Lamounier, P.; Ramos, H.V.L.; Costa, C.C. Aberrant common and internal carotid arteries and their surgical implications: A case report. *Braz. J. Otorhinolaryngol.* **2020**, *87*, 366–369. Available online: https://www.ncbi.nlm.nih.gov/pmc/articles/PMC9422460/ (accessed on 29 August 2023). [CrossRef]

24.   Mahmodlou, R.; Sepehrvand, N.; Hatami, S. Aberrant Right Subclavian Artery: A Life-threatening Anomaly that should be considered during Esophagectomy. *J. Surg. Tech. Case Rep.* **2014**, *6*, 61–63. Available online: https://www.ncbi.nlm.nih.gov/pmc/articles/PMC4290042/ (accessed on 29 August 2023).

25.   Andreou, A.Y.; Iakovou, I.; Vasiliadis, I.; Psathas, C.; Prokovas, E.; Pavlides, G. Aberrant left internal thoracic artery origin from the extrascalenic part of the subclavian artery. *Exp. Clin. Cardiol.* **2011**, *16*, 62–64. Available online: https://www.ncbi.nlm.nih.gov/pmc/articles/PMC3126687/ (accessed on 29 August 2023).

26.   Nanthakumar, H.; Iwanaga, J.; Dumont, A.S.; Tubbs, R.S. A rare cadaveric case of a duplicated internal thoracic artery. *Anat. Cell Biol.* **2020**, *53*, 366–368. Available online: https://www.ncbi.nlm.nih.gov/pmc/articles/PMC7527130/ (accessed on 29 August 2023). [CrossRef]

27.   Sajja, L.R.; Mannam, G. Rare variants of internal thoracic artery in patients with coronary artery disease. *Indian J. Thorac. Cardiovasc. Surg.* **2009**, *25*, 68–70. [CrossRef]

28. Burton, K.R.; Ditkofsky, N. Incidental discovery of a new and clinically important anatomic variant of the internal thoracic artery in a young male trauma patient. *BMJ Case Rep.* **2018**, *2018*, bcr-2017. Available online: https://www.proquest.com/docview/2010536726/abstract/AFCF8082BDB3481APQ/1 (accessed on 29 August 2023). [CrossRef]

29. Bilateral Internal Thoracic Artery Harvesting; Anatomical Variations to Be Considered | SpringerLink. Available online: https://link.springer.com/article/10.1007/s12055-007-0036-3 (accessed on 29 August 2023).

30. Odo, Y.; Iwanaga, J.; Tabira, Y.; Watanabe, K.; Saga, T.; Tubbs, R.S.; Fujishima, Y.; Yamaki, K.-I. Bilateral Lateral Costal Branches of the Internal Thoracic Arteries: A Case Report. *Kurume Med. J.* **2018**, *65*, 105–108. Available online: https://www.jstage.jst.go.jp/article/kurumemedj/65/3/65_MS653003/_article (accessed on 29 August 2023). [CrossRef]

31. Bax, M.; Romanov, V.; Junday, K.; Giannoulatou, E.; Martinac, B.; Kovacic, J.C.; Liu, R.; Iismaa, S.E.; Graham, R.M. Arterial dissections: Common features and new perspectives. *Front. Cardiovasc. Med.* **2022**, *9*, 1055862. Available online: https://www.frontiersin.org/articles/10.3389/fcvm.2022.1055862 (accessed on 29 August 2023). [CrossRef]

32. Vural, Ü.; Aglar, A.A.; Sahin, S.; Kizilay, M. Lateral Costal Artery: Clinical Importance of an Accessory Thoracic Artery. *Braz. J. Cardiovasc. Surg.* **2018**, *33*, 626–630. Available online: https://www.ncbi.nlm.nih.gov/pmc/articles/PMC6326454/ (accessed on 29 August 2023). [CrossRef] [PubMed]

33. Tubbs, R.S.; Loukas, M. Common Carotid and Cervical Part of the Internal Carotid Arteries. In *Bergman's Comprehensive Encyclopedia of Human Anatomic Variation*; John Wiley & Sons, Ltd.: New York, NY, USA, 2016; pp. 475–476. ISBN 978-1-118-43030-9. Available online: https://onlinelibrary.wiley.com/doi/abs/10.1002/9781118430309.ch46 (accessed on 29 August 2023).

34. Westrych, K.; Ruzik, K.; Zielinska, N.; Paulsen, F.; Georgiev, G.; Olewnik, Ł.; Łabętowicz, P. Common trunk of the internal thoracic artery, inferior thyroid artery and thyrocervical trunk from the subclavian artery: A rare arterial variant. *Surg. Radiol. Anat.* **2022**, *44*, 983–986. [CrossRef] [PubMed]

35. Knipe, H. Internal Thoracic Artery | Radiology Reference Article | Radiopaedia.org. Available online: https://radiopaedia.org/articles/internal-thoracic-artery (accessed on 29 August 2023).

36. Singh, K.; Anderson, E.; Harper, J.G. Overview and Management of Sternal Wound Infection. *Semin. Plast. Surg.* **2011**, *25*, 25–33. Available online: https://www.ncbi.nlm.nih.gov/pmc/articles/PMC3140234/ (accessed on 29 August 2023). [CrossRef]

37. Deere, M.; Singh, A.; Burns, B. Central Venous Access of the Subclavian Vein. In *StatPearls*; StatPearls Publishing: Treasure Island, FL, USA, 2023. Available online: http://www.ncbi.nlm.nih.gov/books/NBK482224/ (accessed on 29 August 2023).

38. Willner, D.A.; Grossman, S.A. Pericardiocentesis. In *StatPearls*; StatPearls Publishing: Treasure Island, FL, USA, 2023. Available online: http://www.ncbi.nlm.nih.gov/books/NBK470347/ (accessed on 29 August 2023).

39. Carrier, M.; Grégoire, J.; Tronc, F.; Cartier, R.; Leclerc, Y.; Pelletier, L.C. Effect of internal mammary artery dissection on sternal vascularization. *Ann. Thorac. Surg.* **1992**, *53*, 115–119. [CrossRef]

40. Samak, M.; Fatullayev, J.; Sabashnikov, A.; Zeriouh, M.; Schmack, B.; Ruhparwar, A.; Karck, M.; Popov, A.-F.; Dohmen, P.M.; Weymann, A. Total Arterial Revascularization: Bypassing Antiquated Notions to Better Alternatives for Coronary Artery Disease. *Med. Sci. Monit. Basic Res.* **2016**, *22*, 107–114. Available online: https://www.ncbi.nlm.nih.gov/pmc/articles/PMC5063431/ (accessed on 30 August 2023). [CrossRef] [PubMed]

41. Borović, M.L.; Borović, S.; Perić, M.; Vuković, P.; Marinković, J.; Todorović, V.; Radak, D.; Lacković, V. The internal thoracic artery as a transitional type of artery: A morphological and morphometric study. *Histol. Histopathol.* **2010**, *25*, 561–576. [PubMed]

42. Resilience of the Internal Mammary Artery to Atherogenesis: Shifting From Risk to Resistance to Address Unmet Needs | Arteriosclerosis, Thrombosis, and Vascular Biology. Available online: https://www.ahajournals.org/doi/10.1161/ATVBAHA.121.316256 (accessed on 9 October 2023).

43. Ibrahim, A.; Atiyeh, B.; Karami, R.; Adelman, D.M.; Papazian, N.J. The Deltopectoral Flap Revisited: The Internal Mammary Artery Perforator Flap. *J. Craniofac. Surg.* **2016**, *27*, e189–e192. [CrossRef]

44. Vesely, M.J.J.; Murray, D.J.; Novak, C.B.; Gullane, P.J.; Neligan, P.C. The internal mammary artery perforator flap: An anatomical study and a case report. *Ann. Plast. Surg.* **2007**, *58*, 156–161. [CrossRef]

45. Abdelmonem, K.; Elshahat, A.; Abol-Atta, H.; Abou-Gamrah, S.; Abd Eltawab, R.; Massoud, K. Breast reconstruction using internal mammary artery perforator (IMAP) flap. *Ain Shams J. Surg.* **2013**, *6*, 173–190. Available online: https://asjs.journals.ekb.eg/article_179840.html (accessed on 30 August 2023). [CrossRef]

46. Vesely, M.; Murray, D.; Novak, C.; Gullane, P.; Neligan, P. The Internal Mammary Artery Perforator Flap. *Ann. Plast. Surg.* **2007**, *58*, 156–161. [CrossRef]

Case Report

Duplicated Inferior Vena Cava in a 69-Year-Old White Female Donor

Joanna Klansek [1], Keiko Meshida [2], Elizabeth Maynes [3], Maria Ximena Leighton [3], Gary Wind [3] and Guinevere Granite [3,*]

1 F. Edward Hebert School of Medicine, Uniformed Services University of Health Sciences, Bethesda, MD 20814, USA
2 The Henry M. Jackson Foundation for the Advancement of Military Medicine, Inc., Bethesda, MD 20814, USA
3 Department of Surgery, Uniformed Services University of the Health Sciences, Bethesda, MD 20814, USA
* Correspondence: guinevere.granite@usuhs.edu; Tel.: +1-301-295-1500

Abstract: While relatively uncommon, a duplication of the inferior vena cava is moderately well-discussed in the literature. This anatomical variation was noted in a 69-year-old white female donor. This variation is typically asymptomatic; however, it can be associated with complications, such as confusion with a mediastinal mass, increased risk for thromboembolism, and hemorrhage during surgery. It is also associated with a handful of comorbidities, including, but not limited to, congenital renal anomalies such as horseshoe kidney or fused crossed kidney. Research supports that the variation of a duplicated IVC (DIVC) can be due to a failure of the left supracardinal vein to regress during embryonic development.

Keywords: duplicated inferior vena cava; venous embryonic development; supracardinal veins; anatomical variation; inferior vena cava variations

Citation: Klansek, J.; Meshida, K.; Maynes, E.; Leighton, M.X.; Wind, G.; Granite, G. Duplicated Inferior Vena Cava in a 69-Year-Old White Female Donor. *Anatomia* **2023**, *2*, 117–123. https://doi.org/10.3390/anatomia2020011

Academic Editors: Francesco Cappello, Gianfranco Natale and Francesco Fornai

Received: 8 November 2022
Revised: 22 February 2023
Accepted: 7 April 2023
Published: 10 April 2023

1. Introduction

Malformations and/or variations in venous vasculature may affect any number of structures. They can be classified as extratruncular or truncular based on when they occur during embryonic development [1]. Lee (2012) describes the distinction as extratruncular variations occurring during early embryogenesis and truncular variations occurring later in life [1]. In general, truncular lesions are limited to a vessel trunk, and as such, may have more serious consequences for patients. Lee (2012) also states that truncular venous malformations can be the result of a persistent fetal remnant vein that did not involute or regress normally, which correlates with the variation presented in this case report (Figures 1a,b and 2) [1]. Throughout embryonic development of the inferior vena cava (IVC), there is a cycle of development and regression of venous pairs. These three venous pairs include the postcardinal, supracardinal, and subcardinal veins [1]. The segments must fuse and develop in proper chronology, which affords many opportunities for anomalies [1]. Failure of the left supracardinal vein to regress leads to a duplication of the IVC. Venous embryological anomalies may be associated with additional anomalies, such as renal anomalies, including horseshoe kidney and crossed fused kidney. Even though many of these anomalies, including a DIVC, typically present asymptomatically for the majority, if not all, of a patient's life, increased for thromboembolism and side effects of incidentally found comorbidities are essential to identify prior to any sort of abdominal surgery in order to improve surgical outcomes.

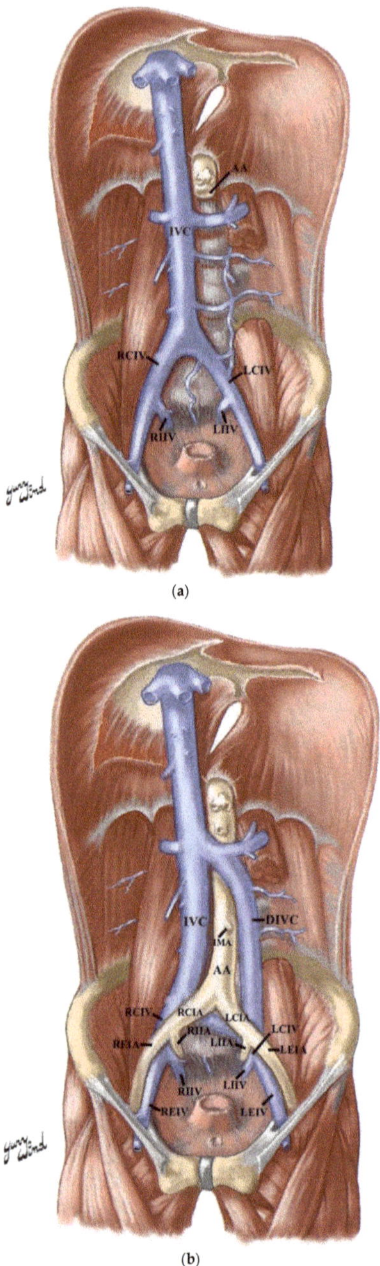

Figure 1. (**a**) Illustrative schematic of normal anatomy and tributaries of the inferior vena cava. (**b**) Illustrative schematic of a duplicated inferior vena cava and its tributaries. Here you can see the duplicated IVC draining into the left renal vein, then merging with the IVC. AA = abdominal aorta; DIVC = duplicated inferior vena cava; IMA = inferior mesenteric artery; IVC = inferior vena cava; LCIA = left common iliac artery; LCIV = left common iliac vein; LEIA = left external iliac artery; LEIV = left external iliac vein; LIIA = left internal iliac artery; LIIV = left internal iliac vein; RCIA = right common iliac artery; RCIV = right common iliac vein; REIA = right external iliac artery; REIV = right external iliac vein; RIIA = right internal iliac artery; RIIV = right internal iliac vein.

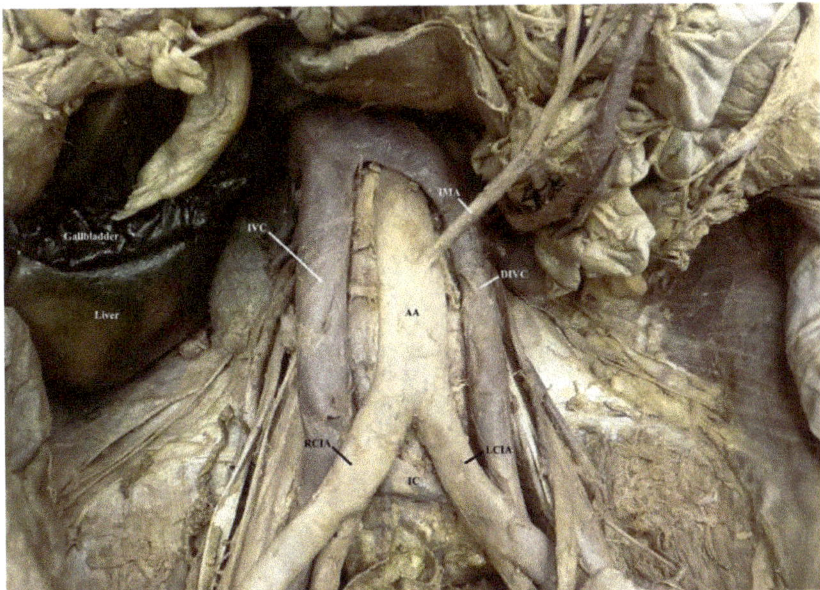

Figure 2. Facilitated display highlighting the duplicated inferior vena cava merging with the inferior vena cava anterior to the abdominal aorta and superior to the inferior mesenteric artery. Also highlighted is the additional venous connection at the base of the duplicated inferior vena cava and the inferior vena cava near the iliac veins. AA = abdominal aorta; DIVC = duplicated inferior vena cava; IMA = inferior mesenteric artery; IC = inferior connection (between the main and duplicated IVC); IVC = inferior vena cava; LCIA = left common iliac artery; RCIA = right common iliac artery.

2. Case Description

During routine anatomical dissection of sixty-five human cadaveric donors during the 2021 first-year medical gross anatomy course and 2021 graduate nursing advanced anatomy course at the Uniformed Services University of the Health Sciences, a duplicated IVC was observed in a 69-year-old white female donor with a listed cause of death of Corticobasilar Degeneration (Neurodegenerative Disease). There was an additional venous connection at the base of the two IVCs near the iliac veins (Figure 2). All cadaveric images depicted in this article were vetted and approved by the USUHS Human Anatomical Specimens Review Committee (HAMRC).

3. Discussion

3.1. Duplicated or Double Inferior Vena Cava

In the majority of cases, the IVC directs deoxygenated blood flow upwards from the inferior and middle body, draining into the right atrium. To achieve this goal, the IVC is located on the right side of the body (Figure 1a). In comparison, a person with a duplicated/double IVC has an IVC on both the left and right sides of the body (Figure 1b). To drain deoxygenated blood into the right atrium of the heart, the left IVC must cross over the abdominal aorta (AA) and merge with the right IVC. This is demonstrated in this particular case as the two IVCs merge anterior to the AA and superior to the inferior mesenteric artery (IMA) (Figure 2). Prior to merging directly inferior to the right atrium, the left and right IVCs drain from the left and right iliac veins (Figure 3). There is a second merge between the left and right common iliac veins posterior to the AA and inferior to the aortic bifurcation into the left and right common iliac arteries, creating a rectangular venous structure in the abdomen (Figures 2 and 3). Measurements of the DIVC were conducted through ImageJ software, using a baseline of 4.10 mm for the diameter of the IMA at its origin, according to a study conducted by Sinkeet et al. (2012) [2]. Four measurements of the DIVC width were conducted along the vessel to average an estimated width of 11.83 mm. To measure the length, the outer and inner lengths were computed to be 11.71 cm and 10.12 cm, respectively, averaging an estimated DIVC length of 10.91 cm.

Figure 3. Facilitated display highlighting the second merge of the duplicated inferior vena cava and the inferior vena cava posterior to the abdominal aorta and inferior to the aortic bifurcation between the left and right common iliac veins, creating a rectangular venous structure in the abdomen. AA = abdominal aorta; DIVC = duplicated inferior vena cava; IMA = inferior mesenteric artery; IC = inferior connection (between the main and duplicated IVC); IVC = inferior vena cava; LCIA = left common iliac artery; LEIA = left external iliac artery; LIIA = left internal iliac artery; RCIA = right common iliac artery; REIA = right external iliac artery; REIV = right external iliac vein; RIIA = right internal iliac artery; RIIV = right internal iliac vein.

3.2. Embryonic Development

The IVC develops between weeks 6 and 8 of embryonic development [3]. During these three weeks, three pairs of veins emerge: postcardinal veins, subcardinal veins, and supracardinal veins (Figure 4) [1]. Regression of the left supracardinal vein allows for the right-sided nature of the mature IVC in the majority of humans. Failure of this regression can result in a duplicated IVC, as noted in this donor [1]. If the right supracardinal vein regresses instead of the left, this can present as a circumcaval ureter, where the ureter passes posterior to the IVC [4]. Circumcaval ureters are clinically significant because they can lead to renal complications such as ureteral obstructions and hydronephrosis [4].

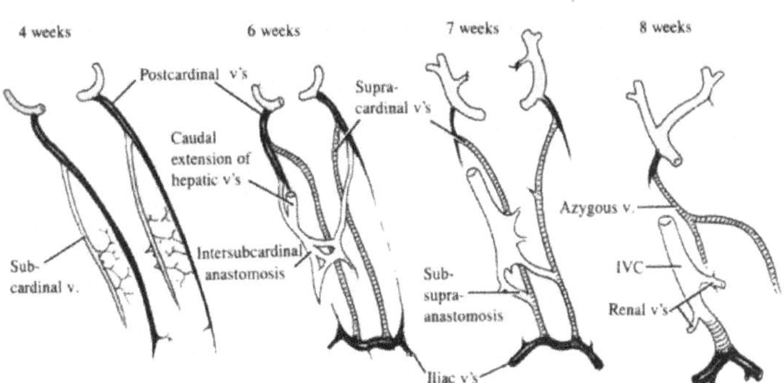

Figure 4. Embryological schematics created by Dr. Gary Wind illustrating the abdominal venous system at four weeks, six weeks, seven weeks, and eight weeks.

3.3. Associated Comorbidities

Investigating other literature reviews of this rare anomaly revealed multiple possible interesting comorbidities. For example, Shaha et al. (2016) documented a case report of a patient who presented with DIVC, crossed fused kidney, and gut malrotation [5]. A possible explanation for these comorbidities would be adjustments to the body's vasculature during embryological development. In particular, renal relocation could be impaired by the presence of a second DIVC occupying abdominal space. This could be a possible explanation for the association with a crossed-fused kidney, where both kidneys are fused on one side of the abdomen, or a horseshoe kidney, where the kidneys are bound by their inferior poles. Both of these anomalies typically present asymptomatically; however, throughout the literature, they have been demonstrated to have associations with an increased risk of developing renal cancer. Renal biopsy for diagnosis of renal cancer is controversial and can indicate a nephrectomy based on imaging results alone. Even then, renal biopsy is usually contraindicated (unless extreme circumstances arise) for evaluation of chronic kidney disease (CKD) if a patient only has one kidney. This contraindication exists because if excessive bleeding were to occur during the biopsy, the ultimate treatment to stop the bleeding is a nephrectomy. In both horseshoe kidney and fused kidney, renal cancer may also involve a complete nephrectomy. In both of these situations, a nephrectomy would place the patient on permanent dialysis as they are no longer able to filter their own blood. Shaha et al. (2016)'s findings also cite sources that support associations between DIVC and polycystic kidney disease, horseshoe kidney, and ectopic kidney [5]. These associations with renal anomalies are important to discover prior to any abdominal surgery in order to preserve renal vasculature during surgery and particularly avoid the possibility of a nephrectomy of a horseshoe or fused kidney if at all feasible. Shaha et al. (2016)'s conclusions encourage surgeons to complete a thorough exploration for other congenital anomalies should a DIVC or other rare finding appear on routine examination or preoperative imaging [5].

Gut malrotation is also mentioned as a comorbidity in Shaha et al. (2016)'s article, although research indicates it is a less common comorbidity than renal malformations [5]. Intestinal malrotation may lead to volvulus, as the intestine fails to rotate correctly during embryological development. This situation requires urgent surgical correction. If not corrected, parts of the bowel may become ischemic and die. The association between gut malrotation and fused crossed kidney both involve failure of proper positioning during embryogenesis. Their association with a duplicated IVC could possibly be explained by a mutation in gene signaling that leads to incorrect completion of embryogenesis, although further studies would be needed to investigate this hypothesis.

3.4. Clinical Significance

Duplicated IVCs are typically incidental findings and present asymptomatically. Research, however, supports a handful of associated complications and surgical relevance. For example, on imaging, the IVC duplication may appear as a mediastinal mass [6]. On computerized tomography (CT), this may be confused with an abdominal aneurysm or aortic dissection. This misinterpretation can lead a patient to undergo unnecessary tumor work-up and imaging [7]. Additionally, due to the increased surface area of vasculature, patients with a duplicated IVC have a higher risk of chronic congestion and hypertension [1]. Patients also present with increased risk for thromboembolism, as a thrombus could become lodged at the extra bifurcation. This could also lead to life-threatening pulmonary implications, such as pulmonary thromboembolism [3]. This risk for IVC thrombosis increases significantly for patients with any type of congenital IVC anomaly. Li et al. (2022) state a prevalence of 60–80% for these patients [8]. Taking into account Virchow's Triad of hypercoagulable state, stasis, and endothelial damage, increased venous stasis from insufficient blood return and drainage can explain this increased risk [8]. Treatment with an IVC filter can increase blood flow, but knowledge of the duplication is essential for proper placement [8]. In particular, Rao et al. (2011) suggest the placement of an IVC filter in both the left and right IVCs [9]. Filter placement may have its own complications, including filter tilting, fracture, perforation, or malposition [9]. These risks increase in filter placement in a DIVC as there are two filters to consider—one for each IVC. This addition makes imaging and precision that much more important. Surgically, knowledge of a DIVC is important prior to any abdominal procedure as it impacts access to neighboring organs and vasculature. Failure to acknowledge a DIVC in a patient could lead to unintentional hemorrhage and death [10]. For example, Effler et al. (1951) describe a potentially preventable fatality after ligation of the anomalous bilateral IVC [10]. This complication arose after an exploratory thoracotomy with right total pneumonectomy following identification of a bronchial neoplasm [10]. Failure to take note of the duplicated IVC after identifying the hepatic vein led to a total ligation of the anomalous IVC during surgery. Ligation resulted in massive retroperitoneal hemorrhage and, ultimately, the patient's death [10]. Identification of this vessel prior to the procedure and plans to reconstruct the vessel during surgery may have avoided the patient's outcome. CT is currently the gold standard for identifying duplication of the IVC and should be utilized prior to abdominal surgery.

4. Conclusions

During embryological development, persistence of both supracardinal veins can lead to a DIVC, as noted in this case. Although this particular patient died of corticobasilar degeneration, symptoms of mediastinal widening on CT and thromboembolism could have appeared if the patient was symptomatic. Associations with gastrointestinal or renal anomalies may have been present asymptomatically as well but were not noted in this particular case. Future research could explore possible genetic associations with such comorbidities and congenital anomalies; perhaps there is an association with a mutation in gene signaling. In conclusion, a DIVC is important to identify via CT in patients prior to undergoing abdominal surgery since it can pose a hemorrhagic threat as well as decreased visibility of left-sided vasculature and anatomy.

Author Contributions: Conceptualization, J.K. and G.G.; methodology, J.K., G.G., K.M., E.M. and M.X.L.; validation, J.K., G.G., K.M., E.M. and M.X.L.; formal analysis, J.K., G.G. and K.M.; investigation, J.K., G.G., K.M. and M.X.L.; resources, J.K., G.G., K.M. and M.X.L.; data curation, J.K., G.G., K.M. and M.X.L.; writing—original draft preparation, J.K. and G.G.; writing—review and editing, J.K., G.G., K.M., E.M., G.W. and M.X.L.; visualization, J.K., G.G., K.M., G.W. and M.X.L.; supervision,

G.G.; project administration, G.G. All authors have read and agreed to the published version of the manuscript.

Funding: This research received no external funding.

Institutional Review Board Statement: Not applicable.

Informed Consent Statement: Not applicable.

Data Availability Statement: Not applicable.

Acknowledgments: We would like to thank the family of our donor for their beneficent contribution. Without their generosity, this article would not have been possible. We would also like to thank Sara Chae for assisting us in our literature review.

Conflicts of Interest: The authors declare no conflict of interest.

References

1. Lee, B.B. Venous embryology: The key to understanding anomalous venous conditions. *Phlebolymphology* **2012**, *19*, 161–204.
2. Sinkeet, S.; Mwachaka, P.; Muthoka, J.; Saidi, H. Branching Pattern of Inferior Mesenteric Artery in a Black African Population: A Dissection Study. *ISRN Anat.* **2013**, *2013*, 962904. [CrossRef] [PubMed]
3. Aaditya, A.; Neelam, D.; Nageswar Rao, J.; Deepak, G. Congenital Anomalies of Inferior Vena Cava, Review of Embryogenesis, Presentation, Associated Congenital Anomalies and Surgical Importance. *Cardiovasc. Thorac. Surg.* **2018**, *3*, 1–6.
4. Lesma, A.; Bocciardi, A.; Rigatti, P. Circumcaval Ureter: Embryology. *Eur. Assoc. Urol.* **2006**, *5*, 444–448. [CrossRef]
5. Shaha, P.; Garg, A.; Sahoo, K.; Kothari, N.; Garg, P. Duplication of Inferior Vena Cava with Associated Anomalies: A Rare Case Report. *J. Clin. Diagn. Res.* **2016**, *10*, TD01. [CrossRef] [PubMed]
6. Cohen, M.; Gore, R.M.; Vogelzang, R.L.; Rochester, D.; Neiman, H.L.; Crampton, A.R. Accessory Hemiazygos Continuation of Left Inferior Vena Cava: CT Demonstration. *J. Comput. Assist. Tomogr.* **1984**, *8.4*, 777–779. [CrossRef] [PubMed]
7. Geley, T.; Unsinn, K.M.; Auckenthaler, T.M.; Fink, C.J.; Gassner, I. Azygos Continuation of the Inferior Vena Cava: Sonographic Demonstration of the Renal Artery Ventral to the Azygos Vein as a Clue to Diagnosis. *AJR* **1999**, *172*, 1659–1662. [CrossRef] [PubMed]
8. Li, W.; Feng, H.; Jin, L.; Chen, X.M.; Zhang, Z.W. Duplication of the Inferior Vena Cava: A Series. *J. Int. Med. Res.* **2022**, *50*, 03000605221100771. [CrossRef] [PubMed]
9. Rao, B.; Duran, C.; Steigner, M.L.; Rybicki, F.J. Inferior Vena Cava Filter—Associated Abnormalities: MDCT Findings. *Am. Roentgen Ray Soc.* **2012**, *198*, 605–610. [CrossRef] [PubMed]
10. Effler, D.B.; Greer, A.E.; Sifers, E.C. Anomaly of the Vena Cava Inferior: Report of Fatality After Ligation. *JAMA* **1951**, *146*, 1321–1322. [CrossRef] [PubMed]

Case Report

Presumed Presence of Extensor Indicis et Digiti Medii Communis Muscle in a 70-Year-Old White Male Donor

Isabella Penkwitz [1], Gary Wind [2], Elizabeth Maynes [2], Maria Ximena Leighton [2] and Guinevere Granite [2,*]

[1]   F. Edward Hebert School of Medicine, Uniformed Services University of the Health Sciences, Bethesda, MD 20814, USA
[2]   Department of Surgery, Uniformed Services University of the Health Sciences, Bethesda, MD 20814, USA
*   Correspondence: guinevere.granite@usuhs.edu; Tel.: +1-301-295-1500

Abstract: Forearm extensor muscle variations can be diverse and, in some instances, rare. During a routine anatomical dissection of human cadaveric donors during the 2021 first-year medical gross anatomy course and 2021 graduate nursing advanced anatomy course at the Uniformed Services University of the Health Sciences, bilateral agenesis of the extensor carpi ulnaris muscle was noted in one 70-year-old white male donor. This variation is described as extremely rare in the literature. The presence of an extensor indicis et digiti medii tendon, a variant of the extensor indicis tendon, appeared to be evident in post-dissection photographs on the left hand. The presence of a duplicated extensor indicis proprious tendon appears to be evident on the right hand. However, further inspection of this region was impeded as the body was sent for cremation prior to the variation being identified. The presence of various juncturae tendinum was also noted bilaterally. Reported prevalence of extensor indicis muscle variants ranges from 0.75% to 13%, depending on the specific type or grouping of variations observed. Knowledge of variations in the extensor compartment of the forearm and wrist is crucial for orthopedic surgeons and specialists. Alteration of surgical approaches may be necessary if such a variation is present. Such variations can be options for grafts, resulting in minimal functional change to the grafted area due to the continued existence of other muscles performing similar functions. Knowledge of such variations, and alternative, synonymous names for them, is also important for anatomy instructors, who may need to assist students in identifying these rare variations during anatomical dissection.

Keywords: extensor indicis et digiti medii muscle; extensor indicis proprius muscle; extensor pollicis longus muscle; extensor indicis et medii communis muscle; extensor indicis et digiti minimi muscle; extensor compartment of the forearm and wrist variations

Citation: Penkwitz, I.; Wind, G.; Maynes, E.; Leighton, M.X.; Granite, G. Presumed Presence of Extensor Indicis et Digiti Medii Communis Muscle in a 70-Year-Old White Male Donor. *Anatomia* **2023**, 2, 109–116. https://doi.org/10.3390/anatomia2010010

Academic Editors: Gianfranco Natale and Francesco Fornai

Received: 11 January 2023
Revised: 24 February 2023
Accepted: 14 March 2023
Published: 16 March 2023

1. Introduction

The extensor compartment of the forearm and wrist in humans can have multiple anatomical variations. Many of these variations are asymptomatic and incidentally found during surgery, trauma response, or assessments of the hand in instances of pain. The extensor indicis et digiti medii muscle (EIMe) originates from the extensor indicis proprius muscle (EIP) (Figure 1a). It plays a role in the extension of the second finger. The origin of the EIP varies between the ulna, radius, carpus, or interosseous membrane [1]. More unusual origins for the EIP include the lunate, scaphoid, and capitate bones with termination at the head of the proximal phalanx of the index finger or second finger [1]. The muscle belly of the EIP runs deep to the extensor digitorum communis tendons (EDCs) and is found rather invariably among humans, gorillas, and chimpanzees [1,2]. Variations in size, origin, insertion, duplications, and/or supernumerary tendons, however, can be found [1,2]. Another variant of the EIP is a duplicated EIP (DEIP), which is a complete duplication of the EIP muscle and should not be confused with a distal splitting to have multiple tendons (Figure 1b).

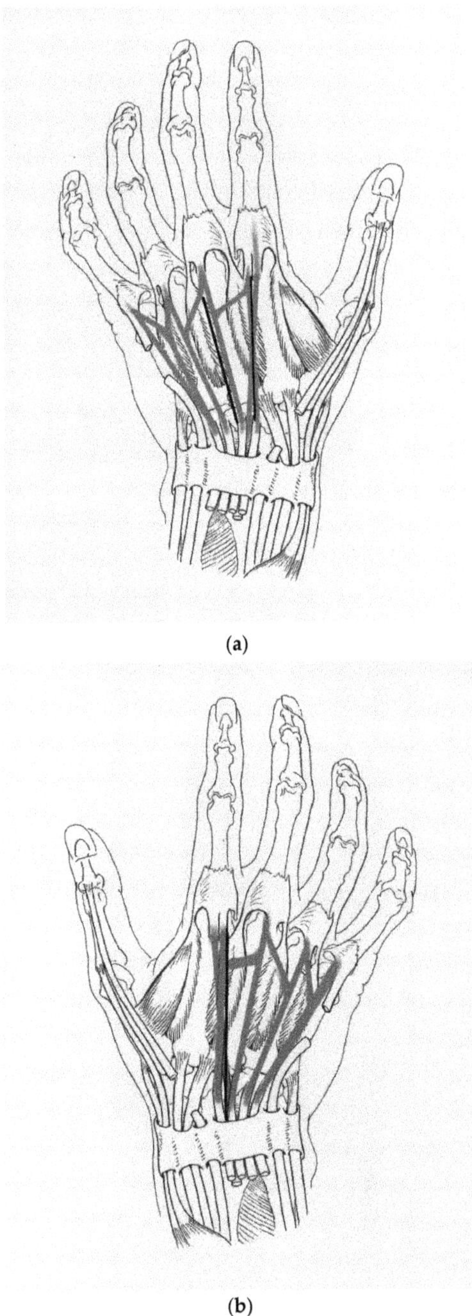

(a)

(b)

Figure 1. (**a**) Schematic of the left hand with the extensor indicis et digiti medii muscle (EIMe) indicated by the thick black lines. (**b**) Schematic of the right hand with the duplicated extensor indicis proprius muscle (DEIP) indicated by the thick black line.

The EIMe and DEIP are categorized as "supernumerary tendons, with normal muscle origins" [1]. In the literature, supernumerary tendons are classified as: complete duplication of the EIP tendon, separate muscles coexisting with normal extensor pollicis longus (EPL) and EIP while providing tendons to the first and second fingers, the extensor medii proprius muscle (EMP) and extensor indicis et medii communis muscle (EIMC), and special tendons to the second, third, and fourth fingers [1]. Specifically, duplicated tendons that insert at the second and third fingers have been observed, with the combined muscle referred to as the EIMC and the muscle inserting on the middle finger as the EMP [1]. It should be noted that the extensor digiti medii and EMP are synonymous terms, but throughout this work, EMP will be used. Similarly, the EIMC can also be referred to as the extensor indicis et medii proprius or extensor indicis et medii accessorius [3–8]. We will use EIMC for consistency. Throughout this work, EIMe refers to the muscle that splits off from the EIMC, which then extends onto the second finger. It should be noted that because the EIMe has a different origin and insertion points, it is considered a separate anatomical variant from a tendinous slip. As a muscular variant of the EIP, the EIMe also plays a role in independent digit extension. Specific rates for the presence of the EIMe were difficult to find. Multiple works, however, have reported the prevalence of other extensor tendon variants, including the EIP, EMP, EIMC, and extensor pollicis et indicis muscle tendon (EPI). Yammine (2015) describes the systematic review and meta-analysis of 21 studies and reports a pooled prevalence estimate of 1.6% for the EIMC, specifically [2]. The article further states that the EIMC was most often found in North American populations when compared to European, Indian, and Japanese populations [2]. We presume the EIMC was present in the left hand of this donor if the observed variant in the first finger was correctly identified as the EIMe. Because of the limited visibility in the photos taken for Granite et al. (2022) [8] and the subsequent cremation of the donor's remains, it is not possible to reflect the superficial structures in order to obtain an unobstructed view of the muscle for more definitive identification as the EIMe. With respect to the DEIP, because it is a duplication, this variant serves the same function as a singular EIP. Historically, the prevalence of the DEIP has varied from 2.7% to 16%. Cauldwell et al. (1943) reported a prevalence of 2.7%, whereas Wood (1868) reported 9.3%, and von Schroeder (1995) reported 16% [1,6,9]. A DEIP involves a completely duplicated and robust muscle belly, which should not be confused with splitting into an additional distal slip.

The purpose of this case report is to alert surgeons and anatomists of the existence of the EIMe anatomical variation and the possibility that this variation may also be present in future patients or donors. It also serves to clarify any alternative, synonymous names for select variants that exist in the literature in order to reduce confusion.

2. Case Description

During routine anatomical dissection of human cadaveric donors during the 2021 first-year medical gross anatomy course and 2021 graduate nursing advanced anatomy course at the Uniformed Services University of the Health Sciences, Granite et al. (2022) found a 70-year-old White male donor with bilateral agenesis of the extensor carpi ulnaris muscle (ECU) [10]. The donor's cause of death was pneumonia.

After the case report on bilateral agenesis of the ECU was published in February 2022, we noted an unfamiliar muscle and tendon variant present on the left hand of the donor [10]. The variant in question was located on the dorsal aspect of the left hand and appeared to be a separate tendon from both the extensor digitorum muscle (ED) and EIP tendons. Because the donor was cremated, further dissection of the limb to remove the superficial tendons and determine the origin of the tendon in question was not possible. There are, however, other works describing the presence of a variant matching what was observed in this donor [1–8,11]. After reviewing descriptions of variants affecting the extensor compartment of the forearm and wrist, we have reason to believe that the tendon in question for this donor is the EIMe (Figure 2).

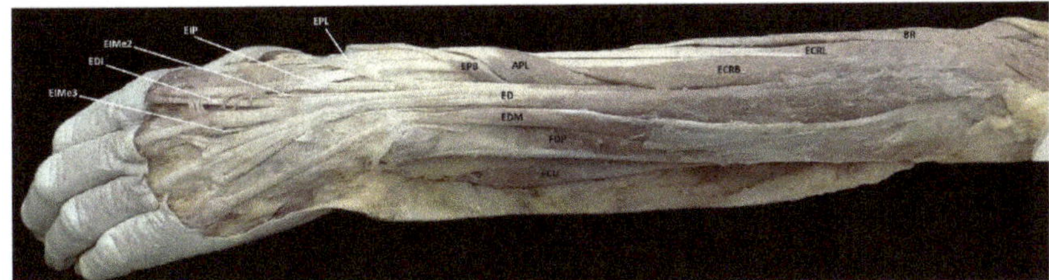

Figure 2. Facilitated display highlighting the following left forearm muscles: APL = Abductor Pollicis Longus Muscle, BR = Brachioradialis Muscle, ECRB = Extensor Carpi Radialis Brevis Muscle, ECRL = Extensor Carpi Radialis Longus Muscle, ED = Extensor Digitorum Muscle, EDi = Extensor Digitorum Muscle Tendon to the Index Finger, EDM = Extensor Digiti Minimi Muscle, EIP = Extensor Indicis Proprius Muscle, EIMe2 = Extensor Indicis et Digiti Medii Muscle Tendon of the Second Digit, EIMe3 = Extensor Indicis et Digiti Medii Muscle Tendon of the Third Digit, EPB = Extensor Pollicis Brevis Muscle, EPL = Extensor Pollicis Longus Muscle, and FCU = Flexor Carpi Ulnaris Muscle.

We noted that the EIMe is located radial to the ED tendon of the second finger on the left hand, but a different variant was present ulnar to the ED tendon of the second finger on the right hand (Figure 3). Because it is fairly robust, we believe that DEIP is a more appropriate designation of this variant rather than simply an additional distal slip.

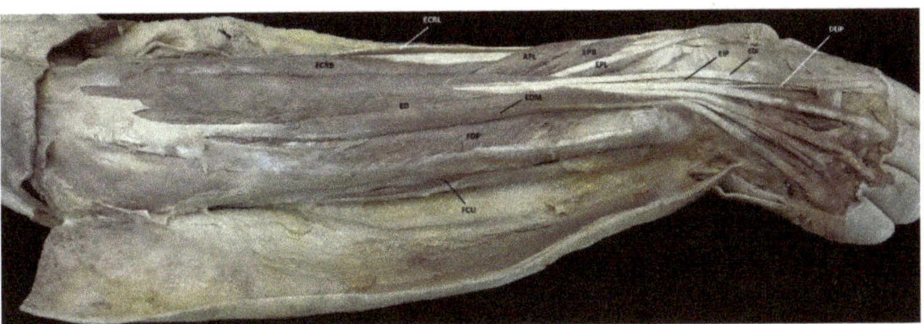

Figure 3. Facilitated display highlighting the following right forearm muscles: APL = Abductor Pollicis Longus Muscle, ECRB = Extensor Carpi Radialis Brevis Muscle, ECRL = Extensor Carpi Radialis Longus Muscle, ED = Extensor Digitorum Muscle, EDi = Extensor Digitorum Muscle Tendon to the Index Finger, EDM = Extensor Digiti Minimi Muscle, EIP = Extensor Indicis Proprius Muscle, DEIP = Duplicated Extensor Indicis Proprius Muscle Tendon, EPB = Extensor Pollicis Brevis Muscle, EPL = Extensor Pollicis Longus Muscle, and FCU = Flexor Carpi Ulnaris Muscle.

The presence of juncturae tendinum (JT) was also noted bilaterally (Figure 4a,b). Based on the classifications outlined by Abdel-Hamid et al. (2013), we believe there are at least six cumulative JTs in this donor [12]. Specifically, three JTs were noted in the left hand: a type 2 JT between the second and third digits, a type 3r JT between the third and fourth digits, and a type 3y JT between the fourth and fifth digits (Figure 4a). Three JTs were noted in the right hand: a type 2 JT between the second and third digits, a type 3r between the third and fourth digits, and a type 3y JT between the fourth and fifth digits (Figure 4b).

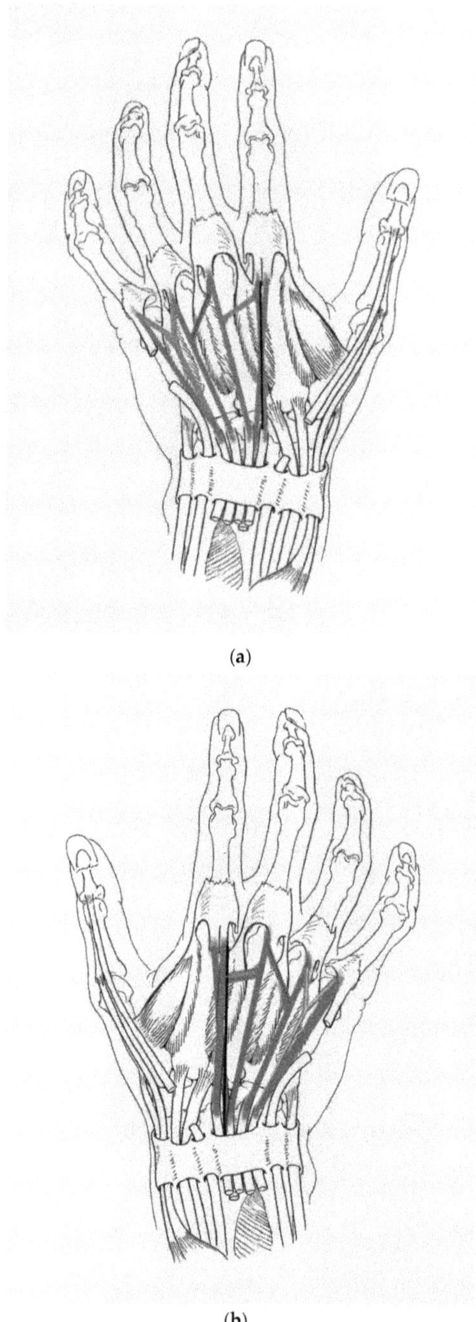

(a)

(b)

Figure 4. (**a**) Schematic of the left hand highlighting the juncturae tendinum (JT): a type 2 JT between the second and third digits (highlighted in green), a type 3r JT between the third and fourth digits (highlighted in red), and a type 3y JT between the fourth and fifth digits (highlighted in blue). (**b**) Schematic of the right hand highlighting the juncturae tendinum (JT): a type 2 JT between the second and third digits (highlighted in green), a type 3r between the third and fourth digits (highlighted in red), and a type 3y JT between the fourth and fifth digits (highlighted in light purple).

3. Discussion

Reported variations of the EIP and its tendon include one or more distal slips and attachment of slips to other digits, including the EMP, EIMC, and EPI. Yammine (2015) reported an EMP prevalence of 3.7%, with a bilateral variation being rarer, while prevalence of the EIMC tendon was 1.6%, and the prevalence of the EPI was 0.75% [2]. Yammine (2015) has cited variations in frequencies based on gender, ancestry, laterality, and side [2].

With respect to slips of the EIP, single-slip EIPs are the most common, whereas the prevalence steeply decreases with the increased number of slips [2,6]. Yammine (2015) reports a pooled prevalence estimate of 92.6%, 7.2%, and 0.3% for the single-, double-, and triple-slip EIP, respectively [2]. The single-slip variation frequently inserts on the ulnar side of the EDC of the second finger (EDC-index) in 98.3% [2]. The attachment location of these single-, double-, or triple-slip EIPs may also vary. Yammine (2015) reports a prevalence of 53.5% for additional EIP slips on the ulnar side of the ED to the second digit [2]. The ulnar location of the observed right-hand variant remains consistent with Yammine's data. However, we believe that this observed variant is more robust than a typical extra slip of the EIP and is more likely to be a DEIP. The prevalence of a DEIP ranges from 2.7% to 16% [1,6,9].

Although the actions or functions may be similar for the tendinous slips of the ED to the second digit and the EIMe, the EIMe has its own muscle belly and origin and, as a result, is referenced as a distinctly different muscle in the anatomical variation literature [1–8,11].

Komiyama et al. (2019) reported variations of the EIP by classification of supernumerary muscles or tendons into four types, each with two subtypes [4]. However, they did not observe any cases of EIMC or EMIe and, therefore, did not include these variations in their subtype classifications. In general, it is agreed that, for the second finger, the most frequent variation of the EIP is the presence of a single supernumerary tendon, mostly occurring on the ulnar side of the ED tendon [4,6].

Embryological development of the upper limb begins at approximately 26 days after fertilization with the presence of the apical ectodermal ridge [13]. Of the four embryological developmental zones, the three that are relevant to this work are the zeugopod (future forearm), mesopod (future wrist), and autopod (future hand) [13]. After zones are established, cartilage and other cells begin to form joints, bones, tendons, and muscles. Muscle fibers can be found in the limb bud by week 7 of development [13]. Specific limb development occurs along three axes: proximodistal, anteroposterior, and dorsoventral. The proximodistal axis influences limb growth proximally to distally. The anteroposterior axis is important for developing ulnar and radial structures, with the Sonic Hedgehog gene (SHH) specifically affecting the development of the second and third fingers via chemical growth and differentiation signaling [13]. The dorsoventral axis is important for development of both the dorsal and ventral structures, with the Wnt family member 7A gene (WNT7A) in the dorsal ectoderm and LIM homeobox transcription factor 1 beta (LMX1B) in the dorsal mesoderm being responsible for the development of extensor tendons in the hand [13].

The EIP can be involved in pain syndromes, trauma, and tendon transfers during reconstructive surgery. Examples include external trauma, extensor indicis proprius syndrome [4], rheumatoid arthritis, and osteoarthritis [11]. In cases of reconstruction involving trauma, options for grafts depend on which ligaments are still present [14]. Grafts sourced from the extensor retinaculum or flexor digitorum superficialis are often used in finger reconstruction requiring joint stabilization [14]. The EIP is frequently used for tendon transfer following extensor rupture due to rheumatoid arthritis [14]. In cases where a DEIP is present, this could allow for tendon transfer of one EIP with preservation of the alternate EIP.

Similar to the EIP, the relevance for the EIMe likely revolves around the surgical possibilities. Currently recognized extra tendon sources aside from the EDC include the EIP, EMP, EIMC, EPI, and EDM [15]. Although clinical practice currently revolves around utilizing the EIP, further research could evaluate the legitimacy of using the EIMe for tendon transfer, if it is a robust graft source, in place of the EIP. Yammine (2019) determined that there is a 40% probability of finding extra slips and an 8.5% probability of an EIP variant tendon (EMP, EIMC, or EPI) being present [15]. Extra slips and EIP variant tendons can be used in reconstructive surgery [15]. If slips are to be used as replacement tendons, surgeons should remember that there are variations in their frequency based on ethnicity [15]. There is also the possibility of retaining second finger extension despite traditional EIP tendon transfer if the EIMe is present. Use of MRI to gain an understanding of each patient's musculature within the hand can allow surgeons to alter approaches as needed.

Other considerations for the identified muscle and tendon variant of the left hand include the EMP, extensor digiti brevis manus muscle (EDBM), and extensor indicis brevis manus muscle (EIBM). The EMP originates at the ulna below the EIP and either inserts at the extensor aponeurosis of the third finger or splits to supply slips to both the second and third fingers [1]. Von Schroeder and Botte (1991) describe the EMP as a distinctly different muscle belly from the EIP that inserts into the third finger [5]. It is possible for the EMP to be fused to part of the EIP, making its identification difficult in some cases [1]. The EDBM originates from the ulna or ulnar half of the carpal bones, which is notably narrower when compared to the wider range of origins for the EIP [1]. It is described as a "short carpal extensor" that inserts either into the metacarpals or other extensor tendons of a respective digit [1]. Similar in name and origin to the EDBM, the EIBM is an EDBM variant with at least one muscle belly and/or tendon that is distinguishable from the EIP [1,16,17]. The EIBM inserts at either the proximal or distal phalanx of the second digit [16,17]. Like the EDBM, the EIBM can join extensor tendons, such as the EIP [1]. Interestingly, the EIBM and EIP coexistence has been reported by Esakkiammal et al. (2021), while EDBM and EIP presence may be mutually exclusive [16].

Bilateral presence of JT was also noted in the donor. Abdel-Hamid et al. (2013) defines JT as "short bands of connective tissue present between the adjacent extensor tendons on the dorsum of the hand" [12]. There are various classifications of JT based on the thickness of the connective tissue, whether it is filamentous or tendinous, and the shape or angle of the band's connections [12]. Variations in incidence exist based on the type and location of the JT in the second, third, or fourth intermetacarpal spaces of the hand [12].

4. Conclusions

Although the extensor compartment of the forearm and hand can have extensive anatomical variations, not all variations are equally common. While the prevalence of the EMP is 3.7% and the EPI is 0.75%, the prevalence of the EIMC is 1.6% [2]. These variations are all quite rare in the literature. This may be due to a general decreased awareness of authors of their existence in patients or donors.

A major limitation of this case report was identifying the possible EIMe after the donor had been sent for incineration. Our hypothesis that the tendon and muscle observed inserting into the second and third digits of the left hand is the EIMe was based on the available images of the donor and the descriptions of the EIMe in other works. We also believe it is possible that the EIMC is also present if the identified variant is the EIMe, but no longer have the means to verify this suspicion. We believe the observed variant of the second digit on the right hand is a DEIP. In the same way, because the donor was cremated, this case report was limited in verifying a complete duplication of the EIP. However, because it is fairly robust, we believe that this is a more appropriate designation rather than simply an additional distal slip.

The importance of identifying anatomical variations is twofold: to alert anatomists of these possible variations that may have embryonic development origins, and to ensure surgeons are aware of alternate tendons that may prove useful during select procedures.

Author Contributions: Conceptualization, I.P. and G.G.; methodology, I.P.; formal analysis, I.P., M.X.L. and G.G.; investigation, I.P., M.X.L. and G.G.; resources, I.P.; writing—original draft preparation, I.P.; writing—review and editing, G.G. and E.M.; visualization, I.P., G.W. and G.G.; supervision, G.G. All authors have read and agreed to the published version of the manuscript.

Funding: This research received no external funding.

Institutional Review Board Statement: Not applicable.

Informed Consent Statement: Not applicable.

Data Availability Statement: Not applicable.

Acknowledgments: We would like to thank the families of our donors for their beneficent contribution. Without their generosity, this article would not have been possible. We would also like to thank Sara Chae for assisting us in our literature review.

Conflicts of Interest: The authors declare no conflict of interest.

Disclaimer: The opinions or assertions contained herein are the private ones of the author/speaker and are not to be construed as official or reflecting the views of the Department of Defense, the Uniformed Services University of the Health Sciences or any other agency of the U.S. Government. The contents of this presentation are the sole responsibility of the author(s) and do not necessarily reflect the views, opinions or policies of Uniformed Services University of the Health Sciences (USUHS), The Henry M. Jackson Foundation for the Advancement of Military Medicine, Inc., the Department of Defense (DoD) or the Departments of the Army, Navy, or Air Force. Mention of trade names, commercial products, or organizations does not imply endorsement by the U.S. Government.

References

1. Anatomy Atlases: An Anatomy Digital Library—Curated by Ronald A. Bergman, Ph.D. Available online: http://www.anatomyatlases.org/AnatomicVariants/MuscularSystem/Text/E/25Extensor.shtml (accessed on 10 January 2023).
2. Yammine, K. The prevalence of the extensor indicis tendon and its variants: A systematic review and meta-analysis. *Surg. Radiol. Anat.* **2015**, *37*, 247–254. [CrossRef] [PubMed]
3. Gonzalez, M.H.; Weinzweig, N.; Kay, T.; Grindel, S. Anatomy of the Extensor Tendons to the Index Finger. *J. Hand Surg.* **1996**, *21*, 988–991. [CrossRef] [PubMed]
4. Komiyama, M.; New, T.M.; Toyota, N.; Shimada, Y. Variations of the Extensor Indicis Muscle and Tendon. *J. Hand Surg. (Br. Eur. Vol.)* **1999**, *24*, 575–578. [CrossRef] [PubMed]
5. Von Schroeder, H.P.; Botte, M. The extensor medii proprius and anomalous extensor tendons to the long finger. *J. Hand Surg.* **1991**, *16*, 1141–1145. [CrossRef] [PubMed]
6. Von Schroeder, H.P.; Botte, M.J. Anatomy of the Extensor Tendons of the Fingers: Variations and Multiplicity. *J. Hand Surg.* **1995**, *20*, 27–34. [CrossRef] [PubMed]
7. Yalcine, B.; Kutoglu, T.; Ozan, H.; Gurbuz, H. The Extensor Indicis Et Medii Communis. *Clin. Anat.* **2006**, *19*, 112–114. [CrossRef] [PubMed]
8. Yoshida, Y. Anatomical Study on the Extensor Digitorum Profundus Muscle in the Japanese. *Okajimas Folia Anat. Jap.* **1990**, *66*, 339–354. [CrossRef] [PubMed]
9. Cauldwell, E.W.; Anson, B.J.; Wright, R.R. The extensor indicis proprius muscle: A study of 263 consecutive specimens. *Q. Bull. Northwestern Univ. Med. Sch.* **1943**, *17*, 267–279.
10. Granite, G.; Maynes, E.; Leighton, M.X.; Wind, G.; Nesti, L. Bilateral Agenesis of the Extensor Carpi Ulnaris Muscle of a 70 Year-Old White Male Donor. *J. Surg.* **2022**, *7*, 1470. [CrossRef]
11. Fischer, K.; Breitfeld, T.; Damert, H.G.; Rothkotter, H.J. Multiple Variations of Extensor Muscles in a Single Hand. *Int. J. Anat. Var.* **2016**, *9*, 32–34, eISSN: 1308-4038. Available online: https://www.pulsus.com/scholarly-articles/multiple-variations-of-extensor-muscles-in-a-single-hand.pdf (accessed on 10 January 2023).
12. Abdel-Hamid, G.A.; El-Beshbishy, R.A.; Abdel Aal, I.H. Anatomical variations of the hand extensors. *Folia Morphol.* **2013**, *72*, 249–257. [CrossRef] [PubMed]
13. Al-Qattan, M.M.; Kozin, S.H. Update on Embryology of the Upper Limb. *J. Hand Surg.* **2013**, *38*, 1835–1844. [CrossRef] [PubMed]
14. UpToDate—Surgical Reconstruction of the Upper Extremity by Kevin C. Chung and Hidemasa Yoneda. Available online: https://www-uptodate-com.usu01.idm.oclc.org/contents/surgical-reconstruction-of-the-upper-extremity (accessed on 22 July 2022).
15. Yammine, K. Predicting Tendon Tissue Grafting Source From the extensors of Long Fingers: A Systematic Review of Cadaveric Studies. *HAND* **2019**, *14*, 651–657. [CrossRef] [PubMed]
16. Esakkiammal, N.; Chauhan, R.; Sharma, R. Clinical Significance of Presence of Extensor Indicis Brevis Manus—A Case Report. *J. Clin. Diagn. Res.* **2017**, *11*, 5–6. [CrossRef]
17. Solomon, D.; Atlaw, D.; Gezahegn, H. Extensor Indicis Brevis Muscle: A Case Report. *Int. Med. Case Rep. J.* **2021**, *14*, 323. [CrossRef] [PubMed]

Case Report

Clinicopathologic Features of Neuroblastoma-like Schwannoma: A Case Report of Unusual Morphologic Variant

Samira Mortazavi [1,*], **Kambiz Kamyab Hesari** [1], **Atieh Khorsand** [2] and **Maryam Ardalan** [3]

[1] Department of Pathology, Razi Hospital, Tehran University of Medical Sciences, Tehran 1983969411, Iran
[2] Department of Pathology, Shariati Hospital, Tehran University of Medical Sciences, Tehran 1983969411, Iran
[3] Department of Physiology, Institute of Neuroscience and Physiology, Sahlgrenska Academy, University of Gothenburg, 40530 Gothenburg, Sweden
* Correspondence: samira.mortazavi@cshs.org

Abstract: Neuroblastoma-like schwannoma is known as a rare unusual variant of schwannoma with difficulties of differential diagnosis with neuroblastoma, Ewing sarcoma/peripheral neuroectodermal tumor and other cutaneous small round cell tumors. Herein, we describe a neuroblastoma-like schwannoma that was presented as a painless lesion on the dorsal side of the left hand in a 39-year-old woman. Composed collagen fibers in the central core of rosettes and diffuse expression of S100 protein in the tumor cells found in the biopsy specimens confirmed the Schwann cell origin of the tumor.

Keywords: neuroblastoma; schwannoma; small round cell; rosettes

Citation: Mortazavi, S.; Kamyab Hesari, K.; Khorsand, A.; Ardalan, M. Clinicopathologic Features of Neuroblastoma-like Schwannoma: A Case Report of Unusual Morphologic Variant. *Anatomia* **2022**, *1*, 217–221. https://doi.org/10.3390/anatomia1020022

Academic Editors: Gianfranco Natale and Francesco Fornai

Received: 3 August 2022
Accepted: 2 December 2022
Published: 9 December 2022

Publisher's Note: MDPI stays neutral with regard to jurisdictional claims in published maps and institutional affiliations.

1. Introduction

Schwannoma is a benign peripheral nerve sheath tumor that is identified by a biphasic histologic appearance composed of antoni A and antoni B areas [1]. Several histologic variants of schwannoma including "ancient", plexiform, epithelioid, and neuroblastoma-like schwannoma have been identified [2]. Neuroblastoma-like schwannoma is a rare variant of this entity. It was first described by Goldblum JR et al. [3] in 1994, and by 2019, 24 cases of neuroblastoma-like schwannoma had been reported [4]. This tumor is characterized by a typical immunophenotyping profile accompanied by areas that resemble neuroblastoma [5]. Neuroblastoma is predominantly composed of small round hyperchromatic Schwann cells surrounding perivascular and giant rosettes [6]. Herein, we report a case of neuroblastoma-like schwannoma that was presented as a lesion on the dorsal side of the left hand.

2. Case Report

A 39-year-old woman presented with a 3-year history of a slowly growing painless lesion of 2 cm in size, located on the dorsal side of the left hand. No accompanying symptoms were reported. There was no family history of schwannoma or any kind of cutaneous malignancy. The patient did not report a history of smoking and comorbidity.

The lesion was completely excised by an expert dermatologist at the hospital. Microscopically, an encapsulated creamy whitish round mass with a size of 2 × 2 cm, homogenous on cutting, without hemorrhage or necrosis, was observed.

Histopathological investigation revealed a neoplastic tissue surrounded by a thin fibrous capsule and consisting of uniform small lymphocyte-like round cells with scant cytoplasm and small hyperchromatic nuclei. The cells were radially arranged around eosinophilic collagenous-core-forming giant rosettes. No cellular atypia, necrosis or mitosis was evident in this specimen. The classic pattern of schwannoma was notably absent in all representative sections.

Masson's trichrome staining confirmed that the central core of the rosettes was composed of collagen fibers. Subsequent immunohistochemical study revealed diffuse expression of S100 protein (as the most commonly used marker to detect tumoral cells) [6] in the tumoral cells, confirming the Schwann cell origin of the tumor (Figures 1 and 2). Additionally, staining for the neural markers showed focal positive expression for Neuron Specific Enolase (NSE) and negative for synaptophysin which confirmed our diagnosis (Figures 3 and 4).

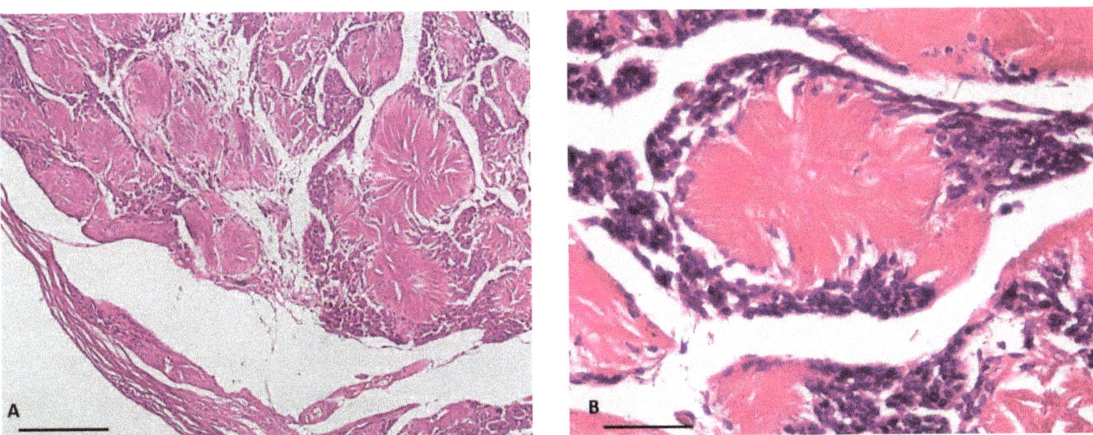

Figure 1. Neuroblastoma-like schwannoma: (**A**) Encapsulated neoplasm consists of small round cells around an eosinophilic core on H&E-stained tissue (10× objective lens, scale bar = 200 µm); (**B**) 40× small Schwann cells with scant cytoplasm and hyperchrome nuclei with a radial arrangement around a collagen core that formed giant rosettes on H&E-stained tissue (40× objective lens, scale bar = 50 µm).

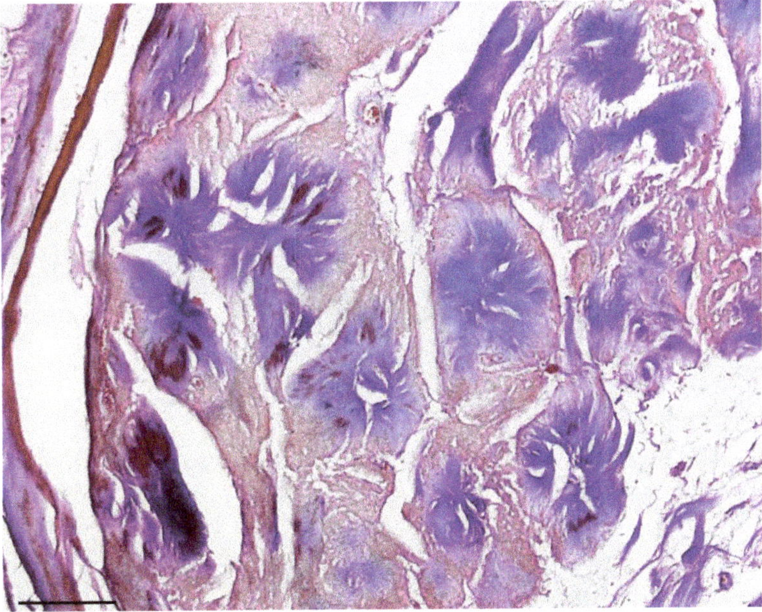

Figure 2. Masson Trichrome, collagen core of rosettes stained blue (40× objective lens, scale bar = 50 µm).

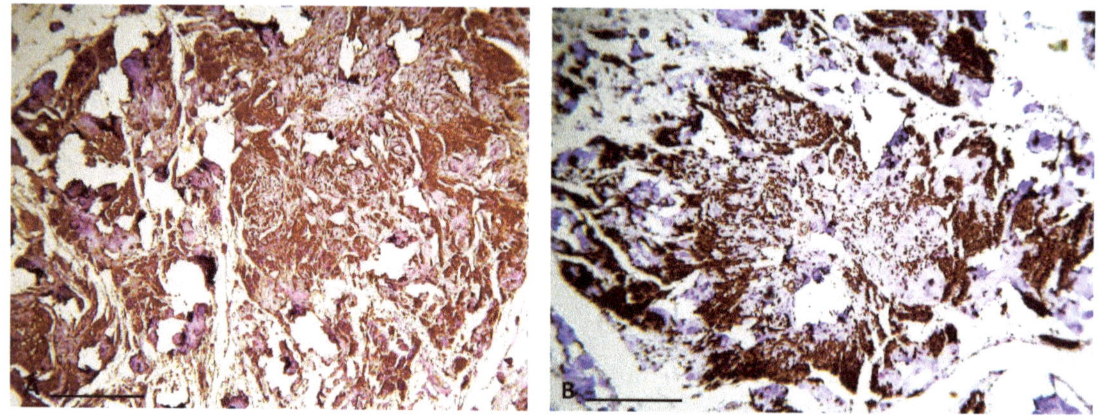

Figure 3. Immunohistochemical profile of neuroblastoma-like schwannoma (10× objective lens): (**A**) Schwann cells show strong and diffuse positivity for S100 marker; (**B**) Tumor cells are focally stained by Neuron Specific Enolase (scale bar = 100 μm).

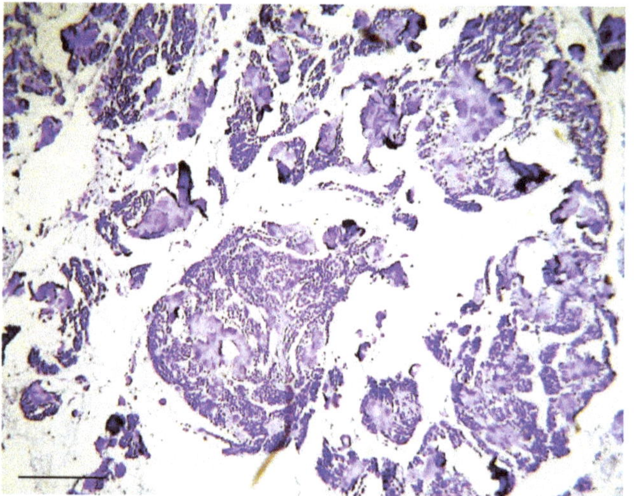

Figure 4. Negative immunostaining of tumor cells with Synaptophysin. (10× objective lens, scale bar = 100 μm).

3. Discussion

Neuroblastoma-like schwannoma is a benign nerve sheath tumor as it grows slowly and usually does not represent an immediate risk to health. To date, a limited number of cases with a diagnosis of neuroblastoma-like schwannoma, as an exceptional case of schwannoma, have been described in the literature. Regarding the location of lesion, it has been shown that neuroblastoma-like schwannoma lesions are mostly located on the neck, trunk and extremities [7]. However, lumbar spine nerve root [8], lower labial mucosa [9], orbit [10] and pleura [11] were also reported as sites of neuroblastoma-like schwannoma lesions. Interestingly, the location of lesion in our case was on the dorsal side of the left hand and that has not been reported previously.

Prevalence of neuroblastoma-like schwannoma is known to be higher in women than men and is usually reported as a painful mass [12,13]. Our patient was also a woman, however had a painless lesion. Higher prevalence has been suggested as a reason for an increased clinicopathological variation of schwannoma in women compared to men [13].

Microscopically, neuroblastoma-like schwannoma can mimic other small round cell tumors, as originally described by Goldblum JR et al., and the presence of these small round cells along with rosette formation may cause confusion with primitive neuroectodermal tumors (PNETs) and neuroblastoma. Zachary T. Lewis believes that the name "neuroblastoma-like schwannoma" may overestimate these similarities [13]. Neuroblastoma-like schwannoma can be differentiated from mentioned entities based on the presence of areas of conventional schwannoma, encapsulation, lack of mitoses and atypia, along with diffuse positivity for S100 protein [2].

Difference in the size of rosettes is considered an important histological criterion to distinguish neuroblastoma from neuroblastoma-like schwannoma [2,5]. In agreement with previously reported observations, rosettes in our neuroblastoma-like schwannoma case were trichrome positive [12].

Dendritic cell neurofibroma is another dermal lesion with pseudorosettes that might be confused in diagnosis with neuroblastoma-like schwannoma. The differentiation between dendritic cell neurofibroma and neuroblastoma-like schwannoma is based on the smaller size of the pseudorosettes, with a histopathological characterization including lagers cells with pale eosinophilic cytoplasm and vesicular nuclei, and strong CD57 positivity [14,15]. The other distinguishing feature between dendritic cell neurofibroma and neuroblastoma-like schwannoma is positive staining with S100 protein and positive/absent CD34 staining [14].

In our case, the histopathological investigation on Hematoxylin and Eosin (H&E) stained sections revealed a small cell tumor with the cellular characteristics of neuroblastoma described earlier, along with areas of conventional schwannoma. Moreover, no cellular atypia, necrosis or mitosis was evident in the specimen of our patient. These findings are consistent with the findings of previous neuroblastoma-like schwannoma case reports [16].

Immunohistochemical methods helped us to reach a definite diagnosis. Indeed, in our case, the tumor cells were strongly and diffusely positive for S100 protein and focally positive for NSE. Neuroblastoma-like schwannoma showed focal areas with typical immunophenotyping of conventional schwannoma and all tumor cells were strongly and diffusely positive for S100 protein, but typically negative for other markers of neural differentiation, such as synaptophysin [3]. However, there are reports in which the tumor cells show positivity for other neural markers such as NSE [9].

4. Conclusions

Neuroblastoma-like schwannoma is an unusual variant of schwannoma which can be misinterpreted as malignancy. Therefore, performing more specific histopathological characterization, such as examining the expression of S100 protein in tumoral cells, in combination with histological investigation, including H&E staining, is of fundamental importance for the pathologists.

Author Contributions: Conceptualization, S.M., K.K.H., A.K.; investigation, S.M.; resources, S.M., K.K.H., A.K., M.A.; writing—original draft preparation S.M., K.K.H., A.K.; writing—review and editing, M.A., S.M. All authors have read and agreed to the published version of the manuscript.

Funding: This research received no external funding.

Institutional Review Board Statement: The study was conducted in accordance with the Declaration of Helsinki, and approved by This research was approved by the ethics committee of Tehran University of Medical Sciences (2017/675, March 2017).

Informed Consent Statement: Informed consent was obtained from subject involved in the study.

Data Availability Statement: Not applicable.

Conflicts of Interest: The authors declare no conflict of interest.

References

1. Rodriguez, F.J.; Folpe, A.L.; Giannini, C.; Perry, A. Pathology of peripheral nerve sheath tumors: Diagnostic overview and update on selected diagnostic problems. *Acta Neuropathol.* **2012**, *123*, 295–319. [CrossRef] [PubMed]
2. Somerhausen, N.D.S.A.; Valaeys, V.; Geerts, M.; André, J. Neuroblastoma-like schwannoma: A case report and review of the literature. *Am. J. Dermatopathol.* **2003**, *25*, 32–34. [CrossRef] [PubMed]
3. Goldblum, J.R.; Beals, T.F.; Weiss, S.W. Neuroblastoma-like neurilemoma. *Am. J. Surg. Pathol.* **1994**, *18*, 266–273. [CrossRef] [PubMed]
4. Mahjoub, W.K.; Jouini, R.; Khanchel, F.; Ben Brahim, E.; Llamas-Velasco, M.; Helel, I.; Khayat, O.; Chadli, A.; Badri, T.; Mentzel, T. Neuroblastoma-like schwannoma with giant rosette: A potential diagnostic pitfall for hyalinizing spindle cell tumor. *J. Cutan. Pathol.* **2019**, *46*, 234–237. [CrossRef] [PubMed]
5. Sharma, P.; Chatterjee, D.; Das, A. Neuroblastoma like schwannoma: A diagnostic challenge. *BMJ Case Rep.* **2017**. [CrossRef] [PubMed]
6. Sulhyan, K.R.; Deshmukh, B.D.; Gosavi, A.V.; Ramteerthakar, N.A. Neuroblastoma-like schwannoma in a case of schwannomatosis: Report of a rare case. *Int. J. Health Sci.* **2015**, *9*, 478–481. [CrossRef]
7. Suchak, R.; Luzar, B.; Bacchi, C.E.; Maguire, B.; Calonje, E. Cutaneous neuroblastoma-like schwannoma: A report of two cases, one with a plexiform pattern, and a review of the literature. *J. Cutan. Pathol.* **2010**, *37*, 997–1001. [CrossRef]
8. Sharma, G.K.; Eschbacher, J.M.; Uschold, T.D.; Theodore, N. Neuroblastoma-like schwannoma of lumbar spinal nerve root. *J. Neurosurg. Spine* **2010**, *13*, 82–86. [CrossRef]
9. Sedassari, B.T.; Lascane, N.A.D.S.; Gallottini, M.H.C.; de Sousa, S.C.O.M.; Júnior, D.D.S.P. Neuroblastoma-like schwannoma of the lower labial mucosa: A rare morphologic variant of peripheral nerve sheath tumor. *Oral. Surg. Oral. Med. Oral. Pathol. Oral. Radiol.* **2014**, *118*, 579–582. [CrossRef] [PubMed]
10. Kukreja, M.; Gupta, R.; Julka, A.; Sharma, M.C. Neuroblastoma-like schwannoma as a rare cause of proptosis: A case report. *Can. J. Ophthalmol.* **2007**, *42*, 624–625. [CrossRef]
11. Adams, K.; Liu, X.S.; Akhtar, I.; Flowers, R.; Baliga, M. Pleural-based neuroblastoma-like schwannoma: A case report with cytologic findings and review of literature. *Diagn. Cytopathol.* **2015**, *43*, 650–653. [CrossRef]
12. Bhatnagar, S.; Banerjee, S.S.; Mene, A.R.; Prescott, R.J.; Eyden, B.P. Schwannoma with features mimicking neuroblastoma: Report of two cases with immunohistochemical and ultrastructural findings. *J. Clin. Pathol.* **1998**, *51*, 842–845. [CrossRef]
13. Lewis, Z.T.; Geisinger, K.R.; Pichardo, R.; Sangueza, O.P. Schwannoma with neuroblastoma-like rosettes: An unusual morphologic variant. *Am. J. Dermatopathol.* **2005**, *27*, 243–246. [CrossRef]
14. Michal, M.; Fanburg–Smith, J.C.; Mentzel, T.; Kutzner, H.; Requena, L.; Zamecnik, M.; Miettinen, M. Dendritic cell neurofibroma with pseudorosettes: A report of 18 cases of a distinct and hitherto unrecognized neurofibroma variant. *Am. J. Surg. Pathol.* **2001**, *25*, 587–594. [CrossRef]
15. Lerman, M.A.; Li, C.-C.; Woo, S.-B. Dendritic cell neurofibroma with pseudorosettes: A clinicopathologic and immunohistochemical study of 5 intraoral cases. *Oral Surg. Oral Med. Oral Pathol. Oral Radiol.* **2014**, *117*, 221–226. [CrossRef]
16. Van Nguyen, A.; Argenyi, Z.B. Cutaneous neuroblastoma. Peripheral neuroblastoma. *Am. J. Dermatopathol.* **1993**, *15*, 7–14. [CrossRef] [PubMed]

Case Report

A Rare Configuration origin of the Superior Thyroid, Lingual and Facial Arteries in a Pentafurcated Common Carotid Artery

Fabrizio Zaccheo [1,*], Francesco Mariotti [2], Alessandro Guttadauro [2], Alfredo Passaretti [2,*],
Maria Elisabetta Campogrande [1], Giulia Petroni [2] and Andrea Cicconetti [2]

[1] Independent Researcher, 00199 Rome, Italy
[2] Department of Oral and Maxillofacial Sciences, Sapienza University of Rome, 00161 Rome, Italy
* Correspondence: fabrizio.zaccheo85@gmail.com (F.Z.); alfredo.passaretti@uniroma1.it (A.P.)

Abstract: Detailed knowledge about the normal anatomy and its variations is a necessity in good clinical and surgical practice. The case we are reporting here shows a rare configuration origin of the superior thyroid, lingual and facial arteries arising directly from a pentafurcated common carotid artery. A fresh frozen latex-infiltrated cadaver was dissected. The superior thyroid artery, the lingual artery and the facial artery were found to be directly detached from the common carotid artery as terminations of the five terminal rami variations. The current classifications of anatomical variants are discussed in detail in relation to this case.

Keywords: common carotid artery; carotid bifurcation; superior thyroid artery; lingual artery; facial artery; external carotid artery; anatomical variation

Citation: Zaccheo, F.; Mariotti, F.; Guttadauro, A.; Passaretti, A.; Campogrande, M.E.; Petroni, G.; Cicconetti, A. A Rare Configuration origin of the Superior Thyroid, Lingual and Facial Arteries in a Pentafurcated Common Carotid Artery. *Anatomia* **2022**, *1*, 204–209. https://doi.org/10.3390/anatomia1020020

Academic Editors: Gianfranco Natale and Francesco Fornai

Received: 13 October 2022
Accepted: 24 November 2022
Published: 1 December 2022

Publisher's Note: MDPI stays neutral with regard to jurisdictional claims in published maps and institutional affiliations.

1. Introduction

Surgical procedures and clinical diagnoses require a deep understanding of the common and less common anatomical variations of structures located in the surgical field, particularly in relation to blood vessels. The concept of "normal anatomy" is based on repeated observations [1]. Detailed knowledge is now more important than ever because of present and future improvements in surgical procedures. Dissection—although not a surgical act—is essential and preparatory for an oral and maxillofacial surgeon, whose aim is to deepen the knowledge of anatomical organization beyond the limits of the operating field, thus guaranteeing greater safety and a less traumatic surgery. Thus, in situations where normal anatomical variation could be confusing, a deep understanding of vessel anatomy is mandatory. During radical neck surgery, the carotid bifurcation (CB) is a significant anatomical landmark, which helps to define the plane of dissection [2], and neck surgery must be planned according to the anatomical pattern of the CB in order to minimize complications [3].

2. Case Report

A rare configuration of the origin of the carotid artery anterior branch, which cannot be included in current classifications, was found on the left side of the neck of a 72-year-old man during a routine didactical neck dissection. The educational focus was on vessels supplying the oral cavity for students at the Sapienza University of Rome, and the dissection was performed at the ICLO Teaching and Research Center of Verona, Italy.

The cadaver specimen was treated using the "fresh frozen" technique ($-20\,°C$). No formalin fixation solution was used in order to preserve the tissue features for the in vivo presentation. A vascular injection fluid (latex) was used to achieve vascular patterns.

In this rare case, the superior thyroid artery (STA), the lingual artery (LA) and the facial artery (FA) were found to be detached directly from the CCA at the terminal ramus just before the carotid bifurcation (CB). Thus, we obtained a pentafurcation of the common carotid artery with figures resembling a hand (see Figure 1). This pattern has never been previously reported in the literature as far as we know. The contralateral carotid artery configuration of the analyzed cadaver specimen is a Natsis type I.

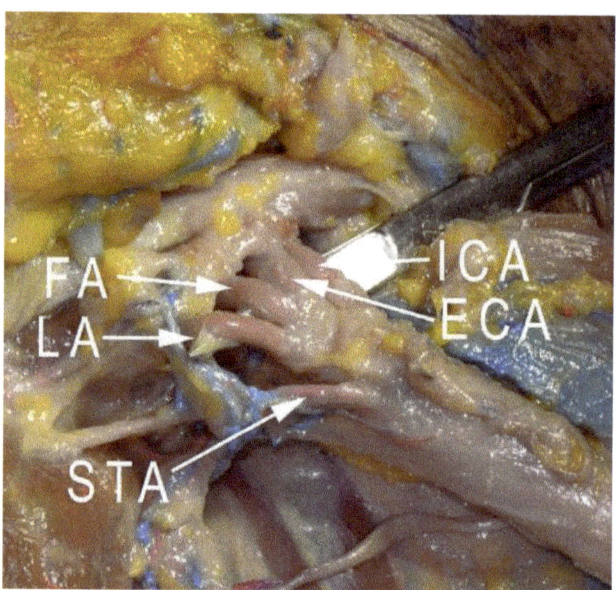

Figure 1. Pentafurcation variant of the common carotid artery: superior thyroid artery (STA); lingual artery (LA); facial artery (FA); common carotid artery (CCA); external carotid artery (ECA); internal carotid artery (ICA); and carotid bifurcation (CB).

Figures 1–4 show the abnormal left CCA and the course of its ventral branches. A detailed description of the cadaver's anatomy follows this paragraph.

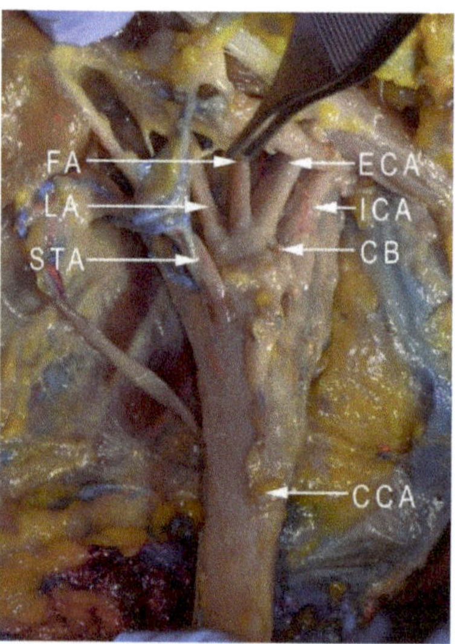

Figure 2. Dissection of the neck: superior thyroid artery (STA); lingual artery (LA); facial artery (FA); common carotid artery (CCA); external carotid artery (ECA); internal carotid artery (ICA); and carotid bifurcation (CB).

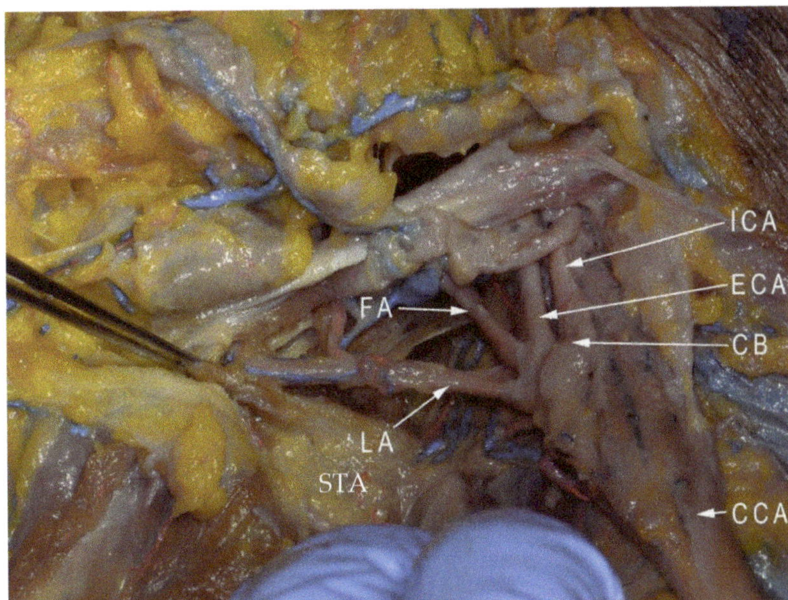

Figure 3. Dissection of the neck: lingual artery (LA); facial artery (FA); common carotid artery (CCA); external carotid artery (ECA); internal carotid artery (ICA); and carotid bifurcation (CB).

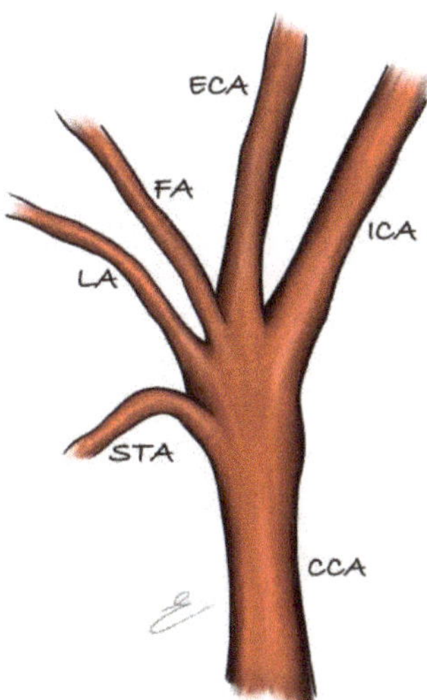

Figure 4. Explanatory illustration of the observed anatomical variant.

The left common carotid artery (CCA) is 9.1 mm in width. It arises from the aortic arch and continues upwards.

The superior thyroid artery (STA) is 1.8 mm in width and arises from the common carotid artery (CCA). The distance between the STA and the origin of the left CCA is 118 mm and 28 mm below the CB. The origin of the STA is located at the height of the C3 vertebral body, 3 mm above the posterior margin of the greater horn of the hyoid bone.

The lingual artery (LA) is 2.1 mm in width and represents the second branch medially arising from the CCA. It is located between the STA and the facial artery (FA). The distance between the STA and LA is 10 mm.

The FA and external carotid artery (ECA) are located above the LA. The lingual artery is anteriorly and superiorly directed, continuing as a frontal course until reaching the posterior margin of the hyoglossus muscle above the hyoid bone. The LA on the right side presents the same level of origin.

The facial artery (FA) is 2.9 mm in width and medially detaches from the CCA at a level of about 2.7 mm upon the lingual artery. Some differences between the right and left side of the facial artery are noted, especially in the distribution. The origin of the left FA is at the same level of the right side, but the distribution of the peripheral branches is different: the left FA crosses the submandibular gland and continues along the internal wall of the mandible, dividing into two main branches. The absence of the left superior labial artery and left angular artery is offset by the contralateral rami of the FA.

The external carotid artery (ECA) can be defined as a common trunk of the posterior auricular, maxillary and superficial temporal arteries. ECA originates at the height of C2 and starts from the glomus caroticum, above the origin of FA. It presents a width of 3.4 mm. From the CB, it runs parallel to the ICA. The posterior auricular artery is 0.9 mm in width and continues superiorly along the back of the head. The maxillary artery and the superficial temporal artery originate 47 mm above the ECA origin from the CCA and present a width of 2.6 mm and 2.4 mm, respectively. We did not find anomalies of the posterior auricular, superficial temporal and internal maxillary arteries.

The internal carotid artery (ICA) shows a width of about 4.1 mm. Its course can be divided into two tracts: cervical and intracranial. The cervical tract does not present collateral branches and continues upward, posteriorly to the ECA. The styloglossus muscle and stylopharyngeus muscle separate the ICA from the ECA. Once the cranial cavity is reached, the intracranial tract starts. No abnormalities were observed on the ICA entrance in the cranial cavity either.

3. Discussion

As a very rare occurrence, the carotid artery can ascend in the neck without dividing into the two usual branches, and either the external or internal carotid artery can be absent [4]. Julius A Ogeng'o [5] described some cases where, instead of bifurcating, the common carotid presents with the variant terminations of trifurcation (31.7%), quadrifurcation (5.4%) and pentafurcation (1.4%). The common carotid artery may provide one or more of the branches, usually derived from the external carotid. In this rare case, the pentafurcation of the CCA comprises the STA, LA, FA, ECA and ICA. This pattern has never been described before.

According to the classifications proposed by Natsis et al. and Ozgur et al., the anterior branches of the external carotid artery (ECA) are classified into four types [6,7]:

I   The STA, the LA and the FA originate from separate branches;
II   The STA and the LA share a common thyrolingual trunk;
III   The LA and the FA share a common linguofacial trunk;
IV   The STA, the LA and the FA originate from a common thyrolinguofacial trunk.

Moreover. Vazquez et al. divided the classification into three subtypes based on the STA origin [8]:

1. From the carotid bifurcation (50.2%);
2. From the common carotid artery (26.6%);
3. From the external carotid artery (23.2%).

Once the CCA gives rise to the CB at the level of the thyroid cartilage, the CCA divides into the ECA and the ICA. Following the CB, the ECA gives rise to its anterior branches: the STA, the LA and the FA [7]. The anterior branches of the ECA have already been described by several authors (STA, LA and FA) [7,9–15]. Natsis 2011 and Zumre

2005 have also described divergences from the ECA in terms of linguofacial trunk (LFT), thyrolingual trunk (TLT) and thyrolinguofacial trunk (TLFT). A variety of origins for these anterior branches have been reported in the literature, including their occurrence either from other carotid arteries or from carotid arterial trunks, but never for the carotid bifurcation before [16–19]. Kaneko has previously described a case report where the ECA branches arose directly from the CCA, but the arrangement differed from the one presented here; there was no external carotid artery, and all the external carotid artery branches were derived from the internal carotid artery instead [4].

4. Conclusions

This case report documents a very rare variant termination of the CCA and an unusual arrangement configuration of the anterior branches of the ECA, which cannot be recognized in the existing classifications.

Due to the wide range of surgical and radiological procedures performed in this area, the distribution of the ECA, the ICA and their branches are clinically relevant.

Possible anatomical variations are documented and classified by various authors and are crucial not only in routine planned surgeries, but in emergencies too. Furthermore, the vascular pattern is important for the evolution and prognosis of oncologic diseases.

Author Contributions: Dissections and writing—original draft preparation, F.Z., F.M., A.G., A.P., G.P. and M.E.C.; supervision, A.C.; original drawing, M.E.C. All authors have read and agreed to the published version of the manuscript.

Funding: This research received no external funding.

Institutional Review Board Statement: This study was conducted in accordance with the Declaration of Helsinki. No additional approval of an ethics board was required for this study.

Informed Consent Statement: All human cadaveric specimens were derived from donors who gave their written consent for the use, postmortem, of their body for educational and scientific purpose.

Acknowledgments: The authors sincerely thank those who donated their bodies to science so that anatomical research could be performed. The results from such research can potentially increase humankind's overall knowledge, which can then improve patient care. Therefore, these donors and their families deserve our highest gratitude.

Conflicts of Interest: The authors declare no conflict of interest.

Abbreviations

CB	carotid bifurcation
STA	superior thyroid artery
CCA	common carotid artery
LA	lingual artery
FA	facial artery
ECA	external carotid artery
ICA	internal carotid artery
Arteries' dimensions (Width in mm)	
9.1	left common carotid artery
1.8	superior thyroid artery
2.1	lingual artery
2.9	facial artery
3.4	external carotid artery
0.9	posterior auricular artery
2.6	maxillary artery
2.4	superficial temporal artery
4.1	internal carotid artery

References

1. Żytkowski, A.; Tubbs, R.S.; Iwanaga, J.; Clarke, E.; Polguj, M.; Wysiadecki, G. Anatomical normality and variability: Historical perspective and methodological considerations. *Transl. Res. Anat.* **2020**, *23*, 100105. [CrossRef]
2. Klosek, S.K.; Rungruang, T. Topography of carotid bifurcation: Considerations for neck examination. *Surg. Radiol. Anat.* **2008**, *30*, 383–387. [CrossRef] [PubMed]
3. Michalinos, A.; Chatzimarkos, M.; Arkadopoulos, N.; Safioleas, M.; Troupis, T. Anatomical Considerations on Surgical Anatomy of the Carotid Bifurcation. *Anat. Res. Int.* **2016**, *2016*, 1–8. [CrossRef] [PubMed]
4. Kaneko, K.; Akita, M.; Murata, E.; Imai, M.; Sowa, K. Unilateral anomalous left common carotid artery; a case report. *Ann. Anat. Anat. Anz.* **1996**, *178*, 477–480. [CrossRef] [PubMed]
5. Ogeng'o, J.A.; Misiani, M.; Malek, A.; Inyimili, M.; Murunga, A.; Ongeti, K. Variant termination of the common carotid artery: Cases of quadrifurcation and pentafurcation. *Anat. J. Afr.* **2014**, *3*, 386–392.
6. Natsis, K.; Raikos, A.; Foundos, I.; Noussios, G.; Lazaridis, N.; Njau, S.N. Superior thyroid artery origin in Caucasian Greeks: A new classification proposal and review of the literature. *Clin. Anat.* **2011**, *24*, 699–705. [CrossRef] [PubMed]
7. Ozgur, Z.; Govsa, F.; Ozgur, T. Assessment of Origin Characteristics of the Front Branches of the External Carotid Artery. *J. Craniofacial Surg.* **2008**, *19*, 1159–1166. [CrossRef] [PubMed]
8. Vazquez, T.; Cobiella, R.; Maranillo, E.; Valderrama, F.J.; McHanwell, S.; Parkin, I.; Sañudo, J.R. Anatomical variations of the superior thyroid and superior laryngeal arteries. *Head Neck* **2009**, *31*, 1078–1085. [CrossRef] [PubMed]
9. Lo, A.; Oehley, M.; Bartlett, A.; Adams, D.; Blyth, P.; Al-Ali, S. Anatomical variations of the common carotid artery bifurcation. *ANZ J. Surg.* **2006**, *76*, 970–972. [CrossRef] [PubMed]
10. Zümre, Ö.; Salbacak, A.; Çiçekcibaşi, A.E.; Tuncer, I.; Seker, M. Investigation of the bifurcation level of the common carotid artery and variations of the branches of the external carotid artery in human fetuses. *Ann. Anat. Anat. Anz.* **2005**, *187*, 361–369. [CrossRef]
11. Hayashi, N.; Hori, E.; Ohtani, Y.; Ohtani, O.; Kuwayama, N.; Endo, S. Surgical anatomy of the cervical carotid artery for carotid endarterectomy. *Neurol. Med.-Chir.* **2005**, *45*, 25–30. [CrossRef] [PubMed]
12. Fazan, V.P.S.; da Silva, J.H.N.; Borges, C.T.; Ribeiro, R.A.; Caetano, A.G. An anatomical study on the lingual–facial trunk. *Surg. Radiol. Anat.* **2009**, *31*, 267–270. [CrossRef] [PubMed]
13. Thwin, S.S.; Soe, M.M.; Myint, M.; Than, M.; Lwin, S. Variations of the origin and branches of the external carotid artery in a human cadaver. *Singap. Med. J.* **2010**, *51*, e40–e42.
14. Troupis, T.; Michalinos, A.; Dimovelis, I.; Demesticha, T.; Vlasis, K.; Skandalakis, P. Bilateral Abnormal Origin of the Anterior Branches of the External Carotid Artery. *Ann. Vasc. Surg.* **2013**, *28*, 494.e5-7. [CrossRef] [PubMed]
15. Murlimanju, B.V.; Prabhu, L.V.; Pai, M.M.; Jayaprakash, D.; Saralaya, V.V. Variant Origins of Arteries in the Carotid Triangle. *Chang Gung Med. J.* **2012**, *35*, 281–284. [PubMed]
16. Poisel, S.; Golth, D. Variability of the large arteries in the carotid triangle. *Wien. Med. Wochenschr.* **1974**, *15*, 229–232.
17. Czerwiński, F. Variability of the course of external carotid artery and its rami in man in the light of anatomical and radiological studies. *Folia Morphol.* **1981**, *40*, 449–453.
18. Lučev, N.; Bobinac, D.; Maric, I.; Dreščik, I. Variations of the great arteries in the carotid triangle. *Otolaryngol. Neck Surg.* **2000**, *122*, 590–591. [CrossRef]
19. Demirtaş, I.; Ayyıldız, B.; Demirbaş, A.T.; Ayyıldız, S.; Topcu, F.S.; Kuş, K.C.; Kurt, M.A. Geometric morphometric study of anterior branches of external carotid artery and carotid bifurcation by 3D-CT angiography. *Surg. Radiol. Anat.* **2022**, *44*, 1029–1036. [CrossRef] [PubMed]

Technical Note

Orofacial Anatomy Discrepancies and Human Identification—An Education Forensic Approach

Ana Corte-Real

Faculty of Medicine, University of Coimbra, 3000-548 Coimbra, Portugal; atgoncalves@fmed.uc.pt

Abstract: The objectives of this study were to identify and correlate orofacial anatomy discrepancies as biometric data and the impact of its rehabilitation as an educational, forensic approach to human identification. An observational and retrospective cohort study was performed on cephalography and photographic exams of a clinical database. The clinical reports were randomly selected according to inclusion and exclusion criteria, reviewed, and interpreted according to the anatomical features. The anatomic features analyzed for human identification compare pre- and post-medical intervention data. The challenging anatomical area was the lower one-third of the face. In an extreme forensic condition, identification should consider the medical history for a positive and accurate conclusion.

Keywords: forensic anatomy; gross anatomy education; oral anatomy; human identification

1. Introduction

Orofacial anatomy is a pillar of knowledge for medical and dental education [1,2]. It provides the framework for future health professionals to develop their clinical skills and performance. Additionally, it supports human identification issues in the forensic context [3].

The orofacial region has a particular property of resilience and resistance to extreme forensic conditions [1]. Orofacial anatomy, namely dental anatomy, is recording along individual life by health professionals into the clinical record. The goal of orofacial practice requires the understanding of the two-dimensional (2D) and three-dimensional (3D) morpho-anatomy of the stomatognathic system [1–5]. Normal anatomy allows for the comparative assumption to define the abnormal condition [1,3]. Orofacial discrepancies and their heterogeneous anatomical traits include different and distinct orofacial and dental characteristics. For example, orofacial cleft (OFC) is the leading congenital disability in the craniofacial region [6]. The OCF phenotypes are broadly divided into three types: isolated cleft lip, isolated cleft palate, and simultaneous cleft lip and palate [7]. OFC is associated with the supernumerary maxillary primary lateral incisor or agenesis, which can also occur in permanent dentition [6]. This pathology prevalence differs by geographical region, ranging from 2.9 to 23.9 per 10,000 [8], concerning its discriminating power in anatomical identification [9,10]. The discriminatory power for each piece of evidence is referring the number of possible candidates (identities) that can be excluded on that basis [10]. Anormal features correlated to a lower frequency have significantly more discriminatory power than normal or usual features [9–11]. This assumption in transversal is used in human identification for forensic purposes under genotyping methodology [9,11]. Orofacial surgery aims to correct facial skeletal discrepancies. Dental rehabilitation's goal is to correct dental disharmony and pathological functions. These rehabilitations are based on objective evaluation and complementary diagnostic exams, such as photography and radiological exams. These exams are a database of individual anatomical information for prospective analysis.

The identification procedure in a forensic scenario includes comparing and analyzing data [1]. The first line of the procedure's orofacial analysis consists of details of tissues, soft and hard, and clinical features. The facial soft tissues analysis corresponds to the definition

Citation: Corte-Real, A. Orofacial Anatomy Discrepancies and Human Identification—An Education Forensic Approach. *Anatomia* **2022**, *1*, 170–176. https://doi.org/10.3390/anatomia1020017

Academic Editors: Gianfranco Natale and Francesco Fornai

Received: 22 September 2022
Accepted: 31 October 2022
Published: 3 November 2022

Publisher's Note: MDPI stays neutral with regard to jurisdictional claims in published maps and institutional affiliations.

of shape, proportionality, and symmetry. The facial hard tissues analysis corresponds to the record of morphological and anatomical features. These analyses should be performed concerning standard guidelines and anatomical references [1]. Positive identification is the goal of the global process, with an impact on legal issues, such as death certificates, insurance proceeds, estate transfers, settlement of probate, execution of wills, remarriage, and child custody [1,2,10,11].

Anatomy teaching is associated with studying human identification within the scope of the missing and unidentified person. The study of human identification has its first line of action: dactyloscopy, visual or visual, anthropological, and dental methods. They are immediate and more economical than genotyping. They integrate the general medical examiner's knowledge of human anatomy and mass disaster team skills. Both areas are part of the Professional Qualifications Directive's requirement and the Bologna Declaration's guidelines [12–15].

This manuscript intends to correlate anatomical changes following orofacial iatrogenic interventions and their impact on forensic human identification within an educational approach.

2. Materials and Methods

An observational and retrospective cohort study was performed on cephalography and photographic exams of patient records of the Clinical Academic Center of Coimbra (CACC), in line with the Strengthening the Reporting of Observational Studies in Epidemiology (STROBE) methodology. The CACC is a consortium between the University and Hospital Center of Coimbra and the Faculty of Medicine at the University of Coimbra (FMUC), Portugal.

The inclusion criteria were as follows: clinical record in national health service; Class III Angle diagnosis as the record of orofacial, dental, and/or facial anatomy discrepancies; orthognathic surgery the scope of orofacial rehabilitation; events between January 2017 and May 2019; and an adult population. Medical reports with systemic diseases and aesthetic surgery goals were excluded.

Informed consent was provided in compliance with the Faculty of Medicine Ethics Committee (process number 23-CE-2017).

A team dentist performed the victim's exam with a forensic odontologist from the Laboratory of Forensic Dentistry at the University of Coimbra, Portugal. The research team had up to 15 years of experience in medico-legal expertise and anatomy education. The sample selected in the CACC database included patients (n = 30) aged between 18 and 26 years.

The horizontal subdivision and proportionality of the face were performed in photography and cephalometry. Following Arnett's study [16], there are several landmarks in face analysis procedure: (1) upper one-third of the face, between triquio (Tri) and glabella (Gl) soft tissue points; (2) medium one-third of the face, between Gl and sub nasale (SN) soft tissue points; and (3) lower one-third of the face, between Sn and menton (Me) soft tissue points. The horizontal proportionality is achieved when the three portions have the same size. The vertical subdivision of the face was performed regarding the DMV Atlas [17] regarding the medium sagittal line, allowing for analyzing the left and right proportion. For example, the breadth of the nasal bridge in the middle portion of the nasal bridge in frontal view: narrow, average, or broad; or the form of the chin contour or its shape: round, square, or pointed [17].

Cone-Beam Computed Tomography (CBCT) guidelines are well defined following clinical procedure [2]. Photographic standards are well defined and discussed in several medical fields, for example, plastic surgery and orthodontics. Digital photography was performed following institutional guidelines and widespread use in clinical documentation. The patient's head should be aligned to the Frankfort horizontal line for the frontal view. The focal point and center of the picture is the intersection between the Frankfort horizontal line and the midline of the face. The patient should look straight, and the interpupillary line should be horizontal. No rotation in the vertical axis should occur. Lips should be relaxed with a visible inter-labial gap if it exists. By enlargement, the following close-up views can be achieved of the full-face front view: front view of eyes, front view of lower face and jaw, front view of the nose, and front view of auricles.

The reports were randomly selected according to previous inclusion, and exclusion criteria were reviewed and interpreted according to the anatomical features in the thirds of the face compared to pre- and post-surgery intervention in the oral rehabilitation plan for identification purposes. Descriptive analysis was performed, and intra- and inter-observer errors were analyzed by the Technical Measurement Error test.

3. Results

The anatomical analysis was performed in two periods: before and after orthognathic surgery regarding the Class III Angle previous diagnostic of 30 patients in homogeneous sex distribution. Intra- and inter-observer errors were nonsignificant.

Surgical interventions caused differences in the individual's facial features after the intervention. Orthognathic surgery was the cause for these differences detected on the lower one-third of the face. The anatomical measure (SN and Me), or the height of the lower one-third height, is highlighted as a change measurement (67%) (Figure 1 and Table 1), detected on frontal and lateral views on photographic and cephalometric analysis. The anteroposterior analysis (between the craniometric and cutaneous points) performed on cephalometric exams was also highlighted as a changed measurement (Table 1).

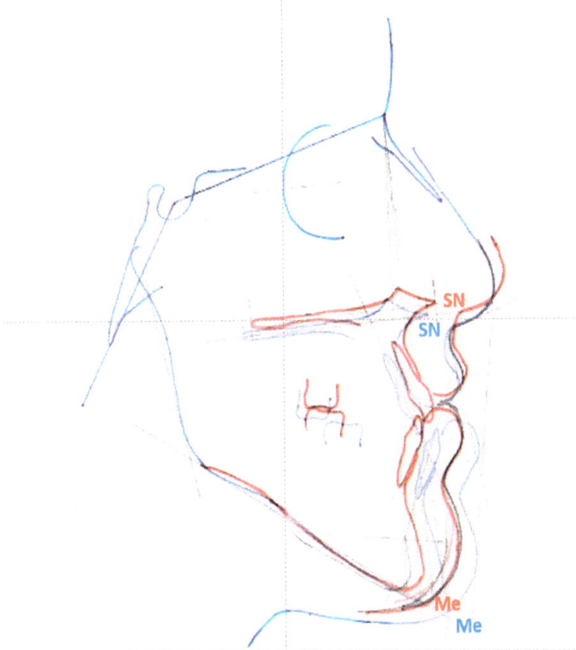

Figure 1. Cephalometric analysis highlighting cutaneous point sub nasale (SN) and menton (Me), before (red) and after (blue) orthognathic surgery.

Table 1. Horizontal and vertical data analysis between pre and post-surgery (n = 30).

N	%	Face *	Horizontal Proportionality				Vertical Proportionality
			Upper	Medium	Lower		Pre-Post Outcome
					Height	Antero-Posterior	
					Pre-Post Value (mm)	Pre-Post Value (mm)	
20	67	Frontal	*	*	+1 ± 0.4	/	Achieved
		Lateral				+1.5 ± 0.4	
4	13	Frontal			*		*
		Lateral					
6	20	Frontal			*		Achieved
		Lateral					

Footer: (*) no change in measurement; (/) no information added in this view.

Regarding the differences with the professional intervention, a positive value indicates the decrease between the two records for the same individual, and a negative value indicates the opposite (Table 1). A positive result means that the starting value is greater than the ending value. Differences were highlighted in horizontal proportionally, in height, and in anteroposterior analysis.

Left and right asymmetry, or vertical proportionality analysis, was performed in the frontal analysis of the face, and a symmetry in 87% of the sample was achieved under the initial diagnosis of facial disharmony. No changes were detected in 13% of the sample.

4. Discussion

Professional interventions that interfere with hard and soft tissue orofacial anatomical features have potential outcomes in biometric data and should be discussed regarding their impact in the forensic scope, such as identity. Asserting one's identity is a universal right.

Orthognathic surgery introduces changes in the anatomy of the face and, consequently, in the individual's visual aspect and identity profile. Our findings highlight anatomical changes in the lower one-third of the face. The results relate interdental relations between lower and upper dental arches under maxilla and mandible surgery [18]. These relations within the stomatognathic system emphasize the bone and dental support of the soft tissues, impacting the anatomy of the lips and chin.

The challenge of human identification consists of undocumented person situations. Science must follow the moment's needs in the practical impact of its outcomes. This is a current and relevant topic in the context of natural disasters and armed conflicts and human migration flows [9]. This technical note intends to adjust the visual recognition methodology, namely orofacial data comparison, as it corresponds to the first line of action in human identification. International forensic experts use morphological or anatomic analysis to compare missing and unidentified data as an identification procedure [11]. Biometric data in visual and dental methodologies lead to a consistent evaluation of anatomical features [11] based on integrated and complementary analysis using photo and radiological exams.

Orofacial identification, according to Pretty and Sweet, 2001, is based on anatomical features like the maxillary sinus (size, shape, foreign bodies, and relationship to teeth), anterior nasal spine, mandibular canal, coronoid and condylar processes, and temporomandibular joint [1]. Atlas of these morphological details must be a standard procedure, and it must comprise part of professional education, because the selection of reference planes is the most relevant in anatomical evaluation. The evaluation of hard and soft tissues and their anatomy can be performed by bi- or three-dimensional analysis, namely CBCT technology [19]. Soft tissue landmarks, namely Sn (the point at which columella merges with the upper lip in the midsagittal plane) and Me (the lowest median landmark on the lower border of the chin) [19], can be identified in photography as well as in CBCT reconstruction. The digital technologies associated with photography and radiological software allow for a realistic metrics analysis (related with minimum distortion) of the horizontal and vertical, proportionally allowing for a positive coordinate evaluation of the sample.

In the present study, the results (Table 1) follow and reinforce the differences in facial profiles following the surgical intervention. The vertical proportionally was performed using a vertical plane for a qualitative analysis of facial asymmetry [17,19–21]. The photographic CBCT records highlight hard tissue landmarks to construct the reference plane following the Hwang study [19]. The advantage of using soft tissue landmarks under hard tissue landmarks is that the hard and soft tissue can be evaluated simultaneously, enabling us to assess the hard and soft tissue interrelationship more accurately and overcome the potential difference in thickness under specific population groups following anatomical regions, such as the lips [19]. The standardized anatomical planes for hard and soft tissue evaluation will simplify the analysis procedure. This general procedure in clinical practice further contributes to the broader use of 3D image analysis in everyday practice. Concerning the right and left bone anatomical discrepancies, it is not fully reflected in soft tissue because it is compensated or masked by muscle and/or skin tissue, which significantly impacts achieving symmetry after bone surgery. The present study emphasizes midline landmarks to overcome the discrepancy between right and left anatomical landmarks. Regarding a methodology for asymmetry evaluation, it highlights that the mandible is a mobile structure compared to the maxilla, connecting to the adjacent structures and adjusting to outcomes of surgical intervention following Severt and Proffit [20].

Anatomical discrepancies compared to regular anatomy can be an essential feature in identity. In the scope of human identification, an orofacial anatomy study (radiographic study of orofacial and dental anomalies) is a population-based feature. The findings specific to the population overcome unbiased comparison in forensic procedures—these clinical conditions spotlight clinical and complementary exams as tools for a forensic report and prospective identity procedure.

The purpose of health professional intervention is to correct abnormalities and functions, such as orthognathic surgery's purpose to overcome the anatomical discrepancies to achieve a stable or normal condition. The technological advances to improve the patient's esthetic and functional clinical condition can alter anatomical relationships and deviate from the predictable anatomy of the face, regarding the present results. The match between clinical data is a condition for overlapping the anatomical region in the comparison between pre- and post-surgery for identification purposes. Positive identification is achieved under the match between orofacial features, and negative identification is achieved under a no match. There is a medico-legal impact of a no match condition. The legal impact of the result is related with legal, civil, religious, and cultural reasons.

The victim's medical history should be considered in this procedure to overcome an incorrect result and its medico-legal impact. The clinical record assessment is relevant knowledge for the discriminating diagnosis and has real value for correct identification.

Standards for individual records in clinical practice should be in accordance with identification analysis for a forensic purpose [21]. Anatomical features discussed in photographs and complementary exams are essential for pre- and post-operative documentation [21]. These planning data illustrate the purpose of the surgical intervention for the patient and its consistent documentation, which is demanded in medico-legal cases. In addition, these records support anatomical education as well as a scientific study.

This analysis is included in the multimodal pedagogical process in anatomy education, performed using two tools, photography and imagology, and namely CBCT technology, that sequentially complement the student's skills [2]. The integration of CBCT technology corresponds to the anatomical interpretation of multiplanar medical images and multiplied 2D slices in 3D reconstructions or 3D models, spotlighting external and internal anatomical details and their relations, whose dynamics positively attracted the students' attention [2,5]. The Master's degree corresponds to integrated education, and anatomy knowledge should be related to legal medicine or forensic issues with an impact on hot topics, such as identification in human traffic and migration flow [9].

Plastic and reconstructive surgery is a particular intervention and topic related to the identity process that was not considered in this manuscript.

5. Conclusions

The analysis of anatomical points is a methodological procedure and must follow the technological advances in its detection and recording. For identification purposes, clinical records are the basis of expert analysis and must be based on medical records and standardized anatomical references. The soft tissue references used in identification concerning the hard tissue references must have as a presupposition a complementary reading for the accuracy of the process. The surgical procedures to which the individual is submitted during his life can alter the interpretation of the aforementioned anatomical references and interfere with the conclusions of the identification process. For a favorable and accurate judgment in an extreme forensic condition, the professional should know the individual clinical data with the intervention records. The present study focused on identifying the hot points based on anatomical references—that is, the soft tissues of the facial region, namely the Sn and Me—and highlights that the reduction of the lower third of the face (1.4 mm) can occur after orthognathic surgery when we are facing a Class III Angle diagnosis. The author conducted this study with the objective to analyze anatomical diagnosis and found that the lower third of the face is a challenging anatomical area for human identification after orthodontic surgery. It is related to the changes in the interrelationship between the maxilla and the mandible, with the relationship between the bone anatomy and dental arches.

Funding: This research received no external funding.

Institutional Review Board Statement: The study was conducted in accordance with the Declaration of Helsinki, and approved by the Institutional Review Board (or Ethics Committee) of the Faculty of Medicine of the University of Coimbra (protocol code 23-CE-2017 and date of approval).

Informed Consent Statement: Informed consent was provided in compliance with the Faculty of Medicine Ethics Committee (process number 23-CE-2017). Written informed consent has been obtained from the patients to publish this paper.

Acknowledgments: The author would like to thank the Laboratory of Forensic Dentistry. She also would like to thank all those who helped carry out the research, namely the surgeons, oral professionals, and the patients.

Conflicts of Interest: The author declares no conflict of interest.

References

1. Pretty, I.; Sweet, D. A look at forensic dentistry–Part 1: The role of teeth in the determination of human identity. *Br. Dent. J.* **2001**, *190*, 359–366. [CrossRef] [PubMed]
2. Corte-Real, A.; Nunes, T.; Caetano, C.; Almiro, P.A. Cone Beam Computed Tomography (CBCT) Technology and Learning Outcomes in Dental Anatomy Education: E-Learning Approach. *Anat. Sci. Educ.* **2021**, *14*, 711–720. [CrossRef] [PubMed]
3. Coelho, J.; Armelim Almiro, P.; Nunes, T.; Kato, R.; Garib, D.; Miguéis, A.; Corte-Real, A. Sex and age biological variation of the mandible in a Portuguese population- a forensic and medico-legal approaches with three-dimensional analysis. *Sci. Justice J. Forensic Sci. Soc.* **2021**, *61*, 704–713. [CrossRef] [PubMed]
4. Field, J.C.; Walmsley, A.D.; Paganelli, C.; McLoughlin, J.; Szep, S.; Kavadella, A.; Manzanares Cespedes, M.C.; Davies, J.R.; DeLap, E.; Levy, G.; et al. The graduating European dentist: Contemporaneous methods of teaching, learning and assessment in dental undergraduate education. *Eur. J. Dent. Educ.* **2017**, *21*, S28–S35. [CrossRef] [PubMed]
5. Elgreatly, A.; Mahrous, A. Enhancing student learning in dental anatomy by using virtual three-dimensional models. *J. Prosthodont.* **2020**, *29*, 269–271. [CrossRef] [PubMed]
6. Yow, M.; Hermann, N.V.; Wei, Y.; Karsten, A.; Kreiborg, S. Deep orofacial phenotyping of population-based infants with isolated cleft lip and isolated cleft palate. *Sci. Rep.* **2020**, *10*, 21666. [CrossRef] [PubMed]

7. Kernahan, D.A.; Stark, R.B. A new classification for cleft lip and cleft palate. *Plast. Reconstr. Surg. Transplant. Bull.* **1958**, *22*, 435–441. [CrossRef] [PubMed]

8. Grosen, D.; Chevrier, C.; Skytthe, A.; Bille, C.; Mølsted, K.; Sivertsen, Å.; Murray, J.C.; Christensen, K. A cohort study of recurrence patterns among more than 54000 relatives of oral cleft cases in Denmark: Support for the multifactorial threshold model of inheritance. *J. Med. Genet.* **2010**, *47*, 162–168. [CrossRef] [PubMed]

9. Presciuttini, S.; Ciampini, F.; Alù, M.; Cerri, N.; Dobosz, M.; Domenici, R.; Peloso, G.; Pelotti, S.; Piccinini, A.; Ponzano, E.; et al. Allele sharing in first-degree and unrelated pairs of individuals in the Ge F IAmpFlSTR Profiler Plus database. *Forensic Sci. Int.* **2003**, *131*, 85–89. [CrossRef]

10. *The Forensic Human Identification Process: An Integrated Approach*; Forensic Unit of the International Committee of the Red Cross (ICRC): Geneva, Switzerland, 2022.

11. Ritz-Timme, S.; Gabriel, P.; Obertovà, Z.; Boguslawski, M.; Mayer, F.; Drabik, A.; Poppa, P.; De Angelis, D.; Ciaffi, R.; Zanotti, B.; et al. A new atlas for the evaluation of facial features: Advantages, limits and applicability. *Int. J. Legal Med.* **2011**, *125*, 301–306. [CrossRef] [PubMed]

12. Gale, N.K.; Heath, G.; Cameron, E.; Rashid, S.; Redwood, S. Using the framework method for the analysis of qualitative data in multi-disciplinary health research. *BMC Med. Res. Methodol.* **2013**, *13*, 117. [CrossRef]

13. European Union Parliament. *Reconhecimento das Qualificações Profissionais e o Regulamento n.o 1024/2012*; Official Journal of the European Union: Brussels, Belgium, 2013. Available online: https://wwwcdn.dges.gov.pt/sites/default/files/celex_3201310055_pt_txt.pdf (accessed on 2 December 2020).

14. EHEA. European Higher Education Area. In *ECTS User's Guide 2015*, 1st ed.; Publications Office of the European Union: Luxembourg, 2015; 105p. Available online: https://ec.europa.eu/assets/eac/education/ects/users-guide/docs/ects-use (accessed on 2 December 2020).

15. Kravitz, A.S.; Bullock, A.; Cowpe, J.; Barnes, E. Council of European Dentists: Manual of Dental Practice. 2015, pp. 287–298. Available online: https://www.omd.pt/content/uploads/2017/12/ced-manual-2015-completo.pdf (accessed on 2 December 2020).

16. Arnett, G.W.; Jelic, J.S.; Kim, J.; Cummings, D.R.; Beress, A.; Worley, C.M., Jr.; Chung, B.; Bergman, R. Soft tissue cephalometric analysis: Diagnosis and treatment planning of dentofacial deformity. *Am. J. Orthod. Dentofac. Orthop.* **1999**, *116*, 239–253. [CrossRef]

17. Olivieri, L.; Mazzarelli, D.; Bertoglio, B.; De Angelis, D.; Previderè, C.; Grignani, P.; Cappella, A.; Presciuttini, S.; Bertuglia, C.; Di Simone, P.; et al. Challenges in the identification of dead migrants in the Mediterranean: The case study of the Lampedusa shipwreck of October 3rd 2013. *Forensic Sci. Int.* **2018**, *285*, 121–128. [CrossRef]

18. Katz, M.I. Angle classification revisited 2: A modified Angle classification. *Am. J. Orthod. Dentofac. Orthop.* **1992**, *102*, 277–284. [CrossRef]

19. Hwang, H.S.; Yuan, D.; Jeong, K.H.; Uhm, G.S.; Cho, J.H.; Yoon, S.J. Three-dimensional soft tissue analysis for the evaluation of facial asymmetry in normal occlusion individuals. *Korean J. Orthod.* **2012**, *42*, 56–63. [CrossRef] [PubMed]

20. Severt, T.R.; Proffit, W.R. The prevalence of facial asymmetry in the dentofacial deformities population at the University of North Carolina. *Int. J. Adult Orthod. Orthognath. Surg.* **1997**, *12*, 171–176.

21. Ettorre, G.; Weber, M.; Schaaf, H.; Lowry, J.C.; Mommaerts, M.Y.; Howaldt, H.P. Standards for digital photography in cranio-maxillo-facial surgery-Part I: Basic views and guidelines. *J. Cranio-Maxillofac. Surg.* **2006**, *34*, 65–73. [CrossRef] [PubMed]

Opinion

Introduction of a New Classification of the Cleidooccipital Muscle

Sandeep Silawal [1,*], Shikshya Pandey [2] and Gundula Schulze-Tanzil [1]

[1] Institute of Anatomy and Cell Biology, Paracelsus Medical University, Nuremberg, Prof. Ernst Nathan Str. 1, 90419 Nuremberg, Germany

[2] Department of Anesthesiology and Intensive Care Medicine, Kliniken des Landkreises Neumarkt i.d.OPf, Nürnberger Straße 12, 92318 Neumarkt in der Oberpfalz, Germany

* Correspondence: sandeep.silawal@pmu.ac.at

Abstract: The "cleidooccipital branches" are integral muscular branches of the sternocleidomastoid muscles (SCM), as well as the trapezius muscles, which construct the anterior and posterior borders of the posterior triangle of the neck, respectively. The term "cleidooccipital muscle", in the literature, generally describes the accessory muscle, which is proximally attached to the middle portion of the clavicle, separate from the clavicular attachment of the SCM or trapezius muscle, and crosses the posterior triangle of the neck obliquely. With proximity to either the trapezius or the SCM, the accessory cleidooccipital muscles can be divided into posterior and anterior accessory cleidooccipital muscles, respectively. At present, most of the descriptions in the literature associated with the accessory cleidooccipital muscles concern the posterior accessory cleidooccipital muscles. The anterior accessory cleidooccipital muscles are mostly recognized as a proximal clavicular-head-sided supernumerary variation of the SCM. We propose a new classification of these muscles, with nomenclatures to help researchers to differentiate the cleidooccipital branches or muscles from one another. Introducing this classification, we hope that more clarity can be achieved when addressing the so-called "cleidooccipital muscle" in the future.

Keywords: cleidooccipital; sternocleidomastoid; trapezius

Citation: Silawal, S.; Pandey, S.; Schulze-Tanzil, G. Introduction of a New Classification of the Cleidooccipital Muscle. *Anatomia* **2022**, *1*, 148–151. https://doi.org/10.3390/anatomia1020015

Academic Editors: Gianfranco Natale and Francesco Fornai

Received: 31 August 2022
Accepted: 27 September 2022
Published: 1 October 2022

Publisher's Note: MDPI stays neutral with regard to jurisdictional claims in published maps and institutional affiliations.

1. Introduction

During vertebrate neck development, the posterior triangle of the neck is formed after the division of a single muscle into posterior and anterior parts, leading to the development of the trapezius muscle and the sternocleidomastoid muscle (SCM), respectively [1–3]. In cases where the splitting of the muscles fails completely, there is an absence of the formation of the posterior triangle of the neck [4]. However, the separation process can also be incomplete in individuals where a separate supernumerary muscle remains in the posterior triangle of the neck. The term "cleidooccipital muscle", in the literature, generally describes this supernumerary muscle, present either unilaterally or bilaterally. This muscle is proximally attached to the middle portion of the clavicle, separate from the clavicular attachment of the SCM or trapezius muscle, and crosses the posterior triangle of the neck obliquely. It is also important to understand that "cleidooccipital branches", formed during the normal development of the neck, are integral parts of the SCM, as well as the trapezius muscles, which construct the anterior and posterior borders of the posterior triangle of the neck, respectively. As the cleidooccipital branch of the SCM is proximally attached to the sternal end of the clavicle, the branch of the trapezius muscle is inserted at the acromial end of the clavicle. The distal insertion of the cleidooccipital branches, as well as the accessory cleidooccipital muscles, is in the superior nuchal line between the mastoid process and the external occipital protuberance on the occipital bone. However, with proximity to either the trapezius or the SCM, the accessory cleidooccipital muscles can be divided into posterior and anterior accessory cleidooccipital muscles, respectively.

2. Relevant Section

Classification of the cleidooccipital muscles with their nomenclatures (Figure 1).

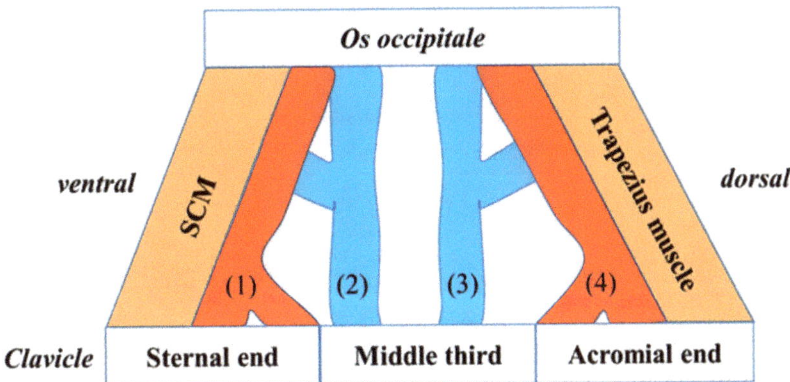

Figure 1. The schematic description of the posterior triangle of the neck with a new classification and nomenclature of the cleidooccipital muscular branches of the sternocleidomastoid muscle (SCM) or trapezius muscles, as well as the supernumerary cleidooccipital muscles. (1) Cleidooccipital branch of the SCM (*R. cleidooccipitalis musculi sternocleidomastoidii*). (2) Anterior accessory cleidooccipital muscle (*M. cleidooccipitalis accessorius anterior*). (3) Posterior accessory cleidooccipital muscle (*M. cleidooccipitalis accessorius posterior*). (4) Cleidooccipital branch of the trapezius muscle (*R. cleidooccipitalis musculi trapezii*).

(1) One of the four anatomical branches of the SCM connecting the sternal end of the clavicle to the occipital bone [5,6]. Cleidooccipital branch of the SCM (*R. cleidooccipitalis musculi sternocleidomastoidii*).

(2) An accessory muscular slip connecting the middle portion of the clavicle to the occipital bone with a close proximity or attachment to the SCM [7]. Anterior accessory cleidooccipital muscle (*M. cleidooccipitalis accessorius anterior*).

(3) An accessory muscular slip connecting the middle portion of the clavicle to the occipital bone with a close proximity or attachment to the trapezius muscle [8–11]. Posterior accessory cleidooccipital muscle (*M. cleidooccipitalis accessorius posterior*).

(4) One of the anatomical branches of the trapezius muscle connecting the acromial end of the clavicle to the occipital bone [12]. Cleidooccipital branch of the trapezius muscle (*R. cleidooccipitalis musculi trapezii*).

3. Discussion

At present, most of the descriptions in the literature associated with the accessory cleidooccipital muscles concern the posterior accessory cleidooccipital muscles [8–11]. Even though the anterior accessory cleidooccipital muscles are available for observation, these muscles are instead defined as clavicular-sided, proximal head variations of the SCM, as described in a literature review [13]. In one of the case reports, the author suggested that the additional head of the SCM could be named as a cleidooccipital muscle, as it was found to originate from the middle one third of clavicle, 1.2 cm lateral to the usual clavicular head [14]. We suggest that the same muscle and many other similar variants qualify to be nomenclated as the anterior accessory cleidooccipital muscles according to our proposed new classification. A swelling in the right posterior triangle of the neck was addressed in a case report of a 48-year old man with pain and sensory impairment on the right-hand side that radiated from the neck to the radial border of the forearm, thumb and index finger [15]. The symptoms were aggravated particularly during the movement of the head or while lifting heavy objects. The magnetic resonance imaging (MRI) showed a soft tissue in the posterior triangle of the neck, detached from the trapezius muscle midway and extending towards the middle part of the clavicle. This muscle was termed as an accessory part of the trapezius muscle in the posterior triangle of the neck, which was surgically excised, resulting in the total relief of the symptoms of the patient. Our proposed terminology for the same muscle would define it as the posterior accessory cleidooccipital muscle.

Table 1 offers the representation of such examples, where new nomenclatures could be implemented in selected case reports in terms of our proposed new classification.

Table 1. A tabular illustration with examples of newly proposed terminologies in comparison to the already existing ones in selected case reports. SCM = sternocleidomastoid muscle.

Literature	Applied Description in the Literature	Proposed Terminology (*English*)
Rahman et al. 1994 [11]	Anomalous cleidooccipitalis muscle	Posterior accessory cleidooccipital muscle
Hug et al. 2000 [15]	Accessory part of the trapezius muscle	Posterior accessory cleidooccipital muscle
Sarikcioglu et al. 2001 [7]	Cleidooccipital muscle	Cleidooccipital branch of the SCM
Kwak et al. 2003 [8]	Cleidooccipitalis cervicalis	Posterior accessory cleidooccipital muscle
Rao et al. 2007 [16]	Additional slip in the origin of the clavicular head of SCM	Anterior accessory cleidooccipital muscle
Cherian et al. 2008 [17]	Additional third head originated from the middle third of the clavicle	Anterior accessory cleidooccipital muscle
Natsis et al. 2009 [18]	Three additional clavicular heads, four in total C1–C4	C2–C3: Cleidooccipital branch of the SCM C4: Anterior accessory cleidooccipital muscle
Mehta et al. 2012 [19]	The clavicular head of the muscle exhibited two bellies, one medial and one lateral	Lateral belly (Anterior accessory cleidooccipital muscle)
Paraskevas et al. 2013 [9]	Accessory cleidooccipitalis muscle	Posterior accessory cleidooccipital muscle
Sabnis et al. 2013 [20]	Third head of the SCM	Anterior accessory cleidooccipital muscle
Kaur et al. 2017 [14]	Additional head of SCM +cleidooccipital muscle	Anterior accessory cleidooccipital muscle
Maslowski, D.J. et al. 2019 [10]	Cleidooccipitalis cervicalis muscle	Posterior accessory cleidooccipital muscle

4. Conclusions and Future Direction

At present, different nomenclatures are used by authors for the accessory cleidooccipital muscles or the cleidooccipital branches of the SCM or the trapezius muscles. Introducing this proposed classification into the anatomical literature, we hope that more clarity could be achieved when addressing the so called "cleidooccipital muscle" in the future. Proper descriptions of these muscle variations are relevant in clinical, surgical and radiological contexts, as such variations can be encountered in daily practice. The use of a common description could optimize the description of these muscles in general.

Author Contributions: Conceptualization, S.S. and G.S.-T.; investigation, S.S. and G.S.-T.; resources, G.S.-T.; data curation, S.S. and S.P.; writing—original draft preparation, S.S.; writing—review and editing, S.P. and G.S.-T.; visualization, S.S.; supervision, G.S.-T.; project administration, S.S. All authors have read and agreed to the published version of the manuscript.

Funding: This research received no external funding.

Institutional Review Board Statement: Not applicable.

Informed Consent Statement: Not applicable.

Data Availability Statement: Not applicable.

Conflicts of Interest: The authors declare no conflict of interest.

References

1. Kuratani, S. Evolutionary developmental studies of cyclostomes and the origin of the vertebrate neck. *Dev. Growth Differ.* **2008**, *50* (Suppl. S1), S189–S194. [CrossRef] [PubMed]
2. Mekonen, H.K.; Hikspoors, J.P.; Mommen, G.; Köhler, S.E.; Lamers, W.H. Development of the epaxial muscles in the human embryo. *Clin. Anat.* **2016**, *29*, 1031–1045. [CrossRef] [PubMed]
3. Cho, K.H.; Morimoto, I.; Yamamoto, M.; Hanada, S.; Murakami, G.; Íguez-Vázquez, J.F.R.; Abe, S. Fetal development of the human trapezius and sternocleidomastoid muscles. *Anat. Cell Biol.* **2020**, *53*, 405–410. [CrossRef] [PubMed]
4. Singh, S. Absence of Posterior Triangle: Clinical and Embryological Perspective. *J. Clin. Diagn. Res.* **2017**, *11*, AD01–AD02. [CrossRef] [PubMed]
5. Bordoni, B.; Varacallo, M. *Anatomy, Head and Neck, Sternocleidomastoid Muscle, in StatPearls*; StatPearls Publishing LLC.: Treasure Island, FL, USA, 2022.
6. Kennedy, E.; Albert, M.; Nicholson, H. The fascicular anatomy and peak force capabilities of the sternocleidomastoid muscle. *Surg. Radiol. Anat.* **2017**, *39*, 629–645. [CrossRef] [PubMed]
7. Sarikcioglu, L.; Donmez, B.O.; Ozkan, O. Cleidooccipital muscle: An anomalous muscle in the neck region. *Folia Morphol.* **2001**, *60*, 347–349.
8. Kwak, H.H.; Kim, H.J.; Youn, K.H.; Park, H.D.; Chung, I.H. An Anatomic Variation of the Trapezius Muscle in a Korean: The Cleido-occipitalis Cervicalis. *Yonsei Med. J.* **2003**, *44*, 1098–1100. [CrossRef] [PubMed]
9. Paraskevas, G.K.; Natsis, K.; Ioannidis, O. Accessory cleido-occipitalis muscle: Case report and review of the literature. *Rom. J. Morphol. Embryol.* **2013**, *54* (Suppl. S3), 893–895. [PubMed]
10. Maslowski, D.J.; Snyder, S.B.; Pellis, Z.; Petrone, A.B.; Zdilla, M.J.; Lambert, H.W. Clinical implications of the cleido-occipitalis cervicalis muscle: A muscular variant of the trapezius muscle. *FASEB J.* **2019**, *33*, 616.16. [CrossRef]
11. Rahman, H.; Yamadori, T. An Anomalous Cleido-Occipitalis Muscle. *Cells Tissues Organs* **1994**, *150*, 156–158. [CrossRef] [PubMed]
12. Johnson, G.; Bogduk, N.; Nowitzke, A.; House, D. Anatomy and actions of the trapezius muscle. *Clin. Biomech.* **1994**, *9*, 44–50. [CrossRef]
13. Silawal, S.; Schulze-Tanzil, G. The sternocleidomastoid muscle variations: A mini literature review. *Folia Morphol.* **2022**. [CrossRef] [PubMed]
14. Kaur, A.; Sharma, A.; Sharma, M. A Case of Unusual Unilateral Accessory Clavicular Head of Sternocleidomastoid Muscle. *North States J. Anat.* **2017**, *2*, 25–28.
15. Hug, U.; Burg, D.; Meyer, V.E. Cervical outlet syndrome due to an accessory part of the trapezius muscle in the posterior triangle of the neck. *J. Hand Surg.* **2000**, *25*, 311–313. [CrossRef] [PubMed]
16. Rao, T.R.; Vishnumaya, G.; Shetty, K.P.; Suresh, R.; Prakashchandra, S. Variation in the Origin of Sternocleidomastoid Muscle: A Case Report. *Int. J. Morphol.* **2007**, *25*, 621–623. [CrossRef]
17. Cherian, S.B.; Nayak, S. A Rare Case of Unilateral Third Head of Sternocleidomastoid Muscle. *Int. J. Morphol.* **2008**, *26*, 99–101. [CrossRef]
18. Natsis, K.; Asouchidou, I.; Vasileiou, M.; Papathanasiou, E.; Noussios, G.; Paraskevas, G. A rare case of bilateral supernumerary heads of sternocleidomastoid muscle and its clinical impact. *Folia Morphol.* **2009**, *68*, 52–54.
19. Mehta, V.; Arora, J.; Kumar, A.; Nayar, A.K.; Ioh, H.K.; Gupta, V.; Suri, R.K.; Rath, G. Bipartite clavicular attachment of the sternocleidomastoid muscle: A case report. *Anat. Cell Biol.* **2012**, *45*, 66–69. [CrossRef] [PubMed]
20. Sabnis, A.; Shaikh, S.; More, R. Third head of sternocleidomastoid muscle. *Natl. J. Clin. Anat.* **2013**, *2*, 218–220. [CrossRef]

Case Report

Anastomosis between Median and Musculocutaneous Nerve: Presentation of a Very Rare Anatomical Variation in Comparison to Classical Divisions

Rosario Barone [1,*,†], Agata Grazia D'Amico [2,†], Noemi Di Lorenzo [1], Grazia Laura Di Grado [1], Egle Matranga [1], Giulio Spinoso [1], Leonardo Luca Bavuso [1], Antonella Marino Gammazza [1], Francesca Rappa [1], Fabio Bucchieri [1], Francesco Cappello [1,3], Weronika Piotrowska [4], Jan Henryk Spodnik [4], Edyta Spodnik [4] and Sławomir Wójcik [4]

[1] Department of Biomedicine, Neurosciences and Advanced Diagnostic, University of Palermo, Via del Vespro n. 129, 90127 Palermo, Italy; noemidilorenzo96@gmail.com (N.D.L.); grazialauradigrado@gmail.com (G.L.D.G.); eglematranga95@gmail.com (E.M.); giulio.spinoso96@gmail.com (G.S.); leoluca94@gmail.com (L.L.B.); antonella.marinogammazza@unipa.it (A.M.G.); francesca.rappa@unipa.it (F.R.); fabio.bucchieri@unipa.it (F.B.); francesco.cappello@unipa.it (F.C.)
[2] Department of Drug Sciences, University of Catania, 95129 Catania, Italy; agata.damico@unict.it
[3] Euro-Mediterranean Institutes of Science and Technology (IEMEST), 90139 Palermo, Italy
[4] Department of Anatomy and Neurobiology, Medical University of Gdansk, 80-210 Gdansk, Poland; wspodnik@gumed.edu.pl (W.P.); jan.spodnik@gumed.edu.pl (J.H.S.); edyta.spodnik@gumed.edu.pl (E.S.); slawomir.wojcik@gumed.edu.pl (S.W.)
* Correspondence: rosario.barone@unipa.it; Tel.: +39-09123865823
† These authors contributed equally to this work.

Citation: Barone, R.; D'Amico, A.G.; Di Lorenzo, N.; Di Grado, G.L.; Matranga, E.; Spinoso, G.; Bavuso, L.L.; Marino Gammazza, A.; Rappa, F.; Bucchieri, F.; et al. Anastomosis between Median and Musculocutaneous Nerve: Presentation of a Very Rare Anatomical Variation in Comparison to Classical Divisions. *Anatomia* 2022, 1, 68–74. https://doi.org/10.3390/anatomia1010007

Academic Editor: Alessandro Castorina

Received: 12 May 2022
Accepted: 14 June 2022
Published: 19 June 2022

Publisher's Note: MDPI stays neutral with regard to jurisdictional claims in published maps and institutional affiliations.

Abstract: The musculocutaneous nerve (MCN) is the terminal branch of the lateral cord of the brachial plexus, and emerges at the inferior border of pectoralis minor muscle. The nerve can interact with the median nerve (MN), adhering to the nerve and sharing fibers with it. During anatomical dissection of twelve cadavers, we have detected a rare variation of the anastomosis between MCN and MN. The knowledge of this anatomical variation could be of great relevance during surgical and clinical practices.

Keywords: musculocutaneous nerve; median nerve; anatomical variation; dissections

1. Introduction

The musculocutaneous nerve (MCN) and the median nerve (MN) derive from the brachial plexus, that is usually formed by the union of the ventral rami of the four lower cervical spinal nerves and the first thoracic spinal nerve (C5-T1) [1]. These nerves are two of the four terminal nerves of the brachial plexus; communications between MCN and the MN are often observed due to failed nerve fiber segregation during embryological development [1].

In its most common anatomic form, the MCN arises from the lateral cord of the brachial plexus, pierces and innervates the coracobrachialis muscle (CBM), descending between the biceps brachii and the brachialis muscles. It then continues into the forearm as the lateral antebrachial cutaneous nerve [2]. The MCN runs laterally to the MN and innervates the anterior arm muscles and the skin covering the lateral surface of the forearm.

The lateral and medial roots of the MN originate from the lateral and medial cords of the brachial plexus, respectively. In the arm, the MN has no branches and is located medially to the coracobrachialis muscle, entering the forearm through the cubital fossa, passing posteriorly to the flexor retinaculum at the wrist, and through the carpal tunnel to reach the palmar aspect of the hand. The MN supplies most of the anterior forearm and thenar muscles, in addition its sensory fibers laterally innervating the palmar surface of the hand and the dorsal surface of digits I-III at the level of the distal phalanx [1,2]. In Figure 1, we have presented the classical localization of the above-mentioned nerves in the axillary region.

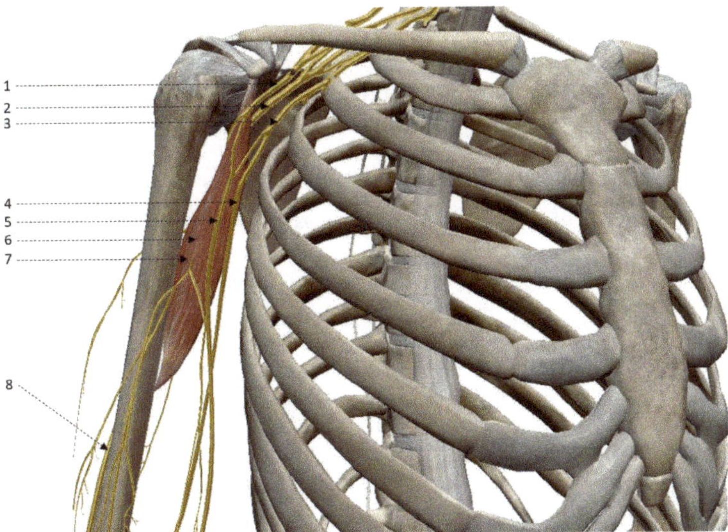

Figure 1. Anatomical relationship between the musculocutaneous nerve and median nerve in the axillary region. Right upper limb, front vision. 1, lateral cord of brachial plexus; 2, posterior cord of brachial plexus; 3, medial cord of brachial plexus; 4, ulnar nerve; 5, median nerve; 6, coracobrachialis muscle; 7, musculocutaneous nerve; 8, radial nerve.

Anatomical variations of the communication between the MCN and the MN are relatively common as a consequence of different factors during embryological development. Anastomoses between the MNC and the MN have been described in literature [1]; in particular, the most commonly used classification proposed by Le Minor divides the existing topographic relations between the MCN and the MN into five categories [3]. Other classifications have been suggested by Guerri-Guttenberg and Inglotti [4] and Venieratos and Anangnostopolou [5].

During the 2018 summer school of anatomical dissection at the laboratory of the Medical University of Gdansk, we had the opportunity to dissect twelve cadavers. Here, we report a case of a rare anatomical variation of an anastomosis between the MCN and MN, which does not correspond to the standard classifications.

2. Materials and Methods

Dissections were performed on 12 adult donor corpses (eight males and four females) of Caucasian race. In particular, 12 bilateral upper limbs (24 limbs) were studied. Age of cadavers at time of death varied between 56–87 years (mean—79.56 years, median—78 years). Cadavers were dissected at the laboratory of the Department of Anatomy and Neurobiology, Medical University of Gdansk, Poland, during a summer school of anatomical dissection. Causes of death were unknown. All cadavers did not present lesions and had no history upper limb surgery. All human cadaveric tissues used in the study were derived from donors who gave their written consent, premortem, to the use of their body for educational and scientific purposes. Gratitude to the donor cadavers and their families is expressed within the Acknowledgments section, in accordance with the formula proposed recently by 20 editors-in-chief of 17 anatomical journals [6]. The Institutional Ethics Committee gave approval for conducting this study (Institutional Review Board number: Ordinance

No. 26/2016 of the Rector of the Medical University of Gdańsk of 6 June 2016 on the implementation of the "Program of Conscious Donation of Corpses").

Deep dissection of upper limbs was performed in formalin-fixed adult human cadavers. The upper limbs were removed from the body, the arm positioned in a supine position, the skin incised and every fat layer was carefully removed not to damage vessels and nerves. Subsequently, the brachial fascia was separated from the underlying muscles and intermuscular septa, and three muscles of the anterior arm compartment (coracobrachialis, brachialis and biceps brachii) were separated. Finally, the musculocutaneous nerve and median nerve were identified and separated.

The following 3D image (Figure 1) depicting the region of interest was created with the "Human Anatomy Atlas" virtual anatomy software, supplied by the laboratory of the Human Anatomy section of the University of Palermo.

The morphological analysis of the MCN and MN consisted of the assessment of: (a) relationship of the MCN and CBM (observation if the MCN goes through the CBM or not); (b) lateral root of the MN (measurement of length and width); (c) medial root of the MN (measurement of length and width); (d) MN on the arm (measurement of length until the cubital fossa and width); (e) MCN (measurement of length and width of an upper part—till CBM, and lower part—till the point where muscular branches to biceps brachii m. emerge from the MCN). The measurements, length and width, have been taken with line ruler and Digital Digimatic Vernier Caliper (Mitutoyo, Japan), respectively. All measurements were carried out by the same investigator. Each measurement was taken twice, and accurate to the 1 mm (length) and 0.1 mm (width). As the final result, the average of both measurements was taken.

3. Results

In 23 out of the 24 cases studied, the MCN and the MN showed a typical morphology as per the classical anatomical description. The results obtained for individual parameters in the studied cases are summarized in Table 1. A representative case of the anatomical relationship between the musculocutaneous nerve and median nerve is shown in Figure 2. In all cases observed, except case 4 bilaterally, the MCN pierces the coracobrachialis muscle (Table 1). In one of the studied cases (case 9, see Table 1), the right upper limb of a male donor, we observed an anastomosis between the MCN and the MN (Figure 3). The MCN, after piercing the coracobrachialis muscle, continued its course descending laterally between the brachialis muscle and the biceps brachii muscle, innervating all three of them. 19 cm from its origin, at the lateral cord of the brachial plexus and 10 cm above the cubital fossa, the MCN anastomosed with the MN (Figure 3, point 7). At the anastomotic site, the two nerves were attached to each other, side-to-side, for a span of 1.7 cm, and some fibers were shared between the MCN and the MN. The roots of the MN originated from the lateral and medial cords of the brachial plexus, joined in a single trunk in the axilla and extended along the forearm. On the arm, just 10 cm above the cubital fossa, 15 cm from the union of the medial and lateral root of the MN, the anastomosis between the MN and the MCN was observed (Figure 3, point 7). This anastomosis between the MCN and the MN represents a new anatomical subvariant, since it does not fall neither within the classical Le Minor [3], nor within the Venieratos and Anangnostopolou [5] classifications. No vascular variations were observed in this case. Dissection using the same methods was also performed on the left upper limb of the same donor to study if the variation was present also in the other arm. Upon dissection of the left upper limb (axilla, arm, forearm), no anatomical variants were observed.

Table 1. Measurements of the observed structures.

Case	Gender [M/F]	Side [R/L]	MCN/ CBM [Y/N]	Lateral Root MN		Medial Root MN		MN		MCN Upper		MCN Lower	
				Length [mm]	Width [mm]	Length [mm]	Width [mm]	Length [mm]	Width [mm]	Length [mm]	Width [mm]	Length [mm]	Width [mm]
1	F	L	Y	42	6.1	39	6.7	254	5.4	38	2.2	90	2.4
1	F	R	Y	49	3.7	60	2.7	240	4.4	62	2.2	100	2.6
2	F	L	Y	115	4.6	96	2.4	186	4.3	110	3.2	120	1.9
2	F	R	Y	128	3.0	144	2.6	156	4.4	73	2.0	173	2.4
3	F	L	Y	20	6.9	40	7.0	277	4.3	41	3.5	42	3.9
3	F	R	Y	20	3.2	12	3.3	292	3.6	69	3.2	64	3.4
4	F	L	N	9	4.9	21	4.2	232	3.9	-	-	142	2.4
4	F	R	N	135	3.7	135	2.3	109	5.8	-	-	165	3.8
5	M	L	Y	18	4.2	17	5.3	251	4.9	46	3.1	49	3.7
5	M	R	Y	16	3.0	11	2.8	254	3.5	61	1.8	29	1.9
6	M	L	Y	61	3.3	45	4.0	265	3.6	65	2.9	128	2.1
6	M	R	Y	64	4.9	62	3.1	238	4.0	75	2.7	89	3.3
7	M	L	Y	12	4.9	1	2.8	297	5.3	32	2.5	115	3.3
7	M	R	Y	53	2.5	56	3.0	247	3.9	94	2.5	78	2.4
8	M	L	Y	13	4.3	9	4.2	295	4.5	37	2.5	80	3.0
8	M	R	Y	3	5.6	4	4.0	264	5.2	35	3.0	65	2.5
9	M	L	Y	23	1.8	24	3.0	238	4.2	35	2.6	72	2.9
9	M	R	Y	14	3.5	7	6.9	250	3.7	76	6.7	164	4.1
10	M	L	Y	10	4.4	19	3.9	264	4.3	78	3.1	120	1.1
10	M	R	Y	9	10.0	19	2.2	265	5.0	71	2.9	92	2.3
11	M	L	Y	146	3.2	105	3.2	106	5.4	49	2.5	60	3.4
11	M	R	Y	34	3.8	40	5.6	190	2.2	57	4.0	85	3.3
12	M	L	Y	23	4.9	30	4.4	283	3.9	14	2.4	111	2.1
12	M	R	Y	42	3.2	45	3.2	268	4.0	95	3.2	110	1.8

Abbreviations: M, male; F, female; R, right; L, left; MCN, musculocutaneous nerve; CBM, coracobrachialis muscle; Y, yes; N, no; MN, median nerve.

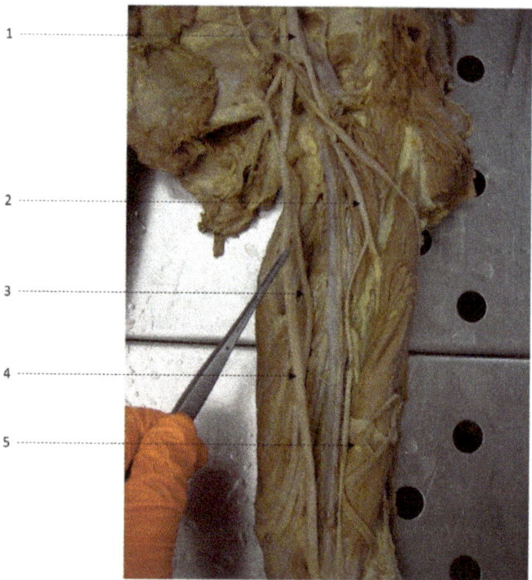

Figure 2. Typical anatomical relationship between the musculocutaneous nerve and median nerve. Left upper limb, anterior view. 1, lateral cord; 2, musculocutaneous nerve, 3, ulnar nerve; 4, median nerve; 5, biceps brachii muscle.

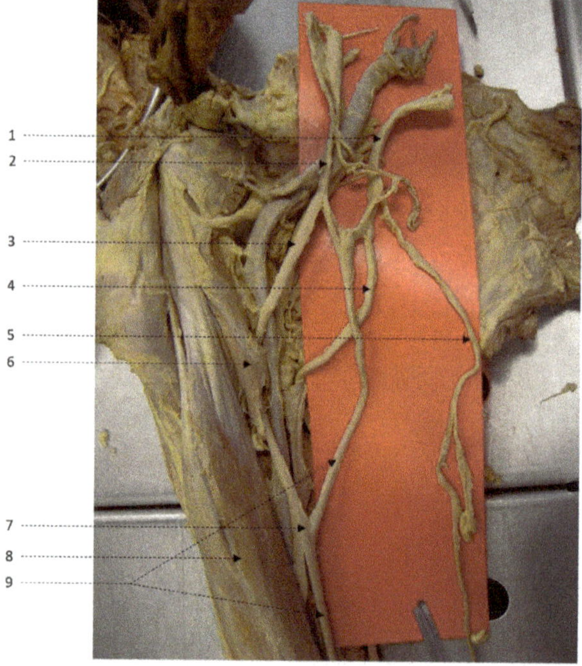

Figure 3. Anatomical variation of musculocutaneous nerve and median nerve communication. Right upper limb, anterior view. 1, medial cord; 2, lateral cord; 3, musculocutaneous nerve; 4, ulnar nerve; 5, medial antebrachial cutaneous nerve; 6, coracobrachialis muscle; 7, communicating branch (MCN-MN anastomosis); 8, biceps brachii muscle; 9, median nerve.

4. Discussion

Over the years, many efforts have been made to increase the knowledge on anatomical variants, since their importance is recognized in both clinical and surgical fields [7]. Among the elements of the peripheral nervous system, the brachial plexus seem to be the most variable one [1]; anatomical variations of the MCN, including its total absence, being among the most significant and present in 10 to 54% of the population according to past studies [1,5,8]. Anatomical studies of embryos suggesting that the MCN is derived from the MN may provide some explanation [1,9] for this high recurrence. The MCN and the lateral root of MN represent two terminal nerves of the lateral cord of the brachial plexus [1]. In literature, many papers have described variants in the communication between the MCN and the MN, that are actually the most frequent anatomical variations in the infraclavicular region of the brachial plexus [10,11]. The first description of communicating branch between the MCN and the MN was reported by Gegenbaur et al. [12], and after this study many other classifications based on such anatomical variations have been drafted. The most specific and most commonly used classifications today are those proposed by Le Minor [3] and Venieratos and Anangnostopolou [5].

According to the Le Minor classification, the large number of variations of the musculocutaneous nerve are classified into five different types in relation to the course of the nerve:

(i)   type I: classic description without anastomoses between the MCN and the MN, with the former passing through the coracobrachialis muscle;

(ii)   type II: fibers link both nerves in the distal section of the arm;

(iii)   type III: lateral root of the MN is merged with the MCN;

(iv)   type IV: all fibers of the MCN are merged with the lateral root of the MN, they follow the MN for some distance and then separate from it as a lateral branch;

(v)   type V: the MCN is absent and branches are located in the upper arm [3].

According to the Venieratos and Anangnostopolou classification, variations are classified into three different types in relation to the sites of communication:

(i)   type I: communications are proximal to the entry point of the MCN into the coracobrachialis muscle;

(ii)   type II: communications are distal to the muscle;

(iii)   type III: the MCN doesn't pierce the coracobrachialis muscle [5].

In the presented study, anatomical dissections were conducted on more than twenty upper limbs, and in 95% of them, type I of MCN-MN variation according to the Le Minor classification was observed. The previously undescribed anatomical variant of MCN-MN anastomosis was found exclusively in one of the studied cadavers, unilaterally. This anatomical variant identified by us does not directly fit within the above-reported classifications. The only variation that shows some similarities is type II of MCN-MN communication according to the Le Minor classification. However, in our case there was no separate branch connecting the MCN and the MN, and the anastomosis was located not at the distal portion, but at the midpoint of the arm. The anastomosis found was in the form of a side-to-side adjacent 1.7 cm section between the two nerves (with visible sharing of fibers between them). In accordance with our findings, Nascimento et al. also reported an anatomical anastomosis between the MCN and the MN localized in a comparable position [13]. In our case, it should also be noted that the anatomical variant was present only on one arm of the donor. This seems to be quite common: Venieratos and Anangnostopolou [5] reported bilateral presence of variations only in 37.5% of cases with detected MCN-MN communication variants. However, further studies and dissections are needed to verify the recurrence of this rare anastomosis between the MCN and the MN.

5. Conclusions

In conclusion, we consider the variation we found as rare, since it is not directly included in any classifications present in literature. In our opinion, this anatomical variation of the anastomosis between the median and the musculocutaneous nerve might assume significant relevance during surgical and clinical practices. A very unusual morphology—a communicating segment between the nerves—may suggest that the structures can be separated during surgery, which could be potentially detrimental to the patient. We hope that our case report will increase the awareness of clinicians and contribute to the improvement of patient safety.

Author Contributions: Dissections—writing—original draft preparation, R.B., A.G.D., N.D.L., G.L.D.G., E.M., G.S., L.L.B., A.M.G., W.P., F.R., E.S. and J.H.S.; supervision, S.W., F.B. and F.C. All authors have read and agreed to the published version of the manuscript.

Funding: This research received no external funding.

Institutional Review Board Statement: The study was conducted in accordance with the Declaration of Helsinki, and approved by the Institutional Ethics Committee of the Medical University of Gdańsk (Ordinance No. 26/2016 of 6 June 2016).

Informed Consent Statement: Premortem informed consent was obtained from all subjects involved in the study.

Data Availability Statement: Not applicable.

Conflicts of Interest: The authors declare no conflict of interest.

References

1. Bergman, R.A.; Tubbs, R.S.; Shoja, M.M.; Loukas, M. *Bergman's Comprehensive Encyclopedia of Human Anatomic Variation*; John Wiley & Sons, Inc.: Hoboken, NJ, USA, 2016; pp. 1069–1089.
2. Standring, S. Pectoral girdle, shoulder region and axilla. In *Gray's Anatomy: The Anatomical Basis of Clinical Practice*, 39th ed.; Elsevier Churchill Livingstone: London, UK, 2005; pp. 817–849.
3. Le Minor, J.M. A rare variant of the median and musculocutaneous nerves in man. *Arch. Anat. Histol. Embryol.* **1990**, *73*, 33–42. [PubMed]
4. Guerri-Guttenberg, R.A.; Ingolotti, M. Classifying musculocutaneous nerve variations. *Clin. Anat.* **2009**, *22*, 671–683. [CrossRef] [PubMed]
5. Venieratos, D.; Anagnostopoulou, S. Classification of communications between the musculocutaneous and median nerves. *Clin. Anat.* **1998**, *11*, 327–331. [CrossRef]
6. Iwanaga, J.; Singh, V.; Ohtsuka, A.; Hwang, Y.; Kim, H.J.; Moryś, J.; Ravi, K.S.; Ribatti, D.; Trainor, P.A.; Sañudo, J.R.; et al. Acknowledging the use of human cadaveric tissues in research papers: Recommendations from anatomical journal editors. *Clin. Anat.* **2021**, *34*, 2–4. [CrossRef] [PubMed]
7. Haviarova, Z.; El Falougy, H.; Killingerova, A.; Matejcik, V. Variation of the median nerve course and its clinical importance. *Biomed. Pap. Med. Fac. Univ. Palacky Olomouc. Czech Repub.* **2009**, *153*, 303–306. [CrossRef] [PubMed]
8. Pedrini, F.A.; Mariani, G.A.; Orsini, E.; Quaranta, M.; Ratti, S.; Cocco, L.; Manzoli, L.; Billi, A.M. Unilateral absence of Casserio's nerve and a communicating branch to the median nerve. An additional variant of brachial flexors motor innervation. *Ital. J. Anat. Embryol.* **2016**, *124*, 16–25.
9. Chitra, R. Various types of intercommunications between musculocutaneous and median nerves: An analytical study. *Ann. Indian Acad. Neurolog.* **2007**, *10*, 100–104. [CrossRef]
10. Radunovic, M.; Vukasanovic-Bozaric, A.; Radojevic, N.; Vukadinovic, T. A new anatomical variation of the musculocutaneous and the median nerve anastomosis. *Folia Morphol.* **2013**, *72*, 176–179. [CrossRef] [PubMed]
11. El Falougy, H.; Selmeciova, P.; Kubikova, E.; Stenova, J.; Haviarova, Z. The variable communicating branches between musculocutaneous and median nerves: A morphological study with clinical implications. *Bratisl Lek List.* **2013**, *114*, 290–294. [CrossRef]
12. Gegenbaur, C. Uber das Verhaltniss des *N. musculocutaneous* zum *N. medianus*. *Jena. Zeitschr F Med. U Nat.* **1867**, *3*, 258–263.
13. Nascimento, S.R.; Ruiz, C.R.; Pereira, E.; Andrades, L.; de Souza, C.C. Rare anatomical variation of the musculocutaneous nerve-case report. *Rev. Bras. Ortop.* **2016**, *51*, 366–369. [CrossRef]

Case Report

Distribution and Appearance of Arrector Pili Muscle in the Skin of the Rhesus Monkey Face

Inga May [1], Kerstin Mätz-Rensing [2] and Christian-Albrecht May [1,*]

[1] Department of Anatomy, Medical Faculty Carl Gustav Carus, TU Dresden, 01099 Dresden, Germany; ingamay@icloud.com

[2] Pathology Unit, German Primate Center, Leibniz Institute for Primate Research, 37077 Göttingen, Germany; kmaetz@dpz.eu

* Correspondence: albrecht.may@tu-dresden.de

Abstract: Although the presence of an arrector pili smooth muscle is documented in many monkey species, its regional peculiarities are hitherto not well documented. We aimed to study this aspect in the face of rhesus monkeys with different areas of hair coat. Eight different regions of six monkeys (male and female) were studied using light microscopy and immunohistochemistry (antibody against smooth muscle alpha actin). We identified two regions (lips and eyelids) with vellus hairs that did not show an arrector pili muscle. In the eyelids, the hairs are rather small and short; in the lips, the vellus hairs were surrounded by striated muscle fibers from the orbicularis oris muscle. In all other regions (frontal region, forehead, cheek, chin), the vellus hairs contained an arrector pili muscle with comparable morphology. Only in the chin region, where additional striated muscles from the face muscles were present, the arrector pili muscles were thinner. All vibrissae showed a close relation to striated muscle fiber bundles of the facial muscles. They never developed smooth muscle bundles assigned as arrector pili equivalent.

Keywords: *Macacca mulatta*; smooth muscle actin; vellus hair; vibrissae; morphology

Citation: May, I.; Mätz-Rensing, K.; May, C.-A. Distribution and Appearance of Arrector Pili Muscle in the Skin of the Rhesus Monkey Face. *Anatomia* 2022, 1, 33–40. https://doi.org/10.3390/anatomia1010004

Academic Editors: Gianfranco Natale and Francesco Fornai

Received: 10 March 2022
Accepted: 8 April 2022
Published: 8 April 2022

Publisher's Note: MDPI stays neutral with regard to jurisdictional claims in published maps and institutional affiliations.

1. Introduction

Mammalian hair follicles are characterized by the presence of an arrector pili muscle. In an extensive study comparing the skin of more than 50 monkeys (a series of individual publications entitled 'The skin of primates' in the American Journal of Physical Anthropology by H. Machida, E. Perkins and co-workers), most species showed the presence of arrector pili muscles, while Lemuriformes and Tarsiiformes seemed to lack these muscles and Lorisiformes seemed to have them inchoate [1]. All these studies did not focus on regional differences which are known, e.g., in the dog, where well developed arrector pili muscles are present in the dorsal back while virtually absent ventrally [2]. A specific regional situation seems to occur in the face, where numerous monkeys develop no fur and only small and thin hairs. In addition, facial muscles insert into the dermal layer [3]. Most of the eyelid seems to lack arrector pili muscles in humans and monkeys [4–7], while a nasal-temporal difference was noted in human Japanese fetuses [8]. The aim of the present study was to specifically investigate the presence and appearance of the arrector pili muscle in different regions of the face of rhesus monkeys.

2. Materials and Methods

Tissue preparation. Skin specimens of selected areas in the face were provided from six rhesus monkeys (*Macacca mulatta*), three males aged 7, 7 and 22 years of age, and three females aged 14, 16 and 20 years of age. The animals were inbreeds from the German Primate Center Göttingen (governmental registration # 122910.3311900, PK 36674) and sacrificed due to other protocols (consultation of the Animal Welfare Committee and the Animal Welfare Officer #E3-22). The skin specimens were fixed in 4% formalin and shipped to Dresden. There, the specimens were transferred to 70% ethanol, trimmed to appropriate size, dehydrated in an ascending series of ethanol (80%, 96%, 96%, 100%, 100%) and embedded in paraffin (standard protocol). The selected areas included the frontal region, the forehead, the medial and lateral upper eyelid, the cheek, the upper and lower lips at the level of the left ala of the nose, and the medial chin.

Histology and immunohistochemistry. Serial sections (5–10 μm thick) of each specimen were performed in sagittal and frontal planes and selected sections dewaxed with xylol, rehydrated (100%, 96%, 70%, 40% ethanol) and stained with hematoxylin and eosin (H&E), Goldner trichrome, or Sirius red solution. For immunohistochemistry, consecutive sections were dewaxed, rehydrated and irradiated with microwaves in 0.01 M sodium citrate buffer (pH 6.0) for 2×5 min at 800 W to unmask the antigens. After washing in PBS, the sections were treated with 0.3% hydrogen peroxide for 10 min and blocked in normal mouse serum for 15 min at 37 °C, followed by washing in PBS. Two primary antibodies raised against smooth muscle alpha-actin (SMA; immunotech PN IM 1144 and Sigma A5228, dilution 1:200) were incubated over night at 4 °C. After washing in PBS, the VECTASTAIN® Elite ABC mouse kit (PK 6101, PK 6102 Vector Laboratories Inc., Burlingame, USA) was applied. Visualization of peroxidase activity was realised by adding 3,3-diaminobenzidine for 8 min.

The sections were examined on a Zeiss Jenamed2 microscope (Carl Zeiss AG, Oberkochen, Germany) and images were recorded by using a Digital Sight DS-Fi1 camera (Nikon AG, Tokio, Japan). Both primary antibodies used revealed identical results.

Statistical analysis. At least five sagittal sections stained with antibodies against SMA of each region and of all animals were used for quantitative measurements. Only proper sections without disturbing artefacts were evaluated. To estimate the density of vellus hairs in the different regions, a length of 5 mm was chosen from each sagittal section and the number of vellus hairs counted within this length. The whole sections were evaluated further, since not all vellus hairs could be used for analysis. Measurements included the length of the arrector pili muscle in mm, which were not always completely present on one section. In this case, consecutive sections (partly with different stains) were used to estimate the total length of the muscle. A min-to-max range was calculated. The maximal width and the depth of muscle origin was measured in μm. For each region, the data of all six animals was pooled and the mean values including their standard deviation calculated.

Due to the low number of cases, statistical analysis was limited. A U-test (Mann–Whitney–Wilcoxon) was applied to compare the male and female regions, but there were no differences, and even no tendencies of difference present. Therefore, the data of both sexes was matched. The U-test was also used to compare the different measurements across the different regions of the face.

The number of associated hair follicles to each arrector pili muscle was counted on the frontal plane sections, since sagittal sections might not show all follicles in one section. Even if possible, a mean number was not statistically appropriate. We therefore noted the majority of association, which showed a clear difference between the frontal region and the other facial regions. A U-test confirmed significance.

3. Results

Vellus hairs were present in all regions investigated, but their number/density varied in different regions. The frontal region, forehead, cheek, and the medial chin showed comparable numbers of vellus hairs within a length of 5 mm: frontal region 11 ± 1, forehead 10 ± 0.5, cheek 12 ± 0.6, and chin 10 ± 0.9. A similar count was also present in the upper lip towards the nose (11 ± 0.8), while in the region with vibrissae, the number of vellus hairs was slightly reduced (8 ± 0.7). In the lower lip, the region of hairy skin always contained vibrissae; the count of vellus hairs (7.4 ± 1) was similar to the comparable region of the upper lip. The number of vellus hairs in the upper eyelid (3.8 ± 0.7) was significantly reduced compared to all other regions.

3.1. Regions Showing Absence of an Arrector Pili Muscle

Upper and **lower lips** (Figure 1). Right after the lip vermilion zone, both vibrissae and vellus hair roots were embedded in dense connective tissue (min–max thickness of this layer: 1.8–2.7 mm) intermingled with numerous bundles of striated muscle fibers. These muscle fibers showed a close relation to the hair roots of both hair types, and reached the dermis close to the epidermal layer. At these places, the muscle fibers showed finger-like endings in contact with the dermal collagen network without a specific orientation of the surrounding collagen. While the striated muscle fibers of the orbicularis oris muscle (min–max total thickness between 1.0–1.4 mm) showed diameters between 40 and 70 μm, the radiating muscle fibers had diameters of 30 to 50 μm, thinning towards 20 μm prior to their finger-like endings.

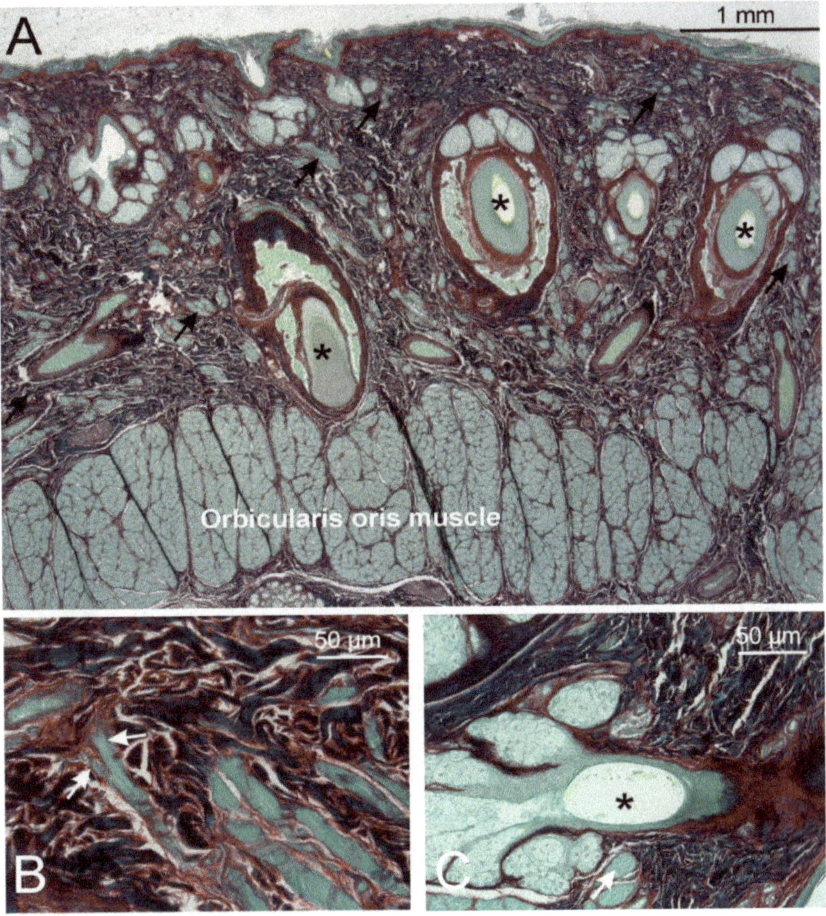

Figure 1. (**A**). In the lip regions, numerous striated muscle bundles (arrows) separated from the orbicularis oris muscle and ran along the vibrissae hair roots (asterisks) to reach the superficial part of the dermis right next to the epidermis. (**B**). At places, the striated muscle fibers formed finger-like endings (arrows). (**C**). The striated muscle fibers (arrow) were also in close contact with the short vellus hair roots (asterisks). No smooth muscle bundles were detected in this part of the skin. (**A**–**C**): Sirius red staining.

The **upper eyelids** (Figure 2). The outer skin of the eyelids was characterized by very small vellus hair roots with no bundles of smooth muscle cells. Within the lid, the orbicularis oculi muscle, min–max 300–500μm in thickness, clearly separated from the skin dermis, which showed a min–max thickness between 160–200μm. At the tip of the eyelid, the striated muscle fibers surrounded the roots of the vellus hairs and the lashes. No difference was seen between the medial and lateral part of the upper lid.

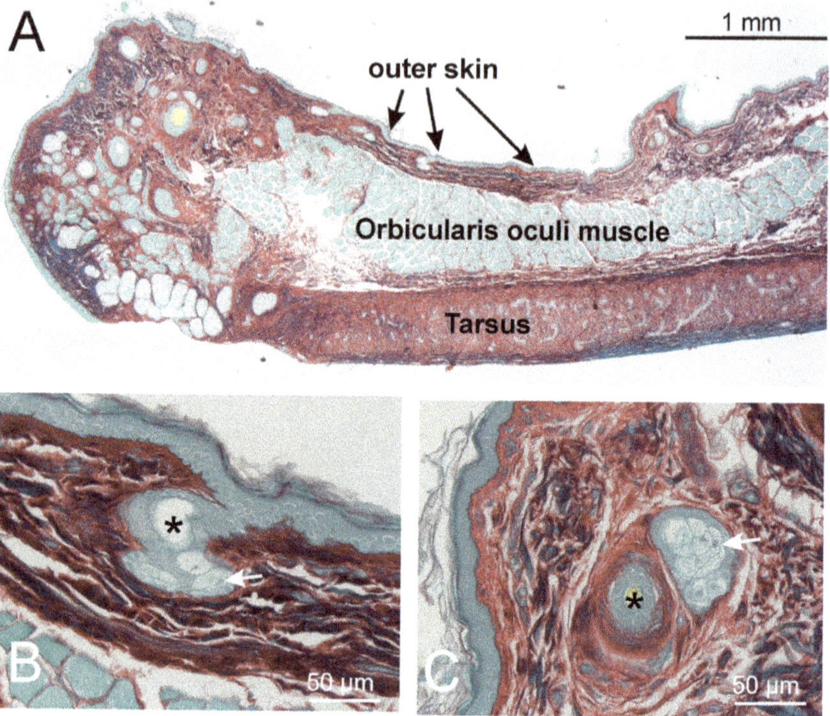

Figure 2. Appearance of the eyelid. (**A**). Overview: the orbicularis oris muscle is clearly separated from the tarsus and the thin connective tissue towards the outer skin. At the tip of the lid (left side), single muscle bundles communicate with the lashes. (**B,C**) show examples of small vellus hairs (asterisks) of the outer lid skin. At the level of the sebaceous glands (arrows), no smooth muscles are present. (**A–C**): Sirius red staining.

3.2. Regions Showing Presence of an Arrector Pili Muscle

Frontal region (Figure 3). The frontalis muscle was 300–500 μm thick and was located at a depth between min-max 1.1 and 1.5 mm to the epidermis. In this dermal connective tissue layer, which contained almost no fat cells, only vellus hair roots were present forming often groups of 2–3 hair roots which showed a shared and well developed arrector pili muscle. The smooth muscle cells formed one strand, 50–80 μm thick, which originated always on one side of the hair root and thinned towards the skin. The length of the arrector pili muscle was 0.8–1.0 mm (Table 1).

The **forehead** and **eyebrow** region (Figure 4). Both vibrissae and vellus hair roots were present. The striated muscle fiber bundles of the mimic muscles were at some distance to the dermis, but at places showed some contact to the deep vibrissae roots. They never reached the roots of the vellus hairs. Only the vellus hairs showed an arrector pili muscle containing of smooth muscle cells. This muscle originated at one side of the vellus hair root, covered the sebaceous gland and reached almost up to the epidermis.

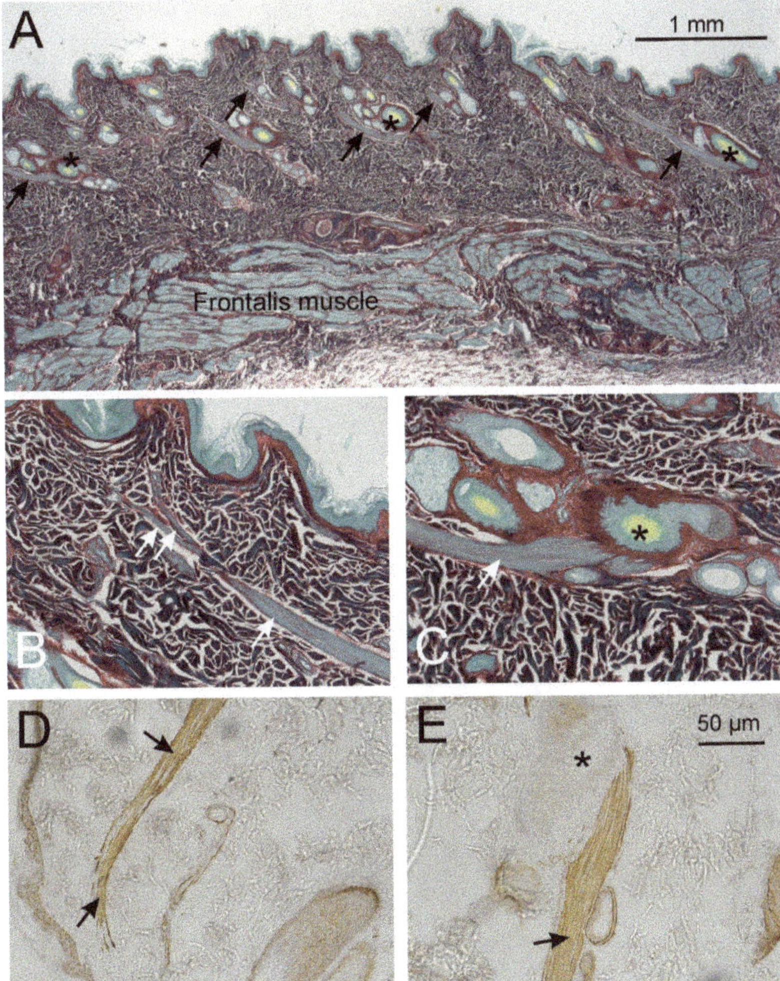

Figure 3. Appearance of the arrector pili muscle (arrows) in the frontal region. (**A**). Overview: each vellus hair complex, mainly containing of 2–3 hairs (asterisks), showed a well developed arrector pili muscle. There was no direct connection to the muscle fibers of the frontalis muscle. (**B,D**). Towards the epithelium, the arrector pili became smaller and showed few and short splitting. (**C,E**). The origin of the arrector pili was at one side of the connective tissue sheath of the vellus hair roots superficial to the bulb. (**A–C**): Sirius red staining. (**D,E**): Smooth muscle alpha actin staining. (**B–E**): Same magnification.

In the **cheek** region, the striated muscle fibers of the mimic muscles were at some distance to the skin (min–max 1.8–2.5 mm), showing no bundles towards the dermis. At places, the dermis showed a fully developed layer of fat tissue (subcutis). There were only vellus hair roots which showed a developed arrector pili muscle as described above (Table 1).

The **chin** region. The skin of the chin region was quite similar to the outer lips. Here too, numerous striated muscle fibers intermingled the connective tissue almost up to the epidermis. Mainly vellus hair roots and only single vibrissae were present. In contrast to the lips, the vellus hairs in the chin region showed an arrector pili muscle, which showed the same topography but was thinner compared to other regions in the face (Table 1).

Table 1. Measuring parameters of the arrector pili muscle at different locations.

Location	Length in mm (Min–Max)	Mean Maximal Width in µm ± Standard Deviation	Mean Depth of Muscle Origin in µm ± Standard Deviation	Number of Associated Hair Follicles
Frontal region	0.8–1.0	64 ± 29	54 ± 18	occasionally 1, mostly 2–3
Forehead	0.9–1.2	64 ± 19	62 ± 21	1, rarely 2
Cheek	0.8–1.0	58 ± 21	64 ± 18	1, rarely 2
Chin	0.7–1.1	38 ± 14	61 ± 20	1, rarely 2

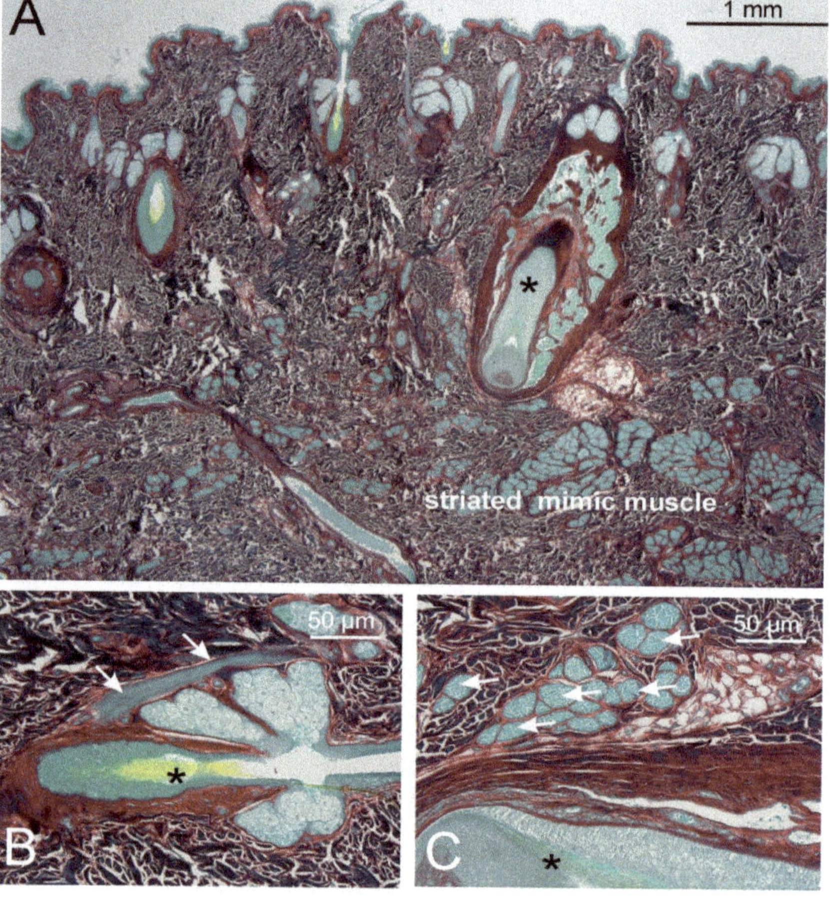

Figure 4. Appearance of the arrector pili muscle in the forehead and eyebrow region. (**A**). Overview: there were only few vibrissae hairs (asterisk) and numerous individual vellus hairs. (**B**). The vellus hair (asterisk) showed a well developed arrector pili muscle (arrows). (**C**). Around the connective tissue sheath of the vibrissae hair (asterisk) there were no smooth muscle bundles, but numerous striated muscle fibers (arrows). These striated muscle fibers did not reach the outer dermis. (**A**–**C**): Sirius red staining.

4. Discussion

The facial skin of the rhesus monkey showed regional differences in the distribution of hair follicles and associated arrector pili muscles.

In all our samples, vibrissae were characterized by large central hair follicles, covered by a thick basal lamina (glassy membrane) and enveloped by a blood sinus encased in a dense connective tissue capsule. They were never accompanied by an arrector pili muscle. However, their bulb and deep root was always in close contact to bundles of the facial striated muscles. In the lips and chin regions, striated muscle bundles oriented perpendicular to the main muscle belly and almost reached the epidermis. In contrast, in the eyelid tip and the forehead regions, the muscle bundles did not branch off the main muscle belly but still surrounded the deep part of the vibrissae. In this respect, our findings are in accordance with that of many mammals [9,10], but in some contrast to the prior description in rhesus monkeys stating a lack of skeletal muscle attachments [11]. It remains to be determined if the sensory nerves around the vibrissae [12] form some neuronal circuits with the specific striated muscle fiber bundles surrounding their bulb and root.

For the vellus hairs there were two regions, the lips and the eyelid, where no arrector pili smooth muscle was formed. The lack of smooth muscle bundles in the eyelid might be caused by the very distinct and rudimentary appearance of these hairs. Interestingly, in this region there were not even remnants of smooth muscle cells present which would indicate a kind of regression. The sheath of the hair root also did not show any signs of smooth muscle alpha actin staining. In contrast, the vellus hairs of the lips were surrounded by fine bundles of striated muscle fibers, presumably adopting the arrector pili function, since fully developed sebaceous glands were present in this region. Although the intense connection between the facial muscles and the skin continued from the lips to the chin, and numerous striated muscle fibers were located in the dermis of the latter region, the chin vellus hairs developed an arrector pili muscle. Interestingly, this muscle was somewhat weaker than in the other regions of the face studied.

In all regions where vellus hairs developed an arrector pili muscle, the muscles inserted on one side of the hair root. This is in accordance with most presentations in text books on human anatomy; the more recently discussed circular smooth muscle cell arrangements around the hair follicle in the human [13–15] were never observed in our rhesus monkey samples. The length of the arrector pili muscle and the depth of origin were similar in all regions studied. The muscle was always located in the dermis and surrounded by dense connective tissue. This might stabilize the muscle bundles for effective support of sebaceous secretion, but might also limit its movement.

Author Contributions: Conceptualization, I.M. and C.-A.M.; tissue samples, K.M.-R.; investigation, I.M.; resources, C.-A.M.; writing—original draft preparation, C.-A.M.; writing—review and editing, I.M. and K.M.-R. All authors have read and agreed to the published version of the manuscript.

Funding: This research received no external funding.

Institutional Review Board Statement: The study was conducted in accordance with the Declaration of Helsinki, and approved by the Institutional Review Board of the German Primate Center Göttingen (#E3-22).

Informed Consent Statement: Not applicable.

Data Availability Statement: Not applicable.

Acknowledgments: The authors thank Daniel Aschoff for tissue preparation and Silvia Bramke for supporting the sectioning and staining of the tissue samples.

Conflicts of Interest: The authors declare no conflict of interest.

References

1. Perkins, E.M., Jr. Phylogenetic significance of the skin of New World monkeys (order Primates, infraorder Platyrrhini). *Am. J. Physic. Anthropol.* **1975**, *42*, 395–423. [CrossRef] [PubMed]
2. Lovell, J.E. Histological and Histochemical Studies of Growth Changes of Canine Skin. Master's Thesis, Iowa State College, Ames, IA, USA, 1955.
3. Diogo, R.; Wood, B.A.; Aziz, M.A.; Burrows, A. On the origin, homologies and evolution of primate facial muscles, with a particular focus on hominoids and a suggested unifying nomenclature for the facial muscles of the Mammalia. *J. Anat.* **2009**, *215*, 300–319. [CrossRef] [PubMed]
4. Sano, M.; Yoshioka, I. Histological observations of eyelids of the Japanese monkey and the rhesus monkey (Histological studies of Primates 1). *Okajimas Folia Anat. Jpn.* **1967**, *43*, 253–261. [CrossRef] [PubMed]
5. Montagna, W.; Ford, D.M. Histology and cytochemistry of human skin. 3. The eyelid. *Arch. Dermatol.* **1969**, *100*, 328–335. [CrossRef] [PubMed]
6. Thibaut, S.; De Becker, E.; Caisey, L.; Baras, D.; Karatas, S.; Jammayrac, O.; Pisella, P.J.; Bernard, B.A. Human eyelash characterization. *Br. J. Dermatol.* **2010**, *162*, 304–310. [CrossRef] [PubMed]
7. Paus, R.; Burgoa, I.; Platt, C.I.; Griffiths, T.; Poblet, E.; Izeta, A. Biology of the eyelash hair follicle: An enigma in plain sight. *Br. J. Dermatol.* **2016**, *174*, 741–752. [CrossRef] [PubMed]
8. Ikeda, M. Das Vorkommen und die Verteilung des M. arrector pili in der Augenlidhaut bei den japanischen Feten. *Okajimas Folia Anat. Jpn.* **1953**, *25*, 79–83. [CrossRef] [PubMed]
9. Vincent, S.B. The tactile hair of the white rat. *J. Comp. Neurol.* **1913**, *23*, 1–34. [CrossRef]
10. Melaragno, H.P.; Montagna, W. The tactile hair follicles in the mouse. *Anat. Rec.* **1953**, *115*, 129–150. [CrossRef] [PubMed]
11. Van Horn, R.N. Vibrissae Structure in the Rhesus Monkey. *Folia Primatol.* **1970**, *13*, 241–285. [CrossRef] [PubMed]
12. Halata, Z.; Munger, B.L. Sensory nerve endings in rhesus monkey sinus hairs. *J. Comp. Neurol.* **1980**, *192*, 645–663. [CrossRef]
13. Narisawa, Y.; Kohda, H. Arrector pili muscles surround human facial vellus hair follicles. *Br. J. Dermatol.* **1993**, *129*, 138–139. [CrossRef] [PubMed]
14. Barcaui, C.; Piñeiro-Maceira, J.; De Avelar Alchorne, M. Arrector pili muscle: Evidence of proximal attachment variant in terminal follicles of the scalp. *Br. J. Dermatol.* **2002**, *146*, 657. [CrossRef] [PubMed]
15. Poblet, E.; Ortega, F.; Jiménez, F. The Arrector Pili Muscle and the Follicular Unit of the Scalp: A Microscopic Anatomy Study. *Dermatol. Surg.* **2002**, *28*, 800–803. [CrossRef] [PubMed]

MDPI

St. Alban-Anlage 66

4052 Basel

Switzerland

www.mdpi.com

Anatomia Editorial Office

E-mail: anatomia@mdpi.com

www.mdpi.com/journal/anatomia